PLANT CELL AND TISSUE CULTURE

Plant Cell and Tissue Culture

Edited by

INDRA K. VASIL
Laboratory of Plant and Molecular Biology, University of Florida, Gainesville, FL, U.S.A.

and

TREVOR A. THORPE
Plant Physiology Research Group, Department of Biological Sciences, The University of Calgary, Alberta, Canada

Springer

Library of Congress Cataloging-in-Publication Data

ISBN 0-7923-2493-5

Vasil and Thorpe: Plant Cell and Tissue Culture

First Indian Reprint 2005
Second Indian Reprint 2010

ISBN 978-81-8128-371-9

This edition is manufactured in India for sale in India, Pakistan, Bangladesh, Nepal and Sri Lanka only.

This edition is published by Springer (India) Private Limited, (a part of Springer Science+Business Media), Registered Office: 3rd Floor, Gandharva Mahavidyalaya, 212, Deen Dayal Upadhyaya Marg, New Delhi – 110 002, India.

Printed in India by Brijbasi Art Press Limited.Noida

Table of Contents

PART B: *IN VITRO* CULTURES

Preface

The earliest organized efforts to induce sustained growth of plant cells in culture were made in the 1930s. They were followed rapidly by the development of aseptic techniques and complex nutrient media. These advances, combined with the discovery of plant growth regulators like auxins and cytokinins, and their profound effects on morphogenesis, led to the regeneration of plants from cultured tissues. Further refinement of techniques made it possible to recover whole plants from isolated single cells, through organogenesis as well as somatic embryogenesis. Today, plant regeneration can be obtained from cell and tissue cultures of a wide variety of plants, including most of the economically important species. Indeed, micropropagation has become an increasingly important and successful industry in all parts of the world.

A new phase in the development of plant cell culture techniques started around 1970 with the regeneration of plants from cultured protoplasts (and protoplast fusion products), anthers and microspores. At the same time, newly emerging recombinant DNA technology provided a powerful new tool for the study of the molecular basis of plant development, as well as their genetic manipulation and improvement. A wide variety of experimental procedures were developed for the delivery and stable integration of alien genes into the germplasm of plants. The synergism of plant cell culture and molecular biology has led to remarkable advances in our understanding of plant development and in the production of transgenic plants with valuable agronomic characteristics. Many such transgenic crops are now undergoing field trials, and are expected to be commercially available before the end of this decade.

Eleven chapters in Part A of this volume provide generic accounts of all of the important aspects of plant cell and tissue culture, from regeneration to genetic transformation. An equal number of chapters in Part B describe advances made in the *in vitro* culture of various economically important groups of plants. Together, these accounts represent the state of the art of plant cell and tissue culture.

Each of the chapters in this volume was prepared by distinguished individuals who have themselves made pioneering and significant contributions to their respective fields of study. We are grateful to them for their diligence in preparing their comprehensive and insightful accounts. It is our hope that this volume will be useful not only to the newly initiated, but will also be a source of much valuable information and new ideas for our colleagues interested in the theory as well as the practical utility of plant cell and tissue cultures.

INDRA K. VASIL
TREVOR A. THORPE

List of Contributors

CONSTABEL, Friedrich, *Plant Biotechnology Institute, National Research Council of Canada, 110 Gymnasium Place, Saskatoon, Saskatchewan, Canada S7N 0W9*
(Chapter 1 with Jerry P. Shyluk)
(Chapter 11 with R.T. Tyler)

DAVEY, M.R., *Plant Genetic Manipulation Group, Department of Life Science, The University of Nottingham, University Park, Nottingham NG7 2RD, United Kingdom*
(Chapter 13 with V. Kumar and N. Hammatt)

DEBERGH, Pierre, *Laboratory of Horticulture, State University of Ghent, Coupure 653, B-9000 Ghent, Belgium*
(Chapter 22)

DIX, Philip J., *Department of Biology, St. Patrick's College, Maynooth, County Kildare, Ireland*
(Chapter 5)

ENGELMANN, F., *ORSTOM, Institut Français de Recherche Scientifique pour le Développement et Coopération, Montpellier, France*
(Co-author of Chapter 9)

FEHÉR, Attila, *Institute of Plant Biology, Biological Research Centre, Hungarian Academy of Sciences, Temesvári Krt 62, P.O. Box 521, H-6701 Szeged, Hungary*
(Chapter 4 with Dénes Dudits)

GROSSER, J.W., *IFAS, Citrus Research and Education Center, University of Florida, Lake Alfred, FL 33850, USA*
(Chapter 19)

HAMMATT, N., *Horticulture Research International, East Malling, Kent ME19 6BJ, United Kingdom*
(Co-author of Chapter 13)

HARRY, Indra S., *Plant Physiology Research Group, Department of Biological Sciences, The University of Calgary, 2500 University Drive NW, Calgary, Alberta, Canada T2N 1N4*
(Chapter 21 with Trevor A. Thorpe)

HINCHEE, Maud A.W., *Monsanto Plant Protection Improvement Institute, 700 Chesterfield VP, St. Louis, MO 63198, USA*
(Chapter 10 with David R. Corbin, Charles L. Armstrong, Joyce E. Fry, Shirley S. Sato, David L. Deboer, William L. Petersen, Toni A. Armstrong, Danette V. Connor-Ward, Jeanne G. Layton, and Robert B. Horsch)

JONES, M.G.K., *Plant Sciences, School of Biological and Environmental Science, Murdoch University, Perth, WA 6150, Australia*
(Chapter 15)

KARP, Angela, *Department of Agricultural Sciences, University of Bristol, AFRC Institute of Arable Crops Research, Long Ashton Research Station, Bristol, BS18 9AF, United Kingdom*
(Chapter 6)

KARTHA, Kutty K., *Plant Biotechnology Institute, National Research Council of Canada, 110 Gymnasium Place, Saskatoon, Saskatchewan, Canada S7N 0W9*
(Co-author of Chapter 3)
(Chapter 9 with F. Engelmann)

KELLER, W.A., *Plant Biotechnology Institute, National Research Council of Canada, 110 Gymnasium Place, Saskatoon, Saskatchewan, Canada S7N 0W9*
(Co-author of Chapter 17)

KRIKORIAN, Abraham D., *Department of Biochemistry and Cell Biology, State University of New York at Stony Brook, Stony Brook, NY 11794-5215, USA*
(Chapters 16 and 20)

KUMAR, V., *Plant Genetic Manipulation Group, Department of Life Science, University of Nottingham, University Park, Nottingham NG7 2RD, United Kingdom*
(Co-author of Chapter 13)

NEHRA, Narender S., *Plant Biotechnology Institute, National Research Council of Canada, 110 Gymnasium Place, Saskatoon, Saskatchewan, Canada S7N 0W9*
(Chapter 3 with Kutty K. Kartha)

PALMER, C.E., *Department of Plant Science, University of Manitoba, Winnipeg, Manitoba, Canada R3T 2N2*
(Chapter 17 with W.A. Keller)

RAGHAVAN, V., *Department of Plant Biology, The Ohio State University, 1735 Neil Avenue, Columbus, OH 43210-1293, USA*
(Chapter 8)

REYNOLDS, John F., *Experimental Plant Genetics, The Upjohn Company, Kalamazoo, MI 49002, USA*
(Chapter 14)

SWARTZ, Harry S., *Department of Horticulture, University of Maryland, College Park, MD 20742, USA*
(Co-author of Chapter 18)

THORPE, Trevor A., *Plant Physiology Research Group, Department of Biological Sciences, The University of Calgary, 2500 University Drive NW, Calgary, Alberta, Canada T2N 1N4*
(Chapter 2)
(Co-author of Chapter 21)

TYLER, R.T., *Department of Applied Microbiology and Food Science, University of Saskatchewan, Saskatoon, Saskatchewan, Canada S7N 0W0*
(Co-author of Chapter 11)

VASIL, Indra K., *Laboratory of Plant Cell and Molecular Biology, Department of Horticultural Sciences, 1143 Fifield Hall, University of Florida, Gainsville, FL 32611-0690, USA*
(Chapter 12 with Vimla Vasil)

WENZEL, G., *Federal Biological Research Centre for Agriculture and Forestry, Institute of Resistance Genetics, W-8059 Grünbach, Germany*
(Chapter 7 with B. Foroughi-Wehr)

ZIMMERMAN, Richard H., *US Department of Agriculture, Agricultural Research Service (USDA/ARS), Plant Sciences Institute, Fruit Laboratory, Bldg 004, Room 120, 10300 Baltimore Avenue, Beltsville, MD 20705-2350, USA*
(Chapter 18 with Harry J. Swartz)

PART A

Methodology

1. Initiation, Nutrition, and Maintenance of Plant Cell and Tissue Cultures

Friedrich Constabel and Jerry P. Shyluk

Contents

1. Introduction

Plant cell and tissue culture has served as methodology of physiology and biochemistry in a quest to increase our knowledge of cell biology for half a century. Today, tissue culture has largely been integrated in biotechnology and permits the regeneration of plants as clones and as transgenics. Callus and cell suspension cultures, the pride of laboratories for years, are being phased out; cell phenomena are being studied *in planta* by employing molecular probes. Also, cells cultured *in vitro* carry a factor of uncertainty in somaclonal variation. Callus and cell suspensions, thus, become reduced in importance to source material for certain enzymes, chemicals, and experimental material for special aspects of differentiation. Changes in goals require changes in treatment of methodology of plant tissue culture.

In general, tissue culture methodology is best studied through apprenticeship in a recognized laboratory. Failing this, many laboratory manuals (see listing of general introductions to plant tissue culture methods) provide exact information of step by step procedures to follow. A manual of micropropagation by Kyte (1983) could be particularly helpful. Tobacco is probably one

Indra K. Vasil and Trevor A. Thorpe (eds.), Plant Cell and Tissue Culture, 3–15.

Milk (liquid endosperm) of coconuts can be obtained by drilling a hole through one of the pores for germination of nuts bought at a local market. Former preference for milk of immature nuts has not been substantiated. Surplus milk can safely be stored in a freezer.

2.5. *Nutrient Media*

Success in plant cell culture is largely determined by the quality of nutrient media. No other factor has received as much attention and, as a result, numerous formulations have been published leaving confusion for any beginner. A systematic approach of nutritional requirements of tissues cultured *in vitro* in the 60's has led to acceptance of the fact that employment of one to three media formulations will permit to at least initiate a culture of plant tissue *in vitro*. Optimization of growth and plant regeneration from cultured cells and tissues may require modifications rather than novel formulations of nutrient media. Formulations designed by Murashige and Skoog (1962) and revised by Linsmaier and Skoog (1965), Gamborg et al. (1968), and by Schenk and Hildebrandt (1972) can be regarded as standard. Still today, the first one is preferred by 9 out of 10 labs.

The MS medium was designed to test the effect of organic supplements on growth and development of tissue cultures. The medium, therefore, was standardized with regard to inorganic nutrients and formulated for tobacco pith tissue. It proved to become widely accepted because replicate cultures were 'astonishingly alike in size and external appearance, the reproducibility was most satisfactory and optimal yields of new growth were obtained' (Murashige and Skoog, 1962). The B_5 medium formulated for growing soybean tissues has been employed with a large variety of plant tissues and offers itself for an initial test with experimental material. In general, concentrations of inorganic nutrients are lower than in the MS medium, 'a feature which apparently is preferred by the cells of some species' (Gamborg et al., 1968). The SH medium for growth of friable callus to be feed stock for cell suspensions and production of protoplasts has exceeded the efficacy of the MS medium with many of the 37 mono- and dicotyledons tested (Schenk and Hildebrandt, 1972). The composition of the three major nutrient media is given in Table 1.

Differences in the concentration of ammonium, potassium, nitrate and phosphate ions are remarkable. High levels of ammonium and nitrate have generally been preferred for callus initiation and culture. Shoot and root development may gain from lower concentrations of mineral salts; the use of half strength MS or B_5 media for plant regeneration is not uncommon.

Indole-3-acetic acid (IAA) is a natural auxin, weaker than naphthalene acetic acid (NAA). Media containing IAA deteriorate faster than those containing NAA. The most potent auxin is 2,4-dichlorophenoxyacetic acid (2,4-D), a commonly used herbicide for broad-leaved weeds. The most active cytokinin is N_6-isopentenyladenine (2iP), kinetin and N_6-benzyladenine are the most common ones, some formulations require zeatin or zeatin riboside (Seabrook, 1980).

Special plant groups like conifers have nutritional requirements which appear not to be met by any of the three standard media. Table 1 lists formulations by von Arnold and Eriksson (AE, 1981) and by Litvay as cited in Verma et al. (LM, 1982). Differences are significant and suggest where to expect and implement modifications when adjusting media to tissues in

Table 1. Composition of nutrient media after Murashige and Skoog (MS, 1962), Gamborg et al. (B_5, 1968), Schenk and Hildebrandt (SH, 1972), von Arnold and Eriksson (AE, 1981), and Litvay (LM, 1982, in Verma et al., 1983) for plant tissue and cell culture

	MS	B_5	SH	AE*	LM
Macronutrients (mg/l)					
NH_4NO_3	1650	–	–	1200	1650
$NH_4H_2PO_4$	–	–	300	–	–
KNO_3	1900	2500	2500	1900	1900
$CaCl_2{\cdot}2H_2O$	440	150	200	180	22
$MgSO_4{\cdot}7H_2O$	370	250	400	370	1850
KH_2PO_4	170	–	–	340	340
$(NH_4)_2SO_4$	–	134	–	–	–
$NaH_2PO_4{\cdot}H_2O$	–	150	–	–	–
Micronutrients (mg/l)					
KI	0.83	0.75	1.0	0.75	4.15
H_3BO_3	6.2	3.0	5.0	0.63	31
$MnSO_4{\cdot}4H_2O$	22.3	–	–	2.20	–
$MnSO_4{\cdot}H_2O$	–	10.0	10.0	–	21
$ZnSO_4{\cdot}7H_2O$	8.6	2.0	1.0	4.0	43
$Na_2MoO_4{\cdot}2H_2O$	0.25	0.25	0.1	0.025	1.25
$CuSO_4{\cdot}5H_2O$	0.025	0.025	0.2	0.002	0.5
$CoCl_2{\cdot}6H_2O$	0.025	0.025	0.1	0.002	0.125
Na_2EDTA	43.0	43.0	20.0	19	37
$+FeSO_4{\cdot}7H_2O$	–	–	15.0	14.0	27
Organics (mg/l)					
Inositol	100	100	1000	100	100
Nicotinic acid	0.5	1.0	5.0	2.0	0.5
Pyridoxine-HCl	0.5	1.0	0.5	1.0	0.1
Thiamine-HCl	0.1	10.0	5.0	5.0	0.1
Carbohydrates (g/l)					
Sucrose	30	20	30	34	30
Hormones (mg/l)					
Auxin	0.1–5.0	0.1–5.0	0.5	0.1–5.0	2.5
Cytokinin	0.01–2.0	0.01–2.0	0.1	0.01–2.0	0.1
Coconut milk (%)	–	–	6	–	–

* Add (in mg/l): D-glucose 180, D-xylose 150, L-arabinose 150, L-glutamine 0.4, L-alanine 0.05, L-cysteine 0.02, L-arginine 0.01, L-leucine 0.01, L-phenylalanine 0.01, L-tyrosine 0.01, glycine 2.0.

question; Mg, Ca, and SO_4 ions, micronutrients, as well as carbohydrate and amino acid supplements deserve attention.

3. Methods

3.1. *Media Preparation*

Two procedures in plant tissue culture are most labor intensive and strenuous: the preparation of media and the transfer of cultured cells and tissues to fresh media. As concerns the preparation of media, this task can be scheduled to once a week and such that concentrates are available for quick response to any demand. Manuals and lab books, again, should be consulted when starting tissue culture for the first time and for comparison and subsequent rationalization of procedures even when experienced.

According to a formula chosen, chemicals are dissolved in water of about half the final volume of the medium. Once all ingredients have been added, the medium is brought to near volume and the pH is adjusted. Finally, the medium is brought to its precise volume.

The actual procedure of preparing the media can substantially be simplified by using stock solutions as recommended by Gamborg and Shyluk (1981; see Table 2). Some labs store 100 ml of concentrate (10×) of medium, i.e. a combination of dissolved mineral salts, vitamins, and carbohydrates in plastic bags (Whirl-Pak) in a freezer. Such concentrates are ready to be used for 1 l preparations within a few minutes of thawing.

Table 2. Stock solutions for preparation of MS and B_5 media (Gamborg and Shyluk, 1981)

Compounds (mg/100 ml)	MS	B_5
Micronutrients (store in freezer)		
$MnSO_4$-H_2O	–	1000
$MnSO_4 \cdot 7H_2O$	2230	–
H_3BO_3	620	300
$ZnSO_4 \cdot 7H_2O$	860	200
$Na_2MoO_4 \cdot 2H_2O$	25	25
$CuSO_4 \cdot 5H_2O$	8.5	2.5
$CoCl_2 \cdot 6H_2O$	2.5	2.5
Vitamins (store in freezer)		
Nicotinic acid	50	100
Thiamine·HCl	10	1000
Pyridoxine·HCl	50	100
Myo-Inositol	10000	10000
Other		
$CaCl_2 \cdot 2H_2O$ (g/100 ml)	44	15
KI (mg/100 ml)	83	75

Table 3. Media preparation (1 l) using stock solutions

	MS	B_5
Double distilled water		
$NaH_2PO_4 \cdot H_2O$ (mg)	–	150
KH_2PO_4	170	–
KNO_3	1900	2500
$(NH_4)_2SO_4$	–	134
NH_4NO_3	1650	–
$MgSO_4 \cdot 7H_20$	370	250
Ferric EDTA	43	43
Sucrose (g)	30	20
$CaCL_2 \cdot 2H_2O$ (ml stock solution)	1	1
Micronutrients (ml stock solution)	1	1
Kl (ml stock solution)	1	1
Vitamins (ml stock solution)	1	1

3.1.1. *Hormones*

Dissolve 50 mg 2,4-D or NAA in ethanol, heat slightly, and dilute to 100 ml with water. Store in refrigerator. For a final concentration of 1 mg/l add 2 ml of stock solution to medium. Dissolve 25 mg kinetin in small volume of 0.5 N HCl by heating slightly and gradually diluting to 100 ml with distilled water. Store in refrigerator. For a final concentration of 0.5 mg/l add 2 ml of stock solution to medium.

3.1.2. *pH*

Adjust pH to 5.5–5.8 with 0.2 N NaOH or HCl.

3.1.3. *Solid Media*

Agar (0.6–1.0%), agarose (0.6–0.8%), or gelrite (0.1–0.3%) are added to nutrient media and heated to boiling once before dispensing in jars and autoclaving, or before autoclaving and subsequent dispensing in petri dishes.

3.1.4. *Sterilization*

Steam sterilization or autoclaving is useful for nutrient media, distilled water, paper products, glassware and metal instruments. A pressure of 1.05 kg/cm^2 (121 °C) over 10–20 min would suffice for up to 2 l of medium. Heat sensitive liquids may be sterilized using membrane filters of 0.45 or 0.2 μm pore size. Chemicals like auxins and cytokinins insoluble in water may be dissolved in alcohol or dimethyl sulfoxide and subsequently added to sterilized media.

3.2. *Explantation*

The process of dissection and culture of small organs or tissue sections is referred to as explantation. Explant choice, the timing of excision, and pretreatment are important determinants of culture success (Donnelly and Vidaver, 1988). Healthy, vigorously growing plants will render suitable explants. For reasons of simplicity, seedlings and plantlets grown *in vitro* from sterilized seeds kept on a moist substrate in a sterile container are preferred. The excision of anthers and of immature embryos for purposes of production of secondary embryos and plant regeneration may require pretreatment of donor plants with appropriate regimes of light, temperature and fertilizer and precise timing.

Origin and size of explanted tissue determine the development of a culture. Number and physiological status of parenchyma cells subtending the cut surface of the explant will give rise to a proportional amount of callus. By means of scalpels and forceps tissues of ca. 0.5 cm^3 pith, 0.5–1 cm^2 leaf, 1 cm hypocotyl, or entire embryos can easily be removed from sterilized donor material in a sterile work area and transferred to nutrient media *in vitro*.

Explanted cells, tissues, and organs as well as their environment must be sterile. Working surfaces are best treated with disinfectants and preferably 70% ethanol. Metal instruments for handling sterile tissue are best sterilized by immersion in absolute ethanol and then flamed. Precise sterilization procedures have to be established for each kind of explant. For example, seeds are sterilized by immersion in 20% chlorox®, a disinfectant containing 52% sodium hypochlorite, with 1–2 drops of wetting agent, i.e. TWEEN 80, for 10 min, followed by a short rinse in sterile water, an immersion in 70% ethanol for 2 min, and 3 thorough rinses in sterile distilled water. Plant parts destined for explantation tolerate surface sterilization with various chemicals as shown in Table 4.

Table 4. Sterilization of plant parts for explantation (Constabel, 1984)

Disinfectant	Concentration (%)	Duration of treatment (min)
Calcium hypochlorite	9–10	10–40
Sodium hypochlorite	0.9–2.0	10–40
Mercury chloride	0.1	10
Ethanol	70	2–3

A number of antibiotics and fungicides have been tested with plant tissue cultures as means to effect sterilization of explants and elimination of contamination. As a result, many of these compounds have been discarded as cytotoxic (Pollock et al., 1983). Indeed, antibiotic toxicity is being employed to select for transformed cells with acquired resistance to antibiotics kanamycin or neomycin for instance. Cefotaxime (60–100 mg/l) and carbenicillin (50–400 mg/l), on the other hand, have been used to eliminate *Agrobacterium* sp. from plant tissues upon co-cultivation for transformation in many instances with great success. These compounds may be recommended for elimination of contamination with gram negative bacteria. Cefotaxime (60 and 100 mg/l) even stimulated callus growth of various wheat and barley varieties (Mathias and Boyd, 1986). Of fungicides, carbendazim (30 mg/l), fenbendazole (30 mg/l), and imazalil (20 mg/l) have been employed relatively safely and against a wide spectrum of fungal contaminants. Also, tissues have been freed from unidentified yeasts by short incubations with 10 mg/l nystatin (Shields et al., 1984). For contaminants of plant tissue cultures and criteria for choosing an antibiotic for control of contamination see Leifert and Waites (1990) and Falkiner (1990).

3.3. *Explant Culture*

The majority of explants are maintained on solid media, solidification being achieved by media supplements of 0.6–0.8% agar. The higher purity of agarose compared to agar may make for better response of an explant to *in vitro* culture. This observation has been particularly noticeable with delicate material like protoplasts; only media with agarose may permit the formation of microcalli. And, for example, media with agarose (0.6–0.8%) have been found to yield 3 times the number of buds with cotyledon explants of Douglas fir. Gelrite (0.1–0.3%) yielded twice as many buds as explants on media with agar. Differences in the source of agar may cause differences in growth response by the explanted tissue (Debergh, 1983).

In a number of cases, explants have been cultured on floats of impregnated nylon fabric on liquid media in order to allow for greatest aeration while maintaining a flow of nutrients by capillary forces. Anthers and immature embryos have been cultured in this manner with advantage. Due to problems with aeration immersion in liquid media is not recommended for explants (Kao et al., 1991).

Petri dishes are still preferred as containers for explant cultures due to accessibility of material to a variety of manipulations. Parafilm can easily be applied to seal the dishes. Such seal permits sufficient ventilation of the atmosphere surrounding the explants and prevents contamination and desiccation over 3–4 weeks of culture.

Explant cultures may be stored in growth chambers in the dark or under any light regime, commonly in a 16 h/day light of 10–40 W m^{-2} of cool white fluorescent sources and at 22–28 °C.

3.4. *Callus Culture*

Within 2–3 weeks of culture, explants show new growth across the surface of the explant depending on the distribution and mitotic activity of the parenchyma residing in the excised tissue. Continued growth may leave the core of the explants fairly undisturbed, or the explants may disintegrate as the callus grows. In light, the callus may turn green. Callus formation may be accompanied by the formation of roots. A twofold increase in the level of auxins in the medium should prevent rhizogenesis.

When cultured for several weeks, any callus will show signs of aging, noted as deceleration of growth, necrosis or browning, and finally desiccation. Transfer of healthy, vigorous callus pieces about 5 mm in diameter to 30 ml fresh medium (subculture) in 120-ml jars at intervals of 4–6 weeks will maintain the callus. Several lines of callus have been maintained for over 15 years. Callus of various species has been stored at 8–10 °C or under mineral oil overlay for 4–6 months without subculturing. After transfer to regular culture conditions growth resumed readily (Augereau et al., 1986).

The establishment of permanent callus collections has been considered by several laboratories. It appears, however, that laboratory collections have been found unsatisfactory in terms of both supply and demand. Maintenance of callus cultures by way of monthly subcultures may lead to slow change of the material and is labor intensive. Exchange of material with other laboratories or re-establishment of a given culture is preferred.

Mailing callus requires the transfer of well-growing callus, about 1–2 cm in diameter, to 50-ml tubes or plastic containers with 5–10 ml liquid rather than agar-medium and the fastest air and courier services. Cross-border regulations covering biological materials must be observed.

3.5. *Suspension Culture*

Initiation of suspension cultures can be achieved by subculture of callus pieces in liquid media. Transfer of 0.5 g of callus to 50 ml of liquid medium in a 200 ml Erlenmeyer flask closed with cotton plugs and kept on a rotary shaker at 100–150 rpm for 1–3 weeks is a common procedure. Over time the inoculum will break up and produce freely floating aggregates of new cells. Replacement of half of the spent medium by fresh medium from sedimented cells will allow for continued growth. Once the suspension of cell aggregates is fine enough, subcultures can be obtained by transfer of about 10 ml suspension by wide-mouthed pipette to 40 ml of fresh medium. Later, the growth rate of the suspension will dictate the size of the inoculum. Aliquots of suspension sampled over a period of time and filtered, dried, and weighed will permit determination of cell growth. Except for cells cultured for light-induced product synthesis and photoautotrophic cells, suspensions do not require special light regimes. Cell suspensions can be maintained for many years. The procedure described here would over time lead to selection of fastest growing cells and a culture quite different in physiological behavior from the initial culture. Cryopreservation of samples and resuspension after a period of time would prevent such changes.

3.6. *Single Cell Culture*

Culturing single cells *in vitro* has been desirable in order to simulate the culture of micro-organisms for industrial production of phytochemicals in fermenters (bioreactors) and for plant development from cells following genetic transformation. Single cell suspensions, however, have never been achieved, because daughter cells tend not to dissociate after a division. The application of cellulases and hemi-cellulases has led to deterioration of cultures over time (Street 1973). The culture of a single isolated cell, on the other hand, requires rather complex conditions which cannot easily be met. The best approach can be deduced from experiments of Kao and Michayluk (1975) with cells (protoplasts) at very low density and from micro-culture conditions suggested by Schaffler and Koop (1990). Here, cell (protoplast) culture was assisted by nurse (feeder) cells and implementation of a dialysis membrane as a physical barrier between target and feeder cells. Using a very low melting point agarose was part of the single cell embedding system.

The concept of engineering transgenic plants *in planta* or with the least implication of cell and tissue culture for reasons of reducing the possibility of somaclonal variation may lead to renewed interest in single cell culture in the future. Short term *in vitro* culture of embryo sacs, egg, and sperm cells would be the objective (cf. Kranz et al., 1990).

4. Outlook

The methodology of initiation, nutrition, and maintenance of plant cell and tissue cultures will continue to be a major concern for purposes of micropropagation, production of phytochemicals and the creation of transgenic plants. Newsletters, journals, and conferences remain the best media to keep abreast with new developments. Manipulation of fertilization with genetically engineered gametes would appear to be the greatest challenge for *in vitro* culture technology (Heberle-Bors, 1991).

5. References

Augereau, J.M., Courtois, D. and Pétiard, V. (1986). Long term storage,of callus cultures at low temperatures or under mineral oil layer. Plant Cell Rep. 5:372–376.

Beyer, E.M. Jr. (1976). A potent inhibitor of ethylene action in plants. Plant. Physiol. 58 268–271.

Chi, G.L., Barfield, D.G., Sim, G.E. and Pua, E.C. (1990). Effect of $AgNO_3$ and aminoethoxyvinylglycine on *in vitro* shoot and root organogenesis from seedling explants of recalcitrant Brassica genotypes. Plant Cell Rep. 9:195–198.

Constabel, F. (1984). Callus culture: induction and maintenance. In: Cell Culture and Somatic Cell Genetics of Plants, Vol. 1. I.K. Vasil (ed.), Academic Press, New York, pp. 27–35.

Debergh, P.C. (1983). Effects of agar brand and concentration on the tissue culture medium. Physiol. Plant. 59:270–276.

Falkiner, F.R. (1990). The criteria for choosing an antibiotic for control of bacteria in plant tissue culture. IAPTC Newsletter 60: 13–23.

Gamborg, O.L. Miller, R.A. and Ojima, K. (1968). Nutrient requirements of suspension cultures of soybean root cells. Exp. Cell Res. 50:151–158.

Gamborg, O.L. and Shyluk, J.P. (1981). Nutrition, media and characteristics of plant cell and tissue cultures. In: Plant Tissue Culture, Methods and Applications in Agriculture. T.A. Thorpe (ed.), Academic Press, New York, pp. 21–44.

Heberle-Bors, E. (1991). Germ line transformation in higher plants. IAPTC Newsletter 64:2–10.

Kao, K.N. and Michayluk, M.R. (1975). Nutritional requirements for growth of Vicia hajastana cells and protoplasts at a very low population density in liquid media. Planta 26:105–110.

Kao, K.N., Saleem, M., Abrams, S., Pedras, M., Horn, D. and Mallard, C. (1991). Culture conditions for induction of green plants from barley microspores by anther culture methods. Plant Cell Rep. 9:595–601.

Kranz, E., Bautor, J. and Lörz, H. (1990). *In vitro* fertilization of single, isolated gametes, transmission of cytoplasmic organelles and cell reconstitution of maize (*Zea mays* L.). In: Progress in Plant Cellular and Molecular Biology. H.J.J. Nijkamp, L.H.W. van der Plas and J. van Aartrijk (eds.), Kluwer Academic Publishers, Dordrecht, pp. 252–257.

Leifert, C. and Waites, W.M. (1990). Contaminants of plant tissue cultures. IAPTC Newsletter 60:2–13.

Linsmaier, E.M. and Skoog, F. (1965): Organic growth factor requirements of tobacco tissue cultures. Physiol. Plant. 18:100–127.

Lynch, D.R., Coleman, M.C. and Lyon, G.D. (1991). Effect of *Alternaria solani* culture filtrate on adventitious shoot regeneration in potato. Plant Cell Rep. 9:607–610.

Murashige, T. and Skoog, F. (1962). A revised medium for rapid growth and bioassays with tobacco tissue cultures. Physiol. Plant. 15:473–479.

Mathias, R.J. and Boyd, L.A. (1986) Cefotaxime stimulates callus growth, embryogenesis, and regeneration in hexaploid bread wheat (*Triticum aestivum* L. em. Thell). Plant Sci. 26:217–223.

Pollock, K., Barfield, D.G. and Shields, R. (1983), The toxicity of antibiotics to plant cell cultures. Plant Cell Rep. 2:36–39.

Roustan, J.P., Latche, A. and Fallot, J. (1989). Stimulation of *Daucus carota* somatic embryogenesis by inhibitors of ethylene synthesis: cobalt and nickel. Plant Cell Rep. 8:182–185.

Schenk, R.U. and Hildebrandt, A.C. (1972) Medium and techniques for induction and growth of monocotyledonous and dicotyledonous plant cell cultures. Can. J. Bot. 50:199–204.

Schäffler, and Koop, H.U. (1990). Single cell nurse culture of tobacco protoplasts. Physiological analysis of conditioning factors. J. Plant Physiol. 137:95–101.

Seabrook, J.E.A. (1980). Laboratory Culture. In: Plant Tissue Culture as a Source of Biochemicals. Staba, E.J. (ed.), CRC Press, Boca Raton, FL, pp. 1–20.

Shields, R., Robinson, S.J. and Anslow, P.A. (1984). Use of fungicides in plant tissue culture. Plant Cell Rep. 3:33–36.

Songstad, D.D., Armstrong, CL. and Petersen, W.L. (1991). $AgNO_3$ increases type II callus production from immature embryos of maize inbred B73 and its derivatives. Plant Cell Rep. 9:699–702.

Verma, D.C., Litvay, LD., Johnson, M.A. and Einspahr, D.W. (1983). Media development for cell suspensions of conifers. In: 5th Intl. Cong. Plant Tissue and Cell Culture. A. Fujiwara (ed.), Maruzen Co., Tokyo, pp. 59–62.

von Arnold, S. and Eriksson, T. (1981). *In vitro* studies of adventitious shoot formation in *Pinus contorta*. Can J. Bot. 50:870–874.

Wake, H., Umetsu, H., Ozeki, Y., Shimomura, K. and Matsunaga, T. (1991). Extracts of marine cyanobactria stimulated somatic embryogenesis of *Daucaus carota* L. Plant Cell Rep. 9:655–658.

Yu, K., Christie, B.R. and Pauls, K.P. (1990). Effects of *Verticillium albo-atrum* culture filtrate on somatic embryogenesis in alfalfa. Plant Cell Rep. 8: 509–511.

General Introduction to Plant Tissue Culture Methods

Coventry, J., Kott, L. and Beversdorf, W.D. (1988). Manual for microspore culture technique for Brassica *napus*. Dept. Crop Science Techn. Bulletin, OAC Publication 0489, University of Guelph, Guelph, Ont., Canada.

de Fossard, R.A. (1981). Tissue Culture for Plant Propagation, University of New England Printery, Armidale, NSW, Australia.

Dixon, R.A. (1985). Plant Cell Culture, a Practical Approach. IRL Press, Oxford, UK.

Dodds, J.H. and Roberts, L.W. (1985). Experiments in Plant Tissue Culture, 2nd ed. Cambridge University Press, Cambridge, UK.

Donnelly, D.J. and Vidaver, W.E. (1988). Glossary of Plant Tissue Culture. Dioscorides Press, Portland, OR, USA.

Gautheret, R.J. (1942). Manuel Technique de Culture des Tissus Végétaux. Masson et Cie, Paris, France.

George, E.F. and Sherrington, P.D. (1984). Plant Propagation by Tissue Culture. Handbook and directory of commercial laboratories. Exegetics Ltd., Eversley, Basingstoke, Hants, UK.

Kyte, L. (1983). Plants from Test Tubes. An introduction to micropropagation. Timber Press, Beaverton, OR, USA.

Lindsay, K. (ed.) (1991). Plant Tissue Culture Manual. Kluwer Academic Publishers, Dordrecht.

Pollard, J.W. and Walker, J.M. (eds.) (1990). Methods in Molecular Biology, Vol. 6. Plant Cell and Tissue Culture. Humana Press, Clifton, NJ, USA.

Seitz, H.U., Seitz, U. and Alfermann, W. (1985). Pflanzliche Gewebekulturen, Ein Praktikum. Gustav Fischer Verlag, Stutgart, New York.

Street, H.E. (ed.) (1973). Plant Tissue and Cell Culture. Blackwell Scientific Publ., Oxford.

Thorpe, T.A. (ed.) (1981). Plant Tissue Culture, Methods and Applications in Agriculture. Academic Press, New York.

Vasil, I.K. (ed.) (1984). Cell Culture and Somatic Cell Genetics of Plants, Vol. 1. Academic Press, New York.

Wetter, L.R. and Constabel, F. (1982). Plant Tissue Culture Methods, 2nd rev. ed., National Research Council of Canada, Ottawa.

2. Morphogenesis and Regeneration

Trevor A. Thorpe

Contents

1. Introduction

All normal living cells within the plant body possess the potential capacity to regenerate the entire organism, i.e. to express their totipotency. This potentiality has been exploited through the culture of protoplasts, cells, tissues and organs *in vitro*. In cultured material it has been possible to study such processes as cytodifferentiation, and organ and somatic embryo formation. Morphogenesis, or the origin of form, can be examined through manipulation approaches, descriptively, physiologically, biochemically or at the molecular level. In addition to such fundamental studies, the capacity to form organs and embryos *de novo* can be exploited to regenerate plantlets for clonal propagation. In this chapter the emphasis will be on the manipulation of tissue cultures, i.e. an examination of the factors leading to the regeneration of plantlets *in vitro* via organogenesis and somatic embryogenesis. In this process, cells and tissues which are mitotically quiescent, or already committed to some function or pathway of development can be (re)directed into organ or embryo formation.

Organogenesis is the process by which cells and tissues are forced to undergo changes which lead to the production of a unipolar structure, namely a shoot or root primordium, whose vascular system is often connected to the

Indra K. Vasil and Trevor A. Thorpe (eds.), Plant Cell and Tissue Culture, 17–36.

parent tissues. In contrast, somatic embryogenesis leads to the production of a bipolar structure containing a root/shoot axis, with a closed independent vascular system. Both of these can occur directly on the explant or indirectly via callus, which is wound parenchyma tissue. Both organogenesis and somatic embryogenesis occur in nature in several species; the tissue culture approach simply dramatically increases the number of species and the plant parts from which de novo organized development can arise (Murashige, 1974).

The earliest reports on controlled organogenesis *in vitro* were by White (1939), who obtained shoots on callus of a tobacco hybrid, and by Nobécourt (1939), who observed root formation in carrot callus. The finding of White was confirmed and extended by Skoog (1944), who showed that auxin could stimulate rooting and inhibit shoot formation, but that the inhibitory effects of auxin could be reduced by increasing the concentration of other medium components, such as inorganic phosphate and adenine. Further studies by Skoog and his associates culminated in the now classical conclusions that organ formation resulted from quantitative interactions between growth regulators, particularly auxin and cytokinin, and other metabolites (Skoog and Miller, 1957). Studies with tobacco callus showed that a high ratio of auxin to cytokinin in the medium favoured root formation, the reverse shoot formation, and that intermediate ratios promoted callus proliferation. Although this approach cannot be demonstrated universally, the concept played a pivotal role in shifting research away from the search for specific morphogens for different types of morphogenesis (although the quest for florigen has not yet been completely abandoned), and placed emphasis on the chemical interaction of substances in the regulation of growth and development (Brown and Thorpe, 1986). As a result of this approach, several hundred plant species have been reported to form shoots and/or roots *in vitro*.

The earliest report on controlled somatic embryogenesis *in vitro* was in 1958 with carrot (Reinert, 1958; Steward et al., 1958). The process occurs naturally in a wide range of species from both reproductive and somatic tissues (Tisserat et al., 1979). Somatic embryogenesis can be induced in callus, cell suspension, and protoplast cultures, or directly from cells of organized structures such as a stem segment or zygotic embryo. As an approach for plantlet regeneration, somatic embryogenesis has several advantages over organogenesis, although not as readily achieved (Vasil and Vasil, 1980; Thorpe, 1988a). The advantages include the efficiency of the process (the formation of plantlets in fewer steps, with a concomitant reduction in labour, time and cost), the potential for the production of much higher number of plantlets, and the morphological and cytological uniformity of the plantlets (Vasil and Vasil, 1988, 1986). Today about 150 species from both angiosperms and gymnosperms have been reported to undergo somatic embryogenesis.

Although plantlet formation *in vitro* was reported as early as the mid 1940s (Ball, 1946), it was only after Morel's work with orchids (1960, 1964) that commercial exploitation began. Today, this use of tissue culture technology is its widest application, with over 300 commercial operations worldwide producing ornamentals (Murashige, 1990). In addition to ornamentals, food crops, vegetable and condiment plants, fruit and nut crops, medicinal plants, and forest trees can all be regenerated *in vitro* (Murashige, 1978; Vasil and Vasil, 1980; Conger, 1981; Debergh and Zimmerman, 1991). There are three ways by which micropropagation can be achieved, namely by enhancing axillary bud breaking, by production of adventitious buds and by somatic embryogenesis (Murashige, 1974, 1978, 1979; Vasil and Vasil, 1980). Even though axillary bud breaking using shoot tips is the major method for micropropagation of ornamentals it will not be discussed here, as the emphasis will be on de novo organized development and regeneration *in vitro* by organogenesis and embryogenesis.

2. Organized Development *in vitro*

Organized development leading to plantlet regeneration via de novo organogenesis is a multi-staged process consisting of at least three distinct stages, namely shoot bud formation, shoot development (and multiplication), and rooting of developed shoots. There are many factors involved in the manipulation of these stages. Plant regeneration via somatic embryogenesis for many species can be divided into two phases, namely the selection or induction of cells with embryogenic competence, and the development of these cells into embryos (Wetherell, 1979; Thorpe, 1982). In general, organized development is successfully achieved through judicious selection of the inoculum, proper choice of medium, including the growth active substances, and the control of the physical environment (Murashige, 1974; Thorpe, 1980, 1982; Evans et al., 1981; Ammirato, 1983).

2.1. *Selection of the Inoculum*

Several factors influence the behaviour of the inoculum in culture (Murashige, 1974). These include (a) the organ that is to serve as tissue source, (b) the physiological and ontogenetic age of the organ, (c) the season in which the explant is obtained, (d) the size of the explant, and (e) the overall quality of the plant from which explants are taken. Genotype is also important, particularly for somatic embryogenesis, e.g. in alfalfa (Brown and Atanassov, 1985) and maize (Hodges et al., 1986). Some of these variables are controlled easily, whereas others are more difficult, thus making the reproducible selection of identical material more problematic. Virtually any plant part can serve as inoculum for organized development (Thorpe, 1982). These include stem segments and apices, root sections, leaf pieces, and petioles, inflorescence sections, flower petals, ovular tissue, seedling parts such as cotyledons, hypocotyls, etc., and seed embryos. Such explants can give rise to organs and embryos directly or indirectly via callus. Another useful explant in some species is thin cell layers, made up of 3–5 layers of epidermal and sub-epidermal cells (Tran Thanh Van, 1980, 1981). Such layers are capable of direct de novo formation of floral buds, vegetative buds, roots and callus. However, for any given species or variety, a particular explant may be necessary for successful regeneration, e.g. embryonic tissue in some cereals (Evans et al., 1981; Vasil and Vasil, 1986). Explants consisting of shoot tips or isolated meristems, which contain mitotically active cells have been very useful for callus initiation and subsequent plantlet regeneration. Although somatic embryogenesis can be induced from a range of tissues, immature and young explants are generally more responsive, with the developmental stage of the explant often being the prime factor (Vasil and Vasil, 1986; Thorpe et al., 1991).

Callus can be maintained more or less indefinitely in semi-solid and/or liquid conditions by subculturing it at regular intervals onto new medium under aseptic conditions. However, in spite of its external morphological appearance, callus is not a uniform tissue. Variation in the size, degree of vacuolation and cytoplasmic content, cell wall characteristics, and shape of cells, from elongated to spherical, are found (Thorpe, 1982). As well, physiological and biochemical differences exist and are often maintained in the cultured tissues. However, the typical callus cell is a highly vacuolated parenchyma cell having a thin peripheral layer of cytoplasm, a nucleus that is usually appressed against the cell wall and one that stains only faintly. Callus cells are therefore not undifferentiated, but rather unorganized (Torrey, 1966). Also tracheary elements are often found in subcultured callus. In conifers a true callus is not induced, rather embryogenic tissue, which consists of highly vacuolated parenchymatous cells, interspersed with single or small groups of densely cytoplasmic cells and filamentous immature embryos. The latter is made up of a small densely plasmatic head and highly vacuolated suspensor cells (Dunstan, 1988). Changes usually occur with continued subculture of callus (Thorpe, 1980; 1982). These include (a) habituation or the loss of a phytohormone requirement for growth, (b) the loss of morphogenic potential, and (c) the spontaneous appearance of tissue with an altered morphology or texture, e.g. friable callus. Generally the above changes reduce the capacity of the tissue to undergo organized development. However, friable callus is needed for induction of cell suspensions. Such tissue can be induced by using higher concentrations of auxin in the medium (Torrey and Reinert, 1961), or by lowering the cytokinin concentration, with or without the addition of gibberellic acid (GA_3) to the medium (Lance et al., 1976).

2.2. *Selection of the Medium*

The major constituents of the medium can be divided into 3 classes (Murashige, 1974). These are (a) inorganic salts, (b) organic substances, and (c) natural complexes. The natural complexes represent a group of constituents, complex in nature and often of unknown and variable composition. These include hydrolysed protein preparations, brewer's by-products, endosperm fluids, fruit pulp and juice, and animal by-products. The most frequently used of these complexes is coconut water (2–15%, v/v). The 5 classes of compounds listed by Earle and Torrey (1965) as being required for successful growth and organogenesis in *Convolvulus* fall into the first 2 classes. These are (a) inorganic macro- and micro-nutrients, (b) carbon and energy source, (c) vitamins, (d) reduced nitrogen, and (e) phytohormones. These five classes of compounds are usually sufficient for most plant species.

2.2.1. *Inorganic Salts*

The earliest widely used inorganic salt formulations were those of White (1943) and Heller (1953). While White's mineral salt medium was used world-wide, Heller's formulation tended to be used in Europe. Since the early 60s more and more researchers have been using Murashige and Skoog's high-salt formulation (1962) or derivatives of this, e.g. Eriksson's (1965), B5 (Gamborg et al., 1968) and Schenk and Hildebrandt's (1972) formulations. The major differences in composition of these high-salt media lie in the amount and form of nitrogen, plus the relative amounts of some of the microelements (Gamborg et al., 1976).

2.2.2. *Organic Substances*

The energy requirement of callus cultures is usually met with sucrose (2–4% w/v), although glucose can sometimes replace sucrose. In addition, carbohydrates have been shown to have an osmotic role in organogenesis (Brown et al., 1979). Vitamin additions to the culture medium may enhance callus growth and differentiation, with thiamine most often added followed by nicotinic acid and pyridoxine. The sugar alcohol, inositol, has often aided the growth and differentiation of numerous tissues in culture.

The addition of organic nitrogen is most often used during callus initiation, but may be beneficial during subculture and organized development. Casein hydrolysate (0.02–0.1%) is a common non-specific organic nitrogen source, while the compounds glutamate, asparagine, tyrosine and adenine are the most frequent specific reduced nitrogen additions. These organic additives can have quantitative as well as qualitative effects on organized development (Dougall, 1981).

Auxins and cytokinins are the two types of phytohormones most often needed in culture. The concentration and ratio of cytokinins and auxins in the medium often control the type and amount of growth which occurs in culture. Both naturally occurring and synthetic compounds are used. The most frequently used auxins are 2,4-dichlorophenoxyacetic acid (2,4-D), indole acetic acid (IAA), naphthalene acetic acid (NAA) and indole butyric acid (IBA). Kinetin and N^6-benzyladenine (BA) are commonly used cytokinins, with isopentenyl adenine (2iP) and zeatin used less frequently.

2.3. *Culture Environment*

There are many aspects of the culture environment that can influence growth and organized development. These include (a) the physical form of the medium, (b) pH, (c) humidity, (d) light, (e) temperature, and (f) the gaseous atmosphere.

2.3.1. *Physical Form of the Medium*

Whether the medium is solidified or liquid can play an important role in growth and differentiation. For gellation, agar and gelrite, both complex polysaccharides, are the most commonly used agents. Callus maintained on medium solidified with agar (0.6–1.0%) grows slowly, with the new cells formed mainly on the periphery of the existing callus mass. In addition, nutrient gradients from the medium into the tissue occur. Cell suspension cultures, on the other hand, tend to grow more rapidly, as the single cells or small cell clumps are constantly exposed to the nutrient medium. Such cell cultures often show true sigmoidal growth curves. In general, most success in organogenesis is achieved with callus or plated cell suspensions on solid medium. On the other hand, somatic embryogenesis can occur in both liquid and solid cultures.

2.3.2. *pH*

The optimum pH for growth and organized development for most species is not critically assessed, as usually the pH of the medium is set between 5.0 and 6.5 during medium preparation. However, pH drifts do occur during culture (Street, 1969; Minocha, 1987). The change in pH tends to be tissue specific and the time at which stability is achieved also varies. Nutrient precipitation, uptake of nutrients and phytohormones are all influenced by medium pH. However, very little is known about the influence of the actual pH value of the medium on organized development *in vitro*. The capacity of cell cultures to utilize NH_4^+ as sole N source was found to depend on keeping the pH of the medium above 5.0 (Martin et al., 1977; Dougall and Verma, 1978). As well, proembryonic cell clusters of carrot could be maintained in auxin-free medium at low pH (~4) but at pH 6–7 development occurred (Smith and Krikorian, 1990). It has also been found that the initial pH affected the long-term viability, growth and differentiation of cotton cells (Shang et al., 1991).

2.3.3. *Humidity*

Virtually no studies on the effects of varying humidity levels on organized development have been made. Part of the reason is that the relative humidity (RH) of the culture environment is usually close to 100%. However, Bouniols (1974) has clearly shown that the RH surrounding the cultures influenced the pattern of organized development. Raising the RH of the atmosphere above agar-solidified medium from 45 to 98% inhibited floral bud formation in endive root segments and favoured vegetative buds, which were preferentially formed in liquid medium on filter paper supports (Margara and Bouniols, 1967; Bouniols and Margara, 1968). Thus the regulation of RH in certain situations may be necessary for a specific form of differentiation. Controlled RH dehydration is used as a tool in somatic embryo maturation and conversion (e.g. Roberts et al., 1990; Compton et al., 1992).

2.3.4. *Light*

This is a major factor of the culture environment, and has been shown clearly to have an effect on organized development *in vitro*. Light requirements for differentiation involve a combination of several components, including intensity, daily light period and quality. The radiant energy requirements are different for tissue cultures than for autotrophic plants, since in the former an adequate supply of carbohydrate, usually sucrose – the sugar of transport – is a normal component of a defined medium, as noted earlier. Nevertheless, light may be required for certain photomorphogenic events (Murashige, 1974). Most researchers do not critically evaluate the light requirements for optimum growth and differentiation. However, there are reports which clearly indicate the importance of this factor (Thorpe, 1980; Hughes, 1981). While maximum callus growth often occurs in darkness, low light intensity (90 nE cm^{-2} sec^{-1}) may enhance organogenesis and embryogenesis. For example, conifer cultures generally require light for morphogenesis, and a light-cytokinin interaction has been observed with bud formation in radiata pine (Villalobos et al., 1984). Furthermore, photoperiod/intensity interactions also occur, leading to the idea that the total radiant energy of specified quality is important. With respect to the spectral quality of light, both blue light and phytochrome effects have been reported (Hughes, 1981).

2.3.5. *Temperature*

Although temperature controls overall plant growth and development *in vivo* and *in vitro*, this factor is not usually thoroughly examined. Generally, cultures are kept at a constant temperature between 20 and 30 °C. However, the optimum temperature for growth and differentiation for a particular species should be determined, as different species have different optima (Hughes, 1981; Chalupa, 1987). For example, bud formation occurred earlier, and a higher percentage of the cultured embryos of *Pinus contorta* responded at 27 °C than at 20 °C (Patel and Thorpe, 1984). Furthermore, thermoperiod and temperature pretreatments, including a chilling temperature, have been shown to affect morphogenesis. Even within the same tissue, different optimum temperatures for shoot formation and rooting have been observed (Rumary and Thorpe, 1984).

2.3.6. *Gaseous Atmosphere*

Another factor affecting the performance of plant culture systems is the gas phase above the cultures (Hughes, 1981). These gases include C_2H_4, O_2, CO_2, ethanol and acetaldehyde. Ethylene is produced by cultured tissues and this gaseous phytohormone has different morphogenic effects, stimulating or inhibiting organized development, depending on the system, and the nature of the culture vessel closures used (Dalton and Street, 1976; Huxter et al., 1979, 1981; Kumar et al. 1987; Thomas and Murashige 1979a, b). Interactions between C_2H_4, CO_2 and O_2 have been observed (e.g., Kumar et al., 1987; Righetti and Facini, 1992). Finally in some urban environments, it may be necessary to filter the air entering the culture room. Such filters, often of charcoal, will also remove dust and spores.

2.4. *Manipulation of Organized Development*

Subcultured callus and explants can be induced to form organs or embryos through manipulation of the above factors. Both embryogenesis and organogenesis may occur in the same culture, e.g., as in pumpkin (Jelaska, 1974). However, this is relatively rare and usually one or the other pattern of reorganization dominates. Similarly, although roots and shoots can be formed in the same tissue, usually one or the other is formed. Both shoots and roots often will be formed in the same tissue piece from bulky initial explants, but callus derived from these explants rapidly loses this capacity and will reproducibly form either roots or shoots under defined conditions. Quite often in longer term cultures, rooting will occur following the development of the shoots. The tendency for organogenesis vs. embryogenesis to occur in culture tends to be species specific. Plant growth regulating substances play a most important role in these processes.

2.4.1. *Organogenesis*

Many growth-active substances, phytohormones as well as other types of compounds, have been included in the culture medium to manipulate organogenesis *in vitro* (Thorpe, 1980; Chandler and Thorpe, 1986). As earlier indicated, a large number of plant species respond to a suitable auxin/cytokinin balance by forming shoots and roots. Evans et al. (1981) found that for 75% of the species forming shoots, either kinetin or BA was used in a concentration range of 0.05–46 μM. Auxins such as IAA and NAA were used in concentrations of 0.06–27 μM. Graminaceous species tend to have a lower requirement for cytokinin for shoot formation than other species. IBA and NAA are the most commonly used auxins for rooting. In some cases, a mixture of two cytokinins or two auxins has proven superior to a single cytokinin or auxin. In a number of cases, however, the permissive phytohormone balance leads to the induction of organogenetic tissue only. This tissue will then develop into organs in a medium with an altered phytohormonal balance.

Either exogenous auxin or cytokinin may be sufficient to bring about organogenesis in some cases. This is generally true for conifers, where a cytokinin alone is sufficient for shoot bud formation (Thorpe et al., 1991). Finally, apparently anomalous situations have also been reported, e.g., by Walker et al. (1978). They found that high levels of auxin (2,4-D) and low levels of cytokinin (kinetin) resulted in shoot formation in alfalfa callus, while the reverse favoured root formation. However, such findings can be reconciled easily, when it is realized that it is the endogenous auxin/cytokinin balance rather than exogenous auxin/cytokinin balance which is important for the initiation of organogenesis (Thorpe, 1980).

A variety of substituted purines, pyrimidines and ureas have been used successfully in place of cytokinins to bring about organogenesis (Thorpe, 1980; Chandler and Thorpe, 1986). Similarly, various auxin-like compounds can satisfy the auxin requirement *in vitro*. Other phytohormones, including gibberellins and abscisic acid added to the medium have been shown to play a role in organogenesis. No generalizations can be made with respect to the effects of these substances in shoot and root formation, as they have been shown to repress, enhance or be without effect in different plant species (Thorpe, 1980).

In addition to phytohormones, other metabolites including adenine, guanine, uracil, uridine, amino acids, various phenolic acids and nicotine, antiauxins, etc. have been shown to stimulate organogenesis in different species (see Thorpe, 1980; Chandler and Thorpe, 1986). The interaction between these different substances is in agreement with the basic ideas of Skoog and Miller (1957). It is possible that many of these metabolites alter the level of endogenous phytohormones. Lastly polyamines (Bagni and Biondi, 1987; Torrigiani et al., 1987; 1989), oligosaccharides (Tran Thanh Van et al.,

1985) and the primary cell wall itself (Fry, 1990) have been implicated in organogenesis.

2.4.2. *Embryogenesis*

In manipulating embryogenesis, it seems clear that two media components, auxin and nitrogen, play crucial roles (Kohlenbach, 1978). The importance of auxin was first recognized by Halperin and Wetherell (1964). Further studies showed that the process of somatic embryogenesis normally takes place in two stages – first, the induction of cells with embryogenic competence (referred to as embryogenic masses or clumps, proembryos, proembryonic tissue, etc.) in the presence of high concentrations of auxin; and second, the development of the embryogenic masses into embryos in the absence of, or in the presence of a lowered concentration of, auxin (e.g. see Nadar et al., 1978; Fujimura and Komamine, 1979; Sung et al., 1979; Vasil and Vasil, 1986). 2,4-D is most commonly used. In a survey of media used to bring about somatic embryogenesis, Evans et al. (1981) found that Murashige-Skoog medium or a modification was used in 70% of the successful cases and that 2,4-D in a concentration of 0.5–27.6 μM was used in 57% of the primary cultures (induction of embryogenic tissue). NAA was used in about 25% of the cases and other auxins much less frequently. Although in carrot and some other species other phytohormones are not required and may even inhibit embryogenesis, Evans et al. (1981) found that in about 50% of the crop species, low levels of cytokinin (0.5–5.0 μM) were required during primary culture. During secondary culture, they found that while 46% of crop species required no growth regulators, some 38% required either IAA or NAA plus cytokinin in lower concentrations than the primary culture. Normal embryo development in caraway required a balance between abscisic acid (ABA), zeatin and GA_3 (Ammirato, 1977); while grapevine required GA_3 and 2iP (Mullins and Srinivasan, 1976). Fujimura and Komamine (1975) found that exogenous GA_3 did not effect the number of globular-stage embryos formed, but decreased the heart and torpedo stages in carrot. Activated charcoal, which acts as an adsorbent, has been shown to stimulate embryogenesis in some species (Ammirato, 1983). Also polyamine involvement in embryogenesis has been reported (Montague et al., 1978; 1979; Feirer et al., 1984).

Halperin (1966) was the first to recognize the importance of reduced nitrogen in the form of NH_4^+ for somatic embryogenesis. Wetherell and Dougall (1976) showed that as little as 0.1 mM ammonium chloride added to nitrate significantly enhanced embryogenesis in carrot, the optimum being KNO_3 (40 mM) and NH_4Cl (30 mM). Glutamine and alanine can serve as sole N-sources. Dougall and Verma (1978) showed that NH_4Cl (10 mM) could serve as the sole source of N provided KCl was present and the pH controlled at 5.4. Kamada and Harada (1979) found that alanine (5–10 mM) was effective in increasing embryo frequency during secondary culture. Furthermore, they found that the reduced nitrogen was needed during embryo development and not during induction.

Osmotic stress can enhance somatic embryogenesis (Litz, 1986; Kamada et al., 1986), which could account for the observed effects of high carbohydrate concentrations in the medium (Lu et al., 1983). A direct correlation between the density of embryogenic cells in suspension and the degree of embryo maturation has been observed, but conditioned medium or the addition of growth regulators and amino acids can overcome the effects of low initial densities (Ammirato, 1983).

3. Determinative Events in Organized Development

As indicated above in all cases of organized development *in vitro*, there is an interplay between the explant and the culture medium and culture conditions. To achieve optimum responses, as is often required in clonal propagation, the interactions of the above factors must be determined empirically. It is clear that although exogenous growth regulators play a pivotal role in the manipulation of organized development, they do so in concert with a variety of other factors. Even so the tendency of a tissue placed in culture to form an organ or an embryo is apparently species dependent. For example, tobacco usually forms organs and rarely embryos, while carrot shows the reverse tendency. An exception appears to be embryonic tissue of conifers which can be made to differentiate either by organogenesis or embryogenesis (Thorpe et al., 1991; see also Chapter 21). In either case, it appears that the process of organized development begins with changes in individual cells, in response to some inductive signal(s), which causes cells which are quiescent or committed to proliferation to undergo the altered path of differentiation. There are at least two other requirements for this altered pathway (Thorpe, 1980). The cells must be able to undergo dedifferentiation (Gautheret, 1966), and to have cellular interaction (Street, 1976). The net result is that the cells undergo precise patterns of cell division leading to a meristemoid or a proembryonic cell cluster and ultimately to a primordium or an embryo, respectively. Examination of these histological events shows many features in common between organogenic and embryogenic tissues (Thorpe, 1982).

The morphogenic signals set up the conditions which presumably allow competent cells to undergo their intrinsically controlled program of differentiation. Such a program results from selective gene action with the subsequent cellular processes of replication, transcription, processing and translation. The net result would be biophysical and biochemical changes in the target cells, including a shift in metabolism. These precede the cytological and histological events associated with the formation of the organ or embryo. The metabolic shift includes the appearance of new enzymes or the alteration (often enhancement) in the activity of metabolic pathways. Further discussion on these aspects is beyond the scope of this chapter, but evidence supporting these ideas can be found elsewhere (e.g. Thorpe, 1980; 1982; 1988a, b; 1993; Brown and Thorpe, 1986; Nomura and Komamine, 1986; Terzi and Loschiavo, 1990; Tran Thanh Van and Trinh, 1990; Komamine et al., 1992).

Although there is evidence to support the idea that organized development de novo begins with changes in a single cell in a cell mass, this does not mean, however, that the entire organ or embryo arises from the derivatives of that single activated cell. Surrounding cells are often induced to divide and become incorporated in the new structure, as has been shown, for example, in *Torenia* (Chlyah, 1974) and radiata pine (Smith and Thorpe, 1975), and through the use of plant chimeras (Norris et al., 1983). In somatic embryogenesis, it appears that single versus multiple-cell proliferation represent varying expressions of the same underlying morphogenic phenomenon (Williams and Maheswaran, 1986). They further suggested that the origin from one to a number of adjacent cells is possible, and was dependent on whether the group of cells can establish and maintain coordinated behaviour as an embryogenic unit. Factors which disrupt intercellular communication contribute to the breakup of the coordination, and lead to further proliferation by single cells or smaller cell groups. Indeed, Komamine and his associates were able to establish a high frequency and synchrönous somatic embryogenic system in carrot by selecting competent single cells (Nomura and Komamine, 1985). These cells were small, round and cytoplasm-rich, and went on to produce embryogenic cell clusters by treatment with auxin, zeatin, mannitol and high O_2; thus confirming the single cell origin of the embryos.

As indicated earlier organized development can occur directly or indirectly on the explant. During indirect organogenesis the primary explant is induced to form a callus prior to reorganization (Hicks, 1980). The tobacco callus system would be an example of this. In contrast direct organogenesis occurs without an intervening callus phase, as in tobacco thin layer explants (Tran Than Van, 1980) or in conifers (Thorpe et al., 1991). Although viewed as distinct pathways of reorganization, they represent a continuum, as is indicated by tissues which differentiate after a small degree of callusing. This can be interpreted as the time taken to establish competence to be able to respond to the inductive signals (Christianson, 1987). This view is expressed as the predetermination theory for somatic embryogenesis (Tisserat et al., 1979). This theory holds that cells undergoing embryo initiation are embryogenic to begin with, and that the *in vitro* culture conditions simply provide the opportunity for embryogenesis to occur. This theory was refined and extended to offer a more unifying view of de novo morphogenesis (Sharp et al., 1980; Evans et al., 1981). It is suggested that explanted tissues contain cells that are already determined for embryogenic development, namely pre-embryogenic determined cells (PEDCs), and those which require redetermination through a period in culture, i.e. induced embryogenic determined cells (IEDCs). There are several examples to support these ideas (Evans et al., 1981). The differences between cells identified as PEDCs and IEDCs can be interpreted in relation to the epigenetic state of the explant (Merkle et al., 1990). Somatic cells which are themselves embryonic or not far removed from the embryonic state are generally more easily induced to undergo somatic embryogenesis than differentiated vegetative cells. In some cases, e.g. in cereals and conifers these embryonic cells are the only ones that can be made to undergo somatic embryogenesis, to date. In contrast highly differentiated cells appear to require major epigenetic changes. The implication of these views is that the program for organized development is an intrinsic one, in which precise external conditions to which the cells are subjected are needed to allow for this differentiation. This idea is supported indirectly by the finding that the transformation of cells by T-DNA coding for auxin or cytokinin allows for organogenesis (Schell et al., 1982), as if the auxin/cytokinin balance was exogenously provided.

A role for physical and/or physiological isolation of cells in a cell or tissue mass from the correlative influences has been suggested in the initiation of organized development (Steward et al., 1958). Although it has been pointed out that there are *in vivo* precedents for this suggestion, the evidence for this requirement *in vitro* is at best equivocal (Williams and Maheswaran, 1986). There is little direct evidence to support the need for physical isolation of cells; although the breakage or wounding of zygotic embryos at the time of explanting led to somatic embryogenesis (Chuang and Chang, 1987; Smith and Krikorian, 1989). Whether the differentiation in these cases occurred as a result of physical rather than physiological isolation remains undetermined. In contrast, some indirect evidence for physiological isolation, namely thickened cell walls and blocked plasmodesmata (Button et al., 1974), severance of plasmodesmata by plasmolysis (Wetherell, 1984), is available. The removal of inter-organ, inter-tissue and inter-cellular correlations has been suggested as a determining factor in the ability of thin cell layers to undergo precise patterns and forms of organogenesis *in vitro* (Than Thanh Van and Trinh, 1978).

As indicated in this chapter, it is clear that we have good information on the factors and requirements for manipulating organized development in a variety of cultured tissues. Nevertheless, some tissues of some species do not respond, when the principles and approaches outlined above are used. Why are some cells of explants or subcultures incapable of regeneration? As a conceptual framework, Halperin (1986) suggests that the answer involves determining the nature of the block to regeneration. He considers that there are three broad possibilities, namely genetic, epigenetic, and physiological. The block may be genetic, involving a lack of totipotency. Mature plant cells *in vivo* and *in vitro* undergo changes at the nuclear level such as polyploidy (including aneuploidy), polyteny, differential DNA synthesis (under replication, gene amplification, etc.), somaclonal variation (arising from point mutations, chromosome structural rearrangements, transpositional events, somatic crossing-over, gene amplification, etc.). Also the block may be genetic in the sense that a high degree of heritability is involved in regeneration of organs *in vitro* (e.g. see Brown and Thorpe, 1986). The implication of this fact is that cultured cells may have a genome functionally identical to that of the fertilized egg, but may lack some genetically determined feature which would allow response to the *in vitro* environment (Halperin, 1986).

The block to regeneration may be epigenetic, involving stable, but potentially reversible constraints in the functioning of genes required for growth and organized development (Halperin, 1986). Such cells are said to lack competence. Habituation and other stable physiological adaptations (e.g. often based on *in vitro* selection for various traits), which are not passed on through meiosis to succeeding generations, fall into this category. An explanation based on stable transcription complexes that repress and activate genes has been proposed for this phenomenon (Brown, 1984). An example of this phenomenon at work is provided by *Hedera helix*, which exhibits distinct juvenile and adult forms. While callus from juvenile tissue formed shoots, callus from adult tissue formed somatic embryos (Banks, 1979). This phenomenon is probably also the basis of our inability to produce organs and embryos in tissues and callus from adult woody plants (see Chapter 21). Lastly, the block to regeneration may be physiological, in the sense that the cells may have no genetic or epigenetic constraints to regeneration, but the requisite environmental signals may be lacking (Halperin, 1986). Such factors as the absence of particular phytohormones in particular ratios and concentrations, or interference with such signals by inhibitors present in the medium or tissue, or improper physical culture conditions, etc., may be at work. As Halperin (1986) points out, this is the basic assumption under which we work, namely, that explanted tissues are capable of regeneration, and that it is only a matter of finding the right combination of chemical and physical stimuli for achieving the goal. Indeed, success in the 1980s in our ability to manipulate tissues from legumes (see Chapter 13), cereals and grasses (see Chapter 12), and woody plants (see Chapter 21), all considered recalcitrant in culture, would attest to this view.

4. Concluding Thoughts

The capacity of cells in culture to undergo organogenesis and somatic embryogenesis has allowed for the widest application of tissue culture technology, namely micropropagation via organogenesis. While this application is mainly with ornamental plants commercially at present, use in agriculture, horticulture and forestry is imminent, as can be gleaned from the various chapters in Part 2 of this volume. In the not-too-distant future, one can expect use of somatic embryogenesis for mass propagation, as bioreactor and other scale-up methods, and artificial seed technology are improved (Levin et al., 1988; Thorpe 1988a; Vasil, 1991). Also embryo cloning or recurrent embryogenesis is poised for exploitation for mass propagation, metabolite production and genetic transformation (Merkle et al., 1990).

As indicated in this chapter, the proper empirical selection of the nutrient medium and the culture environmental conditions allows competent cells to demonstrate their intrinsic capacity for organized development, which is ultimately a reflection of selective gene activity. Both processes of organogenesis and embryogenesis begin with changes in a single cell (or small group of cells) which then divides to produce a globular mass of cells – a meristemoid or an embryogenic cell cluster, and ultimately a primordium (shoot or root) or an embryo, respectively. However, not all cells are competent to respond to the morphogenic signals and a major effort is needed to understand the blocks to the acquisition of this state, as the determinative events are not yet completely understood. The thin cell layer system (Tran Thanh Van, 1980) is proving to be very valuable in understanding cellular and tissue interactions during organogenesis. Similarly, the development of a single cell system for carrot (Nomura and Komamine, 1985) is allowing for an in-depth examination of the entire process of somatic embryogenesis (Nomura and Komamine, 1986; Komamine et al., 1992). Coupled with molecular approaches, a dramatic improvement in our understanding of de novo organized development, the process that best exemplifies the totipotency of the plant cell, can be expected.

5. References

Ammirato, P.V. (1977). Hormonal control of somatic embryo development from cultured caraway. Plant Physiol. 59:579–586.

Ammirato, P.V. (1983). Embryogenesis. In: Handbook of Plant Cell Culture, Vol. 1. D.A. Evans, W.R. Sharp, P.V. Ammirato, and Y. Yamada (eds.), Macmillan, New York, pp. 82–123.

Ball, E. (1946). Development in sterile culture of stem tip and adjacent regions of *Tropaeolum majus* L. and of *Lupinus majus* L. Am. J. Bot. 33:301–318.

Bagni, N. and Biondi, S. (1987). Polyamines. In: Cell and Tissue Culture in Forestry, Vol. 1. J.M. Bonga and D.J. Durzan (eds.), Martinus Nijhoff Publishers, Dordrecht, pp. 113–124.

Banks, M.S. (1979). Plant regeneration from callus from two growth phases of English ivy, *Hedera helix* L. Z. Pflanzenphysiol. 92:349–353.

Bouniols, A. (1974). Néoformation de bourgeons floraux *in vitro* à partir de fragments de racine d'endive *Cichorium intybus* L.: Influence du degré d'hydration des tissus, et ses consequences sur la composition en acides aminés. Plant Sci. Lett. 2:363–371.

Bouniols, A. and Margara, J. (1968). Recherches expérimentales sur la néoformation de bourgeons inflorescentiels ou végétatifs *in vitro* à partir d'explantats d'Endive (*Cichorium intybus* L). Ann. Physiol. Veg. 10:69–81.

Brown, D. (1984). The role of stable complexes that repress and activate eucaryotic genes. Cell 37:359–365.

Brown, D.C.W. and Atanassov, A.I. (1985). Role of genetic background in somatic embryogenesis in *Medicago*. Plant Cell Tissue Org. Cult. 4:111–122.

Brown, D.C.W. and Thorpe, T.A. (1986). Plant regeneration by organogenesis. In: Cell Culture and Somatic Cell Genetics in Plants, Vol. 3. I.K. Vasil (ed.), Academic Press, New York, pp. 49–66.

Brown, D.C.W., Leung, D.W.M. and Thorpe, T.A. (1979). Osmotic requirement for shoot formation in tobacco callus. Physiol. Plant. 46:36–41.

Button, J., Kochba, J. and Bornman, C.H. (1974). Fine structure of and embryoid development from embryogenic ovular callus of 'Shamouti' orange (*Citrus sinensis* Osb.). J. Exp. Bot. 25:446–457.

Chalupa, V. (1987). Temperature. In: Cell and Tissue Culture in Forestry, Vol. 1. J.M. Bonga and D.J. Durzan (eds.), Martinus Nijhoff Publishers, Dordrecht, pp. 142–151.

Chandler, S.F. and Thorpe, T.A. (1986). Hormonal regulation of organogenesis *in vitro*. In: Hormonal Regulation of Plant Growth and Development, Vol. 3. S.S. Purohit (ed.), Agro Botanical Publishers, India, pp. 1–27.

Chlyah, H. (1974). Formation and propagation of cell division centres in the epidermal layers of internodal segments of *Torenia fournieri* grown *in vitro*. Can. J. Bot. 52:867–872.

Christianson, M.L. (1987). Causal events in morphogenesis. In: Plant Tissue and Cell Culture. C.E. Green, D.A. Somers, W.P. Hackett and D.O. Biesboer (eds.), Alan Liss, New York; pp. 45–55.

Chuang, M.-J. and Chang, W.-C. (1987). Somatic embryogenesis and plant regeneration in callus culture derived from immature seeds and mature zygotic embryos of *Dysosma pleiantha* (Hance) Woodson. Plant Cell Rep. 6:484–485.

Compton, M.E., Benton, C.M., Gray, D.J. and Songstad, D.D. (1992). Plant recovery from maize somatic embryos subjected to controlled relative humidity dehydration. *In vitro* Cell. Dev. Biol. 28P:197–201.

Conger, B.V. (ed.) (1981). Principles and Practices of Cloning Agricultural Plants via *in vitro* Techniques. CRC Press, Boca Raton, Florida, 273 pp.

Dalton, C.C. and Street, H.E. (1976). The role of the gas phase in the greening and growth of illuminated cell suspension cultures of spinach (*Spinacea oleracea* L). *In vitro* 12:485–494.

Debergh, P.L. and Zimmerman, R.H. (eds.) (1991). Micropropagation. Kluwer Academic Publishers, Dordrecht, 484 pp.

Dougall, D.K (1981). Media factors affecting growth. Env. Exptal. Bot. 21:277–280.

Dougall, D.K. and Verma, D.C. (1978). Growth and embryo formation in wild-carrot suspension cultures with ammonium ion as a sole nitrogen source. *In vitro* 14:180–182.

Dunstan, D.I. (1988). Prospects and progress in conifer biotechnology. Can. J. For. Res. 18:1497–1506.

Earle, E.D. and Torrey, J.G. (1965). Morphogenesis in cell colonies grown from *Convolvulus* cell suspensions plated on synthetic media. Am. J. Bot. 52: 891–899.

Eriksson, T. (1965). Studies on growth requirements and growth measurements of cell cultures of *Haplopappus gracilis*. Physiol. Plant. 18:976–993.

Evans, D.A., Sharp, W.R. and Flick, C.E. (1981). Growth and behavior of cell cultures. In: Plant Tissue Culture – Methods and Applications in Agriculture. T.A. Thorpe (ed.), Academic Press, New York, pp. 45–113.

Feier, R.P., Mignon, G. and Litvay, J.D. (1984). Arginine decarboxylase and polyamines required for embryogenesis in wild carrot. Science 223:1433–1435.

Fry, S.C. (1990). Roles of the primary cell wall in morphogenesis. In: Progress in Plant Cellular and Molecular Biology. H.J.J. Nijkamp, L.H.W. Van Der Plas and J. Van Aartrijk (eds.), Kluwer Academic Publishers, Dordrecht, pp. 504–5123.

Fujimura, T. and Komamine, A. (1975). Effects of various growth regulators on the embryogenesis in a carrot cell suspension culture. Plant Sci. Lett. 5:359–364.

Fujimura, T. and Komamine, A. (1979). Involvement of endogenous auxin in somatic embryogenesis in a carrot cell suspension culture. Z. Pflanzenphysiol. 95:13–19.

Gamborg, O.L., Miller, R.A. and Ojima, K 1968. Nutrient requirements of suspension cultures of soybean root cells. Exp. Cell Res. 50. 151–158.

Gamborg, O.L., Murashige, T., Thorpe, T.A. and Vasil, I.K. (1976). Plant tissue culture media. *In vitro* 12:473–478.

Gautheret, R.J. (1966). Factors affecting differentiation of plant tissues grown *in vitro*. In: Cell Differentiation and Morphogenesis. W. Beerman (ed.), North Holland Publishig Co., Amsterdam, pp. 55–71.

Halperin, W. (1966). Alternative morphogenetic events in cell suspensions. Am. J. Bot. 53:443–453.

Halperin, W. (1986). Attainment and retention of morphogenetic capacity *in vitro*. In: Cell Culture and Somatic Cell Genetics of Plants, Vol. 3. I.K. Vasil (ed.), Academic Press, New York, pp. 3–47.

Halperin, W. and Wetherell, D.F. (1964). Adventive embryony in tissue cultures of the wild carrot, *Daucus carota*. Am. J. Bot. 51:274–283.

Heller, R. (1953). Recherches sur la nutrition minérale des tissus végétaux cultivés *in vitro*. Ann. Sci. Natl. Biol. Veg. 14:1–223.

Hicks, G.S. (1980). Patterns of organ development in plant tissue culture and the problem of organ determination. Bot. Rev. 46:1–23.

Hodges, T.K., Kamo, K.K., Imbrie, C.W. and Becwar, M.R. (1986). Genotype specificity of somatic embryogenesis and regeneration in maize. Bio/Tech. 4:218–223.

Hughes, KW. (1981). *In vitro* ecology: exogenous factors affecting growth and morphogenesis in plant tissue cultures. Env. Exptal. Bot. 21:281–288.

Huxter, T.J., Reid, D.M. and Thorpe, T.A. (1979). Ethylene production by tobacco (*Nicotiana tabacum*) callus. Physiol. Plant. 46:374–380.

Huxter, T.J., Thorpe, T.A. and Reid, D.M. (1981). Shoot initiation in light and dark-grown tobacco callus: The role of ethylene. Physiol. Plant. 53:319–326.

Jelaska, S. (1974). Embryogenesis and organogenesis in pumpkin explants. Physiol. Plant. 31:257–261.

Kamada, H. and Harada, H. (1979). Studies on organogenesis in carrot tissue culture, II: Effects of amino acids and inorganic nitrogenous compounds on somatic embryogenesis. Z. Pflanzenphysiol. 91:453–463.

Kamada, H., Saga, H., and Harada, H. (1986). The induction of somatic embryogenesis by osmotic stress in *Daucus carota*. In D.A. Somers, B.G. Gengenbach, D.D. Biesboer, W.P. Hackett, C.E. Green (eds.), Abstr., VI. Int. Congr. Plant Tissue & Cell Culture. University of Minnesota, Minneapolis, p. 446.

Kohlenbach, H.W. (1978). Comparative somatic embryogenesis. In: Frontiers of Plant Tissue Culture 1978. T.A. Thorpe (ed.), University of Calgary Printing Services, Calgary, pp. 59–66.

Komamine, A., Kawahara, R., Matsumoto, M., Sunabori, S., Toya, T., Fujiwara, A., Tsukahara, M., Smith, J., Ito, M., Fukuda, H., Nomura, K. and Fujimura, T. (1992). Mechanisms of somatic embryogensis in cell cultures: Physiology, biochemistry, and molecular biology. *In vitro* Cell. Dev. Biol. 28 P:11–14.

Kumar, P.P., Reid, D.M. and Thorpe, T.A. (1987). The role of ethylene and carbon dioxide in differentiation of shoot buds in excised cotyledons of *Pinus radiata in vitro*. Physiol. Plant. 69:244–252.

Lance, B., Reid, D.M. and Thorpe, T.A. (1976). Endogenous gibberelins and growth of tobacco callus cultures. Physiol. Plant. 36:287–292.

Levin, R., Gaba, V., Tal., B., Hirsch, S., DeNola, D. and Vasil, I.K. (1988). Automated plant tissue culture for mass propagation. Bio/Tech. 6:1035–1040.

Litz, R.E. (1986). Effect of osmotic stress on somatic embryogenesis in *Carica papaya* suspension cultures. J. Am. Soc. Hort. Sci. 111:969–972.

Lu, C.-Y., Vasil, V. and Vasil, I.K. (1983). Improved efficiency of somatic embryogenesis and plant regeneration in tissue cultures of maize (*Zea mays* L.). Theor. Appl. Genet. 66:285–289.

Margara, J. and Bouniois, A. (1967). Comparison *in vitro* de l'influence due milieu liquide ou gélosé, sur l'initiation florale chez *Cichoium intybus* L. C.R. Acad. Sci., Sér. D. 264:1166–1168.

Martin, S.M., Rose, D. and Hui, V. (1977). Growth of plant cell suspension cultures with ammonium as the sole source of nitrogen. Can. J. Bot. 55:2838–2843.

Merkle, S.A., Parrott, W.A. and Williams, E.G. (1990). Applications of somatic embryogenesis and embryo cloning. In: Plant Tissue Culture: Applications and Limitations. S.S. Bhojwani (ed.), Elsevier, Amsterdam, pp. 67–101.

Minocha, S.C. (1987). PH of the medium and the growth and metabolism of cells in culture. In: Cell and Tissue Culture in Forestry, Vol. 1. J.M. Bonga and D.J. Durzan (eds.), Martinus Nijhoff Publishers, Dordrecht, pp. 125–141.

Montague, M.J. Koppenbrink, J.W. and Jaworski, E.G. (1978). Polyamine metabolism in embryogenic cells of *Daucus carota*, I: Changes in intracellular content and rates of synthesis. Plant Physiol. 62:430–433.

Montague, M.J., Armstrong, T.A. and Jaworski, E.G. (1979). Polyamine metabolism in embryogenic cells of *Daucus carota*, II: Changes in arginine decarboxylase activity. Plant Physiol. 63:341–345.

Morel, G.M. (1960). Producing virus-free Cymbidium. Am. Orchid Soc. Bull 29:495–497.

Morel, G.M. (1964). Tissue culture – A new means of clonal propagation of orchids. Am. Orchid Soc. Bull. 33:473–478.

Mullins, M.G. and Srinivasan, C. (1976). Somatic embryos and plantlets from an ancient clone of the grape vine (cv. Cabernet-Sauvignon) by apomixis *in vitro*. J. Exp. Bot. 27:1022–1030.

Murashige, T. (1974). Plant propagation through tissue culture. Annu. Rev. Plant Physiol. 25:135–166.

Murashige, T. (1978). The impact of plant tissue culture on agriculture. In: Frontiers of Plant Tissue Culture 1978. T.A. Thorpe (ed.), University of Calgary, Printing Services, Calgary, pp. 15–26, 518–524.

Murashige, T. (1979). Principles of rapid propagation. In: Propagation of Higher Plants Through Tissue Culture: A Bridge Between Research and Application. K.W. Hughes, R. Henke, and M. Constantin (eds.), Conf. 780411. US Tech. Inf. Center, DOE, Springfield, VA, pp. 14–24.

Murashige, T. (1990). Plant propagation by tissue culture: Practice with unrealized potential. In: Handbook of Plant Cell Culture, Vol. 5. P.V. Ammirato, D.A. Evans, W.R. Sharp and Y.P.S. Bajaj (eds.), McGraw-Hill, New York, pp. 3–9.

Murashige, T. and Skoog, F. (1962). A revised medium for rapid growth and bioassays with tobacco tissue cultures. Physiol. Plant. 15:473–497.

Nadar, H.M., Soepraptoto, S., Heinz, D.J. and Ladd, S.L. (1978). Fine structure of sugarcane (*Saccharum* sp.) callus and the role of auxin in embryogenesis. Crop Sci. 18:210–216.

Nobécourt, P. (1939). Sur les radicelles naissant des cultures de tissus du tubercule de carotte. C.R. Seances Soc. Biol. Ses Fil. 130:1271–1272.

Nomura, K and Komamine, A. (1985). Identification and isolation of single cells that produce somatic embryos at a high frequency in a carrot suspension culture. Plant Physiol. 79:988–991.

Nomura, K., Komamine, A. (1986). Somatic embryogenesis in carrot cells. Dev. Growth Differ. 28:511–517.

Norris, R., Smith, R.H. and Vaughn, KC. (1983). Plant chimeras used to establish *de novo* origin of shoots. Science 220:75–76.

Patel, K.R. and Thorpe, T.A. (1984). *In vitro* differentiation of plantlets from embryonic explants of lodgepole pine (*Pinus contorta* Dougl. ex Loud). Plant Cell Tissue Org. Cult. 3:31–142.

Reinert, J. (1958). Morphogenese und ihre Kontrolle an Gewebekulturen aus Karotten. Naturwissenschaffen 45:344–345.

Righetti, B. and Facini, O. (1992). Headspace gas composition in four *Prunus avium* cultivars with differing photosynthetic capabilities. *In vitro* Cell. Dev. Biol. 28P:179182.

Roberts, D.R., Sutton, B.C.S. and Flinn, B.S. (1990). Synchronous and high frequency germination of interior spruce somatic embryos following partial drying at high relative humidity. Can. J. Bot. 68-:1086–1090.

Rumary, C. and Thorpe, T.A. (1984). Plantlet formation in black and white spruce, I: *In vitro* techniques. Can. J. For. Res. 14:10–16.

Schell, J., Van Montague, M., Holsters, M., Hernalsteens, J.P., Dhaese, P., De Greve, H.,

Leemans, J. Joos, H., Inzel, D., Willmitzer, L., Otten, L., Wostemeyer, A. and Schroeder, J. (1982). Plant cells transformed by modified Ti plasmids: A model system to study plant development. In: Biochemistry of Differentiation and Morphogenesis. L. Jaenicke (ed.), Springer-Verlag, Berlin, pp. 65–73.

Schenk, R.U. and Hildebrandt, A.C. (1972). Medium and techniques for induction and growth of monocotyledonous and dicotyledonous plant cell cultures. Can. J. Bot. 50.:199–204.

Shang, X.M., Huang, J.Y., Haigler, C.H. and Trolinder, N.L. (1991). Buffer capacity of cotton cells and effects of extracellular pH on growth and somatic embryogenesis in cotton cell suspensions. *In vitro* Cell. Dev. Biol. 27P:147–152.

Sharp, W.R., Sondhal, M.R., Caldas, L.S. and Maraffa, S.B. (1980). The physiology of *in vitro* asexual embryogenesis. Hortic. Rev. 2:268–310.

Skoog, F. (1944). Growth and organ formation in tobacco tissue cultures. Am. J. Bot. 31:19–24.

Skoog, F. and Miller, C.O. (1957). Chemical regulation of growth and organ formation in plant tissues cultured *in vitro*. Symp. Soc. Exp. Biol. 11:118–131.

Smith, D.L. and Krikorian, A.D. (1989). Release of somatic embryogenic potential from excised zygotic embryos of carrot and maintenance of proembryonic cultures in hormone-free medium. Am. J. Bot. 76:1832–1843.

Smith, D.L. and Krikorian, A.D. (1990). Somatic proembryo production from excised, wounded zygotic carrot embryos on hormone-free medium: Evaluation of the effects of pH, ethylene and activated charcoal. Plant Cell. Rep. 9:34–37.

Smith, D.R. and Thorpe, T.A. (1975). Root initiation in cuttings of *Pinus radiata* seedlings, I: Developmental sequence. J. Exp. Bot. 26:184–192.

Steward, F.C., Mapes, M.O. and Mears, K. (1958). Growth and organized development of cultured cells, II: Organization in cultures growth from freely suspended cells. Am. J. Bot. 45:705–708.

Street, H.E. (1969). Growth in organized and unorganized systems – knowledge gained by culture of organs and tissue explants. In: Plant Physiology, Vol. 5B. F.C. Steward (ed.), Academic Press, New York, pp. 3–224.

Street, H.E. (1976). Cell cultures: a tool in plant biology. In: Cell Genetics in Higher Plants. D. Dudits, G.L. Farkas and P. Maliga (eds.), Akademiai Kiado, Budapest, pp. 7–38.

Sung, Z.R., Smith, R., and Horowitz, J. (1979). Quantitative studies of embryogenesis in normal and 5-methyltryptophan-resistant cell lines of wild carrot-effects of growth regulators. Planta 147:236–240.

Terzi, M. and Loschiaivo, F. (1990). Somatic embryogenesis In: Plant Tissue Culture: Applications and Limitations. S.S. Bhojwani (ed.), Elsevier, Amsterdam, pp. 54–66.

Thomas, D.C. and Murashige, T. (1979a). Volatile emissions of plant tissue cultures, I: Identification of the major components. *In vitro* 15:651–658.

Thomas, D.S. and Murashige, T. (1979b). Volatile emissions of plant tissue cultures, II: Effects of the auxin 2,4-D on production of volatiles in callus cultures. *In vitro* 15:659–663.

Thorpe, T.A. (1980). Organogenesis *in vitro*: Structural, physiological and biochemical aspects. Int. Rev. Cytol. Suppl. 11A:71–111.

Thorpe, T.A. (1982). Callus organization and *de novo* formation of shoots, roots and embryos *in vitro*. In: Techniques and Applications of Plant Cell and Tissue Culture to Agriculture and Industry. D.T. Tomes, B.E. Ellis, P.M. Harney, K.J. Kasha, and R.L. Peterson (eds.), University of Guelph, Ontario, pp. 115–138.

Thorpe, T.A. (1988a). *In vitro* somatic embryogenesis. ISI Atlas of Science – Animal and Plant Sciences 1:81–88.

Thorpe, T.A. (1988b). Physiology of bud induction in conifers *in vitro*. In: Genetic Manipulation of Woody Plants. J.W. Hanover and D.E. Keathley (eds.), Plenum Press, New York, pp. 167–184.

Thorpe, T.A. (1993). In vitro organogenesis and somatic embryogenesis: Physiological and biochemical aspects. In: Markers of Plant Morphogenesis. K.A. Roubelakis-Angelakis and K. Tran Thanh Van (eds.), Plenum Press, New York, in press.

Thorpe, T.A., Harry, I.S. and Kumar, P.P. (1991). Application of micropropagation to forestry. In: Micropropagation. P.C. Debergh and R.H. Zimmerman (eds.), Kluwer Academic Publishers, Dordrecht, pp. 311–336.

Tisserat, B., Esan, E.B. and Murashige, T. (1979). Somatic embryogenesis in angiosperms. Hortic. Rev. 1:1–78.

Torrey, J.G. (1966). The initiation of organized development in plants. Adv. Morphogen. 5:39–91.

Torrey, J.G. and Reinert, J. (1961). Suspension cultures of higher plant cells in synthetic medium. Plant Physiol. 36:483–491.

Torrigiani, P., Altamura, M.M., Capitani, F., Serafini-Fracassini, D. and Bagni, N. (1989). *De novo* root formation in thin layers of tobacco: Changes in free and bound polyamines. Physiol. Plant. 77:294–301.

Torrigiani, P., Altamura, M.M., Pasqua, G., Monacelli, B., Serafini-Fracassini, D. and Bagni, N. (1987). Free and conjugated polyamines during *de novo* floral and vegetative bud formation in thin cell layers of tobacco. Physiol. Plant. 70:453–460.

Tran Thanh Van, K. (1980). Control of morphogenesis by inherent and exogenously applied factors in thin cell layers. Int. Rev. Cytol. Suppl. 11A:175–194.

Tran Thanh Van, K. (1981). Control of morphogenesis in *in vitro* cultures. Annu. Rev. Plant Physiol. 32:291–311.

Tran Thanh Van, K and Trinh, H. (1978). Morphogenesis in thin cell layers: Concept, methodology and results. In: Frontiers of Plant Tissue Culture 1978. T.A. Thorpe (ed.), University of Calgary, Printing Services, Calgary, pp. 37–48.

Tran Thanh Van, K. and Trinh, T.H. (1990). Organogenic differentiation. In: Plant Tissue Culture: Applications and Limitations. S.S. Bhojwanil (ed.), Elsevier, Amsterdam, pp. 34–53.

Tran Thanh Van, K., Toubart, P., Cousson, A., Darvill, A.J., Gollin, D.G., Chelf, P. and Albersheim, P. (1985). Manipulation of morphogenetic pathways of tobacco explants by oligosaccharins. Nature 314:615–617.

Vasil, I.K. (ed.) (1991). Cell Culture and Somatic Cell Genetics of Plants, Vol. 8. Scale-up and Automation in Plant Propagation. Academic Press, San Diego.

Vasil, I.K and Vasil, V. (1980). Clonal propagation. Int. Rev. Cytol. Suppl. 11A:145–173.

Vasil, I.K. and Vasil, V. (1986). Regeneration in cereal and other grass species. In: Cell Culture and Somatic Cell Genetics of Plants, Vol. 3. I.K. Vasil (ed.), Academic Press, New York, pp. 121–150.

Villalobos, V.M., Leung, D.W.M. and Thorpe, T.A. (1984). Light-cytokinin interaction in shoot formation in cultured cotyledon explants of radiata pine. Physiol. Plant. 61:497–504.

Walker, K.A., Yu, P.C., Sato, S.J. and Jaworski, E.G. (1978). The hormonal control of organ formation in callus of *Medicago sativa* L. cultured *in vitro*. Am. J. Bot. 65:654–659.

Wetherell, D.F. (1979). *In vitro* embryoid formation in cells derived from somatic plant tissues. In: Propagation of Higher Plants Through Tissue Culture – A Bridge Between Research and Application. K.W. Hughes, R. Henke and M. Constantin (eds.), Conf. – 780411 U.S. Tech. Inf. Center, DOE, Springfield, VA, pp. 102–124.

Wetherell, D.F. (1984). Enhanced adventive embryogenesis resulting from plasmolysis of cultured wild carrot cells. Plant Cell Tissue Org. Cult. 3:221–227.

Wetherell, D.F. and Dougall, D.K. (1976). Sources of nitrogen supporting growth and embryogenesis in cultured wild carrot tissue. Physiol. Plant. 37:37–103.

White, P.R. (1939). Controlled differentiation in a plant tissue culture. Bull. Torrey Bot. Club 66:507–513.

White, P.R. (1943). Nutrient deficiency studies and an improved inorganic nutrient for cultivation of excised tomato roots. Growth 7:53–65.

Williams, E.G. and Maheswaran, G. (1986). Somatic embryogenesis: Factors influencing coordinated behaviour of cells as an embryogenic group. Ann. Bot. 57:443–462.

3. Meristem and Shoot Tip Culture: Requirements and Applications*

Narender S. Nehra and Kutty K. Kartha

Contents

Introduction

Among the prevalent tissue culture techniques of agricultural and horticultural promise, meristem and shoot tip culture has been exploited at a much wider scale primarily due to its application in diverse areas such as rapid clonal multiplication of vegetatively propagated crop plants, virus elimination and germplasm preservation of both vegetatively and seed propagated crops (Kartha, 1981; 1986; George and Sherrington, 1984; Hussey, 1986), and more recently in the development of genetic transformation protocols for crop improvement through gene transfer (Ulian, 1988; Gould et al., 1991).

The shoot apical meristem, characterized as a dome of totipotent cells situated at the tip of the shoot, is the centre of activity for various developmental programs in the life of a higher plant. During its post-embryonic development, the apical meristem passes through distinct vegetative and reproductive phases. In the vegetative phase, just below the apical meristem there are several ridges of progressively increasing size which represent the newly differentiated leaf primordia. The apical meristems in their juvenile vegetative phase are considered to be the most appropriate explants to

* NRCC Publication No. 32483.

Indra K. Vasil and Trevor A. Thorpe (eds.), Plant Cell and Tissue Culture, 37–70.

initiate meristem culture. Most often confusion arises when we encounter the use of terms meristem culture and meristem tip culture in literature to describe the explants used for virus elimination. In reality, the meristem culture refers to the culture of true meristematic dome, whereas the meristem tip culture comprises the culture of meristematic dome along with two or three rudimentary leaf primordia. However, in practice most often meristem tips are used as explants to obtain virus free plants (George and Sherrington, 1984). Shoot tip explants are commonly used for rapid *in vitro* propagation and are distinguished from meristem tips primarily by the size of the explants. Meristem tips are relatively smaller in size (0.1 to 0.5 mm) than shoot tips (0.5 to 5.0 mm). However, these size dimensions are subjective and exact size of the explant used in culture depends on the plant species and the purpose of the investigation. Since there is a considerable overlap in the use of meristem and shoot tip explants for various applications of the technique, the literature pertaining to both type of explants will be discussed in this chapter.

The history of meristem culture essentially began with the first successful culture of meristem tips of Nasturtium (*Tropaeolium majus*) and formation of rooted plants (Ball, 1946). Since then meristem tip culture has attracted much attention from plant pathologists for its potential in production of virus-free plants. In this context, the observation of Limmaset and Cornuet (1949) that viruses are not uniformly distributed in a plant and are often absent or not detectable in the apical meristem was significant. This observation prompted Morel and Martin (1952) to experiment with Dahlias and postulate the hypothesis that it might be possible to obtain virus-free plants by culturing isolated apical meristems. Subsequently, the hypothesis was confirmed by producing virus-free plants of six potato cultivars using meristem tip culture (Morel and Martin, 1955). With several modifications and improvements, the technique has now become a common practice for virus elimination in a large number of commercial crop species (Table 3). Another chapter in the history of meristem tip culture, which revolutionized the use of this technique as a tool for efficient and rapid clonal multiplication, was once again added by Morel (1960) when he succeeded in inducing multiple protocorm formation in *Cymbidium* orchid. The discovery of cytokinin and development of improved tissue culture media (Murashige and Skoog, 1962) provided further impetus to meristem tip culture technique to such an extent that it is now viewed as a commercially viable technology for mass propagation of numerous crop plants (Debergh and Zimmerman, 1991). The advances in the field of cryobiology have extended the use of meristem tip culture in germplasm storage (Kartha, 1985a). With recent developments in the field of molecular biology and gene manipulations, the meristem tip culture has also been adopted as a tool for gene transfer in higher plants (Ulian, 1988; Gould et al., 1991).

In brief, the meristem tip culture technique requires the aseptic excision of the meristematic region of a shoot, its establishment under sterile con-

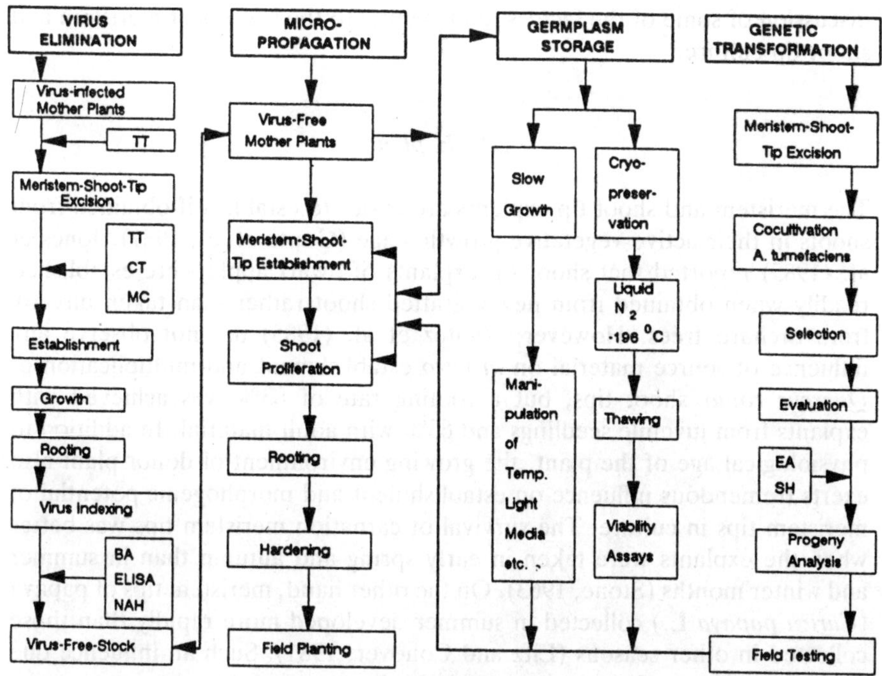

Figure 1. A schematic representation of methodology used for various applications of meristem and shoot tip culture. Abbreviations: TT = Thermotherapy; CT = Chemotherapy; MC = Meristem culture; BA = Bioassay; NAH = Nucleic Acid Hybridization; EA = Enzyme Assay; SH = Southern Hybridization.

ditions on growth medium and subsequent manipulations of growth hormones and culture conditions for obtaining complete plantlets. The details of the general technique are available in previous review articles and book chapters (Kartha, 1981, 1984a; George and Sherrington, 1984). Therefore, in this review only the relevant factors influencing meristem and shoot tip culture are discussed followed by an appraisal of potential applications of the technique with emphasis on virus elimination and mass propagation. The other applications are summarized only briefly as they have been extensively covered elsewhere in this volume.

2. Factors Affecting Meristem and Shoot Tip Culture

The *in vitro* culture of meristem and shoot tip passes through several phases such as initiation of culture and establishment of explant, growth and differentiation, proliferation of shoots, and formation of roots (Fig. 1). Each phase is influenced by several factors. The following section provides a general discussion of some of the factors fundamental to the success of meristem and shoot tip culture.

2.1. *Explant*

The meristem and shoot tip explants are easier to establish if obtained from shoots in their active vegetative growth stage (Gupta et al., 1981). Jones et al. (1985) reported that shoot tip explants of dwarf apple were established readily when obtained from newly grafted shoot rather than taken directly from orchard trees. However, Vieitez et al. (1985) did not observe any influence of source material on *in vitro* establishment and multiplication of *Quercus robur* shoot tips, but a rooting rate of 83% was achieved with explants from juvenile seedlings and 63% with adult material. In addition to physiological age of the plant, the growing environment of donor plant also exerts tremendous influence on establishment and morphogenic potential of meristem tips in culture. The survival of carnation meristem tips was better when the explants were taken in early spring and autumn than in summer and winter months (Stone, 1963). On the other hand, meristem tips of papaya (*Carica papaya* L.) collected in summer developed more rapidly than those collected in other seasons (Litz and Conover, 1981). Such an influence due to seasonal fluctuations can be minimized to some extent by growing the donor plants, wherever possible, under controlled environmental conditions. However, even in the situation of controlled environment one has to establish a set of conditions optimal for donor plant that would give desired response from meristem tip explants in culture. For example, in *Brassica oleracea*, the donor plants grown at 16 °C were successfully established in culture, whereas those taken from plants grown at 21 °C died in culture (Anderson and Carstens, 1977). In crop plants showing strong apical dominance, the meristems excised from terminal position exhibit better *in vitro* response than those taken from lateral position (Hollings and Stone, 1968; Hasegawa, 1979). However, in some other crop species, the explants prepared from lateral buds have been found to be desirable (Dore Swamy et al., 1982, Gupta et al., 1981). These studies indicate that physiological state of donor plant, growing season, and even the position of the explant on donor plant influence the *in vitro* response of meristem and shoot tip culture. Such responses are generally believed to be mediated by differences in endogenous levels of growth hormones which seems to be a plausible explanation, but remains purely speculative in the absence of documented evidence in literature in support of such an explanation.

The importance of size of explant has been emphasized in several review articles on meristem and shoot tip culture (Kartha, 1981; Styer and Chin, 1983). The general consensus appears to be that larger explants give better survival and have higher regeneration capacity than smaller explants. However, the smaller explants have the advantage of greater chances of virus elimination. Hence, the size of the explant has to be empirically determined for each species keeping in view the objective of the study. If the aim is mass propagation, comparatively larger explants would be beneficial for culture initiation. On the other hand, the culture of smallest possible explants consisting of only the meristematic dome or with 2 to 3 leaf primordia would be ideal if the objective is virus elimination. However, there are exceptions to these generalizations where larger explants have been effectively used for virus elimination (Vine and Jones, 1969) and smaller explants successfully established and proliferated for mass propagation (Hunter et al., 1984; Pink and Walkey, 1984) by manipulation of media and culture conditions.

2.2. *Culture Media*

In early attempts on meristem culture, White's medium (1943), originally developed for root culture, was extensively used (see review Hu and Wang, 1983). Stace-Smith and Mellor (1968) tested four different media formulations for potato meristem culture and found that MS (Murashige and Skoog, 1962) out-performed other tissue culture media in terms of meristem survival and development of shoots. Since then MS medium has been invariably used in meristem and shoot tip culture studies (Table 1; Kartha, 1981). However, there are reports where other media formulations have been found to be suitable for meristem and shoot tip culture of some species. Hunter et al. (1984) compared four different basal media for strawberry meristem culture and found that Adams' (1972) medium was superior to that of Boxus (1974) for shoot establishment, but shoot proliferation was similar on both media indicating that media requirements vary at different stages of meristem culture. Recently, Samartin (1989) studied the effect of six different macroelement formulae in *Camellia japonica* and found that MS medium supported highest growth rate when explants were prepared from juvenile seedlings. However, the MS medium proved toxic to the explants taken from mature trees where a low salt medium (Heller, 1953) gave satisfactory results. The toxicity of MS medium has also been observed in some other plant species (Sommer, 1982; Vieitez and Vieitez, 1983).

Among the various macroelements of culture media, the nitrogen component is critical and has been shown to influence the success of meristem culture. Welander (1985) reported that meristem growth of six raspberry genotypes improved if the concentration of NH_4NO_3 and KNO_3 in MS medium was reduced to half strength, whereas rooting was enhanced by one fifth dilution of macroelements. A reduction in the level of mineral salts has also been reported to be beneficial for *in vitro* rooting of shoots in several other crop species (Kartha et al., 1974a; Hasegawa, 1980; Murashige, 1977; Drew, 1987). Hyndman et al. (1982) observed improved rooting of *in vitro* rose shoots by reducing the concentration of total nitrogen in the MS salt formulation suggesting that beneficial effect of lowering the salt concentration

Table 1. Culture media and growth hormones used for meristem and shoot tip culture for different crop species

Crop species	Culture media	Growth regulators (μM)						References
		Initiation (I)		Proliferation (II)		Rooting (III)		
Allium cepa L. (Onion)	BDS	BAP	0.1–10	–		BAP	0.1	Havel and Novak (1985)
			NAA	0.1–10		NAA	0.1	
Alstroemeria species	MS	IBA	2.5–25	–		None		Halkaart and Versluijs (1988)
Arachis hypogaea L. (Peanut)	MS+	BA	0.1	–		–		Kartha et al. (1981b)
	B5 Vitamins	NAA	10					
Camellia japonica	MS/Heller	BAP	4.4	BAP	4.4	None		Samartin (1989)
Carica papava L. (Papaya)	MS	NAA	1.0	–		IBA	10	Drew (1987)
		pCPA	1.0					
Cicer arietinum L. (Chickpea)	MS+	BA	0.1–1.0	–		IBA	1.0	Kartha et al. (1981b)
	B5 Vitamins	NAA	0–10					
Citrus species	MS	BA	0.5–2.0	BA	0.5–2.0	NAA	1.0–5.0	Barlass and Skene (1986)
Coffea arabica L.(Coffee)	MS+	BA	0.1–1.0	BA	5.0–10	–		Kartha et al. (1981a)
	B5 Vitamins							
Cucurbita pepo L. (Pumpkin)	MS	KIN	11.4	BA	4.4	IAA	45	Pink and Walkey (1984)
		IAA	45					
Ficus carica L. (Fig)	MS	BA	0.5	BA	0.5	NAA	2.7	Muriithi et al. (1982)
		NAA	1.0	NAA	1.0	IBA	2.5	
		GA_3	0.1	GA_3	0.1			
Fragaria × *ananassa*	Knops (major)	IBA	4.9	BA	4.4	IBA	4.9	Boxus et al. (1977)
Duch. (strawberry)	MS (minor)		IBA	4.9				
	MS+	BA	1.0	BA	10	BA	1.0	Kartha et al. (1980)
	B5 Vitamins	IBA	1.0	NAA	1.0			
		GA_3	0.1					
	MS	None		BA	4.4	IBA	5.0	Watkins et al. (1990)
				IBA	0.5			
Glycin max (L.) Merr.	MS+	BA	0.05–0.1	–		–		Kartha et al. (1981b)
	B5 Vitamins							

Table 1. Continued

Crop species	Culture media	Growth regulators (μM)						References
		Initiation (I)		Proliferation (II)		Rooting (III)		
Malus species (Crabapple)	MS	BA	4.4	BA	4.4–8.8	NAA	0.5–1.0	Singha (1982)
(Ottawa-3 rootstock)	MS	BA	2.2	BA	2.2	IBA	30	Chong and Pua (1985)
		IBA	2.8	IBA	2.8			
		GA_3	8.6	GA_3	8.6			
(Dwarf mutant apple)	MS	BA	5.6	BA	3.0–6.0	–		Lane and McDougald (1982)
	MS	BA	1.0–10	BA	1.0–10	NAA	1.0–10	Lane and McDougald (1982)
Manihot esculenta (Cassava)	MS+	BA	0.5	–		NAA	1.0	Kartha et al. (1974b)
	B5 Vitamins	NAA	1.0			GA_3	0.1	
		GA_3	0.1					
Musa acuminata L.	MS	BA	22–44	BA	44	IBA	25	DoreSwamy et al. (1982)
		IBA	25					
Musa species	MS	BA	22	BA	22	NAA	5.4	Cronauer and
(Banana and Plantain)								Krikorian (1984)
	MS	BA	3.0	BA	3.0	BA	3.0	Gupta (1986)
		KIN	3.4	KIN	3.4	KIN	3.4	
Phaseolus species	MS+	–		BA	10	NAA	1.0	Rubluo and Kartha (1985)
	B5 Vitamins			IAA	10			
Phoenix dactylifera L.	MS	NAA	53.7	BA	44.4	NAA	0.5	Tisserat (1984)
(Datepalm)				NAA	0.5			
Pisum sativum L. (Pea)	B5	BA	0.5	–		NAA	1.0	Kartha et al. (1974a)
		NAA	1.0					
	MS	BA	10	–		NAA	0.5	Kallak and Koiveer (1990)
		NAA	0.5					
Pogostemon cablin Benth.	MS	BA	8.8	BA	8.8	IAA	5.7	Kukreja et al. (1990)
(Patchouli)	IAA	5.7	IAA	5.7				
Prunus cerasus L.	MS	BA	2.2	BA	2.2	IBA	5.0	Baumann et al. (1984)
(Sour cherry)								
Prunus persica L.	MS	BA	0.9	BA	0.9	–		Hammerschlag (1982)
Batsch. (Peach)		IBA	0.05	IBA	0.05			
Psidium guajava L. (Guava)	MS	BA	1.0	BA	1.0	BA	1.0	Jaiswal and Amin (1987)

Table1. Continued

Crop species	Culture media	Growth regulators (μM)						References
		Initiation (I)		Proliferation (II)		Rooting (III)		
Ouercus robur L. (Oak)	MS	BA	4.4	BA	0.4	None		Vieitez et al. (1985)
Ribes rubrum L. (Red Current)	MS	BA	0.4	BA	0.4	None		Krogstrup and Thomsen (1985)
		IBA	4.9	IBA	4.9			
Rubus ideaus L. (Raspberry)	MS	BA	0.9	BA	4.4	IBA	0.05	Welander (1985)
		IBA	0.05					
Solanum tuberosum L. (Potato)	MS	KIN	18.6	–		KIN	0.05	Cassells and Long (1982)
		IAA	11.4			IAA	0.6	
		GA_3	2.9					
	MS	KIN	1.4	KIN	1.4	IAA	5.7	Lozoya-Saldana and
		IAA	1.7	IAA	1.7			Madrigal-Vargas (1985)
		2iP	49.2	2iP	49.2			
	MS	KIN	0.2	–		None		Mellor and Stace-Smith (1987)
Vigna unguiculata L. (Cowpea)	MS+	IBA	0.005–0.1	–		–		Kartha et al. (1981b)
	B5 Vitamins	NAA 0	0.05					
Vitis species (Grapevine)	MS	BA	10	BA	10	IBA	0.1	Novak and Jovova (1983)
Zingiber officinale Rosc. (Ginger)	MS	BA	2.2	BA	18–22	–		Bhagyalakshmi and Singh (1988)
		IBA	2.0					

Abbreviations: MS = Murashige and Skoog (1962); BS = Gamborg et al. (1968); BDS = Dunstan and Short (1977); BA = benzylamino purine; KIN = kinetin; ZEA = zeatin; 2iP = 2-isopentyladenine; IAA = indoleacetic acid; IBA = indolebutyric acid; NAA = naphthaleneacetic acid; pCPA = p-chlorophenoxyacetic acid; GA_3 = gibberellic acid.

of MS medium on rooting may be predominantly due to achieving a more favourable nitrogen concentration.

Carbon source is another important constituent of the culture medium used for meristem and shoot tip culture. Typically all media contain sucrose in the range of 1–3% as a carbon source. However, there are examples where alternative energy sources have been used for meristem culture of some species. Pua and Chong (1984) showed that meristem tip explants of *Malus robusta* No. 5 apple rootstock required sorbitol for better shoot initiation and multiplication. In a subsequent study, Chong and Pua (1985) investigated the effect of sucrose, glucose, sorbitol and fructose in the range of 1–7% on growth and rooting of Ottawa 3 apple rootstock. Interestingly, normal growth and 100% rooting occurred only when media contained 3% sucrose as a carbon source. All other carbon sources exhibited abnormal growth and poor rooting response. Rubluo and Kartha (1985) observed reduction in growth rate and plant regeneration of *Phaseolus* meristems in the presence of sucrose and mannitol at higher concentrations than 3%. However, the higher concentrations of sucrose were more toxic than mannitol. These studies emphasize the importance of carbon nutrition during different stages of *in vitro* meristem and shoot tip culture.

In addition to chemical composition of culture medium, its physical form can also influence the growth and multiplication rate of cultured meristems and shoot tips (Mellor and Stace-Smith, 1969; Hammerschlag, 1982; Cronauer and Krikorian, 1984; Bhagyalakshmi and Singh, 1988). Most often an agar-gelled medium is used, but in some crop species liquid medium, both in static and agitated form, has also been successfully employed. For example, Mellor and Stace-Smith (1969) reported that liquid medium was superior to solid medium for potato meristem culture. Similarly, growth of peach shoot was found to be better on liquid medium than solid medium (Hammerschlag, 1982). In contrast, ginger meristem cultures performed better on solid than in liquid static or agitated media (Bhagyalakshmi and Singh, 1988).

2.3. *Growth Regulators*

Plant growth regulator requirement for meristem and shoot tip culture depends on plant species and culture stage. Styer and Chin (1983) have extensively reviewed the growth regulator requirement for meristem and shoot tip culture of diverse dicot and monocot plant species. It appears that plant species belonging to families such as Leguminosae and Solanaceae require lower concentrations of exogenous plant growth regulators as compared to those belonging to Brassicaceae and Compositeae. Although there is no experimental data available, it is highly probable that plants belonging to families requiring higher doses of exogenous plant growth regulators may be low in endogenous levels of respective growth hormones. This once again points to the fact that a critical balance of endogenous and exogenous plant growth regulators is required to induce the desired response from meristem and shoot tips in culture.

An extensive review of plant growth regulators used at critical stages in meristem and shoot tip culture is available (Hu and Wang, 1983). Although in some species the stage I (culture initiation) cultures can be initiated without the use of exogenous growth regulators (Kartha et al., 1981b; Mellor and Stace-Smith, 1987), most species require low level of cytokinins to support the growth and development of meristems. Auxins are not essential for stage I culture medium, but beneficial if added at low concentrations. The most critical component of stage II (proliferation) culture medium is high level of cytokinins to promote axillary shoot proliferation. Among various cytokinins used, BA is most commonly employed in stage II culture medium for different species. However, BA has been found toxic for shoot proliferation of some species belonging to family Ericaceae (McCown and Lloyd, 1983; Norton and Norton, 1985). Although, auxins do not promote shoot proliferation, they are required in stage II culture medium to promote growth of shoots by counteracting the suppressive effect of high cytokinin concentrations on shoot elongation (Lundergan and Janick, 1980). However, very high concentration of auxins may induce callus formation (Hasegawa, 1980) which is undesirable for maintaining genetic stability among propagules. Gibberillins are also added to stage I and II culture medium (Kartha et al., 1974b; 1980; Wochok and Sluis, 1980), but their role in *in vitro* shoot elongation and proliferation is controversial (Lundergan and Janick, 1980). The stage III (rooting) culture medium is invariably supplemented with auxins to promote root development in stage II shoots. However, there are reports where rooting has been successfully achieved on hormone free medium (Krogstrup and Thomsen, 1985; Mellor and Stace-Smith, 1987; Hakkaart and Versluijs, 1988; Bhagyalakshmi and Singh, 1988). These generalizations of culture stage-dependent growth regulator requirements for meristem and shoot tip culture are also supported by the review of recent literature summarized in Table 1. The growth regulator requirements for meristem and shoot tip culture of tropical and sub-tropical fruits and vegetable crops have been recently reviewed (Litz and Jaiswal, 1991; Seckinger, 1991).

2.4. *Culture Environment*

It has been realized over the years that various components of tissue culture environment such as light, temperature and photoperiod can influence the process of *in vitro* differentiation from various explant tissues including the meristem and shoot tip explants (Hughes, 1981). In spite of this, in most studies very little attention has been paid to defining the culture environment for various stages of meristem and shoot tip culture. For example, the literature surveyed on meristem tip culture of strawberry, the crop for which the technique is in commercial use since early 1970s, revealed that a tempera-

Table 2. Culture environments for strawberry meristem culture

Species/ cultivar	Photoperiod (light)	Light intensity (lux)	Temperature (°C)	References
Fragaria × *ananassa* Duch. 3 cultivars	16h	4200	19 or 25	Adams (1972)
74 cultivars	12–16h	1000–1500	28	Boxus (1974) Boxus et al. (1977)
cv. Red Gauntlet	dark 12h	24000	25	Lee and de Fossard (1977)
cv. Gento	12h	4200	25	James and Newton (1977)
cv. Redcoat	16h	4000	26	Kartha et al. (1980)
3 cultivars	16h	1800–2000	26	Scott and Zanzi (1980)
3 cultivars	16h	1000–2000	26	Swartz et al. (1981)
cv. Cambridge Favourite	10h	0–10000	12–28	Hunter et al. (1983)
cv. Cambridge Favourite	16h	4400–4200	20–25	Hunter et al. (1984)
6 cultivars	16h	1000–1500	22	Simpson and Bell (1989)

ture between 12–28 °C and a light intensity in the range of 1000–24000 lux has been used in various studies (Table 2). Surprisingly, sometimes many different cultivars are being maintained at a constant temperature and light intensity even at different stages of meristem tip culture (Boxus et al., 1977). Hunter et al. (1983) studied the influence of different light and temperature regimes on *in vitro* propagation of strawberry cv. Cambridge Favourite and concluded that 28 °C was an optimum temperature, but light intensity requirement varied at different stages of meristem culture. Light intensities of 4000, 6000, and 7000 lux were found optimal for meristem tip establishment, shoot proliferation and rooting, respectively. We have observed an interaction between light intensity and temperature on shoot proliferation of strawberry cv. Redcoat (Nehra and Kartha, unpublished). A light intensity of 2500 lux was optimal for shoot proliferation at 26 °C, whereas shoot proliferation was considerably suppressed at 15 °C temperature at the same light intensity (Fig. 2) indicating the need to standardize conditions optimal for individual genotypes.

There are only few reports where the effect of different constituents of culture environment has been examined in conjunction with standardization of culture media and hormone composition. Hammerschlag (1982) studied the effect of temperature on survival of peach shoot tips *in vitro* and found that survival was significantly greater at 21–24 °C than at 28 or 30 °C. Kartha et al. (1981b) reported higher frequency of plant regeneration from meristems of soybean and cowpea at high light intensity (7500 lux) and a constant

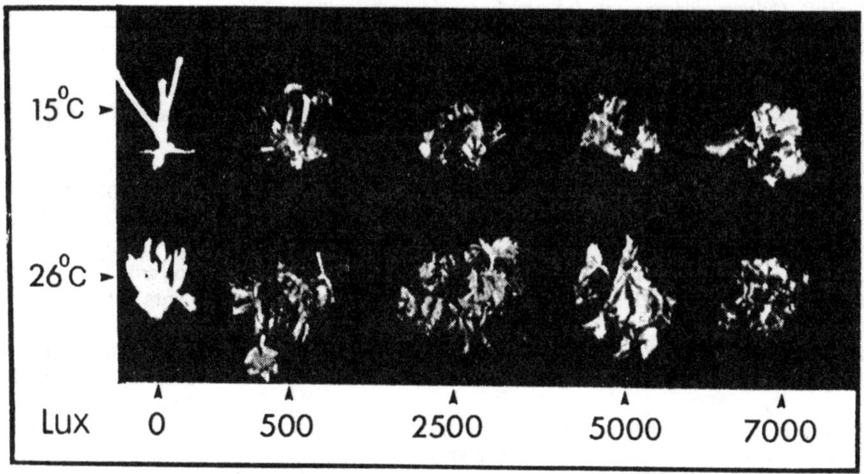

Figure 2. Shoot proliferation response of strawberry cv. Redcoat under various temperature and light intensity regimes after 4 weeks on MS medium containing 10 μM BA.

temperature of 26 °C than at alternating temperature (20 °C day: 15 °C night) and low light intensity (1200 lux). However, the reverse was true for differentiation of pea meristems (Kartha et al., 1979). Hildebrandt and Harney (1988) observed suppression of shoot proliferation in geranium at low light intensity (1350 lux) and temperature (7 °C) than at high light intensity (4000–11200 lux) and an elevated temperature (27 °C). Drew (1987) has recently reported, enhanced shoot proliferation and growth by exposing the foliage of *Carica papaya* shoot tip cultures to increased irradiance and photoperiod. All these studies indicate that the requirement for culture conditions varies according to the stage of meristem and shoot tip culture. Since there exists a strong and complex interaction between growth regulators and culture conditions (Baraldi et al., 1988), adequate attention should be given to optimization of culture environment, along with hormonal regimes, in a strategy aimed at developing efficient meristem and shoot tip culture technique.

2.5. *Genotype*

In plant tissue culture, it is now well known that no two genotypes give similar response under a given set of culture conditions (Nehra et al., 1989; 1990a). However, quite often the meristem and shoot tip culture technique developed for one genotype is used for many different genotypes of the same species which not only decreases the efficiency of plant production but also results in production of abnormal phenotypes among regenerants. The shoot tip survival of 11 commercial peach genotypes was tested on a common medium by Hammerschlag (1982). Significant differences were noted in survival response between cultivars with only 3 cultivars showing more than 50% survival rate. Similarly, Hakkaart and Versluijs (1988) reported that survival of *Alstroemeria* meristems was cultivar dependent and different cultivars required varying levels of IBA in the culture medium. Although there are some studies to indicate that diverse cultivars have a similar growth regulator optimum for shoot proliferation (Lane et al., 1982; Welander, 1985; Marcotrigiano et al., 1984), most other reports point to the contrary (Anderson et al., 1982; Norton and Norton, 1985; Simpson and Bell, 1989; Kallak and Koiveer, 1990). A positive correlation was reported between cytokinin (BA) concentrations and shoot proliferation of several strawberry cultivars (Anderson et al., 1982; Simpson and Bell, 1989). The morphogenetic response of pea meristems has also been shown to be genotype specific (Kartha et al., 1974a; Kallak and Koiveer, 1990). However, the genotype dependent behaviour of pea meristems was also strongly influenced by the exogenous growth regulator content of the culture medium. Similarly, Norton and Norton (1985) reported that effective cytokinin concentration for shoot proliferation in 20 cultivars of family Ericaceae varied with each cultivar. The interesting aspect of this study was that for all species belonging to Ericaceae, cytokinin 2iP was more effective than BA which is contrary to most other studies where BA has been found more effective for shoot proliferation (Table 1). This was suspected to be due to the presence of an enzyme system in family Ericaceae for either effective metabolization of 2iP or degradation of BA which may be lacking in other species. However, the enzyme system responsible for such differences has yet not been identified. Nevertheless, these studies suggest that considerable differences exist for morphogenetic response of meristem and shoot tip culture between different genotypes. Since the differences observed in morphogenetic response can be modified by manipulation of exogenous growth regulators or varying the culture conditions, it would appear that morphogenetic competence of different genotypes is under the control of physiological or epigenetic factors rather than true genetic factors.

3. Applications of Meristem and Shoot Tip Culture

3.1. *Virus Elimination*

Viruses invade almost all major food crops and are difficult to control by conventional control measures used for other plant pathogens. Cultivation of virus resistant genotypes has been considered an effective approach to combat viral diseases. However, plant breeding efforts aimed at developing virus resistant genotypes are often restricted by the availability of a limited germplasm pool for a source of virus resistance as opposed to wide host range of viruses. In addition, the breeding process is laborious and time consuming, especially for crops with long generation cycles and highly polyploid genomes. Furthermore, every breeding step results in the gain of some new traits but the loss of other desirable traits is inevitable. Recently, the use of genetic engineering has been advanced as an attractive and potential alternative for incorporating viral resistance into crop plants. Although considerable progress has been made in cloning of coat protein genes for some viruses and genetic engineering of crop plants for viral resistance (Hoekema et al., 1989; Lawson et al., 1990), the technology has yet to be perceived as a practical means for controlling viral diseases. Consequently, the most effective measure available so far for virus eradication is the use of virus-free propagation material obtained by employing the meristem tip culture alone or in combination with thermo or chemotherapy. The application of meristem and shoot tip culture for virus elimination has been evaluated in several reviews (Walkey, 1980; Monette, 1983; Kartha, 1984b; 1986). The most recent examples of virus elimination are compiled in Table 3.

The technique of virus elimination pioneered by Morel and Martin (1952) was based on the assumption that virus particles are unevenly distributed in plants and their titer decreases (or disappears) as the meristematic dome of vegetative shoot apex is approached (Limmaset and Cornuet, 1949). The apical meristematic domes were thus considered free from virus particles. It would therefore appear that the smaller the size of the explant, the greater will be the chances of virus elimination. In some instances positive linear correlation has been observed between the explant size and virus elimination (Stone, 1982; Converse and Tanne, 1984; Green and Lo, 1989). However, there are examples where elimination of some viruses has been difficult irrespective of the size of the meristem tip (Theiler-Hedtrich and Baumann, 1989, suggesting that in addition to size of the explant, the host-virus relationship plays an important role in success of virus elimination.

The observation of partial virus elimination in most cases (Table 3) indicates that some viruses are probably replicated and transmitted rapidly to the apical growing points. The electron microscopic studies in the recent past have in fact confirmed the presence of high concentration of virus particles in the apical meristematic domes of infected crop plants of some species (Toussaint et al., 1984). However, despite the presence of high concentration of virus particles in the apical meristems, virus elimination does occur in some plants by the use of meristem tip culture. This suggests that virus elimination is probably a result of a combination of several factors such as an inhibitory effect of various components of the culture medium and culture conditions or stress due to excision of meristem tips on virus multiplication. Since thermo and chemotherapy improves the success rate of virus elimination (Table 3), the stress factor seems to play an important role in inactivation of virus particles.

3.1.1. *Thermotherapy and Virus Elimination*

Heat treatment in conjunction with meristem tip culture has been used in many instances for improving the success rate of virus elimination in many crop species (Walkey, 1980; Kartha, 1986; Mellor and Stace-Smith, 1987). It is especially useful for elimination of those viruses that are often difficult to eliminate by meristem tip culture alone (Brown et al., 1988). Another advantage of combining meristem tip culture with heat therapy is that relatively larger explants can be used for elimination of viruses as opposed to very small size of explants needed when meristem culture alone is used. For example, Green and Lo (1989) were able to eliminate sweet potato yellow dwarf virus from sweet potato tissues by excision of small meristem tips (0.3 mm), but the meristem tips of 1–2.5 cm were also satisfactory for virus elimination provided the parent plants were subjected to heat treatment of 37 °C for 1 or 2 months. Similar results were reported for elimination of cassava mosaic virus by meristem culture alone or in combination with heat treatment (Kartha and Gamborg, 1975).

For thermotherapy, the infected parent plants or portions of the parent plant are subjected to elevated temperatures before excision of meristem tips (Fig. 1). The temperature is often gradually increased to reach a critical range effective for inactivation or suppression of viruses in the active growing regions without adversely affecting the growth of plants. The preconditioned plants are held at the critical temperature for a certain duration and then the meristem tips are excised and cultured *in vitro* for obtaining virus-free plants. Alternatively, the exposure of *in vitro* shoots before excision of meristems (Brown et al., 1988), and excised meristems (Janeckova, 1988) to high temperature have also been successfully used for virus elimination (Fig. 1).

The temperature requirement and duration of the treatment depends upon the plant genotype and the virus being eliminated. Converse and Tanne (1984) demonstrated that elimination of mild yellow edge virus from strawberry was directly related to length of time the stock plants were kept at 38 °C. The virus was eliminated from 50% of the regenerants when the meristems were excised from stock plants after 15 days at 38 °C and from 100% after 48 days at the same temperature. Lozoya-Saldana and Merlin-Lara (1984) exposed 6 potato cultivars uniformly infected with potato virus X to gradual temperature increase from 37 to 40 °C for two weeks and observed differences in tolerance to heat and subsequent virus elimination. The cultivars that were most tolerant to heat also produced largest number of virus-free plants suggesting that temperature regimes and duration of treatment should be determined empirically for each genotype.

In most cases the maintenance of infected host plants or excised meristems at constant high temperature is effective for virus elimination. However, sometimes the exposure to constant high temperature is damaging to the plant tissues and some viruses are difficult to eliminate by this procedure

Table 3. Virus and viroid elimination in crop plants by meristem and shoot tip culture

Crop species	Virus/viroid	Success rate (%)	Treatment	Virus indexing	Reference
Allium sativum L.	Onion yellow dwarf	85	TT	EM	Walkey et al. (1987)
Allium ascalonicum	Leek yellow stripe	85	TT	ISEM	Walkey et al. (1987)
	Shallot latent				
Alstroemeria species	Alstroemeria mosaic	–	None	ELISA	Hakkaart and Versluijs (1988)
Citrus species	Citrus tristeza	–	CT	ELISA	Greno et al. (1990)
	Citrus infectious vareigation				
	Exocortis viroid				
Chrysanthemum species	Chrysanthemum stunt viroid	18–80	Low temperature	Bioassay	Paduch-Cichal and Kryczynski (1987)
	Chrysanthemum chlorotic mottle viroid				
	Cucumber pale fruit viroid				
Fragaria × *ananassa*	Strawberry mild yellow edge	100	TT	Leaf graft	Converse and Tanne (1984)
	June yellow complex	0	TT	–	Watkins et al. (1990)
	Strawberry mottle	74	CT	Bioassay	Kondakova and Schuster (1991)
	Strawberry crincle				
Ipomoea batatas L.	Sweet potato yellow dwarf	100	TT	ELISA	Green and Lo (1989)
Manihot esculenta	Cassava latent	100	TT	Visual and bioassay	Adejare and Coutts (1981)
	Cassava mosaic	100	TT	Bioassay	Kartha and Gamborg (1975)
Nicotiana tobacum L.	Tobacco mosaic	0	TT	Bioassay	Lozoya-Saldana and Dawson (1982)

Table 3. Continued

Crop species	Virus/viroid	Success rate (%)	Treatment	Virus indexing	Reference
Pisum sativum	Pea seed-borne mosaic	90–100	None	Bioassay	Kartha and Gamborg (1978)
Pogostemon cablin (Blanco) Benth.	Patchouli mosaic	100	None	Bioassay	Kukreja et al. (1990)
Prunus cerasus L.	Prunus dwarf	29	None	ELISA	Baumann et al. (1984)
Ribes rubrum L.	Raspberry ring spot	100	None	Bioassay	Krogstrup and Thomsen (1985)
Rubus ideaus L.	Nine sap transmissible	–	TT	Bioassay/ELISA	Janeckova (1988)
	Apple mosaic Raspberry bushy dwarf	85–100	None	ELISA	Theiler-Hedtrich and Baumann (1989)
	Raspberry bushy dwarf	100	CT	ELISA	Kudell and Buchenaur (1989)
Solanum tuberosum	Potato virus X (PVX)	33–62	TT	ELISA	Lozoya-Saldana and Merlin-Lara (1984)
	PVX	15–40	TT	ELISA	Lozoya-Saldana and Madrigal-Vargas (1985)
	Potato virus Y (PVY) Potato virus S (PVS)	20	CT	ELISA	Wambugu et al. (1985)
	PVS	90	TT	ELISA	Brown et al. (1988)
	Potato spindle tuber viroid	18–80	TT	Bioassay	Paduch-Cichal and Kryczynski (1987)
Ullucus tuberosus	Tobamovirus Papaya mosaic Ullucus mosaic Ullucus virus C	6–25	CT	Bioassay/EM/ISEM	Stone (1982)
Vigna unguiculata (L >) Walp.	Cowpea chlorotic mottle	100	IT	Bioassay	Lozoya-Saldana and Dawson (1982)
Vitis vinifera L.	Grapevine fan leaf	–	TT	ELISA	Monette (1986)

Treatment: TT = Thermotherapy; CT = Chemotherapy.
Virus indexing: ELISA = Enzyme-linked immunosorbent assay; EM = Electron microscopy; ISEM = Immunosorbent electron microscopy.

because of their ability to replicate at high temperature (Dawson, 1976). In such instances the temperature regimes alternating between optimum and restricting temperature have been found effective for virus elimination. Using such an approach, Lozoya-Saldana and Dawson (1982) were able to eliminate cowpea chlorotic mosaic virus from infected cowpea plants, but the procedure was not effective for eliminating tobacco mosaic virus from tobacco plants. Recently, Monette (1986) reported successful elimination of grapevine fan leaf virus and arabis mosaic virus from grapevine using alternating temperature regimes. In addition, a combination of low temperature therapy and meristem tip culture has been successfully used for eliminating viroids from infected potato and chrysanthemum plants (Paduch-Cichal and Kryczynski, 1987).

3.1.2. *Chemotherapy and Virus Elimination*

Chemotherapy has been tested for several years for its usefulness in virus elimination in combination with meristem tip culture. However, the technique has met with only limited success and often the role of antiviral compounds in virus suppression or inactivation has been doubtful (see reviews by Cassells, 1987; Kartha, 1986). Cassells and Long (1980) reported production of virus-free plants from cucumber mosaic virus and potato virus X infecting tobacco explants by inclusion of a synthetic broad-spectrum antiviral compound Ribavirin (Virazole) in the culture medium. Since then this compound has been successfully used for eradication of several viral complexes from *in vitro* meristem cultures of potato cultivars (Cassells and Long, 1982; Klein and Livingstone, 1982). In a subsequent study, Wambugu et al. (1985) reported that Ribavirin was phytotoxic to potato axillary bud tips at all concentrations tested, but exhibited considerable inhibitory effect on virus replication (72%) even at the lowest concentration (5 mg/l). In contrast, Simpkins et al. (1981) reported that Ribavirin was not effective in production of virus-free plants from meristem tip cultures of *Nicotiana rustica* infected with cucumber mosaic virus even at a concentration of 100 mg/l.

In the light of contradictory reports on the effectiveness and phytotoxicity of antiviral compounds in the past, chemotherapy has not been generally recognized as an effective method for virus elimination from infected plants. However, recent reports are very encouraging and suggest that some antiviral compounds have potential in improving the efficiency of virus elimination when used in combination with meristem and shoot tip culture. Ribavirin has been effectively used for elimination of chlorotic leaf spot virus from apple (Hansen and Lane, 1985) and green ring mottle virus-like agent from *Prunus* species (Hansen, 1984). The antiviral activity of Ribavirin against virus and viroids has also been proven in several other studies on fruit tree species (Belles et al., 1986; Hansen, 1988). Raspberry bushy dwarf virus is considered difficult to eliminate by thermotherpy and meristem culture due to its heat stability and meristem tip colonization. However, this virus was effectively eliminated by addition of Ribavirin and DMEB (dedecyl-N-methylephedrinecin bromide) in the culture medium at non-phytotoxic concentrations (Kudell and Buchenauer, 1989). Recently, Greno et al. (1990) determined the effect of five antiviral compounds on three distinct virus and virus-like agents of citrus species and concluded that Ribavirin was most affective when used in the culture medium.

In addition to Ribavirin, some other antiviral compounds have also been found to have beneficial effect on virus elimination. Stone (1982) succeeded in obtaining virus-free plants of *Ullucus* by using 1 mg/l of an adenine derivative antiviral compound, Vidarabine. Recently, Kondakova and Schuster (1991) demonstrated that rate of elimination of two strawberry viruses can be considerably enhanced by the addition of 0.2 mg/l DHT (2,4-dioxohexahydro-1,3,5-triazine) in the culture medium. However, in both studies it was not conclusive if the virus elimination was due to true antiviral properties of these compounds or due to their favourable effect on growth of explants because of their kinetin-like properties. Nevertheless, the above discussion suggests the possibility of successful incorporation of chemotherapy in virus elimination schemes in conjunction with meristem tip culture, especially for those viruses or virus-like agents that are difficult to control by heat therapy and meristem tip culture (Fig. 1). However, since the antiviral compounds are often toxic to plant tissues and may not result in complete virus elimination, more rigorous testing of these compounds is warranted before they can be routinely employed in virus elimination programs.

3.1.3. *Virus Indexing*

The availability of a rapid, effective and sensitive procedure for virus indexing is of paramount importance for success of a program aimed at production of virus-free plant material (Fig. 1). The accurate and rapid diagnosis of viruses in meristem tip culture regenerants is also relevant to the international exchange of plant material. In the past, several biological assays have been employed for virus indexing (George and Sherrington, 1984). The biological assays are reasonably accurate, but too slow and difficult to use for screening of large number of plants. Alternatively, the techniques involving electron microscopy, serology or a combination of both were developed for virus detection in plant tissues. The basic features of these techniques and their potential in virus indexing have been discussed in previous reviews (Van Regenmortel, 1982; George and Sherrington, 1984; Kartha, 1986).

Among various serological techniques, the enzyme-linked immunosorbent assay (ELISA) has been extensively used for virus indexing (Table 3). With the advent of recombinant DNA methodology, nucleic acid hybridization techniques have also been successfully applied for detection of viruses and viroids with equal or high degree of sensitivity than the conventional ELISA method (Sela et al., 1984; Varveri et al., 1988; Polston et al., 1989). Since these two techniques are in forefront for virus indexing in current literature, a brief discussion of their application is given below. More detailed information on the development and use of both techniques can be obtained from a number of recent reviews (Lawson, 1986; Martin, 1987; Miller and Martin, 1988).

3.1.3.1. *Enzyme-Linked Immunosorbent Assay (ELISA)*. In recent past, ELISA has become a popular immuno-detection system for virus indexing of mature and tissue cultured plantlets (Gallenberg and Jones, 1985; Table 3) because of its sensitivity, simplicity and speed with which the viruses can be detected and quantified in plant tissues. The technique involves binding of an antigen (virus) to a specific enzyme-labelled antibody (conjugate). The enzyme acts upon its substrate added to the bound complex and results in a coloured product. The intensity of the colour can be quantified spectrophotometrically and is proportional to the amount of substrate bound or the amount of antigen present in the test sample. Two versions of this technique, namely direct and indirect ELISA, are commonly used for virus indexing (George and Sherrington, 1984; Miller and Martin, 1988). In direct or double antibody sandwich (DAS) ELISA, an unlabelled antibody is bound to a solid phase to which antigen is added. The antigen-antibody complex is then detected by binding to a specific enzyme labelled antibody. On the other hand, indirect ELISA uses an enzyme labelled anti-immunoglobulin as a second antibody to detect the antigen-antibody complex bound to the solid phase. Lommel et al. (1982) compared both direct and indirect ELISA for detection of two plant viruses in carnations and concluded that direct ELISA was superior for quantitative virus detection in crude extracts, but was highly strain specific and required purified antiserum for each antigen coupled to an enzyme. On the other hand, indirect ELISA avoided the limitations of direct ELISA and proved to be a sensitive and useful tool for virus detection in crude extracts when accurate quantization is not required.

Several modifications of the above two methods such as a 'two step' or 'cocktail' ELISA, in which both samples and conjugate are incubated simultaneously (Kanaiewski and Thomas, 1988), cocktail ELISA combined with enzyme amplification (Van den Heuvel and Peters, 1989), and the use of mixed antisera (Etienne et al., 1991) have been tested for improving the speed and sensitivity of plant virus detection. Recently, a simplified DAS-ELISA using dry reagent carriers immobilized on filter papers has also been developed for detection of plant viruses under field condition (Reichenbacher et al., 1990).

3.1.3.2. *Nucleic Acid Hybridization*. Nucleic acid hybridization techniques, based on the formation of DNA/RNA or RNA/RNA duplexes between the complementary nucleotide sequences, is fast becoming an attractive method for detecting virus and viroid particles in plant tissues (Varveri et al., 1988; Roy et al., 1988; Polston et al., 1989; Fuchs et al., 1991). The technique makes use of a cloned DNA sequence as a probe and a dot blot hybridization method. The technical details for dot blot assay, and development and labelling of nucleic acid probes are available (Ausubel et al. 1989). The basic concepts of these techniques in relation to their use in plant disease diagnosis have been discussed in a recent review article (Miller and Martin, 1988).

The nucleic acid based techniques are especially useful for detection of viroid particles because viroids lack the protein component and can not be detected by ELISA (Mohamed and Imperial, 1984). Since these techniques are several fold more sensitive than conventional ELISA, they may also become a useful complement to ELISA in those situations where the latter is unable to detect virus particles in plant tissues (Robinson et al., 1984). Recently, Polston et al. (1989) developed a nucleic acid dot hybridization assay for detection of cucurbit geminivirus and found that the technique was able to detect 300–400-fold less amount of virus in plant tissues than ELISA. Similarly, Varveri et al. (1988) reported that dot hybridization based detection of plum pox virus using a RNA probe was 250 times more sensitive than ELISA. The cloned hybridization probes are commonly labelled with radioactive ^{32}P. However, Roy et al. (1988) have demonstrated that a non-radioactive biotin labelled probe was as effective as ^{32}P labelled probe for the detection of papaya mosaic potex virus. The development of such non-radioactive probes is expected to augment the use of nucleic acid hybridization techniques for virus indexing.

3.2. *Mass Propagation*

Besides virus elimination, the second most important field of application of meristem and shoot tip culture has been the rapid clonal multiplication of vegetatively propagated plant species (Fig. 1). In general, three different modes of *in vitro* plant regeneration, namely organogenesis, embryogenesis and axillary shoot proliferation, have been recognized (Murashige, 1977; George and Sherrington, 1984). Although the rate of plant regeneration is usually higher through organogenesis or embryogenesis, the meristem and shoot tip culture, which allows shoot proliferation through axillary branching, is still considered a method of choice for *in vitro* mass propagation because the method is less prone to the risk of genetic instability. Despite the obvious advantages of rapidity, time and space saving, and higher plant production potential compared to conventional methods, the *in vitro* mass production of plants using meristem and shoot tip culture is not on sound footing, on the practical front, mainly due to labour intensive and highly specialized nature of operation leading to higher cost of plant production per unit (Vasil, 1991). As a consequence, it seems that the use of this technique may remain justified only in those situations where either the conventional propagation techniques do not work or a new variety is to be released for commercial production.

According to an estimate (Donnan, 1986), labour accounts for over 60%

of the cost involved in a tissue culture operation for plant propagation. Keeping this in view, attempts are being made to introduce automation and robotization in various phases of *in vitro* plant production to bring down the cost, and make meristem and shoot tip culture as an attractive and profitable enterprise for plant production (DeBry, 1986; Aitken-Christie, 1991; Vasil, 1991). Alternatively, efforts are also afoot to minimize the cost of *in vitro* plant production by improving the rate of multiplication, growth and quality of shoots, and reducing the rate of mortality by improving the conditions for rooting and acclimatization (see reviews in Debergh and Zimmerman, 1991).

Following the success of meristem and shoot tip culture with orchids and other ornamental plants (Morel, 1972; Murashige, 1974), the concept was pursued much more vigorously and, as a result, many new crop species were quickly brought into the domain of this technology (Vasil and Vasil, 1980). During the last decade, tremendous progress has been made in developing meristem and shoot tip culture protocols with potential application for mass propagation (Table 1; Nehra and Singh, 1982; Styer and Chin, 1983; George and Sherrington, 1984; Debergh and Zimmerman, 1991). The question is that how many of this plethora of reports on meristem and shoot tip culture are pragmatic. The answer will probably narrow it down to a few known examples of fruit and ornamental plants. The reason for this is that most published reports only indicate the potential for micropropagation and do not provide a complete package that could be put to use for large scale clonal propagation. While this may justify the need for further research for improving the existing protocols, it also reflects on the incomplete nature of published protocols which is probably responsible for widening the gap between research and the commercial application of the technology.

The most attractive feature of meristem and shoot tip culture technique, that provides this method an edge over other *in vitro* methods, is its ability to maintain a high degree of genetic fidelity among propagules (Murashige, 1974; Samyn et al., 1984; Ahmed et al., 1987). However, there are reports to indicate that even this method of micropropagation is not free from the risk of genetic instability (Vajrabhaya, 1977; Swartz et al., 1981; 1983). Is the risk of genetic instability genuine and serious enough to cause a concern among those who are using the meristem and shoot tip culture as a plant production system and those who are going to use the end product? Unfortunately, the answer is affirmative as indicated by the closure of many small tissue culture nurseries and decline in production of micropropagated plants of certain species by big nurseries in Europe and Asia, mainly due to the realization that they were not producing true to type plants (Schaeffer et al., 1980; Pierik, 1991; Gavinlertvatana and Prutpongse, 1991). In some countries, even the farmers are reluctant to accept tissue cultured plants for the fear of ending up with something other than what they are planting (Gavinlertvatana and Prutpongse, 1991). Thus the fear of genetic instability is genuine and serious, but not to the extent that it can not be dispelled. The problem generally arises due to use of inappropriate culture medium, growth hormones and culture conditions, and repeated proliferation of the explants once established in culture (Karp, 1989). By paying attention to these variables and by avoiding callusing of explants, the problem can be greatly minimized, if not circumvented.

3.3. *Germplasm Storage*

Preservation of elite and rare germplasm is essential for maintaining sufficient genetic diversity in the gene pool of present day crop species that can be utilized for continued crop improvement. Traditionally seeds are used for germplasm preservation of grain crops, whereas vegetatively propagated crops are preserved by field maintenance. The storage of germplasm in the form of seed is efficient and economical for most crop species, but becomes a problem in those crops that produce sterile or highly heterozygous seed or loose seed viability in a short time period. In addition, the seeds are also vulnerable to deterioration by seed borne diseases and pathogens. On the other hand, the field maintenance of asexually propagated crops is extremely expensive, labour intensive, and beset with pathogen and pest problems. A panacea to these problems is often sought in the use of *in vitro* techniques for germplasm storage (Kartha, 1985a). Among various types of explants used for *in vitro* germplasm storage, meristem and shoot tips are considered ideal for both short and long-term storage needs because they possess certain specific attributes that are not shared by other tissues and organs. As discussed above, the meristems and shoot tips are capable of producing virus-free plants at high frequency. In addition, meristem tips are less dedifferentiated and often result in production of genetically identical progeny. Since meristematic cells are densely cytoplasmic and non-vacuolated, theoretically, they should better withstand freezing injury. However, there are reports of difficulties encountered in cryopreservation of meristems and shoot tips for long term storage (Withers et al., 1990). The problem often arises due to recovery of incomplete meristematic domes following freezing which results in poor survival rate (Haskins and Kartha, 1980; Fukai and Oe, 1990). In addition, the freeze damaged meristem and shoot tips often show greater propensity to callusing and adventitious shoot development which enhances the risk of producing abnormal phenotypes (Withers et al.. 1990: Fukai and Oe. 1990).

Two approaches, namely slow growth and cryopreservation (Fig. 1), are generally used for germplasm preservation by meristem and shoot tip culture Kartha, 1985a). Slow growth is suitable for short to medium-term storage, whereas cryopreservation is a preferred method for long-term storage. The plant species that have inherent slow proliferation rate *in vitro* can be maintained for a considerable time period even at room temperature by subculturing at a regular interval. For example, Kartha et al. (1981a) successfully maintained coffee genotypes by *in vitro* shoot tip cultures for over two years with only a single transfer. There are also examples in literature for *in vitro* storage of various other plant species for several years by periodic transfer of shoots at 1–2 years interval (Kartha, 1985b; Wanas et al., 1986). In our laboratory about 25 different genotypes of strawberry have been maintained at optimal temperature for more than 6 years by subculturing every 3–4 months on fresh medium (Kartha, unpublished). However, in most plant species with vigorous growth and proliferation rate, the slow growth is achieved by maintaining cultures at reduced temperature and light, by induction of osmotic and hormonal stress, or by addition of growth retardants in the culture medium (see reviews by Kartha, 1985b; Sakai, 1985). In general, growth limitation can be achieved by maintaining tropical plant species at relatively higher temperature and light intensity as compared to temperate plants. In recent studies with slow growth storage of cassava at Centro Internacional de Agricultura Tropical (CIAT), Cali, Colombia (Roca et al., 1989), it has been shown that shoot elongation could be considerably reduced by decreasing temperature to 22 °C from 28–30 °C. However, temperatures lower than 18 °C were detrimental to shoot growth. The addition of increased concentration of BA (40 to 110 μM) and sucrose (0.05 to 0.12 M) and reduction of nitrogen content (20 mM) in the culture medium were also effective in containing the *in vitro* growth of cassava shoots without affecting the shoot viability when retrieved to normal growth conditions. Similarly, banana shoot tip cultures have been stored for more than one year at low temperature (15 °C) and reduced light (1000 Lux) on media containing 10 μM BA (Banerjee and De Langhe, 1985). In contrast, temperature regimes between 0 to 15 °C have been found satisfactory for *in vitro* storage of various temperate forest and fruit tree species (Aitken-Christie and Singh, 1987).

In spite of the information and technology available for *in vitro* storage of several crop species, *in vitro* active gene banks exist so far only for potato (International Potato Centre, Lima, Peru) and cassava (CIAT) (Roca et al., 1989). Recently, efforts have been geared up under the auspices of International Board for Plant Genetic Resources (IBPGR) to exploit the full potential of *in vitro* storage methods employing meristem and shoot tip cultures. In this context, the initiation of a pilot project by collaborative efforts of IBPGR and CIAT, using cassava as a model system, for assessing the potential and feasibility of establishing and running an *in vitro* active gene bank is a first significant step (Roca et al., 1989; Withers et al., 1990). Under the project more than 4000 clones of cassava are being maintained *in vitro* under minimal growth conditions. The study is expected to provide guidelines for operation and maintenance of *in vitro* active gene banks for cassava and may serve as a model for establishing similar gene banks for other vegetatively propagated crop species. Although the slow growth storage approach is a promising method for short to medium-term storage of crop plants in *in vitro* active gene banks for immediate utilization, the practice is labour intensive and often results in loss of morphogenetic potential of stored cultures, increased risk of producing abnormal phenotypes and mutants, and loss or deterioration of material due to contamination during subculture. The limitations of slow growth method of germplasm storage can be overcome by the use of cryopreservation method of germplasm storage. The subject matter of cryopreservation will not be discussed here as it has been extensively evaluated elsewhere in this volume (see Chapter 9).

3.4. *Genetic Transformation*

The application of meristem and shoot tip culture as a tool for genetic transformation of crop plants is of relatively recent origin. Many different strategies, such as electroporation, PEG induced DNA uptake, particle bombardment, and *Agrobacterium*-mediated gene transfer, have been used in recent past for delivery of foreign DNA into host plant cells of both dicot and monocot plant species (Chapter 10). Among these methods, the *Agrobacterium*-mediated gene transfer has now become routine for genetic transformation of some dicot plant species (Gasser and Fraley, 1989). Besides other advancements in the field of molecular biology, a major share of the credit for success of *Agrobacterium*-mediated gene transfer system goes to Horsch et al. (1985) for developing a leaf disk transformation system that helped overcome the problems associated with the previously used protoplast transformation system (Fraley et al., 1984). However, adventitious shoot regeneration from leaf disks is often associated with an intermediate callus phase which enhances the risk of somaclonal variation (Karp, 1989). Although shoot regeneration from leaf disks without an intervening callus phase is possible in some crop species (Nehra et al., 1989), such systems are highly genotype specific limiting their use in genetic transformation of only few genotypes (Nehra et al., 1990b). Since in most cases the totipotent meristematic cells are capable of developing into a fertile plant without an intermediate callus phase and the system is generally applicable to a range of cultivars within a single species, the use of meristem and shoot tip culture has been recently proposed as an alternative to leaf disk system for *Agrobacterium*-mediated genetic transformation (Ulian et al., 1988).

Until recently, the meristem and shoot tips were considered insensitive to *Agrobacterium* infection, and inconvenient to use in transformation studies. The first successful *Agrobacterium*-mediated genetic transformation procedure (Fig. 1) using shoot tips isolated from seedlings was reported in petunia (Ulian et al, 1988). Since then the procedure has been used for transformation of meristematic cells in shoot apices of pea (Hussey et al., 1989) and production of transgenic sunflower by cocultivating shoot apical meristems with *Agrobacterium tumefaciens* (Schrammeijer et al., 1990). Recently, the method has also been used for production of transgenic corn (*Zea mays*) plants (Gould et al., 1991). These studies suggest the potential of meristem and shoot tip culture in *Agrobacterium*-mediated genetic transformation, but the general application of the procedure is limited, at present, due to very low transformation frequencies and occurrence of chimeric transgenic plants (Schrammeijer et al., 1990; Gould et al., 1991).

4. Conclusions and Future Prospects

Since the inception of the concept of meristem and shoot tip culture about 40 years ago, the technology has advanced to such an extent that it is now possible to culture almost all cultivated crop species by this method. The review of various requirements for meristem and shoot tip culture revealed that often there exists a strong and complex interaction between the explant, growth regulators, culture conditions, and genotype. There are very few instances where all these factors have been collectively taken into account for developing a protocol for meristem and shoot tip culture, rather most studies have capitalized on standardization of growth regulator requirements. The most neglected among the various factors has been the optimization of culture environment. Since culture conditions are as critical as other factors, there is a need to pay appropriate attention to various components of culture environment in a strategy aimed at devising efficient meristem and shoot tip culture techniques.

Among various applications of meristem and shoot tip culture technology, its use in virus elimination has already become a reality for some crop species. With further evaluation of the effectiveness of thermo and chemotherapy in conjunction with meristem culture and development of sensitive and rapid virus indexing procedures, the technique will be used much more extensively. This will not only boost agricultural production by cultivation of healthy virus-free plant material, but also facilitate the international exchange of elite germplasm. However, the application of meristem and shoot tip culture technology for commercial mass propagation of cultivated crop species is declining mainly due to the high production cost and the fear of producing abnormal phenotypes. Both these problems need to be addressed for revitalizing meristem and shoot tip culture as a commercially viable technology for mass propagation. In the wake of increased awareness and efforts to preserve valuable and rare germplasm, the *in vitro* storage of meristem and shoot tips is emerging as a viable and attractive alternative to conventional methods for medium and long-term storage of sexually and asexually propagated crop plants. However, to realize the full potential of the system, there is a need to develop efficient procedures for culture maintenance, pathogen indexing and monitoring the genetic stability of crop plants being maintained in *in vitro* active gene banks. In addition, more concerted research efforts are needed in the direction of developing suitable cryopreservation techniques for meristem and shoot tips as a means for long-term storage. The use of meristem and shoot tip culture technique for *Agrobacterium*-mediated genetic transformation is encouraging and may be useful for genetic modification of those crop species that are not amenable to other modes of regeneration.

5. References

Adams, A.N. (1972). An improved medium for strawberry meristem culture. J. Hort. Sci. 47:263–264.

Adejare, G.O. and Coutts, R.H.A. (1981). Eradication of cassava mosaic disease from Nigerian cassava (*Manihot esculenta*) clones by meristem tip culture. Plant Cell Tissue Org. Cult. 1, 25–32.

Ahmed, R., Gupta, S.D. and Ghosh, P.D. (1987). The cytological status of plants regenerated from shoot meristem culture of *Pisum sativum* L. Plant Breeding 98:306–311.

Aitken-Christie, J. (1991). Automation. In: Micropropagation – Technology and Application. P.C. Debergh and R.H. Zimmerman (eds.), Kluwer Academic Publishers, Dordrecht, pp. 363–388.

Aitken-Christie, J. and Singh, A.P. (1987). Cold storage of tissue cultures. In: Tissue Culture in Forestry. J. Bonga, and D. Durzan (eds.), Martinus Nijhoff Publishers, The Hague.

Anderson, W.C. and Carstens J.B. (1977). Tissue culture propagation of broccoli, *Brassica oleracea* (Italica group), for use in F_1 hybrid seed production. J. Am. Soc. Hort. Sci. 102:69–73.

Anderson, H.M., Abbott, A.J. and Wiltshire, S. (1982). Micropropagation of strawberry plants *in vitro* – Effect of growth regulators on incidence of multi-apex abnormality. Sci. Hort. 16:331–341.

Ausubel, F.M., Brent, R., Kingston, R.E., Moore, D.D., Seidman, J.G., Smith, J.A. and Struhl, K. (1989). Current Protocols in Molecular Biology, Vol. 1. Greene Publishing, New York.

Ball, E. (1946). Development of sterile cultures of stem tips and adjacent regions of *Tropaeolum majus* L. and *Lupinus majus* L. Am. J. Bot. 33:301–318.

Banerjee, N. 1 and De Langhe, E. (1985). A tissue culture technique for rapid clonal propagation and storage under minimal growth conditions of *Musa* (banana and plantain). Plant Cell Rep. 4:351–354.

Baraldi, R., Rossi, F. and Lercari, B. (1988). *In vitro* shoot development of *Prunus* GF665–2: interaction between light and benzyladenine. Physiol. Plant. 74:440–443.

Barlass, M. and Skene, K.G.M. (1986). Citrus (*Citrus* species). In: Biotechnology in Agriculture and Forestry, Vol. 1: Trees. Y.P.S. Bajaj (ed.), Springer-Verlag, Heidelberg, pp. 207–219.

Baumann, G., Casper, R. and Kornkamhaeng, P. (1984). Detection of prune dwarf virus by ELISA in meristem-propagated sour cherry plants during *in vitro* culture. Phytopath. Z. 110:168–171.

Belles, J.M., Hansen, A.J., Granell, A. and Conejero, V. (1986). Antiviroid effects of ribavirin on citrus exocortis viroid infection in *Gynura aurantiaca* DC. Physiol Mol. Plant Pathol. 28:61–65.

Bhagyalakshmi and Singh, N.S. (1988). Meristem culture and micropropagation of a variety of ginger (*Zingiber officinale* Rosc.) with a high yield of oleoresin. J. Hort. Sci. 63:321–327.

Boxus, P. (1974). The production of strawberry plants by *in vitro* micropropagation. J. Hort. Sci. 49:209–210.

Boxus, P., Quoirin, M. and Laine, J.M. (1977). Large scale propagation of strawberry plants from tissue culture. In: Applied and Fundamental Aspects of Plant Cell Tissue and Organ Culture. J. Reinert and Y.P.S Bajaj (eds.), Springer-Verlag, New York, pp. 130–143.

Brown, C.R., Kwiatkowski, S., Martin, M.W. and Thomas, P.E. (1988). Eradication of PVS from potato clones through excision of meristems from *in vitro*, heat treated shoot tips. Am. Potato J. 65:633–638.

Cassells, A.C. (1987). In vitro induction of virus-free potatoes by chemotherapy. In: Biotechnology in Agriculture and Forestry, Vol. 3: Potatoes. Y.P.S. Bajaj (ed.), Springer-Verlag, New York, pp. 40–50.

Cassells, A.C. and Long, R.D. (1980). The regeneration of virus-free plants from cucumber mosaic virus and potato virus Y infected tobacco explant cultures in the presence of virazole. Z. Naturf 35:350–351.

Cassells, A.C. and Long, R.D. (1982). The elimination of potato virus X, Y, S, and M in meristem and explant cultures of potato in the presence of virazole. Potato Res. 25:165–173.

Chong, C. and Pua, E.C. (1985). Carbon nutrition of Ottawa 3 apple rootstock during stages of *in vitro* propagation. J. Hort. Sci. 60:285–290.

Converse, R.H. and Tanne, E. (1984). Heat therapy and stolon apex culture to eliminate mild yellow-edge virus from Hood strawberry. Phytopathology 74:1315–1316.

Cronauer, S.S. and Krikorian, A.D. (1984). Rapid multiplication of bananas and plantains by *in vitro* shoot tip culture. Hort. Sc. 19:234–235.

Dawson, W.O. (1976). Synthesis of TMV RNA at restrictive high temperatures. Virology 73:319–326.

Debergh, P.C. and Zimmerman, R.H. (1991). Micropropagation – Technology and Application. Kluwer Academic Publishers, Dordrecht.

De Bry (1986). Robots in plant tissue culture: an insight. IAPTC News Lett. 49:2–22.

Donnan, A. (1986). Determining and minimizing production costs. In: Tissue Culture as a Plant Production System for Horticultural crops. R.H. Zimmerman, R.H. Griesbach, R.A. Hammerschlag and R.H. Lawson, (eds.), Martinus Nijhoff Publishers, Dordrecht, pp. 167–173.

DoreSwamy, R., Srinivasa Rao, N.K. and Chacko, E.K. (1982). Tissue culture propagation of banana. Sci. Hort. 18:247–252.

Drew, R.A. (1987). The effect of media composition and cultural conditions on *in vitro* root initiation and growth of papaya (*Carica papaya* L.). J. Hort. Sci. 62:551–556.

Dunstan, D.I. and Short, K.C. (1977). Improved growth of tissue culture of the onion, *Allium cepa*. Physiol. Plant. 41:70–72.

Etienne, L., Clauzel, J.M. and Fuchs, M. (1991). Simultaneous detection of several nepoviruses infecting grapevine in a single DAS-ELISA test using mixed antisera. J. Phytopathology 131:89–100.

Fraley, R.T., Horsch, R.B. and Matzke, A. (1984). *In vitro* transformation of Petunia cells by an improved method of co-cultivation with *A. tumefaciens* strains. Plant Mol. Biol. 3:371–378.

Fukai, S. and Oe, M. (1990). Morphological observations of chrysanthemum shoot tips cultured after cryoprotection and freezing. J. Japan Soc. Hort. Sci. 59:383–387.

Fuchs, M., Pinck, M., Etienne, L., Pinck, L. and Walter, B. (1991). Characterization and detection of grapevine fan leaf virus by using cDNA probes. Phytopathology 81:559–565.

Gallenberg, D.J. and Jones, E.D. (1985). Detection of potato virus X and S in tissue culture plantlets. Am. Potato J. 62:111–118.

Gamborg, O.L., Miller, R.A. and Ojima, K. (1968). Nutrient requirements of suspension cultures of soybean root cells. Expt. Cell Res. 50:151–158.

Gasser, C.S. and Fraley, R.T. (1989). Genetically engineering plants for crop improvement. Science 244:1293–1299.

Gavinlertvatana, P. and Prutpongse, P. (1991). Commercial micropropagation in Asia. In: Micropropagation – Technology and Application. P.C. Debergh and R.H. Zimmerman (eds.), Kluwer Academic Publishers, Dordrecht, pp. 181–190.

George, E.F. and Sherrington, P.D. (1984). Plant Propagation by Tissue Culture – Handbook and Directory of Commercial Laboratories. Eastern Press, Reading.

Gould, J., Devey, M., Hasegawa, O., Ulian, E.C., Peterson, G. and Smith, R.H. (1991). Transformation of *Zea mays* L. using *Agrobacterium tumefaciens* and the shoot apex. Plant Physiol. 95:426–434.

Green, S.K. and Lo, C.Y. (1989). Elimination of sweet potato yellow dwarf virus (SPYDV) by meristem tip culture and by heat treatment. J. Plant Dis. Protection 96:464–469.

Greno, V., Cambra, M., Navarro, L. and Duran-Vila, N. (1990). Effect of antiviral chemicals on the development of virus content of citrus buds cultured *in vitro*. Scientia Hort. 45:75–87.

Gupta, P.P. (1986). Eradication of mosaic disease and rapid clonal multiplication of bananas and plantains through meristem tip culture. Plant Cell Tissue Org. Cult. 6:33–40.

Gupta, P.K., Mascarenhas, A.F. and Jagannathan, V. (1981). Tissue culture of forest trees – clonal propagation of mature trees of *Eucalyptus citriodora* Hook. by tissue culture. Plant Sci. Lett. 20:195–201.

Hakkaart, F.A. and Versluijs, J.M.A. (1988). Virus elimination by meristem-tip culture from a range of *Alstroemeria cultivars*. Neth. J. Plant Path. 94:49–56.

Hammerschlag, F. (1982). Factors affecting establishment and growth of peach shoots *in vitro*. Hort. Sc. 17:85–86.

Hansen, A.J. (1984). Effect of ribavirin on green ring mottle causal agent and necrotic ring spot virus in *Prunus* species. Plant Dis. 68:216–218.

Hansen, A.J. (1988). Chemotherapy of plant virus infections. Applied Virology Res. 1:285–299.

Hansen, A.J. and Lane, W.D. (1985). Elimination of apple chlorotic leaf spot virus from apple shoot cultures by ribavirin. Plant Dis. 69:134–135.

Hasegawa, P.M. (1979). *In vitro* propagation of rose. Hort. Sc. 14:610–612.

Hasegawa, P.M. (1980). Factors affecting shoot and root initiation from cultured rose shoot tips. J. Am. Soc. Hort. Sci. 105:216–220.

Haskins, R.H. and Kartha, K.K. (1980). Freeze-preservation of pea meristems: Cell survival. Can J. Bot. 58:833–840.

Havel, L. and Novak, F.J. (1985). Meristem tip culture of *Allium cepa* L. Sci. Hort. 27:209–214.

Heller, R. (1953). Recherches sur la nutrition minerale des tissus vegetaux cultives *in vitro*. Ann. Sci. Natl. Biol. Veg. 36:31–38.

Hildebrandt, V. and Harney, P.M. (1988). Factors affecting the release of phenolic exudate from explants of *Pelargonium* × *hortorum*, Bailey 'Sprinter Scarlet'. J. Hort. Sci. 63:651–657.

Hoekema, A., Huisman, M.J., Molendijk, L., Van den-Elzen, P.J.M. and Cornelissen, B.J.C. (1989). The genetic engineering of two commercial potato cultivars for resistance to potato virus X. Bio/Technology 7:273–278.

Hollings, M. and Stone, O.M. (1968). Techniques and problems in the production of virus-tested planting material. Sci. Hort. 20:57–72.

Horsch, R.B., Fry, S.E., Hoffman, N.L., Eicholtz, D., Rogers, S.G. and Fraley, R.T. (1985). A simple and general method for transferring genes into plants. Science 227:1229–1231.

Hu, C.Y. and Wang, P.J. (1983). Meristem shoot tip and bud cultures. In: Handbook of Plant Cell Cultures. A. Evans, W.R. Sharp, P.V. Ammirato and Y. Yamada (eds.), Vol. 1. Macmillan, New York, p. 177–227.

Hughes, K.W. (1981). *In vitro* ecology: exogenous factors affecting growth and morphogenesis in plant culture systems. Env. Expt. Bot. 21:281–288.

Hunter, S.A., Foxe, M.J. and Hennerty, M.J. (1983). The influence of temperature and light intensity on the *in vitro* propagation of strawberry (*Fragaria* × *ananassa* Duch.) cv. Cambridge Favourites. Acta Hort. 131:153–161.

Hunter, S.A., Hannon, M., Foxe, M.J. and Hennerty, M.J. (1984). Factors affecting the *in vitro* production of strawberry (*Fragaria* × *ananassa*) meristems. cv. Cambridge Favourite. J. Life Sci. R. Dubl. Soc. 5:13–19.

Hussey, G. (1986). Vegetative propagation of plants by tissue culture. In: Plant Cell Culture Technology, M.M. Yeoman (ed.), Botanical Monographs, Vol. 23. Blackwell Scientific Publications, Oxford, pp. 29–66.

Hussey, G., Johnson, R.D. and Warren, S. (1989). Transformation of meristematic cells in the shoot apex of cultured pea shoots by *Agrobacterium tumefaciens* and *A. rhizogenes*. Protoplasma 148:101–105.

Hyndman, S.E., Hasegawa, P.M. and Bressan, R.N. (1982). Stimulation of root initiation from cultured rose shoots through the use of reduced concentrations of mineral salts. Hort. Sc. 17:82–83.

Jaiswal, V.S. and Amin, M.N. (1987). *In vitro* propagation of guava from shoot cultures of mature trees. J. Plant Physiol. 130:7–12.

James, D.J and Newton, B. (1977). Auxin: cytokinin interactions in the *in vitro* micropropagation of strawberry. Acta Hort. 78:321–331.

Janeckova, M. (1988). The raspberry virus elimination program in Czechoslovakia. Acta Hort. 236:45–50.

Jones, O.P., Zimmerman, R.H., Fordham, I.M. and Hopgood, M.E. (1985). Propagation *in vitro* of some dwarf apple trees. J. Hort. Sci. 60:141–144.

Kallak, H. and Koiveer, A. (1990). Induction of morphogenesis in meristems of different cultivars of *Pisum sativum* L. Plant Sci. 67:221–226.

Kanaiewski, W.K. and Thomas, P.E. (1988). A two step ELISA for rapid, reliable detection of potato viruses. Am. Potato J. 65:561–571.

Karp, A. (1989). Can genetic instability be controlled in plant tissue cultures? IAPTC News Lett. 58:2–11.

Kartha, K.K. (1981). Meristem culture and cryopreservation – methods and applications. In: Plant Tissue Culture – Methods and Applications. T.A. Thorpe (ed.), Academic Press, New York, pp. 118–211.

Kartha, K.K. (1984a). Culture of shoot apical meristems. In: Cell Culture and Somatic Cell Genetics of Plants, Vol. 1. I.K. Vasil (ed.), Academic Press, New York, pp. 106–110.

Kartha, K.K. (1984b). Tissue culture techniques for virus elimination and germplasm preservation. In: Plant Breeding Reviews, Vol. 2. J. Janick (ed.), AVI Publishing Co., Connecticut, pp. 265–279.

Kartha, K.K. (1985a). Cryopreservation of Plant Cells and Organs. CRC Press Inc., Boca Raton, Florida.

Kartha, K.K. (1985b). Meristem culture and germplasm preservation. In: Cryopreservation of Plant Cells and Organs. K.K. Kartha (ed.), CRC Press Inc., Boca Raton, Florida, pp. 115–134.

Kartha, K.K. (1986). Production and indexing of disease free plants. In: Pant Tissue Culture and its Agricultural Applications. L.A. Withers and P.G. Anderson (eds.), Butterworths, London, pp. 219–238.

Kartha, K.K. and Gamborg, O.L. (1975). Elimination of cassava mosaic disease by meristem culture. Phytopathology 65:826–828.

Kartha, K.K. and Gamborg, O.L. (1978). Meristem culture techniques in the production of disease-free plants and freeze-preservation of germplasm of tropical tuber crops and grain legumes. In: Diseases of Tropical Food Crops. H. Maraite and J.A. Mayer (eds.), Université Catholique, Louvain-la-Neuve, pp. 267–283.

Kartha, K.K., Gamborg, O.L. and Constabel, F. (1974a). Regeneration of pea (*Pisum sativum* L.) plants from shoot apical meristems. Z. Pflanzenphysiol. 72:172–176.

Kartha, K.K., Gamborg, O.L. and Constabel, F. (1974b). Regeneration of cassava plants from apical meristems. Plant Sci. Lett. 2:107–113.

Kartha, K.K., Leung, N.L. and Gamborg, O.L. (1979). Freeze-preservation of pea meristems in liquid nitrogen and subsequent plant regeneration. Plant Sci. Lett. 15:7–15.

Kartha, K.K., Leung, N.L. and Pahl, K. (1980). Cryopreservation of strawberry meristems and mass propagation of plantlets. J. Am. Soc. Hort. Sci. 105:481–484.

Kartha, K.K., Mroginski, L.A., Pahl, K. and Leung, N.L. (1981a). Germplasm preservation of coffee (*Coffea arabica* L.) by *in vitro* culture of shoot apical meristems. Plant Sci. Lett. 22:301–307.

Kartha, K.K., Pahl, K., Leung, N.L. and Mroginski, L.A. (1981b). Plant regeneration from meristems of grain legumes: soybean, cowpea, peanut, chickpea, and bean. Can. J. Bot. 59:1671–1679.

Klein, R.E. and Livingstone, C.H. (1982). Eradication of potato virus X from potato by ribavirin treatment of cultured potato shoot tips. Am. Potato J. 59:359–365.

Kondakova, V. and Schuster, G. (1991). Elimination of strawberry mottle virus and strawberry crincle virus from isolated apices of three strawberry varieties by the addition of 2,4-dioxohexa-hydro-1,3,5-triazine (S-Azadihydrouracil) to the nutrient medium. J. Phytopathology 132:84–86.

Krogstrup, P. and Thomsen, A. (1985). Use of meristem culture *in vitro* for elimination of raspberry ringspot virus (RRSV) in red current (*Ribesn* L. cv. Random). In: *In vitro* Techniques-Propagation and Long Term Storage. A. Schaefer-Menuhr (ed.), Martinus Nijhoff Publishers, Dordrecht, pp. 79–83.

Kudell, A.R. and Buchenauer, H. (1989). Elimination of raspberry dwarf virus from axillary bud cultures of red raspberry cv. Lloyd George by antiviral compounds. J. Phytopathology 124:332–336.

Kukreja, A.K., Mathur, A.K. and Zaim, M. (1990). Mass production of virus-free patchouli plants, *Pogostemon cablin* (Balnco) Benth., by *in vitro* culture. Trop. Agric. 67:101–104.

Lane, W.D., Looney, N.E. and Mage, F. (1982). A selective tissue culture medium for growth of compact (dwarf) mutants of apple. Theor. Appl. Genet. 61:219–223.

Lane, W.D. and McDougald, J.M. (1982). Shoot tip culture of apple: comparative response of five cultivars to cytokinin and auxin. Can. J. Plant Sci. 62:689–694.

Lawson, R.H. (1986). Pathogen detection and elimination. In: Tissue Culture as a Plant Production System for Horticultural Crops. R.H. Zimmerman (ed.), Martinus Nijhoff Publishers, Dordrecht, pp. 97–117.

Lawson, C., Kaniewski, W., Haley, L., Rozman, R., Newell, C., Sanders, P. and Tumer, N.E. (1990). Engineering resistance to mixed virus infection in a commercial potato cultivar: resistance to potato virus X and potato virus Y in transgenic Russet Burbank. Bio/Technology 8:127–134.

Lee, E.C.M. and de fossard, R.A. (1977). Some factors affecting multiple bud formation of strawberry (*Fragaria* × *ananassa* Duch.). Acta. Hort. 78:187–195.

Limmaset, P. and Cornuet, P. (1949). Recherche de virus de la mosaique du tabac (Marmor Tabaci Holmes) dans les meristemes des plantes infectées. Compt. Rend. 228:1971–1972.

Litz, R.E. and Conover, R.A. (1981). Effect of sex type, season, and other factors on *in vitro* establishment and culture of *Carica papava* L. explants. J. Am. Soc. Hort. Sci. 106:792–794.

Litz, R.E. and Jaiswal, V.S. (1991). Micropropagation of tropical and subtropical fruits. In: Micropropagation – Technology and Application. P.C. Debergh and R.H. Zimmerman (eds.), Kluwer Academic Publishers, Dordrecht, pp. 247–264.

Lommel, S.A., McCain, A.H. and Morris, T.J. (1982). Evaluation of indirect enzyme-linked immunosorbent assay for the detection of plant viruses. Phytopathology 72:1018–1022.

Lozoya-Saldana, H. and Dawson, W.O. (1982). Effect of alternating temperature regimes on reduction or elimination of viruses in plant tissues. Phytopathologyy 72:1059–1064.

Lozoya-Saldana, H. and Merlin-Lara, O. (1984). Thermotherapy and tissue culture for elimination of potato virus X (PVX) in Mexican potato cultivars resistant to late blight. Am. Potato J. 61:735–739.

Lozoya-Saldana, H. and Madrigal-Vargas, A. (1985). Kinetin, thermotherapy, and tissue culture to eliminate potato virus (PVX) in potato. Am. Potato J. 62:339–345.

Lundergan, C.A. and Janick, J. (1980). Regulation of apple shoot proliferation and growth *in vitro*. Hort. Res. 20:19–24.

Marcotrigiano, M., Swartz, H.J., Gray, S.E., Tokaric, D. and Popenoe, J. (1984). The effect of benzylaminopurine on the *in vitro* multiplication rate and subsequent field performance of tissue culture propagated strawberry plants. Recent Adv. Strawberry Prod. 3:23–25.

Martin, R.R. (1987). The use of monoclonal antibodies for virus detection. Can. J. Plant Pathol. 9:177–181.

McCown, B.H. and Lloyd, G.B. (1983). A survey of response of Rhododendron to *in vitro* culture. Plant Cell Tissue Org. Cult. 2:77–85.

Mellor, F.C. and Stace-Smith, R. (1969). Development of excised potato buds in nutrient medium. Can. J. Bot. 47:1617–1621.

Mellor, F.C. and Stace-Smith, R. (1987). Virus-free potatoes through meristem culture. In: Biotechnology in Agriculture and Forestry, Vol. 3: Potatoes. Y.P.S. Bajaj (ed.), Springer-Verlag, New York, pp. 30–39.

Miller, S.A. and Martin, R.R. (1988). Molecular diagnosis of plant disease. Ann. Rev. Phytopathol. 26:409–432.

Mohamed, N.A. and Imperial, J.S. (1984). Detection and concentration of coconut cadang-cadang viroid in coconut leaf extracts. Phytopathology 74:165–169.

Monette, P.L. (1983). Virus eradication through *in vitro* techniques. Int. Plant Propag. Soc. Comb. Proc. 33:90–100.

Monette, P.L. (1986). Elimination *in vitro* of two grapevine nepoviruses by an alternating temperature regime. J. Phytopathology 116:88–91.

Morel, G. (1960). Producing virus-free cymbidium. Am. Orchid Soc. Bull. 29:495–497.

Morel, G.M. (1972). Morphogenesis of stem apical meristem cultivated *in vitro*: Application to clonal propagation. Phytomorphology 22,:265–277.

Morel, G. and Martin, C. (1952). Guerison de dahlias atteints d'une malade à virus. C.R. Acad. Sci., Paris 235:1324–1325.

Morel, G. and Martin, C. (1955). Guerison de pommes de terre atteintis de malades à virus. C.R. Acad. Agric. Fr. 41:471–474.

Murashige, T., and Skoog, F. (1962). A revised medium for rapid growth and bioassays with tobacco tissue cultures. Physiol. Plant. 15:473–497.

Murashige, T. (1974). Plant propagation through tissue culture. Ann. Rev. Plant Physiol. 25:135–166.

Murashige, T. (1977). Clonal crops through tissue culture. In: Plant Tissue Culture and its Biotechnological Application. W. Barz, E. Reinhard and M.H. Zenk (eds.), Springer-Verlag, New York, pp. 392- 403.

Muriithi, L.M., Rangan, T.S. and Waite, B.H. (1982) *In vitro* propagation of fig through shoot tip culture. Hort. Sc. 17:86–87.

Nehra, N.S. and Singh, K. (1982). Propagation of fruit crops through tissue culture – a review. Haryana J. Hort. Sci. 11:13–16.

Nehra, N.S., Stushnoff, C. and Kartha, K.K. (1989). Direct shoot regeneration from strawberry leaf disks. J. Am. Soc. Hort. Sci. 114:1014–1018.

Nehra, N.S., Stushnoff, C. and Kartha, K.K. (1990a). Regeneration of plants from immature leaf-derived callus of strawberry (*Fragaria* × *ananassa*). Plant Sci. 66:119–126.

Nehra, N.S., Chibbar, R.N., Kartha, K.K., Datla, R.S.S., Crosby, W.L. and Stushnoff, C. (1990b). Genetic transformation of strawberry by *Agrobacterium tumefaciens* using a leaf disk regeneration system. Plant Cell Rep. 9:293–298.

Norton, M.E. and Norton, C.R. (1985). *In vitro* propagation of Ericaceae: a comparison of the activity of the cytokinins N^6-benzyladenine and N^6-isopentyladenine in shoot proliferation. Sci. Hort. 27:335–340.

Novak, F.J. and Jovova, Z. (1983). Clonal propagation of grapevine through *in vitro* axillary bud culture. Sci. Hort. 18:231–240.

Paduch-Cichal, E. and Kryczynski, S. (1987). A low temperature therapy and meristem-tip culture for eliminating four viroids from infected plants. J. Phytopathology 118:341–346.

Pierik, R.L.M. (1991). Commercial propagation in Western Europe and Israel. In: Micropropagation – Technology and Application. P.C. Debergh and R.H. Zimmerman (eds.), Kluwer Academic Publishers, Dordrecht, pp. 155–166.

Pink, D.A.C., and Walkey, D.G.A. (1984). Rapid propagation of *Cucurbita pepo* L. by culture of meristem tips. Sci. Hort. 24:107–114.

Polston, J.E., Dodds, A.J. and Perring, T.M. (1989). Nucleic acid probes for detection and strain discrimination of cucurbita geminiviruses. Phytopathology 79:1123–1127.

Pua, E-C. and Chong, C. (1984). Requirement for sorbitol (D-glucitol) as carbon source for *in vitro* propagation of *Malus robusta* No. 5. Can. J. Bot. 62:1545–1549.

Reichenbacher, D., Reiss, E., Hunger, H.-D. and Richter, J. (1990). Detection of plant viruses by ELISA using different reagent strips. J. Phytopathology 128:333–339.

Robinson, D.J., Harrison, B.D., Sequeira, J.C. and Duncan, G.H. (1984). Detection of strains of African cassava mosaic virus by nucleic acid hybridization and some effects of temperature on their multiplication. Ann. Appl. Biol. 105:483–493.

Roca, W.M., Chavez, R., Martin, M.L., Arias, D.I., Mafla, G. and Reyes, R. (1989). *In vitro* methods of germ-plasm conservation. Genome 31:813–817.

Roy, B.P., AbouHaidar, M.G., Sit, T.L. and Alexander, A. (1988). Construction and use of cloned cDNA biotin and ^{32}P-labeled probes for the detection of papaya mosaic potex virus RNA in plants. Phytopathology 78:1425–1429.

Rubluo, A. and Kartha, K.K. (1985). *In vitro* culture of shoot apical meristems of various *Phaseolus* species and cultivars. J. Plant Physiol. 119:425–433.

Sakai, A. (1985). Cryopreservation of shoot tips of fruit trees and herbaceous plants. In: Cryopreservation of Plant Cells and Organs. K.K. Kartha (ed.), CRC Press, Boca Raton, Florida, pp. 135–158.

Samartin, A. (1989). A comparative study of effects of nutrient media and culture conditions on shoot multiplication of *in vitro* cultures of *Camellia japonica* explants. J. Hort. Sci. 64:73–79.

Samyn, G.L., Debergh, P.C. and Vermaerke, D. (1984). Field performance and phenotypic stability of virus-free tissue cultured *Begonia* × *tuberhybrida* 'Multiflora'. Sci. Hort. 24:185–191.

Schaeffer, W.G., Damiano, C., Scott, D.H., McGrew, J.R., Krul, W.R. and Zimmerman, R.H. (1980). Transcription of panel discussion on the genetic stability of tissue culture propagated plants. In: Proceedings of the Conference on Nursery Production of Fruit Plants Through Tissue Culture-Applications and Feasibility. USDA-SEA, Agric. Res. Results, ARE-NE-11, pp. 64–79.

Schrammeijer, B., Sijmons, P.C., van den Elzen, P.J.M. and Hoekema, A. (1990). Meristem transformation of sunflower via *Agrobacterium*. Plant Cell Rep. 9:55–60.

Scott, D.H. and Zanzi, C. (1980). Rapid propagation of strawberry from meristems. In: The Strawberry-Cultivars to Marketing. N.F. Childers (ed.), Horticultural Publications, Gainesville, Florida, pp. 213–221.

Seckinger, G.R. (1991). Micropropagation of vegetable crop species. In: Micropropagation – Technology and Application. P.C. Debergh and R.H. Zimmerman (eds.), Kluwer Academic Publishers, Dordrecht, pp. 265–284.

Sela, I., Reichman, M. and Weissbach, A. (1984). Comparison of dot molecular hybridization and enzyme-linked immunosorbent assay for detecting tobacco mosaic virus in plant tissues and protoplasts. Phytopathology 74:385–389.

Simpkins, I., Walkey, D.G.A. and Neely, H.A. (1981). Chemical suppression of virus in cultured plant tissues. Ann. Appl. Biol. 99:161–169.

Simpson, D.W. and Bell, J.A. (1989). The response of different genotypes of *Fragaria* × *ananassa* and their seedling progenies to *in vitro* micropropagation and the effect of varying the concentrations of 6-benzylaminopurine in the proliferation medium. Plant Cell Tissue Org. Cult. 17:225–234.

Singha, S. (1982). *In vitro* propagation of crabapple cultivars. Hort. Sc. 17:191–192.

Sommer, H.E. (1982). Organogenesis in angiospermous trees. Application to vegetative propagation. Bulletin de la Societé Botanique de la France 72:128–130.

Stace-Smith, R. and Mellor, F.C. (1968). Eradication of potato virus X and S by thermotherapy and axillary bud culture. Phytopathology 58:199–203.

Stone, O.M. (1963) Factors affecting the growth of carnation plants from shoot apices. Ann. Appl. Biol. 52:199–209.

Stone, O.M. (1982). The elimination of four viruses from *Ullucus tuberosus* by meristem tip culture and chemotherapy. Ann. Appl. Biol. 101:79–83.

Styer, D.J. and Chin, C.K. (1983). Meristem and shoot tip culture for propagation, pathogen elimination and germplasm preservation. Hort. Rev. 5:221–277.

Swartz, H.J., Galletta, G.J. and Zimmerman, R.H. (1981). Field performance and phenotypic stability of tissue-culture-propagated strawberries. J. Am. Soc. Hort. Sci. 106:667–673.

Swartz, H.J., Galletta, G.J. and Zimmerman, R.H. (1983). Field performance and phenotypic stability of tissue-culture-propagated thornless blackberries. J. Am. Soc. Hort. Sci. 108:285–290.

Tisserat, B. (1984). Propagation of date palms by shoot tip cultures. Hort. Sc. 19:230–231.

Theiler-Hedtrich, R. and Baumann, G. (1989). Elimination of apple mosaic virus and raspberry bushy dwarf virus from infected red raspberry (*Rubus idaeus* L.) by tissue culture. J. Phytopathology 127:193–199.

Toussaint, A., Dekegel, D. and Vanheule, G. (1984). Distribution of *Odontoglossum* ringspot virus in apical meristems of infected Cymbidium cultivars. Physiological Plant Pathology 25:297–305.

Ulian, E.C., Smith, R.H., Gould, J.H. and McKnight, T.D. (1988). Transformation of plants via the shoot apex. *In vitro* Cell. Dev. Biol. 24,:951–954.

Vajrabhaya, T. (1977). Variation in clonal propagation. In: Orchid Biology: Review and Perspectives. J. Arditti (ed.), Columbia University Press, New York, pp. 179–201.

Van den Heuvel, J.F.J.M. and Peters, D. (1989). Improved detection of potato leafroll virus in plant material and in aphids. Phytopathology 79:963–967.

Van Regenmortel, M.H.V. (1982). Serology and Immunochemistry of Plant Viruses. Academic Press, New York.

Varveri, C., Candresse, T., Cugusi, M., Ravelonandro, M. and Dunez, J. (1988). Use of ^{32}P-labeled transcribed RNA probe for dot hybridization detection of plum pox virus. Phytopathology 78:1280–1283.

Vasil, I.K. (1991). Cell Culture and Somatic Cell Genetics of Plants, Vol. 8. Scale-up and Automation in Plant Propagation. Academic Press, San Diego.

Vasil, I.K. and Vasil, V. (1980). Clonal propagation. In: Perspectives in Plant Cell and Tissue Culture. I.K. Vasil (ed.), Int. Rev. Cytol., Supplement 11A, Academic Press, New York, pp. 145–174.

Vieitez, A.M., Carmen San-Jose, M. and Vieitez, E. (1985). *In vitro* plantlet regeneration from juvenile and mature *Ouercus robur* L. J. Hort. Sci. 60:99–106.

Vieitez, A.M. and Vieitez, M.L. (1983). *Castanea sativa* plantlets proliferated from axillary buds cultivated *in vitro*. Sci. Hort. 18:343–351.

Vine, S.J. and Jones, O.P. (1969). The culture of shoot tips of hop (*Humulus lupulus*) to eliminate viruses. J. Hort. Sci. 44:281–284.

Walkey D.G.A (1980) Production of virus-free plant by tissue culture. In: Tissue Culture Methods for Plant Pathologist. D.S. Ingram and J.P. Helgeson (eds.), Blackwell Scientific Publications, Oxford, pp. 109–117.

Wakley, D.G.A., Webb, M.J.W., Bolland, C.J. and Miller, A. (1987). Production of virus-free garlic (*Allium sativum* L.) and shallot (*A. ascalonicum* L.) by meristem-tip culture. J. Hort. Sci. 62:211–220.

Wambugu, F.M., Secor, G.A. and Gudmestad, N.C. (1985). Eradication of potato virus Y and S from potato by chemotherapy of cultured axillary bud tips. Am. Potato J. 62:667–672.

Wanas, W.H., Callow, J.C. and Withers, L.A. (1986). Growth limitation for the conservation of pear genotypes. In: Plant Tissue Culture and its Agricultural Applications. L.A. Withers. and P.G. Alderson (eds.), Butterworths, London, pp. 285–290.

Watkins, C.A., McNicol, R.J., Young, K. and Jones, A.T. (1990). The effect of heat treatment and meristem tip culture on June yellows in strawberry. Ann. Appl. Biol. 116:489–492.

Welander, M. (1985). *In vitro* culture of raspberry (*Rubus ideaus*) for mass propagation. Hort. Sci. 60:493–500.

White, P.R. (1943). A Handbook of Plant Tissue Culture. Jaques Cattell Press, Lancaster.

Withers, L.A., Wheelans, S.K. and Williams, J.T. (1990). *In vitro* conservation of crop germplasm and IBPGR databases. Euphytica 45:9–22.

Wochok, R.J. and Sluis, C.J. (1980). *In vitro* propagation and establishment of wax current (*Ribes inebrians*). J. Hort. Sci. 55:355–357.

4. Plant Protoplasts for Cell Fusion and Direct DNA Uptake: Culture and Regeneration Systems

Attila Fehér and Dénes Dudits

Contents

1. Plant Protoplasts as Unique Experimental Objects

1.1. *Definition*

The plasma membrane of plant cells is surrounded by a cellulose wall and adjacent cells are connected with a pectin-rich matrix. Disconnection of plant cells and removal of the cell wall experimentally, by either a mechanical or an enzymatic process, results in the production of single, 'naked' plant cells termed 'isolated protoplasts'. Protoplasts thus can be considered as special individual plant cells with fully exposed outer plasma membrane which is the only barrier between the external environment and the interior of the living

Indra K. Vasil and Trevor A. Thorpe (eds.), Plant Cell and Tissue Culture, 71–118.

cells. However, plant protoplasts cannot serve as the botanical analog of animal cells, but they should be viewed as injured cells (see Section 2.3.2) that have to go through a repair process before they are capable of sustained division.

1.2. *Protoplasts as Tools in Plant Sciences*

Plant cells are normally connected to each other via many plasmodesmata in multicellular tissues. Isolated protoplasts represent actually the only source of single-cell system obtainable in higher plants. Protoplasts forming millions of individual cells in culture, can be applied for microbial techniques (e.g. mutant isolation, cell cloning, cell fusion, genetic transformation and complementation etc.) and their totipotency allows the regeneration of fully mature, flowering plants.

In several aspects, protoplasts represent an unique experimental system as 'naked' plant cells. They can be used to study *de novo* cell wall formation, although the consequences of their stressed nature have to be considered (see Section 2.3.2). As the plasma membrane is directly exposed to the culture medium, protoplasts are very useful objects of membrane studies (e.g. membrane transport or other uptake processes, membrane structure etc.). Protoplasts provide advantageous experimental material for ultrastructural investigations (e.g. microtubule array and rearrangement) as well as for the isolation of cell organelles (nuclei, metaphase chromosomes, chloroplasts, mitochondria, vacuoles).

The above special features have made plant protoplasts highly valuable and irreplaceable experimental objects in many fields of plant science (for more details see the reviews by Fowke and Gamborg, 1980; Galun, 1981).

1.3. *Key Events in History of Protoplast Research*

Mechanical isolation of protoplasts from plasmolysed tissues of higher plants has been reported as early as 1892 by Klerckel (1892). The real start of plant protoplast research, however, can be dated back to the 1960's when the enzymatic removal of plant cell walls was demonstrated (Cocking, 1960). It became a widely used technique with the production of commercially available enzymes, first used by Takabe et al. (1968). Only three years later, the totipotency of plant protoplasts was affirmed by the successful regeneration of whole tobacco plants from mesophyll protoplasts (Takabe et al., 1971). During this period, simultaneously with the investigation of the specific features and behavior of protoplasts in culture (as reviewed by Cocking, 1972) the elaboration of techniques for genetic manipulation of plants through the use of protoplasts was started. After the regeneration of the first interspecific somatic hybrid plants (Carlson et al., 1972), the protoplast fusion experiments were stimulated by the introduction of polyethylene-glycol (PEG) as a potent fusogen (Kao et al., 1974; Wallin et al., 1974). The improved fusion methods have resulted in the production of several somatic hybrids including the first intergeneric hybrid (Melchers et al., 1978) and the first cybrid (Zelcer et al., 1978) plants, and the achievement of gene transfer by asymmetric hybridization (Dudits et al., 1979, 1980a). During these years, the induction of protoplast fusion by electrical stimulus was also achieved by Senda et al. (1979). Further improvement of this technique by Zimmermann and Scheurich (1981) provided an efficient alternative to chemically-induced fusion methodology.

Although the first report on the detection of DNA uptake by plant protoplasts using radioactive bacterial DNA can be dated back to 1972 (Ohyama et al., 1972), the real development of direct protoplast transformation techniques only started with the introduction of plasmid vector molecules (Lurquin and Kado, 1977) and expression of plasmid DNA-coded characters (Davey et al., 1980) in plant protoplasts. During the 1980s, the elaboration of PEG-mediated (Krens et al., 1982; Paszkowsky et al., 1984) and electroporation-induced (Fromm et al., 1985, 1986) DNA uptake procedures and the use of selectable marker genes in the transformation experiments (Paszkowsky et al., 1984) resulted in the regeneration of numerous transgenic plants (as reviewed e.g. by Weising et al., 1988; see Chapter 10). The development of micromanipulation and microculture techniques by Koop and Schweiger (1985a, b) during this period, can be considered another significant contribution to protoplast research.

In spite of this great progress in protoplast manipulation techniques, some agronomically significant taxa including cereal crops, grasses, legumes, woody trees, conifers etc. proved to be notoriously recalcitrant in protoplast culture and regeneration. The concentrated efforts of several laboratories have resulted in a real breakthrough in this respect only during the late 1980's and at the beginning of the 1990's (Vasil, 1988). This breakthrough has been achieved due to the choice of proper genotypes and tissues as the source of starting material, to the efforts to minimize stress during the isolation procedure and to the improvement of culture and regeneration systems.

2. Plant Regeneration from Cultured Protoplasts

2.1. *Differences between Species, Role of Genotype*

Genotype of the donor plant has considerable influence on every step of protoplast isolation and culture as the yield, viability and even more, the regeneration capability is concerned. The composition of the cell wall and the plasmolytic behaviour of the cells (determining the sensitivity of protoplasts against isolation procedures) can be responsible for the differences between species in response to the isolation process. Cell wall components of monocots, especially *Gramineae*, are different from those of dicots that can cause significant differences in requirement for use of defined enzyme combinations for effective protoplast isolation (see the review of Ishii, 1989). Comparison between amenable and recalcitrant species to protoplast isolation and culture emphasized the significance of plasmolitic response (Smith et al., 1989a).

The effect of the genotype on the competence of plant protoplasts for colony formation and plant regeneration is evident from large number of studies with several cultivars/genotypes under the same protoplast isolation and culture conditions (e.g. Vardi et al., 1982; Haberlach et al., 1985; Kyozuka et al., 1988; Loudon et al., 1989; Poulsen and Nielsen, 1989; Jenes and Pauk, 1989). Based on the analysis of the genetic background of potato cell and tissue culture response, Coleman et al. (1990) have suggested that protoplast plating efficiency and regeneration from leaf discs is under separate, and relatively simple genetic control with 'blocks of genes' located on specific chromosomes. In *Solanum phureja*, 'protoplast culturability' as the capability of a genotype to develop calli from cultured protoplasts, was proposed to be controlled by two independent loci with complete dominance (Cheng and Veilleux, 1991). It has to be mentioned, however, that the genotype with the best culture response may not necessarily exhibit the best regeneration capability (e.g. Myers et al., 1989).

Recently, the successes in the use of breeding strategies to combine valuable tissue culture traits in synthetic genotypes of tomato (Koornneef et al., 1986) and maize (Mórocz et al., 1990) have been reported.

2.2. *Sources of Plant Material*

Besides the genotype, the source of explant material and its actual physiological state is one of the most important factors during the entire process of protoplast isolation and culture.

2.2.1. *Growth Conditions of the Donor Plants*

Plants grown in the field or grown in greenhouse are often readily usable for protoplast work. Light and temperature regimes, nutrient supply, developmental stage of the donor plants, however, as sources of variation can often not be well enough controlled to give optimal yield (Cassells and Cocker, 1982; Cassells and Tamma, 1986). Furthermore, the surface sterilization of the explants can result in additional problems.

The use of *in vitro* shoot cultures ensures juvenil growth, provides maximum control on the growth conditions and axenic explant material for protoplast isolation. During optimalization of the culture conditions for *in vitro* grown plants several factors should be considered. The age of the culture, the temperature and light regimes, the composition of the culture medium have been shown among others to influence protoplast yield and stability. In *in vitro* cultures, the accumulation of ethylene can be a crucial factor affecting the yield and quality of released protoplasts (Perl et al., 1988; Rethmeier et al. 1991). In this respect, the significance of the length of the *in vitro* culture period of the donor plants before protoplast isolation has been demonstrated by Rethmeier et al. (1991). The presence of ethylene antagonists such as silver ions (silver thiosulphate or nitrate) or acetylsalicylic acid in the medium was beneficial for improving protoplast yield and plating efficiency in potato (Perl et al., 1988) and *Lycopersicon pennellii* (Rethmeier et al., 1991).

2.2.2. *Protoplast Source*

In general, actively growing, young and healthy cells or tissues provide protoplasts with high yield and good quality. The most convenient and widely used source of plant protoplasts is the leaf. Protoplasts have been isolated, however, from a variety of other organs and tissues. Seedling tissues (first of all hypocotyl and cotyledon) can be an alternative source of material for isolation (see Table 1) if the culture of protoplasts from leaves is unsuccessful. Embryonic or meristematic tissues may provide protoplasts with high proliferation potential but the efficiency of isolation can be relatively low.

Cell suspension cultures are also frequently used for protoplast isolation. Since the cultured cells are already conditioned to *in vitro* growth, they can tolerate stresses related to isolation. They provide a population of uniform protoplasts at a high yield. Moreover, the division capability of protoplast derived cells can be easily insured in the protoplast culture medium. Genetic instability of long term cultures, however, can result in a high rate of aneuploids and in the loss of morphogenic potential. The age of the suspension cultures influences the complexity of the cell wall structure and thus protoplast yield and quality vary with time in culture and with the frequency of subculture of the cell suspension. Cell cultures as protoplast sources gained remarkable significance with the discovery and exploitation of embryogenic cell suspension cultures (Vasil, 1987, 1988). The use of these cultures has contributed to the development of procedures for protoplast culture and plant regeneration of many formerly recalcitrant species with economic importance (e.g. cereals and certain conifers).

2.3. *Isolation of Plant Protoplasts – a Stressful Procedure*

2.3.1. *General Guidelines for Isolation*

(*a*) *Enzymatic cell wall digestion.* Enzymes used for plant protoplast isolation and sold under different commercial names (listed e.g. by Ishii, 1989), fall into two main categories in respect to their action on plant cell walls. Pectin-

Table 1. Selected papers on recent achievements in plant regeneration from cultured protoplasts

Species	Pp. source[a]	Reg.[b]	Remarks	References
CEREALS				
Oryza sativa	ECS(ant.c)	FP,HP		Toriyama et al., 1986
var. *japonica*	ECS(imm.e)	P	NC	Kyozuka et al., 1987
	ECS(mat.s)	P		Wang et al., 1989a
	ECS(germ.e)	P,S	DG	Masuda et al., 1989
	ECS(imm.e)	FP	AE	Jenes and Pauk, 1989
	ECS(sec.scut)	FP	NC	Li and Murai,1990
var. *indica*	ECS(mat.s)	P	NC	Kyozuka et al., 1988
	ECS(mat.s)	P		Wang et al., 1989a
	ECS(imm.e,mat.s)	P	NC	Lee et al., 1989
	ECS(micr.sp)	FP	AE	Datta et al., 1990a
	lf.b	P	NC	Gupta and Pattanayak, 1993
Zea mays	ECS(imm.e)	SP	NC	Rhodes et al., 1988a
	ECS(imm.e)	FP	NC	Prioli and Söndahl, 1989
	ECS(imm.e)	FP	NC,AE	Shillito et al., 1989
	ECS(micr.sp)	HP		Sun et al., 1989
	ECS(imm.e)	FP		Mórocz et al., 1990
	ECS(micr.sp)	HP		Mitchell and Petolino, 1991
Triticum aestivum	ECS(imm.e)	P	AE	Vasil et al., 1990
	ECS(CS)	P	AE	Chang et al., 1991
	ECS(imm.e)	P,FP	AE	Ahmed and Sági, 1993
Hordeum vulgare	ECS(imm.e)	APl	AE	Luhrs and Lörz, 1988
	ECS(ant.c.)	AP,FP	AE	Jahne et al., 1991
Sorghum vulgare	ECS(imm.inf)	FP		Wei and Xu, 1990
GRASSES				
Dactylis glomerata	ECS(lf.b)	P	AE,CM	Horn et al., 1988a
Lolium perenne	ECS(mat.e)	P,AP		Dalton et al., 1988
	ECS(imm.e/inf)	P	AE	Creemers-Molenaar et al., 1989
Festuca arundinacea	ECS(mat.e)	P,AP		Dalton et al., 1988
	ECS(imm.e,lf.b)	P	AE,NC	Takamizo et al., 1990
Fagopyrum esculentum	hypoc.	P		Adachi et al., 1989
Agrostis alba	ECS(mat.s)	P	AE	Asano and Sugiura, 1990
VEGETABLES				
Brassica oleracea				
var. *italica*	hypoc.	P		Kao et al., 1990
	hypoc.,leaf	P	NC	Walters and Earle, 1990
var. *capitata*	leaf	P		Loudon et al., 1989
var. *botrytis*	hypoc.,leaf	P	NC	Walters and Earle, 1990
	leaf	P		Jourdan et al., 1990
Cucumis sativus	cotyl.,leaf	P	AE	Colijn-Hooymans et al., 1988
	cotyl.,leaf	P	AE	Punja et al., 1990
Asparagus officinalis	EC	P		Kunitake and Mii, 1990a
Capsicum annuum	leaf	FP		Diaz et al., 1988
TUBER AND ROOT CROPS				
Ipomoea batatas	pet.,stem	P		Sihachakr and Ducreux, 1987
Beta vulgaris	leaf	P		Krens et al., 1990
OIL CROPS				
Brassica napus				
var. *oleifera*	leaf	P		Loudon et al., 1989

Table 1. Continued

Species	Pp. source[a]	Reb.[b]	Remarks	References
	hypoc.	FP		Poulsen and Nielsen, 1989
Brassica carinata	cotyl.	P	AE	Jaiswal et al., 1990
Helianthus annuus	hypoc.	FP	AE	Burrus et al., 1991
LEGUMES				
Glycine max	imm.cotyl.	FP		Wei and Xu, 1988
	imm.cotyl.	P	AE	Dhir et al., 1991
Pisum sativum	e.ap.,lat.b.	FP		Lehminger-Mertens and Jacobsen, 1989
Stylosanthes sp.	cotyl.	P	AlgE	Vieira et al., 1990
Trifolium pratense	leaf,CS	P		Myers et al., 1989
Lotus corniculatus	root hairs	FP	AE	Rasheed et al., 1990
Lotus tenuis	cotyl.,root	P		Piccirilli et al., 1988
Lotus pedunculatus	leaf,cotyl.	P		Pupilli et al., 1990
WOODY TREES				
Larix eurolepis	ECS(imm.e.)	P	DG	Klimaszewska, 1989
Picea glauca	ECS	Pl		Attree et al., 1989
Populus sp.	leaf	P		Russel and McCown, 1988
Malus pumila	leaf	Pl	AE	Wallin and Johansson, 1989
Malus × *domestica*	leaf	P	AE	Patat-Ochatt et al., 1988
Prunus cerasus	root callus	P	AE	Ochatt, 1990
Prunus avium × *pseudocerasus*	CS(root)	P	ES	Ochatt et al., 1988a, b
Pyrus communis	leaf	P		Ochatt and Power, 1988
Actinidia deliciosa	pet. callus	P		Oliviera and Pais, 1991
Coffea canephora	som.e.	Pl		Shöpke et al., 1987
ORNAMENTALS				
Rosa sp.	ECS(root,int.n)	Pl	AE	Matthews et al., 1991.
Pelargonium sp.	CS(leaf,pet.)	Pl		Yarrow et al., 1987
Centaurea cyanus	leaf	P	AE	Pillai et al., 1990
Senecio x hybridus	leaf	P	AE	Pillai et al., 1990
Callistephus chinens.	leaf	P	AE	Pillai et al., 1990
Chrysanthemum sp.	leaf	P		Sauvadet et al., 1990
Limonium perezii	CS(leaf)	P	GgE	Kunitake and Mii, 1990b
MISCELLANEOUS				
Fragaria × *Ananassa*	leaf	P		Nyman and Wallin, 1988
Arabidopsis thaliana	leaf	FP	AlgE	Damm and Willmitzer, 1988
	leaf	FP	AlgE	Karesch et al., 1991a

[a]Source of protoplasts: the original source tissue of callus or cell cultures are indicate between the parentheses.

[b]Reported achievements in plant regeneration.

Abbreviations: EC = embryogenic callus; CS = cell suspension; ECS = embriogenic cell suspension; hypoc = hypocotyl; cotyl = cotyledon; pet = petiole; imm = immature; mat = mature; germ = germinating; som = somatic; e = embryo; ap = apex; s = seed; inf = inflorescence; ant.c = anther culture; micr.sp = microspore; lat.b = lateral shoot bud; lf.b = leaf base; P = plant; Pl = plantlet; F = fertile; H = haploid; S = sterile; A = albino; S = shoot; AE = agarose embedding; AlgE = alginate embedding; GgE = gellan gum embedding; ES = electrostimulation; DG = density gradient fractionation; NC = nurse culture; CM = conditioned medium.

ases dissolve the middle lamella and in this way separate cells from each other, while cellulases and hemicellulases degrade cellulose cell walls and eventually release protoplasts. Cellulase Onozuka R-10 and Macerozyme R-10 (and Pectolyase Y-23 if cell wall material is diffucult to degrade) are the most widely used enzymes that provide a suitable yield of protoplasts in many cases. Most often an empirical approach is required, however, to develop the proper enzyme mixture for protoplast isolation.

To facilitate the penetration of enzyme solution into the intracellular spaces of plant tissues, that is essential for effective digestion, various methods can be applied. The most simple and commonly practiced method is the cutting of the leaf (or other explant) into small pieces or strips. Excessive damage of the cells, however, can result in a low yield of healthy protoplasts. Infiltration of the enzyme into the tissues by vacuum can further enhance the digestion efficiency.

The duration of the enzyme treatment also has to be optimized to get the suitable yield and quality. Some cells can tolerate a long treatment period (up to 20–24 h) with lower enzyme concentration, while in other cases, a short but extensive cell wall digestion gives better results in respect of protoplast viability. Protoplast yield and survival rate often are inversely affected by the incubation time and enzyme concentration (e.g. Dupuis et al., 1990).

Dai et al. (1987) found that the adjusment of the appropriate osmotic potential in the enzyme solution close to the isotonic level of potato hypocotyl cells is one of the most important factors in the isolation of stable protoplasts from this tissue. The significance of the type of the osmoticum in diminishing of stresses during oat mesophyll protoplast isolation has been reported by Mussell et al. (1986).

For standard treatment, a mixture of 'cellulase' and 'pectinase' at concentrations around 0.5–2% w/v is dissolved in an osmoticum (giving 400–800 mOsm kg^{-1}) at pH 5.2–5.6. The digestion time can be varied from 0.5 up to 20 h at 24 °C.

(*b*) *Purification and harvest of protoplasts.* Protoplasts are usually freed of the hydrolyzing enzymes and the debris by a combination of filtration, centrifugation, and washing. The enzyme solution with the protoplasts is filtered through a stainless steel or nylon mesh (50–100 μm) to remove larger portions of undigested tissues and cell clumps. For further purification one of the following two methods has been generally followed. In situations, where sugar alcohols and/or salts are used as osmotica in the enzyme and washing solutions, the protoplasts can be pelleted by gentle centrifugation (75–100 × g for 2–5 min) and washed at least twice by repeated resuspention and sedimentation. Alternatively, if dense solutions like Percoll or sucrose are applied, the highly vacuolated protoplasts can be floated to the top of the solution or to the interface between solutions with different densities. In this way a more homogeneous protoplast population can be obtained even in the case of very fragile protoplasts. Furthermore, it is possible to separate different protoplast types by discontinuous density gradient centrifugation (Harms and Potrykus, 1978) and to enrich protoplast populations with higher division (Masuda et al., 1989) or embryogenic (Klimaszewska, 1989) potential. These steps are usually also followed by washing with a low density solution. During protoplast purification, the maintenance of the same osmotic pressure during entire procedure, is of significant importance. The presence of Ca^{2+} ions in the washing solution is often necessary for membrane stability (Chen et al., 1988).

The use of protoplasts in further studies often requires the check of their viability after accomplishing the isolation procedures. A number of viability tests are available for this purpose (Gahan, 1989).

2.3.2. *Attempts to Diminish Stress Incurred by Plant Protoplasts During Isolation*

The stressful procedure of the enzymatic plant cell wall removal is often accompained by increased ethylene synthesis (e.g. Cassells et al., 1980, Facciotti and Pilet, 1981; Anderson et al. 1982; Mussell et al., 1986; Cassells and Tamma, 1986), the production of polyphenols (e.g. Glimelius, 1984; Chen et al., 1988), phytoalexins (Mieth et al., 1986) and superoxide radicals (e.g. Ishii, 1988; Cutler et al., 1989). All these factors can often cause a lowered protoplast survival. The stress incurred during isolation of protoplasts, however, is not necessarily reflected by viability test, therefore the subsequent performance in culture should also be analysed (Smith et al., 1989a).

Tobacco mesophyll protoplasts have been shown to synthetise proteins (Meyer et al., 1984a, b; Grosset et al., 1990a) and mRNAs (Grossett et al., 1990b) that were undetectable in the leaf tissue. Some of them could be characterized as pathogenesis related (PR) proteins. The accumulation of these proteins as well as the disappearance of leaf specific proteins in this case could be ascribed to the effect of wounding independently of the enzyme treatment (Grossett et al., 1990b). Additionally, according to several reports (e.g. Asamizu and Nishi, 1980; Pilet et al., 1984; Mock et al., 1990) the incipient cell walls of plant protoplasts are very different from that of normal plant cells. One of the most important differences is the synthesis of callose. This polysaccharide is otherwise typically synthesized in response to injury and elicitation of plant cells instead of cellulose (see the review of Bowles, 1990). Recently, Jouanneau et al. (1991) have reported that in plant protoplasts these spontaneous defense reactions are under auxin control.

These observations indicate that plant protoplasts exhibit typical stress responses similarly to injured or pathogen attacked cells. Furthermore, enzymatic plant cell wall digestion has been found to inhibit nuclear division (Weber et al., 1986; Wang et al., 1989b), change the protoplast membrane lipid composition (Browse et al., 1988) and plasmamembrane protein phosphorylation (Blowers et al., 1988). These have to be always considered when one intends to use protoplasts as analogoues of the original differentiated cells.

The above mentioned stress responses can be ascribed to several factors associated with the protoplast isolation procedure: to the removal of the cells from their natural tissue environment and placing them under unphysiological conditions, to wounding effect before and during the enzyme treatment, to osmotic shock, to the action of mostly microbial enzymes used for cell wall digestion, and to the impurities of enzyme solutions.

Recognition of the fact that protoplasts are subjected to a range of stresses during isolation and culture which finally result in low viability, has led to the search for possible ways to diminish these stresses. On the one hand, the proper physiological state of the donor plants/cells has been found to be very important (see also Section 2.2.1). In this respect, the application of some pretreatments of plants or tissues before isolation has a beneficial effect on protoplast yield and stability. Furthermore, the removal of impurities from commercial enzyme preparations and the use of inhibitors of stress response pathways as well as the application of optimal culture conditions can be considered as possible approaches to diminish the above mentioned stresses.

(*a*) *Preconditioning of donor plants/tissues*. The exposure of the donor plants or explants to low light intensities and/or shortened photoperiods or even dark for a defined period of time can result in consistently higher yield of viable protoplasts, for example in *Solanum* (e.g. Shepard and Totten, 1977; Haberlach et al., 1985; Masson et al., 1987; Fehér et al., 1989; Chang and Loescher, 1991) and *Lycopersicon* (Tan et al., 1987; Niedz and Sink, 1988) species. Explanations for these observations could be the synthesis of thinner cell walls with lower concentrations of pectate at low light intensities (Cassels and Barlass, 1976) and/or the mobilization of starch deposited in starch grains which otherwise distrupt fragile protoplasts during plasmolysis (Lee-Stadelmann et al., 1985) or centrifugation (Watts et al., 1974; Chang and Loescher, 1991). Stability of protoplasts can be increased by cold (4–10 °C for 6–24 hours) or a combination of cold/dark pretreatment as well (Haberlach et al., 1985, Tan et al., 1987; Niedz and Sink, 1988; Fehér et al., 1989), but the basis of this effect is not yet clearly understood. A change in the lipid composition of the plasma membrane or as suggested by Tan et al. (1987) a decreased activity of enzymes involved in harmful oxidation processes can be responsible factors in the better adaptation of the cells to stress conditions during protoplast release. Excision and preculture of plant tissues prior to protoplast isolation is sometimes recommended (e.g. Berry et al., 1982; Haberlach et al., 1985).

(*b*) *Preplasmolysis*. Isolation of plant protoplasts should be carried out in hypertonic solutions otherwise the removal of the cell wall would result in protoplast bursting due to the internal turgor of the plant cells. In order to minimize the detrimental effect of the enzyme treatment, plant tissues are often plasmolysed before cell wall digestion. Preplasmolysis reduces the amount of enzyme uptake by endocytosis during protoplast contraction and the shrinking of protoplasts seals plasmodesmata probably reducing the leakage of cell contents. Preplasmolysis time was found to be critical in respect of protoplast yield and viability in sugarbeat (Krens et al., 1990) and apple (Wallin and Johansson, 1989). The protoplasmic properties of some plants (high cell sap osmolarity or strong adherence of the plasma membrane to the cell wall) can inhibit plasmolysis, with negative influence on the success of isolation and the quality of protoplasts (Lee-Stadelmann et al., 1985). Use of octylguanidine in plasmolysing solution has been shown to decrease the adherence of the plasmamembrane to the cell wall and to enhance plasmolysis in soybean leaf cells (Lee-Stadelmann et al., 1985). Abscisic acid treatment of apple leaf tissue during preplasmolysis has been found to improve division ability of protoplasts by Wallin and Johansson (1989).

(*c*) *Cell wall hydrolysing enzymes as stress factors*. Crude commercial preparations of enzymes often contain impurities (proteinaceous as well as low molecular weight compounds) that may be harmful to protoplasts. Gel filtration of enzyme solution by elution through Sephadex G-25 or G-75 or BioGel P-6 columns has often resulted in improved yield of viable protoplasts (e.g. Patnaik et al., 1981; Chand et al., 1990). Shea et al. (1989) reported the synthesis of cell walls of carrot protoplasts with very little callose formation (callose synthesis is a characteristic stress response) similar to the original carrot cells, if a purified enzyme preparation providing high viability of protoplasts (75% for over 7 days) was used.

In a sense, enzymatic cell wall hydrolysis mimics certain aspects of fungal attack on the plant tissue. From plant cell walls elicitors activating plant defense genes are known to be liberated by microbial hydrolases (see e.g. the review of Bowles, 1990). Furthermore, oligosaccharides isolated from plant cell walls were shown to regulate defense responses and morphogenetic events in plants (Ryan and Farmer, 1991). The release of phytotoxic factors (endogenous elicitors) from plant cell walls also has been observed as a result of enzymatic cell wall removal during protoplast isolation (Hahne and Lörz, 1988). Generally, pectinases and hemicellulases appeared to produce more toxic products than cellulases. Macerase and Pectolyase frequently used for isolation of protoplasts also have been reported to produce pectic cell wall fragments eliciting ethylene synthesis in leaves (Tong et al., 1986; Baldwin and Biggs, 1988).

Anderson et al. (1982) reported the induction of ethylene synthesis by Cellulysin treatment of tobacco leaf discs, and suggested that Cellulysin preparations contain a specific factor responsible for this induction. Rethmeier et al. (1991) also reported increased ethylene generation during protoplast isolation with Cellulysin as compared to Cellulase R-10. Fuchs et al. (1989) characterized the factor in Cellulysin that is responsible for this ethylene generation to be an endoxylanase. Amongst the products of this xylanase, however, no oligosaccharides with capability for induction of ethylene biosynthesis were found (Dean et al., 1991).

Various fractionations of Cellulase R-10 also were analyzed according to their influence on the viability of barley protoplasts and it was found that only a small part of the original enzyme was beneficial and necessary for protoplast isolation (Krogaard and Andersen, 1990).

On the basis of these observations it is likely that cell wall fragments liberated by wounding and enzyme treatment can be at least partially responsible for the physiological changes which are common in elicited cells and plant protoplasts: ethylene generation, lipid peroxidation and ethane formation, polyphenol production (necrosis) and callose synthesis (see the review of Ryan and Farmer, 1991). Besides the purification of commercial enzyme preparations used for isolation of protoplasts, several attempts have been made to diminish these stress responses during protoplast isolation and culture. We have mentioned before that the proper physiological state of the donor plants/tissues has a significant role to avoid extreme ethylene generation, or alternatively ethylene antagonists can be applied to the plants/tissues prior to or during protoplast isolation (see Section 2.2.1.). Lipid peroxidation and subsequent ethane generation can be considerably decreased by the application of n-propyl gallate, a known lipoxygenase inhibitor (Saleem and Cutler, 1987; Cutler et al., 1989; Mórocz et al., 1990; Krens et al., 1990). Its effects increase protoplast stability. Protoplast viability and plating efficiency have been reported to be improved in the presence of enzymes detoxyfying and reducing lipid hydroperoxides in membranes (e.g. glutathion peroxidase and phospholipase A2) (Ishii, 1988; Creemers-Molenaar et al., 1990). Moreover, Schnabl and his coworkers have proved the usefulness of alginate immobilization in the maintenance of protoplast membrane integrity through the inhibition of enzymatic lipid degradation (as recently thoroughly discussed by Schnabl and Zimmermann, 1989). Polyphenol production by cultured protoplasts or protoplast derived cells leading to necrosis often reflects unsuitable culture conditions. The proper cell number/medium volume ratio, the appropriate timing of decreasing the cell density and the osmotic pressure during culture have been frequently found to be crucial to avoid this culture problem (e.g. Glimelius, 1984; Lenee and Chupeau, 1986; Klimaszewska, 1989; Loudon et al., 1989). Embedding (Tan et al., 1987; Lee and Wetzstein, 1988) as well as the changing of the auxin concentration during culture (Tan et al., 1987; Chen et al., 1988) has also been reported to result in decreased polyphenol production by the cells. Alternatively, ascorbic acid, polyvinyl pyrrolidon (PVP) and activated charcoal can be added to the culture medium (e.g. Wei and Xu, 1990).

Isolation of protoplasts in enzyme mixture in the presence of a culture medium with all essential components was found to be very beneficial especially during prolonged isolation procedures (see Kao and Michayluk, 1974).

2.4. *Culture of Protoplasts – Special Requirements and Culture Systems*

Media used for protoplast culture are basically the same as those employed for tissue and cell cultures. However, protoplasts are fragile, therefore, cannot be shaken too much and there is a need for the proper adjustment of the osmolarity in the culture medium. Moreover, as it was discussed above, protoplasts are exposed to severe stresses during isolation and their recovery needs special care avoiding additional stresses during culture.

2.4.1. *General Guidelines in Protoplast Culture*

(*a*) *Osmotic pressure*. Isolated protoplasts require osmotic protection in the culture medium until they regenerate a strong wall. Osmolarity in the medium is adjusted to the same level as in the enzyme and washing solutions (400–800 mOsm kg^{-1}). The proper type of the osmoticum has to be experimentally determined (e.g., Wallin and Johansson, 1989). Prolonged culture at high osmotic pressure can result in browning of the cultures and inhibition of callus growth (e.g. Kao and Michayluk, 1980; Klimaszewska, 1989) that is why after cell wall synthesis and few divisions the osmolarity of the medium should be gradually decreased. The proper timing of this decrease is of significant importance (e.g. Chand et al., 1990; Loudon et al., 1989).

(*b*) *Nutritional requirements*. Generally, the basic constituents in protoplast culture media are similar to those used for cell cultures (for example many of them are based on MS or B5 media). However, some modifications (most often the change of carbon or nitrogen sources, vitamins) can have beneficial effects on protoplast survival and colony formation. For example, NH_4^+ ions proved to be toxic to protoplasts of potato (Shepard and Totten, 1977; Bokelmann and Roest, 1983), fruit tree species (e.g. Ochatt and Power, 1988; Ochatt and Caso, 1986; Patat-Ochatt et al., 1988), and some species of *Compositae* (Okamura et al., 1984, Lenee and Chupeau, 1986; Sauvadet ct al., 1990).

The actual requirement of protoplasts in terms of the composition and nature of the culture medium is complex and the physiological bases are poorly understood. Frequently the socalled complex ('rich') media (e.g. KM8p and its derivatives) developed for culturing protoplasts at low cell densities (Kao and Michayluk, 1975) provide sufficient growth conditions. However, the organic components of these 'rich' media can have negative side effects in certain protoplast cultures (Ochatt and Power, 1988a; Ochatt, 1990; Patat-Ochatt et al., 1988; Hunt and Helgeson, 1989).

Recently, it has been shown that ultrafiltration of the protoplast culture media can result in improved protoplast division (Davies et al., 1989). It was likely due to the removal of inhibitory high molecular weight compounds present in glucose, coconut water, casein hydrolysate and minor carbohydrates.

(*c*) *Growth regulators*. While nuclear RNA synthesis of freshly isolated protoplasts is not dependent upon exogenously applied hormones during the first 18h of culture, replicative DNA synthesis (entering of the cells into S phase) requires the application of auxin and cytokinin (e.g. Cook and Meyer, 1981; Bergounioux et al., 1988). Thus the majority of protoplast culture media contain (at concentrations of 0.05–5 mg l^{-1}) one or more auxins (e.g. 2,4-dichlorophenoxyacetic acid, 1-naphthaleneacetic acid, indoleacetic acid) plus one or more cytokinins (e.g. kinetin, benzyladenine, zeatin) to stimulate protoplast division and growth. Only a few protoplast systems can be cultured without the requirement for exogeneous hormones, as e.g. *Citrus* (Vardi et al., 1982; Sim et al; 1988) or *Petunia* tumor (Phillips and Darrel, 1988) protoplasts.

It is recommended to change hormone supply, after the first divisions, as the phytohormone requirement of protoplasts and protoplast-derived cells may be very different (e.g. Tan et al., 1987; Masson et al., 1987; Zhongyi et al., 1990).

(*d*) *Environmental factors*. High light intensities often inhibit protoplast division (e.g. Harris et al., 1988; Dhir et al., 1991; Krens et al., 1990) and the culture of protoplasts in complete dark or at dim light at the initial stages may be beneficial. Further development of protoplast derived cells, however, is frequently supported by light.

Protoplasts are generaly cultured at 24–26 °C. Cold treatment of freshly isolated protoplasts is sometimes recommended (e.g. Mühlback and Thiele, 1980; Damm and Willmitzer, 1988). Recently, heat shock (45 °C 5 min) applied to rice (Thompson et al., 1987), wheat (Chang et al., 1991) and *Pennisetum* (Gupta et al., 1988) cell culture protoplasts has been shown to promote division.

Prolonged low-voltage electric treatment protoplasts resulted in enhanced division in *Trifolium* (Zhongyi et al., 1990) and stimulation of direct embryogenesis in *Medicago* (Dijak et al., 1986). Alternatively, conditions of electroporation treatments with short high voltage pulses (250–1500 V for 10–50 μsec) have been found to have a long-term effect on the protoplasts of several species. The treatment resulted in the stimulation of protoplast division, colony growth and morphogenesis (Gupta et al., 1988; Ochatt et al., 1988a, b; Chand et al., 1988b).

Protoplasts from most species grow well at pH 5.5–5.7. Mórocz et al. (1990), however, reported the culture of maize cell culture protoplasts in a medium without pH adjustment. Low pH values (pH 3.5–4) caused by organic acid constituents of the medium can influence division and morphogenetic response.

Successfull culture of plant protoplasts often strongly depends on the appropriate medium volume per cell ratio (plating density). The optimal protoplast density usually has to be experimentally determined, but in most of the cases it is in the range of $0.5–2 \times 10^5$ protoplasts ml^{-1} culture medium.

Protoplasts of some species, however, prefer much higher plating density (1–2×10^6 cells ml^{-1}) (e.g. Wallin and Johansson, 1989; Mórocz et al., 1990, Chang et al., 1991). Most often the density of protoplast derived cells can be serially diluted during culture with simultaneous lowering of the osmotic pressure in the medium.

In special cases, there is a need for the culture of protoplasts at very low plating densities or even for the culture of individual protoplasts. For this purpose special culture media (e.g. Kao and Michayluk, 1975; Hunt and Helgeson, 1989) or culture techniques (as reviewed by Schweiger et al., 1987; Schaffler and Koop, 1990) are required.

2.4.2. *Special Culture Systems*

(*a*) *Immobilization of plant protoplasts.* Embedding of plant protoplasts into solid or semi-solid media is said to improve plating efficiency (Shillito et al., 1983; Lörz et al., 1983; Draget et al., 1988). This technique has resulted in a significant improvement in the culture of protoplasts of formerly recalcitrant species as well (see Table 1). Immobilization of protoplasts was reported to be absolutely necessary for sustained divisions for example in maize (Imbrie-Milligan et al., 1987), barley (Lührs and Lörz, 1988; Jahne et al., 1991), and sunflower (Moyne et al., 1988; Schmitz and Schnabl, 1989). In several other instances, the division of protoplasts was markedly enhanced in solidified versus liquid medium (e.g. soybean, Dhir et al., 1991; cucumber, Colijn-Hooymans et al., 1988; Punja et al., 1990; sunflower, Burrus et al., 1991. etc). Genotypic difference has been described in respect of the response to embedding in barley by Lührs and Lörz (1988). It has to be mentioned, however, that other important factors such as the genotype and the culture conditions may be more essential for protoplast division than immobilization itself, and the usefulness of protoplast embedding should not be overestimated. For example, the protoplasts from maize genotype HE/89 can be very efficiently cultured in liquid culture medium (Mórocz et al., 1990).

The positive effect of immobilization can be ascribed to several factors. The gel matrix provides mechanical support to the fragile protoplasts. The effect of gel entrapment of protoplasts on their viability and membrane integrity through the inhibition of enzymatic lipid peroxidation has been recently reviewed by Schnabl and Zimmermann (1989). The diffusion of metabolites and large molecules necessary for cell wall synthesis and division (e.g. 'viability' and 'conditioning' factors; see part (b) of this Section) is reduced by the gel matrix. The same holds true for proteolytic enzymes released from dying cells. Immobilization allows to combine the adventages of high and low density cultures if the embedded protoplasts are cultured in a high volume of liquid medium. Via immobilization of protoplasts, aggregation is prevented and individual protoplasts can be positioned to facilitate cloning or micromanipulation. Moreover, these cultures are easy to handle with the dilution or replacement of the surrounding liquid medium without interferring with the cells. Deleterious effect, of embedding, however, have been also reported in some cases (Oliviera and Pais, 1991; Vasil and Vasil, 1987).

Because of the fragility of protoplasts, only a mild immobilization technique can be used. In the majority of experiments, plant protoplasts were embedded in agarose or alginate matrices. The first report of embedding of tobacco protoplasts in a semi-solid medium containing 0.6% agar was reported by Nagata and Takabe as early as 1971. The superiority of agarose, a chemically modified and purified agar with neutral characteristics and lowered gelling temperature, versus agar has been clearly demonstrated by Lörz et al. (1983). Methodology and applications of agarose plating have been recently reviewed by Dons and Colijn-Hooymans (1989).

Alginates are linear copolymers of 1–4 linked D-mannuronic acid and L-guluronic acid. Entrapment in alginate gels by the help of multivalent cations (e.g. Ca^{2+}) is one of the mildest immobilization procedures known (Brodelius and Nilsson, 1980) because the gelling agent itself is non-toxic and the gelling process is thermo-independent. An additional advantage of alginate versus agarose is that the gel can be easily dissolved using chelating agents such as citrate or phosphate without arising the temperature. The characteristics of alginate as an immobilization matrix for cells have been throughly discussed by Smidsrod and Skjak-Braek (1990). Methodology and applications of alginate entrapment of plant protoplasts have been reviewed in detail by Schnabl and Zimmermann (1989).

(*b*) *Nurse cultures and conditioned media.* As a result of protoplast isolation, plant cells are removed from their tissue environment and their plasmodesmatal connections are disrupted. Moreover, they are cultured under non-balanced conditions. Crossfeeding by 'nurse' or 'feeder' cells or using media previously 'conditioned' by fast growing cells has proven to be useful during the initial period of culture in the promotion of recovery and division of protoplasts. Similar systems could be used efficiently to support the growth of protoplasts at very low densities or even as single cells (e.g. Schaffler and Koop, 1990).

'Nurse' cultures rely on the simultaneous culture of target cells and feeder cells. However, these cells have to be somehow distinguishable from each other. This can be based for example on morphological differences between the target and nurse cells (e.g. Menczel et al., 1978) or on the inactivation of feeder cells by irradiation (e.g. Vardi et al., 1982). In most of the cases, however, feeder cells are physically separated from target cells by different types of filters (e.g. Shneyour et al., 1984; Rhodes et al., 1988a; Prioli and Söndahl 1989; Shillito et al., 1989; Sun et al., 1989; Walters and Earle, 1990), nylon mesh (Eigel and Koop, 1989), by immobilization of target cells culturing them in liquid medium with the feeders ('mixed nurse culture'; e.g. Kyozuka et al., 1987, 1988; Hayashi and Shimamoto, 1988; Zaghmout and Torello, 1990) or the reverse (Hahne et al., 1990) or via the immobilization of both kind of cells labelling 'nursing beads' with charcoal (Larkin et al. 1988).

Nurse cultures were found to be absolutely necessary in some of the laboratories for the success in culture of protoplasts of rice (Kyozuka et al., 1987, 1988; Lee et al., 1989; Shimamoto et al., 1989), wheat (Hayashi and Shimamoto, 1988), *Festuca* sp. (Takamizo et al., 1990), *Brassica oleracea* (Walters and Earle, 1990) and it was reported to significantly improve plating efficiency in maize (Rhodes et al., 1988a; Shillito et al., 1989; Sun et al., 1989) protoplasts.

The nursing effect is 'dosage' dependent: under the optimal density of nurse cells there is not enough supply of growth stimulating substances, while at supraoptimal density target and nurse cells compete for nutrients (Li and Murai, 1990; Schaffler and Koop, 1990). The timing of nurse culture is also an important parameter (Rhodes et al., 1988a; Li and Murai, 1990; Schaffler and Koop, 1990). One of the most significant factors, however, is the combination of nurse and target cells. Combinations of cells from different species often suffer from the differences in the nutritional requirements of the respective cells. This can drastically reduce the efficiency of nursing in heterologous combinations (Kyozuka et al., 1987; Hayashi and Shimamoto, 1988; Schaffler and Koop, 1990; Walters and Earle, 1990; Hahne et al., 1990). Interestingly, it was also observed in some cases that cells of the same cell line, genotype, or species is not suitable for nursing their own protoplasts (Hayashi and Shimamoto, 1988; Lee et al., 1989; Takamizo et al., 1990). Importance of cell type rather than genotype of feeder cells was emphasized by Hayashi and Shimamoto (1988), Schaffler and Koop (1990), and Takamizo et al. (1990).

Character and amount of substances added to or removed from the medium by nurse cells remain unknown. Schaffler and Koop (1990) reported the use of dialysis membrane as barrier between target and feeder cells. The molecular weight of 'nursing factor(s)' (NF) was in the range of 1000 daltons, and the nursing efficiency (diffusion of the factor(s)) correlated with agarose spacer thickness between the cells.

There are some successes, however, in the purification and characterization of 'conditioning' factors (CF) from conditioned media. Since the conditioned medium frequently can not substitute for feeder cells (Rhodes et al., 1988a; Shillito et al., 1989; Walters and Earle, 1990; Schaffler and Koop, 1990) it is likely that 'conditioning' and 'nursing' factors are chemically different (Bellincampi and Morpurgo, 1989). While NF can be chemically unstable, CF isolated from maize cell suspension culture was found to be stable, polar but neutral compound with a nominal molecular weight of 1200, resembling oligosaccharides in these respects (Birnberg et al., 1988). Chemically different CFs have been isolated from barley anther culture (Köhler and Wenzel, 1985) and dicot suspension cultures (Bellincampi and Morpurgo, 1987, 1989).

The effect of conditioned medium on colony formation of maize protoplasts has been throughly investigated by Somers et al. (1987) and a 10-fold increase could be ascribed to the use of this medium. Colony formation frequency has been described as a linear function of the amount of conditioned medium in other species as well (Kitamura et al., 1989; Horn et al., 1988a; Birnberg et al., 1988).

The effect of CFs is reported to rely on the promotion of cell divisions irrespectively of protoplast viability and cell wall regeneration (Somers et al., 1987). Recently, in tobacco cell cultures so called viability factors (VF) have also been identified and characterized as likely oligosaccharides or their derivatives (Steinbrenner et al., 1989; Schröder et al., 1989).

2.4.3. *Recent Achievements in Plant Regeneration from Protoplasts*

In a majority of cases, regeneration of plants from protoplasts proceeds through a callus phase. In some instances, somatic embryos can develop directly from protoplast-derived cells via an unequal first division initiating a polar growth of cell colonies (e.g. in *Medicago* sp., Dijak et al., 1986, Dijak and Brown, 1987; Song et al., 1990, in *Brassica juncea*, Pua, 1990; Kirti and Chopra, 1990, in *Citrus* sp., Sim et al., 1988). Once protoplasts develop into calli, plant regeneration can be achieved as in callus developed from other explants. In general, plant regeneration *in vitro* occurs via two developmental pathways: embryogenesis and organogenesis. During somatic embryogenesis the ontogenesis starts from a single cell with bipolar structure established through cell division. In the particularly recalcitrant *Poaceae* and in woody tree species the differentiation of embryogenic cells triggers plant redifferentiation. The recent achievements in the regeneration of plants from protoplasts of these species (see Table 1) mainly rely on the establishement and use of suspension cultures from embryogenic cells (as reviewed by Vasil, 1987, 1988; see also Chapter 12). Protoplasts isolated from these cells can maintain their embryogenic capability through their development into microcolonies. The major factor controlling embryogenesis *in vitro* is unquestionably the genotype dependent auxin (2,4-dichlorophenoxyacetic acid) responses that cause the reactivation of the cell cycle. The embryogenic program and germination generally occurs under hormone free conditions. The possible molecular and cellular mechanisms underlaying somatic embryo induction have been recently reviewed by Dudits et al. (1991).

As compared to embryogenesis, shoot organogenesis (Brown and Thorpe, 1986) is characterized by the formation of an unipolar bud primordium with subsequent development into a shoot with leaves. Root formation is a separate step during organogenesis. The control of organogenesis in the majority of dicot cultures is largely a function of auxin/cytokinin ratio within a particular range of concentrations. As plant regeneration from protoplasts through organogenesis is a complex response, a sequential nutrient and hormone treatment is required. The hormone levels applied during colony growth have been shown to influence regeneration: auxin requirement for bud regeneration depended on hormone level in the cultured tissues and in the medium during the previous culture stage (Renaudin et al., 1990). Roots can be induced by a cytokinin-deprived medium after addition of auxins (indol butyric acid). This process is dependent upon the previous culture of calli in the presence of a high cytokinin concentration (Renaudin et al., 1990).

Recently the long-term effect of electrostimulation on embryo as well as shoot formation has been reported in several cases (Dijak et al., 1986; Ochatt et al., 1988a; Chand et al., 1988b, see also Section 2.4.1).

Table 1 summarizes significant results in plant regeneration from protoplasts of some of the agronomically important species. Although there have been several important achievements during the past few years, there are still many recalcitrant species. In these cases, plant regeneration has failed, in spite of considerable progress in the culture of protoplasts has been made (e.g. in grape, Theodoropoulos and Roubelakis-Angelakis, 1990; and forest trees, Tremblay, 1988; Lang and Kohlenbach, 1988; Brison and Lamant, 1990; Tautorus et al., 1990; Teulieres and Boudet, 1991).

3. Chemical and Physical Methods in Induction of Protoplast Fusion–Somatic Hybridization

In vivo, the specific interaction between egg cell and sperm initiates the start of the developmental program and the formation of the zygote with a new combination of genetic determinants. Fertilization mediated by haploid gametes restores the diploid chromosome set in the newly formed individual. Sexual crossing is one of the major sources of genetic variablity. Both prezygotic and postzygotic incompatibilities strictly restrict the hybridization of evolutionarily distant species. Artificial manipulation of the fertilization processes and *in vitro* culture of hybrid embryos with reduced vitality, makes it possible to overcome the phylogenetic barriers and to a limited extent, interspecific and intergeneric sexual hybrids can be produced between related species. These sexual hybrids with important agronomic traits can serve as a source of useful breeding material for crop improvement. The need for integration of traits from species that can not be crossed, has encouraged speculation and experimentation to find asexual ways for hybrid production. The so called 'vegetative hybridization' by grafting of tissues can result in chimeric structures, without the formation of hybrid cells and tissues and production of hybrid plants with characters inherited in the subsequent generations. The hybrid nature of a cell involves the presence of organelles (chromosomes, chloroplasts, mitochondria) from both partners in a single cell, that can be the starting point for the formation of a new individual. In the case of higher plants, somatic cell hybridization requires the removal of the cell wall and the unification of the cells through fusion of protoplasts. The very first observation of such event was made by Küster (1909), who proposed protoplast fusion as a tool for production of hybrids. Michel (1937) demonstrated protoplat fusion by treatment with sodium nitrate. After Küster's and Michel's pioneer discoveries, significant improvement was necessary in methodology of plant cell and tissue culture until the dreams could become true. Introduction of the use of enzyme digestion to release protoplasts, optimalization of culture conditions, especially in plant regeneration from single cells, has significantly contributed to the extensive use of protoplast fusion for somatic hybridization. After the first successful regeneration of hybrid plants in the early 1970's (Carlson et al., 1972), somatic hybridization has been used both in basic plant science and experiments for production of new variants for plant breeding (reviewed by Gleba and Sytnik, 1984; Scheider and Kohn, 1986; Dudits, 1987; Glimelius et al., 1991). The complex procedure of parasexual hybridization includes the isolation of protoplasts, the induction of membrane fusion, the mixing of cytoplasms and organelles, the formation of synkaryons during the culture of hybrid cells, the selection of the fusion products, the regeneration of hybrid plants, molecular and biochemical characterization of primary regenerants and plants from the subsequent generations.

3.1. *Membrane Fusion as an Aspecific Biophysical and Biochemical Event*

The recent models of the plant plasmalemma describe this outer membrane bilayer as a flexible asymmetric structure of membrane proteins integrated into phospholipid molecules. The prerequisite for fusion of plant protoplasts from somatic cells is the substantial alteration of physico-chemical parameters of the outer membranes that allows the direct contact between the membrane surfaces and the reorganization of the membrane structure. The adhesion of protoplasts depends on surface charge. According to electrophoretic studies, plant protoplasts possess a negatively charged outer membrane surface (Grout and Coutts, 1974; Nagata and Melchers, 1978; Abe and Takeda, 1988). The P-potentials of protoplast surface (from -10 mV to -50 mV) can be significantly reduced by Ca^{2+} (Nagata and Melchers, 1978). Positively charged protoplasts can be obtained in the presence of poly-L-lysin, poly-L-ornithin or synthetic phospholipids (Grout and Coutts, 1974; Nagata and Melchers, 1978). In addition to chemical treatment, close membrane contact can be achieved by dielectrophoresis and pearl-chain formation in a non-uniform AC field (Zimmermann and Vienken, 1982). After establishment of protoplast-protoplast contact, the membrane fusion is a consequence of the subsequent structural changes that are triggered by external physical or chemical stimuli. The mutual restructuring between membranes with tight contact requires membrane fluidity that makes possible the interchange of lipid molecules. The lipid matrices play a significant role in the structural reorganization of biomembranes leading to fusion. Membrane fusion is preceded by the translocation of protein molecules along the contact region of membranes. The effective shape of lipid molecules is changed in the external monolayers, during the formation of contact bilayer (Stenger and Hui, 1986). Cullis and Hope (1978) proposed a model that is based on the formation of an intermediate hexagonal H_{II} phase through combining the two outer monolayers. During the completion of fusion process the bilayer structure is restabilized. The presence of Ca^{2+} is required for cell-cell contact, the processes of protein aggregation (Poste and Allison, 1973) as well as for the formation of the H_{II} phase (Cullis and Hope, 1978).

Kinetic studies on animal cells (Prives and Shinitzky, 1977) and plant protoplasts (Yamada et al., 1979; Senda et al., 1980) clearly showed the crucial role of membrane fluidity in fusion. The transient fluidisation required for fusion can be regulated by alteration of fatty acid composition or change of temperature. Exposure of myoblasts to fatty acids (stearic or elaidic acids) that increases membrane microviscosity reduced fusion, while the decrease of microviscosity by application of oleic or linoleic acids stimulated fusion (Prives and Shinitzky, 1977). The relation between membrane fluidity and fusion potential can be concluded from experiments with plant protoplasts as well. Senda et al. (1980) showed an increase in number of fused protoplasts and shortened time required for fusion at elevated temperature. In agreement with the above theoretical considerations, various methods based on either chemical or physical treatments are designed to generate conditions that promote physical contact between the outer membranes and stimulate the phase transition of phospholipid bilayers.

The lack of specificity in events of membrane fusion is clearly demonstrated by the fact that fusion can be induced between any cell or protoplast types studied regardless of their origin. There is no selectivity or incompatibility at the stage of fusion. Protoplasts from diverse species, e.g. dicot and monocot species can be fused (Dudits et al., 1976a; Kao et al., 1974; Niizeki, 1989). Plant protoplasts can be fused with protoplasts of filamentous fungi (Lynch et al., 1989), or even with insect (*Drosophila*) cells (Hadlaczky et al., 1980), amphibian cells (Davey et al., 1978) and mammalian cells (Jones et al., 1976; Dudits et al., 1976b). Fusion between protoplasts from haploid tetrads and cultured cells (Lee and Power, 1988) or from microspores and somatic cells (Pental et al., 1988) as well as from pollen or generative cells (Ueda et al., 1990) provides further evidence for the cell type independent nature of fusion events.

3.2. *Polyethylene Glycol (PEG) as one of the Most Effective Fusogenic Chemical Agents*

The primary observations by Kao and Michayluk (1974) and Wallin et al. (1974) about the aggregation and fusion of plant protoplasts in solutions of the water-soluble polymer, PEG, with Ca^{2+} have considerably stimulated research in cell fusion and somatic hybridization. The basic protocols for

PEG-mediated fusion have been described in detail (Kao et al., 1974; Kao, 1975; Menczel et al., 1981; Evans, 1983; Kao and Michayluk, 1989).

The use of highly basic solutions (pH 10.5) containing high concentration of Ca^{2+} (0.05M $CaCl_2$) resulted in fusion with high frequency (Melchers and Labib, 1974). Later this method was combined with PEG treatment to improve the efficiency increasing the number of fusion products (Kao et al., 1974; Kao and Saleem, 1986).

Beneficial effect of dimethyl sulfoxide (DMSO) on PEG mediated fusion was first shown by Haydu et al. (1977) in protoplasts from cultured cells. Menczel et al. (1981) have further improved this approach by including DMSO in the wash solution. From the studies by Chand et al. (1988a) it becomes evident that the commercial preparations of PEG contain contaminants that can considerably alter the fusogenecity and the capability of treated protoplasts for cultivation. The authors suggest the use of PEG products with low carbonyl content, omitting of the Ca^{2+} from the PEG solution and the use of hypotonic mannitol solution for elution. Frequently protoplasts are treated by PEG solution after settlement onto glass coverslip or polystyrene Petri dish. Kao (1986) has developed a protocol for fusion of protoplasts at the interphase of a glucose and a sucrose-PEG solution. As far as the mode of action of PEG is concerned, Kao and Michayluk (1989) suggest bridge formation mediated by Ca^{2+} and PEG through its slightly negative polarity. Osmotic shock can also has a role in fusion. In addition to PEG, other polymers such as polyvinyl alcohol and dextran sulfate, have been also used to induce fusion between plant protoplasts (Nagata, 1978; Kameya, 1975).

3.3. *Electrofusion of Plant Protoplasts*

Close membrane contact, transient membrane breakdown and reorganization of membrane components as basic requirements for cell fusion can be achieved also by exposing the cells to electrical fields. Electro-stimulated fusion (Senda et at., 1979; Zimmermann and Scheurich, 1981; Zimmermann et al., 1981) is widely used in studies on membrane fusion and for production of somatic hybrid colonies or plants (Bates and Hasenkamp, 1985; Kohn et al., 1985; Tempelaar and Jones, 1985; Chapel et al., 1986). As emphasized by Zimmermann (1982), electrofusion is a biophysical process that involves two steps. In the first, the protoplasts are exposed to a high frequency alternating electric field (0.5–1.5 MHz) that generates dipoles through dielectrophoresis. In the non-uniform field, the dipoles, in this case the protoplasts due to dielectrophoresis, move in the direction of the increasing field as a function of the field strength, cell size and the mobility of charge within the protoplast and in the surrounding solution. As a result of mutual dielectrophoresis (Zimmermann, 1982; Zacharison and Bornman, 1986) in non-uniform as well as uniform field the protoplasts clump, and bind together to form pearl chains. The second step of electrofusion is the application of one or more short (10–100 μs) direct current (DC) (1–3 kV cm^{-1}) pulses that cause reversible membrane breakdown resulting in pores in the aligned membranes. The contacting membranes can be fused and this fusion opens the way for hybrid cell formation. To insure the maintenance of close membrane contact during fusion, the AC field is briefly reapplied. Frequently, the fusion pulses are repeated. So far, considerable efforts were focused on the determination of optimum conditions for induction of cell-to-cell fusion (Zimmermann and Scheurich, 1981; Watts and King, 1984; Zacharison and Bornman, 1986; Tempelaar et al., 1987; Mehrle et al., 1990; Naton et al., 1991). In these respects both the physical parameters (field strength, pulse duration) and the protoplast characters (average diameter, origin of protoplasts such as cell suspension or leaf mesophyll cells, presence or absence of vacuole) should be considered to achieve high fusion and survival rate. Various electrofusion apparates have been designed for large scale fusion (Watts and King 1984; Zacharison and Bornman, 1984; Hibi et al., 1988; Mehrle et al., 1989). Progress in the development of techniques for micromanipulation and microculture of plant protoplasts has also provided the background for microfusion of preselected protoplast pairs (Spangenberg et al., 1990). This approach allows the control of the number and type of protoplasts involved in fusion event, and eliminates the need for selective markers. However, it should be mentioned that the application of selective pressure can be essential in production of asymmetric hybrids with the required traits. The microfusion system has been also successfully used for organelle transfer (Spangenberg et al., 1991).

3.4. *Hybrid Cell Lines, Somatic Hybrid Plants and their Potential Values in Genetic Manipulation of Higher Plants*

Since the majority of the described fusion methods are based on handling of protoplast populations, there is a need to identify and select the fusion products either at the single cell level or at the stage of callus tissue, or alternatively after regeneration of plants. Gleba and Shlumukov (1990) have recently published an excellent comprehensive review about the various approaches applied in large number of somatic hybridization experiments. There are also several publications with the list of somatic hybrids and description of their characteristic features (Evans, 1983; Binding et al., 1986; Gleba and Shlumukov, 1990). Here, we outline the main general trends in the field of somatic hybridization research.

We recognize at least three major directions of research during the past twenty years.

3.4.1. *Establishment of Basic Methodology and Providing Evidence for Hybrid Production*

During the 1970's a variety of intra- and inter-speicfic hybrids were produced mainly with species of the *Solanaceae*. Analysis of fusion products of *Nicotiana*, *Datura* and *Petunia* species have proven the possibility of parasexual hybridization and regeneration of amphidiploid hybrid plants with intermediate or additive characters of parental species.

3.4.2. *Somatic Hybrids with New Cytoplasmic Organelle Populations and Cybrids*

From the very first period of experimentation with protoplast fusion up to today, selection and analysis of cytoplasmic hybrids has attracted the most attention. The unique potential of this approach has been clearly demonstrated by new discoveries that could not be achieved by classical genetics (see reviews by Gleba, 1978; Hanson et al., 1985; Galun and Aviv, 1986; Kumar and Cocking, 1987; Kemble and Barsby, 1988; Ichikawa et al., 1989; Medgyesy, 1990). Based on the use of chloroplast encoded mutants and molecular markers, it was shown that there is a continuous segregation of the mixed chloroplast population. The non-parental combination of nuclei and chloroplasts can originate from spontaneous segregation (Belliard et al., 1978) or from irradiation stimulated elimination of donor chromosomes (Zelcer et al., 1978). Under effective selection pressure, recombination between chloroplast DNAs from different origin can occur with low frequency (Medgyesy et al., 1985). In contrast, frequent and extensive recombination of the mitochondrial DNAs was detected in a variety of somatic hybrids (Belliard et al., 1979; Smith et al., 1989b; reviewed by Hanson et al., 1985). The 'donor-recipient' protoplast fusion system has been also applied to transfer chloroplasts and mitochondria in several crop plants (Barsby et al., 1987; Vardi et al., 1987, 1989; Thomzik and Hain, 1988; Akagi et al., 1989; Yang et al., 1988a; Kyozuka et al., 1989). Fusion experiments were carried out to produce useful breeding material as well (Tanno-Suenaga et al., 1988; Perl et al., 1990; Sakai and Imamura, 1990; Kao et al., 1991).

3.4.3. *Success in Wide Somatic Hybridization: Asymmetry in Hybrid Genomes*

As a consequence of physico-chemical nature of fusion events without selectivity, hybrid cells and cell lines could be produced regardless of phylogenetic distance between the parents. In most cases, these remote cell hybrids showed instability in their chromosome constitution and failed to regenerate plants. Several such non-morphogenic cell lines have been described: soybean + *Nicotiana glauca* (Kao, 1977); soybean + *N. tabacum* (Chien et al., 1982); carrot + tobacco (Hauptmann et al., 1983); carrot + rice (Sala et al., 1985); *Duboisia* + tobacco (Endo et al., 1987); soybean + rice (Niizeki, 1989). In selected intergeneric fusion combinations, hybrid plants with additive chromosome number were regenerated, e.g. after fusion between potato and tomato (Melchers et al., 1978; Shepard et al., 1983), *Arabidopsis* and *Brassica* (Gleba and Hoffmann, 1980), *Datura* and *Atropa* (Krumbiegel and Schieder, 1979), *Lycopersicon* and *Solanum* (Handley et al., 1986; O'Connel and Hanson, 1987; Guri et al., 1988) and *Eruca* and *Brassica* (Sikdar et al., 1990). In these instances, certain degrees of taxonomic relation is reflected by the fact that the two parental genomes could cooperatively coexist in the regenerated plants. However, morphological abnormalities and sterility occurred in several of these intergeneric hybrids. In addition, somatic incompatibility as a limiting factor was clearly indicated (see review by Dudits et al., 1980b; Harms, 1983; Dudits and Praznovszky, 1985; Dudits, 1987). In attempts to reduce incompatibility reactions in the fusion products, an enforced and directed chromosome elimination was proposed as a requirement to reach gene transfer between species (Dudits et al., 1980b). Genomic asymmetry through preferential loss of chromosomes belonging to one of the fusion partners could be established through application of irradiation with lethal doses of X or gamma rays (Dudits et al., 1980a; Gupta et al., 1984; Bates et al., 1987; Dudits et al., 1987; Imamura et al., 1987; Sjödin and Glimelius, 1989; Famelaer et al., 1989; Gleba et al., 1988; Yamashita et al., 1989; Bates, 1990).

Both cytological and molecular evidence have been provided for chromosomal rearrangement, integration of chromosome fragments or the presence of only very few chromosomes (Piastuch and Bates, 1990; Itoh et al., 1991; Hinnisdaels et al., 1991). Frequently, there was no direct relation between the degree of chromosome loss and the irradiation doses applied (Dudits et al., 1987; Imamura et al.; 1987; Famelaer et al., 1989). Complementation of mutation was detected in highly asymmetric hybrids lacking cytologically detectable donor chromosomes (Somers et al., 1986; Agoudgil et al., 1990). Extensive chromosome elimination can occur spontaneously, as e.g. in hybrids between *Nicotiana* and *Petunia* (Pental et al., 1986). Alternatively, despite the application of high doses of irradiation only a limited chromosome loss was detected in defined fusion combinations mainly with related species (Wijbrandi et al., 1990; Fehér et al., 1992). These cases clearly indicate that irradiation is not the only factor that influences chromosomal changes during asymmetric hybrid formation. The efficiency can be variable according to the repair potential of the irradiated cells. Spontaneous chromosome loss is strongly determinated by the taxonomic relation between the partners. A back-crossing program with asymmetric hybrids can provide a chance for a more extensive loss of the donor chromosomes in the subsequent generation as shown by Famelaer et al. (1990) and Bates (1990). These considerations emphasize the need to search alternative ways in addition to irradiation for control of chromosome behavior, especially if the fusion is carried out with related species. One of the possible approaches can be based on fusion of protoplasts from various cell cycle stages. In interpretation of results of fusion experiments with non-dividing leaf mesophyll and actively dividing cell suspension protoplasts, we proposed the role of premature chromosome condensation (PCC) (Dudits et al., 1979). Later we obtained cytological evidence for PCC that can occur in fused plant cells (Szabados and Dudits, 1980). Fusion between protoplasts with mitotic stage and S phase causes extensive fragmentation of the S phase nuclei. This pulverized chromatin is lost during the subsequent divisions. Finally, it is important to mention that production of useful asymmetric hybrids postulates the control not only of chromosome loss but stimulation of intergenomic recombination events as well.

3.4.4. *The Present Trends and Potentials in Somatic Hybridization*

We can see a significant increase in the number of somatic hybridization programs based on important crop species and the transfer of agronomic traits. Extensive studies are in progress with potato and its relatives (Helgeson, 1989; Jones et al., 1990; Preiszner et al., 1991). Somatic hybridization with *Brassica* species has also provided examples for production of breeding material (Glimelius et al., 1991). As a consequence of establishment of plant regeneration from cereal protoplasts, the somatic hybridization could be extended to these agronomically important crops, mainly to rice (Terada et al., 1987; Hayashi et al., 1988; Toriyama and Hinata, 1988). Several rice cybrids have been produced with this method (Yang et al., 1988a; Akagi et al., 1989; Kyozuka et al., 1989). Because of the progress in protoplast culture system in wheat (Vasil et al., 1990) and maize (Mórocz et al., 1990) similar fusion work is expected to start with these crops as well.

In the future, the manipulation of cytoplasmic traits will be continued as a preferred, major area of activities in somatic hybridization work. In addition, more efforts will be concentrated into the use of asymmetric genomic hybrids. This approach can have a complementary function in transfer of gene complexes. It has a potential in introduction of those genes that encode important traits, but the DNA transformation can not be accomplished because of the lack of the cloned gene. This method is less specific, it requires the application of selective pressure and production and characterization of hybrids in large numbers. At the same time, the produced asymmetric hybrids can serve as a source to clone the gene of interest. In general, there is a tendency to use simultaneously the molecular and cellular techniques. We can expect further considerable progress from combination of these two methods in both plant biology and applied studies.

Table 2. Recent advances in the production of transgenic plants via DNA uptake into protoplasts

Species	Promoter/gene	Method	Reg	References
Oryza sativa	Adh1/I1/GUS	PEG	P	Zhang and Wu, 1988
	35S/neo	EP	P	Toriyama et al., 1988
	35S/neo	EP	P	Zhang et al., 1988
	35S/hph	EP	FP	Shimamoto et al., 1989
	NOS/hph;35S/hph	PEG	FP	Li et al.,1989
	35S/GUS;35S/hph	EP	FP	Tada et al., 1990
	35S/hph	PEG	FP	Datta et al., 1990b
	mas/GUS;35S/hph 35S/dhfr;wNOS	EP,PEG	P	Meijer et al., 1991
Zea mays	35S/neo	EP	SP	Rhodes et al., 1988b
	35S/pat;35S/neo	PEG	FP	Omirulleh et al., 1993
	35S/dhfr	PEG	FP	Golovkin et al., 1993
Dactylis glomerata	35S/hph	EP,PEG	P	Horn et al., 1988b
Citrus sp.	mas/CAT;mas/neo	PEG	P	Vardi et al., 1990
Solanum tuberosum	NOS/neo	PEG	P	Feher et al., 1991
Lactuca sativa	35S/neo	EP	FP	Chupeau et al., 1989
Arabidopsis thaliana	35S/hph	PEG	FP	Damm et al., 1989
	35S/hph	PEG	FP	Karesch et al., 1991b

Abbreviations: Adhl/I1 = promoter and the first intron of the alcohol dehidrogenase gene of maize; 35S = cauliflower mozaic virus 35S gene promoter; NOS = nopaline synthase gene promoter of *Agrobacterium tumefaciens*; mas = mannopine synthase gene promoter of *Agrobacterium tumefaciens*; neo = neomycin phosphotransferase gene; hph = hygromycin phosphotransferase gene; GUS = β-glycuronidase gene; dhfr = mutated mouse dihydrofolate reductase gene; CAT = cloramphenicol acetyltransferase gene; pat = phosphinotrycine acyltransferase gene; wNOS = wild type nopaline synthase gene; PEG = polyethylene glyccol; EP = electroporation; P = plant; FP = fertile plant; SP = sterile plant.

4. Direct DNA Uptake into Plant Protoplasts–Transient Expression and Integrative Transformation

Due to the fluid mosaic nature of the outer membranes, transient aqueous pores in the bilayer portion can make the membrane permeable for macromolecules. The formation of pores and reversible membrane discontinuities opens the way for introduction of nucleic acid molecules into the cytoplasm and the nucleus. The reversible, nondestructive permeabilization of membranes can be achieved by both chemical treatments and action of short-duration high electric field pulses. Uptake of DNA molecules into a variety of plant protoplasts was successfully promoted by application of polyethylene glycol (PEG) treatment or electroporation (Davey et al., 1980; Krens et al., 1982, Paszkowski et al., 1984; Shillito et al., 1985; Fromm et al., 1986; Riggs and Bates, 1986; Negrutiu et al., 1987).

Recently, cationic methods (using polybrene or lipofectin) have also been reported as an efficient alternative of the above mentioned and most frequently used techniques (Antonelli and Stadler, 1989, 1990). The methodology, efficiency and applications of DNA-mediated gene transfer into protoplasts have been recently discussed by several authors (e.g. Davey et al., 1989; Weising et al., 1988; Paszkowsky et al., 1989; Potrykus, 1991; Joersbo and Brunstedt, 1991). According to the comparative studies of Negrutiu et al. (1990), the chemical transfection method with optimized parameters resulted in a higher efficiency than electroporation in the case of leaf mesophyll protoplasts. There are experiments with other conclusions (Yang et al., 1988b). Therefore, each experimental system based on various species and/or protoplast types require the choice of the most effective way of transformation.

Direct DNA uptake into protoplasts as one of the methods out of several transformation approaches (see the review by Potrykus, 1991) exhibits specific features and potentials. It is extensively used in transient gene expression studies. Assay of gene activity within few days after DNA introduction allows to optimalize the parameters for DNA uptake (Vasil et al., 1989; Maas and Werr, 1989; Wilson et al., 1989; Oliviera et al., 1991; Teulieres et al., 1991).

Furthermore, the regulatory elements in control of expression of the introduced genes can be studied and optimized by this interval. Recently, for example, the enhancement of heterologous gene expression (10–1000 fold) by the inclusion of intron or intron/exon sequences of Adh-1 and Shrunken-1 genes of maize into chimeric gene constructs has been demonstrated in transient expression experiments (Vasil et al., 1989; Maas et al., 1991; Mascarenhas et al., 1990; Oard et al., 1989). Moreover, tissue specificity seems to be retained within a certain range in the transformed protoplasts, therefore transient expression studies may be used for investigation of tissue and developmental control of genes in homologous and heterologous systems. This possibility has been demonstrated via the introduction of reporter genes linked to the promoters of the α-amylase gene of barley (Lee et al., 1989; Gopalokrishan et al., 1991; Jacobsen and Close, 1991; Salmenkallio et al., 1990) and zein or sucrose synthase genes of maize (Junker et al., 1987; Thompson et al., 1990; Ueda and Messing 1991; Schwall and Feix, 1988) into somatic and/or endosperm protoplasts. However, soybean 7S seed storage protein gene promoter and potato patatin class II gene promoter have been found to be expressed in an unregulated fashion in protoplasts of different tissue origin (Jones et al., 1989; Fujiwara et al., 1991). In general, it is safe to propose that the transient expression data provide preliminary information about the characteristics of the promoters that should also be checked with analysis of stable transformants.

Production of transgenic tissues and plants requires the stable integration of the introduced DNA molecules into the genome of recipient species used as protoplast source. Selection and identification of stable transformants can be based on various marker genes such as the bacterial neomycin phosphotransferase (NPTII) gene = kanamycin or G418 resistance; hygromycin phosphotransferase (HPH) = hygromycin resistance; dihydrofolate reductase (DHFR) = methotreste resistance; phosphinothricin acetyltransferase (PAT) = Basta resistance, and so on. The use of various reporter genes as chloramphenicol acetyl transferase (CAT); β-glucuronidase (GUS) can also help the identification of transformed microcolonies and they serve as a detection system to quantify of level of gene expression in different organs, cell types etc. (see review by Paszkowsky et al., 1989; Davey et al., 1989).

At present, the available experimental data indicate that the integration of the introduced plasmid molecules into the host genome is generally based on non-homologous recombination under influence of several factors (Negrutiu et al., 1987). However, homologous recombination can also occur (Paszkowsky et al., 1988). As far as the structure of vector DNA is concerned, several studies show a more complex pattern of integration with linear DNA than with application of supercoiled DNA without any carrier DNA (see Kartzke et al., 1990). Concatemerization and truncation of the introduced DNA was frequently observed (e.g. Riggs and Bates, 1986).

Furner et al. (1989) have found no differences in transformation efficiency by using single or double stranded DNA. The authors suggested that the introduced ssDNA became double stranded rapidly through extrachromosomal processes. In contrast, Rodenburg et al. (1989) reported 3–10-fold higher frequency of stable transformants when ssDNA was used.

From both technical and applied aspects there is a need to carry out cotransformation. Frequently, the selectable marker gene and the gene of interest is cloned in un-linked plasmid molecules. In these cases the transformation with a plasmid mixture can result in the coexpression of the genes in transformed tissues or plants selected for the presence and expression of the resistance marker gene. We can see a wide range of frequency data (20–50%) as far as the efficiency of cotransformation is concerned (Uchimiya et al., 1986; Schocher et al., 1986; Tagu et al., 1990; Damm et al., 1989; Lyznik et al., 1989; Christou and Swain, 1990). Studies on the nature of integration of the non-selected gusA gene into corn genomes indicated the presence of tandem head-to-tail repeats. These structures are supposed to be formed by homologous recombination between plasmid molecules (Lyznik et al., 1989). The process of integration of foreign DNAs or the efficiency of transformation can be modified by specific genomic sequences (so called transformation booster sequence, see Meyer et al., 1988) or by highly repeated nuclear DNA sequences (Marchesi et al., 1989).

The fate of the introduced DNA molecules and their integration into the genome is under the influence of susceptibility, that is, the competence of the recipient cells to transformation. Genotype dependent transient expression or transformation frequency was reported by several authors (Tyagi et al., 1989; Negrutiu et al., 1990; Damm et al., 1989).

Both the frequency of transformants and the pattern of integration can be considerably different in recipients with various cell cycle stages (Meyer et al. 1988; Kartzke et al., 1990). The implication of repair processes in the integration of foreign DNA was indicated by the increase of transformation events by application of X-ray irradiation (Köhler et al., 1989) or UV light and 3-aminobenzamide treatment (Gharti-Chherti et al., 1990).

As a consequence of the significant progress in methodology of protoplast culture and regeneration, extensive experimentation can now be carried out not only with dicot plants but several monocot cereal crops as well. The direct delivery of DNA into protoplasts can be one of the key methods in the production of transgenic plants. In addition to its high efficiency, the origin of transformants from single cells is guaranteed.

5. References

Abe, S. and Takeda, J. (1988). Effects of La^{3+} on surface charges, dielectrophoresis, and electrofusion of barley protoplasts. Plant Physiol. 87:389–394.

Adachi, T., Yamaguchi, A., Miike, Y. and Hoffmann, F. (1989). Plant regeneration from protoplasts of common buckwheat (*Fagopyrum esculentum*). Plant Cell Rep. 8:247–250.

Agoudgil, S., Hinnisdaels, S., Mouras, A., Negrutiu, I. and Jacobs, M. (1990). Metabolic complementation for a single gene function associated with partial and total loss of donor DNA in interspecific somatic hybrids. Theor. Appl. Genet. 80:337–342.

Ahmed, K. Z. and Sági, F. (1993). Culture of and fertile plant regeneration from regenerable embryogenic suspension cell derived protoplasts of wheat (*Triticum aestivum* L.). Plant Cell Rep. 12:175–179.

Akagi, H., Sakamoto, M., Negishi, T. and Fujimura, T. (1989). Construction of rice cybrid plants. Mol. Gen. Genet. 215:501–506.

Anderson, J.D., Mattoo, A.K. and Lieberman, M. (1982). Induction of ethylene biosynthesis in tobacco leaf discs by cell wall digesting enzymes. Biochem. and Biophys. Res. Comm. 107:588–596.

Antonelli, N.M. and Stadler, J. (1989). Chemical methods for efficient direct gene transfer to maize cells: treatment with polyethylene glycol or polybrene. J. Genet. Breed. 43:113–122.

Antonelli, N.M. and Stadler, J. (1990). Genomic DNA can be used with cationic methods for highly efficient transformation of maize protoplasts. Theor. Appl. Genet. 80:395–401.

Asamizu, T. and Nishi, A. (1980). Regenerated cell wall of carrot protoplasts isolated from suspension-cultured cells. Physiol. Plant. 48:207–212.

Asano, Y. and Sugiura, K. (1990). Plant regeneration from suspension culture-derived protoplasts of *Agrosis alba* L. (Redtop). Plant Sci. 72:267–273.

Attree, S.M., Dunstan, D.I. and Fowke, L.C. (1989). Plantlet regeneration from embryogenic protoplasts of white spruce (*Picea glauca*). Biotechnology 7:1060–1062.

Baldwin, E.A. and Biggs, R.H. (1988). Cell-wall lysing enzymes and products of cell-wall digestion elicit ethylene in citrus. Physiol. Plant. 73:58–64.

Barsby, T.L., Yarrow, S.A., Kemble, R.J. and Grant, I. (1987). The transfer of cytoplasmic male sterility to winter-type oilseed rape (*Brassica napus* L.) by protoplast fusion. Plant Sci. 53:243–248.

Bates, G.W. (1990). Asymmetric hybridization between *Nicotiana tabacum* and *N. rapanda* by donor recipient protoplast fusion: transfer of TMV resistance. Theor. Appl. Genet. 80:481–487.

Bates, G.W. and Hasenkamp, C.A. (1985). Culture of plant somatic hybrids following electrical fusion. Theor. Appl. Genet. 70:227–233.

Bates, G.W., Hasenkamp, C.A., Contolini, C.L. and Piastuch, W.C. (1987). Asymmetric hybridization in *Nicotiana* by fusion of irradiated protoplasts. Theor. Appl. Genet. 74:718–726.

Belliard, G. and Pelletier, G. (1978). Cytoplasmic hybridization by protoplast fusion in *Nicotiana tabacum*. Physiol. Veg. 16:441–448.

Belliard, G., Vedel, F. and Pelletier, G. (1979). Mitochondrial recombination in cytoplasmic hybrids of *Nicotiana tabacum* by protoplast fusion. Nature 281:401–403.

Bellincampi, D. and Morpurgo, G. (1987). Conditioning factor affecting growth in plant cells in culture. Plant Sci. 51:83–91.

Bellincampi, D. and Morpurgo, G. (1989). Evidence for the presence of a second conditionary factor in plant cell cultures. Plant Sci. 65:125–130.

Bergounioux, C., Perennes, C., Brown, S.C. and Gadal, P. (1988). Cytometric analysis of growth-regulator-dependent transcription and cell-cycle progression in *Petunia* protoplast cultures. Planta 175:500–505.

Berry, S.F., Lu, D.Y., Pental, D. and Cocking, E.C. (1982). Regeneration of plants from protoplasts of *Lactuca sativa* L. Z. Pflanzenphysiol. 108:31–38.

Binding, H., Krumbiegel-Schroeren, G. and Nehls, R. (1986). II. Protoplast fusion and early development of fusants. In: Results and Problems in Cell Differentiation. J. Reinert and H. Binding (eds.), Differentiation of Protoplasts and of Transformed Plant Cells, Vol. 12. Springer-Verlag, Berlin, Heidelberg, pp. 37–66.

Birnberg, P.R., Somers, D.A. and Brenner, M.L. (1988). Characterization of conditioning factors that increase colony formation from 'Black Mexican Sweet Corn' protoplasts. J. Plant Physiol. 123:316–321.

Blowers, D.P., Boss, W.F. and Trewavas, A.J. (1988). Rapid changes in plasma membrane protein phosphorylation during initiation of cell wall digestion. Plant Physiol. 86:505–509.

Bokelmann, G.S. and Roest, S. (1983). Plant regeneration from protoplasts of potato (*Solanum tuberosum* cv. Bintje). Z. Pflanzenphysiol. 109:259–265.

Bowles, D.J. (1990). Defense-related proteins in higher plants. Annu. Rev. Biochem. 59:873–907.

Brison, M.L. and Lamant, A. (1990). Callus formation from root protoplasts of *Quercus rubra* L. (red oak). Plant Cell Rep. 9:139–142.

Brodelius, P. and Nilsson, K. (1980). Entrapment of plant cells in different matrices. FEBS Letters 122:312–316.

Brown, D.C.W. and Thorpe, T.A. (1986). Plant regeneration by organogenesis. In: Cell Culture and Somatic Cell Genetics of Plants. I.K. Vasil (ed.), Plant Regeneration and Genetic Variability, Vol. 3. pp. 49–65.

Browse, J., Sommerville, C.R. and Slack, C.R. (1988). Changes in lipid composition during protoplast isolation. Plant Sci. 56:15–20.

Burrus, M., Chanabe, C., Alibert, G. and Bidney, D. (1991). Regeneration of fertile plants from protoplasts of sunflower (*Helianthus annuus* L.) Plant Cell Rep. 10:161–166.

Carlson, P.S., Smith, H.H. and Dearing, R.D. (1972). Parasexual interspecific plant hybridization. Proc. Natl. Acad. Sci. USA 69:2292–2294.

Cassells, A.C. and Barlass, M. (1976). Environmentally induced changes in the cell walls of tomato leaves in relation to cell and protoplast release. Physiol. Plant. 37:239–246.

Cassells, A.C., Cocker, F.M and Austin, S. (1980). Ethylene release during tobacco protoplast isolation and subsequent protoplast survival. Plant Sci. Lett. 19:169–173.

Cassells, A.C. and Cocker, F.M. (1982). Seasonal and physiological aspects of the isolation of tobacco protoplasts. Physiol. Plant. 56:69–79.

Cassells, A.C. and Tamma, L. (1986). Ethylene and ethane release during tobacco protoplast isolation and protoplast survival potential *in vitro*. Physiol. Plant. 66:303–308.

Chand, P.K., Davey, M.R., Power, J.B. and Cocking, E.C. (1988a). An improved procedure for protoplast fusion using polyethylene glycol. J. Plant Physiol. 133:480–485.

Chand, P.K, Ochatt, S.J., Rech, E.L., Power, J.B and Davey, M.R. (1988b). Electroporation stimulates plant regeneration from protoplasts of the woody medicinal species *Solanum dulcamara* L. J. of Exp. Bot. 39:1267–1274.

Chand, P.K., Davey, M.R. and Power, J.B. (1990). Efficient plant regeneration from cell suspension protoplasts of the woody medicinal plant *Solanum dulcamara* L. (bittersweet, woody nightshade). Plant Cell Tissue Org. Cult. 22:119–125.

Chang, M-M. and Loescher W.H. (1991). Effects of preconditioning and isolation conditions on potato (*Solanum tuberosum* L. cv. Russet Burbank) protoplast yield for shoot regeneration and electroporation. Plant Sci. 73:103–109.

Chang, Y.F., Wang, W.C., Warfield, C.Y., Nguyen, H.T. and Wong, J.R. (1991). Plant regeneration from protoplasts isolated from long-term cell cultures of wheat (*Triticum aestivum* L). Plant Cell Rep. 9:611–614.

Chapel, M., Montane, M.H., Ranty, B., Teissie, J. and Alibert, G. (1986). Viable somatic hybrids are obtained by direct current electrofusion of chemically aggregated plant protoplasts. FEBS Letters 196:79–85.

Chen, W.H., Davey M.R., Power J.B. and Cocking, E.C. (1988). Sugarcane protoplasts: factors affecting division and plant regeneration. Plant Cell Rep. 7:314–347.

Cheng, J. and Veilleux, R.E. (1991) Genetic analysis of protoplast culturability in *Solanum phureja*. Plant Sci. 75:257–265.

Chien, Y.C., Kao, K.N. and Wetter, L.R. (1982). Chromosomal and isoenzyme studies of *Nicotiana tabacum – Glycine max* hybrid cell lines. Theor. Appl. Genet. 62:301–304.

Christou, P. and Swain, W.F. (1990). Cotransformation frequencies of foreign genes in soybean cell cultures. Theor. Appl. Genet. 79:337–341.

Chupeau, M-C., Bellini C., Guerche P., Maisonneuve B., Vastra G. and Chupeau Y. (1989) Transgenic plants of lettuce (*Lactuca sativa*) obtained through electroporation of protoplasts. Biotechnology 7:503–508.

Cocking, E.C. (1960). A method for the isolation of plant protoplasts and vacuoles. Nature 187:927.

Cocking E.C. (1972) Plant cell protoplasts – Isolation and development. Annu. Rev. Plant Physiol. 23:29–50.

Coleman M., Waugh R. and Powell W. (1990) Genetic analysis of *in vitro* cell and tissue culture response in potato. Plant Cell Tissue Org. Cult. 23:181–186.

Colijn-Hooymans, C.M., Bouwer, R., Orczyk, W. and Dons, J.J.M. (1988). Plant regeneration from cucumber (*Cucumis sativus*) protoplasts. Plant Sci. 57:63–71.

Cooke, R. and Meyer, Y. (1981). Hormonal control of tobacco protoplast nucleic acid metabolism during *in vitro* culture. Planta 152:1–7.

Creemers-Molenaar, J. and Van Oort, Y. (1990). Antioxidants influence the plating efficiency and microcallus-growth of protoplasts in *Lolium perenne* L. In: Progress in Plant Cellular and Molecular Biology. H.J.J. Nijkamp, L.H.W. Van der Plas, and J. Aartrijk (eds.), Kluwer Academic Publishers, Dordrecht, pp. 44–49.

Creemers-Molenaar, J., Van der Valk, P., Loeffen, J.P.M. and Zaal, M.A.C.M. (1989). Plant regeneration from suspension cultures and protoplasts of *Lolium perenne* L. Plant Sci. 63:167–176.

Cullis, P.R. and Hope, M.J. (1978). Effects of fusogenic agent on membrane structure of erythrocyte ghosts and the mechanism of membrane fusion. Nature 271:672–674

Cutler, A.J., Saleem, M., Coffey, M.A. and Loewen, M.K. (1989). Role of oxidative stress in cereal protoplast recalcitrance. Plant Cell. Tissue Org. Cult. 18:113–127.

Dai, C., Mertz, D. and Lambeth, V. (1987). Improved procedures for the isolation and culture of potato protoplasts. Plant Sci. 50:79–84.

Dalton, S.J. (1988). Plant regeneration from cell suspension protoplasts of *Festuca arundinacea* Schreb. (tall fescue) and *Lolium perenne* L. (perennial ryegrass). J. Plant Physiol. 132:170–175.

Damm, B. and Willmitzer, L. (1988). Regeneration of fertile plants from protoplasts of different *Arabidopsis thaliana* genotypes. Mol. Gen. Genet. 213:15–20.

Damm, B., Schmidt, R. and Willmitzer, L. (1989). Efficient transformation of *Arabidopsis thaliana* using direct gene transfer to protoplasts. Mol. Gen. Genet. 217:6–12.

Datta, S.K., Datta, K. and Potrykus, I. (1990a). Fertile indica rice plants regenerated from protoplasts isolated from microspore derived cell suspensions. Plant Cell Rep. 9:253–256.

Datta, S.K., Peterhans, A., Datta, K. and Potrykus, I. (1990b). Genetically engineered fertile indica rice recovered from protoplasts. Biotechnology 8:736–740.

Davey, M.R., Clothier, R.H., Balls, M. and Cocking, E.C. (1978). An ultrastructural study of the fusion of cultured amphibian cells with higher plant protoplasts. Protoplasma 96:157–172.

Davey, M.R., Cocking, E.C., Freeman, J., Pearce, N. and Tudor, I. (1980). Transformation of *Petunia* protoplasts by isolated *Agrobacterium* plasmids. Plant Sci. Lett. 18:307–313.

Davey, M.R., Rech, E.L. and Mulligan, B.J. (1989). Direct DNA transfer to plant cells. Plant Mol. Biol. 13:273–285.

Davies, P.A., Larkin, P.J. and Tanner, G.J. (1989). Enhanced protoplast division by media ultrafiltration. Plant Sci. 60:237–244.

Dean, J.F.D., Gross, K.C. and Anderson, J.D. (1991). Ethylene biosynthesis-inducing xylanase. III. Product characterization. Plant Physiol. 96:571–576.

Dhir, S.K., Dhir, S. and Widholm, J.M. (1991). Plantlet regeneration from immature cotyledon protoplasts of soybean (*Glycine max* L.). Plant Cell Rep. 10:39–43.

Diaz, I., Moreno, R. and Power, J.B. (1988). Plant regeneration from protoplasts of *Capsicum annuum*. Plant Cell Rep. 7:210–212.

Dijak, M., Smith, D.L., Wilson, T.J. and Brown, D.C.W. (1986). Stimulation of direct embryogenesis from mesophyll protoplasts of *Medicago sativa*. Plant Cell Rep. 5:468–470.

Dijak, M. and Brown, D.C.W. (1987). Patterns of direct and indirect somatic embryogenesis from mesophyll protoplasts of *Medicago sativa*. Plant Cell Tissue Org. Cult. 9:121–130.

Dons, J.J.M. and Colijn-Hooymans, C.M. (1989). Agarose plating of protoplasts and its applications. In: Biotechnology in Agriculture and Forestry. Y.P.S. Bajaj (ed.), Plant Protoplasts and Genetic Engineering, Vol. 8. Springer, Berlin Heidelberg, pp. 50–62.

Draget, K.I., Myhre, S., Skjak-Braek, G. and Ostgaard, K. (1988). Regeneration and differentiation of plant protoplasts immobilized in Ca-alginate beads. J. Plant Physiol. 132:552–556.

Dudits, D. (1987). New approaches to genetic manipulation of plants. In: Cell Culture and Somatic Cell Genetics of Plants. F. Constabel and I.K. Vasil (eds.), Cell Culture and Phytochemistry, Vol. 4, Academic Press, Orlando, pp. 139–152.

Dudits, D., Kao, K.N., Constabel, F. and Gamborg, O.L. (1976a). Fusion of carrot and barley protoplasts and division of heterokaryocytes. Can. J. Genet. Cytol. 18:263–269.

Dudits, D., Raskó, I. and Hadlaczky, Gy. (1976b). Fusion of human cells with carrot protoplasts induced by polyethylene glycol. Hereditas 82:121–124.

Dudits, D., Hadlaczky, Gy., Bajszár, Gy., Koncz, C., Lázár, G. and Horváth, G. (1979). Plant regeneration from intergeneric cell hybrids. Plant Sci. Lett. 15:101–112.

Dudits, D., Fejer, O., Hadlaczky, G., Koncz, C. and Horvath, G. (1980a). Intergeneric gene transfer mediated by plant protoplast fusion. Mol. Gen. Genet. 179:283–288.

Dudits, D., Hadlaczky, Gy., Lázár, G. and Zs. Haydú (1980b). Increase in genetic variability through somatic cell hybridization of distantly related plant species. In: Plant Cell Culture: Results and Perspectives. F. Sala et al. (eds.), Elsevier/North Holland, Amsterdam, pp. 207–214.

Dudits, D. and Praznowsky, T. (1985). Intergeneric gene transfer by protoplast fusion and uptake of isolated chromosomes. In: Biotechnology in Plant Science. M. Zaitlin, P. Day, and A. Hollaender (eds.), Vol. 4. Academic Press, New York, pp. 115–127.

Dudits, D., Maróy, E., Praznovszky, T., Oláh, Z., Györgyey, J. and Cella, R. (1987). Transfer of resistance traits from carrot into tobacco by asymmetric somatic hybridization: regeneration of fertile plants. Proc. Natl. Acad. Sci. USA 84:8434–8438.

Dudits, D., Bögre, L. and Györgyei, J. (1991). Molecular and cellular approaches to the analysis of plant embryo development from somatic cells *in vitro*. J. Cell Sci. 99:474–484.

Dupuis, J.M., Pean, M. and Chagvardieff, P. (1990). Plant donor tissue and isolation procedure effect on early formation of embryoids from protoplasts of *Helianthus annuus* L. Plant Cell Tissue Org. Cult. 22:183–189.

Eigel, L. and Koop, H-U. (1989). Nurse culture of individual cells: regeneration of colonies from single protoplasts of *Nicotiana tabacum*, *Brassica napus* and *Hordeum vulgare*. J. Plant Physiol. 134:577–581.

Endo, T., Komiya, T., Masumitsu, Y., Morikawa, H. and Yamada, Y. (1987). An intergeneric hybrid cell line of *Duboisia hopwoodii* and *Nicotiana tabacum* by protoplast fusion. J. Plant Physiol. 129:453–459.

Evans, D.A. (1983). Protoplast fusion. In: Handbook of Plant Cell Culture. D.A. Evans, W.R.

Sharp, P.V. Ammirato, and Y. Yamada (eds.) Techniques for Propagation and Breeding, Vol. 1. Macmillan Publishing Company, New York, pp. 291–321.

Facciotti, D. and Pilet, P.E. (1981). Ethylene release during haploid and diploid protoplast isolation and viability. Z. Pflanzenphysiol. 104:401–407.

Fehér, A., Preiszner, J. and Dudits, D. (1989). Differentiation of potato (*Solanum tuberosum* L.) plants from cultured leaf protoplasts. Acta Biologica Hungarica 40:369–380.

Fehér, A., Felföldi, K., Preiszner, J. and Dudits, D. (1991). PEG-mediated transformation of leaf protoplasts of *Solanum tuberosum* L. cultivars. Plant Cell Tissue Org. Cult. 27:105–114.

Fehér, A., Preiszner, J., Litkey, Z., Csánadi, G. and Dudits, D. (1992). Characterization of chromosome instability in interspecific somatic hybrids obtained by X-ray fusion between potato (*Solanum tuberosum* L.) and *S. brevidens* Phil. Theor. Appl. Genet. 85:880–890.

Famelaer, I., Gleba, Y.Y., Sidorov, V.A., Kaleda, V.A., Parokonny, A.S., Boryshuk, N.V., Cherep, N.N., Negrutiu, I. and Jacobs, M. (1989). Intergeneric asymmetric hybrids between *Nicotiana plumbaginifolia* and *Nicotiana sylvestris* obtained by 'gamma-fusion'. Plant Sci. 61:105–117.

Famelaer, I., Negrutiu, I., Mouras, A., Vaucheret, H. and Jacobs, M. (1990). Asymmetric hybridization in *Nicotiana* by 'gamma-fusion' and progeny analysis of self fertile hybrids. Theor. Appl. Genet. 79:513–520.

Fowke, L.C. and Gamborg, O.L. (1980). Applications of protoplasts to the study of plant cells. Int. Rew. Cytol. 68:9–51.

Fromm, M., Taylor, L. and Walbot, V. (1985). Expression of genes transferred into monocot and dicot plant cells by electroporation. Proc. Natl. Acad. Sci. USA 82:5824–5828.

Fromm, M.E., Taylor, L.P. and Walbot, V. (1986). Stable transformation of maize after electroporation. Nature 319:791–793.

Fujiwara, T., Naito, S., Chino, M. and Nagata, T. (1991). Electroporated protoplasts express seed specific gene promoters. Plant Cell Rep. 9:602–606.

Furner, I.J., Higgins, E.S. and Berrington, A.W. (1989). Single-stranded DNA transforms plant protoplasts. Mol. Gen. Genet. 220:65–68.

Fuchs, Y., Saxena, A., Gamble, H.R. and Anderson, J.D. (1989). Ethylene biosynthesis-inducing protein from cellulysin is an endoxylanase. Plant Physiol. 89:138–143.

Gahan, P.B. (1989). Viability of plant protoplasts. In: Biotechnology in Agriculture and Forestry. Y.P.S. Bajaj (ed.), Plant Protoplasts and Genetic Engineering, Vol. 8. Springer-Verlag, Berlin, pp. 34–49.

Galun, E. (1981). Plant protoplasts as physiological tools. Annu. Rev. Plant Physiol. 32:237–266.

Galun, E. and Aviv, D. (1986). Organelle transfer. In: Methods in Enzymology. A. Weisbach and H. Weisbach (eds.), Plant Molecular Biology, Vol. 118. Academic Press, Orlando, pp. 595–611.

Gharti-Chhetri, G.B., Cherdshewasart, W., Dewulf, J., Paszkowski, J., Jacobs, M., and Negrutiu, I. (1990). Hybrid genes in the analysis of transformation conditions: 3. Temporal/spatial fate of NPTII gene integration, its inheritance and factors affecting these processes in *Nicotiana plumbaginfolia*. Plant Mol. Biol. 14:687–696.

Gleba, Y.Y. (1978). Extranuclear inheritance investigated by somatic hybridization. In: Frontiers of Plant Tissue Culture 1978. T.A. Thorp (ed.), International Assoc. Plant Tissue Culture, Calgary, pp. 95–102.

Gleba, Y.Y. and Hoffmann, F. (1980). 'Arabidobrassica': a novel plant obtained by protoplast fusion. Planta 149:112–117.

Gleba, Y.Y. and Sytnik, K.M. (1984). Protoplast Fusion: Genetic Engineering in Higher Plants. Springer-Verlag, Berlin and New York.

Gleba, Y.Y. and Shlumukov, L.R. (1990). Selection of somatic hybrids. In: Plant Cell Line Selection. P.J. Dix (ed.), VCH Verlagsgesellschaft, Weinheim, pp. 257–286.

Gleba, Y.Y., Hinnisdaels, S., Sidorov, V.A., Kaleda, V.A., Parokonny, A.S., Boryshuk, N.V., Cherep, N.N., Negrutiu, I. and Jacobs, M. (1988). Intergeneric asymmetric hybrids between

Nicotiana plumbaginifolia and *Atropa belladonna* obtained by 'gamma-fusion'. Theor. Appl. Genet. 76:760–766.

Glimelius, K. (1984). High growth rate and regeneration capacity of hypocotyl protoplasts in some *Brassicaceae*. Physiol. Plant. 61:38–44.

Glimelius, K., Fahlesson, J., Landgren, M., Sjödin, C. and Sundberg, E. (1991). Gene transfer via somatic hybridization in plants. Trends in Biotechnology 9:24–30.

Golovkin, M.V., Ábrahám, M., Mórocz, S., Bottka, S., Fehér, A. and Dudids, D. (1993). Production of transgenic maize plants by direct DNA uptake into plant protoplasts. Plants Sci. 90:41–52.

Gopalakrishnan, B., Sonthayanon, B., Rahmatullah, R. and Muthukrishnan, S. (1991). Barley aleuron layer cell protoplasts as a transient expression system. Plant Mol. Biol. 16:463–467.

Grosset, J., Meyer, Y., Chartier, Y., Kauffmann, S., Legrand, M. and Fritig, B. (1990a). Tobacco mesophyll protoplasts synthesize 1,3-β-glucanase, chitinases, and osmotins during *in vitro* culture. Plant Physiol. 92:520–527.

Grosset, J., Marty, I., Chartier, Y. and Meyer, Y. (1990b). mRNAs newly synthetised by tobacco mesophyll protoplasts are wound-inducible. Plant Mol. Biol. 15:485–496.

Grout, B.W.W. and Coutts, R.H.A. (1974). Additives for the enhancement of fusion and endocytosis in higher plant protoplasts: an electrophoretic study. Plant Sci. Lett. 2:397–403.

Gupta, H.S. and Pattanayak, A. (1993). Plant regeneration from mesophyll protoplasts of rice (*Oryza sativa* L.). Biotechnology 11:90–94.

Gupta, H.S., Schieder, O. and Gupta, M. (1984). Intergeneric nuclear gene transfer between somatically and sexually incompatible plants through asymmetric protoplast fusion. Mol. Gen. Genet. 197:30–35.

Gupta, H.S., Rech, E.L., Cocking, E.C. and Davey M.R. (1988). Electroporation and heat shock stimulate division of protoplasts of *Pennisetum squamulatum*. J. Plant Physiol. 133:457–459.

Guri, A., Levi, A. and Sink, K.C. (1988). Morphological and molecular characterization of somatic hybrid plants between *Lycopersicon esculentum* and *Solanum nigrum*. Mol. Gen. Genet. 212:191–198.

Haberlach, G.T., Cohen, B.A., Reichert, N.A., Baer, M.A., Towill, L.E. and Helgeson, J.P. (1985). Isolation, culture and regeneration of protoplasts from potato and several related *Solanum* species. Plant Sci. 39:67–74.

Hadlaczky, Gy., Burg, K., Maroy, P. and Dudits, D. (1980). DNA synthesis and division in interkingdom heterokaryons. *In vitro* 16:647–650.

Hahne, G. and Lörz, H. (1988). Release of phytotoxic factors from plant cell walls during protoplast isolation. J. Plant Physiol. 132:345–350.

Hahne, B., Lörz, H. and Hahne, G. (1990). Oat mesophyll protoplasts: Their response to various feeder cultures. Plant Cell Rep. 8:590–593.

Handley, L.W., Nickels, R.L., Cameron, M.W., Moore, P.P. and Sink, K.C. (1986). Somatic hybrid plants between *Lycopersicon esculentum* and *Solanum lycopersicoides*. Theor. Appl. Genet. 71:691–697.

Hanson, M.R., Rothenberg, M., Boeshore, M.L. and Nivison, H.T. (1985). Organelle segregation and recombination following protoplast fusion: analysis of sterile cytoplasms. In: Biotechnology in Plant Science. Relevance to Agriculture in the Eighties. M. Zaitlin, P. Day and A. Hollaender (eds.), Academic Press, Orlando, pp. 129–144.

Harms, C.T. (1983). Somatic incompatibility in the development of higher plant somatic hybrids. Quarterly Rev. Biol. 58:325–353.

Harms, C.T. and Potrykus, I. (1978). Fractionation of plant protoplast types by iso-osmotic density gradient centrifugation. Theor. Appl. Genet. 53:57–63.

Harris, R., Wright, M., Byrne, M., Varnum, J., Brightwell, B. and Schubert, K. (1988). Callus formation and plantlet regeneration from protoplasts derived from suspension cultures of wheat (*Triticum aestivum* L.). Plant Cell. Rep. 7:337–340.

Hauptmann, R., Kumar, P. and Widholm, J. (1983). Carrot (x) tobacco somatic cell hybrids selected by amino acid analog resistance complementation. In: Protoplasts 1983. I. Potrykus,

C.T. Harms, A. Hinnen, R. Hutter, P.J. King and R.D. Shillito (eds.), poster proc. Birkhauser, Basel, pp. 92–93.

Hayashi, Y., Kyozuka, J. and Shimamoto, K. (1988). Hybrids of rice (*Oryza saltiva* L.) and wild *Oryza* species obtained by cell fusion. Mol. Gen. Genet. 214:6–10.

Hayashi, Y. and Shimamoto, K. (1988). Wheat protoplast culture: embryogenic colony formation from protoplasts. Plant Cell Rep. 7:414–417.

Haydu, Zs., Lazar, G. and Dudits, D. (1977). Increased frequency of polyethylene glycol induced protoplast fusion by dimethyl-sulphoxide. Plant Sci. Lett. 10:357–360.

Helgeson, J.P. (1989). Somatic hybridization of wild Solanum species with potato: a potential source of diversity for breeders. In: Parental Line Breeding and Selection in Potato Breeding. Pudoc, Wageningen, pp. 87–94.

Hibi, T., Kano H., Suguiura, M., Kazami T. and Kimura, S. (1988). High-speed electro-fusion and electro-transfection of plant protoplasts by a continuous flow electro-manipulator. Plant Cell Rep. 7:153–157.

Hinnisdaels, S., Bariller, L., Mouras, A., Sidorov, V.A., Del-Favero, J., Veuskens, J., Negrutiu, I. and Jacobs, M. (1991). Highly asymmetric intergeneric nuclear hybrids between *Nicotiana* and *Petunia*: evidence for recombinogenic and translocation events in somatic hybrid plants after 'gamma-fusion'. Theor. Appl. Genet. 82:609–614.

Horn, M.E., Conger, B.V. and Harms, C.T. (1988a). Plant regeneration from protoplasts of embryogenic suspension cultures of orchardgrass (*Dactylis glomerata* L.). Plant Cell Rep. 7:371–374.

Horn, M.E., Shillito, R.D., Conger, B.V. and Harms, C.T. (1988b). Transgenic plants of orchardgrass (*Dactylis glomerata* L.) from protoplasts. Plant Cell Rep. 7:469–472.

Hunt, G.J. and Helgeson, J.P. (1989). A medium and simplified procedure for growing single cells from *Solanum* species. Plant Sci. 60:251–257.

Ichikawa, H., Tanno-Suenaga, L. and Imamura, J. (1989). Transfer of mitochondria through protoplast fusion. In: Biotechnology in Agriculture and Forestry. Y.P.S. Bajaj (ed.), Plant Protoplasts and Genetic Engineering, Vol. 9. Springer-Verlag, Berlin, pp. 360–375.

Imamura, J., Saul, M.W. and Potrykus, I. (1987). X-ray irradiation promoted asymmetric somatic hybridization and molecular analysis of the products. Theor. Appl. Genet. 74:445–450.

Imbrie-Milligan, C., Kamo, K.K. and Hodges, T.K. (1987). Microcallus growth from maize protoplasts. Planta 171:58–64.

Ishii, S. (1988). Factors influencing protoplast viability of suspension-cultured rice cells during isolation process. Plant Physiol. 88:26–29.

Ishii, S. (1989). Enzymes for the isolation of protoplasts. In: Biotechnology in Agriculture and Forestry. Y.P.S. Bajaj (ed.), Plant Protoplasts and Genetic Engineering, Vol. 8. Springer-Verlag, Berlin, pp. 23–33.

Itoh, K., Iwabuchi, M. and Shimamoto, K. (1991). *In situ* hybridization with species-specific DNA probes gives evidence for asymmetric nature of *Brassica* hybrids obtained by X-ray fusion. Theor. Appl. Genet. 81:356–362.

Jacobsen, J.V. and Close, T.J. (1991). Control of transient expression of chimaeric genes by gibberellic acid and abscisic acid in protoplasts prepared from mature barley aleurone layers. Plant Mol. Biol. 16:713–724.

Jahne, A., Lazzeri, P.A. and Lörz, H. (1991). Regeneration of fertile plants from protoplasts derived from embryogenic cell suspension of barley (*Hordeum vulgare* L.). Plant Cell Rep. 10:1–6.

Jaiswal, S.K., Hammatt, N., Bhojwani, S.S., Cocking, E.C. and Davey, M.R. (1990). Plant regeneration from cotyledon protoplasts of *Brassica carinata*. Plant Cell Tissue Org. Cult. 22:159–165.

Jenes, B. and Pauk, J. (1989). Plant regeneration from protoplast derived calli in rice (*Oryza sativa* L.) using dicamba. Plant Sci. 63:187–198.

Joersbo, M. and Brunstedt, J. (1991). Electroporation: Mechanism and transient expression, stable transformation and biological effects in plant protoplasts. Physiol. Plant. 81:256–264.

Jones, C.W., Mastrangelo, I.A., Smith, H.H., Liu, H.Z. and Meck, R.A. (1976). Interkingdom fusion between human (HeLa) cells and tobacco hybrid (GGLL) protoplasts. Science 193:401–403.

Jones, H., Ooms, G. and Jones, M.G.K. (1989). Transient gene expression in electroporated *Solanum* protoplast. Plant Mol. Biol. 13:503–511.

Jones, M.G.K., Dunckley, R., Steele, S., Karp, A., Gibson, R., Fish, N., Valkonen, J., Poutala, T. and Pehu, E. (1990). Transfer of resistance to PLRV, PVX and PVY from *S. brevidens* to potato by somatic hybridization: characterization and field evaluation. In: Progress in Plant Cellular and Molecular Biology. H.J.J. Nijkamp, L.H.W. Van der Plas and J. Van Aartrijk (eds.), Kluwer Academic Publishers, Dordrecht, Boston, pp. 286–304.

Jouanneau, J.P., Lapous, D. and Guern, J. (1991). In plant protoplasts, the spontaneous expression of defense reactions and the responsiveness to exogenous elicitors are under auxin control. Plant Physiol. 96:459–466.

Jourdan, P.S., Earle, E.D. and Mutschler, M.A. (1990). Improved protoplast culture and stability of cytoplasmic traits in plants regenerated from leaf protoplasts of cauliflower (*Brassica oleracea* ssp. *botrytis*). Plant Cell Tissue Org. Cult. 21:227–236.

Junker, B., Zimny, J., Luhrs, R. and Lorz, H. (1987). Transient expression of chimaeric genes in dividing and non-dividing cereal protoplasts after PEG-induced DNA uptake. Plant Cell Rep. 6:329–332.

Kameya, T. (1975). Induction of hybrids through somatic cell fusion with dextran sulphate and gelatin. Jap. J. Genet. 50:235–246.

Kao, H.M., Keller, W.A., Gleddie, S. and Brown, G.G. (1990). Efficient plant regeneration from hypocotyl protoplasts of broccoli (*Brassica oleracea* L. ssp. *italica* Plenck). Plant Cell Rep. 9:311–315.

Kao, H.M., Brown, G.G., Scoles, G. and Seguin-Swartz, G. (1991). Ogura cytoplasmic male sterility and triazine tolerant *Brassica napus* cv. Westar produced by protoplast fusion. Plant Sci. 75:63–72.

Kao, K.N., Constabel, F., Michayluk, M. and Gamborg, O.L. (1974). Plant protoplast fusion and growth of intergeneric hybrids cells. Planta 120:215–227.

Kao, K.N. (1975). A method for fusion of plant protoplasts with polyethylene glycol. In: Plant Tissue Culture Methods. O.L. Gamborg and L.R. Wetter (eds.), National Research Council of Canada, pp. 22–27.

Kao, K.N. (1977). Chromosomal behavior in somatic hybrids of soybean-*Nicotiana glauca*. Mol. Gen. Genet. 150:225–230.

Kao, K.N. (1986). Fusion of plant protoplasts at the interface of a glucose and a sucrose-polyethylene glycol solution. J. Plant Physiol. 126:55–58.

Kao, K.N. and Michayluk, M.R. (1974). A method for high-frequency intergeneric fusion of plant protoplasts. Planta 115:355–367.

Kao, K.N. and Michayluk, M. (1975). Nutritional requirements for growth of *Vicia hajstana* cells and protoplasts at a very low population density in liquid media. Planta 126:105–110.

Kao, K.N. and Michayluk, M.R. (1980). Plant regeneration from mesophyll protoplasts of alfalfa. Z. Pflanzenphysiol 96:135–141.

Kao, K.N. and Saleem, M. (1986). Improved fusion of mesophyll and cotyledon protoplasts with PEG and high pH-Ca^{2+} solutions. J. Plant Physiol. 122:217–225.

Kao, K.N. and Michayluk, M. (1989). Fusion of plant protoplasts – techniques. In: Biotechnology in Agriculture and Forestry. Y.P.S. Bajaj (ed.), Plant Protoplasts and Genetic Engineering, Vol. 8. Springer-Verlag, Berlin, pp. 277–285.

Karesch, H., Bilang, R. and Potrykus, I. (1991a). *Arabidopsis thaliana*: protocol for plant regeneration from protoplasts. Plant Cell Rep. 9:575–578.

Karesch, H., Bilang, R., Scheid, O.M. and Potrykus, I. (1991b). Direct gene transfer to protoplasts of *Arabidopsis thaliana*. Plant Cell Rep. 9:571–574.

Kartzke, S., Saedler, H. and Meyer, P. (1990). Molecular analysis of transgenic plants derived from transformations of protoplasts at various stages of the cell cycle. Plant Sci. 67:63–72.

Kemble, R.J. and Barsby, T.L. (1988). Use of protoplast fusion systems to study organelle genetics in a commercially important crop. Biochem. Cell Biol. 66:665–676.

Kirti, P.B. and Chopra, V.L. (1990). Rapid plant regeneration through organogenesis and somatic embryogenesis from cultured protoplasts of *Brassica juncea*. Plant Cell Tissue Org. Cult. 20:65–67.

Kitamura, Y., Morikawa, T. and Miura, H. (1989). Isolation and culture of protoplasts from cell suspension cultures of *Duboisia myoporoides* with subsequent plant regeneration. Plant Sci. 60:245–250.

Klercker, J. (1982). Methode zur Isolterung lebender Protoplasten. Ofvers Vetensk Akad. Forh. (Stockholm) 9:463–475.

Klimaszewska, K. (1989). Recovery of somatic embryos and plantlets from protoplasts cultures of *Larix* × *eurolepis*. Plant Cell Rep. 8:440–444.

Köhler, F. and Wenzel, G. (1985). Regeneration of isolated barley microspores in conditioned media and trials to characterize the responsible factor. J. Plant Physiol. 121:245–257.

Köhler, F., Cardon, G., Pohlman, M., Gill, R. and Schieder, O. (1989). Enhancement of transformation rates in higher plants by low-dose irradiation: Are DNA repair systems involved in the incorporation of exogenous DNA into the plant genome. Plant Mol. Biol. 12:189–199.

Kohn, H.R., Schieder, R. and Schieder, O. (1985). Somatic hybrids in tobacco mediated by electrofusion. Plant Sci. 38:121–128.

Koop, H.U. and Schweiger, H.G. (1985a). Regeneration of plants from individually cultivated protoplasts using an improved microculture system. J. Plant Physiol. 121:245–257.

Koop, H.U. and Schweiger, H.G. (1985b) Regeneration of plants after electrofusion of selected pairs of protoplasts. Eur. J. Cell Biol. 39:46–49.

Koornneef, M., Hanhart, C.J., Jonsma, M., Tomay, I., Weide, R., Zabel, P. and Hill, I. (1986). Breeding of a tomato genotype readily accessible to genetic manipulation. Plant Sci. 45:201–208.

Krens, F.A., Molendijk, L., Wullems, G.J. and Schilperoort, R.A. (1982). *In vitro* transformation of plant protoplasts with Ti-plasmid DNA. Nature 296:72–74.

Krens, F.A., Jamar, D., Rouwendal, G.J.A. and Hall, R.D. (1990). Transfer of cytoplasm from new *Beta* CMS sources to sugar beet by asymmetric fusion. Theor. Appl. Genet. 79:390–396.

Krogaard, H. and Andersen, A.S. (1990). Effect of cellulase fractionation on the viability of cultured barley (*Hordeum vulgare*) protoplasts. Physiol. Plant. 80:119–125.

Krumbiegel, G. and Schieder, O. (1979). Selection of somatic hybrids after fusion of protoplasts from *Datura innoxia* Mill. and *Atropa belladonna* L. Planta 145:371–375.

Kumar, A. and Cocking, E.C. (1987). Protoplast fusion: A novel approach to organelle genetics in higher plants. Am. J. Bot. 74:1289–1303.

Kunitake, H. and Mii, M. (1990a). Somatic embryogenesis and plant regeneration from protoplasts of asparagus (*Asparagus officinalis* L.). Plant Cell Rep. 8:706–710.

Kunitake, H. and Mii, M. (1990b). Plant regeneration from cell culture-derived protoplasts of statice (*Limonium perezii* Hubbard). Plant Sci. 70:115–119.

Küster, E. (1909). Über die Verschmelzung nachter Protoplasten. Ber. Dtsch. Bot. Ges. 27:589–598.

Kyozuka, J., Hayashi, Y. and Shimamoto, K. (1987). High frequency plant regeneration from rice protoplasts by novel nurse culture methods. Mol. Gen. Genet. 206:408–413.

Kyozuka, J., Otoo, E. and Shimamoto, K. (1988). Plant regeneration from protoplasts of indica rice: genotypic differences in culture response. Theor. Appl. Genet. 76:887–890.

Kyozuka, J., Kaneda, T. and Shimamoto, K (1989). Production of cytoplasmic male sterile rice (*Oryza sativa*) by cell fusion. 7:1171–1174.

Lang, H. and Kohlenbach, H.W. (1988). Callus formation from mesophyll protoplasts of *Fagus sylvatica* L. Plant Cell Rep. 7:485–488.

Larkin, P.J., Davies, P.A. and Tanner, G.J. (1988). Nurse culture of low numbers of *Medicago* and *Nicotiana* protoplasts using calcium alginate beads. Plant Sci. 58:203–210.

Lee, B., Murdoch, K., Topping, J., Kreis, M. and Jones, M.G.K. (1989). Transient gene expression in aleurone protoplasts isolated from developing caryopses of barley and wheat. Plant Mol. Biol. 13:21–29.

Lee, C.H. and Power, J.B. (1988). Interspecific gametosomatic hybridization in *Petunia hybrida* Plant Cell Rep. 7:17–18.

Lee, L., Schroll, R.E., Grimes, H.D. and Hodges, T.K. (1989). Plant regeneration from indica rice (*Oryza sativa* L.) protoplasts. Planta 178:325–333.

Lee, N. and Wetzstein, H.Y. (1988). Protoplast isolation and callus production from leaves of tissue cultured *Vitis* spp. Plant Cell Rep. 7:531–534.

Lee-Stadelmann, O., Chung, I. and Stadelmann, E.J. (1985). Plasmolysis of *Glycine max* mesophyll cells: the use of octylguanidine and its implications in protoplast isolation. Plant Sci. 38:1–7.

Lehminger-Mertens, R. and Jacobsen, H.J. (1989). Plant regeneration from pea protoplasts via somatic embryogenesis. Plant Cell Rep. 8:379–382.

Lenee, P. and Chupeau, Y. (1986). Isolation and culture of sunflower protoplasts (*Helianthus annuus* L.): factors influencing the viability of cell colonies derived from protoplasts. Plant Sci. 43:69–75.

Li, Z. and Murai, N. (1990). Efficient plant regeneration from rice protoplasts in general medium. Plant Cell Rep. 9:216–220.

Li, Z., Burow, M.D. and Murai, N. (1989). High frequency generation of fertile transgenic rice plants after PEG-mediated protoplast transformation. Plant Mol. Biol. Rep. 8:276–291.

Lörz, H., Larkin, P.J., Thomson, J. and Scowcroft, W.R. (1983). Improved protoplast culture and agarose media. Plant Cell Tissue Org. Cult. 2:217–226.

Loudon, P.T., Nelson, R.S. and Ingram, D.S. (1989). Studies of protoplast culture and plant regeneration from commercial and rapid-cycling *Brassica* species. Plant Cell Tissue Org. Cult. 19:213–223.

Lührs, R. and Lörz, H. (1988). Initiation of morphogenic cell-suspension and protoplast cultures of barley (*Hordeum vulgare* L.). Planta 175:71–81.

Lurquin, P.F. and Kado, C.I. (1977). *Eschericia coli* plasmid pBR313 insertion into protoplasts and into their nuclei. Mol. Gen. Genet. 154:113–121.

Lynch, P.T., Isaac, S. and Collin, H.A. (1989). Electrofusion of protoplasts from celery (*Apium graveolens* L.) with protoplasts from the filamentous fungus *Aspergillus nidulans*. Planta 178:207–214.

Lyznik, L.A., Kamo, K.K., Grimes, H.D., Ryan, R., Chang, L.L. and Hodges, T.K. (1989). Stable transformation of maize: the impact of feeder cells on protoplast growth and transformation efficiency. Plant Cell Rep. 8:292–295.

Maas, C. and Werr, W. (1989). Mechanism and optimized conditions for PEG mediated DNA transfection into plant protoplasts. Plant Cell Rep. 8:148–151.

Maas, C., Laufs, J., Grant, S., Korfhage, C. and Werr, W. (1991). The combination of a novel stimulatory element in the first exon of the maize *Shrunken*-1 gene with the following intron 1 enhances reporter gene expression up to 1000-fold. Plant Mol. Biol. 16:199–207.

Marchesi, M.L., Castiglione, S. and Sala, F. (1989). Effect of repeated DNA sequences on direct gene transfer in protoplasts of *Nicotiana plumbaginifolia*. Theor. Appl. Genet. 78:113–118.

Mascarenhas, D., Mettler, I.J., Pierce, D.A. and Lowe, H.W. (1990). Intron-mediated enhancement of heterologous gene expression in maize. Plant Mol. Biol. 15:913–920.

Masson, J., Lecerf, M., Rousselle, P., Perennec, P. and Pelletier, G. (1987). Plant regeneration from protoplasts of diploid potato derived from crosses of *Solanum tuberosum* with wild *Solanum* species. Plant Sci. 53:167–176.

Masuda, K., Kudo-Shiratori, A. and Inoue, M. (1989). Callus formation and plant regeneration from rice protoplasts purified by density gradient centrifugation. Plant Sci. 62:237–246.

Matthews, D., Mottley, J., Horan, I. and Roberts, A.V. (1991). A protoplast to plant system in roses. Plant Cell Tissue Org. Cult. 24:173–180.

Medgyesy, P. (1990). Selection and analysis of cytoplasmic hybrids. In: Plant Cell Line Selection.

P.J. Dix (ed.), VCH Verlagsgesellschaft, Weinheim, New York, Basel, Cambridge, pp. 287–316.

Medgyesy, P., Fejes, E. and Maliga, P. (1985). Interspecific chloroplast recombination in a *Nicotiana* somatic hybrids. Proc. Natl. Acad. Sci. USA 82:6960–6964.

Mehrle, W., Hampp, R., Naton, B. and Grothe, D. (1989). Effects of microgravitation on electrofusion of plant cell protoplasts. Plant Physiol. 89:1172–1177.

Mehrle, W., Naton, B. and Hampp, R. (1990). Determination of physical membrane properties of plant cell protoplasts via the electrofusion technique: prediction of optimal fusion yields and protoplast viability. Plant Cell Rep. 8:687–691.

Meijer, E.G.M., Schilperoort, R.A., Rueb, S., Van Os-Ruygrok and Hensgens, L.A.M. (1991). Transgenic rice cell lines and plants: Expression of transferred chimeric genes. Plant Mol. Biol. 16:807–820.

Melchers, G. and Labib, G. (1974). Somatic hybridization of plants by fusion of protoplasts I. Selection of light resistant hybrids of 'haploid' light sensitive varieties of tobacco. Mol. Gen. Genet. 135:277–294.

Melchers, G., Sacristan, M.D. and Holder, A.A. (1978). Somatic hybrid plants of potato and tomato regenerated from fused protoplasts. Carlsberg Res. Commun. 43:203–218.

Menczel, L., Lazar, G. and Maliga, P. (1978). Isolation of somatic hybrids by cloning *Nicotiana* heterokaryons in nurse cultures. Planta 143:29–32.

Menczel, L., Nagy, F., Kiss, Zs. and Maliga, P. (1981). Streptomycin resistant and sensitive somatic hybrids of *Nicotiana tabacum* (×) *Nicotiana knightiana*: correlation of resistance to *Nicotiana tabacum* plastids. Theor. Appl. Genet. 59:191–195.

Meyer, Y., Aspart, L. and Chartier, Y. (1984a). Auxin-induced regulation of protein synthesis in tobacco mesophyll protoplast cultivated *in vitro*, 1: Characteristics of auxin-sensitive proteins. Plant Physiol. 75:1027–1033.

Meyer, Y., Aspart, L. and Chartier, Y. (1984b). Auxin-induced regulation of protein synthesis in tobacco mesophyll protoplast cultivated *in vitro*, 2: Time course and level of auxin control. Plant Physiol. 75:1034–1039.

Meyer, P., Kartzke, S., Niedenhof, I., Heidmann, I., Bussmann, K. and Saedler, H. (1988). A genomic DNA segment from *Petunia hybrida* leads to increased transformation frequencies and simple integration patterns. Proc. Natl. Acad. Sci. USA 85:8568–8572.

Michel, W. (1937). Über die experimentelle Fusion pflanzlicher Protoplastsen. Arch. Exp. Zellforsch. 20:230–252.

Mieth, H., Speth, V. and Ebel, J. (1986). Phytoalexin production by isolated soybean protoplasts. Z. Naturforsch. 41:193–201.

Mitchell, J.C. and Petolino, J.F. (1991). Plant regeneration from haploid suspension and protoplast cultures from isolated microspores of maize. J. Plant Physiol. 137:530–536.

Mock, H.P., Emmerling, M. and Seitz, H.U. (1990). Cell wall synthesis in carrot cells: comparison of suspension-cultured cells and regenerating protoplasts. Physiol. Plant. 79:347–353.

Mórocz, S., Donn, G., Nemeth, J. and Dudits, D. (1990). An improved system for the obtention of fertile regenerants via maize protoplasts isolated from a highly embryogenic suspension culture. Theor. Appl. Genet. 80:121–726.

Moyne, A. L., Thor, V., Pelissier, B., Bergounioux, C., Freyssinet, G. and Gadal, P. (1988). Callus and embryoid formation from protoplasts of *Helianthus annuus*. Plant Cell Rep. 7:437–440.

Mühlback, H.P. and Thiele, H. (1980). Response to chilling of tomato mesophyll protoplasts. Planta 151:399–401.

Mussel, H., Earle, E., Campbell, L. and Batts, L.A. (1986). Ethylene synthesis during protoplast formation from leaves of *Avena sativa*. Plant Sci. 47:207–214.

Myers, J.R., Grosser, J.W., Taylor, N.L. and Collin, G.B. (1989). Genotype-dependent whole plant regeneration from protoplasts of red clover (*Trifolium pratense* L.). Plant Cell Tissue Org. Cult. 19:113–127.

Nagata, T. (1978). A novel cell-fusion method of protoplasts by polyvinyl alcohol. Naturwissenschaften 65:263–264.

Nagata, T. and Melchers, G. (1978). Surface charge of protoplasts and their significance in cell-cell interaction. Planta 142:235–238.

Nagata, T. and Takabe, I. (1971). Plating of isolated tobacco mesophyll protoplasts on agar medium. Planta 99:12–20.

Naton, B., Hoffmann, E.M., Hampp, R. and Vasil, I.K. (1991). Improved electrofusion of protoplasts of varied fusibility by selective pairing: application of asymmetric breakdown of plasma membranes. Plant Sci. 75:93–105.

Negrutiu, I., Shillito, R.D., Potrycus, I., Biasini, G. and Sala, F. (1987). Hybrid genes in the analysis of transformation conditions. I. Setting up a simple method for direct gene transfer to protoplasts. Plant Mol. Biol. 8:363–373.

Negrutiu, I., Dewulf, J., Pietrzak, M., Botterman, J., Rietveld, E., Wurzer-Figurelli, E.M., Ye, D. and Jacobs, M. (1990). Hybrid genes in the analysis of transformation conditions: II. Transient expression vs stable transformation – analysis of parameters influencing gene expression levels and transformation efficiency. Physiol. Plant. 79:197–205.

Niedz, R.P., and Sink, K.C. (1988). Multifactor analysis of environmental preconditioning of tomato seedlings on protoplasts culture and development. J. Plant Physiol. 133:385–391.

Niizeki, M. (1989). Somatic hybridization in rice × soybean. In: Biotechnology in Agriculture and Forestry. Y.P.S. Bajaj (ed.), Plant Protoplasts and Genetic Engineering, Vol. 8. Springer-Verlag, Berlin, pp. 410–434.

Nyman, M. and Wallin, A. (1988). Plant regeneration from strawberry (*Fragaria* × *Ananassa*) mesophyll protoplasts. J. Plant Physiol. 133:375–377.

O'Connel, M.A. and Hanson, M.R. (1987). Regeneration of somatic hybrid plants formed between *Lycopersicon esculentum* and *L. pennelli*. Theor. Appl. Genet. 75:83–89.

Oard, J.H., Paige, D. and Dvorak, J. (1989). Chimeric gene expression using maize intron in cultured cells of breadwheat. Plant Cell Rep. 8:156–160.

Ochatt, S.J. (1990). Plant regeneration from root callus protoplasts of sour cherry (*Prunus cerasus* L.). Plant Cell Rep. 9:268–271.

Ochatt, S.J. and Caso, O.H. (1986). Shoot regeneration from leaf mesophyll protoplasts of wild pear (*Pyrus Communis* var. *pyraster* L.). J. Plant Physiol. 122:243–249.

Ochatt, S.J., Chand, P.K., Rech, E.L., Davey, M.R., and Power, J.B. (1988a). Electroporation-mediated improvement of plant regeneration of colt cherry (*Prunus avium* × *pseudocerasus*) protoplasts. Plant Sci. 54:165–169.

Ochatt, S.J. and Power, J.B. (1988). Plant regeneration from mesophyll protoplasts of Williams' Bon Chretien (syn. Bartlett) pear (*Pyrus communis* L.). Plant Cell Rep. 7:587–589.

Ochatt, S.J., Rech, E.L., Davey, M.R. and Power, J.B. (1988b). Long-term effect of electroporation on enhancement of growth and plant regeneration of colt cherry (*Prunus avium* × *pseudocerasus*) protoplasts. Plant Cell Rep. 7:393–395.

Ohyama, K., Gamborg, O.L. and Miller, R.A. (1972). Uptake of exogenous DNA by plant protoplasts. Can. J. Bot. 50:2077–2080.

Okamura, M., Hayashi, T. and Miyasaki, S. (1984). Inhibiting effect of ammonium ion in protoplast culture of some *Asteraceae* plants. Plant and Cell Physiol. 25:281–286.

Oliviera, M. and Pais, M.S. (1991). Plant regeneration from protoplasts of long-term callus cultures of *Actinidia deliciosa* var. *deliciosa* cv. Hayword (kiwifruit). Plant Cell Rep. 9:643–646.

Oliviera, M., Barroso, J. and Pais, M.S. (1991). Direct gene transfer into *Actidinia deliciosa* protoplasts: analysis of transient expression of the *CAT* gene using TLC autoradiography and a GC-MS-based method. Plant Mol. Biol. 17:235–242.

Omirulleh, S., Ábrahám, M., Golovkin, M., Stefanov, I., Karabaev, M. A., Mustárdy, L., Mórocz, S. and Dudits, D. (1993). Activity of a chimeric promoter with doubled CaMV 35S enhancer element in protoplast-derived cells and transgenic plants in maize. Plant. Mol. Biol. 21:415–428.

Paszkowski, J., Shillito, R.D., Saul, M., Mandak, V., Hohn, T., Hohn, B. and Potrykus, I. (1984). Direct gene transfer to plants. EMBO J. 3:2717–2722.

Paszkowski, J., Baur, M., Bogucki, A. and Potrykus, I. (1988). Gene targeting in plants. EMBO J. 8:4021–4026.

Paszkowski, J., Saul, M.W. and Potrykus, I. (1989). Plant gene vectors and genetic transformation: DNA-mediated direct gene transfer to plants. In: Cell Culture and Somatic Cell Genetics of Plants. I.K. Vasil and J. Schell (eds.), Vol. 6. Academic Press, San Diego, New York, Berkeley, pp. 51–68.

Patat-Ochatt, E.M., Ochatt, S.J. and Power, J.B. (1988). Plant regeneration from protoplasts of apple rootstocks and scion varieties (*Malus* × *domestica* Borkh.). J. Plant Physiol. 133:460–465.

Patnaik, G., Wilson, D. and Cocking, E.C. (1981). Importance of enzyme purification for increased plating efficiency and plant regeneration from single protoplasts of *Petunia parodii*. Z. Pflanzenphysiol. 102:199–205.

Pental, D., Hamill, D.J., Pirrie, A. and Cocking, E.C. (1986). Somatic hybridization of *Nicotiana tabacum* and *Petunia hybrida*. Recovery of plants with *P. hybrida* nuclear genome and *N. tabacum* chloroplast genome. Mol. Gen. Genet. 202:342–347.

Pental, D., Mukhopadhyay, A., Grover, A. and Pradhan, A.K. (1988). A selection method for the synthesis of triploid hybrids by fusion of microspore protoplasts (n) with somatic cell protoplasts (2n). Theor. Appl. Genet. 76:237–243.

Perl, A., Aviv, D. and Galun, E. (1988). Ethylene and *in vitro* culture of potato: suppression of ethylene generation vastly improves protoplast yield, plating efficiency and transient expression of an alien gene. Plant Cell Rep. 7:403–406.

Perl, A., Aviv, D. and Galun, E. (1990). Protoplast-fusion-derived *Solanum* cybrids: Application and phylogenetic limitations. Theor. Appl. Genet. 79:632–640.

Phillips, R. and Darrell, N.J. (1988). A simple technique for single-cell cloning of crown gall tumor tissue: *Petunia* protoplast regeneration without exogenous hormones. Plant Physiol. 133:447–451.

Piastuch, W.C. and Bates, G.W. (1990). Chromosomal analysis of *Nicotiana* asymmetric somatic hybrids by dot blotting and *in situ* hybridization. Mol. Gen. Genet. 222:97–103.

Piccirilli, M., Pupilli, F. and Arcioni, S. (1988). *Lotus tenuis* Wald. & Kit.: *in vitro* conditions for plant regeneration from protoplasts and callus of various explants. Plant Sci. 55:77–82.

Pilet, P.E., Blaschek, W., Senn, A. and Franz, G. (1984). Comparison between maize root cells and their respective regenerating protoplasts: Wall polysaccharides. Planta 161:465–469.

Pillai, V., Davey, M.R. and Power, J.B. (1990). Plant regeneration from mesophyll protoplasts of *Centaurea cyanus*, *Senecio* × *hybridus* and *Callistephus chinensis*. Plant Cell Rep. 9:402–405.

Poste, G. and Allison, A.C. (1973). Membrane fusion. Biochim. Biophys. Acta 300:421–465.

Potrykus, I. (1991). Gene transfer to plants: Assessment of published approaches and results. Ann. Rev. Plant Physiol. Plant Mol. Biol. 42:205–225.

Poulsen, G.B. and Nielsen, S.V.S. (1989). Regeneration of plants from hypocotyl protoplasts of rapeseed (*Brassica napus* L. var. *oleifera*) cultivars. Plant Cell Tissue Org. Cult. 17:153–158.

Preiszner, J. Fehér, A., Veisz, O., Sutka, J. and Dudits, D. (1991). Characterization of morphological variation and cold resistance in interspecific somatic hybrids between potato (*Solanum tuberosum* L.) and *S. brevidens* Phil. Euphytica 57:37–49.

Prioli, L.M. and Söndahl, M.R. (1989). Plant regeneration and recovery of fertile plants from protoplasts of maize (*Zea mays* L.). Biotechnology 7:589–594.

Prives, J. and Shinitzky, M. (1977). Increased membrane fluidity precedes fusion of muscle cells. Nature 268:761–763.

Pua, E.C. (1990). Somatic embryogenesis and plant regeneration from hypocotyl protoplasts of *Brassica juncea* (L.) Czern & Coss. Plant Sci. 68:231–238.

Punja, Z.K., Tang, F.A. and Sarmento, G.G. (1990). Isolation, culture and plantlet regeneration from cotyledon and mesophyll protoplasts of two pickling cucumber (*Cucumis sativus* L.) genotypes. Plant Cell Rep. 9:61–64.

Pupilli, F., Arcioni, S., Damiani F. and Pezzotti M. (1990). Plant regeneration from callus and protoplast cultures of *Lotus pedunculatus* Cav. Plant Cell Tissue Org. Cult. 23:193–199.

Rasheed, J.H., Al-Mallah, M.K., Cocking, E.C. and Davey, M.R. (1990). Root hair protoplasts of *Lotus corniculatus* L. (birdsfoot trefoil) express their totipotency. Plant Cell Rep. 8:565–569.

Renaudin, J.P., Tournaire, C., Brillat, M. and De La Serve, B.T. (1990). Sequential hormone requirement for growth and organogenesis of *Petunia hybrida* protoplasts-derived calli. Plant Sci. 71:239–250.

Rethmeier, N.O.M., Jansen, C.E., Snel, E.A.M., Nijkamp, H.J.J. and Hille, J. (1991). Improvement of regeneration of *Lycopersicon pennellii* protoplasts by decreasing ethylene production. Plant Cell Rep. 9:539–543.

Rhodes, C.A., Lowe, K.S. and Ruby, K.L. (1988a). Plant regeneration from protoplasts isolated from embryogenic maize cell cultures. Biotechnology 6:56–60.

Rhodes, C.A., Pierce, D.A., Mettler, I.J., Mascarenhas, D. and Detmer, J.J. (1988b). Genetically transformed maize plants from protoplasts. Science 240:204–206.

Riggs, C.D. and Bates, G.W. (1986). Stable transformation of tobacco by electroporation: evidence for plasmid concatenation. Proc. Natl. Acad. Sci. USA 83:5602–5606.

Rodenburg, K.W., De Groot, J.A., Schilperoort, R.A. and Hooykaas, P.J.J. (1989). Single-stranded DNA used as an efficient new vehicle for transformation of plant protoplasts. Plant Mol. Biol. 13:711–719.

Russell, J.A. and McCown, B.H. (1988). Recovery of plants from leaf protoplasts of hybrid-poplar and aspen clones. Plant Cell Rep. 7:59–62.

Ryan, C.A. and Farmer, E.E. (1991). Oligosaccharide signals in plants: A current assessment. Ann. Rev. Plant Physiol. Plant Mol. Biol. 42:651–674.

Sakai, T. and Imamura, J. (1990). Intergeneric transfer of cytoplasmic male sterility between *Raphanus sativus* (cms line) and *Brassica napus* through cytoplast-protoplast fusion. Theor. Appl. Genet. 80:421–427.

Sala, C., Biasini, M.G., Morandi C., Nielsen, E., Parisi B. and Sala, F. (1985). Selection and nuclear DNA analysis of cell hybrids between *Daucus carota* and *Oryza sativa*. J. Plant Physiol. 118:409–419.

Saleem, M. and Cutler, A.J. (1987). Stabilizing corn leaf protoplasts with n-Propyl Gallate. J. Plant Physiol. 128:479–484.

Salmenkallio, M., Hannus, R., Teeri, T. H. and Kauippinen, V. (1990). Regulation of α-amylase promoter by gibberellic acid and absisic acid in barley protoplasts transformed by electroporation. Plant Cell Rep. 9:352–355.

Sauvadet, M.A., Brochard, P. and Boccon-Gibod, J. (1990). A protoplast-to-plant system in chrysanthenum: differential responses among several commercial clones. Plant Cell Rep. 8:692–695.

Schaffler, E. and Koop, H.U. (1990). Single cell nurse culture of tobacco protoplasts: physiological analysis of conditioning factors. J. Plant Physiol. 137:95–101.

Schieder, O. and Kohn, H. (1986). Protoplast fusion and generation of somatic hybrids. In: Cell Culture and Somatic Cell Genetics of Plants. I.K. Vasil (ed.), Plant Regeneration and Genetic Variability, Vol. 3. pp. 569–588.

Schmitz, P. and Schnabl, H. (1989). Regeneration and evacuolation of protoplasts from mesophyll, hypocotyl and petioles from *Helianthus annuus* L. J. Plant Physiol. 135:223–227.

Schnabl, H. and Zimmermann, U. (1989). Immobilization of plant protoplasts. In: Biotechnology in Agriculture and Forestry. Y.P.S. Bajaj (ed.), Plant Protoplasts and Genetic Engineering, Vol. 8. Springer-Verlag, Berlin, pp. 63–96.

Schocher, R.J., Shillito, R.D., Saul, M.W., Paszkowski, J. and Potrykus, I. (1986). Co-transformation of unlinked foreign genes into plants by direct gene transfer. Biotechnology 4:1093–1096.

Schröder, R., Gartner, F., Steinbrenner, B., Knoop, B. and Beiderbeck, R. (1989). Viability factors in plant suspension cultures – some properties. J. Plant Physiol. 135:422–427.

Schwall, M. and Feix, G. (1988). Zein promoter activity in transiently transformed protoplasts from maize. Plant Sci. 56:161–166.

Schweiger, H.G., Dirk, J., Koop, H.U., Kranz, E., Neuhaus, G., Spangenberg, G. and Wolff, D. (1987). Individual selection, culture and manipulation of higher plant cells. Theor. Appl. Genet. 73:769–783.

Senda, M., Takabe, J., Abe, S. and Nakamura, T. (1979). Induction of cell fusion of plant protoplasts by electrical stimulation. Plant and Cell Physiol. 20:1441–1443.

Senda, M., Morikawa, H., Katagi, H., Takada, T. and Yamada, Y. (1980). Effect of temperature on membrane fluidity on protoplast fusion. Theor. Appl. Genet. 57:33–35.

Shea, E.M., Gibeaut, D.M. and Carpita, N.C. (1989). Structural analysis of the cell walls regenerated by carrot protoplasts. Planta 179:293–308.

Shepard, J.F. and Totten, R.E. (1977). Mesophyll cell protoplasts of potato. Plant Physiol. 60:313–316.

Shepard, J.F., Bidney, D., Barsby, T. and Kemble, R. (1983). Genetic transfer in plants through interspecific protoplast fusion. Science 219:683–688.

Shillito, R.D., Paszkowski, J. and Potrykus, I. (1983). Agarose plating and a bead type culture technique enable and stimulate development of protoplast-derived colonies in a number of plant species. Plant Cell Rep. 2:244–247.

Shillito, R.D., Saul, M.W., Paszkowski, J., Muller, M. and Potrykus, I. (1985). High frequency direct gene transfer to plants. Biotechnology 3:1099–1103.

Shillito, R.D., Carswell, G.K., Johnson, C.M., DiMaio, J.J. and Harms, C.T. (1989). Regeneration of fertile plants from protoplasts of elite inbred maize. Biotechnology 7:581–587.

Shimamoto, K., Terada, R., Izawa, T. and Fujimoto, H. (1989). Fertile transgenic rice plants regenerated from transformed protoplasts. Nature 338:274–276.

Shneyour, Y., Zelcer, A., Izhar, S. and Beckmann, J.S. (1984). A simple Feeder-Layer technique for the plating of plant cells and protoplasts at low density. Plant Sci. Lett. 33:293–302.

Shöpke, C., Müller, L.E. and Kohlerbach, H-W. (1987). Somatic embryogenesis and regeneration of plantlets in protoplast cultures from somatic embryos of coffee (*Coffea canephora* P. ex Fr.). Plant Cell Tissue Org. Cult. 8:243–248.

Sihachakr, D. and Ducreux, G. (1987). Plant regeneration from protoplast culture of sweet potato (*Ipomoea batatas* Lam.) Plant Cell Rep. 6:326–328.

Sikdar, S.R., Chatterjee, G., Das, S. and Sen, S.K. (1990). '*Eurissica*' the intergeneric fertile somatic hybrid developed through protoplasts fusion between *Eruca sativa* Lam. and *Brassica juncea* (L.) Czern. Theor. Appl. Genet. 79:561–567.

Sim, G.E., Loh, C.S. and Goh, C.J. (1988). Direct somatic embryogenesis from protoplasts of *Citrus mitis* Blanco. Plant Cell Rep. 7:418–420.

Sjödin, C. and Glimelius, K. (1989). Transfer of resistance against *Phoma lingam* to *Brassica napus* by asymmetric somatic hybridization combined with toxin selection. Theor. Appl. Genet 78:513–520.

Smidsrod, O. and Skjak-Braek, G. (1990). Alginate as immobilization matrix for cells. Trends in Biotechnology 8:71–78.

Smith, M.A.L., Palta, J.P., McCrown, B.H. and Steffen, K.L. (1989a). Plasmolytic behavior of the donor cell may affect protoplast response. Physiol. Plant. 76:201–204.

Smith, M.A.L., Pay, A. and Dudits, D. (1989b). Analysis of chloroplast and mitochondrial DNAs in asymmetric somatic hybrids between tobacco and carrot. Theor. Appl. Genet. 77:641–644.

Somers, D.A., Narayanan, K.R., Kleinhofs, A., Cooper-Bland, S. and Cocking, E.C. (1986). Immunological evidence for transfer of the barley nitrate reductase structural gene to *Nicotiana tabacum* by protoplasts fusion. Mol. Gen. Genet. 204:296–301.

Somers, D.A., Birnberg, P.R., Petersen, W.L. and Brenner, M.L. (1987). The effect of conditioned medium on colony formation from 'Black Mexican Sweet' corn protoplasts. Plant Sci. 53:249–256.

Song, J., Sorensen, E.L. and Liang, G.H. (1990). Direct embryogenesis from single mesophyll protoplasts in alfalfa (*Medicago sativa* L.). Plant Cell Rep. 9:21–25.

Spangenberg, G., Osusky, M., Oliviera, M.M., Freydl, E., Nagel, J., Pais, M.S. and Potrykus, I. (1990). Somatic hybridization by microfusion of defined protoplast pairs in *Nicotiana*: morphological, genetic, and molecular characterization. Theor. Appl. Genet. 80:577–587.

Spangenberg, G., Freydl, E., Osusky, M., Nagel, J., and Potrykus, I. (1991). Organelle transfer by microfusion of defined protoplast-cytoplast pairs. Theor. Appl. Genet. 81:477–486.

Steinbrenner, B., Schröder, R., Knoop, B. and Beiderbeck, R. (1989). Viability factors in plant suspension cultures – a novel bioassay. J. Plant Physiol. 134:582–585.

Stenger, D.A. and Hui, S.W. (1986). Kinetics of ultrastructural changes during electrically-induced fusion of human erythrocytes. J. Membr. Biol. 93:43–53.

Sun, C.S., Prioli, L.M. and Sondahl, M.R. (1989). Regeneration of haploid and dihaploid plants from protoplasts of supersweet (sh2sh2) corn. Plant Cell Rep. 8:313–316.

Szabados, L. and Dudits, D. (1980). Fusion between interphase and mitotic plant protoplasts. Exp. Cell Res. 127:441–446.

Tada, Y., Sakamoto, M. and Fujimura, T. (1990). Efficient gene introduction into rice by electroporation and analysis of transgenic plants: Use of electroporation buffer lacking chloride ions. Theor. Appl. Genet. 80:475–480.

Tagu, D., Bergounioux, C., Perennes, C. and Gadal, P. (1990). Inheritance of two foreign genes cointroduced *Petunia hybrida* by direct gene transfer. Plant Cell Tissue Org. Cult. 21:259–266.

Takabe, I., Otsuki, Y. and Aoki, S. (1968). Isolation of tobacco mesophyll cells in intact and active state. Plant and Cell Physiol. 9:115–124.

Takabe, I., Labib, G. and Melchers, G. (1971). Regeneration of whole plants from isolated mesophyll protoplasts of tobacco. Naturwissenschaften 58:318–320.

Takamizo, T., Suginobu, K.I. and Ohsugi, R. (1990). Plant regeneration from suspension culture derived protoplasts of tall fescue (*Festuca arundinacea* Schreb.) of a single genotype. Plant Sci. 72:125–131.

Tan, M.L.M.C., Rietveld, E.M., Van Marrewijk, G.A.M. and Kool, A.J. (1987). Regeneration of leaf mesophyll protoplasts of tomato cultivars (*L. esculentum*): Factors important for efficient protoplast culture and plant regeneration. Plant Cell Rep. 6:172–175.

Tanno-Suenaga, L., Ichikawa, H. and Imamura, J. (1988). Transfer of the CMS trait in *Daucus carota* L. by donor-recipient protoplast fusion. Theor. Appl. Genet. 76:855–860.

Tautorus, T.E., Attree, S.M., Fowke, L.C. and Dunstan, D.I. (1990). Somatic embryogenesis from immature and mature zygotic embryos and embryo regeneration from protoplasts in black spruce (*Picea mariana* Mill.). Plant Sci. 67:115–124.

Tempelaar, M.J. and Jones, M.G.K. (1985). Fusion characteristics of plant protoplasts in electric fields. Planta 165:205–216.

Tempelaar, M.J., Duyst, A., De Vlas, S.Y., Krol, G., Symmonds, C. and Jones, M.G.K (1987). Modulation and direction of the electrofusion response in plant protoplasts. Plant Sci. 48:99–105.

Terada, R., Kyozuka, J., Nishibayashi, S. and Shimamoto, K. (1987). Plantlet regeneration from somatic hybrids of rice (*Oryza sativa* L.) and barnyard grass (*Echinocloa oryzicola* Vasing). Mol. Gen. Genet. 210:39–43.

Teulieres, C. and Boudet, A.M. (1991). Isolation of protoplasts from different *Eucalyptus* species and preliminary studies on regeneration. Plant Cell Tissue Org. Cult. 25:133–140.

Teulieres, C., Grima-Pettenati, J., Curie, C., Teissie, J. and Boudet, A.M. (1991). Transient foreign gene expression in polyethylene/glycol treated or electropulsated *Eucalyptus Gunnii* protoplasts. Plant Cell Tissue Org. Cult. 25:125–132.

Theodoropoulos, P.A. and Roubelakis-Angelakis, K.A. (1990). Progress in leaf protoplast isolation and culture from virus-free axenic shoot cultures of *Vitis vinifera* L. Plant Cell Tissue Org. Cult. 20:15–23.

Thompson, J.A., Abdullah, R., Chen, W.-H. and Gartland, K.M.A. (1987). Enhanced protoplast division in rice (*Oryza sativa* L.) following heat shock treatment. J. Plant Physiol. 127:367–370.

Thompson, G.A., Boston, R.S., Lyznik, L.A., Hodges, T.K. and Larkins, B.A. (1990). Analysis of promoter activity from an α-zein gene 5′ flanking sequence in transient expression assays. Plant Mol. Biol. 15:755–764.

Thomzik, J.E. and Hain, R. (1988). Transfer and segregation of triazine tolerant chloroplasts in *Brassica napus* L. Theor Appl. Genet. 76:165–171.

Tong, C.B., Labavitch, J.M. and Yang, S.F. (1986). The induction of ethylene production from pear cell culture by cell wall fragments. Plant Physiol. 81:929–930.

Toriyama, K., Hinata, K. and Sasaki, T. (1986). Haploid and diploid plant regeneration from protoplasts of anther callus in rice. Theor. Appl. Genet. 73:16–19.

Toriyama, K., Arimoto, Y., Uchimiya, H. and Hinata, K. (1988). Transgenic rice plants after direct gene transfer into protoplasts. Biotechnology 6:1072–1074.

Toriyama, K. and Hinata, K. (1988). Diploid somatic-hybrid plants regenerated from rice cultivars. Theor. Appl. Genet. 76:665–668.

Tremblay, F.M. (1988). Callus formation from protoplasts of *Betula papyrifera* March. cell suspension culture. J. Plant Physiol. 133:247–251.

Tyagi, S., Spörlein, B., Tyagi, A.K., Herrmann, R.G. and Koop, H.U. (1989). PEG-mediated electroporation-induced transformation in *Nicotiana tabacum*: Influence of genotype on transformation frequencies. Theor. Appl. Genet. 78:287–292.

Uchimiya, H., Hirochika, H., Hashimoto, H., Hara, A., Masuda, T., Kasumimoto, T., Harada, H., Ikeda, J.F. and Yoshioka, M. (1986). Co-expression and inheritance of foreign genes in transformants obtained by direct DNA transformation of tobacco protoplasts. Mol. Gen. Genet. 205:1–8.

Ueda, T., Miyamoto, Y. and Tanaka, I. (1990). Fusion studies of pollen protoplasts and generative cell protoplasts in *Lilium longiflorum*. Plant Sci. 72:259–266.

Ueda, T. and Messing, J. (1991). A homologous expression system for cloned zein genes. Theor. Appl. Genet. 82:93–100.

Vardi, A., Spiegel-Roy, P. and Galun E. (1982). Plant regeneration from citrus protoplasts: variability in methodological requirements among cultivars and species. Theor. Appl. Genet. 62:171–176.

Vardi, A., Breiman, A. and Galun, E. (1987). *Citrus* cybrids: production by donor-recipient protoplastfusion and verification by mitochondrial-DNA restriction profiles. Theor. Appl. Genet. 75:51–58.

Vardi, A., Arzee-Gonen, P., Frydman-Shani, A., Bleichman, S. and Galun, E. (1989). Protoplastfusion-mediated transfer of organelles from *Microcitrus* into *Citrus* and regeneration of novel alloplasmic trees. Theor. Appl. Genet. 78:741–747.

Vardi, A., Bleichman, S. and Aviv, D. (1990). Genetic transformation of *Citrus* protoplasts and regeneration of transgenic plants. Plant Sci. 69:199–206.

Vasil, I.K. (1987). Developing cell and tissue culture systems for the improvement of cereal and grass crops. J. Plant Physiol. 128:193–218.

Vasil, I.K. (1988). Progress in the regeneration and genetic manipulation of cereal crops. Biotechnology 6:397–402.

Vasil, V., Clancy, M., Ferl, R.J., Vasil, I.K. and Hannah, L.C. (1989). Increased gene expression by the first intron maize *shrunken*-1 locus in grass species. Plant Physiol. 91:1575–1579.

Vasil, V., Redway, F. and Vasil, I.K. (1990). Regeneration of plants from embryogenic suspension culture protoplasts of wheat (*Triticum aestivum* L.). Biotechnology 8:429–432.

Vasil, V. and Vasil, I.K. (1987). Formation of callus and somatic embryos from protoplasts of a commercial hybrid of maize (*Zea mays* L.). Theor. Appl. Genet. 73:793–798.

Vieira, M.L.C., Jones, B., Cocking, E.C. and Davey, M.R. (1990). Plant regeneration from protoplasts isolated from seedling cotyledons of *Stylosanthes guianansis*, *S. macrocephala* and *S. scabra*. Plant Cell Rep. 9:289–292.

Wallin, A., Glimelius, K. and Eriksson, T. (1974). The induction of aggregation and fusion of *Daucus carrota* protoplasts by polyethylene glycol. Z. Pflanzenphysiol. 74:64–80.

Wallin, A. and Johansson, L. (1989). Plant regeneration from leaf mesophyll protoplasts of *in vitro* cultured shoots of a columnar apple. J. Plant Physiol. 135:565–570.

Walters, T.W. and Earle, E.D. (1990). A simple, versatile feeder layer system for *Brassica oleracea* protoplast culture. Plant Cell Rep. 9:316–319.

Wang, D., Miller, P.D. and Sandahl, M.R. (1989a). Plant regeneration from protoplasts of Indica type rice and CMS rice. Plant Cell Rep. 8:329–332.

Wang, H., Cutler, A.J., Saleem, M. and Fowke, L.C. (1989b). Treatment of soybean cells with cell wall degrading enzymes inhibits nuclear division but not DNA synthesis. J. Plant Physiol. 135:404–408.

Watts, J.W., Motoyoshi, F. and King, J.M. (1974). Problems associated with the production of stable protoplasts of cells of tobacco mesophyll. Annu. Bot. 38:667–671.

Watts, J.W. and King, J.M. (1984). A simple method for large-scale electrofusion of plant protoplasts. Biosci. Rep. 4:335–342.

Weber, G., De Groot, E. and Schweiger, H.G. (1986). Synchronization of protoplasts from *Glycine max* (L.) Merr. and *Brassica napus* (L.). Planta 168:273–280.

Wei, Z.M. and Xu, Z.H. (1988). Plant regeneration from protoplasts of soybean (*Glycine max* L.). Plant Cell Rep. 7:348–351.

Wei, Z.M. and Xu, Z.H. (1990). Regeneration of fertile plants from embryogenic suspension culture protoplasts of *Sorghum vulgare*. Plant Cell Rep. 9:51–53.

Weising, K, Schell, J. and Kahl, G. (1988). Foreign genes in plants: transfer, structure, expression and applications. Annu. Rev. Genet. 22:421–477.

Wijbrandi, J., Zabel, P. and Koornneef, M. (1990). Restriction fragment length polymorphism of somatic hybrids between *Lycopersicon esculentum* and irradiated *Lycopersicon peruvianum*: evidence for limited donor genome elimination and extensive chromosome rearrangement. Mol. Gen. Genet. 222:270–277.

Wilson, S.M., Thorpe, T.A. and Moloney, M.M. (1989). PEG-mediated expression of GUS and CAT genes in protoplasts from embryogenic suspension cultures of *Picea glauca*. Plant Cell Rep. 7:704–707.

Yamada, Y., Hara, Y., Senda, M., Nishihara, M. and Kito, M. (1979). Phospholipids of membranes of cultured cells and the products of protoplast fusion. Phytochemistry 18:423–426.

Yamashita, Y., Terada, R., Nishibayashi, S. and Shimamoto, K. (1989). Asymmetric somatic hybrids of *Brassica*: partial transfer of *B. campestris* genome into *B. oleracea* by cell fusion. Theor. Appl. Genet. 77:189–194.

Yang, Z.Q., Shikanai, T. and Yamada, Y. (1988a). Asymmetric hybridization between cytoplasmic male-sterile (CMS) and fertile rice (*Oryza sativa* L.) protoplasts. Theor. Appl. Genet. 76:801–808.

Yang, H., Zhang, H.M., Davey, M.R., Mulligan, B.J. and Cocking, E.C. (1988b). Production of kanamycin resistant rice tissues following DNA uptake into protoplasts. Plant Cell Rep. 7:421–425.

Yarrow, S.A., Cocking, E.C. and Power, J.B. (1987). Plant regeneration from cultured cell-derived protoplasts of *Pelargonium andum*, *P.* × *hortorum* and *P. peltatum*. Plant Cell Rep. 6:102–104.

Zacharison, A. and Bornman, C.H. (1984). Application of electric field fusion in plant tissue culture. Physiol. Plant. 61:314–320.

Zacharison, A. and Bornman, C.H. (1986). Electromanipulation of plant protoplasts. Physiol. Plant. 67:507–516.

Zaghmout, O.M. and Torello, W.A. (1990). Isolation and culture of protoplasts from embryogenic suspension cultures of red fescue (*Festuca rubra* L.). Plant Cell Rep. 9:340–343.

Zelcer, A., Aviv, D. and Galun, E. (1978). Interspecific transfer of cytoplasmic male sterility by fusion between protoplasts of normal *Nicotiana sylvertris* and X-ray irradiated protoplasts of male sterile *N. tabacum*. Z. Pflanzenphysiol. 90:397–407.

Zhang, H.M., Yang, H., Rech, E.L., Golds, T.J., Davis, A.S., Mulligan, B.J., Cocking, E.C., and Davey, M.R. (1988). Transgenic rice plants produced by electroporation-mediated plasmid uptake into protoplasts. Plant Cell Rep. 7:379–384.

Zhang, W. and Wu, R. (1988). Efficient regeneration of transgenic plants from rice protoplasts and correctly regulated expression of the foreign gene in the plants. Theor. Appl. Genet. 76:835–840.

Zhongyi, L., Tanner, G.J. and Larkin, P.J. (1990). Callus regeneration from *Trifolium subterraneum* protoplasts and enhanced protoplast division by low-voltage treatment and nurse cells. Plant Cell Tissue Org. Cult. 21:67–73.

Zimmermann, U. (1982). Electrical field-mediated fusion and related electrical phenomena. Biochim. Biophys. Acta 694:227–277.

Zimmermann, U. and Scheurich, P. (1981). High frequency fusion of plant protoplasts by electric fields. Planta 151:26–32.

Zimmermann, U., Scheurich, P., Pilwat, G. and Benz, R. (1981). Cells with manipulated functions: New perspectives for cell biology, medicine and technology. Angewandte Chemie 20:325–344.

Zimmermann, U. and Vienken, J. (1982). Electric field-induced cell-to-cell fusion. J. Membrane Biol. 67:165–182.

5. Isolation and Characterisation of Mutant Cell Lines

Philip J. Dix

Contents

1. Introduction

It is perhaps dangerous to try to pinpoint the nascence of efforts at selection or isolation of modified phenotypes *in vitro*. Rather, awareness of the genetic potential of tissue culture dawned as researchers became conscious of the nature and scope of these cultures from numerous fundamental studies throughout the 1950s and 1960s. Those early to appreciate these possibilities deserve credit and this reviewer intends no disrespect to them in placing the watershed in the development of this field in the early 1970s with the reports on amino acid analogue resistant cell lines (Widholm, 1972a, b) and the selection of a streptomycin resistant mutant (Maliga et al., 1973). The former is noteworthy in that a rational selection strategy was devised to obtain a defined phenotype, namely tryptophan accumulation. Not only was the strategy carried through successfully, but characterisation of the resistant lines confirmed the predicted basis of the change, namely, reduced feedback sensitivity of the enzyme anthranilate synthetase.

The streptomycin resistant mutant, SR1, of tobacco is even more significant. It is the first *in vitro* selected trait for which sexual inheritance (ma-

Indra K. Vasil and Trevor A. Thorpe (eds.), Plant Cell and Tissue Culture, 119–138.

ternal) was unequivocally demonstrated (Maliga et al., 1973, 1975), and the first for which the exact site and nature of the genetic lesion has been identified, a single base substitution in the 16SrRNA gene in the chloroplast DNA (Etzold et al., 1987). Both these mutant types have since proved useful for fundamental investigations and as selectable markers for genetic manipulation.

There have of course subsequently been numerous studies on selection of a wide range of variants *in vitro*. These show a strong emphasis on traits of agronomic value such as disease, herbicide and environmental stress resistance, but include mutations of more fundamental interest, such as auxotrophs and hormone mutants. Reviews of some or all of these areas have been frequent, culminating in a recent treatise examining the entire field (Dix, 1990). The number and diversity of relevant reports precludes an exhaustive review here. Rather, the emphasis will be placed on evaluating the progress in a few of the key areas, describing some of the successes, particularly highlighting the problems, and those mutants which have been extensively characterised genetically and/or biochemically.

2. Principles of Selection

2.1. *Approaches*

The principles behind the different selection strategies are very simple, and will be outlined here. Unfortunately, the practise is a different matter, usually requiring a great deal of patience and a number of trials to establish the most appropriate selection conditions. Detailed evaluation of the problems of *in vitro* selection can be found elsewhere (Dix, 1986; Collin and Dix, 1990). A large majority of the mutant cell lines described have been obtained by resistance selection, in which a bulk population is subjected to a selection pressure (e.g. toxic drug, microbial culture filtrate, salt, low temperature) and resistant cell lines recovered from surviving cells. The ideal starting material for selection is protoplast-derived colonies, but more frequently callus or cell suspension cultures provide a more easily manipulated inoculum. The use of regenerating culture systems (e.g. McCabe et al., 1989) is attractive and may be on the increase.

Negative selection approaches, employing counterselective agents to kill wild type cells, have been successfully used for the selection of auxotrophs (e.g. Negrutiu et al., 1985) but the technique is extremely rigorous to apply and is dependent on an efficient procedure for culture of haploid protoplasts. It has therefore been largely restricted to *Nicotiana* species, and very few laboratories.

Most other selection strategies are based on screening. At its simplest this involves visual selection of colonies for the production or loss of a pigment, an approach which may be automated and brought down to the cell level by flow cytometry and cell sorting (Lister, 1990), but then it becomes much less simple! At its most tedious, screening involves individual chemical analysis of colonies, which can sometimes be accelerated by the development of a suitable immunoassay (Zenk et al., 1977). Equally time-consuming is use of total selection, by testing the nutritional requirements of haploid protoplast-derived colonies, to obtain auxotrophs, which has been successfully applied in several laboratories (Pythoud and King, 1990).

Occasionally a novel selection strategy can be devised, as in the elegant procedure developed to obtain temperature sensitive mutants of carrot (Breton and Sung, 1982 – described in Section 4.3), which is really a variation on the principle of counterselection.

2.2. *Mutagenesis*

In vitro mutagenesis is another topic which merits, and has recently received (Negrutiu, 1990), a lengthy review in its own right. Here, a few general observations can be made about the use of mutagens to generate variation *in vitro*. Firstly, when using a strong selection pressure (i.e. resistance selection) for a dominant trait, or in haploid cultures, it is usually not essential. Secondly, mutagenic treatments can be very effective, as demonstrated by the (extremely rare) systematic, quantitative studies on the use of physical (Grandbastien et al., 1985) and chemical (Negrutiu et al., 1984) mutagens on protoplast populations. Thirdly, for certain purposes mutagenesis is important. Total selection for auxotrophs requires mutagenesis to bring the number of colonies to be screened down to manageable levels, and it is also valuable when a limited number of propagating units are to be subjected to selection, as in regenerating explants. Plastome targetted mutagens have also been valuable in several protocols for the selection of plastid mutations (see Section 4.2). Finally, the optimal system to which to apply mutagens, is freshly isolated mesophyll protoplasts. These cells are predominantly in G1, and can give rise to single cell derived colonies, improving the chances of obtaining non-chimeral mutant lines. If haploid, they enhance the prospects for selecting recessive mutants, and establishing homozygosity.

The most widely used mutagens *in vitro* are ethyl-methane sulphomate (EMS), 1-methyl-3-nitro-1-nitrosoguanidine (MNNG), and the nitrosoureas, and physical agents: X-, gamma-, and UV-irradiation.

Physical mutagens, where appropriate, should be favoured for two simple reasons: they are less hazardous to the user, and don't require additional washing steps for their removal. The latter is not a trivial consideration when handling delicate protoplast populations. UV-irradiation is particularly attractive as it is of proven effectiveness in inducing gene mutations (Grandbastien et al., 1985), and is relatively cheap and available to most laboratories.

3. Mutants of Agricultural Interest

3.1. *Herbicide Resistance*

This character is the one for which arguably the greatest amount of progress has been made, in terms both of selection of mutants, and their detailed molecular characterisation. The reasons for this success probably relate to the relatively high level of understanding of the biochemical basis of phytotoxicity of many (but not all) of the chemicals in question, the severe and reproducible symptoms produced by them, and in some cases the clear genetic potential for resistance, as demonstrated by the regular occurrence of resistant weed populations. Here, we shall restrict our attention to those herbicides for which the resistant mutants have been extensively and recently characterised. For a broader review the reader is referred to Smith and Chaleff (1990).

The broad spectrum herbicide glyphosate blocks aromatic amino acid biosynthesis by inhibiting the enzyme 5-enolpyruvyl-shikimate 3-phosphate synthase (EPSPS). *In vitro* selected, resistant lines have been the target of several laboratories and have been obtained for carrot (Nafziger et al., 1984), tomato (Smith et al., 1986), *Petunia* (Steinrucken et al., 1986), *Corydalis sempervivens* (Hollander-Czytko et al., 1988), and tobacco (Dyer et al., 1988; Goldsbrough et al., 1990). While there remains a paucity of inheritance data on these resistant lines, the mechanism of resistance has been clearly established. In all cases resistant lines exhibited elevated levels of EPSPS, in contrast to the bacterial mode of resistance: altered EPSPS insensitive to the inhibitor (Comai et al., 1983; Kishore et al., 1986). Generally EPSPS overproduction in plants is due to gene amplification, but in *C. semperivivens* alterations in gene expression are implicated as there is no change in EPSPS gene copy number. Glyphosate is not the only herbicide for which resistance is associated with gene amplification. Amplification of the glutamine synthetase gene confers resistance to phosphinotricin in alfalfa cell lines (Donn et al., 1984).

Another group of herbicides which has generated considerable interest is the sulfonylureas. These herbicides, which include chlorsulfuron, sulfometuron methyl and primisulfuron inhibit the enzyme acetolactate synthase, ALS (aka. acetohydroxy acid synthase, AHAS), the first enzyme in the synthesis of the branched chain amino acids.

In vitro selection for resistance to these herbicides was first reported by Chaleff and Ray (1984) who selected lines resistant to chlorsulfuron and sulfometuron methyl in callus cultures of tobacco. Genetic analysis of six mutants indicated that resistance is conferred by single dominant or semidominant alleles, at two unlinked loci. Further characterization (Chaleff and Mauvais, 1984; Chaleff and Bascomb, 1987) revealed that mutation at each locus resulted in half the enzyme being herbicide insensitive, demonstrating that tobacco has at least two distinct ALS isozymes encoded by different genes. Amino acid substitutions resulting in resistance at both loci have now been identified (Lee et al., 1988).

Other reports on *in vitro* selection for sulfonylurea resistance include those for flax (Jordan and McHughen, 1987), *Datura innoxia* (Saxena and King, 1988), *Brassica napus* (Swanson et al., 1988; Saxena et al., 1990), and a further report for tobacco (Harms and DiMaio, 1991), this time using a different sulfonylurea herbicide, primisulfuron. Where characterised these mutants all show resistant ALS, but an additional report (Armour et al., 1990) suggests sulfonylureas should also be added to the list of herbicides for which resistance may result from gene amplification. Selection was generally carried out in callus or cell suspension cultures but one of the procedures used for *B. napus* (Swanson et al., 1988) is notable, being based on microspore mutagenesis (with ethyl nitrosourea) followed by resistance selection among microspore-derived embryos.

The imidazolinones are another group of herbicides for which the primary site of action is ALS. Cell lines of maize resistant to one of these herbicides, imazaquin, were selected in embryogenic callus cultures. The resistance was expressed at the whole plant level, inherited as a semidominant nuclear trait, and correlated with a resistant form of ALS (Shaner and Anderson, 1985).

The *Brassica napus* microspore mutagenesis and selection system of Swanson et al., 1988) proved equally successful when applied to the imidazolinone 'Pursuit' (Swanson et al., 1989). The mutants showed semidominant inheritance of at least two unlinked genes which could be combined to give a highly tolerant hybrid. Of the two mutants which were extensively characterised, one has an ALS 50 times more tolerant than the wild type, but the other only has a marginal improvement in ALS, suggesting a different tolerance mechanism. While the sulfonylureas and imidazolinones clearly both act on the same enzyme, there are also important differences since imidazolinone resistant mutants of maize both with and without cross resistance to sulfonylureas have been reported (Anderson and Georgeson, 1990), and Saxena and King (1988), observing absence of cross-resistance in a imidazolinone resistant line of *Datura innoxia* suggest two separate sites of action on ALS.

Cséplő and Medgyesy (1986) were able to obtain photomixotrophic (partially photosynthetic) cultures, in which the herbicides caused bleaching of green callus. By challenging protoplast-derived colonies of *Nicotiana plumbaginifolia*, they were able to select first triazine resistant mutants, and then mutants resistant to several other photosynthetic inhibitors (Cséplő et al., 1985).

Recently similar reports have emerged from Wang et al., (1991) who plated photomixotrophic cell suspension cultures of tobacco into herbicide-containing medium, and Rey et al. (1990) who used fully photoautotrophic protoplast-derived colonies of *N. plumbaginifolia*. Detailed characterization is restricted to the first triazine resistant mutations (Cséplő et al., 1985), selected as resistant to terbutryn. In keeping with its predicted plastome location the mutation is maternally inherited. The mutants are cross-resistant to other triazine and uracil type herbicides, but not to other photosynthetic herbicides. For one of the mutants a point mutation has been identified in the psbA gene, encoding the 32 kDa protein (Páy et al., 1988), at a site already implicated in triazine resistance in algal mutants.

3.2. *Disease Resistance*

In contrast to herbicide resistance, where emphasis has been on a limited range of chemicals, with well characterised modes of action, and a small number of plant species, in the case of disease resistance there is a very diffuse body of work, reflecting the range of pathogens and crops for which there is a clear socioeconomic incentive for developing resistance. A large number of resistant lines have been described and a full review is available (Jones, 1990). Many of the lines reported, however, are poorly characterised, and inheritance data are frequently lacking.

The selection strategy itself has posed problems, generally being less straightforward than selection for resistance to antimetabolites. An understanding of the mechanism of pathogenicity is a clear advantage here, particularly if it is associated with the production of a defined toxin. All too often, however, this understanding is lacking and selection is based on a mechanism which is unknown or merely suspected.

Resistance selection may be to the pathogen itself, or to a toxin produced by the pathogen in culture. The latter approach is more common since maintenance of dual cultures involves careful optimization of the bacterial/fungal inoculum and control of its subsequent growth. The use of toxins is more satisfactory, but again the response of the plant to the toxin must be understood. The toxins may be pathogenicity factors, facilitating the infectivity of the pathogen, or virulence factors influencing the spread of the disease after infection. In practise crude culture filtrates have been far more frequently used as selective agents than the toxin itself. This is because they are much easier to obtain and their use circumvents the need to address the more difficult questions about virulence mechanisms. Selection has usually been carried out in callus or cell suspension cultures, although protoplast derived colonies have also been used for amenable species like tomato (Shahin and Spivey, 1986) and potato (Wenzel and Foroughi-Wehr, 1990).

The efficiency of these selection procedures is called into question by numerous reports on the inordinately high frequency with which somaclonal variants, with comparable improvements in resistance, can be obtained from unselected cultures. This is particularly telling where a direct comparison of the two approaches is available, as for *Phoma lingum* resistance in *Brassica napus* (Sacristan, 1982), *Xanthomonas campestris* resistance in peach (Hammerschlag, 1988, 1990), *Fusarium* wilt (Shahin and Spivey, 1986) and *Pseudomonas solanacearum* (Toyoda et al., 1989b) resistance in tomato.

A sobering assessment of the prospects of toxin selection as a route to crops with improved disease resistance can be found in Wenzel and Foroughi-Wehr (1990). This is one of the few reports on exhaustive progeny tests and deals with three different crop species: barley selected for resistance to *Helminthosporium sativum* extract and fusaric acid, wheat selected for resistance to *H. sativum*, and potato selected for resistance to *Fusarium* species and *Phytophthora infestans*. Despite clear improvements in the first *in vitro*-derived generation, and with the exception of a single *P. infestans* resistant potato clone (only vegetative progeny were assessed for potato), none of the data obtained using natural or artificial infection suggested significant improvements in the field tolerance of the progeny for selected lines. The authors reiterate the need to know about the infective process, and the mechanism of pathogenicity, or the plant response, in order to devise sufficiently powerful and precise selection strategies.

A promising approach to virus resistance is that of Toyada et al. (1989a), who, working on tobacco mosaic virus (TMV), initiated callus cultures from systemically infected tobacco plants and screened the TMV titre of a number of callus sublines to detect those in which a high level of the virus is maintained. They were then able to detect healthy, resistant plants among the regenerants from these calli, and obtained inheritance data suggesting a dominant single gene mutation conferring TMV resistance.

3.3. *Amino Acid Accumulators*

The principle behind the selection of regulatory mutants of amino acid biosynthetic pathways, resulting in accumulation of specific amino acids, is now well established (Dix, 1986), and the pioneering work in this area has already been alluded to in the introduction. The selective agents are either analogues of the amino acids, or amino acids such as valine (Bourgin, 1978) or combinations of amino acids, such as lysine and threonine (Hibberd et al., 1980) on branched or parallel biosynthetic pathways with common control enzymes. The considerable interest in such mutants derives from their value for both fundamental studies on metabolism, and for their potential applications, which are varied, including the direct improvement of nutritional quality of the crop, the possible role of amino acid accumulation in conferring improved stress tolerance, and the shifting of metabolism towards the production of valuable secondary products.

In the first category the phenotype which has attracted most attention is lysine accumulation by mutants selected for resistance either to a combination of lysine and threonine, or to the lysine analogue amino ethyl-cysteine. The nutritional value of lysine rich cereals is illustrated by the emphasis placed on cereal species including maize (Hibberd and Green, 1982), barley (Bright et al., 1982a, b), rice (Schaeffer and Sharpe, 1987), wheat (Kumpaisal et al., 1988), pearl millet (Boyes and Vasil, 1987) and Napier grass (Redway and Vasil, 1990). A number of these mutants have now been extensively characterized, both biochemically and genetically. Typically, increased levels of one or more members of the aspartate group of amino acids (particularly lysine or threonine) are observed, and are correlated to a reduced feedback sensitivity of aspartate kinase (Hibberd et al., 1980; Cattoir-Reynarts et al., 1983; Rognes et al., 1983; Jacobsen, 1986; Boyes and Vasil, 1987; Redway and Vasil, 1990). Elevated levels of the amino acids have also been found in seeds and, in the case of rice (Schaeffer and Sharpe, 1990), modification of the composition of endosperm proteins was reported. Inheritance data have generally supported a single dominant gene, and two unlinked loci are suggested for barley (Bright et al., 1982b) and maize (Diedrick et al., 1990). For one of the latter, genetic analysis by backcrossing to isogenic maize mutants, has shown it to be linked to the 02 gene, which causes a shift from zein to soluble endosperm proteins, on the short arm of chromosome 7 (Azevedo et al., 1990).

In contrast to these findings lysine accumulation in rice is inherited as a recessive character (Schaeffer et al., 1988). These authors also observed that in some genetic backgrounds it was associated with such negative traits as infertility and abnormal seed fill.

Proline is an amino acid frequently invoked in tolerance of environmental stresses (Stewart and Lee, 1974; Greenway and Munns, 1980) and this has provided the impetus for several investigations into the selection of proline accumulating lines, usually through resistance to the proline analogues hydroxyproline or azetidine, carboxylic acid (Cella et al., 1982; Ricardi et al., 1983; Dix et al., 1984; Van Swaai et al., 1986). In most, but not all (Van Swaai et al., 1986; Mori et al., 1989) cases, the expected proline accumulation is found, but is sometimes less pronounced in regenerated plants than in selected cultures. A consideration of stress tolerance of these lines will be included in the following Section 3.4.

Amino acid accumulation as a route to increased production of secondary metabolites has been reviewed recently (Wilson, 1990). Most interest has been in the production of indole alkaloids and serotonin by cell lines selected for resistance to tryptophan analogues. Other products have also been investigated and an interesting recent report is that of increased capsaicin production by p-fluorophenylalanine resistant cell lines of *Capsicum annuum* (Salgado-Garciglia and Ochoa-Alejo, 1990).

3.4. *Environmental Stress Tolerance*

The stress which has generated by far the most attention is salinity. A large number of reports have emerged describing salt tolerant cell lines of a diverse range of species. In most cases selection was applied to callus or cell suspension cultures by including growth inhibitory levels of NaCl in the medium. Unfortunately, thorough characterisation of these lines is rare, and convincing inheritance data even more so. Dracup (1991) recently emphasised the need for an improved understanding of the mechanisms of salt tolerance, in an assessment rather critical of the somewhat simplistic approach to *in vitro* selection adopted by most workers. He suggests that more productive strategies should be based on specific cell-based physiological traits. these might include selection for turgour regulation in photoautotrophic cultures, or for tolerance of high Na^{+}:Ca^{+} ratios.

Despite the perceived problems in selecting for salt tolerance *in vitro*, several recent reports provide much needed encouragement. Inheritance of *in vitro* selected salt tolerance has been reported for rice (Vajrabhaya et al., 1989), sugarbeet (Freytag et al., 1990), *Brassica juncea* (Jain et al., 1991) and *Nicotiana plumbaginifolia* (Sumaryati et al.,1991). The last of these is particularly intriguing as it cuts a swathe across different selection strategies suggesting they all give rather similar results. Colonies derived from UV-mutagenised haploid protoplasts were stressed with NaCl, KCl or polyethylene glycol (PEG). Lines resistant to each stress were obtained and plants regenerated. Inheritance data showed resistance to be due to a single dominant nuclear gene. Furthermore, each mutant was cross-resistant to the other two stresses, and all accumulated proline to 10–15 times the level of wild type cells. This is in fact higher than the level of proline accumulation in proline analogue resistant lines selected by the same team, and also inherited as a single gene dominant trait (Syukur et al., 1991). This work reinforces the link, frequently suggested, between salinity and drought stress, with proline accumulation a major factor in both, and under simple genetic control. The temptation is great to link these results to those of Van Swaai et al. (1986), who established that proline accumulation could confer improved frost tolerance on potato cells, in suggesting proline may have an even greater pivotal role in protection against stresses influencing water relations, than hitherto suspected. Salt tolerant *Brassica juncea* lines also accumulate proline (Jain et al., 1991), and proline analogue selection as an alternative route to salt tolerance continues to be successfully applied, most recently in *Vigna radiata* cultures (Kumar and Sharma, 1989).

Molecular characterization of salt tolerant cell lines is beneficial in helping to identify genes specifically associated with stress adaptation, with obvious implications for future efforts at genetically engineering stress tolerance into crop plants. For example, Winicov et al. (1989) used *in vitro* translation to detect stable changes in gene expression in salt tolerant cell lines of alfalfa. Many of the polypeptides obtained were different from those induced by salt shock in sensitive cells.

Selection for aluminium tolerance was complicated by its complex interaction with pH, and the uptake of other ions from the medium. A great deal of groundwork was needed (Conner and Meredith, 1985a) before a suitable selective medium could be devised. This involved provision of iron in an unchelated form in the aluminium containing medium, and the lowering of levels of calcium and phosphate ions. Using this medium, Conner and Meredith (1985b, c) were able to select *Nicotiana plumbaginifolia* calli with enhanced aluminium tolerance which was inherited as a single dominant allele.

Efforts at selecting for frost tolerance *in vitro* have received a boost from the recent report on 'cryoselection' in Spring wheat callus (Kendall et al., 1990). These workers subjected embryogenic callus to extreme low temperature stress, by immersing, without cryopreservatives, in liquid nitrogen, and were able to obtain a heritable enhancement of tolerance to −12 °C in selected lines. This drastic approach is relatively easy to apply and we should soon know whether it can yield dividends with a broader range of species.

4. Mutants of Fundamental Interest

This section is intended to cover those classes of mutants which, while not of direct agricultural interest, have been extensively studied because they may offer valuable insights into fundamental aspects of plant biology, or because of their potential use as genetic markers. A number of mutants could be included in this category but we will restrict ourselves to three types on which a particularly high level of effort, and in some cases ingenuity, has been focussed.

4.1. *Nitrate Reductase Deficient Mutants*

Muller and Grafe (1978) first exploited chlorate resistance as a strategy for selecting nitrate reductase deficient mutants in tobacco calli. Subsequently the same approach was applied to other species and several groups have made particularly effective use of haploid protoplasts of *Nicotiana plumbaginifolia* (Marton et al., 1982; Negrutiu et al., 1983; Gabard et al., 1987). Well over 200 such mutants have now been described for this species (see review by Pythoud and King, 1990). That review includes a detailed account of the characterization of these mutants, of which only the main points will be summarised here. The mutants fall into two main classes, the *nia* mutants, defective in the nitrate reductase apoenzyme, and the *cnx* mutants, defective in the Molybdenum-cofactor. The latter are selected at a lower frequency and are more severely impaired in their growth and development. Complementation analysis, through sexual crosses and somatic cell fusion, has suggested that the *nia* mutants are all allelic, indicating a single gene for the apoenzyme, but that intragenic complementation can occur (Gabard et al., 1987), an observation supported by recent reconstitution experiments (Pelsy and Gonneau, 1991).

The *cnx* mutants, on the other hand, fall into at least six complementation groups (*cnx*A-*cnx*F). All have immunologically detectable apoenzyme activity, and nitrate reductase activity of *cnx*A, but not the other *cnx*, mutants can be restored by the addition of high levels of molybdate to the culture medium. The *cnx*A gene product is thought to be associated with the incorporation of molybdenum, while *cnx*B, *cnx*C and *cnx*D are involved with synthesis of the molybdopterin moiety of the cofactor, although one of the roles ascribed to the cofactor, namely dimerisation of the nitrate reductase subunits, is not impaired by *cnx*D mutations.

The characterization of this fascinating group of mutants is very much ongoing, and in combination with molecular techniques, including the use of cDNA probes for nitrate reductase, from *Arabidopsis* and other species, should give us a powerful insight into the molecular architecture and functioning of this key enzyme complex.

4.2. *Cytoplasmic Mutants*

The obtention of a number of selectable, plastome-encoded, antibiotic resistant mutants, has become a major objective of those wishing to study the genetics of chloroplasts. Most of these mutants have been resistant to the amino-glycoside antibiotic streptomycin, and selected using a single strategy (green = resistant, against a bleached background) using an extremely diverse range of culture systems, callus (Maliga et al., 1973), protoplast derived colonies (Hamill et al., 1986; To et al., 1989; Jansen et al., 1990), mutagenised seedlings (Fluhr et al., 1985) and morphogenetically competent explants (McCabe et al., 1989). The last two procedures, in particular, leaned heavily on the use of efficient, plastome-targetted, mutagen nitrosomethylurea. Similar approaches have now been used to obtain mutants resistant to other antibiotics such as lincomycin (Cséplő and Maliga, 1982; Dix et al., 1990; Glas et al., 1990) and spectinomycin (Fluhr et al., 1985; Dix et al., 1990). Notably, the species range has now stretched beyond *Nicotiana* to include *Onobrychis viciifolia* (Hamill et al., 1986), *Solanum nigrum* (McCabe et al., 1989), *Lycoperscion peruvianum* (McCabe et al., 1989; Jansen et al., 1990; Glas et al., 1990), and *L. esculentum* (Glas et al., 1990). Streptomycin plus spectinomycin resistant double mutants of tobacco (Svab et al., 1990) and *S. nigrum* (Dix et al., 1990) have also been reported.

While a plastome location for most of the antibiotic resistant mutants has been supported by the maternal inheritance of the trait, detailed molecular characterisation is restricted to several of the *Nicotiana* mutants. In common with earlier mutants of green algae the '912 site' (named from the positional equivalent coordinate of *E. coli*), which is base pair 860 in *Nicotiana tabacum*, of the chloroplast 16SrRNA gene, was found to be the site of a C→A transversion in the SR1 mutant (Etzold et al., 1987) and a C→U transition in another streptomycin resistant tobacco mutant (Fromm et al., 1989). The latter report identifies a C→U transition at base pair 472, in the '530 loop' of the same gene, as the mutational site in another tobacco mutant with a lower level of streptomycin resistance. The same workers have reported a different chloroplast DNA locus conferring streptomycin resistance in yet another tobacco mutant (Galili et al., 1989), this time in the gene encoding the ribosomal protein S12.

Spectinomycin resistance has also been mapped to the 16SrRNA gene. The sites identified for two different tobacco mutants were at base pairs 1012 and 1140, on opposite sides of a conserved stem structure (Fromm et al., 1987).

Point mutations conferring resistance to lincomycin in four *N. plumbaginifolia* mutants have as anticipated, been found in the chloroplast 23SrRNA gene (Cséplő et al., 1988). This type of fine structure mapping of chloroplast mutants will be invaluable for their use in studies on plastome transformation and recombination (see for example Svab et al., 1990; Golds et al., 1993), and needs to be extended to the increasing number of comparable mutants now available for other species.

A number of plastome encoded chlorophyll deficient mutants have also been isolated *in vitro* (McCabe et al., 1989; Malone and Dix, 1990), sometimes in combination with antibiotic resistance mutations (Svab and Maliga, 1986; Toki et al., 1990).

The only selectable mitochondrial markers obtained to date are oligomycin resistant mutants of *Nicotiana sylvestris* (Durand, 1987; Aviv and Galun, 1988). For both mutants the mitochondrial DNA was confirmed as the location of the mutation, by mitochondrial transfer in interspecific protoplast fusion products (Durand and Harada, 1989; Perl et al., 1991).

4.3. *Developmental and Hormonal Mutants*

Perhaps the most aesthetically pleasing exercise in *in vitro* selection in recent years has been the generation of a number of temperature sensitive carrot mutants impaired in somatic embryogenesis. Since the development of somatic embryos parallels that of zygotic embryos, passing through globular, heart and torpedo stages, and since mutants arresting at the different stages can be obtained, these mutants appear to provide an invaluable opportunity for studying the patterns of gene expression associated with normal embryo development.

The principle behind selecting these mutants was devised by Breton and Sung (1982) using haploid suspension cultures, and has since been applied in modified forms in several other laboratories, using both haploid and diploid cultures (for review see Terzi and Lo Schiavo, 1990). Mutagenised cultures are subjected to embryogenesis inductive conditions (by removal of 2,4-D) at a high temperature (usually 31–33 °C). After a suitable period embryos developing under these conditions are removed by sieving and the remaining culture incubated at the permissive temperature (usually 24 °C), at which the temperature sensitive (ts) aggregates or arrested embryos are able to complete their embryogenic development. Generally callus is initiated from these embryos or plantlets to permit checking of the ts phenotype. A number of stable ts mutants have been obtained using this approach, and rapid progress is being made with their characterisation. The most frequent mutants are those displaying temperature sensitivity between globular and torpedo stages, but a few affected at very early stages of embryo development have also been obtained (Terzi and Lo Schiavo, 1990).

Two mutants in particular have been characterized at the molecular level. ts59 is a dominant mutant (as indicated by somatic hybridization – no inheritance data are yet available for these mutants) with two temperature sensitive periods. one early in embryo development, and one at the heart-torpedo transition (Lo Schiavo et al., 1988), as determined using 5 day temperature shocks on different subpopulations of embryos. 2D PAGE separation of heat shock proteins (HSPs) reveals a change in a subset of HSPs (in the 30–50 KDa range), which is thought to be due to a lack of phosphorylation rather than a change in the amino acid sequences. This work supports the involvement of heat shock proteins in developmental processes and, according to Terzi and Lo Schiavo (1990) points to a role of specific kinases in activating a function in plant cells. A second mutant tsllc isolated from a diploid cell line (Giuliano et al., 1984) is arrested in embryogenesis at the end of the globular stage (Lo Schiavo et al., 1990), and is recessive in somatic hybrids. This mutant appears to be defective in glycan processing on the extracellular proteins at non-permissive temperatures. These proteins show reduced fucosylation and their transport to the periphery of the embryos is impaired. The addition of wild type extracellular proteins restores normal embryonic development, and a revertant has been isolated in which normal fucosylation and embryogenesis at non-permissive temperatures are restored. We must await with keen interest the characterisation of more of these mutants, which should allow the gradual piecing together of the events occurring during embryo development, and their genetic control.

Temperature sensitive mutants are not the only kind which can be of value for these investigations. A mutant resistant to the hypomethylating drug 2-amino 5-ethoxy-carbonyl-pyrimidine-4(3H) one (ECP) was found to have a reduced efficiency of embryogenesis and a high incidence of tumors in the plantlets recovered (Lo Schiavo et al., 1989), suggesting that embryogenesis and normal plant development is influenced by changes in the pattern of DNA methylation.

Mention must also be made of another interesting class of mutants, namely those altered in hormone metabolism, for which selection is not straightforward and which have therefore not attracted the attention they deserve. An auxin auxotrophic line was obtained by total selection among haploid protoplast-derived colonies of *Nicotiana plumbaginifolia* (Blonstein et al., 1988), and subsequently a procedure was developed for enriching a cell population for temperature sensitive auxin auxotrophs by negative selection using BUdR (Suter et al., 1988). One of these clones (IVA3) has been further characterised (Fracheboud and King, 1990). Plants regenerated from it are also temperature sensitive, and the trait is transmitted as a monogenic recessive trait. Callus cultures also show resistance to natural and synthetic auxins, and tracer experiments suggest rapid metabolism rather than impaired synthesis, as the basis for the auxotrophy. A more detailed review of hormone mutants can be found in Pythoud and King (1990).

5. Future Prospects

Some of the areas where substantial progress can be anticipated have already been highlighted in the preceding sections, for example, an improved understanding of environmental stress tolerance, and fundamental studies on nitrogen metabolism and embryogenesis. The latter areas serve to illustrate the value of 'loss of function' mutants for the elucidation of biochemical and developmental pathways and, as we become more skilled at developing selection strategies, an expansion of effort directed at selecting such mutants can be anticipated. This author has long felt that mutants sensitive to stresses, such as salinity, drought or frost, might ultimately prove more valuable than resistant mutants, particularly if acquired through transposon 'tagging'. A trend towards this use of 'internal', rather than 'external' mutagenesis can be seen in an increased interest in *Mutator* systems, such as the *Mu* element in maize which has been shown to be active *in vitro* (James and Stadler, 1989). Grandbastien et al. (1989) demonstrated the efficiency of such a mutator gene in tobacco, in increasing the frequency of valine resistant, protoplast-derived, colonies from less than 10^{-6} to greater than 10^{-3}. Transposon tagging is of course a means of fishing for the genes associated with a process, and as the objectives of molecular biologists become yet more ambitious, viz. the manipulation of pathways (biochemical or developmental), rather than single proteins, this will become more important.

Disease resistance will continue to be a prime target for *in vitro* selection, but resistance to agrochemicals has its political problems, and an increased emphasis on breeding resistance to pests, or biological control approaches, can be envisaged. A little lateral thinking might reveal applications of *in vitro* selection in the latter area.

Improved efficiency is a probable future target. While improved photosynthetic efficiency (with its yield implications) still awaits a suitable selection strategy, the large area of mineral nutrition is beginning to attract attention. Recent studies on mineral deficiency *in vitro* include iron deficiency in soybean (Stephens et al., 1990) and zinc deficiency in wheat (He et al., 1991). Also Naik et al. (1990) have described iron efficient plants of sugarcane, obtained after selecting for callus growth on Fe-stress medium.

6. References

Anderson, P.C. and Georgeson, M. (1989). Herbicide-tolerant mutants of corn. Genome 3:994–999.

Armour, S., DiMaio, J.J., Shillito, R.D., Harms, C.T. and Jen, G. (1990). Amplification of a mutant acetohydroxacid synthase gene in a sulfonylurea herbicide-resistant tobacco cell line. J. Cell. Biochem. 14E:295.

Aviv, D. and Galun, E. (1988). Transfer of cytoplasmic organelles form an oligomycin-resistant *Nicotiana* cell suspension into tobacco protoplasts yielding oligomycin-resistant cybrid plants. Mol. Gen. Genet. 215:128–133.

Azevedo, R.A., Arana, J.L. and Arruda, P. (1990). Biochemical genetics of the interaction of the lysine plus threonine resistant mutant Ltr 1 with opacque-2 maize mutant. Plant Sci. 70:81–90.

Blonstein, A.D., Vahala, T., Koorneef, M. and King, P.J. (1988). Plants regenerated from auxin-auxotrophic variants are inviable. Mol. Gen. Genet. 215:58–64.

Bourgin, J.P. (1978). Valine-resistant plants from *in vitro* selected tobacco cells. Mol. Gen. Genet. 161:225–230.

Boyes, C.J. and Vasil, I.K. (1987). *In vitro* selection for tolerance to S-(2-aminoethyl)L-cysteine and overproduction of lysine by embryogenic calli and regenerated plants of *Pennisetum americanum* (L.) K. Schum. Plant Sci. 50:195–203.

Breton, A.M. and Sung, Z.R. (1982). Temperature-sensitive carrot variants impaired in somatic embryogenesis. Dev. Biol. 90:58–60.

Bright, S.W.J., Norburg, P.B. and Miflin, B.J. (1982a). Isolation of a recessive barley mutant resistant to S-(2-aminoethyl)-L-cysteine. Theor. Appl. Genet. 55:1–4.

Bright, S.W.J., Miflin, B.J. and Roynes, S.E. (1982b). Threonine accumulation in the seeds of a barley mutant with an altered aspartate kinase. Biochem. Genet. 20:229–243.

Cattoir-Reynarts, A., Degryse, E., Verbruggen, I. and Jacobs, M. (1983). Isolation and characterization of a lysine + threonine resistant variant in *Daucus carota*. Biochem. Physiol. Pflanz. 178:81–90.

Cella, R., Parisi, B. and Nielsen, E. (1982). Characterization of a carrot cell line resistant to azetidine-2-carboxylic acid. Plant Sci. Lett. 24,125–135.

Chaleff, R.S. and Bascomb, N.F. (1987). Genetic and biochemical evidence for multiple forms of acetolactate synthase in *Nicotiana tabacum*. Mol. Gen. Genet. 210:33–38.

Chaleff, R.S. and Mauvais, C.J. (1984). Acetolactate synthase is the site of action of two sulfonylurea herbicides in higher plants. Science 224:1443–1445.

Chaleff, R.S and Ray, T.B. (1984). Herbicide-resistant mutants from tobacco cell cultures. Science 223:1148–1151.

Collin, H.A. and Dix, P.J. (1990). Culture systems and selection procedures. In: Plant Cell Line Selection. P.J. Dix (ed.), VCH, Weinheim, pp. 3–18.

Comai, L., Sen, L.C. and Stalker, D.M. (1983). An altered aroA gene product confers resistance to the herbicide glyphosate. Science 221:370–371.

Conner, A.J. and Meredith, C.P. (1985a). Simulating the mineral environment of aluminium-toxic soils in plant cell culture. J. Exp. Bot. 36:870–880.

Conner, A.J. and Meredith, C.P. (1985b). Strategies for the selection and characterization of aluminium-resistant variants from cell cultures of *Nicotiana plumbaginifolia*. Planta 136:466–473.

Conner, A.J. and Meredith, C.P. (1985c). Large scale selection of aluminium resistant mutants from plant cell culture: Expression and inheritance in seedlings. Theor. Appl. Genet. 71:159–165.

Cséplő, A., Etzold, T., Schell, J. and Schreier, P.H. (1988). Point mutations in the 23S rRNA genes of four lincomycin resistant *Nicotiana plumbaginifolia* mutants could provide new selectable markers for chloroplast transformation. Mol. Gen. Genet. 214:295–299.

Cséplő, A. and Maliga, P. (1982). Lincomycin resistance, a new type of maternally inherited mutation in *Nicotiana plumbaginifolia*. Curr. Genet. 6:105–109.

Cséplő, A. and Medgyesy, P. (1986). Characteristic symptoms of photosynthesis inhibition by herbicides are expressed in photomixotrophic tissue cultures of *Nicotiana*. Planta 168:24–28.

Cséplő, A., Medgyesy, P., Hideg, E., Demeter, S., Marton, L. and Maliga, P. (1985). Triazine-resistant *Nicotiana* mutants from photomixotrophic cell cultures. Mol. Gen. Genet. 200:508–510.

Diedrick, T.J., Frisch, D.A. and Gengenbach, B.G. (1990). Tissue culture isolation of a second mutant locus for increased threonine accumulation in maize. Theor. Appl. Genet. 79:209–215.

Dix, P.J. (1986). Cell line selection. In: Plant Cell Culture Technology. M.M. Yeoman (ed.), Blackwell Scientific Publications, Oxford, pp. 143–201.

Dix, P.J., McKinley, C.P. and McCabe, P.F. (1990). Antibiotic resistant mutants of *Solanum nigrum*. In: Progress in Plant Cellular and Molecular Biology. H.J.J. Nijkamp, L.H.W. van Der Plas and J. Van Aartrijk (eds.), Kluwer Academic Publishers, Dordrecht, pp. 169–174.

Dix, P.J., McLysaght, U.A. and Pearce, R.S. (1984). The potential of cell cultures for the production of salt tolerant cultivars. In: Efficiency in Plant Breeding. W. Lange, A.C. Zeven and N.G. Hogenboom (eds.), Pudoc, Wageningen, pp. 219–223.

Donn, G., Tischer, E., Smith, J.A. and Goodman, H.M. (1984). Herbicide-resistant alfalfa cells: An example of gene amplification in plants. J. Mol. Appl. Genet. 2:621 -635.

Dracup, M. (1991). Increasing salt tolerance of plants through cell culture requires greater understanding of tolerance mechanisms. Aust. J. Plant Physiol. 18:1–15.

Durand, J. (1987). Isolation of antibiotic resistant variants in a higher plant, *Nicotiana sylvestris*. Plant Sci. 51:113–118.

Durand, J. and Harada, H. (1989). Interspecific protoplast fusion in *Nicotiana* provides evidence for a mitochondrial determinism for oligomycin-resistance. Plant Sci. 62:263–274.

Dyer, W.E., Weller, S.C., Bressan, R.A. and Herrmann, K.M. (1988). Glyphosate tolerance in tobacco (*Nicotiana tabacum* L.). Plant Physiol. 88:661–666.

Etzold, T., Fritz, C.C., Schell, J. and Schreier, P.H. (1987). A point mutation in the chloroplast 16SrRNA gene of a streptomycin resistant *Nicotiana tabacum*. FEBS Lett. 219:343–346.

Fluhr, R., Aviv, D., Galun, E. and Edelman, M. (1985). Efficient induction and selection of chloroplast encoded antibiotic resistant mutants in *Nicotiana*. Proc. Natl. Acad. Sci. USA 82:1485–1489.

Fracheboud, Y. and King, P.J. (1990). A single-gene recessive mutation leading to auxin auxotrophy in *Nicotiana plumbaginifolia*. In: Abstracts 7th International Congress on Plant Tissue and Cell Culture. IAPTC, Amsterdam, p. 278.

Freytag, A.H., Wrather, J.A. and Erichsen, A.W. (1990). Salt tolerant sugarbeet progeny from tissue cultures challenged with multiple salts. Plant Cell Rep. 8:647–650.

Fromm, H., Edelman, M., Aviv, D., and Galun, E. (1987). The molecular basis for rRNA-dependent spectinomycin resistance in *Nicotiana* chloroplasts. EMBO J. 6:3233–3237.

Fromm, H., Galun, E. and Edelman, M. (1989). A novel site for streptomycin resistance in the '530' loop of chloroplast 16S ribosomal RNA. Plant Mol. Biol. 12:499–505.

Gabard, J.A., Marion-Poll, A., Cherel, I., Meyer, C., Muller, A. and Cabocke, M. (1987). Isolation and characterization of *Nicotiana plumbaginifolia* nitrate reductase-deficient mutants: genetic and biochemical analysis of the N1A complementation group. Mol. Gen. Genet. 209:596–606.

Galili, S., Fromm, H., Aviv, D., Edelman, M. and Galun, E. (1989). Ribosomal protein S12 as a site for streptomycin resistance in Nicotiana chloroplasts. Mol. Gen. Genet. 218:289–292.

Giuliano, G., Lo Schiavo, F. and Terzi, M. (1984). Isolation and developmental characterization of temperature sensitive carrot cell variants. Theor. Appl. Genet. 67:179–183.

Glas, C., Kamp, J.C., Jongsma, C., Nijkamp, H.J.J. and Hille, J. (1990). The induction of lincomycin resistance in *Lycopersicon peruvianum* and *Lycopersicon esculentum*. Plant Sci. 70:231–241.

Golds, T., Maliga, P. and Koop, H.V. (1993). Stable plastid transformation in PEG-mediated photoplasts of *Nicotiana tabacum*. Bio/Technology 11:95–97.

Goldsbrough, P.B., Hatch, E.M., Huang, B., Kosinski, W.G., Dyer, W.E., Herrmann, K.M. and Weller, S.C. (1990). Gene amplification in glyphosate tolerant tobacco cells. Plant Sci. 72:53–62.

Grandbastien, M.A., Bourgin, J.P. and Caboche, M. (1985). Valine resistance, a potential marker in plant cell genetics, II: Optimization of UV-mutagenesis and selection of valine-resistant colonies derived from tobacco mesophyll protoplasts. Genetics 109:409–425.

Grandbastien, M.A., Missonier, C., Goujaud, J., Bourgin, J.P., Deshayes, A. and Caboche, M. (1989). Cellular genetic study of a somatic instability in a tobacco mutant: *in vitro* isolation of valine-resistant spontaneous mutants. Theor. Appl. Genet. 777:482–488.

Greenway, H. and Munns, R. (1980). Mechanisms of salt tolerance in nonhalophytes. Ann. Rev. Plant Physiol. 31:139–190.

Gressel, J. (1979). A review of the place of *in vitro* cell culture systems in studies of action, metabolism and resistance of biocides affecting photosynthesis. Z. Naturforsch. 34c:905–913.

Hamill, J.D., Ahuja, P.S., Davey, M.R. and Cocking, E.C. (1986). Protoplast derived streptomycin resistant plants of the forage legume, *Onobrychis viciifolia* Scop. (sainfoin). Plant Cell Rep. 5:439–441.

Hammerschlag, F.A. (1988). Selection of peach cells for insensitivity to culture filtrates of *Xanthomonas campestris* pv. *pruni* and regeneration of resistant plants. Theor. Appl. Genet. 76:865–869.

Hammerschlag, F.A. (1990). Resistance responses of plants regenerated from peach callus cultures to *Xanthomonas campestris* pv. *pruni*. J. Am. Soc. Hort. Sci. 115:1034–1037.

Harms, C.T. and DiMaio, J.J. (1991). Primisulfuron herbicide-resistant tobacco cell lines. Application of fluctuation test design to *in vitro* mutant selection with plant cells. J. Plant Physiol. 137:513–519.

He, D-G., Yang, Y-M. and Scott, K.J. (1991). Zinc deficiency and the formation of white structures in immature embryo cultures of wheat (*Triticum aestivum* L.). Plant Cell Tissue Org. Cult. 24:9–12.

Hibberd, K.A. and Green, C.E. (1982). Inheritance and expression of lysine plus threonine resistance selected in maize tissue culture. Proc. Natl. Acad. Sci. USA 79:559–563.

Hibberd, K.W., Walter, T., Green, C.E. and Gengenbach, B.G. (1980). Selection and characterization of a feedback-insensitive tissue culture of maize. Planta 148:183–187.

Hollander-Czytko, H., Johanning, D., Meyer, H.E. and Amrhein, N. (1988). Molecular basis for the overproduction of 5 enolpyruvylshikimate-3-phosphate synthase in a glyphosate-tolerant cell suspension culture of *Corydalis sempervirens*. Plant Mol. Biol. 11:215–220.

Jacobsen, E. (1986). Isolation, characterization and regeneration of an S-(2-aminoethyl)-L-cysteine resistant cell line of dihaploid potato. J. Plant Physiol. 123:307–315.

Jain, S., Nainawatee, H.S., Jain, R.K. and Chowdhury, J.B. (1991). Proline status of genetically stable salt-tolerant *Brassica juncea* L. somaclones and their parent cv. Prakash. Plant Cell Rep. 9:684–687.

James, M.G. and Stadler, J. (1989). Molecular characterisation of *Mutator* systems in maize embryogenic cultures indicates Mu element activity *in vitro*. Theor. Appl. Genet. 77:383–393.

Jansen, C.E., Snel, E.A.M., Akerboom, M.J.E., Nijkamp, H.J.J. and Hille, J. (1990). Induction of streptomycin resistance in the wild tomato *Lycopersicon peruviallum*. Mol. Gen. Genet. 220:261–268.

Jones, P.W. (1990). *In vitro* selection for disease resistance. In: Plant Cell Line Selection. P.J. Dix (ed.), VCH, Weinheim), pp. 113–150.

Jordan, M.C. and McHughen, A. (1987). Selection for chlorsulfuron resistance in flax (*Linum usitatissimum*) cell cultures. J. Plant Physiol. 131:333–338.

Kendall, E.J., Qureshi, J.A., Kartha, K.K., Leung, N., Chevrier, N., Caswell, K. and Chen, T.H.H. (1990) Regeneration of freezing-tolerant spring wheat (*Triticum aestivum* L.) plants from cryoselected callus. Plant Physiol. 94:1756–1762.

Kishore, G.M., Brundage, L., Kolk, K., Padgette, S.R., Rochester, D., Huynh, K. and dela-

Cioppa, G. (1986). Isolation, purification and characterization of a glyphosate tolerant mutant *E. coli* EPSP synthase. Fed. Proc. 45:1506.

Kumar, V. and Sharma, D.R. (1989). Selection and characterization of an L-thiazolidine-4-carboxylic acid resistant cellus culture of *Vigna radiata* (L.) Wilczek var. *radiata*. Plant Cell Rep. 7:648–651.

Kumpaisal, R., Hashimoto, T and Yamada, Y. (1988). Selection and characterization of S-(2-aminoethyl)-L-cysteine-resistant wheat cell cultures. J. Plant Physiol. 133:608–614.

Lister, A. (1990). Flow cytometry for selection of plant cells *in vitro*. In: Plant Cell Line Selection. P.J. Dix (ed.), VCH, Weinheim, pp. 39–85.

Lo Schiavo, F., Giuliano, G., de Vries, S.C., Genga, A., Bollini, R., Pitto, L., Cozzani, F., Nuti-Ronchi, V. and Terzi, M. (1990). A carrot cell variant temperature sensitive for somatic embryogenesis reveals a defect in the glycosylation of extracellular proteins. Mol. Gen. Genet. 223:385–393.

Lo Schiavo F., Giuliano, G. and Sung, Z.R. (1988). Characterization of a temperature sensitive carrot cell mutant impaired in somatic embryogenesis. Plant Sci. 54:157–164.

Lo Schiavo, F., Pitto, L., Giuliano, G., Torti, G., Nuti-Ronchi, V., Marazziti, D., Vergara, R., Orselli, S. and Terzi, M. (1989). DNA methylation of embryogenic carrot cell cultures and its variations as caused by mutation differentiation, hormones and hypomethylating drugs. Theor. Appl. Genet. 77:325–331.

Maliga, P., Sz. Breznovits, A. and Marton, L. (1973). Streptomycin-resistant plants from callus culture of haploid tobacco. Nature Neur. Biol. 244:29–30.

Maliga, P., Sz. Breznovits, A., Marton, L. and Joo, F. (1975). Non-Mendelian streptomycin-resistant tobacco mutant with altered chloroplasts and mitochondria. Nature 225:401–402.

Malone, R.P. and Dix, P.J. (1990). Mutagenesis and triazine herbicide effects in strawberry shoot cultures. J. Exp. Bot. 41:463–469.

Marton, L., Dung, T.M., Mendel, R. and Maliga, P. (1982). Nitrate reductase deficient cell lines from haploid protoplast cultures of *Nicotiana plumbaginifolia*. Mol. Gen. Genet. 186:301–304.

McCabe, P.F., Timmons, A.M. and Dix, P.J. (1989). A simple procedure for the isolation of streptomycin resistant plants in *Solanaceae*. Mol. Gen. Genet. 216:132–137.

Mori, S., Hasegawa, H., Che, R., Nakanishi, H. and Murakami, M. (1989). Free proline contents in two different groups of rice mutants resistant to hydroxy-L-proline. Theor. Appl. Genet. 77:44–48.

Muller, A.J., and Grafe, R. (1978). Isolation and characterization of cell lines of *Nicotiana tabacum* lacking nitrate reductase. Mol. Gen. Genet. 161:67–76.

Nafziger, E.D., Widholm, J.M., Steinruchen, H.C. and Killmer, J.L. (1984). Selection and characterization of a carrot cell line tolerant to glyphosate. Plant Physiol. 76:571–574.

Naik, G.R., Babu, K.H. and Lingappa, G. (1990). Studies on *in vitro* selection of Fe-efficient lines in sugarcane. Plant and Soil 129:183–186.

Negrutiu, I. (1990). *In vitro* mutagenesis. In: Plant Cell Line Selection. P.J. Dix (ed.), VCH, Weinheim, pp. 19–38.

Negrutiu, I., de Brouuver, D., Dirks, R. and Jacobs, M. (1985). Amino acid auxotrophs from protoplast cultures of *Nicotiana plumbaginifolia*, Viviani, I BUdR enrichment selection, plant regeneration and general characterisation. Mol. Gen. Genet. 199:330–337.

Negrutiu, I., Dirks, R. and Jacobs, M. (1983). Regeneration of fully nitrate reductase-deficient mutants from protoplast culture of *Nicotiana plumbaginifolia* (Viviani). Theor. Appl. Genet. 66:341–347.

Negrutiu, I., Jacobs, M. and Caboche, M. (1984). Advances in somatic cell genetics of higher plants – the protoplast approach in basic studies on mutagenesis and isolation of biochemical mutants. Theor. Appl. Genet. 67:289–304.

Pay, A., Smith, M.A., Nagy, F. and Marton, L. (1988). Sequence of the psbA gene from wild-type and triazine resistant *Nicotiana plumbaginifolia*. Nucleic Acids Res. 16:8176.

Pelsy, F. and Gonneau, M. (1991). Genetic and biochemical analysis of intragenic complementation events among nitrate reductase apoenzyme-deficient mutants of *Nicotiana plumbaginifolia*. Genetics 127:199–204.

Perl, A., Aviv, D. and Galun, E. (1991). Protoplast fusion mediated transfer of oligomycin resistance from *Nicotiana sylvestris* to *Solanum tuberosum*. Mol. Gen. Genet. 225:11–16.

Pythoud, F. and King, P.J. (1990). Auxotrophic, temperature-sensitive and hormone mutants isolated *in vitro*.In: Plant Cell Line Selection. P.J. Dix (ed.), VCH, Weinheim, pp. 233–255.

Redway, F.A. and Vasil, I.K. (1990). Selection of S-(2-aminoethyl)-L-cysteine tolerance in embryogenic calli and regenerated plants of *Pennisetum purpureum* Schum. Plant Sci. 67;203–209.

Rey, P., Eymerg, F. and Peltier, G. (1990). Atrazine and diuron resistant plants from photoautotrophic protoplast-derived cultures of *Nicotiana plumbaginifolia*. Plant Cell Rep. 9:241–244.

Ricardi, G., Cella, L.R., Camerino, G. and Ciferri, O. (1983). Resistance to azetidine-2-carboxylic acid and sodium chloride tolerance in carrot cell cultures and *Spirulina platensis*. Plant Cell Physiol. 24:1073–1078.

Rognes, S.E., Bright, S.W.J. and Miflin, B.J. (1983). Feedback-insensitive asparate kinase isoenzymes in barley mutants resistant to lysine plus threonine. Planta 157:32–38.

Sacristan, M.D. (1982). Resistance responses to *Phoma lingum* of plants regenerated from selected cell and embryogenic cultures of haploid *Brassica napus*. Theor. Appl. Genet. 61:193–200.

Salgado-Garciglia, R. and Ochoa-Alejo, N. (1990). Increased capsaicin content in PFP-resistant cells of chilli pepper (*Capsicum annuum* L.). Plant Cell Rep. 8:617–620.

Saxena, P.K. and King, J. (1988). Herbicide resistance in *Datura innoxia*. Cross-resistance of sulfonylurea-resistant cell lines to imidazolinones. Plant Physiol. 86:863–867.

Saxena, P.K., Williams, D. and King, J. (1990). The selection of chlorsulfuron-resistant cell lines of independent origin from an embryogenic cell suspension culture of *Brassica napus* L. Plant Sci. 69:231–237.

Schaeffer, G.W. and Sharpe, F.T. (1987). Increased lysine and seed storage protein in rice plants recovered from calli selected with inhibitory levels of lysine plus threonine and S-(2-aminoethyl) cysteine. Plant Physiol. 84:509–515.

Schaeffer, G.W. and Sharpe, F.T. (1990). Modification of amino acid composition of endosperm proteins from *in-vitro*-selected high lysine mutants in rice. Theor. Appl. Genet. 80:841–846.

Schaeffer, G.W., Sharpe, F.T. and Dudley, J.T. (1988). Segregation for endosperm lysine in F_2, F_3, and F_4 progeny from a cross of *in-vitro*-selected and unselected cultivar of rice. Theor. Appl. Genet. 77:176–183.

Shahin, E.A. and Spivey, R. (1986). A single dominant gene for *Fusarium* wilt resistance in protoplast-derived tomato plants. Theor. Appl. Genet. 73:164–169.

Shaner, D.L. and Anderson, P.C. (1985). Mechanism of action of the imidazolinones and cell culture selection of tolerant maize. In: Biotechnology in Plant Science. M. Zaitlin, P.R. Day and A. Hollaender (eds.), Academic Press, New York, pp. 287–300.

Smith, C.M., Pratt, D. and Thompson, G.A. (1986). Increased 5-enolpyruvyl-shikimic acid 3-phosphate synthase activity in a glyphosate-tolerant variant strain of tomato cells. Plant Cell Rep. 5:298–301.

Smith, W.A. and Chaleff, R.S. (1990). Herbicide resistance. In: Plant Cell Line Selection. P.J. Dix (ed.), VCH, Weinheim, pp. 151–166.

Steinrucken, H.C., Schulz, A., Amrhein, N., Porter, C.A. and Fraley, R.T. (1986). Overproduction of 5-enolypyruvyl-shikimate 3-phosphate synthase activity in a glyphosate-tolerant *Petunia hybrida* cell line. Arch. Biochem. Biophys. 244:169–178.

Stephens, P.A., Widholm, J.M. and Nickell, C.D. (1990). Iron-deficiency chlorosis evaluation of syoybean with tissue culture. Theor. Appl. Genet. 80:417–420.

Stewart, G.R. and Lee, J.A. (1974). The role of proline accumulation in halophytes. Planta 120:279–289.

Sumaryati, S., Negrutiu, I. and Jacobs, M. (1991). Characterization and regeneration of salt and water stress mutants from protoplast cultures of *Nicotiana plumbaginifolia* (Viviani). Theor. Appl. Genet. (in press).

Suter, M., Schnebli, V. and King, P.J. (1988). The development of a negative selection system for the isolation of plant temperature-sensitive auxin auxotrophs. Theor. Appl. Genet. 75:869–874.

Svab, Z., Hajdnkiewicz, P. and Maliga, P. (1990). Stable transformation of plastids in higher plants. Proc. Natl. Acad. Sci. USA 87:5826–5830.

Svab, Z. and Maliga, P. (1986). *Nicotiana tabacum* mutants with chloroplast encoded streptomycin resistance and pigment deficiency. Theor. Appl. Genet. 72:637643.

Swanson, E.B., Coumans, M.P., Brown, G.L., Patel, J.D. and Beversdorf, W.D. (1988). The characterization of herbicide resistant plants in *Brassica napus* L. after *in vitro* selection of microspores and protoplasts. Plant Cell Rep. 7:83–87.

Swanson, E.B., Herogesell, M.J., Arnoldo, M., Sippell, D.W. and Wong, R.S.C. (1989). Microspore mutagenesis and selection: Canola plants with field tolerance to the imidazolinones. Theor. Appl. Genet. 78:525–530.

Syukur, S., Jacobs, M and Negrutiu, I. (1991). Analysis of mutant plants resistant to salt or water stress and to proline analogues obtained from the protoplasts of *Nicotiana plumbaginifolia* viviani. In: Plant Mutation Breeding for Crop Improvement, Vol. 2. IAEA, Vienna, pp. 265–269.

Terzi, M. and Lo Schiavo, F. (1990). Developmental mutants in carrot. In: Progress in Plant Cellular and Molecular Biology. H.J.J. Nijkamp, L.H.W. Van Der Plas and J. Van Aartrijk (eds.), Kluwer Academic Publishers, Dordrecht, pp. 391–397.

To, K.-Y., Chen, C.-C. and Lai, Y.-K. (1989). Isolation and characterization of streptomycin-resistant mutants in *Nicotiana plumbaginifolia*. Theor. Appl. Genet. 78:81–86.

Toki, S., Kameya, T. and Abe, T. (1990). Production of a triple mutant, chlorophyll-deficient, streptomycin-, and kanamycin-resistant *Nicotiana tabacum*, and its use in intergeneric somatic hybrid formation with *Solanum* melongena. Theor. Appl. Genet. 80:588–592.

Toyoda, H., Chatani, K., Matsuda, Y. and Ouchi, S. (1989a). Multiplication of tobacco mosaic virus in tobacco callus cultures and *in vitro* selection for viral disease resistance. Plant Cell Rep. 8:433–436.

Toyoda, H., Shimizu, K., Chatani, K., Kita, N., Matsuda, Y. and Ouchi, S. (1989b). Selection of bacterial wilt-resistant tomato through tissue culture. Plant Cell Rep. 8:317–320.

Vajrabhaya, M., Thanapaisal, T. and Vajrabhaya, T. (1989). Development of salt tolerant lines of KDML and LPT rice cultivars through tissue culture. Plant Cell Rep. 8:411–414.

Van Swaai, A.C., Jacobsen, E., Kiel, J.A.K.Wt and Feenstra, W.J. (1986). Selection, characterization, and regeneration of hydroxyproline-resistant cell lines of *Solanum tuberosum*: Tolerance to NaCl and freezing stress. Physiol. Plant 68:359–366.

Wang, W.-C., Myers, J.R. and Collins, G.B., (1991). Selection of atrazine-resistant tobacco variants from photomixotrophic cultures. Plant Sci. 73:199–209.

Wenzel, G. and Foroughi-Wehr, B. (1990). Progeny tests of barley, wheat, and potato regenerated from cell cultures after *in vitro* selection for disease resistance. Theor. Appl. Genet. 80:359–365.

Widholm, J.M. (1972a). Anthranilate synthetase from 5-methyltryptophan-susceptible and -resistant cultured *Daucus carota* cells. Biochim. Biophys. Acta 279:48–57.

Widholm, J.M. (1972b). Cultured *Nicotiana tabacum* cells with an altered anthranilate synthetase which is less sensitive to feedback inhibition. Biochim. Biophys. Acta 261:52–58.

Wilson, G. (1990). Screening and selection of cultured plant cells for increased yields of secondary metabolites. In: Plant Cell Line Selection. P.J. Dix (ed.), VCH, Weinheim, pp. 187–213.

Winicov, I., Waterborg, J.H., Harrington, R.E. and McCoy, T.J. (1989). Messenger RNA induction in cellular salt tolerance of alfalfa (*Medicago sativa*). Plant Cell Rep. 8:6–11.

Zenk, M.H., El-Shagi, H., Arens, H., Stockigt, J., Weiler, E.W. and Deus, B. (1977). Formation of the indole alkaloids serpentine and ajinalicine in cell suspension cultures of *Catharanthus roseus*. In: Plant Tissue Culture and its Biotechnological Application. W. Barz, E. Reinhard and M.H. Zenk (eds.), Springer-Verlag, Berlin, pp. 27–43.

6. Origins, Causes and Uses of Variation in Plant Tissue Cultures

Angela Karp

Contents

1. Introduction

Plant cell and tissue culture were not expected to provide a useful source of variation for crop improvement. The growth of plant cells *in vitro* and their regeneration into whole plants is an asexual process, involving only mitotic division of the cell. The expected result was clonal multiplication of genetically uniform plants. This expectation has formed the basis of the micropropagation industry and provided a technical basis for genetic manipulation in plants. The occurrence of uncontrolled variation during the culture process was largely unexpected and undesired.

Early evidence of chromosome instability in cultured cells gained significance when regeneration systems were established from explants and protoplasts of important crops and wide-scale phenotypic and genetic variation were revealed. Interest heightened with widely publicised reports of changes in agronomic traits (e.g., Shepard et al., 1980) and promises of a 'novel source of variation for crop improvement'. In 1981, the subject was reviewed by Larkin and Scowcroft, who named the phenomenon of variation found

Indra K. Vasil and Trevor A. Thorpe (eds.), Plant Cell and Tissue Culture, 139–151.

in plants regenerated from cell cultures 'somaclonal variation' . The term is now universally adopted although alternatives such as 'protoclonal' or 'gametoclonal', describing variation from protoplast and anther culture, respectively, are also used.

Over the past decade, considerable effort has been placed on determining the causes and origins of somaclonal variation (Karp and Bright, 1985). The applications of the variation has been thoroughly addressed and numerous attempts made to isolate useful variants for breeding purposes. Their limited success has meant that somaclonal variation has suffered a decline in popularity as a useful source of agronomic changes. This chapter reviews the origins of variation in plant tissue cultures and the problems and opportunities associated with its application for crop improvement.

2. Origins and Causes of Somaclonal Variation

Regeneration from somatic cells is possible in plants because the cells can retain totipotentcy. During regeneration, the differentiated plant cell becomes undifferentiated and then redifferentiated through a new pathway of development. The degree of change taking place during this process relates to the tissue culture system utilised and is most extensive in protoplast culture where regeneration is achieved from isolated cells after cell division and growth are induced over several months in the culture environment. The causes and origins of somaclonal variation are manifold and relate to both internal and external aspects of plant tissue culture (Karp, 1991a).

2.1. *The Importance of Cellular Organisation*

Plant tissue culture systems vary in their involvement of disorganised growth. In meristem-tip culture the organisation of the meristem remains undisturbed during the culture process. To date, this is the only culture system that can be considered free from somaclonal variation. In other systems such as protoplast and explant culture, regeneration is achieved through the formation of adventitious meristems after a phase of disorganised callus or cell suspension growth. These latter systems are subject to instability, suggesting that cellular organisation is a critical feature and that somaclonal variation is related to disorganised growth. The longer the duration of the disorganised phase and the greater the departure from organised structures the greater the chances of somaclonal variation. This remains true whether regeneration is from organogenic or embryogenic cultures, although the direct formation of such structures from cultured plant tissues (without any intermediate callus phase) will clearly minimise the chances of instability. The stabilising influence of the meristem is lost when plant cells are grown in culture, but whether the resultant instability arises through lack of selective constraints or whether variation is induced by the culture environment is not easy to discern.

2.2. *Variation in the Source Tissue*

Gross changes in the genome including endopolyploidy, polyteny and amplification or diminution of DNA sequences can occur during somatic differentiation in normal plant growth and development (D'Amato, 1989). It is therefore not surprising that differences in both the frequency and nature of somaclonal variation may occur when regeneration is achieved from different tissue sources. Furthermore, the processes of de-differentiation and redifferentiation may involve both qualitative and quantitative changes in the genome and different DNA sequences may be amplified or deleted during these changes in the state of the cell. These in turn are related to the original tissue source and regeneration system, which may explain why certain culture systems can be associated with specific types of change, such as the high frequency of albino plants and the amplification of repetitive sequences in anther culture of cereals and tobacco, respectively.

Where experiments in somaclonal variation have been designed so that the identity of each somaclone and its donor plant are known and several plants regenerated from the same culture can be compared, it has become clear that whilst most changes arise during the culture phase, some can be traced to the donor tissue, indicating that somaclonal variation can arise from somatic mutations already present in the donor plant. The degree to which somatic mutations and changes associated with differentiation of somatic cells account for the origins of somaclonal variation will be better assessed when our knowledge of these processes in plants is improved.

2.3. *Abnormalities in Cell Division in vitro*

Mitosis is induced when differentiated cells or tissues are isolated and cultured *in vitro* in the presence of exogenous growth regulators. Cytological studies have revealed that abnormalities in cell division can occur such as multi-polar spindle formation, bridges and fragments, laggards, micronuclei and chromosome fragmentation (D'Amato, 1989). These errors result in numerical and structural chromosome changes in regenerated plants. Growth regulators, temperature, light, osmolarity and agitation rate of the culture medium are all known to affect the cell cycle *in vivo* in plants (Gould, 1984) suggesting that inadequate control of the cell cycle *in vitro* is one of the causes of somaclonal variation.

In protoplast regeneration, the problems of cell cycle control are accentuated. During the isolation of protoplasts the cell wall is removed. New walls are then synthesised in culture, and division is induced in the protoplast-

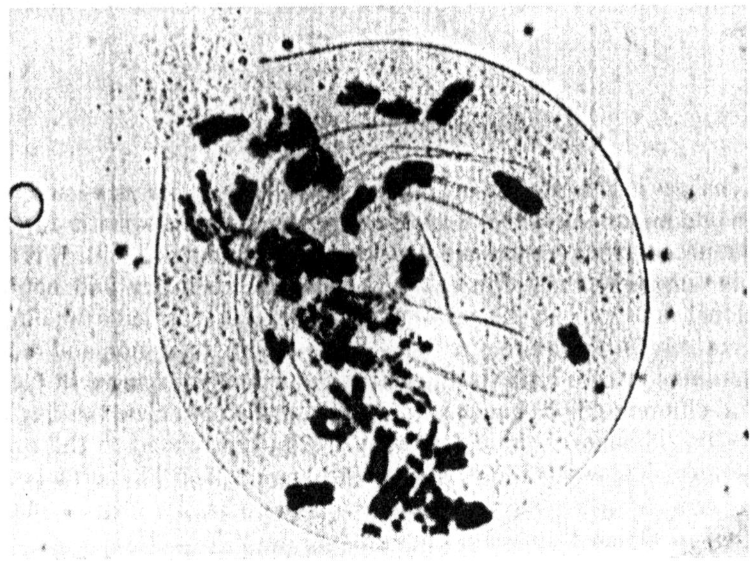

Figure 1. Chromosome fragmentation in protoplast-derived cell of wheat.

derived cell. Cytological studies indicate that during protoplast culture a high frequency of errors in microtubule synthesis, spindle formation, spindle orientation, chromatid segregation and cross wall formation occurs, resulting in extensive variation in chromosome number and structure (Fig. 1). More detailed studies have shown that most errors occur in the earliest cells undergoing division suggesting that in these cells the cytoplasm has had insufficient time to recover from the protoplast isolation. This indicates that protoplasts have unique problems of instability which are a consequence of their isolation and explains why somaclonal variation is most prevalent in plants regenerated from protoplast cultures. In particular, chromosome aberrations might be expected to occur at high frequencies in protoplast-derived plants and, where only stable plants are regenerated, careful growth conditions and stringent selection must be in operation. A further corollary might be that chromosome instability could be reduced if the initial division of the protoplast was delayed and better control of the cell cycle achieved.

2.4. *The Importance of Growth Regulators*

The influence of growth regulators on the cell cycle is one way in which they contribute to the causes of somaclonal variation. The synthetic auxin, 2,4-dichlorophenoxyacetic acid (2,4-D), in particular has been reputed to increase chromosome instability at high concentrations. Evidence for a direct mutagenic action of the growth regulator is somewhat contradictory and most evidence points to a more indirect effect through stimulation of rapid disorganised growth. There is also some evidence that shifts in ploidy levels can be effected through changes in the choice and concentration of growth regulators, suggesting that they may also influence the variation observed in regenerated plants through selective stimulation of cells with specific ploidies.

Growth regulators are also important in somaclonal variation in the way that they can cause transient modifications of phenotype which, although inherited mitotically during plant growth, are not sexually transmitted and are therefore, epigenetic. The requirement for sterility in plant tissue culture necessitates growth in contained vessels where gas exchange may be inadequate and accumulation of growth regulators such as ethylene can result in epigenetic modifications. Although they are not inherited through a sexual cycle such changes are an important aspect of somaclonal variation in micropropagation of ornamentals where the direct product of tissue culture is important and in vegetatively propagated species, or plants with long generation times such as trees.

2.5. *The Responsiveness of Plant Genomes*

The enforced changes in the differentiated state of plant cells during culture and regeneration, and the peculiar conditions of the culture environment, place the plant genome under stress. Somaclonal variation can consequently be viewed as variation induced by the stress of tissue culture. Evidence that this, in part, explains the origins of somaclonal variation comes from the finding that different genomes respond differently and thus that somaclonal variation has a genotypic component. Some genotypes are remarkable for their stability in culture whilst others are consistently unstable irrespective of their ploidy. It seems likely that these differences in stability relate to differences in genetic make-up and that some components of plant genomes render them more unstable during the culture process. Although it is possible that major genes are involved in this 'response' most evidence points to repetitive DNA sequences which can differ in quality and quantity between plant species. There are numerous accounts of amplification (Fig. 2a) and diminuition of repetitive sequences, including ribosomal RNA genes, as a result of tissue culture, indicating that this is a common mode of genome response.

One class of repetitive sequence, those present in heterochromatin, have been shown to play a role in the induction of chromosome aberrations in tissue culture. Numerous studies have demonstrated a correlation between culture-induced chromosome breakage and the presence of heterochromatin. This correlation can be explained by the late replication of heterochromatin resulting in a 'perturbation of the cell cycle' and that 'any perturbation affecting the synchrony between chromosome replication during S-phase and

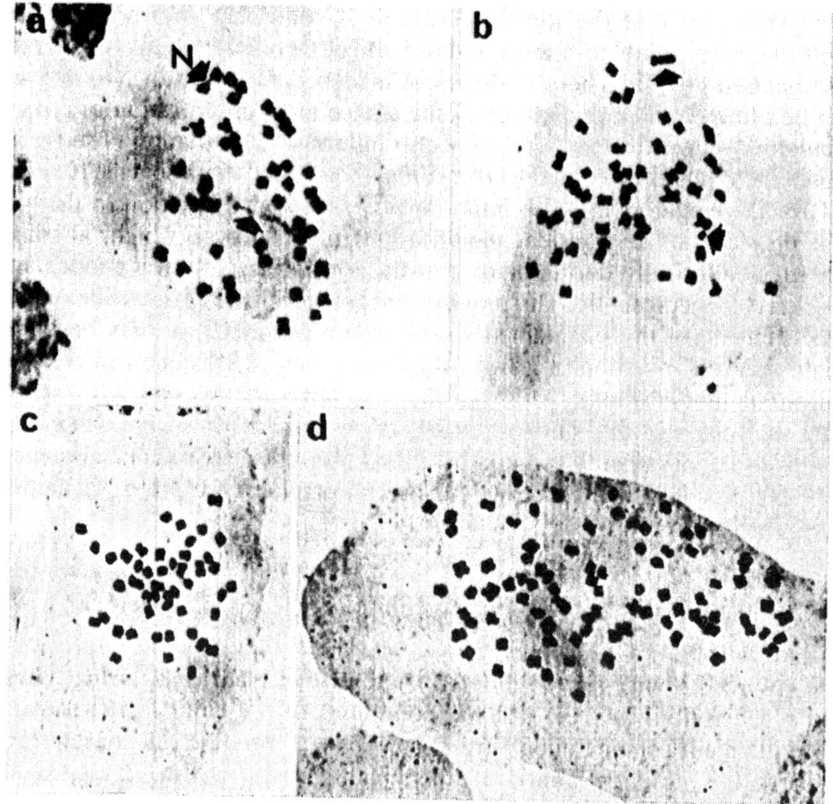

Figure 2. Numerical and structural chromosome variation in regenerated potato plants (a) amplification of nucleolus organiser region (arrow), normal homologue is also marked (arrow-N) (b) translocation (arrows) (c) aneuploid with 47 instead of 48 chromosomes (d) octoploid, after chromosome doubling, also missing 8 chromosomes.

cell division would likely result in chromosome aberration' (Lee and Phillips, 1988).

The heterochromatin theory predicts that chromosome aberrations will occur at higher frequencies in tissue cultures of plants whose chromosomes contain large segments of heterochromatin. Differences in heterochromatic content may, therefore, be one explanation for the differences in stability observed among different species in tissue culture. Other components of the genome must also be involved since differences in somaclonal variation can be observed among closely related species and even among different cultivars of the same species. The presence of transposable elements in some plant genomes clearly predisposes them to this class of instability and the activation of such elements has been reported in tissue cultures of maize. Whether transposons play a role in somaclonal variation of other species has yet to

Table 1. Problems and attractions of somaclonal variation in crop improvement

Problems	Attractions
The variation is cultivar-dependent	A rapid source of variation is available
Frequencies of change vary	Some changes occur at high frequency
Many changes are undesirable	Agronomic traits can change
Some changes are unstable	Some changes are homozygous
Many changes are not novel	Novel variants can arise
Characters of interest may not change	New varieties have been produced

be demonstrated, but it might be supposed that each plant genome may have some capacity for instability relating to specific components of its composition, whether this be transposable elements, repeated sequences or even major genes. A better understanding of this origin of somaclonal variation will come from studies investigating the nature of molecular changes induced by tissue culture (Brown et al., 1990, 1991).

3. Uses of Somaclonal Variation

An understanding of the causes and origins of somaclonal variation is crucial for its realistic application to crop improvement. Earlier promises of a new source of variation for plant breeding have fallen far short of their expectations largely because at the time they were made little was known about the nature of the variation and the way in which it arises. It has become increasingly clear that somaclonal variation provides no easy alternative to conventional breeding and that whilst there are attractions of this source of variability, there are also many problems associated with its application (Table 1).

3.1. *Problems in the Use of Somaclonal Variation for Crop Improvement*

General conclusions as to the usefulness of somaclonal variation are difficult to make, even within plant species, because plant regeneration from different cultivars can give differing amounts of somaclonal variation. This genotype-dependency means that the recovery of useful somaclonal variants in one cultivar does not guarantee that similar success will be achieved in another. Where possible, pilot experiments in which small numbers of regenerated plants are screened for variability, should precede any large scale commitment.

A related problem is that somaclonal variation is not always as extensive as the first widely publicised reports implied. Although there are many instances where changes have occurred at high frequencies, there are many others where frequencies of change are very low. Genotypic differences

Table 2. Factors that influence somaclonal variation

Factor	General influence*
Genotype	Different genotypes give differing amounts of variation
Ploidy	Higher ploidies give more numerical chromosome variation
Regeneration system	Protoplasts give more variation than explants
Time in culture	The longer the time in culture the greater the variation
Tissue source	Some tissues (e.g. older/storage) give more variation
Growth regulators	High concentrations increase variability

* See Karp (1991a) for more detailed discussion of influences.

are just one cause of this disparity. Other factors influencing the extent of somaclonal variation are listed in Table 2. This problem is further complicated by the lack of a universal system for measuring somaclonal variation. In some reports, variations have been assessed from morphological characteristics, whilst in others changes are detected at a cytogenetical, biochemical or molecular level. These approaches differ in sensitivity making direct comparisons of frequency nonsensical.

Even in cases where somaclonal variation has occurred at high frequencies, and changes have been observed in a whole range of characters, some characters do not change. This means that there is no guarantee that any character of specific interest will vary among somaclones.

Depression in vigour and yield are common in regenerated plants, illustrating that when important characters do vary in somaclonal populations, the changes are not always in the right direction. Variation in chromosome number and structure (Fig. 2) are common causes of unwanted upsets in growth, fertility and vigour. Since chromosome instability is more prevalent in regeneration from protoplasts, explant regeneration systems are more suitable in studies where useful variation is sought.

Where useful changes do occur they are not always stable. In fact three classes of variation can be recognised (Karp, 1991a):

- heritable stable variation
- heritable unstable variation
- non-heritable (epigenetic) variation

Heritable stable variation results from changes in the information content of the genome whereas heritable unstable variation results from heritable changes in gene activity. Epigenetic variation has already been referred to as transient phenotypic modifications which most probably result from the influences of exogenous growth regulators.

Much of the variation observed in somaclones is similar to that obtainable by more conventional approaches such as wide crossing and mutation breeding. Whether somaclonal variation really represents a novel source of variability for crop improvement remains a controversial issue. Where direct comparisons have been made, results are inconsistent. In some cases new mutations have been reported whilst, in others, crossing somaclonal mutants with their radiation-induced counterparts has demonstrated that they are allelic. What is clear is that there is considerable overlap between these two sources of variation and it is therefore reasonable to question whether somaclonal variation has anything to offer in cases where mutation breeding has already been applied.

3.2. *Attractions of Somaclonal Variation for Crop Improvement*

Amongst the early attractions of somaclonal variation as a potential source of variability for crop improvement were the extensive nature of the variations and their rapid availability. Even within a species, the range of changes can be broad (Shepard et al., 1980). This aspect of somaclonal variation still retains promise particularly in asexually propagated and apomictic species where alternative breeding approaches are limited, and in crops grown in developing countries where paucity of funds has restricted breeding efforts. In such species, a relatively simple procedure such as regeneration from cultured explants can release variability for breeding.

Another attractive feature is the high frequency at which some characters change which has led to the exploitation of somaclonal variation as a source of useful mutants. For example, numerous somaclones with altered patterns of seed storage proteins on SDS-PAGE (sodium dodecylsulphate polyacrylamide gel electrophoresis) have been identified in plants regenerated from cultured immature embryos of cereals. High frequencies of chromosome doubling have been exploited as a means of obtaining polyploid plants and the high incidences of structural chromosome re-arrangements, in particular translocations (Fig. 3), have been exploited as a means of achieving introgression in inter-specific and inter-generic hybrids (Table 3).

A related attraction is that many changes occur in agronomic traits. Useful variation described in somaclones include changes in disease response, heading and maturation dates, grain yield, protein content and grain quality. Somaclonal variants have also been screened for their tolerance against numerous environmental stresses and changes have been reported in responses to cold, frost, salinity, waterlogging and freezing. In addition, a selective pressure can be imposed during the culture phase to enhance the frequency at which desired changes are recovered. This approach of *in vitro* selection has been used effectively for the isolation of mutants that are resistant to fungal toxins, salt or herbicides (Table 4).

When somaclonal variation was first widely publicised, it was heralded as a new source of variation and, indeed, novel changes have been identified in somaclonal populations. Other rather unusual attributes are that single regenerated plants may have variation in combinations of characters, or have complicated changes involving a succession of events which are difficult to explain or achieve by more conventional approaches. Furthermore, some changes affect the whole population of somaclones rather than single individ-

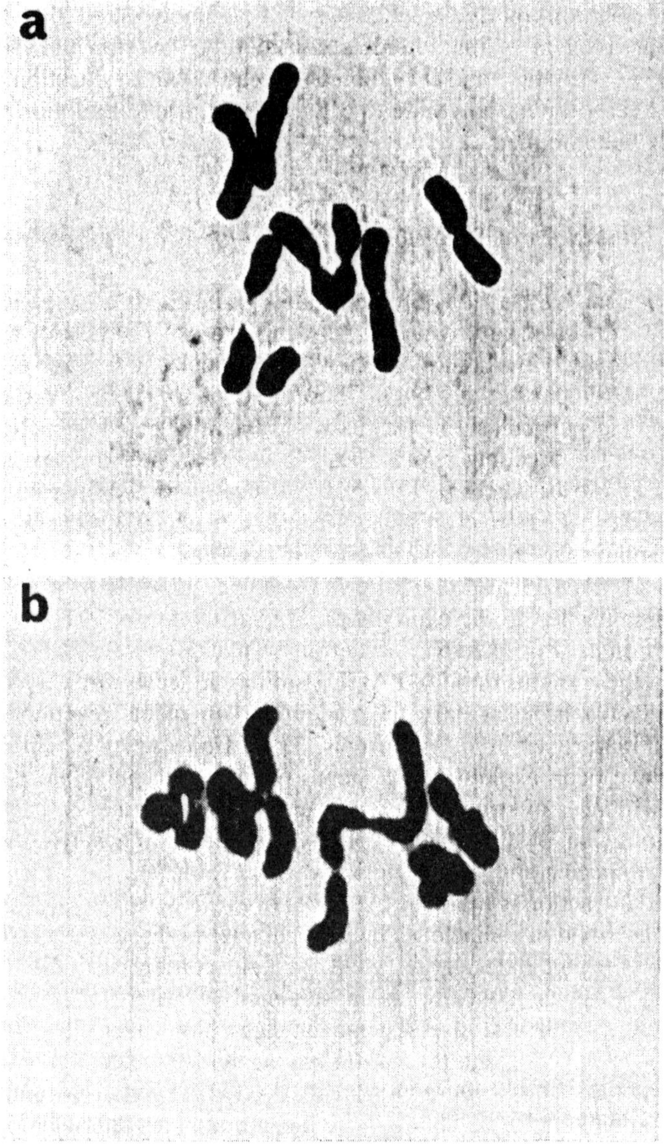

Figure 3. Translocations detected at metaphase-1 of meiosis in regenerated rye plants as multivalent formation (a) trivalent and univalent (b) quadrivalent.

uals as expected from mutation breeding. These findings suggest that somaclonal variation does offer sufficient differences from conventional approaches to be considered an alternative source of variation, and that where the isolation of specific changes from conventional approaches has failed it may be worthwhile screening for the change in somaclonal populations.

Table 3. Exploiting chromosome variation in plant tissue culture

Chromosome variation	Examples of use
Chromosome doubling for production of polyploids	Potato Sugar beet Tobacco Brassicas
Structural chromosome variation for hybrid introgression in inter-specific hybrids	*Lolium multiflorum* × *Lolium perenne* *Hordeum vulgare* × *Hordeum jubatum* *Hordeum vulgare* × *Hordeum bulbosum* *Oryza sativa* × *Oryza latifolia*
Numerical and structural chromosome variation for production of amphidiploids and introgression in inter-generic hybrids	*Triticum crassum* × *Hordeum vulgare* *Triticum aestivum* × *Secale cereale* *Elymus canadensis* × *Secale cereale* *Triticum tauschii* × *Secale cereale* *Hordeum vulgare* × *Triticum aestivum* *Lolium multiflorum* × *Festuca pratensis* *Triticum aestivum* × *Agropyron* sp. *Saccharum officinarum* × *Sclerotachya fuchsa*

'Somaclonal Snowstorm', as it's name suggests, is a new variety of *Paulownia tomentosa* derived from somaclonal variation (Marcotrigiano and Jagannathan, 1988). Also tomato variety with high solids in the fruit (Plant Variety Protection Certificate,NO. 8400146, 1984. USDA Plant Variety Protection Office, Beltsville, MD) has been obtained. These examples illustrate that in some cases the potential of somaclonal variation for crop improvement has been successfully realised.

4. Future Prospects

Somaclonal variation is still a relatively young phenomenon and, given that conventional breeding practices may take ten years to reach fruition, it is possible that more new cultivars, derived from somaclones, will appear in the future. However, there has been limited success so far, despite considerable input and this has led to a decline in the popularity of somaclonal variation as a source of variability for crop improvement.

It is the uncontrolled and essentially random way in which the changes arise that makes somaclonal variation so difficult to use. Considerable progress has been made in the identification of external and internal factors (Table 2) that influence the variation, but the way that these factors interact is still to be determined. The genotype-dependency further complicates the issue since influential factors may have differing effects in different genotypes.

Until somaclonal variation can be controlled it will remain an uncertain approach, whose attractions should be carefully weighed against the problems associated with its utilisation. Whether the approach is adopted or not de-

Table 4. Somaclonal variation in conjunction with *in vitro* selection

in vitro selection for	Somaclone	Heritability
(A) Disease resistance		
Fusarium oxysporium	Alfalfa	√ (dominant)
Fusarium oxysporium j. sp. *apii*	Celery	√ (nuclear)
Phytophthora infestans	Potato	?
Phytophthora parasitica	Tobacco	?
Helminthosporium maydis	Maize	√ (cytoplasmic)
Helminthosporium oryzae	Rice	√ (nuclear dominant)
Helminthosporium victoriae	Oats	√ (nuclear recessive)
Phoma lingam	Oilseed rape	√ (nuclear)
Pseudimonas syringae	Tobacco	√ (nuclear)
Alternaria alternata	Tobacco	√ (nuclear)
Ceratocystis ulmi	Elm	?
(B) Herbicide/drug		
Imidazolinone	Maize	?
Sulfonylurea	Maize, Tobacco	√
Methotrexate	Petunia	?
Atrazine	Tobacco	?
Methomyl	Maize	√ (cytoplasmic)
8-azaguanine	Tomato	?
Amitrole	Tobacco	√
Glyphosate	Carrot, Petunia	√
MCPA, NIDEL	Potato	√ (vegetatively)
Picloram	Tobacco	√ (nuclear dominant)
Paraquat	Tobacco	√ (nuclear dominant)
Hydroxyurea	Tobacco	√ (nuclear dominant)
Amino acid analogues	Rice, Tobacco	√
(C) Salt tolerance	Sugar beet, Rice	
(D) Aluminium tolerance	Alfalfa (cell line only)	

pends on the alternatives available. Unless it can be demonstrated that variants can be obtained more readily from somaclonal variation compared with conventional breeding practices it will not be considered economically viable. But in cases where all other approaches have failed there are sufficient attractions to make somaclonal variation an alternative worth exploring.

To improve our understanding and control of somaclonal variation more work is needed on the mechanisms and key processes involved. In particular, more efforts are required to determine:

1. Why the stabilising influence of the meristem is lost in disorganised growth;
2. What happens to nuclear and cytoplasmic genomes during changes in the differentiated state of the plant cell;
3. How cell cycle control can be improved *in vitro*;

4. What concentrations and combinations of growth regulators are required to maximise control;
5. What components of the genome are responsible for the generation of instability.

A better understanding of these key processes is of relevance to plant biology in general. It is important, not only from the view point of utilising somaclonal variation in a more predictable manner, but also for control of unwanted instability in current approaches of genetic manipulation which rely on plant tissue culture and regeneration and in which somaclonal variation remains a serious problem.

5. References

Brown, P.T.H., Kyozuka, J., Sukekiyo, Y., Ximura, Y., Shimamoto, K. and Lorz, H. (1990). Molecular changes in protoplast-derived rice plants. Mol. Gen. Genet. 223:324–328.

Brown, P.T.H., Gobel, E. and Lorz, H. (1991). RFLP analysis of *Zea mays* callus cultures and their regenerated plants. Theor. Appl. Genet. 81:227–232.

D'Amato, F. (1989). Polyploidy in cell differentiation. Caryologia, 42:183–211.

Gould, A. R. (1984). Control of the cell cycle in cultured plant cells. CRC Critical Reviews in Plant Science 1:315–344.

Karp, A. and Bright, S.W.J. (1985). On the causes and origins of somaclonal variation. In: Oxford Surveys of Plant Molecular and Cell Biology. B.J. Miflin (ed.), Vol. 2. pp. 199–234.

Karp, A. (1991). On the current understanding of somaclonal variation. In: Oxford Surveys of Plant Molecular and Cell Biology. B.J. Miflin (ed.), Vol. 7, pp. 1–58.

Larkin, P.J. and Scowcroft, W.R. (1981). Somaclonal variation – a novel source of variability from cell cultures for plant improvement. Theor. Appl. Genet. 67:443–455.

Lee, M. and Phillips, R.L. (1988). The chromosomal basis of somaclonal variation. Ann. Rev. Plant Physiol. Plant Mol. Biol. 39:413–438.

Marcotrigiano, M. and Jagannathan, L. (1988). Paulownia-tomentosa cultivar 'Somaclonal Snowstorm'. Hort. Sc. 23:226–227.

Shepard, J.F., Bidney, D. and Shahin, E. (1980). Potato protoplasts in crop improvement. Science 208:17–24.

7. Poduction and Use of Isogenic Lines

G. Wenzel and B. Foroughi-Wehr

Contents

1. Introduction

Isogenic lines are defined as individuals that possess the same genotype irrespective of their homo- or heterozygous nature (Johannsen, 1926). In classical plant breeding this means that vegetatively propagated clones are isogenic and lines of inbreeders become isogenic after repeated self-fertilization. In outbreeders, isogeny is achieved in hybrid varieties. The creation of the homozygous lines, from which hybrids are produced, normally requires special breeding techniques in which isogeny is again a central tool. With the exception of population varieties, e.g., rye, *Secale cereale*, isogenic line production is the central breeding step in variety production. This fact has been realized since the beginning of scientific plant breeding and, consequently, great efforts have been undertaken to speed up this process, which normally needs at least six inbreeding generations.

After introducing variability, e.g., in a combination breeding program, selection and selfing steps must always follow. With tissue culture techniques allowing a rapid production of such isogenic lines, breeders should have been enthusiastic to use *in vitro* approaches. The reality was, however, different. Although all the advantages of haploids were defined as soon as the first haploid plant was described in the 1920s, it took 40 years before haploids gained broader scientific interest, and another 10 years until they were taken

Indra K. Vasil and Trevor A. Thorpe (eds.), Plant Cell and Tissue Culture, 153–172.

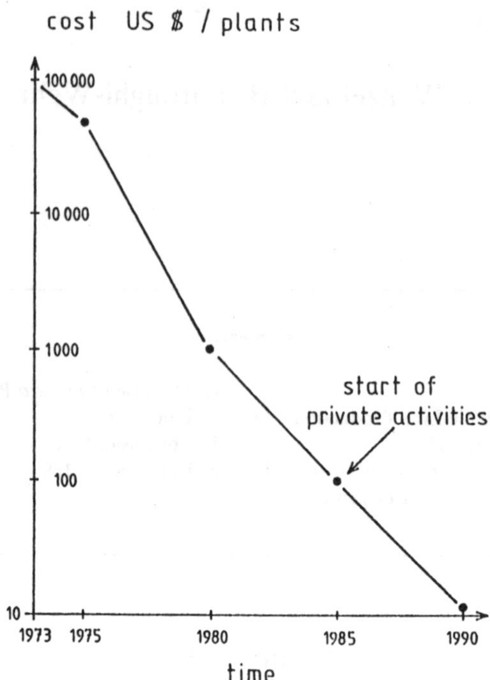

Figure 1. Estimated price of one DH line of barley. The costs include labor and running costs of an established laboratory.

up in private breeding stations. Only today do most breeding companies have access to laboratories where haploids from barley, wheat, potato and rapeseed are produced routinely in numbers sufficient for applied breeding needs. Figure 1 demonstrates the progress in the production and use of isogenic lines by comparing the price of a single doubled haploid and, consequently, the isogenic line. The progress follows a logarithmic scale and the methodology has now reached the stage where private laboratories are using this technique at their own economic risk.

The starting material for the production of homozygous lines in only one generation are the haploid gametes. Consequently, two principal methods exist for the induction of plants with a simpler genome: either starting from the female or the male gametophyte, which means the *in vitro* regeneration of immature pollen or the induction of haploid tissue from an embryo sac cell. In all probability, only the androgenetic procedure will be cheap enough to be of direct importance for economic breeding programs. No other technique (see reviews Yang and Zhou, 1982; Foroughi-Wehr and Wenzel, 1991), although many are of utmost scientific interest, promises to replace microspore androgenesis. This chapter will therefore concentrate on approaches to producing isogenic lines via androgenetic haploids. Of course, the haploids must be diploidized again, in order to result in doubled haploid lines (DH) with the necessary stability for the new genotype.

Parthenogenetic haploids occur sooner or later spontaneously in nature; the first were found in twin seeds. For practical breeding purposes this spontaneous haploidization rate is much too low. Because of this, procedures for increasing the haploid rate were developed, such as using specific mutants capable of inducing increased frequencies of haploids, and pollination with irradiated or chemically treated pollen. However, none of these approaches has attained practical importance, as they were normally rather specific. General progress in tissue culture opened up programmed *in vitro* approaches. When techniques such as embryo culture or isolation of unfertilized egg cells and their regeneration *in vitro* became possible, the artificial induction of parthenogenetic haploids attained an acceptable frequency and was used to a great extent in barley isogenic line production.

A breakthrough occurred in 1964, however, with the successful production of haploids originating from very young pollen in anthers cultured *in vitro* (see Guha and Maheshwari, 1966). The principle of androgenesis is to stop the development of the young pollen whose fate is normally to develop into the male gametephyte, and to force the pollen into a somatic development. The uninucleate state of the microspore is needed as a pathway to a haploid sporophyte. This occurs only when the anther is taken away from its normal environment and placed under artificial *in vitro* conditions.

The main advantage of using haploids is the rapid and complete homozygosity of the offspring, because phenotype selection for quantitative, and particularly for qualitative, inherited characters is much easier, and therefore breeding is more efficient. In polyploids like *Medicago* and potato, haploidization is used to produce diploids, not only for breeding by classical techniques with such diploids but also as a prerequisite for the successful application of somatic hybridization. In most tetraploid crops it has become possible, using this approach, to combine wild species with the cultivated forms and/or to combine complex genotypes.

A rather recent application of isogenic lines is their use in genome mapping. In a population of DH lines the identification of phenotypic markers is much more secure, as most intermediate expressions resulting from heterozygosity are excluded. In both the expression at the plant level as well as at the marker level, a gene will segregate in a 1:1 ratio. Actually, it is very difficult to reproducibly map and correlate DNA polymorphisms when the bands segregate in a 3:1 or a 1:2:1 ratio. This is of particular importance when a polygenically coded character is to be mapped in a QTL analysis.

2. Paths towards Isogenic Line Production

Attempts were made to extract general principles and to clarify the up-to-date knowledge in isogenic line production and use, with the main effort being on androgenesis of important crops.

2.1. *Parthenogenesis*

It has been possible to induce parthenogenesis by *in vitro* culture of unpollinated ovaries and ovules. Haploid embryo sac cells are able to develop a haploid embryo or callus *in vitro*. Attempts have been made to culture unfertilized ovules or ovaries in different plant species, but growth of most of them stopped at the callus stage. Many of the conditions mentioned for other plant tissue culture events are practicable for the ovule or ovary culture. As in all tissue culture procedures, there is a different regeneration response depending on the genotype. In five wheat genotypes, for example, the percentage of ovules producing haploid plants ranged from 1.3 to 10.9% (Zhu et al., 1981). Among 50 breeding lines of onions, four were found to be the most productive (Keller, 1990). From one ovule several embryos can be induced in barley (San Noeum, 1976) and sugar beets (e.g. D'Halluin and Keimer, 1986).

Until now the practical application of parthenogenesis has been rare because of the low frequency of haploid induction in most species, aside from potato. Only in plant species where anther culture has been unsuccessful as in sugar beet or onion are isolated, unpollinated ovules or ovaries of practical interest (Muren, 1989).

In some interspecific crosses, the phenomenon of chromosome elimination occurs during early embryo development. Due to a mechanism not yet understood, the chromosomes of one parent are partially or completely lost during the early divisions of the young hybrid embryo. In nature, such a loss of chromosomes normally coupled to a failure of endosperm development leads to sterility. If all chromosomes of one parent are lost and if the embryo rescue *in vitro* is successful, the procedure often results in haploid plants of the female parent. For barley, this system was studied in crosses of *Hordeum vulgare* × *H. bulbosum*. Kasha and Kao (1970), and Jensen (1975) demonstrated that the method of crossing diploid barley with diploid *H. bulbosum* could be used to produce high frequencies of isogenic barley lines. Although normal fertilization takes place, the *H. bulbosum* chromosomes were eliminnated during the development of the embryo. Since the endosperm fails to develop, the embryo rescue technique is required to prevent starvation and death of the embryos. Snape et al. (1979) also produced wheat haploids by fertilizing wheat with *H. bulbosum*. A similar mechanism of chromosome elimination leads to haploids in wheat crosses with maize (Laurie and Bennett, 1987).

The bulbosum system in barley has been practiced rather extensively to produce isogenic barley lines. The first doubled haploid variety of barley 'Mingo' was licensed in Canada in 1979 (Ho and Jones, 1980). But although the method has been remarkably improved by selecting more compatible *H. bulbosum* genotypes and by optimizing different culture factors, its economic importance is limited. Since it is possible to make use of the greater potential of the androgenetic system for haploid production, the bulbosum method is of decreasing value for practical plant breeding.

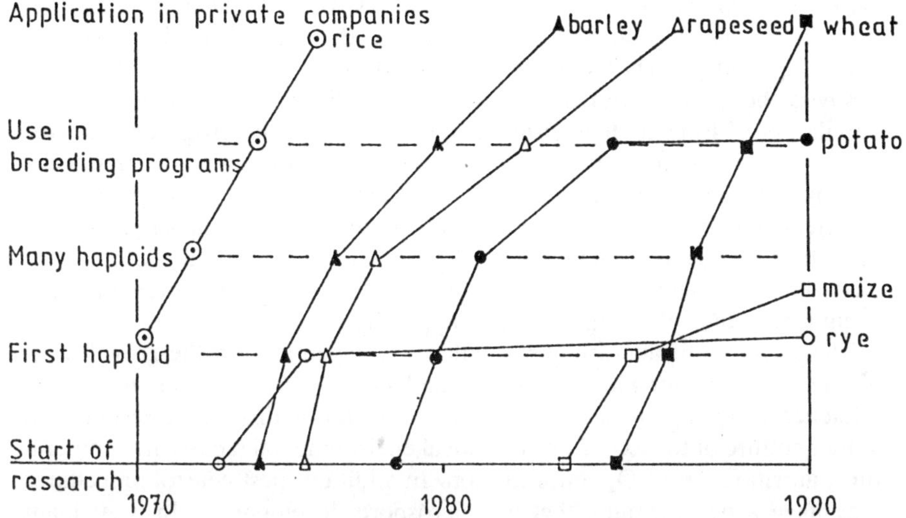

Figure 2. Progress in the use of DH lines.

2.2. *Androgenetic Haploid Production*

Today the most promising and also most successful method of inducing isogenic lines is microspore androgenesis. A microspore, i.e., a uninucleate pollen in an early developmental stage, can be induced *in vitro* to form a plantlet. This technology has been applied successfully to a lot of species in many genera. However, most important is the research progress in crop plant species from the Solanaceae, Gramineae and Cruciferae, where the technique is used nowadays for practical breeding; a time scale is given in Fig. 2. In spite of the application of this method, we know very little about the induction mechanisms responsible for transforming a microspore to a sporophyte, although our knowledge is getting broader. Here, the most important factors that influence success as well as certain general principles, will be summarized.

2.2.1. *Donor Plants*

The genotype-environment interaction not only influences agronomic traits but also any success of *in vitro* cultures. Consequently, there is a need to optimize both, the environment via *in situ* and *in vitro* culture conditions, and the genotype by selecting or constructing most responsive types. Dunwell (1981) demonstrated that optimizing culture conditions for each genotype is possible. It could also be shown that certain genotypes, e.g., in rye, *Secale cereale* × *S. vavilovii* hybrids (Wenzel et al., 1977), in barley the, variety Igri (Foroughi-Wehr and Friedt, 1984) or in winter wheat the cultivar Florida (Foroughi-Wehr and Zeller, 1990), are extraordinarily responsive. If both parameters complement one another, impressive maxima in regeneration frequencies can be obtained - and these are the data normally published. Under applied aspects there are dramatic drawbacks, however. Only very rarely is there an interest in just one isogenic line, but rather usually in a population of isogenic lines originating from, e.g., a heterogeneous microspore population. In such cases, it is unrealistic to maximize the frequency of isogenic line regeneration for each single line. An optimum should be sufficient. Furthermore, it turns out that the regeneration capacity is a heritable trait. Thus, we believe today that it is cheaper to broaden the genetic bases for tissue culture ability by selection and combination breeding than to maximize the culture medium-genotype interaction.

The response of microspores in culture is dependent on the growth conditions of the donor plants. Dunwell and Perry (1973) first pointed to the influence of light (intensity, photoperiod) on the androgenetic response in anther culture of tobacco. Further critical environmental factors are temperature, nutrition, and CO_2 concentration. In addition, pest control procedures may have a detrimental effect on microspore development. In most plant species, field-grown material during the natural season was inferior to greenhouse-grown material. But even under controlled growth conditions there is an endogenous rhythm for success. For applied use it is thus economical to plate anthers during spring and early autumn.

In maize. the anther culture process itself was used to select for genetic improvement of *in vitro* response. Two anther-derived plants were crossed to produce an F_1 hybrid. The anther response of the S_1 families from this cross was almost eight times higher than the original cross. The enhanced anther culture responsiveness of commercial germplasm should broaden the practical applicability (Petolino et al., 1988). Barloy et al. (1989) concluded from their investigations with different isogenic maize lines that it is relatively easy to introduce this ability into elite lines by repeated backcrosses.

In rye (Wenzel et al., 1977), barley (Foroughi-Wehr and Friedt, 1984) and potato (Wenzel, 1980; Uhrig and Salamini, 1987), high-responding genotypes were crossed with agronomically important varieties. The F_1-generation showed an intermediate reaction between both parents, making it feasible that a high response in anther culture is heritable (Wenzel and Foroughi-Wehr, 1984). In winter barley, the high responding genotype 'Igri' (Foroughi-Wehr and Friedt, 1984) produced among eight varieties the highest percentage of green structures in the Scottish Crop Research Institute (Powell, 1988), as well as among 16 varieties in Denmark (Knudsen et al., 1989). But even in such a responding genotype, spring and autumn give much better results than winter and summer (Fig. 3). In wheat, the good regeneration capacity of the variety 'Centurk' could be transferred in reciprocal crosses to the F_1 progeny. The cytoplasmic effects were negligible (Bullock et al., 1982). The genotype of the donor plant is also a strong determinant of the embryogenic capacity in *Brassica napus* (Keller, 1984; Siebel and Pauls, 1989), and this character is under genetic control and is transferable to F_1 hybrids.

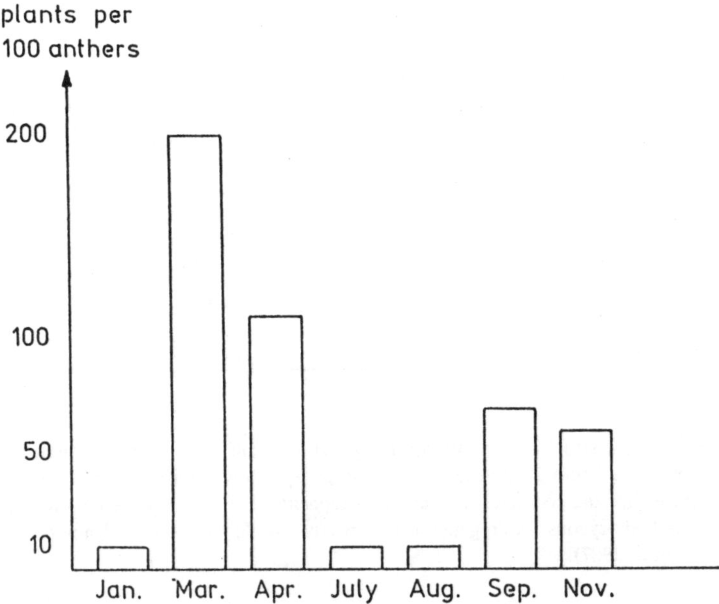

Figure 3. Frequency of regeneration of the high responding barley variety 'Igri' over the year (Kuhlmann, 1991).

Genetic analysis of the heritability of *in vitro* culture response is rare. Henry and de Buyser (1985) stated that haploid production in wheat is controlled by at least three different and independently inherited traits: embryo induction, regeneration ability, and the ratio of green to albino plants. They suggested that regeneration ability could be linked to the 1BL-1RS translocation. These results with 45 winter wheat cultivars obtained by Müller et al. (1989), confirmed the preferential regeneration frequency in 1BL-1RS translocation lines. The results were verified in other investigations (Foroughi-Wehr and Zeller, 1990, Fadel and Wenzel, 1990), but it could be assumed that, in addition to the 1BL-1RS translocations, there are other genetic systems that influence microspore regeneration in wheat. Higgins and Mathias (1987) analyzed the chromosomal effect of regeneration capacity from callus cultures and reported significantly increased morphogenesis of different substitution lines of 4B chromosome in Chinese Spring donor material. For potato, Figure 4 summarizes the heritability of the regeneration response.

2.2.2. *In vitro Culture*

For most plant species, flower buds, inflorescences, or spikes harvested at the early or mid-uninucleate stage of the microspore contain the cells that are most responsive *in vitro*. As soon as starch deposition has begun in

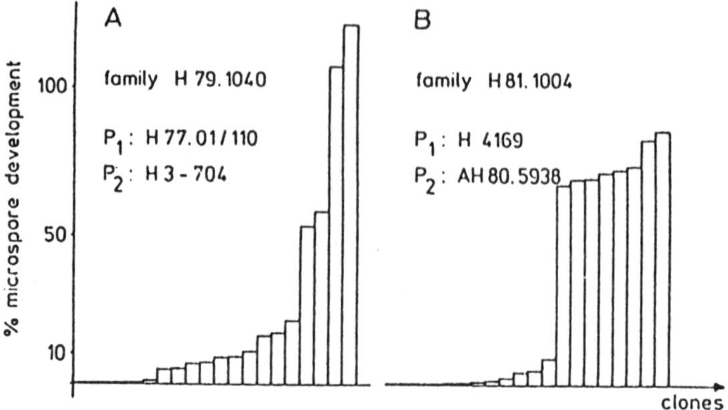

Figure 4. Macroscopic structure formation (embryoids and calli in % of the number of plated anthers) in potato. A From 20 different clones of an F_1 hybrid family. the parents of which were heterozygous for the character 'regeneration capacity'; B From 20 different clones where one parent was homozygous for 'regeneration capacity', while the other did not possess any (Uhrig and Wenzel, 1987).

microspores, no sporophytic development and, subsequently, no macroscopic structure formation occurs. On the other hand, very young microspores, tetrads or spores immediately after release from the tetrads are ineffective for a development towards a sporophyte. The pollen stage in the anther is either determined by rapid staining with acetocarmine or, if the exine is too thick as in potato or rapeseed, by fixation in Carnoy's solution overnight and with subsequent staining. The size of the flower buds that contain anthers in the right stage depends, to some extent, on their position in the inflorescence, on the genotype and on the plant age; thus size is only a vague measure for the development of the microspsores inside. In cereals, the check of the stage from one anther can be used as a marker for the estimation of microspore development in the remaining florets from the same variety during the same season. Pollen in florets from the middle of the inflorescence are further developed than those in apical and basal florets.

Certain pretreatments of the inflorescences can have a positive effect on the development of the microspore. The most effective technique used in anther culture, especially for the cereals, is cold pretreatment. Periods from 72 h up to 4 weeks at 4–10 °C are recommended (barley: Sunderland et al., 1981; potato: Wenzel, 1980; maize: Genovesi and Collins, 1982; wheat: Lazar et al., 1985; rice: Chen et al., 1981; rye: Wenzel et al., 1977). An influence of the ploidy level of regenerated rice plants by cold pretreatment was reported by Yang et al. (1985).

The mechanisms of the positive effect of temperature treatment on macroscopic structure formation are unknown. As a result of cold storage, weak or nonviable anthers and microspores are killed; they then become dark brown in color. Thus, the spikes are enriched in vigorous material. At low temperature, increased numbers of microspores are arrested during the first mitosis, as starch production is blocked. This increases the possibility of subsequent development. The cold pretreatment retards aging of the anther wall and, in particular, the degradation of the tapetum matrix, allowing a higher proportion of microspores to change their developmental pattern from gametophytic to sporophytic. Cold pretreatment may also provide an unspecific shock, resulting in the establishment of endogenous cellular conditions that favor the desired development. This is quite probable, as also other shocks such as pretreatment with high temperature, spraying with ethrel, centrifugation and irradiation sometimes have stimulatory effects (for literature see review by Foroughi-Wehr and Wenzel, 1991). None of these techniques, although effective in several cases during the early experimentation phase, are routinely used today.

On the other hand, several investigations showed a high frequency of callus or embryoid formation for good responding genotypes, such as cv. 'Igri' in barley, without any pretreatment (Powell, 1988). In wheat a negative effect of low temperature on the anther response has also been reported (Marsolais et al., 1984). If the donor plants grow under optimal conditions, the promoting effect of pretreatment is minimized and the whole procedure may be omitted.

The induction of uninucleate microspores to form sporophytic embryoids or calli has been studied in detail for *H. vulgare* (Sunderland et al., 1979; Sunderland and Evans, 1980) and for wheat (Krüger, 1987). It was found that multicellular microspores develop along all pathways, even if the vegetative or the generative nucleus starts to divide first (Sunderland and Evans, 1980). In wheat, the generative nucleus degenerates during the *in vitro* development and is not involved in formation of microstructures. During establishment of an anther culture procedure, it is recommended that some anthers be sacrificed after five days up to three weeks of culture. These should be checked microscopically to find the first multicellular structures, a reliable proof for the desired reaction. In liquid culture media such a check is quite often possible with an inverted microscope without opening the culture vessel. An experienced investigator will recognize responsive microspores after a few days. This is possible in barley even through the very thin anther wall.

The culture of entire florets, tassels, or parts of inflorescences and the induction of haploid plants are reported in barley (Wilson, 1977), rice (Kim and Raghavan, 1988), *Festuca arundinacea* (Kasperbauer et al., 1980) and maize (Pareddy and Greyson, 1985). The results are highly genotype dependent and generally less successful than anther or microspore culture.

The available literature does not allow recommendation of the anther culture medium for general applicability. Nevertheless, for practical application where F_1 hybrids are used containing a heterozygous microspore

Table 1. Successful induction medium for anther culture of potato (Zitzlsperger, unpublished); rape seed (Keller, 1984); rice (Chen et al., 1982); wheat (Fadel and Wenzel, 1990) and barley (Kuhlmann and Foroughi-Wehr, 1989; Foroughi-Wehr, unpublished)

Variety	Basal medium	Additional components	mg/l
Solanum tuberosum	Linsmaier and Skoog (1965); half concentration; without NH_4NO_3 liquid	NH_4NO_3	400
		L-Glutamine	750
		Tiamine HCl	0.2
		BAP	1
		IAA	0.1
		Saccharose	60000
		Maltose	60000
		Activated charcoal	5000
Brassica napus	B5-medium (Gamborg et al., 1968) liquid	$CaCl_2$–$2H_2O$	750
		L-Glutamine	800
		Serine	100
		NAA	0.1
		2,4-D	0.1
		Saccharose	100000
Oryza ssp.	inorganic salts. N-6-medium modified, (Chen et al., 1982) organic substances MS-medium (Murashige and Skoog, 1962) solid	NAA	4
		Kinetin	2
		Saccharose	30000
Triticum aestivum	Potato 2 medium (Chuang et al., 1978; Henry and De Buyser, 1981) semiliquid	L-Glutamine	200
		Ficoll	100000
		Maltose	70000–100000
Hordeum vulgare	inorganic salts: modified MS (Foroughi-Wehr et al., 1976) liquid	Thiamine HCl	0.4
		Myo-inositol	100
		NAA	2
		BAP	1
		Ficoll	200000
		Maltose	30000
		Barley starch	80 mg/dish

population, a general working method is needed; thus, for all important crops there is a most recommended medium that guarantees good response of a sufficient number of genotypes. Table 1 summarizes one successful medium each for barley, wheat, rice, potato and rapeseed. As basal media they contain mainly the inorganic salts of the N6 medium (Chu, 1978) or those of the Murashige and Skoog (1962) medium. In the People's Republic of China, a potato extract medium was developed and further improved by Chuang et al. (1978), on which especially wheat gives a high androgenetic response. As this medium is not completely defined, many trials were made to replace it, but until now all synthetic defined media have been less responsive in wheat anther culture (Fadel and Wenzel, 1990; Han-min et al., 1990).

A carbon source is obligatory for microspore regeneration. In most culture media, sucrose is recommended. Since Raquin (1983) found that maltose improves embryo induction and development in anther culture of petunia, more attention has been paid to this sugar. In potato anther culture, sucrose was found to stimulate embryo or callus induction more than maltose (Batty and Dunwell, 1989), but plant production was significantly increased in maltose-containing induction medium. Keller et al. (1975) found similar results in *Brassica campestris*, where sucrose induced more embryoids than maltose. In wheat (Fadel and Wenzel, 1990; Orchinsky et al., 1990; Last and Brettell, 1990) and barley (Finnie et al., 1989), maltose in the induction medium increased the rate of regeneration.

All early experiments started with agar solidified media. Due to different agar qualities and to the high price of agar, a main research effort was recently made for haploid production by replacing the agar with starch. For barley, barley starch was optimal (Sorvari, 1986), and Foroughi-Wehr and Zeller (1990) were able to verify these results for wheat, where wheat starch gave a better response than agar. Gelrite is also under investigation and often gives more reproducible results than agar.

Liquid media were normally not successful, as the anthers sunk to the bottom of the petri dishes, not allowing sufficient gas exchange. Kao (1981) reported considerable increase in the frequency of pollen regeneration by adding Ficoll to liquid medium. Ficoll prevents small structures such as callus or embryoids from sinking into the medium and thus dying through anaerobiosis. But again Ficoll is so expensive that it should not be used for routine work. Furthermore, the system becomes vulnerable: evaporation and change of culture media must be controlled much more accurately than in culture on solid media.

Kuhlmann and Foroughi-Wehr (1989) were able to regenerate a high frequency of green plants in barley in liquid Ficoll medium containing barley starch and maltose, but they obtained almost the same results in a maltose-free medium. Figure 5 compares the development of barley microspores to macroscopic structures on media with different carbohydrate sources. The addition of maltose accelerates the microspore development more than starch, resulting in a higher regeneration frequency. The best results were gained by addition of both starch and maltose. The effects of incubation conditions such as light (quantity and quality) and temperature have been investigated in a number of studies. Usually the anthers were cultured in the dark in a range between 20 and 30 °C. Ouyang et al. (1983) showed that in wheat the optimal culture temperature ranged from 26 to 30 °C, varying with genotypes. For field-grown material, the optimal culture temperature was generally 2 °C higher than for greenhouse-grown material (Ouyang et al., 1987). The orientation of the anthers on the medium with regard to the

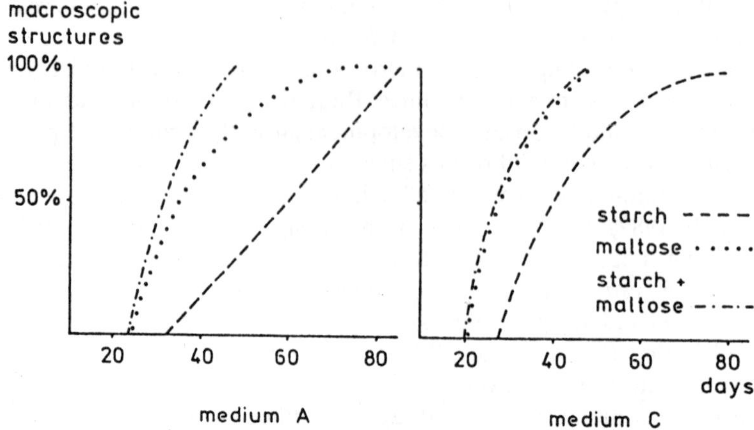

Figure 5. Time needed for the development of embryoids and calli dependent of the carbohydrate source: medium A: modified N6 (Chu,1978); medium C: modified Murashige and Skoog (Kuhlmann. 1991).

loculi has a positive effect on response in barley (Shannon et al., 1985), and in maise (Tsay et al., 1986).

2.2.3. *Isolated Microspore Culture*

Today, the culture of isolated microspores is possible in many crop plants. The first reports were given for tobacco by Nitsch (1974), for rice by Chen et al. (1981), for rapeseed by Lichter (1982), for barley by Sunderland and Xu (1982), for wheat by Wei (1982), for potato by Uhrig (1985) and for maize by Pescitelli et al. (1989). The microspores are either isolated mechanically from the anther or shed initially into liquid culture media. Figure 6 shows a flow diagram on how to proceed in barley isolated microspore culture. To improve the efficiency of isolated pollen culture, it proved to be beneficial to condition the liquid culture medium by adding anthers or ovules (Köhler and Wenzel, 1985). Siebel and Pauls (1989) compared anther culture and microspore culture in *Brassica napus* and demonstrated a two to ten times higher efficiency in embryo production in microspore culture. There were no differences between the two populations with respect to agronomic characters in the field.

It can be concluded that isolated microspore culture is superior to anther culture in *Brassica napus*. In all other cases the culture of isolated microspores is possible but anther culture gives better results and is easier for routine haploid production. Nevertheless, for some high-responding genotypes and under aspects other than only maximal plantlet formation, isolated microspore cultures offer a more effective system of regenerating a random sample from the microspore population than does anther culture. This is

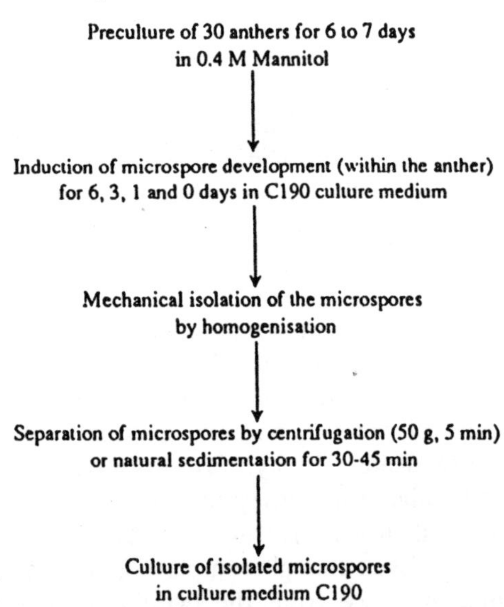

Figure 6. Flow diagram for the preparation of isolated microspores of barley; medium CI90 contains in addition to the MS components 2 g/l inositol and 90 g/l maltose (Kuhlmann, 1991).

particularly important when the desired trait is linked with low plasticity. In addition, these cultures are a good source for *in vitro* selection and gene transfer.

3. From a Haploid Plant to an Isogenic Line

Plants developed from microspores or ovaries are either haploid, diploid or tetraploid; a few are aneuploid or mixoploid. In barley, potato, rapeseed and rye the majority (ca. 70%) of the plants are diploid, whereas in maize and in wheat, most of the regenerants remain haploid. Various factors influence the ploidy level of the regenerated plants. In *Brassica napus*, this character seems to be under genetic control (Chuong et al., 1988) and the ratio of diploids rises with increasing duration of the culture period (Siebel and Pauls, 1989). The same is true for potatoes, where the monoploid ($x = n = 12$) formation is dependent on the genotype; it ranged from 0 to 20% (J. Zitzlperger, personal communication). Epigenetic factors, such as the pollen stage at the time of inoculation, the hormones in the media (particularly high auxin levels) or the culture conditions, may also influence the ploidy level of the plants.

Haploid plants are doubled by colchicine, as described by Jensen (1975). This occurs with some variation in the concentration, without problems in most plant species. In potato and maize, the efficiency of inducing doubled haploid plants is very low. However, Wan et al. (1989) used colchicine treatment of an embryogenic haploid maize callus initiated from anther culture and produced fertile doubled haploid inbred lines at high frequency.

In rice and barley the induction of doubled haploids has been improved to the extent that colchicine treatment of the remaining haploids can be omitted. Seed set could be found in more than 70% of haploid barley plants without colchicine treatment, preferably under field conditions.

A very specific doubling technique is applied in vegetatively propagated crops, particularly in potato, where somatic hybridization offers a very efficient procedure for the return from the 2X to the important (for practical purposes) 4X level. Somatic hybridization allows at the same time the addition of all characters of the two parental genomes, and fertility of the haploid lines is no prerequisite. The hybrids can be vegetatively propagated as clones, despite their highly heterozygous nature. For applied purposes, however, it must be clarified why hybrids from the same parents are often genetically not identical. Even though complete genomes are added, there seems to be a strong influence from the cytoplasm, which segregates in different clones, thus ending up in different plasmon/genome combinations. Furthermore, the fusion technique has showed that different isogenic lines have different genetic strength: it became clear that for somatic combination of isogenic lines the rules of specific and general combining ability may be applied. This technique of hybrid variety production only has a future, however, in vegetatively propagated crops. For other plants the somatic doubling system is not economical.

4. The Use of *in vitro* Produced Isogenic Lines

Since Wark (1977) obtained improved, doubled haploid tobacco lines via anther culture, this technique has worked in practice in potato, rapeseed and the cereals; in Taiwan, e.g., 1500–15000 isogenic lines of rice were produced per season for selection (Huang et al., 1988). It is generally accepted today that a step-wise reduction in the ploidy level of tetraploids offers the advantage of a simpler inheritance and also a better chance to combine qualitatively inherited characters. In potato, it could be demonstrated that field resistance to potato virus Y and potato leaf roll virus is maintained and expressed on all ploidy levels, which means that both monogenically and polygenically inherited resistances are maintained during the successive haploidization process (Wenzel and Uhrig, 1981). When a genome has passed the monohaploid level, a homozygous additive character can be transferred efficiently to the next generation via classical crossing. In a vegetatively propagated crop, such as potato, the F_1 hybrid can be propagated as a new cultivar.

In *Brassica napus*, quality characteristics are emphasized in breeding techniques with haploids. Particularly the glucosinolate content and the oil quality is of interest in terms of an alternative use of this crop in the non-food market. In isogenic line production of *B. napus* it was found that vigor of microspores was positively correlated with high glucosinolate content. More than 70% of androgenetic plantlets tested had a glucosinolate content much higher than did the original material (Hoffmann et al., 1982). Since the aim of breeding was for low glucosinolate content, spontaneous selection for this character worked against the program. Development of microspores with a high level of glucosinolates, being chemically quite similar to phytohormones, may have been favored. Very low or very high glucosinolate levels are diverse qualities needed either for human consumption or for industry. A simple screening for extremes would increase breeding flexibility at an early stage.

Since regeneration is no longer a main obstacle in using *in vitro* derived isogenic lines, other questions are gaining importance. Snape and Simpson (1981) already discussed the genetic effects when starting from different filial generations. The question is: can isogenic lines be extracted more economically from an F_1 generation, or should the haploid step be started from a preselected F_2 or F_3? It could be demonstrated that one haploid step followed by selection in the greenhouse and in the field during the first androgenetic (A_1) and two subsequent selfed generations (A_2; A_3) is the most efficient procedure, if characters from related varieties are to be combined.

Haploids can also increase the efficiency for the most complex breeding problem: the transfer of quantitatively inherited traits from distant genotypes to cultivars. For such complex breeding programs a combination of recurrent selection, which guarantees variability. and haploid selection, which guarantees secure selection, has been developed. This breeding technique is called recurrent selection alternating with haploid steps. The degree of relationship of the two parents determines the number of backcrosses needed. The necessary offspring size of DH population for quantitative characters depends on the degree of linkage and on the number of genes involved. The main advantage of using this method is to speed up the breeding process and to make selection of quantitatively inherited traits more effective and reliable (Foroughi-Wehr and Wenzel, 1990). As soon as methods are available which allow secure selection of characters on a single plant level, e.g., by applying molecular markers such as closely correlated DNA probes (Graner et al., 1990), selection and combination can be done in one year, resulting in the breeding scheme proposed in Figure 7.

5. Conclusion

The use of isogenic lines is a common step in all breeding programs. In principle, it is not relevant whether these lines are produced classically by inbreeding or *in vitro* parthenogenetically or androgenetically. Normally, parthenogenesis is laborious, as it demands greenhouse work and, additionally, *in vitro* culture. An exception is potato, where no *in vitro* culture step

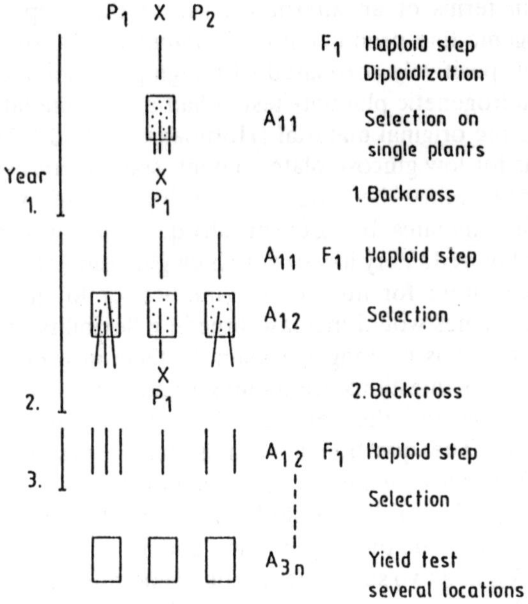

Figure 7. Scheme for recurrent selection alternating with haploid steps using an early selection in the A_1 (Foroughi-Wehr and Wenzel, 1990).

is necessary for the reduction from the 4X to the 2X ploidy level. In most other cases, microspore androgenesis has been improved so much that the numbers of green haploid plants necessary for applied work – normally around 100 per donor genotype – can be produced. This is particularly true for most cereals and for the *Cruciferae*. There are strong genotypic differences but, in principle, the procedure works economically.

There is no doubt that *in vitro* procedures for isogenic line production have shifted from research to application. Unfortunately this statement can still not be generalized, as several major crops are still very recalcitrant. These are nearly all the legumes and maize. Of similar difficulty is the outbreeder rye, where only some genotypes prove that the technique works in principle. Thus, the outbreeders maize and rye, where a rapid production of isogenic lines for broadening the hybrid seed production would be very important, are not yet improved by this *in vitro* methodology. Unfortunately, the knowledge of which parameters have to be altered in which direction is still closer to an art than to a logical science. A basic knowledge of regeneration processes is lacking, although molecular approaches for a better understanding of regeneration processes have been started. Another most elegant approach. to induce meiosis thus converting a sporophytic cell to a gametophytic one – the artificial induction of meiosis – is still wishful thinking.

Nevertheless. it presently looks as if approaches via the genotype of the anther donor plants are more promising than medium alterations.

In those crops where the technical problems of haploid induction are solved, the question of their incorporation into breeding schemes is presently under investigation. For aims such as rapid incorporation of a monogenic trait or combination of monogenic traits, the time savings is interesting, but much more important are strategies that allow the combination of quantitatively inherited characters with a polygenic genetical basis. Here, the *in vitro* procedures have their biggest potential, as they allow a very secure selection of polygenic traits in a very early generation. That an A_1 is equivalent to a F_n is probably the most far-reaching advantage of the ability to produce an isogenic line in one step, and *in vitro* production of these lines is the most recommended technique.

6. References

Barloy, D., Denis, L., and Beckert, M. (1989). Comparison of the aptitude for anther culture in some androgenetic doubled haploid maize lines. Maydica 34:303–308.

Batty, N. and Dunwell, J. (1989). Effect of maltose on the response of potato anthers in culture. Plant Cell Tissue Org. Cult. 18:221–226.

Bullock. W.P. Baenziger, P.S., Schaeffer G.W. and Bottino, P.J. (1982). Anther culture of wheat (*Triticum aestivum* L.) F_1's and their reciprocal crosses. Theor. Appl. Genet. 62:155–159.

Chen, L.J.. Lai, P.C., Liao, C.H. and Tsay, H.S. (1982). Medium evaluation for rice anther culture. In: Plant Tissue Culture. 5th Intl. Cong. Plant Tissue and Cell Culture, Tokyo, pp. 551–552.

Chen. Y., Zuo. Q., Li. S., Lu, D. and Zheng, S. (1981). Green plant regenerated from isolated rice pollen grains *in vitro* and the induction factors. Acta Gen. Sin. 8:158–163.

Chu, C.C. (1978). The N6 medium and its applications to anther culture of cereal crops. In: Proc. Symp. Plant Tissue Culture. Science Press, Peking, pp. 43–45.

Chuang, C.C., Ouyang, T.W., Chia, H., Chou, S.M. and Ching, C.K. (1978). A set of potato media for wheat anther culture. In: Proc. Symp. Plant Tissue Culture. Science Press, Peking, pp. 51–56.

Chuong, P.V., Deslauriers, C., Kott, L.S. and Beversdorf, W.D. (1988). Effects of donor genotype and bud sampling in microspore culture of *Brassica napus*. Can. J. Bot. 66:1653–1657.

D'Halluin. K. and Keimer, B. (1986). Production of haploid sugarbeets (*Beta vulgaris* L.) by ovule culture. In: Genetic Manipulation in Plant Breeding. W. Horn, C.J. Jensen, W. Odenbach, O. Schieder (eds.), Walter de Gruyter, Berlin, pp. 307–309.

Dunwell, J.M. (1981). Influence of genotype and environment on growth of barley embryos *in vitro*. Ann. Bot. (London) (N.S.) 48:535–542.

Dunwell, J.M. and Perry, E. (1973). The influence of *in vivo* growth conditions of *N. tabacum* plants on the *in vitro* embryogenetic potential of their anthers. John Innes Inst. Rep. 64:69–70.

Fadel. F. and Wenzel, G. (1990). Medium-genotype-interaction on androgenetic haploid production in wheat. Plant Breed. 105:278–282.

Finnie, S.J.. Powell, W. and Dyer, A.F. (1989). The effect of carbohydrate composition and concentration on anther culture response in barley (*Hordeum vulgare* L.). Plant Breed. 106:110–118.

Foroughi-Wehr, B., Mix, G., Gaul, H. and Wilson, H.M. (1976). Plant production from cultured anthers of *Hordeum vulgare* L. Z. Pflanzenzücht. 77:198–204.

Foroughi-Wehr, B. and Friedt, W. (1984). Rapid production of recombinant barley yellow mosaic virus resistant *Hordeum vulgare* lines by anther culture. Theor. Appl. Genet. 67:377–382.

Foroughi-Wehr. B. and Wenzel, G. (1990). Recurrent selection alternating with haploid steps a rapid breeding procedure for combining agronomic traits in inbreeders. Theor. Appl. Genet. 80:564568.

Foroughi-Wehr, B. and Wenzel, G. (1993). Andro- and Parthenogenese. In: Plant Breeding – Principle and Prospects. M.D. Hayward, N.O. Bosemark, and I. Romagosa (eds.), Chapman & Hall London, pp. 261–277.

Foroughi-Wehr, B. and Zeller, F.J. (1990). *In vitro* microspore reaction of different German wheat cultivars. Theor. Appl. Genet. 79:77–80.

Gamborg, O.L., Miller, R.A. and Ojima, K. (1968). Nutrient requirement of suspension cultures of soybean root cells. Exp. Cell. Res. 50:151–158.

Genovesi. A.D. and Collins G.B. (1982). *In vitro* production of haploid plants of corn via anther culture. Crop Sci. 22:137–1144.

Graner, A., Siedler, H., Jahoor, A., Herrmann, R.G. and Wenzel, G. (1990). Assessment of the degree and the type of restriction fragment length polymorphism in barley (*Hordeum vulgare*). Theor. Appl. Genet. 80:826–832.

Guha, S. and Maheshwari, S.C. (1966). Cell division and differentiation of embryos in the pollen grains of *Datura* in vitro. Nature 212:97.

Han-min, Y., Keppenne, V.D., Baenziger, P.S., Berke, T. and Liang, G.H. (1990). Effect of genotype and medium on wheat (*Triticum aestivum* L.) anther culture. Plant Cell Tissue Org. Cult. 21:253–258.

Henry, Y. and de Buyser, J. (1981). Float culture of wheat anthers. Theor. Appl. Genet. 60:77–79.

Henry. Y. and de Buyser. J. (1985). Effect of the 1B/1R translocation on anther culture ability in wheat. (*Triticum aestivum* L.). Theor. Appl. Genet. 4:307–310.

Higgins. P. and Mathias, R.J. (1987). The effect of the 4B chromosomes of hexaploid wheat on the growth and regeneration of callus cultures. Theor. Appl. Genet. 74:439–444.

Ho, K.M. and Jones, G.E. (1980). Mingo barley. Can. J. Plant Sci. 60:279–280.

Hoffmann, F., Thomas, E. and Wenzel, G. (1982), Anther culture as a breeding tool in rape. II. Progeny analysis of androgenetic lines and of induced mutants from haploid cultures. Theor. Appl. Genet. 61:225–232.

Huang, C.S., Tsay, H.S., Chern, C.G., Chen, C.C., Yeh, C.C. and Tseng, T.H.(1988). Japonica rice breeding using anther culture. J. Agric. Res. China 37:1–8.

Jensen, C.J. (1975), Barley monoploids and doubled monoploids: technique and experience. In: Barley Genetics. III, H. Gaul (ed.), Karl Thiemig, München, pp. 316–345.

Johannsen,W. (1926).In: Elemente der exakten Erblichkeitslehre. G. Fischer Verlag, Jena.

Kao, K.N. (1981). Plant formation from barley anther cultures with Ficoll-media. Z. Pflanzenphysiol. 103:437–443.

Kasha, K.J. and Kao, K.N. (1970). High frequency haploid production in barley (*Hordeum vulgare* L.) Nature 225:874–876.

Kasperbauer, M.J., Bucknerand, R.C. and Springer, W.D. (1980). Haploid by anther-panicle culture of tall fescue *Festuca arundinacea*. Crop. Science 20:103–107.

Keller, J. (1990). Haploids from unpollinated ovaries of allium cepa-single plant screening, haploid determination, and long term storage. In: Progress in Plant Cellular and Molecular Biology. H.J.J. Nijkamp, L.H.W. Plas, J. van der Aartri (eds.), Kluwer Academic Publishers, Dordrecht, pp. 275–279.

Keller, W.A. (1984). Anther culture of *Brassica*. In: Cell Culture and Somatic Cell Genetics of Plants. I.K. Vasil (ed.), Academic Press, Orlando, pp. 302–310.

Keller. W.A., Rajhathy, F. and Lacapra, J. (1975). *In vitro* production of plants from pollen of *Brassica campestris*. Can. J. Gen. Cytol. 17:655–666.

Kim, M.Z. and Raghavan, V. (1988). Induction of pollen plantlets in rice by spikelet culture. Plant Cell Rep. 7:560- 563.

Knudsen, S., Due, I.K. and Andersen, S.B. (1989). Components of response in barley anther culture. Plant Breed. 103:241–246.

Köhler. F. and Wenzel, G. (1985). Regeneration of isolated barley microspores in conditioned media and trials to characterize the responsible factor. J. Plant. Physiol. 121:181–191.

Krüger, H.-U. (1987). Zytologische Charakterisierung androgenetischer Entwicklungsphasen bei Weizen (*Triticum aestivum* L.). Arch. Züchtungsforch. 17:297–307.

Kuhlmann, U. (1991). Mikrosporenregeneration und Untersuchungen zur Transformation bei Gerste. Dissertation, TU München.

Kuhlmann, U. and Foroughi-Wehr, B. (1989). Production of doubled haploid lines in frequencies sufficient for barley breeding programs. Plant Cell Rep. 8:78–81.

Last, D.I. and Brettell, I.S. (1990). Embryo yield in wheat anther culture is influenced by the choice of sugar in the culture medium. Plant Cell Rep. 9:14–16.

Laurie, D.A. and Bennett, M.D. (1987). The effect of the crossability loci Krl and Kr2 on fertilization frequency in hexaploid wheat × maize crosses. Theor. Appl. Genet. 73:403–409.

Lazar, M.D., Schaeffer. G.W. and Baenziger, P.S. (1985). The physical environment in relation to high frequency callus and plantlet development in anther culture of wheat (*Triticum aestivum* L.) cv. Chris. J. Plant Physiol. 121:103–109.

Lichter. R. (1982). Induction of haploid plants from isolated pollen of *Brassica napus*. Z. Pflanzenphysiol. 105:427–434.

Linsmaier, E.M. and Skoog, F. (1965). Organic growth factor requirements of tobacco tissue cultures. Physiol. Plant. 18:100–127.

Marsolais, A.A., Seguin-Swarts, G. and Kasha, K.J. (1984). The influence of anther cold pretreatments and donor plant genotype on *in vitro* androgenesis in wheat (*Triticum aestivum* L.). Plant Cell Tissue Org. Cult. 3:69–79.

Müller, G., Borschel. H., Vahl, U., Wiberg, A., Härtel, H. und Damisch, W. (1989). Die Nutzung der Antherenkulturmethode im Zuchtprozeß von Winterweizen I. Zur Androgenesefähigekeit von 1B-1R-Weizen-Roggen-Translokationsformen. Plant Breed. 102:196–207.

Muren, R.C. (1989). Haploid plant induction from unpollinated ovaries in onion. Hort. Sc. 24:833–834.

Murashige, T. and Skoog, F. (1962). A revised medium for rapid growth and bio assays with tobacco tissue cultures. Physiol. Plant. 15:473–497.

Nitsch, C. (1974). La culture de pollen isolé sur mllieu synthetique. CR. Acad. Sci., Ser D278:1031–1034.

Orchinsky, B.R., McGregor, L.J., Johnson. G.I.E., Hucl, P. and Kartha, K.K. (1990). Improved embryoid induction and green shoot regeneration from wheat anthers cultured in medium with maltose. Plant Cell Rep. 9:365–369.

Ouyang, J.W., Zhou, S.M. and Jia, S.E. (1983). The response of anther culture to culture temperature in *Triticum aestivum*. Theor. Appl. Genet. 66:101–109.

Ouyang, J.W., He, D.G., Feng, G.H. and Jia, S.E. (1987). The response of anther culture to culture temperature varies with growth conditions of anther-donor plants. Plant Sci. 49:145–148.

Pareddy, D.R. and Greyson. R.J. (1985). *In vitro* culture of immature tassels of an inbred. field variety of *Zea mays* cv.Oh43. Plant Cell Tissue Org. Cult. 5:119–128.

Pescitelli, S.M., Mitchell, J.C., Jones, A.M., Pareddy. D.R. and Petolini, J.F. (1989). High frequency androgenesis from isolated microspore of maize. Plant Cell Rep. 7:673–676.

Petolino, J.F., Jones, A.M. and Thompson, S.A. (1988). Selection for increased anther culture response in maize. Theor. Appl. Genet. 76:157–159.

Powell, W. (1988). The influence of genotype and temperature pre-treatment on anther response in barley (*Hordeum vulgare* L.). Plant Cell Tissue Org. Cult. 12:291–297.

Raquin, C. (1983). Utilization of different sugars as carbon sources for *in vitro* cultures of *Petunia*. Z. Pflanzenphysiol. 111:453–457.

San Noeum, L.H. (1976). Haploides d'Hordeum vulgare 1. par culture *in vitro* d'ovaries non fécondés. Ann. Amélior. Plantes 26:751–754.

Shannon, P.R.M., Nicholson, A.E., Dunwell, J.M. and Davies, D.R. (1985). Effect of anther orientation on microspore-callus production in barley (*Hordeum vulgare* L.). Plant Cell Tissue Org. Cult. 4:271–280.

Siebel, J. and Pauls, K.J. (1989). A comparison of anther and microspore culture as a breeding tool in *Brassica napus*. Theor. Appl. Genet. 78:473–479.

Snape, J.W., Chapman, V., Moss, J., Blanchard, C.E. and Miller, T.E. (1979). The crossabilities of wheat varieties with *Hordeum bulbosum*. Heredity 42:291–298.

Snaper J.P. and Simpson, E. (1981). The genetical expectations of doubled haploid lines derived from different filial generations. Theor. Appl. Genet. 60:123–128.

Sorvari, S. (1986). Comparison of anther cultures of barley cultivars in barley-starch and agar gelatinized media. An. Agr. Fenniae 25:249–254.

Sunderland, N. and Evans L.J. (1980). Multicellular pollen formation in cultured barley anthers, II: The A-pathway, B-pathway and C-pathway. J. Exp. Bot. 31:501–514.

Sunderland. N. and Xu. Z.H. (1982). Shed pollen culture in *Hordeum vulgare*. J. Exp. Bot. 33:1086–1095.

Sunderland, N. and Xu, Z.H. and Huan, B. (1981). Recent advances in barley anther culture. Barley Genet. VI; Proc. 4. Int. Barley Genet. Symp., Edinburg Uni. Press, pp. 699–703.

Sunderland, N., Roberts, M., Evans, L.J. and Wildon, D.C. (1979). Multicellular pollen formation in cultured barley anthers. I. Independent division of the generative and vegetative cells. J. Exp. Bot. 30:1133–1144.

Tsay, H.S., Miao, S.H. and Widholm, J.M. (1986). Factors affecting haploid plant regeneration from maize anther culture. J. Plant Physiol. 126:33–40.

Uhrig, H. (1985). Genetic selection and liquid medium conditions improve the yield of androgenetic plants from diploid potatoes. Theor. Appl. Genet. 71:455–460.

Uhrig, H. and Salamini. F. (1987) Dihaploid plant production from 4X-genotypes of potato by the use of efficient anther plants producing tetraploid strains (4X EAPP-clones) – proposal of a breeding methodology. Plant Breed. 98:228–235.

Uhrig, H. and Wenzel, G. (1987). Breeding for virus and nematode resistance in potato through microspore culture. In: Biotechnology in Agriculture & Forestry, Vol. 3. Y.P.S. Bajaj (ed.), Springer Verlag, Berlin, pp. 346–357.

Wan, Y., Petolino, J.F. and Widholm, J.M. (1989). Efficient production of doubled haploid plants through colchicine treatment of anther-derived maize callus. Theor. Appl. Genet. 77:889–892.

Wark; D.C. (1977). Doubled haploids in Australian tobacco breeding. SABROJ 2:19–23.

Wei, Z.M. (1982). Pollen callus culture of *Triticum aestivum*. Theor. Appl. Genet 63:71–73.

Wenzel, G. (1980). Resent progress in microspore culture of crop plants. In: The Plant Genome. D.R. Davies and D.A. Hopwood, (eds.), John Innes Charity, Norwich, pp. 185–196.

Wenzel, G. and Foroughi-Wehr, B. (1984). Anther culture in cereals and grasses. In: Cell Culture and Somatic Cell Genetics of Plants. I.K. Vasil (ed.), Academic Press, New York, pp. 311–327.

Wenzel, G., Hoffmann, F.L and Thomas, E. (1977). Increased induction and chromosome doubling of androgenetic haploid rye. Theor. Appl. Genet. 51:81–86.

Wenzel, G. and Uhrig, H. (1981). Breeding for virus and nematode resistance in potato via anther culture. Theor. Appl. Genet. 59:333–340.

Wilson, H.M. (1977). Culture of whole barley spikes stimulates high frequencies of pollen calluses in individual anthers. Plant Sci. Lett. 9:233–238.

Yang, H.Y. and Zhou, C. (1982). *In vitro* induction of haploid plants from unpollinated ovaries and ovules. Theor. Appl. Genet. 63:97–104.

Yang, J.S., Sohn, J.K. and Chung, G.S. (1985). Effects of low temperature treatment on culturability and ploidy level in rice anther culture. Korean J. Breed. 17:344–347.

Zhu, Z.C., Wu, H.S., An. Q.K. and Liu, Z.Y. (1981). Induction of haploid plantlets from unpollinated ovaries of *Triticum aestivum* culture *in vitro*. Acta Genet. Sinic. 8:386–389.

8. *In vitro* Methods for the Control of Fertilization and Embryo Development

V. Raghavan

Contents

1. Introduction

An overruling fact about sexual reproduction in angiosperms is that it is the sole means for the production of seeds and fruits. Towards this goal, angiosperms rely upon pollination of the stigma of the flower by pollen grains, fertilization involving genetic recombination between one of the male gametes and the egg and triple fusion between the second male gamete and the polar fusion nucleus. The product of union of the sperm and the egg known as the zygote is the progenitor of the embryo while the product of triple fusion is transformed into a nutritive tissue known as the endosperm. If pollination is compatible, pollen germination, pollen tube growth, fertilization, development of the embryo and endosperm and seed and fruit formation ensue according to a genetic blueprint characteristic of each species. Thus, in a sense, the initial pollen-stigma interaction determines whether a flower will complete its prescribed function of seed and fruit production or succumb at some early stage in its ontogeny.

In traditional plant breeding programs involving incompatible pollinations and the transfer of beneficial alien genes across interspecific or intergeneric barriers, it has been found that pre-fertilization blocks such as the failure of pollen grains to germinate on the stigma or physiological disturbances in the

Indra K. Vasil and Trevor A. Thorpe (eds.), Plant Cell and Tissue Culture, 173–194.

directional growth of the pollen tube through the style towards the vicinity of the egg often thwart seed-set in plants. In some wide hybridizations, despite the occurrence of normal pollination and fertilization, deleterious genes carried forward into the zygote cause disturbances in the equilibrium between the growth of the maternal tissues, embryo and endosperm leading to embryo lethality and seed collapse. A logical deduction from these considerations is that future success in our agricultural practices would depend upon devising techniques to overcome the barriers to pollination, fertilization and embryo and endosperm development in plants.

In vitro methods to achieve pollination, fertilization and embryo growth have attracted a great deal of attention in the last few years. This has in part been due to the ease of handling the reproductive organs of the flower in aseptic culture and the introduction of methods to secure pollen germination and embryo growth under defined conditions. The purpose of this Chapter is to provide an overview of the *in vitro* methods that have been developed to control pollination, fertilization and embryo development in angiosperms. In keeping with the general theme of this volume, primary emphasis will be on the principles and applications with reference to key papers; no attempt has been made to achieve a comprehensive coverage. Excellent and detailed protocols for *in vitro* fertilization and embryo culture are given by Yeung et al. (1981).

2. Control of Pollination and Fertilization

In vitro methods are generally geared to favor the encounter of the sperm and the egg, bypassing stigmatic pollination and pollen tube growth in the style, and to lead to fertilization and seed development. This may be attained in different ways. In one of the first published full investigations of its kind, Bosio (1940) was able to induce fertilization of ovules of *Helleborus* and *Paeonia* by excising the style and depositing pollen grains on the upper part of the ovary. It was not until 1960 when Kanta (1960) attempted to induce seed set in *Papaver rhoeas* that the full potentialities of this technique, christened as 'intraovarian pollination', were realized. The experimental design involved emasculating receptor flowers before anthesis and injecting into the ovary small quantities of a suspension of pollen grains made in boric acid. This led to the germination of pollen grains, growth of pollen tubes and their direct entry into the ovule and finally into the embryo sac leading to fertilization. Results acquired in terms of seed set demonstrated the success of intraovarian pollination in this species and in other members of Papavaraceae such as *Papaver somniferum*, *Argemone mexicana*, *A. ochroleuca* and *Eschscholtzia californica*. By intraovarian pollination, it was also possible to overcome pre-fertilization incompatibility barriers between the diploid *A. mexicana* and the tetraploid *A. ochroleuca* when either species was used as the female parent (Kanta and Maheshwari, 1963a). Later, Kameya et al.

(1966) demonstrated the suitability of intraovarian pollination to obtain seeds from *Brassica oleracea*. An extension of the method described above is to inject sperm directly into the embryo sac; this approach has been tested with limited success but holds promise in plants where the sperm do not reach the embryo sac after intraovarian pollination (Keijzer et al., 1988). For a historical perspective on intraovarian pollination, see Tilton and Russell (1984).

2.1. *Test-Tube Fertilization*

If the intent is to follow fertilization and embryogenesis in the ovule under total *in vitro* conditions, pre-fertilization ovules must be cultured, pollinated and monitored stage by stage. It was in *Papaver somniferum* that this was done first (Kanta et al., 1962). In this technique designated as 'test-tube fertilization', unpollinated ovules were cultured on a nutrient medium and dusted with pollen grains. During the ensuing period of 2–3 weeks, pollen germination, pollen tube growth, fertilization and embryogenesis occur, but because of the microscopic size of the pollen grains and the presence of tissue layers surrounding the embryo sac, one has to depend on fixed and sectioned materials for developmental information. Nonetheless, it has been established that following test-tube fertilization, ovules developed into mature seeds while surviving on the original medium. The success obtained in inducing seed set following test-tube fertilization of ovules of other plants attests to the versatility of the method (Kanta and Maheshwari, 1963b). In some cases whole pistils were cultured and pollinated *in vitro* with results similar to those obtained with cultured ovules. This is analogous to conducting stigmatic pollination *in vitro* and the technique was successful with both self-pollinated species such as *Nicotiana tabacum* (Dulieu, 1966) and cross-pollinated species like *N. rustica* (Rao and Rangaswamy, 1972).

Efficient recovery of seeds following test-tube fertilization is imperative before this technique can be applied in the genetic transformation of plants. From this point of view, several investigators have attempted to identify the conditions for maximizing fertilization and embryo development following *in vitro* fertilization in maize (*Zea mays*). These studies have examined the influence of donor plant genotype (Gengenbach, 1977; Bajaj, 1979; Higgins and Petolino, 1988), the ratio of ovary to cob tissue (Sladky and Havel, 1976; Higgins and Petolino, 1988), culture conditions (Gengenbach, 1977, Jones et al., 1981), pollen quantity (Raman et al., 1980), explant size (Higgins and Petolino, 1988), silk length (Dupuis and Dumas, 1989), stage of spikelet development (Dupuis and Dumas, 1990a) and temperature stress (Dupuis and Dumas, 1990b) on kernel development in *in vitro* pollinated ovules. Based on electron microscopic studies, Schel and Kieft (1986) have called attention to the occasional absence of embryos and frequent abnormalities in endosperm development in cultured ovaries of maize. Although it was not possible to identify a set of conditions as contributing to the low yield of kernels, these studies have shown that the block to kernel development was not in the pollination-fertilization process, but was in the incomplete embryo-endosperm development in the fertilized ovules. The review by Gengenbach (1984) is recommended for additional information.

Test-tube fertilization is an obvious method for raising interspecific and intergeneric hybrids. One of the most convincing examples is the formation of hybrids from the crosses, *Melandrium album* × *M. rubrum* and *M. album* × *Silene schfta* (both Caryophyllaceae) in the test tube and their later transfer to the field to flower (Zenkteler, 1967). Success was however less pronounced with some intergeneric combinations involving *Melandrium* and other genera of Caryophyllaceae. Although most of the cultured ovules in this group were found to favor the alien pollen leading to fertilization, the competence of the zygote to form the embryo was found to fade after a few divisions. Other aberrations observed were bursting of the pollen tubes, degeneration of the egg apparatus and a minimal differentiation of the endosperm (Zenkteler et al., 1975).

The above experiments suggested that one might be able to overcome self-incompatibility in plants by resorting to test-tube fertilization, but this was not achieved until the basic protocol was modified. The modification entailed the culture of placentae with attached ovules together with a short length of the pedicel and pollinating the ovules by dusting them with pollen. This technique was first applied to overcome self-incompatibility in *Petunia axillaris* (Rangaswamy and Shivanna, 1967). The result was prompt germination of pollen grains, rapid growth of the pollen tubes in the style and fertilization of the egg within about two days after pollination. During the next several days, embryo and endosperm development ensued in tandem, leading to the formation of normal seeds. When ovules on one placenta on the ovary were left unpollinated and those on another placenta on the same ovary were pollinated with self- or cross-pollen grains, the unpollinated ovules invariably shriveled and did not set seeds, whereas, both selfed and crossed ovules set seeds. The efficacy of self- and cross-pollination in culture was demonstrated by the absence of differences in seed set when ovules on one placenta were selfed and those on another placenta were crossed (Shivanna, 1971; Rangaswamy and Shivanna, 1971a). In yet another experiment, when stigmatic pollination and placental pollination were performed on the same pistil by simple manipulative techniques, the former was found to lead to seed set only if cross-pollinated, while both self- and cross-pollinations were found to be equally effective with the latter (Rangaswamy and Shivanna, 1971b). These phenomena were collectively termed 'placental pollination' and apart from their intrinsic interest, were considered as evidence that in the absence of the style and stigma, the ovules do not show any preference to the pollen grains they encounter. Placental pollination has also been successfully used to overcome self-incompatibility in *P. hybrida* (Niimi, 1970; Wagner and Hess, 1973). Stewart (1981) has compiled a list of plants in which *in vitro* fertilization and seed development have been achieved by the above methods.

In summary, all of what is known about *in vitro* methods for the control of fertilization in angiosperms comes from experiments performed on eggs enclosed within the ovule. As such they fall short of a true *in vitro* fertilization system as understood in animal embryology. This represents a serious stumbling block to comprehend the process of double fertilization and the physiological reactions in the egg and the sperm at fertilization. However, methods are now at hand to remove embryo sac from angiosperm ovules by enzymatic digestion (Zhou and Yang, 1985; Wagner et al., 1988) and so there is reason to feel optimistic that in a few years we will be able to monitor fertilization in angiosperms using egg and sperm in isolation (see for example, Kranz and Lörz, 1990; Kranz et al., 1991).

3. Embryo Culture

The goal of embryo culture is to rear plants under aseptic conditions from isolated embryos grown in a defined nutrient medium. Although the technique was originally devised to isolate and culture embryos from seeds, in later years, it was extended to the culture of progressively small embryos extracted from immature ovules. There is no organ in an angiosperm more fundamental for the evolution of its sporophyte than the zygote or the embryo derived from it. The transformation of the zygote into an embryo involving cell division, cell enlargement and organ differentiation depends upon the interaction between the newly formed cells on the one hand and the nutrient materials provided by the endosperm and the surrounding tissues of the ovule, on the other. Hence, the importance of the study of factors which influence the metabolism, growth and differentiation of embryos of different ages is unquestioned. This information is best gained by growing embryos outside the environment of the ovule; the continuing quest for this information accounts in a large measure for the wide spread use of embryo culture in experimental embryology. On the practical side, embryo culture method is now extensively used to obtain hybrids from inviable crosses, whose seeds are traditionally condemned and discarded due to their inability to germinate. As alluded to earlier, in plant breeding programs, embryo culture goes hand in hand with *in vitro* control of pollination and fertilization to ensure hybrid production.

Some comments about the technique are in order here, since in many plants, embryo culture is neither simple nor straightforward. It requires isolating the embryo from the ovule without injury, formulating a suitable nutrient medium, and inducing continued embryogenic growth and seedling formation. Since embryos are lodged in the sterile environment of the ovule, their surface-sterilization is not necessary; instead, entire ovules or seeds are sterilized and embryos removed aseptically. Seeds with hard coats are to be soaked in water for a few hours to a few days so that the embryos can be removed with ease. These seeds can be surface-sterilized before or after soaking, although after prolonged soaking they may need to go through another round of sterilization. Splitting open the seeds and transferring embryos to the nutrient medium is perhaps the simplest technique that can be used with seeds. Relatively small embryos are best isolated under a dissecting microscope. Although nutrient media for culture generally vary with the species, immature embryos and proembryos require more complex nutrients than seed embryos. Embryos are generally cultured m vitro up to the plantlet stage, after which they are transferred to sterilized soil or vermiculite and grown to maturity in the greenhouse.

In this section, the results of studies on the fundamental and applied aspects of embryo culture will be highlighted. For recent reviews on this subject, see Bhojwani and Razdan (1983), Raghavan (1986b), Williams et al. (1987) and Pierik (1987); the older literature is summarized by Raghavan (1976).

3.1. *Culture of Mature and Differentiated Embryos*

One aspect of the culture of mature and differentiated embryos of the kind excised from seeds that is often emphasized is their simple nutritional requirement. This has been observed a number of times in the many years since Hannig (1904) cultured embryos excised from seeds of *Raphanus* and *Cochlearia* (Cruciferae) and grew them into plantlets in a medium containing mineral salts and sucrose. However, a rigorous foundation for a medium for culture of seed embryos was laid only after advancements in the general field of plant tissue culture made it possible to culture with relative ease cells, tissues and organs of plants under controlled conditions in defined media. Table 3 in Raghavan (1977a) lists the composition of several media that have been successfully employed to culture seed embryos of various plants; this list has been updated by Williams et al. (1987). Included in the media are such elements as K, Ca, Mg, N, P and S which have been found essential for plant growth. Later, as the role of trace elements such as Fe, B, Mn, Cu, Zn and Mo in the nutrition of plants was recognized, various combinations of these elements were also added. Vitamins are another group of compounds used in embryo culture media, although there is no hard evidence to justify their essentiality. Besides major salts, trace elements and vitamins, it is the general experience that growth and survival of cultured embryos are enhanced by fortifying the medium with a carbon energy source. The superiority of sucrose in this role has been established for embryos of several plants and rarely has any other carbohydrate been as successful (Tables 4, 5 in Raghavan, 1977a). At the morphogenetic level, addition of sucrose to the medium has been shown to promote root growth (Rietsema et al., 1953;

Honma, 1955; Ozsan and Cameron, 1963; del Rosario and de Guzman, 1978) or the growth of both shoot and root primordia (Augusten, 1956; Buffard-Morel, 1968) of cultured embryos.

To the extent that supplementation of the medium with different nitrogen compounds tells us something about the amino acid metabolism of cultured embryos, a few generalizations have emerged. One is that embryos are able to grow moderately well in a medium utilizing nitrates or ammonium salts as the sole source of nitrogen. Second is the demonstration that the amide glutamine is an efficient source of nitrogen for the growth of embryos of a number of plants (Paris et al., 1953; Rijven, 1956; Matsubara, 1964). Third is the existence of mutual antagonism and synergism between different amino acids in the growth of embryos. For example, growth of cultured embryos of barely, wheat (Bright et al., 1978) and maize (Green and Donovan, 1980) is inhibited by the simultaneous addition of lysine and threonine and the inhibition alleviated by methionine, homocysteine or homoserine. Since these observations suggest that embryos are starved for methionine and its metabolic intermediates by feedback inhibition of enzymes of the aspartate pathway, they are of interest for possible selection of feedback-sensitive mutants.

Following the discovery of plant hormones, especially of auxins, gibberellins and cytokinins, there has been a concerted effort to study their effects on the growth and morphogenesis of cultured seed embryos. The results of these studies have been reviewed periodically (Raghavan, 1980, 1986b; Raghavan and Srivastava, 1982) and will not be considered here again. Suffice it to state that although the addition of hormones modifies the growth of cultured embryos, there is no evidence to demonstrate the need for a specific hormone for the growth of embryos of a wide range of plants. As far as the effects of hormones on the growth of embryonic organs are concerned, they are quite similar to those displayed by the corresponding organs of a seedling plant (see, for example, Neal and Topoleski, 1985). If there is a lesson to be learned from these investigations, it is that the basis for hormone action on the embryo is probably no different from that on a seedling plant.

In a recent work, Nyman et al. (1986) found that seed embryos of taro (*Colocasia esculenta*) developed into plantlets only when cultured in the presence of the endosperm including an intact aleurone layer. Indicative of a role for the endosperm in supplying hormonal substances for embryo growth, the authors have subsequently shown that NAA and 6-dimethylaminopurine can replace the endosperm effects on cultured embryos (Nyman et al., 1987).

The involvement of auxin, especially 2,4-dichlorophenoxyacetic acid in the production of callus on cultured embryos is now firmly established. With the demonstration that the callus can be induced to produce somatic embryos, considerable interest has been generated in the possibility of mass production of clonal seedlings of various plants, with the seed embryo as the starting material. For a review of this work, see Williams et al. (1987).

3.2. *Culture of Immature Embryos*

A fundamental feature of the growth of seed embryos described above that may serve as a basis for comparison with embryos at earlier stages of development is the proportionality of growth of the different organs to produce a normally functional seedling. Immature embryos, in contrast, change their pattern of growth drastically in culture and evolve into weak seedlings by a process known as precocious germination. This observation made during the early days of embryo culture investigations by Dieterich (1924) has been repeatedly noticed by later workers. There is also firm evidence that some of the biosynthetic activities such as storage protein accumulation normally associated with late embryogenesis are curtailed in precociously germinating embryos (Dure and Galau, 1981; Crouch and Sussex, 1981). This is no doubt due to the fact that these embryos are deprived of a natural developmental hiatus leading to quiescence or dormancy before reinitiation of growth.

An arresting observation by Norstog (1972) that precocious germination of cultured barley (*Hordeum vulgare*) embryos can be inhibited by the plant hormone abscisic acid (ABA) is relevant at this point. This result which has been confirmed in other plants such as *Brassica napus* (Crouch and Sussex, 1981; Wilen et al., 1991), *Phaseolus vulgaris* (Long et al., 1981), *Triticum aestivum* (Triplett and Quatrano, 1982), *Glycine max* (Ackerson, 1984), *Zea mays* (Rivin and Grudt, 1991) and *Linum usitatissimum* (Wilen et al., 1991) raises the possibility that ABA is a natural factor that suppresses precocious germination during normal embryogenesis. At the molecular level, a manifestation of the control of precocious germination by ABA is the reinstatement of the synthesis of storage proteins (Dure et al., 1981; Long et al., 1981; Triplett and Quatrano, 1982; Rivin and Grudt, 1991; Wilen et al., 1991).

In some species, the embryo begins to germinate while the seed is still attached to the plant, a phenomenon known as vivipary. Since vivipary is analogous to precocious germination *in planta*, it is in some way related to a decreased ABA content of the seed or to a decreased ABA-sensitivity of the embryo. The latter has been found to be the case. In the embryo of viviparous mutants of maize, the genetic lesion apparently confers a greater resistance to growth inhibition by exogenous ABA than in the embryo of the wild type (Robichaud et al., 1980; Robichaud and Sussex, 1986).

In a sense these observations strike at the heart of the question of how ABA acts in plant cells. It is generally noted that embryos germinate precociously when their water content is high and that they lapse into quiescence when their water content is low. A conclusion suggested by these results is that ABA might regulate cell metabolism during embryogenesis by its effects on the water potential of the cells (Walbot, 1978).

Table 1. Progressive embryogenesis in *Capsella*

Developmental stage	Length (μm)	Nutritional requirements
Early globular	0–60	Unknown for embryos <40 μm
Late globular	61–80	Basal medium + indoleacetic acid (0.1 mg/1), kinetin (0.001 mg/1) and *adenine sulfate* (0.001 mg/1)
Early heart-shaped	81–150	
Late heart-shaped	151–250	
Intermediate-stage	251–450	Macronutrient salts + 2% sucrose + vitamins + *trace elements* (basal medium)
Torpedo-shaped	451–700	Macronutrient salts + 2% sucrose + vitamins
Walking-stick-shaped	701–1000	
Inverted-U-shaped	1001–1700	
Mature embryo	>1700	Macronutrient salts + 2% sucrose

Additional nutrients added and/or the synthetic pathways probably becoming active at different stages of embryogenesis are italicized.

Modified from Raghavan (1965).

3.3. *Culture of Proembryos*

Over the years, additional efforts in embryo culture have involved the culture of proembryos – a term applied to embryos at developmental stages that precede cotyledon initiation. In practice, early division phase embryos such as globular and heart-shaped stages are considered as proembryos, although in some cases it is difficult to draw a dividing line between the end of proembryo phase and the beginning of the maturation phase. An implicit assumption in most investigations is that during their growth *in ovulo* proembryos are heterotrophic in nature and are dependent not only upon the metabolites present in their cells but also upon those diffusing from the endosperm and the surrounding maternal tissues. From the embryo culture perspective, this suggests that nutritional requirements for the culture of proembryos are bound to be more exacting than those required for immature and mature and differentiated embryos. A summary of the nutritional requirements for growth in culture of progressively small embryos of *Capsella bursa-pastoris* illustrates this point (Table 1). Although previous workers (Rijven, 1952; Veen, 1963) successfully cultured heart-shaped and older embryos of *C. bursa-pastoris* in an inorganic liquid medium of high osmolality secured by the addition of 12–18% sucrose, later work showed that it is possible to culture heart-shaped embryos in an agar-solidified mineral salt medium supplemented with 2% sucrose. As shown in Table 1, growth of still smaller embryos was secured by fortifying this medium with IAA, kinetin and adenine sulfate (Raghavan and Torrey, 1963). Monnier (1976) found that the high-salt Murashige-Skoog medium was also satisfactory for the culture of proembryos of *C. bursa-pastoris*, although best effects were obtained by modifying the medium by increasing the concentrations of K and Ca salts and decreasing the concentration of NH_4NO_3. In view of the changing nutritional requirements of growing embryos, a culture technique was devised which allows initial growth of proembryos in the revised medium and their later growth in a medium of a slightly different composition. A role for kinetin in inducing growth of proembryos of flax (*Linum usitatissimum*) (Prĕtová, 1986) and for IAA and benzylaminopurine for growth of heart-shaped embryos of *Medicago sativa* and *M. scutella* (Bauchan, 1987) has also been reported.

An idea established from the work of van Overbeek et al. (1942) is the use of nutritionally rich substances of endospermic origin such as coconut milk for the culture of proembryos. From their early studies, there seemed to have been a general agreement that it is possible to grow to plantlet stage torpedo-shaped and heart-shaped embryos of *Datura stramonium* in an inorganic medium enriched with a mixture of vitamins (nicotinic acid, thiamine, pyridoxine, ascorbic acid, pantothenic acid) and an assortment of organic substances (adenine, glycine, succinic acid). Although proembryos failed to grow or grew feebly in this medium before callusing, supplementation of the medium with nonautoclaved coconut milk – the liquid endosperm of coconut – led to a recrudescence of growth in them. This was attributed to the presence of a hormonal factor, designated as 'embryo factor' which was later obtained in a sufficiently pure form to promote growth of embryos in a dilution as low as 1:19,000 on a dry weight basis as compared with 1:110 for the crude milk (van Overbeek et al., 1944). These studies have spawned several successful attempts to culture proembryos of other plants by the use of plant extracts of endospermic and nonendospermic origin (Kent and Brink, 1947; Norstog, 1961; Matsubara, 1962; Nakajima, 1962) or by using the endosperm as a nurse tissue (Stefaniak, 1987).

Some attention has also been bestowed on the physical factors of the milieu of the embryo sac in the growth of proembryos in culture. It has been known for some time that the liquid endosperm bathing the proembryo has a very low (more negative) osmotic potential that substantially decreases (becomes more negative) as the embryo matures (Ryczkowski, 1960; Yeung and Brown, 1982). This has led to the thought that the osmotic pressure of the medium might play a role in promoting the growth of proembryos *in vivo* as well as *in vitro*. In line with this, artificially increasing the osmotic concentration of the culture medium by the addition of high concentrations of sucrose or mannitol lent itself to the culture of proembryos of some plants which did not previously grow even in the most complex media tested (Rietsema et al., 1953; Prĕtová, 1974; Stafford and Davies, 1979; Cook et al., 1988). It is not known how a high osmolality of the medium promotes the growth of proembryos, but its effectiveness supports a possible mechanism that controls the traffic of metabolites and inorganic ions into embryo cells that are starved of them.

A structural feature of the proembryo that is often ignored, but which might provide a clue to its growth potential is the suspensor. It is a filament made up of a variable number of cells and is generally terminated at the chalazal end by the organogenetic part of the embryo and is attached to the embryo sac at the micropylar end. Recent electron microscopic investigations have shown that the suspensor cells possess morphological features of transfer cells serving to facilitate the absorption of metabolites from the endosperm (for a review see Raghavan and Sharma, 1994). This has raised the question whether there is a causal relationship between the presence of the suspensor and growth of the embryo. Some leads have come from studies in which the growth of the embryo severed of its connection with the suspensor is followed. This is no doubt an exacting task and has been conducted only with only a few species, perhaps involving a limited number of embryos. Nonetheless, experiments of this kind have shown that continued growth in culture of proembryos of *Eruca sativa* (Corsi, 1972), *Phaseolus coccineus* (Nagl, 1974; Yeung and Sussex, 1979) and *Capsella bursa-pastoris* (Monnier, 1984), is enhanced in the presence of an attached suspensor than in its absence. The experiments of Yeung and Sussex (1979) are particularly revealing for showing that growth of the embryo is promoted even by the presence of a detached suspensor kept in close vicinity. Other experiments with *P. coccineus* involving supplementation of the medium with growth hormones and determination of growth hormone levels in the embryo and suspensor cells at specified stages of embryogenesis have provided indirect evidence to show that the presumed suspensor function is facilitated by the production of gibberellins and cytokinins (Yeung and Sussex, 1979; Alpi et al., 1975; Cionini et al., 1976; Lorenzi et al., 1978; Bennici and Cionini, 1979). These results are important because they provide a strong argument for a role for hormonal gradients from the suspensor in promoting growth of proembryos. This might serve as base line information for concocting a medium for growing *in vitro* still smaller embryos and even the fertilized egg.

Since small embryos must be isolated in order to be cultured, the difficulty of extracting them from the inner sanctum of the ovule could well be one of the reasons why there are so few reports of successful culture of proembryos. From this perspective, ovule and ovary culture has proved to be particularly rewarding for the analysis of the growth requirements of proembryos of certain plants that have hitherto defied attempts at excision and culture. An excellent summary of the work in this area is provided by Bhojwani and Razdan (1983).

As in embryo culture, much effort in ovule culture has been devoted to devising an appropriate medium. This became evident from the early studies. Maheshwari (1958), Maheshwari and Lal (1961) obtained only limited success in the culture of isolated ovules of *Papaver somniferum* containing the zygote or two-celled proembryo in a mineral salt medium containing 5% sucrose; however, supplementation of the medium with casein hydrolyzate, yeast extract, or kinetin led to a rapid growth of the enclosed embryo. Other examples illustrating this theme for successful growth of early stage embryos in cultured ovules are *Zephyranthes* (Kapoor, 1959), *Capsella bursa-pastoris* (Lagriffol and Monnier, 1985) and *Triticum aestivum* (Zenkteler and Nitzsche, 1985).

Accelerated growth of the ovular tissues in relation to that of the enclosed embryo or vice versa may also determine the success or failure of ovule culture as a means of obtaining seed set. This is clearly shown in attempts to obtain embryo growth in cultured ovules of cotton (*Gossypium*) enclosing the zygote. Stewart and Hsu (1977) found that the key to the successful culture of post-fertilization ovules of cotton was the use of a medium containing low concentrations of IAA, kinetin and GA along with 15 mM NH_4^+. Evaluation of the effects of the components of the medium showed that the addition of the hormones enables the ovule to attain a normal size whereas NH_4^+ promotes continued growth and division of the zygote.

Following upon the success with ovule culture, it is not surprising to see that it is also possible to raise viable seeds starting with excised and cultured ovaries of various plants (Bhojwani and Razdan, 1983). An outgrowth of these studies has been a reconsideration of the role of the calyx and corolla (perianth) in the post-fertilization development of the flower. Based on the work with both monocots and dicots it now appears that proper and complete development of the embryo in cultured ovaries depends upon the presence of the floral envelopes. What substances emanate from the vegetative parts of the flower to cause embryo growth is however obscure.

3.4. *Applications*

The principal applications of techniques directly related to the isolation and culture of embryos appear to be in the area of overcoming embryo abortion in inviable hybrids. As alluded to earlier, because of the transfer of deleterious genes across interspecific and intergeneric barriers, hybrid seeds suffer disturbances during embryogenesis and endosperm formation. In developmental terms, as shown in a cross between *Trifolium ambiguum* and *T. repens*, the embryo did not differentiate beyond the heart-shaped stage in the hybrid seeds and it began to disintegrate soon after attaining that stage. A similar developmental arrest also occurred in the endosperm. Although the division of the endosperm nucleus ceased at different stages in the hybrid seed, in no case did endosperm development proceed beyond the 128-nucleate level. In contrast, the selfed seed had a typical cellular endosperm with characteristic haustoria (Williams and White, 1976). Similar correlations between embryo abortion and the onset of endosperm deterioration have also been described in inviable crosses in a number of other plants to indicate that either endosperm malfunction is one of several events that trigger embryo abortion or that endosperm behavior may stimulate embryo abortion caused by some other events.

Because developing embryos are heterotrophic and depend upon the nutritional supplies of the endosperm, it was naturally thought that hybrid embryos would complete their development if they are grown in an artificial milieu containing nutrients that are normally supplied by the endosperm. As early as 1925, Laibach showed that progeny from crosses between *Linum perenne* and *L. austriacum* can be rescued by excising and culturing embryos before they disintegrate. Since this pioneering work, embryo culture methods have been widely used to obtain transplantable seedlings from seeds of interspecific and intergeneric crosses previously considered incapable of germination. In some cases, growth of the hybrid embryo has been secured by implanting it on a normal endosperm which is then cultured, thus initiating a nurse tissue culture. Embryos of several hybrids have also been rescued by culturing ovules and ovaries.

Numerous examples of hybrid rescue by embryo culture have been described or tabulated in reviews written on this subject (Raghavan, 1976, 1977a, b, 1985, 1986a, b; Collins and Grosser, 1984; Williams, 1987) which have also been updated with references to later papers. Included in these lists are examples of interspecific hybrids in *Abelmoschus*, *Allium*, *Arachis*, *Brassica*, *Carthamus*, *Cerasus*, *Chrysanthemum*, *Corchorus*, *Cucumis*, *Datura*, *Gossypium*, *Hordeum*, *Iris*, *Lathyrus*, *Lilium*, *Lotus*, *Lycopersicon*, *Medicago*, *Melilotus*, *Ornithopus*, *Oryza*, *Phaseolus*, *Solanum*, *Trifolium* and *Vigna* and intergeneric hybrids between *Agropyron tsukushiense* × *Hordeum bulbosum*, *Hordeum* × *Secale*, *Ribes nigrum* × *Grossularia reclinata*, *Tripsacum dactyloides* × *Zea mays* and *Triticum* × *Aegilops*, *Elymus* and *Secale*. Some recent additions include interspecific hybrids between *Phaseolus vulgaris* × *P. angustissimus* (Belivanis and Doré, 1986), *Vigna pubescens* × *V. unguiculata* (Fatokun and Singh, 1987), *V. radiata* × *V. glabrescens* (Chen et al., 1989), *Cucumis metuliferus* × *C. anguria* (Fassuliotis and Nelson, 1988), *Glycine max* × *G. tomentella* (Coble and Schapaugh, 1990; Chung and Kim, 1990) and *Solanum acaule* × *S. tuberosum* (Iwanaga et al., 1991) and intergeneric hybrids between *Hordeum californicum* × *Triticum aestivum* (Gupta and Fedak, 1985a), *Hordeum* × various species of *Agropyron* (Fedak, 1985a, b; Gupta and Fedak, 1985b, c), *Triticum aestivum* × *Leymus innovatus* (Plourde et al., 1989), *Elymus canadensis* × *Psathyrostachys juncea* (Park and Walton, 1990), *T. aestivum* × *P. juncea* (Plourde et al., 1990), *T. aestivum* × *Agropyron intermedium*, *A. elongatum* and *A. trichophorum* (Sharma and Ohm, 1990), *Brassica napus* × *Raphanobrassica* (Agnihotri et al., 1990b) and *Eruca sativa* × *B. campestris* (Agnihotri et al., 1990a).

Transplanting the hybrid embryo into a normal endosperm eliminates the need to formulate a nutrient medium to foster embryo growth in culture, yet the embryo is able to draw upon the nutrients of a healthy endosperm. In the embryo implantation method described by Kruse (1974) to rescue hybrid embryos from intergeneric crosses between *Hordeum* and *Agropyron*, *Secale* and *Triticale*, the embryo is removed from the dehulled grain at the same time as the endosperm is extruded from the pericarp. The endosperm, either singly or in groups of two to four, harboring 5 to 10 hybrid embryos is then aseptically transferred to the surface of a solidified mineral salt medium. Best growth of relatively small hybrid embryos and high yield of hybrids were obtained when the donor endosperm was cultured in a medium containing 25% coconut milk. In the work on interspecific hybrids in pasture legumes such as *Lotus*, *Ornithopus* and *Trifolium*, excised embryos are enveloped in the endosperm dissected from normal seeds and cultured in a medium whose composition depended upon the stage of development of the embryo. Generally, early heart-shaped embryos required a medium containing casein hydrolyzate and vitamins while advanced stage embryos were satisfied with a less complex medium (Williams and de Lautour, 1980).

It was noted earlier that ovules excised at the stage of the fertilized egg or early division phase embryos develop in culture and produce viable seeds. This discovery has prompted some investigators to use ovule culture as an alternate method for rearing embryos from unsuccessful crosses. Two of the widely investigated genera from which interspecific hybrids have been generated by ovule culture are *Gossypium* (Stewart and Hsu, 1978; Gill and Bajaj, 1984; Thengane et al., 1987) and *Nicotiana* (Reed and Collins, 1978; Iwai et al., 1985, 1986; Shintaku et al., 1988). Ovule culture has also been employed to overcome partial cross-incompatibility observed between tissue-culture derived tetraploid tobacco and its diploid progenitor (Contolini and Hughes, 1989). Other examples involving the use of ovule culture to generate desirable hybrids worth mentioning are interspecific hybrids in *Lens* (Cohen et al., 1984), *Impatiens* (Arisumi, 1985), *Medicago* (McCoy, 1985; McCoy and Smith, 1986) and *Cuphea* (Mathias et al., 1990) and intergeneric hybrids between wheat and pearl millet (Ahmad and Comeau, 1990).

Another technique to raise hybrids is based on the culture of ovaries instead of ovules. This has been elegantly demonstrated in several studies involving various species of *Brassica* (Inomata, 1978; Mohapatra and Bajaj, 1988; Sarla and Raut, 1988). In a further extension of this technique, intergeneric hybrids between *Diplotaxis erucoides* and *B. napus* were obtained by ovary culture (Delourme et al., 1989).

Embryo, ovule and ovary culture methods have been used in combination or in combination with test-tube fertilization to bypass pre- and post-fertilization barriers to crossability in some plants. Thus, pre-fertilization barriers to hybridization in *Nicotiana* were removed by combining *in vitro* pollination, placenta culture and ovule culture (DeVerna et al., 1987). In hybridizing *Trifolium repens* with *T. hybridum*, both ovule culture and embryo culture were employed to obtain viable seedings (Przywara et al., 1989) while successful interspecific hybridization of *Brassica napus* and *B. juncea* was possible through ovary, ovule and embryo culture (Bajaj et al., 1986). *Lilium* is another genus in which both *in vitro* pollination and ovary culture have been found to increase seed set in incompatible crosses (van Tuyl et al., 1991).

Several other uses have been attributed to ovary, ovule and embryo culture techniques in horticultural and breeding practices. These include breakage of seed dormancy to shorten breeding cycles, production of monoploids, testing for seed viability, study of host-pathogen interactions, overcoming self-sterility and germination of seeds of parasitic plants. These have been reviewed in earlier publications (Raghavan, 1976, 1977b; Yeung et al., 1981; Williams et al., 1987). In cassava (*Manihot esculenta*), embryo culture has been reported to account for a higher level of recovery of plants than that obtained by conventional seed germination (Biggs et al., 1986). Because embryo culture provides aseptic seedling material, cultured embryos can serve as a host for screening plants for disease resistance (Sharma and Bedi, 1990). A potentially important new use of ovule culture is in raising seedlings from early maturing cultivars of fruit trees or from seeds in which embryo aborts early in development such as in seedless fruits (Cain et al., 1983; Ramming, 1985; Weigel and Hughes, 1986; Emershad et al., 1989). Considerable interest is focused on embryo culture in establishing a direct somatic embryogenesis system capable of yielding embryoids reproducibly and in large numbers, amenable to mechanical harvesting (Maheswaran and Williams, 1984).

4. Conclusions

In vitro methods for the control of pollination, fertilization and embryo development have been empirically developed over the years to gain a better understanding of the physiology and genetics of seed development in angiosperms. In comparison with some of the other *in vitro* methods, those described in this chapter suffer since they are time-consuming and involve the use of a great deal of manipulative skill. Nonetheless, in various agricultural stations around the world, both at the local and national levels, embryo culture has become the choice method to obtain hybrids from wide crosses. In plant breeding stations, *in vitro* pollination and fertilization methods can be of enormous value in overcoming cross-incompatibility in economically important plants. While both methods need refinements to improve efficiency, recognition of their importance in plant breeding affords the hope for significant new advances in our understanding of the reproductive biology of angiosperms.

5. References

Ackerson, R.C. (1984). Abscisic acid and precocious germination in soybeans. J. Expt. Bot. 35:403–413.

Agnihotri, A., Gupta, V., Lakshmikumaran, M.S., Shivanna, K.R., Prakash, S. and Jagannathan, V. (1990a). Production of *Eruca-Brassica* hybrids by embryo rescue. Plant Breeding 104:281–289.

Agnihotri, A., Shivanna, K.R., Raina, S.N., Lakshmikumaran, M., Prakash, S. and Jagannathan, V. (1990b). Production of *Brassica napus* × *Raphanobrassica* hybrids by embryo rescue: An attempt to introduce shattering resistance into *B. napus*. Plant Breeding 105:292–299.

Ahmad, F., and Comeau, A. (1990). Wheat × pearl millet hybridization: Consequences and potential. Euphytica 50:181–190.

Alpi, A., Tognoni, F. and D'Amato, F. (1975). Growth regulator levels in embryo and suspensor of *Phaseolus coccineus* at two stages of development. Planta 127:153–162.

Arisumi, T. (1985). Rescuing abortive *Impatiens* hybrids through aseptic culture of ovules. J. Am. Soc. Hort. 110:273–276.

Augusten, H. (1956). Wachstumsversuche mit isolierten Weizen-Embryonen. Planta 48:24–46.

Bajaj, Y.P.S. (1979). Test-tube fertilization and development of maize (*Zea mays* L.) plants. Indian J. Expt. Biol. 17, 475–478.

Bajaj, Y.P.S., Mahajan, S.K. and Labana, K.S. (1986). Interspecific hybridization of *Brassica napus* and *B. juncea* through ovary, ovule and embryo culture. Euphytica 35:103–109.

Bauchan, G. R. (1987). Embryo culture of *Medicago scutellata* and *M. sativa*. Plant Cell Tissue Org. Cult. 10:21–29.

Belivanis, T. and Doré, C. (1986). Interspecific hybridization of *Phaseolus vulgaris* L. and *Phaseolus angustissimus* A. Gray using *in vitro* embryo culture. Plant Cell Rep. 5:329–331.

Bennici, A. and Cionini, P.G. (1979). Cytokinins and *in vitro* development of *Phaseolus coccineus* embryos. Planta 147:27–29.

Bhojwani, S.S. and Razdan, M.K. (1983). Plant Tissue Culture: Theory and Practice. Elsevier, Amsterdam.

Biggs, B.J., Smith, M.K. and Scott, K.J. (1986). The use of embryo culture for the recovery of plants from cassava (*Manihot esculenta* Crantz) seeds. Plant Cell Tissue Org. Cult. 6:229–234.

Bosio, M.G. (1940). Ricerche sulla fecondazione intraovarica in *Helleborus* et *Poeonia*. Nuovo G. Bot. Ital. 47:591–598.

Bright, S.W.J., Wood, E.A. and Miflin, B.J. (1978). The effect of aspartate-derived amino acids (lysine, threonine, methionine) on the growth of excised embryos of wheat and barley. Planta 139:113–117.

Buffard-Morel, J. (1968). Effets du glucose, du lévulose, du maltose et du saccharose sur le développement des embryons de palmier à huile (*Elaeis guineensis* Jacq. var. *dura* Bec.) en culture *in vitro*. C.R. Acad. Sci. Paris 267:185–188.

Cain, D.W., Emershad, R.L. and Tarailo, R.E. (1983). *In-ovulo* embryo culture and seedling development of seeded and seedless grapes (*Vitis vinifera* L.). Vitis 22:9–14.

Chen, H.K., Mok, M.C., Shanmugasundaram, S. and Mok, D.W.S. (1989). Interspecific hybridization between *Vigna radiata* (L.) Wilczek and *V. glabrescens*. Theor. Appl. Genet. 78:641–647.

Chung, G.H. and Kim, J.H. (1990). Production of interspecific hybrids between *Glycine max* and *G. tomentella* through embryo culture. Euphytica 48:97–101.

Cionini, P.G., Bennici, A., Alpi, A. and D'Amato, F. (1976). Suspensor, gibberellin and *in vitro* development of *Phaseolus coccineus* embryos. Planta 131:115–117.

Coble, C.J. and Schapaugh, W.T., Jr. (1990). Nutrient culture medium for components affecting plant recovery from immature embryos of three *Glycine* genotypes and an interspecific hybrid grown *in vitro*. Euphytica 50:127–133.

Cohen, D., Ladizinsky, G., Ziv, M. and Muehlbauer, F.J. (1984). Rescue of interspecific *Lens* hybrids by means of embryo culture. Plant Cell Tissue Org. Cult. 3:343–347.

Collins, G.B. and Grosser, J.W. (1984). Culture of embryos. In: Cell Culture and Somatic Cell Genetics of Plants, Vol. 1. Laboratory Procedures and Their Applications. I.K. Vasil (ed.), Academic Press, Orlando, pp. 241–257.

Contolini, C.S. and Hughes, K.W. (1989). Reciprocal differences in intraspecific crosses of tobacco result from embryo death. Am. J. Bot. 76:6–13.

Cook, S.K. Adams, H., Hedley, C.L., Ambrose, M.J. and Wang, T.L. (1988). An analysis of seed development in *Pisum sativum*, VII: Embryo development and precocious germination. Plant Cell Tissue Org. Cult. 14:89–101.

Corsi, G. (1972). The suspensor of *Eruca sativa* Miller (Cruciferae) during embryogenesis *in vitro*. G. Bot. Ital. 106:41–54.

Crouch, M.L. and Sussex, I.M. (1981). Development and storage-protein synthesis in *Brassica napus* L. embryos *in vivo* and *in vitro*. Planta 153:64–74.

Delourme, R., Eber, F. and Chevre, A.M. (1989). Intergeneric hybridization of *Diplotaxis erucoides* with *Brassica napus*, I: Cytogenetic analysis of F_1 and BC_1 progeny. Euphytica 41,:123–128.

del Rosario, A.G. and de Guzman, E.V. (1978). The growth of coconut 'makapuno' embryo *in vitro* as affected by mineral composition and sugar level of the medium during the liquid and solid cultures. Philipp. J. Sci. 105:215–222.

DeVerna, J.W., Myers, J.R. and Collins, G.B. (1987). Bypassing prefertilization barriers to hybridization in *Nicotiana* using *in vitro* pollination and fertilization. Theor. Appl. Genet. 73:665–671.

Dieterich, K. (1924). Über Kultur von Embryonen ausserhalb des Samens. Flora 117:379–417.

Dulieu, H.L. (1966). Pollination of excised ovaries and culture of ovules of *Nicotiana tabacum* L. Phytomorphology 16:69–75.

Dupuis, I. and Dumas, C. (1989). *In vitro* pollination as a model for studying fertilization in maize (*Zea mays* L.). Sex. Plant Reprod. 2:265–269.

Dupuis, I. and Dumas, C. (1990a). Biochemical markers of female receptivity in maize (*Zea mays* L.) assessed using *in vitro* fertilization. Plant Sci. 70:11–19.

Dupuis, I. and Dumas, C. (199Ob). Influence of temperature stress on *in vitro* fertilization and heat shock protein synthesis in maize (*Zea mays* L.) reproductive tissues. Plant Physiol. 94:665–670.

Dure, L., III. and Galau, G.A. (1981). Developmental biochemistry of cottonseed embryogenesis and germination, 13: Regulation of biosynthesis of principal storage proteins. Plant Physiol. 68:187–194.

Dure, L., III, Greenway, S.C. and Galau, G.A. (1981). Developmental biochemistry of cottonseed embryogenesis and germination: Changing messenger ribonucleic acid populations as shown by *in vitro* and *in vivo* protein synthesis. Biochemistry 20:4162–4168.

Emershad, R. L., Ramming, D. W. and Serpe, M. D. (1989). *In ovulo* embryo development and plant formation from stenospermic genotypes of *Vitis vinifera*. Am. J. Bot. 76:397–402.

Fassuliotis, G., and Nelson, B.V. (1988). Interspecific hybrids of *Cucumis metuliferus* × *C. anguria* obtained through embryo culture and somatic embryogenesis. Euphytica 37:53–60.

Fatokun, C.A. and Singh, B.B. (1987). Interspecific hybridizatlon between *Vigna pubescens* and *V. unguiculata* Walp through embryo rescue. Plant Cell Tissue Org. Cult. 9:229–233.

Fedak, G. (1985a). Intergeneric hybrids between *Hordeum vulgare* and *Agropyron intermedium* var. *trichophorum*. Z. Pflanzenzücht. 95:45–49.

Fedak, G. (1985b). Intergeneric hybrids of *Hordeum vulgare* with *Agropyron caninum* and *A. dasystachyum*. Can. J. Genet. Cytol. 27:387–392.

Gengenbach, B. G. (1977). Genotypic influences on *in vitro* fertilization and kernel development in maize. Crop Sci. 17:489–492.

Gengenbach, B.G. (1984). *In vitro* pollination, fertilization, and development of maize kernels. In: Cell Culture and Somatic Cell Genetics of Plants, Vol. 1: Laboratory Procedures and Their Applications. I.K. Vasil (ed.), Academic Press, Orlando, pp. 276–282.

Gill, M.S. and Bajaj, Y.P.S. (1984). *In vitro* production of interspecific hybrids in cotton. Curr. Sci. 53:102–104.

Green, C.E. and Donovan, C.M. (1980). Effect of aspartate-derived amino acids and aminoethyl cysteine on growth of excised mature embryos of maize. Crop Sci. 20:358–362.

Gupta, P.K. and Fedak, G. (1985a). Intergeneric hybrids between *Hordeum californicum* and *Triticum aestivum*. J. Hered. 76:365–368.

Gupta, P.K. and Fedak, G. (1985b). Hybrids of *Hordeum californicum* and 2 × *H. brevisubulatum* s.l. with *Agropyron caninum*. Can. J. Genet. 27:380–386.

Gupta, P.K. and Fedak, G. (1985c). New hybrids of *Hordeum parodii* with *Hordeum vulgare*, *H. bogdanii*, *Agropyron caninum* and ×*Triticosecale*. Theor. Appl. Genet. 70:265–270.

Hannig, E. (1904). Zur Physiologie pflanzlicher Embryonen, I: Über die Cultur von Cruciferen-Embryonen ausserhalb des Embryosacks. Bot. Ztg. 62:45–80.

Higgins, R.K. and Petolino, J.F. (1988). *In vitro* pollination fertilization of maize: Influence of explant factors on kernel development. Plant Cell Tissue Org. Cult. 12:21–30.

Honma, S. (1955). A technique for artificial culturing of bean embryos. Proc. Am. Soc. Hort. Sci. 65:405–408.

Inomata, N. (1978). Production of interspecific hybrids in *Brassica campestris* × *B. oleracea* by culture of *in vitro* excised ovaries. I. Development of excised ovaries in the crosses of various cultivars. Jap. J. Genet. 53:161–173.

Iwai, S., Kishi, C., Nakata, K. and Kubo, S. (1985). Production of a hybrid of *Nicotiana repanda* Willd. × *N. tabacum* L. by ovule culture. Plant Sci. 41:175–178.

Iwai, S., Kishi, C., Nakata, K. and Kawashima, N. (1986). Production of *Nicotiana tabacum* × *Nicotiana acuminata* hybrid by ovule culture. Plant Cell Rep. 5:403–404.

Iwanaga, M., Freyre, R. and Watanabe, K. (1991). Breaking the crossability barriers between disomic tetraploid *Solanum acaule* and tetrasomic tetraploid *S. tuberosum*. Euphytica 52:183–191.

Jones, R.J., Gengenbach, B.G. and Cardwell, V.B. (1981). Temperature effects on *in vitro* kernel development of maize. Crop Sci. 21:761–766.

Kameya, T., Hinata, K. and Mizushima, U. (1966). Fertilization *in vitro* of excised ovules treated with calcium chloride in *Brassica oleracea* L. Proc. Jap. Acad. 42:165–167.

Kanta, K. (1960). Intraovarian pollination in *Papaver rhoeas*. Nature 188:683–684.

Kanta, K. and Maheshwari, P. (1963a). Intraovarian pollination in same Papavaraceae. Phytomorphology 13:215–229.

Kanta, K. and Maheshwari, P. (1963b). Test-tube fertilization in some angiosperms. Phytomorphology 13:230–237.

Kanta, K., Rangaswamy, N.S. and Maheshwari, P. (1962). Test-tube fertilization in a flowering plant. Nature 194:1214–1217.

Kapoor, M. (1959). Influence of growth substances on the ovules of *Zephyranthes*. Phytomorphology 9:313–315.

Keijzer, C.J., Reinders, M.C. and Leferink-ten Klooster, H.B. (1988). A micromanipulation method for artificial fertilization in *Torenia*. In: Sexual Reproduction in Higher Plants. M. Cresti, P. Gori and E. Pacini (eds.), Springer Verlag, Berlin, pp. 118–124.

Kent, N. and Brink, R.A. (1947). Growth *in vitro* of immature *Hordeum* embryos. Science 106:547–548.

Kranz, E. and Lörz, H. (1990). Micromanipulation and *in vitro* fertilization with single pollen grains of maize. Sex. Plant Reprod. 3:160–169.

Kranz, E., Bautor, J. and Lorz, H. (1991). *In vitro* fertilization with single, isolated gametes of maize medicated by electrofusion. Sex. Plant Reprod. 4:12–16.

Kruse, A. (1974). An *in vivo/in vitro* embryo culture technique. Hereditas 77:219–224.

Lagriffol, J. and Monnier, M. (1985). Effects of endosperm and placenta on development of *Capsella* embryos in ovules cultivated *in vitro*. J. Plant Physiol. 118:127–137.

Laibach, F. (1925). Das Taubwerden von Bastardsamen und künstliche Aufzucht früh absterbender Bastardembryonen. Z. Bot. 17:417–459.

Long, S.R., Dale, R. M. K. and Sussex, I. M. (1981). Maturation and germination of *Phaseolus vulgaris* embryonic axes in culture. Planta 153:405–415.

Lorenzi, R., Bennici, A., Cionini, P. G., Alpi, A. and D'Amato, F. (1978). Embryo-suspensor relations in *Phaseolus coccineus*: Cytokinins during seed development. Planta 143:59–62.

Maheshwari, N. (1958). *In vitro* culture of excised ovules of *Papaver somniferum*. Science 127:342.

Maheshwari, N. and Lal, M. (1961). *In vitro* culture of excised ovules of *Papaver somniferum* L. Phytomorphology 11:307–314.

Maheswaran, G. and Williams, E.G. (1984). Direct somatic embryoid formation on immature embryos of *Trifolium repens*, *T. pratense* and *Medicago sativa*, and rapid clonal propagation of *T. repens*. Ann. Bot. 54:201–221.

Mathias, R., Espinosa, S. and Röbbelen, G. (1990). A new embryo rescue procedure for interspecific hybridization. Plant Breeding 104:258–261.

Matsubara, S. (1962). Studies on a growth promoting substance, 'embryo factor', necessary for the culture of young embryos of *Datura tatula* in vitro. Bot. Mag. Tokyo 75:10–18.

Matsubara, S. (1964). Effect of nitrogen compounds on the growth of isolated young embryos of *Datura*. Bot. Mag. Tokyo 77:253–259.

McCoy, T.J. (1985). Interspecific hybridization of *Medicago sativa* L. and *M. rupestris* MB using ovule-embryo culture. Can. J. Genet. Cytol. 27:238–245.

McCoy, T.J. and Smith, L.Y. (1986). Interspecific hybridization of perennial *Medicago* species using ovule-embryo culture. Theor. Appl. Genet. 71:772–783.

Mohapatra, D. and Bajaj, Y.P.S. (1988). Hybridization in *Brassica juncea* × *Brassica campestris* through ovary culture. Euphytica 37:83–88.

Monnier, M. (1976). Culture *in vitro* de l'embryon immature de *Capsella bursa-pastoris* Moench. Rev. Cytol. Biol. Vég. 39:1–120.

Monnier, M. (1984). Survival of young immature *Capsella* embryos cultured *in vitro*. J. Plant Physiol. 115:105–113.

Nagl, W. (1974). The *Phaseolus* suspensor and its polytene chromosomes. Z. Pflanzenphysiol. 73:1–44.

Nakajima, T. (1962). Physiological studies of seed development, especially embryonic growth and endosperm development. Bull. Univ. Osaka Pref. Ser. B13:13–48.

Neal, C.A. and Topoleski, L.D. (1985). Hormonal regulation of growth and development of tomato embryos *in vitro*. J. Am. Soc. Hort. Sci. 110:869–873.

Niimi, Y. (1970). *In vitro* fertilization in the self-incompatible plant *Petunia hybrida*. J. Jap. Soc. Hort. Sci. 39:345–352.

Norstog, K. (1961). The growth and differentiation of cultured barley embryos. Am. J. Bot. 48:876–884.

Norstog, K. (1972). Factors relating to precocious germination in cultured barley embryos. Phytomorphology 22:134–139.

Nyman, L.P., Webb, E.L., Gu, Z. and Arditti, J. (1986). Structure and *in vitro* growth of zygotic embryos of taro (*Colocasia esculenta* var. *antiquorum*). Ann. Bot. 57:623–630.

Nyman, L.P., Webb, E.L., Gu, Z. and Arditti, J. (1987). Effects of growth regulators and glutamine on *in vitro* development of zygotic embryos of taro (*Colocasia esculenta* var. *antiquorum*). Ann. Bot. 59:517–523.

Ozsan, M. and Cameron, J.W. (1963). Artificial culture of small *Citrus* embryos, and evidence against nucellar polyembryony in highly zygotic varieties. Proc. Am. Soc. Hort. Sci. 82:210–216.

Paris, D., Rietsema, J., Satina, S. and Blakeslee, A.F. (1953). Effect of amino acids, especially aspartic and glutamic acid and their amides, on the growth of *Datura stramonium* embryos *in vitro*. Proc. Natl. Acad. Sci. USA 39:1205–1212.

Park, C.H. and Walton, P.D. (1990). Intergeneric hybrids and an amphidiploid between *Elymus canadensis* and *Psathyrostachys juncea*. Euphytica 45:217–222.

Pierik, R.L.M. (1987). *In vitro* Culture of Higher Plants. Martinus Nijhoff Publishers, Dordrecht.

Plourde, A., Comeau, A., Fedak, G. and St-Pierre, C.A. 1989. Production and cytogenetics of hybrids of *Triticum aestivum* × *Leymus innovatus*. Theor. Appl. Genet. 78:436–444.

Plourde, A., Fedak, G., St-Pierre, C. A. and Comeau, A. (1990). A novel intergeneric hybrid in the Triticeae: *Triticum aestivum* × *Psathyrostachys juncea*. Theor. Appl. Genet. 79:45–48.

Pretová, A. (1974). The influence of the osmotic potential of the cultivation medium on the development of excised flax embryos. Biol. Plant. 16:14–20.

Pretová, A. (1986). Growth of zygotic flax embryos *in vitro* and influence of kinetin. Plant Cell Rep. 3:21–211.

Przywara, L., White, D.W.R., Sanders, P.M. and Maher, D. (1989). Interspecific hybridization of *Trifolium repens* with *T. hybridum* using *in ovulo* embryo and embryo culture. Ann. Bot. 64:613–624.

Raghavan, V. (1965). Hormonal control of embryogenesis in *Capsella*. In: Proceedings of an International Conference on Plant Tissue Culture. R. White and A.R. Grove (eds.), McCutchan Publishing Corp., Berkeley, pp. 357–369.

Raghavan, V. (1976). Experimental Embryogenesis in Vascular Plants. Academic Press, London.

Raghavan, V. (1977a). Diets and culture media for plant embryos. In: CRC Handbook Series in Nutrition and Food. Section G: Diets, Culture Media, Food Supplements, Vol. IV. Culture Media for Cells, Organs and Embryos. M. Rechcigl, Jr. (ed.), CRC Press, Cleveland, pp. 361–413.

Raghavan, V. (1977b). Applied aspects of embryo culture. In: Applied and Fundamental Aspects of Plant Cell, Tissue, and Organ Culture. J. Reinert and Y.P.S. Bajaj (eds.), Springer Verlag, Berlin, pp. 375–397.

Raghavan, V. (1980). Embryo culture. Int. Rev. Cytol. Suppl. 11B:209–240.

Raghavan, V. (1985). The applications of embryo rescue in agriculture. In: Biotechnology in International Agricultural Research. Int. Rice Res. Inst., Los Banos, Manila, pp. 189–197.

Raghavan, V. (1986a). Variability through wide crosses and embryo rescue. In: Cell Culture and Somatic Cell Genetics of Plants, Vol. 3. Plant Regeneration and Genetic Variability. I.K. Vasil (ed.), Academic Press, Orlando, pp. 613–633.

Raghavan, V. (1986b). Embryogenesis in Angiosperms. A Developmental and Experimental Study. Cambridge University Press, New York.

Raghavan, V. and Sharma, K.K. (1994). Same perspectives on zygotic embryogenesis in gymnosperms and angiosperms. In: *In vitro* Embryogenesis in Plants. T.A. Thorpe (ed.), CRC Press (in press).

Raghavan, V. and Srivastava, P.S. (1982). Embryo culture. In: Experimental Embryology of Vascular Plants. B.M. Johri (ed.), Springer Verlag, Berlin, pp. 195–230.

Raghavan, V. and Torrey, J.G. (1963). Growth and morphogenesis of globular and older embryos of *Capsella* in culture. Am. J. Bot. 50:540–551.

Raman, K., Walden, D.B. and Greyson, R.I. (1980). Fertilization in *Zea mays* L. by cultured gametophytes. J. Hered. 71:311–314.

Ramming, D.W. (1985). *In ovulo* embryo culture of early-maturing *Prunus*. Hort. Sc. 20:419–420.

Rangaswamy, N.S. and Shivanna, K. R. (1967). Induction of gametic compatibility and seed formation in axenic cultures of a diploid self-incompatible species of *Petunia*. Nature 16:937–939.

Rangaswamy, N.S. and Shivanna, K.R. (1971a). Overcoming self-incompatibility in *Petunia axillaris* (Lam.) BSP, 2: Placental pollination. J. Indian Bot. Soc. 50A:286–296.

Rangaswamy, N.S. and Shivanna, K.R. (1971b). Overcoming self-incompatibility in *Petunia axillaris*, 3: Two-site pollinations *in vitro*. Phytomorphology 21:284–289.

Rao, P.S. and Rangaswamy, N.S. (1972). *In vitro* development of the pollinated pistils of *Nicotiana rustica* L. Bot. Gaz. 133:350–355.

Reed, S.M. and Collins, G.B. (1978). Interspecific hybrids in *Nicotiana* through *in vitro* culture of fertilized ovules: *N. stocktonii* × *N. tabacum*; *N. nesophila* × *N. tabacum*; *N. repanda* × *N. tabacum*. J. Hered. 69:311–315.

Rietsema, J., Satina, S. and Blakeslee, A.F. (1953). The effect of sucrose on the growth of *Datura stramonium* embryos *in vitro*. Am. J. Bot. 40:538–545.

Rijven, A.H.G.C. (1952). *In vitro* studies on the embryo of *Capsella pursa-pastoris*. Acta Bot. Neerl. 1:157–200.

Rijven, A.H.G.C. (1956). Glutamine and asparagine as nitrogen sources for the growth of plant embryos *in vitro*: A comparative study of 12 species. Austral. J. Biol. Sci. 9:511–527.

Rivin, C. J. and Grudt, T. (1991). Abscisic acid and the developmental regulation of embryo storage proteins in maize. Plant Physiol. 95:358–365.

Robichaud, C. and Sussex, I.M. (1986). The response of viviparous-l and wild type embryos of *Zea mays* to culture in the presence of abscisic acid. J. Plant Physiol. 126:235–242.

Robichaud, C.S., Wong, J. and Sussex, I.M. (1980). Control of *in vitro* growth of viviparous embryo mutants of maize by abscisic acid. Devel. Genet. 1:325–330.

Ryczkowski, M. (1960). Changes of the osmotic value during the development of the ovule. Planta 55:343–356.

Sarla, N. and Raut, R.N. (1988). Synthesis of *Brassica carinata* from *Brassica nigra* × *Brassica oleracea* hybrids obtained by ovary culture. Theor. Appl. Genet. 76:846–849.

Schel, J.H.N. and Kieft, H. (1986). An ultrastructural study of embryo and endosperm development during *in vitro* culture of maize ovaries (*Zea mays* L.). Can. J. Bot. 64:2227–2238.

Sharma, H.C. and Ohm, H.W. (1990). Crossability and embryo rescue enhancement in wide crosses between wheat and three *Agropyron* sp. Euphytica 49:209–214.

Sharma, N.K. and Bedi, S. (1990). Effects of *Phytophthora palmivora* on zygotic embryos of *Papaya in vitro*. Ann. Bot. 66:597–603.

Shintaku, Y., Yamamoto, K. and Nakajima, T. (1988). Interspecific hybridizatlon between *Nicotiana repanda* Willd. and *N. tabacum* L. through pollen irradiation technique and the egg cell irradiation technique. Theor. Appl. Genet. 76:293–298.

Shivanna, K.R. (1971). Overcoming self-incompatibility in *Petunia*. Differential treatment *in vitro* of the whole placentae. Experientia 2:864–865.

Sladky, Z. and Havel, L. (1976). The study of the conditions for the fertilization *in vitro* in maize. Biol. Plant. 18:469–472.

Stafford, A. and Davies, D.R. (1979). The culture of immature pea embryos. Ann. Bot. 44:315–321.

Stefaniak, B. (1987). The *in vitro* development of isolated rye proembryos. Acta Soc. Bot. Polon. 56:37–42.

Stewart, J.M. (1981). *In vitro* fertilization and embryo rescue. Environ. Expt. Bot. 21:301–315.

Stewart, J.M. and Hsu, C.L. (1977). *In-ovulo* embryo culture and seedling development of cotton (*Gossypium hirsutum* L.). Planta 137:113–117.

Stewart, J.M. and Hsu, C.L. (1978). Hybridization of diploid and tetraploid cottons through *in ovulo* embryo culture. J. Hered. 69:404–408.

Thengane, S., Paranjpe, S.V., Khuspe, S.S. and Mascarenhas, A.F. (1986). Hybridization of *Gossypium* species through *in ovulo* embryo culture. Plant Cell Tissue Org. Cult. 6:209–219.

Tilton, V.R. and Russell, S.H. (1984). Applications of *in vitro* pollination/fertilization technology. Bioscience 34:239–242.

Triplett, B.A. and Quatrano, R.S. (1982). Timing, localization, and control of wheat germ agglutinin synthesis in developing wheat embryos. Dev. Biol. 91:491–496.

van Overbeek, J., Conklin, M.E. and Blakeslee, A.F. (1942). Cultivation *in vitro* of small *Datura* embryos. Am. J. Bot. 29:472–477.

van Overbeek, J., Siu, R. and Haagen-Smit, A.J. (1944). Factors affecting the growth of *Datura* embryos *in vitro*. Am. J. Bot. 31:219–224.

van Tuyl, J.M., van Dien, M.P., van Creij, M.G.M., van Kleinwee, T.C.M., Franken, J. and Bino, R.J. (1991). Application of *in vitro* pollination, ovary culture, ovule culture and embryo rescue for overcoming incongruity barriers in interspecific *Lillium* crosses. Plant Sci. 74:115–126.

Veen, H. (1963). The effect of various growth-regulators on embryos of *Capsella bursa-pastoris* growing *in vitro*. Acta Bot. Neerl. 12:129–171.

Wagner, G. and Hess, D. (1973). *In vitro*-Befruchtungen bei *Petunia hybrida*. Z. Pflanzenphysiol. 69:262–269.

Wagner, V. T., Song, Y., Matthys-Rochon, E. and Dumas, C. (1988). The isolated embryo sac of *Zea mays*: Structural and ultrastructural observations. In: Sexual Reproduction in Higher Plants. M. Cresti, P. Gori and E. Pacini (eds.), Springer Verlag, Berlin, pp. 125–131.

Walbot, V. (1978). Control mechanisms for plant embryogeny. In: Dormancy and Developmental Arrest. Experimental Analysis in Plants and Animals. M.E. Clutter (ed.), Academic Press, New York, pp. 113–166.

Weigel, R.C. and Hughes, K.W. (1986). *In vitro* ovule culture of a seedless persimmon. J. Hered. 77:213.

Wilen, R.W., van Rooijen, G.J.H., Pearce, D.W., Pharis, R.P., Holbrook, L.A. and Moloney, M.M. (1991). Effects of jasmonic acid on embryo-specific processes in *Brassica* and *Linum* oilseeds. Plant Physiol. 95:399–405.

Williams, E. and White, D.D.R. (1976). Early seed development after crossing of *Trifolium ambiguum* and *T. repens*. N. Z. J. Bot. 18:215–220.

Williams, E.G. (1987). Interspecific hybridization in pasture legumes. Plant Breed. Rev. 5:237–305.

Williams, E.G. and de Lautour, G. (1980). The use of embryo culture with transplanted nurse endosperm for the production of interspecific hybrids in pasture legumes. Bot. Gaz. 141:252–257.

Williams, E.G., Maheswaran, G. and Hutchinson, J.F. (1987). Embryo and ovule culture in crop improvement. Plant Breed. Rev. 5:181–236.

Yeung, E.C. and Brown, D.C.W. (1982). The osmotic environment of developing embryos of *Phaseolus vulgaris*. Z. Pflanzenphysiol. 106:149–156.

Yeung, E.C. and Sussex, I.M. (1979). Embryogeny of *Phaseolus coccineus*: The suspensor and the growth of the embryo-proper *in vitro*. Z. Pflanzenphysiol. 91:423–433.

Yeung, E.C., Thorpe, T.A. and Jensen, C.J. (1981). *In vitro* fertilization and embryo culture. In: Plant Tissue Culture. Methods and Applications in Agriculture. T.A. Thorpe (ed.), Academic Press, New York, pp. 253–271.

Zenkteler, M. (1967). Test-tube fertilization of ovules in *Melandrium album* Mill. with pollen grains of several species of the Caryophyllaceae family. Experientia 23:775.

Zenkteler, M., Misiura, E. and Guzowska, I. (1975). Studies on obtaining hybrid embryos in test tubes. In: Form, Structure and Function in Plants. H.Y. Mohan Ram, J.J. Shaw and C.K. Shaw (eds.), Sarita Prakashan, Meerut, pp. 180–187.

Zenkteler, M. and Nitzsche, W. (1985). *In vitro* culture of ovules of *Triticum aestivum* at early stages of embryogenesis. Plant Cell Rep. 4:168–171.

Zhou, C. and Yang, H. Y. (1985). Observations on enzymatically isolated, living and fixed embryo sacs in several angiosperm species. Planta 165:225–231.

9. Cryopreservation and Germplasm Storage*

K.K. Kartha and F. Engelmann

Contents

1. Introduction

Cryopreservation is defined as the viable freezing of biological material and their subsequent storage at ultra-low temperatures, preferably at that of liquid nitrogen. The development of cryopreservation strategy for plant cells and organs has followed the advances made with mammalian systems, albeit several decades later. Even for mammalian systems, the discovery of chemicals with cryoprotective properties was a significant step towards the development and refinement of cryopreservation technology. A major breakthrough in this context was the finding that glycerol was capable of protecting avian spermatozoa from freezing injury (Polge et al., 1949). This generated widespread enthusiasm and renewed interest among people interested in low temperature preservation in such fields as biology and medicine. Since the early 1950's a number of low molecular weight neutral solutes have been identified as potential cryoprotectants, the most commonly recognized ones being dimethylsulfoxide (DMSO or Me_2SO) and glycerol. Dimethylsulfox-

*NRCC Publication No. 32487.

Indra K. Vasil and Trevor A. Thorpe (eds.), Plant Cell and Tissue Culture, 195–230.

ide, originally used to prevent freezing damage to human and bovine red blood cells and bull spermatozoa (Lovelock and Bishop, 1959), has become a universal cryoprotectant. In recent years, considerable progress has been made in the low temperature preservation of red cells and platelets, leucocytes, bone marrow cells, protozoa, and helminth parasites of man and animals, insects and their cells and microorganisms (Ashwood-Smith and Farrant, 1980). Despite all these advances, unlike plants, most attempts to preserve animal organs at ultra-low temperature have met with limited success.

One of the earliest reports on the survival of plant tissues exposed to ultra-low temperature was made by Sakai when he demonstrated that very hardy mulberry twigs, upon induction of dehydration mediated by extracellular freezing, are capable of survival following immersion in liquid nitrogen provided the frozen samples are subsequently rewarmed slowly at an air temperature of 0 °C (Sakai, 1956). Historically, this finding had great significance in the understanding of the mechanism by which plant cells withstand such severe low temperature stress. Later, Sun (1958) achieved partial success when desiccated seedlings of *Pisum sativum* were immersed in liquid nitrogen. Although work continued on the freezing behaviour of tissues of various annual and perennial species, such studies were not extended to cultured plant cells or organs until a decade later when Quatrano (1968) first demonstrated that cultured flax (*Linum usitatissimum*) cells treated with DMSO are capable of survival after freezing to −50 °C. This is the first recorded case on the utilization of a cryoprotectant for the freezing of plant cells. This study was followed by a similar kind with cell cultures of *Ipomoea* and *Daucus carota* in which the former survived freezing to only −40 °C, while the latter survived immersion in liquid nitrogen (Latta, 1971). The experiments carried out with cell cultures of *D. carota* may thus be considered as the first successful attempt to cryopreserve cultured cells of any plant species. Since D. carota was one of the earliest species to be brought into culture, this became a model species to study cryobiology of cultured cells. Nag and Street (1973) successfully regenerated somatic embryos from a cell culture of *D. carota* which had been frozen to the temperature of −196 °C, while Dougall and Wetherell (1974) stored cell cultures of wild carrot in a frozen state. In retrospect, it is fortunate that the early investigators chose *D. carota* cell cultures, albeit unknowingly, for their research since even today certain cell cultures are extremely sensitive to freezing injury (Kartha, 1985). Since the report on the successful cryostorage of *D. carota* cell cultures, several types of material such as callus and protoplasts, meristems/shoot-tips, zygotic and somatic embryos, anthers/pollen and even whole seeds have been studied extensively from a cryopreservation perspective. Examples of successful application of cryopreservation technology to a wide array of plant material can be found in Table 1.

Table 1. List of successfully clyopreserved culture systems

Species	References
A. Meristems/Shoot-tips	
Arachis hypogaea	Bajaj (1979)
Asparagus officicinalis	Uragami et al. (1990)
Brassica napus	Benson et al. (1984)
B. oleracea	Harada et al. (1985)
Carica papaya	Towill (1990b)
Cicer arietinum	Kartha and Gamborg (1978); Bajaj (1979)
Dianthus caryophyllus	Anderson (1979); Seibert (1976) (1977); Seibert and Wetherbee (1977); Uemura and Sakai (1980); Dereuddre et al. (1988); Langis et al. (1990); Towill (1990b)
Fragaria × *ananassa*	Kartha et al. (1980); Sakai et al. (1978)
Lactuca sativa	Seibert (1977)
Lillium sp.	Bouman and de Klerk (1990)
Lycopersicon esculentum	Grout et al. (1978)
Malus domestica	Katano et al. (1983); Sakai and Nishiyama (1978); Katano (1986); Kuo and Lineberger (1985); Tyler and Stushnoff (1988a, b)
Manihot esculenta	Bajaj (1977a; 1983a); Kartha et al. (1982a); Marin et al. (1990a)
Mentha aquatica × *M. spicata*	Towill (1990a)
Pisum sativum	Kartha et al. (1979); McAdams et al. (1991)
Phoenix dactylifera	Bagniol et al. (1990)
Pyrus communis	Dereuddre et al. (1990a, b)
Ribes sp.	Sakai and Nishiyama (1978)
Rubus sp.	Sakai and Nishiyama (1978); Reed (1988)
Sambuscus racemosa	Sakai and Nishiyama (1978)
Solanum etuberosum	Towill (1981a)
S. goniocalyx	Grout and Henshaw (1978)
S. phureja	Fabre and Dereuddre (1990)
S. tuberosum	Standke (1978); Bajaj (1978); Towill (1981b; 1990b); Benson et al. (1989); Manzhulln et al. (1983); Harding (1990); Harding et al. (1991)
Trifolium T. pratense (red clover)	Yamada et al. (1991)
repens (white clover)	Thorn (1990)
Vanda hookeriana	Kadzimin (1988)
Xanthosoma sp.	Zandvoort (1987)
B. Cell suspensions	
Acer pseudoplatanus	Sugawara and Sakai (1974); Nag and Street (1975a, b); Withers (1978)
Atriplex sp.	Goldner et al. (1990)
Atropa belladonna	Nag and Street (1975a, b)
Berberis dictyophylla	Withers (1985)
B. wilsoniae	Reuff (1987)
Brassica napus	Weber et al. (1983)
B. campestris	Langis et al. (1989)
Brunefelsia dentifolia	Pence (1990)
Capsicum annum	Withers and Street (1977)
Catharanthus roseus	Kartha et al. (1982b); Chen et al. (1984a, b); Mannonen et al. (1990); Schrijnemarkers et al. (1990)

Table 1. Continued

Species	References
Citrus sinensis	Kobayashi et al. (1990); Sakai et al. (1990; 1991)
Datura innoxia	Hauptman and Widholm (1982); Weber et al. (1983)
D. stramonium	Bajaj (1976)
Daucus carota	Nag and Street (1973; 1975a, b); Bajaj (1976); Dougall and Whitten (1980); Hauptman and Widholm (1982); Weber et al. (1983); Withers and Street (1977)
Dioscorea deltoidea	Butenko et al. (1984)
Digitalis lanata	Deitrich et al. (1982); Seitz et al. (1983)
Glaucium flavum	Withers (1985)
Glycine max	Bajaj (1976); Weber et al. (1983)
Hyoscyamus muticus	Withers (1985)
Malus domestica	Dereuddre and Kartha (1984)
Musa sp.	Panis et al. (1990)
Myrtillocactus geometrizans	Haffner (1985)
Nicotiana plumbaginigolia	Maddox et al. (1982)
N. sylvestris	Maddox et al. (1982)
N. tabacum	Bajaj (1976); Hauptman and Widholm (1982); Maddox et al. (1982); Withers (1985); Schrijnemakers et al. (1990)
Onobrychis viciifolia	Withers (1985)
Oryza sativa	Cella et al. (1982); Sala et al. (1979); Finkle et al. (1982); Finkle and Ulrich (1982); Ulrich et al. (1984a); Kuriyama et al. (1989); Schrijnemakers et al. (1990); Meijer et al. (1990)
Panax ginseng	Butenko et al. (1984); Seitz and Reinhardt (1987); Mannonen et al. (1990)
Papaver somniferum	Friessen et al. (1991)
Pennisetum americanum	Withers (1985)
Petunia hybrida	Schrijnemakers et al. (1990)
Picea abies	Galerne and Dereuddre (1987); Gupta et al. (1987)
P. glauca	Kartha et al. (1988)
Pinus taeda	Gupta et al. (1987)
Populus sp.	Binder and Zaerr (1980a)
Pseudotsuga menziesii	Binder and Zaerr (1980b)
Puccinella distans	Jekkel et al. (1990)
Rhazya orientalis	Withers (1985)
R. stricta	Withers (1985)
Rosa 'Paul's Scarlet'	Withers and King (1980)
Saccharum spp.	Chen et al. (1979); Finkle and Ulrich (1979); Finkle et al. (1983); Ulrich et al. (1979); Gnanapragasam and Vasil (1990)
Solanum melongena	Withers (1985)
Sorghum bicolor	Withers and King (1980)
Tabernaemontana divaricata	Schrijnermakers et al. (1990)
Triticum aestivum	Chen et al. (1985)
T. monococcum	Withers (1985)
Vinca minor	Caruso et al. (1987)
Zea mays	Withers and King (1980); Shillito et al. (1989)
C. Callus	
Gossypium arboreum	Bajaj (1982)

Table 1. Continued

Species	References
G. hirsutum	Bajaj (1982)
Hordeum vulgare	Hahne and Lorz (1987)
Lavandula vera	Wanatabe et al. (1983, 1990)
Medicago sativa	Finkle et al. (1979, 1983); Cachita et al. (1990)
Oryza sativa	Finkle et al. (1979, 1983)
Phoenix dactylifera	Tisserat et al. (1981); Ulrich et al. (1979)
Populus euramericana	Sakai and Sugawara (1973)
Saccharum spp.	Ulrich et al. (1979); Ling et al. (1987)
Triticum aestivum	Chen et al. (1985)
Trifolium spp.	Cachita et al. (1990)
Ulmus americana	Ulrich et al. (1984b)
D. Protoplasts	
Atropa helladonna	Bajaj (1988)
Bromus inermis	Mazur and Hartman (1978)
Datura innoxia	Hauptman and Widholm (1982)
Daucus carota	Hauptman and Widholm (1982); Mazur and Hartman (1978); Takeuchi et al. (1980, 1982); Weber et al. (1983)
Glycine max	Weber et al. (1983); Takeuchi et al. (1982)
Hordeum vulgare	Takeuchi et al. (1982)
Nicotiana tabacum	Bajaj (1982)
Oryza sativa × *Pisum sativun*	Bajaj (1983b)
Secale cereale	Langis and Steponkus (1990b)
Triticum aestivum	Takeuchi et al. (1982)
T. aestivum × *P. sativum*	Bajaj (1983b)
Zea mays	Withers (1980)
E. Somatic embryos	
Asparagus officinalis	Uragami et al. (1989)
Citrus sinensis	Marin and Duran-Vila (1988)
Coffea arabica	Bertrand-Desbrunais et al. (1999)
Daucus carota	Withers (1979); Dereuddre et al. (1991a, b)
Elaeis guineensis	Engelmann et al. (1995); Engelmann and Duval (1986); Engelmann and Dereuddre (1988)
Manihot esculenta	Sudarmonowati and Henshaw (1990)
Xanthosoma	Zandvoort (1987)
F. Zygotic embryos	
Camellia sinensis	Chaudury et al. (1991)
Carva sp.	Pence and Dresser (1988)
Cocos nucifera	Bajaj (1984); Chin et al. (1989)
Elaeis guineensis	Grout et al. (1983)
Hevea brasiliensis	Normah et al. (1986)
Hordeum vulgare	Withers (1982)
Howea fosteriana	Chin et al. (1988)
Juglans regia	de Boucaud et al. (1991)
Lycopersicon esculentum	Grout (1979)
Manihot esculenta	Marin et al. (1990b)
Musa sp.	Mora et al. (1991)
Oryza sativa	Bajaj (1981)
Picea glauca (cotyledons)	Toivonen and Kartha (1989)

Table 1. Continued

Species	References
Triticum aestivum	Withers (1982)
Veitchia merrillii	Chin et al. (1988)
Zea mays	de Boucault and Cambecedes (1988)
G. Pollen embryos	
Arachis hypogaea	Bajaj (1983c)
A. villosa	Bajaj (1983c)
Atropa belladonna	Bajaj (1977b)
Brassica campestris	Bajaj (1983c)
B. napus	Bajaj (1983c)
Citrus spp.	Bajaj (1984)
Nicotiana tabacum	Bajaj (1977b)
Oryza sativa	Bajaj (1981)
Triticum aestivum	Bajaj (1983c)

2. The Rationale for Cryopreservation

Preservation of both plant and animal germplasm is an integral component of sustainable agriculture systems. From a crop improvement perspective, preservation of all valuable germplasm is being accorded a high priority (IBGR, 1985). It is estimated that of the over 240,000 species of angiosperms in the world, 3000 or more have been used for food at one time or another. Twelve major crops supply the world's population with food indicating man's dependence for food on only a few species of the several thousand angiosperms on earth (Wehner, 1988). Therefore a great deal of emphasis has been placed on the need to preserve plant genetic resources as a means of maintaining biodiversity.

Cultivated crops, on the basis of their mode of propagation, can be broadly grouped into vegetatively propagated and seed-propagated species. Both groups represent an enormous array of annual and perennial species. Amongst the vegetatively propagated species, some exhibit a high degree of heterozygosity and do not produce seeds. Such plants are clonally propagated through various vegetative means such as tubers or cuttings which generally have a limited life span. In addition to being labour intensive and expensive, this practice also increases the accompanying risks associated with field maintenance such as exposure to pathogens, pests, climatic perturbations and human error. Although the most economical means of germplasm storage for seed-propagated species is in the form of seeds, this is not always feasible because of the following reasons: (i) some crops do not produce viable seeds, (ii) some seeds remain viable for a limited duration only and are recalcitrant to storage by reduction of moisture content and temperature, (iii) seeds of certain species deteriorate rapidly due to seed-borne pathogens, and (iv) some seeds are very heterozygous and, therefore, not suitable for maintaining true-to-type genotypes (Kartha, 1982, 1985). Accordingly, an effective approach to circumvent the above problems may be the application of cryopreservation technology. It is now well established that one of the disciplines of plant tissue culture, meristem culture, is an extremely valuable technique with proven record of applications in clonal propagation, retention of genetic fidelity among propagules and elimination of viral pathogens (see chapter 3). These unique attributes make plant meristems ideal candidates for cryogenic preservation and establishment of virus-free germplasm banks and also in their international exchange (Kartha, 1985). In recent years, cryopreservation of zygotic embryos, embryo axes, and even whole seeds is also gaining popularity for the preservation of recalcitrant and tropical species.

Plant cell cultures have also become subjects of intensive research with emphasis on biotechnological applications such as genetic engineering, induction and isolation of mutants, selection of novel variants, screening for resistance against various stress factors, production of compounds of pharmaceutical value and creation of novel species of plants by somatic cell hybridization. As in the past, increasing use of cell cultures is also being made in addressing academic questions in physiology, biochemistry, genetics, and pathology, to name a few. It is now well known that cell lines subjected to any biotechnological manipulations become 'unique germplasm' in their own right and assurance of their uniqueness thus becomes a priority issue. Maintenance of cell or callus cultures in a continued state of division by repeated subculture on appropriate nutrient media is the practice often followed in several laboratories, unfortunately even today. Often, such a practice results in increased ploidy, accumulation of spontaneous mutations, decline and/or loss of morphogenic potential and biosynthetic capacity for product formation, reversion of valuable mutants to wild types, and most importantly, unintentional selection of undesirable phenotypes. Storage of cells in liquid nitrogen (−196 °C) may well alleviate these problems since at such an ultra-low temperature, all metabolic activities of cells are arrested and, consequently, no genetic changes would occur as a function of passage of time. The advantages offered by cryopreservation in preserving both the genetic stability and induced variability in plant cell cultures and organs alone are sufficient justifications to consider cryopreservation as an adjunct activity in any plant cell culture laboratory.

3. Cryopreservation Components

3.1. *Candidates for Cryopreservation*

An array of plant material could be considered for cryopreservation as dictated by the actual needs vis-a-vis preservation. These include meristems, cell, callus and protoplast cultures, somatic and zygotic embryos, anthers, pollen or microspores and whole seeds. Most of the developmental research on cryopreservation was focused on meristems or shoot-tips, cell cultures and somatic and/or zygotic embryos with the objective of preserving clonally propagated species, experimental cell lines and problem crop species (recalcitrant) where conventional storage strategies are inadequate. It is not the intent of this chapter to cover the work carried out with all types of such material since detailed information is available in a number of publications (for a comprehensive treatment of the subject matter, see Withers, 1985; Kartha, 1985). This chapter, would instead concentrate on the technology of cryopreservation and discuss various factors which influence the survival of different culture systems at ultra-low temperature.

3.2. *Pretreatments and Cryoprotection*

Plant cells contain high amounts of cellular water and freezing of plant cells implies conversion of some or all of their liquid water to ice, whereas thawing is reversal of this transition. In plant tissues, the majority of free water is involved in solvation and thus available for freezing, whereas bound water is associated with macromolecular constituents of the cells in a structural and/or functional role (Mazur, 1969; Levitt, 1980). Although freezing of plant cells involves both free and bound water, freezing of the latter has been considered to be more harmful to survival (Levitt, 1980). Therefore, any mechanism by which less of free water is made available for freezing will provide greater chances of survival of the cells or tissues post-freezing. Plant species which are capable of withstanding severe freezing stress in nature have evolved mechanisms to accomplish this by allowing the water to either remain supercooled or by permitting the outflow of free water to extracellular space by exosmosis where they freeze resulting in the promotion of freeze- induced dehydration. The freeze-induced dehydration further depresses the freezing point of the remaining cellular water thus facilitating the cells/tissues to withstand exposure to even lower temperatures including that of liquid nitrogen. On the other hand, in the case of cryopreservation, since most of the experimental systems (meristems, shoot-tips, cultured cells etc.) contain high amount of cellular water and hence are extremely sensitive to freezing injury, protection from freezing and thawing injury has to be imposed artificially. This involves various strategies such as the use of cryoprotectants or other manipulations as explained below.

3.2.1. *Cryoprotectants*

Since cryopreservation involves essentially three steps, specimen treatments and freezing, storage at ultra-low temperature, and thawing, an ideal cryoprotectant should protect cells from all the factors which would affect the viability of the frozen biological sample during all these stages. Since viability loss is not expected to occur at ultra-low temperatures, protection of the cells during freezing and thawing becomes of paramount importance. A number of compounds such as glycerol, DMSO, ethylene glycol, polyethylene glycol, sugars and sugar alcohols either alone or in combination protect living cells against damage during freezing and thawing. Such compounds can lower the temperature at which freezing first occurs and can alter the crystal habit of ice when it separates. Although the exact mechanism of action of the cryoprotective compounds is still poorly understood, the colligative properties of the cryoprotectants can minimize the deleterious action of excessive electrolyte concentration resulting from removal of water and conversion of water to ice (Nash, 1966). The neutral solutes prevent the attainment of critical injurious level of solute concentration in a colligative manner by reducing the solute accumulation at a given level of dehydration to which the biological system is exposed (Lovelock, 1953; Meryman et al., 1977). The neutral cryoprotectants may also interact directly with the membrane systems to counteract the deleterious effect of freeze concentration (Santarius, 1971; Volger and Heber, 1975) or afford protection by eliciting a membrane configuration which is inherently more resistant to freezing stress (Williams and Meryman, 1970). Heber and co-workers (Heber and Santarius, 1964; Heber et al., 1971; Volger and Heber, 1975) and Santarius (1971, 1973) while demonstrating the cryoprotective properties of sugars, amino acids, organic acids and low molecular weight proteins showed that these compounds prevent freeze-induced uncoupling of photophosphorylation. Lineberger and Steponkus (1980) presented evidence to support that differential cryoprotection is afforded to chloroplast thylakoids against freeze-induced uncoupling of cyclic photophosphorylation by equimolar concentrations of glucose, sucrose, and raffinose. This differential protection effect appears to be due to nonideal activity-concentration profiles exhibited by sugars during freezing. For a detailed treatment of the role played by cryoprotective compounds in the viable freezing of plant tissue see Finkle et al. (1985).

The cryoprotectants generally used for freezing biological specimens fall into two categories, permeating and non-permeating. The most commonly used permeating additives are dimethylsulfoxide (DMSO) and glycerol. Dimethylsulfoxide permeates into cells more rapidly than glycerol, and therefore, requires shorter treatment duration. In regard to the toxicity, at equimolar concentration glycerol is less toxic to cells than DMSO while DMSO is superior to glycerol in cryoprotection of plant cells and organs. Here again, the permeation capability of glycerol is species-specific and temperature dependent. For example, at 22 °C, glycerol does not enter the cells of *Haplopappus gracilis*, but does enter *Acer saccharum*, while certain sugars like sucrose do not penetrate either type of cell (Towill and Mazur, 1976). On the other hand, the penetration rate of glycerol at 20 °C was twice that at 0 °C (Mazur and Miller, 1976). The non-permeating compounds include sugars, sugar alcohol and high molecular weight additives, polyvinylpyrolidone, polyethylene glycol (PEG), dextran, hydroxyethyl starch/etc. Most of the cryoprotectants exhibit varying degrees of cytotoxicity at higher concentrations. Generally a concentration of 5–10% for DMSO and 10–20% for glycerol is adequate for most material. In instances where application of a single cryoprotectant does not result in survival, a mixture of cryoprotectants has been beneficial (Chen et al., 1984a, b; Finkle and Ulrich, 1979; Finkle et al., 1985; Withers, 1985). Ulrich et al. (1979) demonstrated the beneficial effect of using a mixture of cryoprotectants involving PEG, glucose, and DMSO (10–8–10%) in the successful cryopreservation of sugarcane callus. In subsequent studies several other cell or callus lines have been cryopreserved using the same mixture (Finkle et al., 1985). Prior to the discovery of a sorbitol – DMSO system (Chen et al., 1984a), the most effective cryoprotectant mixtures applicable to a variety of cell cultures were 0.5 M DMSO + 0.5 M glycerol + 1 M sucrose, and 0.5 M DMSO + 0.5 M glycerol + 1 M proline (Withers, 1985). It is to be borne in mind that the exposure duration of cryoprotectants to cells needs to be such that the concentration applied does not cause sudden plasmolysis which in itself could be a major cause of injury to the osmotic responsiveness of the cells. Although opinions differ as to the method of application of cryoprotectants to the cells, it is preferable to apply them very gradually in an ice bath at 4 °C in order to minimize any injury. Similarly, the removal of cryoprotectants from the thawed samples should also be a gradual process to alleviate the problems associated with deplasmolysis.

3.2.2. *Preculture Strategies*

Where the application of cryoprotectants, singly or in combination, at the nontoxic level becomes inadequate to ensure survival following cryopreservation, certain physiological modifications have been attempted with various types of culture systems to overcome the problem.

Incorporation of cryoprotectants into the culture medium followed by an incubation of isolated shoot-tips or meristems prior to their freezing has been the practice of choice. Pea and strawberry meristems precultured on respective shoot regeneration media supplemented with 5% DMSO for a period of two days and frozen at slow cooling rates exhibited enhanced survival (Kartha et al., 1979, 1980). The same method also resulted in enhanced survival in case of potato meristems (Manzhulin et al., 1983). Substituting 5% glycerol for DMSO only marginally enhanced the survival of strawberry meristems over the controls (Kartha et al., 1980). The DMSO preculture effect on the survival of pea meristems was further studied recently combining the preculture duration and the terminal freezing temperature. It appears that for pea shoot-tips immersed in liquid nitrogen (LN_2), the terminal freezing temperature had no effect whereas the number of days of preculture in 5% DMSO supplemented medium did influence the survival, with a 2-day preculture duration being the optimum. In pea shoot-tips not immersed in LN_2 the number of days of preculture did not influence survival whereas the terminal freezing temperature had an effect (McAdams et al., 1991). Higher rates of survival and plant regeneration were also obtained when shoot meristems of white clover (*Trifolium repens*) were precultured on agar-B5 medium supplemented with 5% DMSO and 5% glucose at 4 °C for two days prior to freezing by slow cooling (Yamada et al., 1991).

Preculturing isolated meristems or shoot-tips for a specific duration on shoot regeneration medium without the addition of cryoprotectants has also been found to enhance survival. For example, *Solanum goniocalyx* (Grout and Henshaw, 1978) and *S. etuberosum* (Towill, 1981a) shoot-tips exhibited enhanced survival if the freshly isolated explants were precultured in light on the regeneration medium for a period of 2 to 3 days prior to cryoprotectant treatment and freezing. Benson et al. (1989) carried out detailed investigations on pre- and post-freeze light regimes on the survival of *S. tuberosum* shoot-tips. They concluded that post-freezing survival was significantly affected by different light combinations and two different cultivars tested exhibited different pre-light requirements. However, other than the physiological alterations brought about by various light combinations, the exact manner in which light conditions influence cryosurvival still remains to be elucidated. Sucrose is the major carbon source in all plant tissue media and it is a well known cryoprotectant. Obviously, it is no surprise that several attempts have been made to utilize sucrose in the preculture media as well as the cryoprotectant during the actual freezing of plant material. Somatic embryos of oil palm (*Elaeis guineensis*) exhibited excellent survival if cryopreserved following 7-day preculture on nutrient medium enriched with 0.75 M sucrose (Engelmann et al., 1985; Engelmann and Duval, 1986; Engelmann and Dereuddre, 1988). Similar effect has also been noted for the cryopreservation of somatic embryos of coffee (Bertrand-Desbrunais et al., 1988) and *Picea abies* (Galerne and Dereuddre, 1987) and sweet orange (Marin and Duran-Vila, 1988). Regrowth rates of close to 100% were obtained with cryopreserved apical shoot tips isolated from 2-month old carnation stems precultured on medium supplemented with 0.75 M sucrose and treated with 5% DMSO or more (Dereuddre et al., 1988). An interesting aspect in this study was the finding that the resistance of axillary shoot tips to freezing decreased as a function of their distance from the apex implying that preculture in sucrose-supplemented medium alone cannot counteract the inherent developmental and physiological differences found in explants derived from the same plant.

Cell suspension cultures have been extensively used for cryobiological research because of their ease of availability, handling and maintenance. However, cells present in any suspension exhibit an array of phenotypes, physiological variations, differences in mitotic oscillations, ploidy and cell cycle. From a cryobiological prospective then, it is always possible to find certain fraction of the cell population in a stage condusive to freezing. In other words, cryopreservation of cell cultures is relatively easier to accomplish as compared to organized structures such as meristems where structural conformity has to be maintained in order to ensure faithful regeneration of plantlets. Although extensive studies have not been carried out to examine the role of heterogeneity brought upon the culture by the variables mentioned above, it is now apparent that cells have to be maintained in an active state of division often accompanied by frequent subculture. Highly vacuolated and quiescent cells are extremely sensitive to cryoprotectant treatment and freezing and such cells are naturally eliminated during the specimen preparation stage (Kartha et al., 1988). In attempts to enrich the cell population most resistant to freezing, Sugawara and Sakai (1974) observed high viability of *Acer pseudoplatanus* cells when cells at their late lag phase or early division phase were used for cryopreservation. In a subsequent study with cell cultures of the same species, Withers (1978) obtained enhanced post-freezing survival and mitotic index from the cells newly entered into Gl phase of the cell cycle. Sala et al. (1979) and Withers and Street (1977) presented evidence to indicate that the viability and regrowth potential of cryopreserved cells are enhanced when cells in late lag and early- to mid-exponential phases are selected for experimentation. The sensitivity of cells in early lag phase or stationary phase to freeze-thaw stress is attributable to the increase in cell size, vacuolar volume and water content whereas the resistance of small sized cells to freezing stress and thereby enhanced survival may relate to their dense cytoplasm. Therefore, one approach to enhance the post-freezing survival of cells is to reduce the cell volume and water content by inducing partial dehydration in prefreezing steps. Supplementation of culture medium with such osmotically active compounds as mannitol, sorbitol, sucrose, and proline increases freezing resistance. Mannitol has been found to be beneficial in reducing the mean cell volume of cells of *Acer pseudoplatanus* and *Capsicum annuum* and increasing their post-freezing survival (Withers and Street, 1977). A preculture strategy of using 3 to 6% mannitol has been successfully used for the cryopreservation of a number of cell cultures such as *Nicotiana tabacum*, *N. plumbaginifolia* (Maddox et al., 1982/83), soybean and sycamore (Pritchard et al., 1986a, b) and *Digitalis lanata* (Diettrich et al., 1982, 1985). Pritchard et al. (1986a, b) examined osmotic stress as a pregrowth procedure for cryopreservation of sycamore and soybean cell suspensions. Addition of 6% mannitol or sorbitol to liquid culture medium decreased the water potential by – 0.93 MPa. Sycamore cells grown to exponential phase in such media exhibited increased levels of total and soluble protein and respiratory activity, but decreased free proline. Soybean cells showed increased respiratory activity and free proline levels, but total protein levels remained unaffected. Water relation studies indicated that sycamore cells are capable of greater osmotic adjustment than soybean cells, and that mannitol uptake does not contribute significantly to that effect. The differential behaviour in osmotic adjustment may be attributable to the differences noticed in responses of cells to mannitol treatment and post freezing survival. For example, preculturing in mannitol supplemented medium did not enhance the survival of *N. sylvestris* (Maddox et al., 1983) or alkaloid-producing cell cultures of *Catharanthus roseus* (Chen et al., 1984a).

Sorbitol has been used either as an osmotic agent in the preculture medium or as the sole cryoprotectant during the freezing process. Weber et al. (1983) successfully cryopreserved cell cultures of *Glycine max*, *Datura innoxia*, *Brassica napus*, and *Daucus carota* and cell culture-derived protoplasts of *G. max*. Similarly, sorbitol has been successfully used as the sole cryoprotectant in the cryopreservation of *Coleus blumei* cell cultures (Reuff et al., 1988). However, cell cultures of alkaloid producing *Catharanthus roseus* could not be cryopreserved using sorbitol as the only cryoprotectant (Chen et al., 1984a), but they could be successfully cryopreserved by preculturing the cells in nutrient medium containing 1 M sorbitol for 6–20 hours followed by freezing using 1 M sorbitol and 5% DMSO. In this study, equimolar concentrations of $CaCl_2$ and KCl, in addition to being toxic to cells, were ineffective in their cryoprotective efficacy. Although considerable survival was noticed subsequent to preculturing the cells in media supplemented with 1 M concentrations of glucose, trehalose, and sucrose, the overall survival and recovery was much superior when sorbitol was used as the preculturing agent.

In order to understand the mode of action of cryoprotectants, the freezing behaviour of DMSO and sorbitol solutions and the alkaloid producing *C. roseus* cells treated with DMSO and sorbitol, alone or in combination, were examined by nuclear magnetic resonance (NMR) and differential thermal analysis (DTA). Incorporation of DMSO or sorbitol into the liquid medium had a significant effect on the temperature range from initiation to completion of ice crystallization. Compared to control, less water crystallized at temperatures below −30 °C in DMSO-treated cells. Similar results were obtained with sorbitol-treated cells, except sorbitol has less effect on the amount of water crystallized below −25 °C. There was a close association between the percent unfrozen water at −40 °C and per cent survival after freezing in liquid nitrogen. These studies indicated that in the alkaloid-producing *C. roseus* cells, the amount of liquid water at −40 °C is critical for successful cryopreservation and the combination of DMSO and sorbitol was effective in preventing this critical level of water from freezing (Chen et al., 1984b).

Attempts to cryopreserve an embryogenic cell culture of white spruce (*Picea glauca*) using DMSO as the only cryoprotectant were unsuccessful, and DMSO was found to cause 35–40% viability loss. However, successful cryopreseration could be accomplished following the sorbitol preculture strategy. Additionally, sorbitol was found to counteract the toxicity caused by DMSO (Kartha et al., 1988). The same procedure has been successfully extended to the cryopreservation of an embryogenic cell culture of a sugarcane (*Saccharum*) hybrid leading to efficient plant regeneration (Gnanapragasam and Vasil, 1990). The sorbitol – DMSO system is also applicable to embryogenic cell cultures of other conifer material such as *Picea glauca* and *Laryx* species (Kartha, unpublished).

Other additives to preculture media to enhance post-cryopreservation survival include various amino acids including proline (Butenko et al., 1984). Withers and King (1979) tested a number of amino acids for cryoprotection, and found that proline is only one of those that can cryoprotect plant cells. They suggested that proline might act as a membrane stabilizer. Hellergen and Li (1981) while trying to elucidate the mode of action of proline on the protection of *Solanum tuberosum* suspension cultures frozen to −14 °C reported that proline might exert its effect by the removal of excessive intracellular water by osmotic gradient. If this is the case, any potent osmoticum should be able to do the same since sucrose at about the same concentration acted in the same way as proline (Hellergren and Li, 1981). From the above discussion it becomes very clear that appropriate preculture strategies coupled with judicious selection and application of cryoprotectants can definitely increase cryosurvival.

3.2.3. *Cold Acclimation*

Plant species with inherent capacity to withstand severe freezing stress have evolved various mechanisms; one such mechanism being cold hardening or cold acclimation. These species achieve cold hardening by responding to such early environmental cues as low temperature, reduced photoperiod or a combination thereof (Levitt, 1980). The underlying physiological and molecular mechanisms of freezing tolerance of plants are not well understood. Hardy plants undergo a series of metabolic and physiological changes during cold acclimation. It is believed that these changes, which are promoted by low temperature, enable the plants to survive severe freezing stress. Preferential synthesis of some metabolites during cold hardening such as proteins, sugars and nucleic acids has been suggested to be induced by low temperature (Brown, 1978; Kacperska-Palacz, 1978; Levitt, 1980). Marked increases were also found in the content of total soluble sugars, reducing sugars and ATP in winter wheat during cold hardening (Perras and Sarhan, 1984). In subsequent studies, Perras and Sarhan (1989) showed that cold hardening induced important changes in the soluble protein patterns depending upon the tissue and cultivar freezing tolerance. For example, a 200 kilodalton protein was induced concomitantly in the leaves, crown and roots of cold acclimated winter wheat. Guy and Haskell (1987) have also found three high molecular weight induced proteins (160,117 and 85 kD) in spinach during cold acclimation. On the other hand, appearance of several small polypeptides (11 to 38 kD) were found in acclimated alfalfa seedlings (Mohapatra et al., 1987). Similarly, during induction of freezing tolerance in cell suspension cultures, low molecular weight polypeptides (20 to 48 kD) were detected in *Brassica napus* (Johnson-Flanagan and Singh, 1987) and *Bromus inermis* (Robertson et al., 1987).

Taken together, all the physiological and metabolic alterations that occur during the cold acclimation process influence the freezing survival of plants either directly or indirectly. In this context, two scenarios have been developed on the exploitation of the cold acclimation process in the context of cryopreservation; the first involves utilization of explants from plants which are naturally acclimated and the other is the artificial induction of cold acclimation provided the experimental plants are capable of doing so, and use such material for cryopreservation. Both approaches have successfully resulted in cryopreservation of various types of plant material. Sakai and Nishiyama (1978) employed winter vegetative buds of apple for cryopreservation. They found that in hardy shoots of apple prefrozen at the temperature range from −30 to −50 °C, little or no injury was observed in the leaf buds and cortex following immersion in LN_2 and that the buds stored for almost two years still remained alive. The same approach, according to them, was applicable to hardy buds of other fruit crop species such as gooseberry, currant, raspberry and pear. Similarly, shoot-tips aseptically isolated from dormant apple buds survived immersion in LN_2 following pre-freezing below −10 °C without the application of cryoprotectants.

The process of cold acclimation is influenced by seasonal changes and, therefore, explants isolated from donor plants at various time periods may have a profound effect on their cryosurvival. Two examples amply illustrate this point. Apices prepared from brussels sprouts (*Brassica oleracea*) from late July to early November showed no tolerance to freezing to −30 °C and LN_2 while those obtained on 25 November or 16 December survived freezing to LN_2 temperature. Apices preconditioned by low temperature in autumn or winter could survive freezing in LN_2, whereas those isolated from plants growing in summer and which had experienced no low temperature could not survive freezing (Harada et al., 1985). In the case of apple, shoot tips suspended in distilled water survived freezing slowly to −40 °C if taken from January to middle of March and their freezing tolerance (FT) decreased to −20 °C from late March to early April. Again from early May to late October, the FT increased to −40 °C (Katano, 1986).

The second approach involves acclimating the donor plants or *in vitro* cultures to low temperature prior to explant isolation for freezing. Cold treatment of donor carnation plants at 4 °C for 3 days or more resulted in a doubling of the percentage of excised, frozen shoot apices which survived freezing and a 6 to 7-fold increase in the percent that formed leaf primordia and shoots (Seibert and Wetherbee, 1977).

Studies on cold acclimating *in vitro* cultures have been attempted. Cultures of apple and Saskatoon berry plants (*Amelanchier alnifolia*) subjected to a 10 week short photoperiod, low temperature hardening treatment, including a −3 °C exposure followed by 5–7 days at 2 °C were 4 to 8 °C hardier than untreated shoot cultures (Caswell et al., 1986). A correlation has been established between cold acclimation and the level of dehydration attained by dormant apple vegetative buds in relation to cryopreservation in studies carried out by Tyler and Stushnoff (1988a, b). They found that the survival of apple buds could be increased if the tissue was sufficiently dehydrated prior to cryopreservation. Buds collected early in the cold-acclimating period suffered injury as a result of dehydration, but the percent survival of the dehydrated buds, after storage in LN_2 was greater than non-dehydrated buds. As cold acclimation progressed, the buds became more resistant to dehydration stress and the survival increased after cryostorage. Other studies on the effect of cold acclimation of *in vitro* plants on the survival of cryopreserved shoot tips are those by Kuo and Lineberger (1985) again with apple, Reed (1988) with *Rubus* spp. and Dereuddre et al. (1990a) with *Purus* spp.

Callus and cell cultures have also been used to study the effect of cold acclimation on cryopreservation. For example, Sakai and Sugawara (1973) showed that callus derived from the cambial area of poplar twigs survived freezing down to −120 °C or even the temperature of liquid nitrogen, after cold acclimation. Similarly growth at 2 °C increased the freezing tolerance of cultured pear cells (Wu and Wallner, 1985). In a study on the cryopreservation of winter wheat suspension culture and regenerable callus, it was found that cold hardening or ABA treatment before cryopreservation increased freezing resistance and improved survival in liquid nitrogen (Chen et al., 1985). While cold acclimation is an excellent way of enhancing cryosurvival, its application is limited to those species which exhibit this trait for low temperature acclimation. It is doubtful if tropical species would respond in a similar way.

3.2.4. *Application of Abscisic Acid*

Abscisic acid (ABA), a naturally occurring plant hormone, is implicated in eliciting an array of physiological and biological responses in plants such as stomatal control, water relations, photosynthesis, growth, dormancy, embryo maturation and adaptation to stress (for a review see Zeevaart and Crelman, 1988). It is also know as a 'stress hormone'. In recent years, ABA has been demonstrated to induce freezing resistance in plants (Irving, 1969), plant parts (Chen et al., 1983) and in plant tissue cultures (Chen and Gusta, 1983). It has been shown that tissue cultures from five species capable of cold hardening increased their cold hardiness upon treatment with ABA while species that could not harden, did not (Chen and Gusta, 1983). ABA was also effective in circumventing the requirements for both low temperature and lengthy acclimation periods in the induction of freezing tolerance in an embryogenic microspore-derived cell suspension culture of winter rape (*Brassica napus* cv. Jet Neuf). A higher level of freezing tolerance was induced when cells were cultured for 7 to 8 days in 50 μM ABA and 13% sucrose at 25 °C (Orr et al., 1986). *Bromus inermis* cell suspension cultures treated with 75 μM ABA for 7 days at 25 °C survived slow cooling to −60 °C and over 80% of the cells in ABA-treated cultures survived immersion in LN_2. As discussed earlier, cold hardening or ABA treatment of winter wheat suspension culture before cryopreservation increased the freezing resistance and improved the survival of the cells in LN_2 (Chen et al., 1985). The exact mechanism by which ABA imparts this effect remains to be elucidated.

3.3. *Freezing Methods*

3.3.1. *Slow Freezing*

Various physicochemical events occur during freezing of biological specimens at slow cooling rates. Mazur (1969) identified the events to which a cell is subjected during freezing and thawing. With temperature reduction the cell and its external medium initially supercool followed by ice formation in the medium. The cell membrane/wall act as a physical barrier and prevent the ice from seeding the cell interior at temperatures above ca −10 °C and thus the cells remain unfrozen but supercooled. As the temperature is further lowered, an increasing fraction of extracellular solution is converted into ice resulting in the concentration of extracellular solutes. Since the cells remain supercooled and its aqueous vapour pressure exceeds that of frozen exterior, the cell equilibrates by loss of water to external ice (dehydration). Slowly cooled cells reach equilibrium with the external ice by efflux of water and remain shrunk provided the cell is sufficiently permeable to water. In such cases, intracellular ice formation, considered to be one of the most important factors responsible for causing freezing injury, will not occur. This phenomenon is utilized in devising cryopreservation techniques by slow freezing.

A number of factors such as cooling rates, pretreatments and cryoprotection, type and physiological state of the experimental material, and the terminal freezing temperature (also called prefreezing temperature) influence the success of slow freezing method. If the freeze preservation equipment does not have the capability to predict the initiation of freezing the solution and mechanisms to override the supercooling of the sample, a linear cooling rate cannot be obtained. In other words, the specimen would be frozen at different cooling rates even though a precise cooling rate was originally selected. This biophysical variable can have a profound effect on the viability of the sample being frozen.

The most commonly used method of freezing meristems and cell cultures is by regulated slow cooling at a rate of 0.5 to 1.0 °C/min down to either −30, −35 or −40 °C followed by storage in liquid nitrogen. In some instances, a modification to slow cooling by holding the samples for a predetermined period of time in the temperature region of −30 to −40 °C is also practised and is referred to as 'step-wise freezing'. Meristems, cell cultures and somatic embryos of a number of species have been successfully cryopreserved following slow freezing methods (see Kartha, 1985; Withers, 1985).

3.3.2. *Rapid Freezing*

Rapid freezing is the simplest form of cryopreservation since the procedure does not require sophisticated and expensive equipment. It is believed that during rapid lowering of temperature, the cells do not have sufficient time to equilibrate with the external ice by efflux of water as explained for the slow freezing method. However, the cells attain equilibrium by intracellular freezing which is lethal for most biological specimens. The extent of intracellular freezing or ice formation is also governed by the extent to which the cells have been dehydrated prior to their quenching in liquid nitrogen. Meristems and somatic embryos of a few plant species have been successfully cryopreserved by rapid freezing. How such cells survive is explained on the basis of the mechanism of Luyet (1937) which suggests that viability of cells may be maintained by preventing the growth of intracellular ice crystals formed during rapid cooling by rapidly passing the tissue through the temperature zone in which lethal crystal (ice) growth occurs (Seibert and Wetherbee, 1977, for cryopreservation of carnation meristems). In recent years, the conventional rapid freezing technique has been modified to evolve 'vitrification' effects resulting in the viable freezing of a number of culture systems (see the section 3.3.4)

3.3.3. *Droplet Freezing*

The basis of droplet freezing is identical to slow freezing. Protective dehydration of the cells is induced by regulated slow cooling avoiding the chances of intracellular ice formation. This technique was originally developed for the cryopreservation of cassava meristems (Kartha et al., 1982a).

In the droplet freezing method, the cryoprotectant solution (15% DMSO and 3% sucrose) is dispensed into droplets of 2 to 3 μl on an aluminum foil contained in a petri dish. Each cassava meristem, treated with the cryoprotectant solution, is then transferred to the droplets and frozen by programmed slow cooling at a rate of 0.5 °C/min to various sub-zero temperatures and stored in liquid nitrogen. This technique resulted in successful cryopreservation of cassava meristems. Before this technique can be adapted for cassava germplasm preservation, the plant regeneration frequency of cryopreserved meristems needs to be improved. The advantage of droplet freezing on aluminum foil could be attributed to the fact that the metal has an efficient thermal conductivity resulting in homogeneous cooling of the sample by uniform dispersion of temperature. Moreover, since a very small amount of cryoprotectant solution is used for each meristem, a uniform cooling rate occurs accompanied by less amount of extracellular ice formation, thus avoiding the excess mechanical pressure exerted on the cell wall as the ice propagates.

3.3.4. *Vitrification*

The development of vitrification techniques for the cryostorage of cultured plant cells and organs is of recent origin although Sakai as early as 1958 succeeded in obtaining the survival of hardy mulberry cortical tissues in LN_2 by vitrification. The conventional cryopreservation techniques so far discussed involved a freeze-induced cell dehydration step prior to LN_2 storage. An alternate approach to cryopreservation is based on the ability of highly concentrated solutions of cryoprotectants to supercool to very low temperatures upon imposition of rapid cooling rates, to become viscous at sufficiently low temperatures, and solidify without the formation of ice. This process is known as vitrification (Rall and Fahy, 1985). If cells are capable of tolerating this severe osmotic stress to allow glass transition to occur, theoretically they should be able to survive the freezing process.

There are both advantages and disadvantages to vitrification. The advantages are that (i) it is a very simple process, and (ii) it does not require regulated cooling so that the need for expensive programmable freezing equipment is eliminated. The main disadvantage is that the highly concentrated levels of many cryoprotective agents required for vitrification may cause extreme toxicity to the cells. Toxicity of the vitrification solution is relative to its osmotic potential and therefore formulation of vitrification solutions differing in osmotic potential may alleviate the problem of toxicity (Langis and Steponkus, 1990a). The most commonly used compounds in vitrification solutions include DMSO, glycerol, ethylene glycol, polyethylene glycol, propylene glycol, sucrose, etc. In all cases where vitrification has been shown to occur, osmotic dehydration of the cells prior to their quenching in LN_2 has been found to be essential. Cryopreservation by vitrification has been successfully so far in a limited number of culture systems, such as cell cultures of *Brassica campestris* (Langis et al., 1989), navel orange (Sakai et al., 1990), somatic embryos of *Asparagus officinalis* (Uragami et al., 1989), *Daucus carota* (Dereuddre et al., 1991a, b), mesophyll protoplasts of *Secale cereale* (Langis and Steponkus, 1990b), shoot tips of mint, potato, papaya (Towill, 1990a, b) and carnation (Towill, 1990b; Langis et al., 1990). A brief description of the methodology of vitrification is given in Table 2.

3.4. *Cryostorage and Thawing Regimes*

Storage temperature should permit total immobilization of metabolic activities of cells and arrest the demand for energy inputs if long-term preservation is desired. This is accomplished at ultra-low temperatures such as that of liquid nitrogen (−196 °C). Ease of availability of liquid nitrogen makes this strategy feasible in most countries and preferable to other systems wherein any defect in the mechanical system or interruption in power supply could seriously endanger specimen viability.

The events occurring during thawing do not simply reflect freezing in reverse. If thawing is carried out at slow rates, structural changes in ice may occur which may or may not affect cell viability. Of primary importance is avoidance of recrystallization and ice crystal growth which may occur during warming. In order to avoid any potential problems, thawing is generally carried out rapidly by immersing the specimen for 1 to 2 minutes in 35 to 40 °C water bath.

Table 2. Examples of cryopreservation by vitrification

Cultures systems and references	Procedure
Cell cultures of *Brassica campestris* L. (Langis et al., 1989)	Equilibration of cells at 0 °C with 1.5 M ethylene glycol (EG); dehydration of cells at 0 °C in a concentrated solution containing 7.0 M EG + 0.88 M sorbitol + 6% (w/v) BSA; transfer of cell suspension into 0.5 ml polypropylene straw and quenching in liquid nitrogen (LN_2). Thawing in air for 10 s followed hy 10 s in an alcohol bath at 20 °C. Viability of 40% based on TTC assay. Note: To attain maximum viability, the cells were unloaded into a 1.5 osmolar sorbitol medium following thawing.
Mesophyll protoplasts of rye (*Secale cereale* L.) (Langis and Steponkus, 1990b)	Essentially the same as above.
Somatic emhryos of *Asparagus officinalis* (Uragami et al., 1989)	The vitrification solution (PVS) contained (w/v) 22% glycerol, 15% EG, 15% propylene glycol, and 7% DMSO in MS medium enriched with 0.5 M sorbitol. After initial cryoprotection with sorbitol-supplemented MS medium containing 12% EG, the cells and somatic emhryos were exposed stepwise to 85% PVS at 0 °C, loaded into 0.5 ml transparent straws and quenched into LN_2. PVS was removed and diluted stepwise after rapid thawing. Survival in the range of 50–65%; embryos developed into plantlets.
Cell cultures derived from nucellar tissues of navel orange (*Citrus sinensis*, var. *brasiliensis* Tanaka) (Sakai et al., 1990)	Cells were dehydrated with highly concentrated 60% cryoprotective solution (PVS2) containing (w/v) 30% glycerol, 15% EG, and 15% DMSO in Murashige and Tucker medium containing 0.15 M sucrose at 25 C for 5 min followed by chilling at 0 °C for 3 min. Aliquots of 0.1 ml treated cells were then loaded into 0.5 ml transparent plastic straws and quenched in LN_2. After rapid warming, the cell suspension was expelled in 2 ml of MT medium containing 1.2 M sucrose. Average survival rate was 80%. The vitrified cells regenerated plantlets.
Same cell culture as above (Sakai et al., 1990)	Procedure same as above except in that 1.8 ml plastic cryotubes were employed instead of plastic straws. The frozen cultures were stored for 40 days retaining an average survival rate of 90%. The method has been extended to six other species or cultivars of *Citrus*.
Shoot tips of Mint (Mentha aquatica × *M. spicata*) (Towill, 1990a).	Mint shoot tips were gradually exposed to a mixture containing 35% EG, 1.0 M DMSO and 10% PEG-8000 and quenched in LN_2. Rapid thawing at 35°C. Cooling and warming rates calculated at about 4800 and 9000 °C/min, respectively.

Table 2. Continued

Cultures systems and references	Procedure
	Survival ranged from 31 to 75% and in many cases, the frozen shoot tips regenerated into shoots without or with only slight callus formation.
Shoot tips of mint, potato, carnation and papaya (Towill, 1990b)	Preliminary report. The above procedure was applied with minor modification. Initial addition of DMSO and PEG was done slowly to facilitate permeation of DMSO into the shoot tips. The vitrification solution caused considerable loss of viability to shoot tips within an hour. Brief treatment was less toxic. The final concentration of between 35–50% (w/w) of either EG or PEG were effective on retaining viability upon exposure to LN. Exact survival rates not given.
Shoot-tip of carnation (*Dianthus caryophyllus*) (Langis et al., 1990)	Shoot tips isolated from either fresh cuttings or cuttings stored at 5 °C in the dark up to 16 weeks (cold acclimated) were equilibrated 60 min with 8 wt% EG (1.53 osm) at 20 °C or 10 wt% DMSO (1.53 osm) at 0 °C. Further dehydration of the shoot tips was carried out in a vitrification solution composed of 50 wt% EG + 15wt%sorbitol + 6 wt% BSA, for an optimal duration of 15 min. Following dehydration, the shoot tips and the solution were frozen in heat sealed polypropylene straw (0.5 ml) in LN_2. The samples were stored in LN_2 for 34 days. The shoot tips were recovered in 1.5 osm sorbitol solution at 20 °C and after 30 min, returned to a hypotonic semi-solid regeneration medium. This procedure resulted in 100% survival (76/76) for shoot tips initially equilibrated with EG.
Alginate-coated somatic embryos of Carrot (*Daucus carota*) (Dereuddre et al., 1991a, b)	Although the authors did not identify this procedure as Vitrification (Dereuddre et al., 1991a), the subsequent study on thermal analysis (Dereuddre et al., 1991b) showed that vitrification has in fact occurred during freezing. Somatic embryos of carrot were encapsulated in calcium alginate using standard procedure. The encapsulated embryos were precultured for 18 h in a nutrient medium enriched with 0.3 M sucrose. Prior to quenching in LN_2, the precultured embryo beads were dehydrated at room temperature, in a laminar air flow cabinet, for 4 h. Survival rates of up to 92% were obtained. Resistance of somatic embryos to LN_2 was correlated with glass transition between −50 and −70 °C and similar glass transitions were obtained with encapsulating material and highly concentrated sucrose solution.

3.5. *Viability Assays and Regrowth Manipulations*

In the past a technique which has proven to be satisfactory for a number of cryopreserved cell cultures involved gradually washing the thawed cells with chilled nutrient medium prior to their return to culture either in liquid or semi-solid medium. It was later found to be deleterious to a number of cell cultures. For example, viability of suspension cultured sugarcane cells protected with a mixture of cryoprotectants was poorer when they were washed with a cold solution than with a 22 °C wash (Finkle and Ulrich, 1982). The washing step has been found to be deleterious to a few cell cultures such as the alkaloid – producing *Catharanthus roseus* (Chen et al., 1984a), *Zea mays*, Paul's Scarlet rose, and *Acer pseudoplatanus* (Withers and King, 1979; 1980). While a special post-thaw culturing technique had to be developed to enhance regrowth of the alkaloid-producing cultures of *C. roseus* (Chen et al., 1984a), the non-producing cell cultures of the same species could be successfully returned to either liquid or semi-solid medium immediately following post-thaw wash (Kartha et al., 1982b). The special growth technique involved plating thawed cells on to filter papers placed over regrowth medium for a specified period of time to facilitate slow diffusion of cryoprotectants and subsequently transferring the filter paper with cells to two changes of nutrient medium. This technique has since been applied to other cell cultures such as *Picea glauca* (Kartha et al., 1988), *Saccharum* sp. (Gnanapragasam and Vasil, 1990) and *Citrus sinensis* (Sakai et al., 1991b). Other manipulations included omission of ammonium ions in post-thaw culture media or incorporation of activated charcoal (Kuriyama et al., 1989).

A number of viability assays are available for rapid estimation of viability of cryopreserved cells. These include fluorescein diacetate staining (FDA), triphenyltetrazolium chloride (TTC) reduction assays and the use of a number of vital stains. The FDA and TTC are generally used to assess viability. The FDA staining technique was originally devised by Widholm (1972) and is based on the observation that only living cells are stained with FDA and emit fluorescence under UV light. In the TTC method (Steponkus and Lanphear, 1967), determination of survival is based on the reduction of TTC by viable cells by mitochondrial activity, to form formazan, a water insoluble red compound. This reduction is considered to be quantitative and the formazan is made soluble by addition of ethanol and the supernatant can thus be examined spectrophotometrically.

Other parameters such as mitotic index, cell number, cell volume, dry and fresh weight, plasmolysis and deplasmolysis, plating efficiency, leakage of electrolytes, could also be used and compared with the unfrozen control in order to assess viability. A combination of various assays is advisable for a true representation of viability.

The most convincing test for viability of frozen-thawed cells would be their ability to regrow upon return to culture without a long lag phase, and exhibit identical growth characteristics as the original culture. A long lag phase is a direct reflection of freezing injury resulting from the use of sub-optimal freezing conditions. In case of meristems, the most reliable and convincing method of viability assessment is the direct differentiation of the cryopreserved meristems into plantlets without intervening callus. This is extremely important from the perspective of germplasm preservation which demands retention of genetic fidelity. Indirect shoot regeneration mediated by a callus phase should be viewed with caution since it could lead to undesirable somaclonal variation. Direct regeneration of plants can only be obtained if the meristem explants do not suffer extensive freezing damage. In this context, some of the earlier studies revealed extensive freezing damage to the meristems and preferential survival of leaf primordia (Haskins and Kartha, 1980; Grout and Henshaw, 1980).

3.6. *Post-Cryopreservation Stability*

Cryopreservation strategy has been proposed as the final answer to the problems of genetic stability on the grounds that storage at liquid nitrogen temperature will not permit the occurrence of any genetic drift in the preserved specimen. A limited number of examples available so far totally endorse this theory. For example, extensive field evaluations extending over 3 asexual generations of strawberry plants regenerated from cryopreserved meristems after 5 years of storage did not reveal any unique variation resulting from cryopreservation *per se*. Tissue culture-induced variations were noted however, these variants were present both in plants derived from unfrozen and cryopreserved meristems (Kartha, unpublished). Similarly plants regenerated from cryopreserved cassava meristems were field tested at CIAT (Cali, Colombia) and a number of well defined genetic markers as well as esterase isozyme profiles were employed to detect variations, if any. The results revealed no variation arising as a consequence of cryopreservation (Kartha and Roca, unpublished). Towill (1983) examined over 1000 regenerants from cryopreserved meristems of two cultivars of potato and did not detect any variation. These few examples are very encouraging and indicate that cryopreservation has the potential of being a safe method for germplasm preservation. Retention of maximum viability and plant regeneration has to be accorded a high priority from a preservation perspective. Extremely low rate of survival following cryopreservation should be viewed with extreme caution since other variables such as cryoselection might operate in this instance.

When special lines of cell cultures are cryopreserved, it is imperative that the cells maintain the same trait after retrieval from liquid nitrogen storage. A number of examples are now available to depict that the cryopreserved cells maintain their embryogenic potential or plant regeneration capability (for a review see Withers, 1985). Cryopreservation technology has also been applied to a number of cell cultures which possess unique biosynthetic potential (for a review see Kartha, 1987). For example, 25 different anthocyanin producing cell cultures of wild carrot produced approximately the same amount of anthocyanin after cryogenic storage as did the unfrozen controls (Dougall and Whitten, 1980). Cryopreserved cells of *Digitalis lanata* also retained biochemical activity relative to the transformation of cardenolides (Dietrich et al., 1982; Seitz et al., 1983). *Lavandula vera* callus recovered after cryopreservation retained not only the biosynthetic capacity for biotin, but also plant regeneration capability as did the original green callus cultures (Watanabe et al., 1983). Cell lines of alkaloid-producing *Catharanthus roseus* also retained their biosynthetic capability for the production of indole alkaloids (Chen et al., 1984a; Mannonen et al., 1990). Other examples of the cryopreservation of such special cell cultures (the name of the natural compound they produce are included in parenthesis) are: *Chenopodium rubrum* (betalaines) (Ziebolz and Forche, 1985); *Coleus blumei* (rosmarinic acid) (Reuff et al., 1988); *Dioscorea deltoidea* (steroids) (Butenko et al., 1984); *Eschscholtzia californica* (anthridines) (Ziebolz and Forche, 1985); *Panax ginseng* (ginsenosides) (Seitz and Reinhard, 1987); *Papaver bracteatum* (chlorophylles) (Ziebolz and Forche, 1985); *P. somniferum* (sanguinarine) (Friessen et al., 1990) and *Thalictrum rugosum* (isoquinolin alkaloids) (Ziebolz and Forche, 1985). These results indicate the potential application of cryopreservation for the long-term preservation of cell cultures in a stable condition.

4. Recent Developments in Cryopreservation

In the recent past, certain new developments have occurred in cryopreservation technology. These include (i) vitrification, (ii) simple freezing method, (iii) encapsulation–dehydration, and (iv) cryoselection. The vitrification procedure has already been discussed elsewhere in this chapter.

4.1. *Simple Freezing Method*

This method developed by Sakai et al. (1991b) for cell cultures of *Citrus sinensis* totally eliminates the need for programmable freezing and the use of DMSO as a cryoprotectant. The nucellar cells are cryoprotected with a mixture of 2 or 3 M glycerol and 0.4 M sucrose for 10 minutes at 25 °C. A 0.2 ml of this sample is loaded to a 0.5 ml transparent straw and frozen spontaneously by placing the straw in a freezer at −30 °C for 20–30 minutes prior to immersion in LN_2. The thawing is carried out using established procedures by placing the straws in a water bath at 40 °C. After thawing, the cell suspension was expelled into 2 ml of a diluent solution containing 1.2 M sucrose in a nutrient medium at 25 °C. This procedure resulted in 90% survival based on FDA and phenosafranin staining. Upon plating, the cryopreserved cells resumed growth within 3 days and subsequently developed into plantlets via embryogenesis.

If this procedure can be successfully extended to other culture systems, it would make cryopreservation technology accessible and affordable throughout the world.

4.2. *Encapsulation-Dehydration*

Encapsulation is now commonly used in the 'synthetic seeds' technology by coating somatic embryos in alginate beads. Some preliminary conservation experiments have been carried out recently using this technique. Mulberry buds and sandalwood somatic embryos encapsulated in alginate could be stored for 45 days at 4 °C and made to resume growth after the storage period (Bapat et al., 1987; Bapat and Rao, 1988). For cryopreservation requirements, this technique was further refined by Dereuddre and his co-workers by combining a dehydration step. In the encapsulation-dehydration step, the alginate-coated shoot-tips or somatic embryos are precultured in a medium supplemented with 0.3 to 0.75 M sucrose, dehydrated in a laminar air flow cabinet for 2–6 hours and quenched in LN_2. This procedure resulted in high survival rates of encapsulated shoot-tips of *Solanum phureja* (Fabre and Dereuddre, 1990), *Pyrus communis* (Dereuddre et al., 1990b) and somatic embryos of carrot (Dereuddre, 1991a, b). Based on thermal analysis during freezing, these authors propose that this technique may allow storage of plant organs at higher temperatures than that of liquid nitrogen without the risk of ice recrystallization since vitrification is in operation during the freezing process.

4.3. *Cryoselection*

Conventional cryopreservation involves cryoprotection with or without predehydration followed by either regulated or rapid freezing and storage in liquid nitrogen. Unless the cells or tissues are derived from fully hardened donor sources and maintained in a hardened or dehydrated state, their exposure to liquid nitrogen temperature invariably results in immediate death of cells. We have developed a novel procedure by which immature embryo-derived callus of non-hardy spring wheat could be frozen to liquid nitrogen temperature without the application of cryoprotectants (Kendall et al., 1990). In this process called 'cryoselection', during the first challenge to LN_2, less than 15% of the calli survived, whereas this survival could be increased to 30 to 40% upon subsequent challenge. Survival during cryoselection was independent of the method of freezing, cooling rates and thawing regimes. Seed progeny derived from five of the 11 regenerant cryoselected lines exhibited significantly enhanced tolerance to freezing at −12 °C thus implying that cryoselection appears to involve, at least in part, selection for genetic rather than epigenetic variants. Analysis of one callus line indicated that cryoselection did not induce significant alterations in lipid composition, adenylate energy charge, or freezing point. An increase in the soluble sugar component was detected. Changes were also detected in the protein complement of microsomal membranes and soluble protein extracts of cryoselected callus. In all, seven unique proteins ranging from 79 to 149 kilodaltons were identified. These results demonstrate that freezing tolerant callus can be isolated from a heterogeneous population by cryoselection, and factors that contribute to hardiness at the callus level are biologically stable and can contribute to tolerance at the whole plant level.

5. Conclusions and Future Prospects

Cryopreservation technology has advanced very rapidly ever since the first demonstration of successful cryopreservation of *Daucus carota* cell cultures by Latta (1971) two decades ago. The reason for this rapid progress is attributable to several factors, the most notable being the realization of the merit of cryopreservation as indispensable for germplasm preservation as well as for the safe storage of valuable experimental cell cultures. During the current decade, the dire need to maintain valuable cell cultures in a genetically stable condition with concomitant retention of organogenic or embryogenic potential has become more transparent with the advent of biotechnology. It is assumed that the future years would further hightlight the need for cryopreservation for the preservation of proprietary transgenic germplasm.

An assessment of the current status of cryopreservation technology as it relates to various culture systems reveals that cell culture cryopreservation is leading the way as compared to other organized structures such as meristems or zygotic embryos. The realization of the fact that the extent to which cells are non-lethally dehydrated prior to freezing by various strategic manipulations (preculture, cold acclimation, etc.) plays a role in improving post-freezing survival has, in fact, led to improving cryopreservation technology as well as developing novel approaches such as vitrification and encapsulation- dehydration. It is expected that these improvements in cryopreservation technology would translate into actually bringing difficult-to-cryopreserve systems into cryostorage.

Advocating cryopreservation strategy for germplasm preservation is theoretically very valid. But before doing so, it should satisfy the following rigid criteria: (i) an evaluation of germplasm should be carried out to determine the extent of inherent variability and chosing a representative sample of the variability of the species in question for cryopreservation, (ii)the regeneration capability of the cryopreserved material should be very high both before and after cryostorage, (iii) the nature of regeneration mode is unaltered after cryopreservation, (iv) the procedure should guarantee retention of genetic stability, and (v) the procedure should not lead to inadvertent selection as a consequence of poor survival post-freezing. However, at the moment, the use of cryopreservation is limited to small laboratory collections and its use on a large scale is only exceptional. Indeed, the successful freezing of a plant material which implies the establishment and optimization of very precise conditions requires extensive research using highly sophisticated equipment. In this context, the search for less complicated freezing techniques such as encapsulation-dehydration, Sakai's simplified freezing method and vitrification is of great interest and may be helpful in certain areas. However, the resort to conventional freezing techniques may still remain obligatory in the majority of cases. Additionally, further exploration of cryoselection strategy as an adjunct to cryopreservation may aid in the genetic improvement of crops which are sensitive to low temperature.

Determination of genetic stability of cryopreserved material based on currently practised methods involving conventional phenotypic, biochemical and cytological markers could reach a higher degree of resolution if complemented with restriction fragment length polymorphisms (RFLP) or random amplified polymorphic DNA (RAPD) analyses.

Over the years, national and international bodies, public research institutes and private firms have shown increasing interest in germplasm storage and cryopreservation and this trend is expected to rise in the future as new generations of germplasm and transgenic germplasm become available.

6. References

Anderson, J.O. (1979). Cryopreservation of apical meristems and cells of carnation (*Dianthus caryophyllus*). Cryobiology 16:583.

Ashwood-Smith, M.J. and Farrant, J. (1980). Low Temperature Preservation in Medicine and Biology. M.J. Ashwood-Smith and J. Farrant (eds.), University Park Press, England, p. 323.

Bagniol, S., Engelmann, F., Monfort, S. and Ferry, M. (1990). First successful cryopreservation of date palm (*Phoenix dactylifera* L.) meristems. In: Abstracts: 7th Intl. Cong. Plant Tissue and Cell Culture, Amsterdam, p. 374.

Bajaj, Y.P.S. (1976). Regeneration of plants from cell suspensions frozen at −20°, −70° and −196 °C. Physiol Plant. 37:263–268.

Bajaj, Y.P.S. (1977a). Clonal multiplication and cryopreservation of cassava through tissue culture. Crop Improv. 4:198–204.

Bajaj, Y.P.S. (1977b). Survival of *Nicotiana* and *Atropa* pollen embryos frozen at −196 °C. Curr. Sci. 46:305–307.

Bajaj, Y.P.S. (1978). Tuberization in potato plants regenerated from freeze-preserved meristems. Crop Improv. 5:137–141.

Bajaj, Y.P.S. (1979). Freeze preservation of meristems of *Arachis hyppogaea* and *Cicer arietinum*. Indian J. Exp. Biol. 17:1405–1407.

Bajaj, Y.P.S. (1981). Growth and morphognesis in frozen (−196 °C) endosperm and embryos of rice. Curr. Sci. 50:947–948.

Bajaj, Y.P.S. (1982). Survival of anther- and ovule-derived cotton callus frozen in liquid nitrogen. Curr. Sci. 51:139–140.

Bajaj, Y.P.S. (1983a). Cassava plants from meristem cultures freeze-preserved for three years. Field Crop Res. 7:161–167.

Bajaj, Y.P.S. (1983b). Survival of somatic hybrid protoplasts of wheat × pea and rice × pea subjected to −196 °C. Indian J. Exp. Biol. 21:120–122.

Bajaj, Y.P.S. (1983c). Regeneration of plants from pollen embryos of *Arachis*, *Brassica* and *Triticum* spp. cryopreserved for one year. Curr. Sci. 52:484–486.

Bajaj, Y.P.S. (1984). Induction of growth in frozen embryos of coconut and ovules of citrus. Curr. Sci. 53:1215–1216.

Bajaj, Y.P.S. (1988). Regeneration of plants from frozen (−196 °C) protoplasts of *Atropa belladonna* L., *Datura innoxia* Mill and *Nicotiana tabacum* L. Indian J. Exp. Biol. 26:289–292.

Bapat, V.A. and Rao, P.S. (1988). Sandalwood plantlets from 'synthetic seeds'. Plant Cell Rep. 7:434–436.

Bapat, V.A., Mathre, M. and Rao, P.S. (1987). Propagation of *Morus indica* L. (mulberry) by encapsulated shoot buds. Plant Cell Rep. 6:393–395.

Benson, E.E., Marshall, H. and Withers, L.A. (1984). A light and electron microscopical study of the cryopreservation of cultured shoot-tips of *Brassica napus* and *Solanum tuberosum*. In: Abstracts: Plant Tissue Culture and Its Agricultural Applications, Univ. of Nottingham School of Agriculture, Nottingham, England, p. 95.

Benson, E.E., Harding, E. and Smith, H. (1989). Variation in recovery of cryopreserved shoot-tips of *Solanum tuberosum* exposed to different pre- and post-freeze light regimes. Cryo-Letters 10:323–344.

Bertrand-Desbrunais, A., Fabre, J., Engelmann, F., Dereuddre, J. and Charrier, A. (1988). Reprise de l'embryogenèse adventive d'embryons somatique de cafeier (*Coffea arabica*) après leur congelation dans l'azote liquide. C.R. Acad. Sci. Paris. Sér. III. 307:795–801.

Binder, W.D. and Zaerr, J.B. (1980a). Freeze preservation of suspension cultured cells of a gymnosperm Douglas fir. Cryobiology 17:624.

Binder, W.D. and Zaerr, J.B. (1990b). Freeze preservation of suspension cultured cells of a hardwood poplar. Cryobiology 17:624–625.

Boucaud (de), M-T. and Cambecedes, J. (1988). The use of 1,2 propanediol for cryopreservation of recalcitrant seeds: The model case of *Zea mavs* imbibed seeds. Cryo-Letters 9:94–101.

Boucaud (de), M-T., Brison, M., Ledoux, C., Germain, E. and Lutz, A. (1991). Cryopreservation of embryonic axes of recalcitrant seed: *Juglans regia* L. *cv. Franquette*. Cryo-Letters 12:163–166.

Bouman, H. and de Klerk, G.J. (1990). Cryopreservation of lily meristems. In: Abstracts: 7th Intl. Cong. Plant Tissue and Cell Culture, Amsterdam, p. 374.

Brown, G.N. (1978). Protein synthesis mechanisms relative to cold hardiness. In: Plant Cold Hardiness and Freezing Stress. Mechanisms and Crop Implications. P.H. Li and A. Sakai (eds.), Academic Press, New York, pp. 153–164.

Butenko, R.G., Popov, A.S., Volkova, L.A., Chernyak, D.N. and Nosov, A.M. (1984). Recovery of cell cultures and their biosynthetic capacity after storage of *Dioscorea deltoidea* and *Panax ginseng* cells in liquid nitrogen. Plant Sci. Lett. 33: 285–292.

Cachita, C.D., Zapirtan, M., Craciun, C. and Vicol, A. (1990). Regeneration from alfalfa and clover calli following their 40 days long preservation in liquid nitrogen – electronmicroscopic aspects. In: Abstracts: 7th Intl. Cong. Plant Tissue and Cell Culture, Amsterdam, p. 375.

Caruso, M., Crespi-Perellino, N., Garofano, L. and Guicciardi, A. (1987). Long term storage of *Vinca minor* and *Panax ginseng* cell cultures. In: Proc. Adv. Stud. on Plant Cell Biol. Albufeira (Algarve). Portugal, 29 March–10 April, pp. 1–4.

Caswell, K.L., Tyler, N.J. and Stushnoff, C. (1986). Cold hardening of *in vitro* apple and saskatoon shoot cultures. Hort. Sc. 210:1207–1209.

Cella, R., Colombo, R., Galli, M.G., Nielsen, E., Rollo, F. and Sala, F. (1982). Freeze preservation of rice cells: A physiological study of freeze-thawed cells. Physiol. Plant. 55: 279–284.

Chaudhury, R., Radhamani, J. and Chandel, K.P.S. (1991). Preliminary observations on the cryopreservation of desiccated embryonic axes of tea (*Camellia sinensis* (L.) O. Kuntze) seeds for genetic conservation. Cryo-Letters 12:31–36.

Chen, H.H., Li, P.H. and Brenner, M.L. (1983). Involvement of abscisic acid in potato cold acclimation. Plant Physiol. 71:362–365.

Chen, T.H.H. and L.V. Gusta. (1983). Abscisic acid induced freezing resistance in cultured plant cells. Plant Physiol. 73:71–74.

Chen, T.H.H., Kartha, K.K., Leung, N.L., Kurz, G.W.G., Chatson, K.B. and Constabel, F.C. (1984a). Cryopreservation of alkaloid-producing cell cultures of periwinkle (*Catharanthus roseus*). Plant Physiol. 75:726–731.

Chen, T.H.H., Kartha, K.K., Constabel, F.C. and Gusta, L.V. (1984b). Freezing characteristics of cultured *Catharanthus roseus* (L.) G. Don cells treated with dimethylsulfoxide and sorbitol in relation to cryopreservation. Plant Physiol. 75:720–725.

Chen, T.H.H., Kartha, K.K. and Gusta, L.V. (1985.). Cryopreservation of wheat suspension culture and regenerable callus. Plant Cell Tissue Org. Cult. 4:101–109.

Chen, W.H., Cockburn, W. and Street, H.E. (1979). Preliminary experiments on freeze-preservation of sugarcane cells. Taiwania 24:70–74.

Chin, H.F., Krishnapillay, B. and Alang, Z.C. (1988). Cryopreservation of *Veitchia* and *Howea* palm embryos: Non-development of haustorium. Cryo-Letters 9:183–186.

Chin, H.F., Krishnapillay. B. and Hor, Y.L. (1989). A note on the cryopreservation of embryos from young embryos of coconuts (*Cocos nucifera* var. *mawa*). Pertanika 12:183–186.

Dereuddre, J. and Kartha, K.K. (1984). Cryopreservation of apple cell suspension cultures. In: Abstracts: First Plant Genetic Engineering Workshop, Canadian IAPTC, Saskatoon, p. 46.

Dereuddre, J., Fabre, J. and Bassaglia, C. (1988). Resistance to freezing in liquid nitrogen of carnation (*Dianthus caryophyllus* L. var. *eolo*) apical and axillary shoot tips excised from different aged *in vitro* plantlets. Plant Cell Rep. 7:170–173.

Dereuddre, J., Scottez, C., Arnaud, Y. and Duron, M. (1990a). Effets d'un endurcissement au froid des vitroplants de poirier (*Pyrus communis* L. cv. Beurré Hardy) sur la résistance à une congélation dans l'azote liquide. C.R. Acad. Sci. Paris 310, Sér. III:265–272.

Dereuddre, J., Scottez, C., Arnaud, Y. and Duron, M. (1990b). Résistance d'apex caulinaire de vitroplants de poirier (*Pyrus communis* L. cv. Beurré Hardy), enrobés dans l'alginate, a une déshydratation puis à une congélation dans l'azote liquide: effet d'un endurcissement préalable au froid. C.R. Acad. Sci. Paris 310, Sér. III:317–323.

Dereuddre, J., Blandin, S. and Hassen, N. (1991a). Resistance of alginate-coated somatic embryos of carrot (*Dacus carota* L.) to desiccation and freezing in liquid nitrogen, 1: Effects of preculture. Cryo-Letters 12:125–134.

Dereuddre, J., Hassen, N., Blandin S. and Kaminski, M. (1991b). Resistance of alginate-coated somatic embryos of carrot (*Daucus carota* L.) to desiccation and freezing in liquid nitrogen, 2: Thermal analysis. Cryo-Letters 12:135–148.

Diettrich, B., Popov, A.S., Pfeiffer, B., Neumann, D., Butenko, R. and Luckner, M. (1982). Cryopreservation of *Digitalis lanata* cell cultures. Planta Medica 46:82–87.

Diettrich, B., Haack, U., Popov, A.S., Butenko, R.G. and Luckner, M. (1985). Long-term storage in liquid nitrogen of an embryogenic cell strain of *Digitalis lanata*. Biochemie und Physiologie der Pflanzenl. 180:33–48.

Dougall, D.K. and Wetherall, D.F. (1974). Storage of wild carrot cultures in the frozen state. Cryobiology 11:410–415.

Dougall, D.K. and Whitten, G.H. (1980). The ability of wild carrot cell cultures to retain their capacity for anthocyanin synthesis after storage at −140 °C. Planta Med. (1980 Suppl.), pp. 129–135.

Engelmann, F. and Dereuddre, J. (1988). Cryoreservation of oil palm somatic embryos: Importance of the freezing process. Cryo-Letters 7:220–235.

Engelmann, F. and Duval, Y. (1986). Cryoconservation des embryons somatiques de palmier à huile (*Elaeis guineensis* Jacq.): Résultats et perspectives d'application. Oléagineux 41:169–173.

Engelmann, F., Duval, Y. and Dereuddre, J. (1985). Survie et prolifération d'embryons somatiques de palmier à huile (*Elaeis guineensis* Jacq.) après congélation dans l'azote liquide. C. R. Acad. Sci. Paris 301, Ser. III:111–116.

Fabre, J. and Dereuddre, J. (1990). Encapsulation dehydration: a new approach to cryopreservation of *Solanum* shoot-tips. Cryo-Letters 11:413–426.

Finkle, B.J. and Ulrich, J.M. (1979). Effect of cryoprotectants in combination on the survival of frozen sugarcane cells. Plant Physiol. 63:598–604.

Finkle, B.J. and Ulrich, J.M. (1982). Cryoprotectant removal as a factor in the survival of frozen rice and sugarcane cells. Cryobiology 19:329–335.

Finkle, B.J., Ulrich, J.M., Rains, D.W., Tisserat, B.B. and Schaeffer, G.W. (1979). Survival of alfalfa, rice, and date palm callus after liquid nitrogen freezing. Cryobiology 16:583.

Finkle, B.J., Ulrich, J.M. and Tisserat, B. (1982). Responses of several lines of rice and date palm callus to freezing at −196 °C. In: Plant Cold Hardiness and Freezing Stress. P.H. Li and A. Sakai (eds.), Academic Press, New York, pp. 643–660.

Finkle, B.J., Ulrich, J.M., Schaeffer, G.W. and Sharpe, F. (1983). Cryopreservation of rice cells. In: Cell and Tissue Culture Techniques for Cereal Crop Improvement. Academia Sinica/International Rice Research Institute, Science Press, Beijing, pp. 343–369.

Finkle, B.J., Zavala, M.E. and Ulrich, J.M. (1985). Cryoprotective compounds in the viable freezing of plant tissues. In: Cryopreservation of Plant Cells and Organs. K.K. Kartha (ed.), CRC Press, Boca Raton, Florida, pp. 75–113.

Friesen, L.J., Kartha, K.K., Leung, N.L., Englund, P., Giles, K.L., Park, J. and Songstad, D.D. (1991). Cryopreservation of *Papaver somniferum* cell suspension cultures. Planta Med. 57:53–55.

Galerne, M. and Dereuddre, J. (1987). Survie de cals embryogènes d'épicéa après congélation a −196 °C. Extrait des Annales AFOCEL 1983, pp. 1–33.

Gnanapragasam, S. and Vasil, I.K. (1990). Plant regeneration from a cryopreserved embryogenic cell suspension of a commercial sugarcane hybrid (*Saccharum* sp.). Plant Cell Rep. 9:419–423.

Goldner, E., Krell, H.W. and Seitz, U. (1990). Cryopreservation of *Atriplex* cell cultures. In: Abstracts. 7th Intl. Cong. Plant Tissue and Cell Culture, Amsterdam, p. 375.

Grout, B.W.W. (1979). Low temperature storage of imbibed tomato seeds: A model for recalcitrant seed storage. Cryo-Letters 15:71–76.

Grout, B.W.W. and Henshaw, G.G. (1978). Freeze-preservatiion of potato shoot-tip cultures. Ann. Bot. (London) 42:1227–1229.

Grout, B.W.W. and Henshaw, G.G. (1980). Structural observations on the growth of potato shoot-tip cultures after thawing from liquid nitrogen. Ann. Bot. (London) 46:243–248.

Grout, B.W.W., Westcott, R.J. and Henshaw, G.G. (1978). Survival of shoot meristems of tomato seedlings frozen in liquid nitrogen. Cryobiology 15:478–483.

Grout, B.W.W., Shelton, K. and Pritchard, H.W. (1983). Orthodox behaviour of oilpalm seed and cryopreservation of the excised embryo for genetic conservation. Ann. Bot. (London) 52:81–384.

Gupta, P.K., Durzan, D. J. and Finkle, B. J. (1987). Somatic polyembryogenesis in embryonic cell masses of *Picea abies* (Norway sruce) and *Pinus taeda* (Loblolly pine) after thawing from liquid nitrogen. Can. J. For. Res. 17:1130–1134.

Guy, C.L. and Haskell, D. (1987). Induction of freezing tolerance in spinach is associated with the synthesis of cold acclimation induced proteins. Plant Physiol. 84:872–878.

Haffner, V. (1985). Mise au point d'un protocole adapte aux celluies de *Myrtillocactus geometrizans* (Mart.) Cons. (T.). Memoire de DEA, Univ. Aix-Marseille III, 32 pp.

Hahne, G.J and Lorz, H. (1987). Cryopreservation of embryogenic callus cultures from barley (*Hordeum vulgare*). Plant Breeding 99:330–332.

Harada, T., Inaba, A., Yakuwa, Y. and Tamura, T. (1985). Freeze-preservation of apices isolated from small heads of brussels sprouts. Hort Sc. 20:678–680.

Harding, K. (1990). Molecular stability of cryopreserved shoot tips of *Solanum tuberosum*. In: Abstracts. 7th Intl. Congr. of Plant Tissue and Cell Culture, Amsterdam, p. 376.

Harding, K., Benson, E.E. and Smith, H. (1991). The effects of *in vitro* culture period on the recovery of cryopreserved shoot-tips of *Solanum tuberosum*. Cryo-Letters 12:17–22.

Haskins, R.H. and Kartha, K.K. (1980). Freeze preservation of pea meristems: Cell survival. Can. J. Bot. 58:833–840.

Hauptman, R.H. and Widholm, J.M. (1982). Cryostorage of cloned amino acid analog-resistant carrot and tobacco suspension cultures. Plant Physiol. 70:30–34.

Heber, U.W. and Santarius, K.A. (1964). Loss of adenosine triphosphate synthesis caused by freezing and its relationship to frost hardiness problems. Plant Physiol. 39:712–719.

Heber, U.W., Tyankova, L. and Santarius, K.A. (1971). Stabilization and inactivation of biological membranes during freezing in the presence of amino acids. Biochim. Biophys. Acta. 241:587–592.

Hellergren, J. and Li. (1981). Survival of *Solanum tuberosum* suspension cultures to −14 °C: the mode of action of proline. Physiol. Plant. 52:449–453.

International Board for Plant Genetic Resources (IBPGR)(1985). Annual Report 1984, IBPGR, Rome.

Irving, R.M. (1969). Influence of growth retardants on the development and loss of hardiness of *Acer regundo*. J. Am. Soc. Hort. Sci. 94:419–422.

Johnson-Flanagan, A.M. and Singh, J. (19873. Alteration in gene expression during the induction of freezing tolerance in *Brassica napus* suspension cultures. Plant Physiol. 85:699–705.

Jekkel, Z.S., Heszky, L.E. and Ali, A.H. (1990). Transfer temperature dependent cryoprotectant and holding time effect in the survival response of cryopreserved cells (*Puccinellia distans* L. Parl). In: Abstracts, 7th Intl. Congr. of Plant Tissue and Cell Culture, Amsterdam, p. 76.

Kacperska-Palacz, A. (1978). Mechanism of cold acclimation in herbaceous plants. In: Plant Cold Hardiness and Freezing Stress. Mechanisms and Crop Implications. P.H. Li and A. Sakai (eds.), Academic Press, New York, pp. 139–153.

Kadzimin, S.B. (1988). Germplasm preservation of orchid through tissue culture. In: The Application of Tissue Culture Techniques in Economically Important Tree Species. R.C. Umaly, I. Umboh, S. Halos and M.N. Normah (eds.), SAMEO-Biotrop, Bogor, Indonesia, Biotrop Special Publ. 35:167–179.

Kartha, K.K. (1982). Cryopreservation of germplasm using meristem and tissue Culture. In: Application of Plant Cell and Tissue Culture to Agriculture and Industry. D.T. Tomes, B.E. Ellis, P.M. Harney, K.J. Kashaj and R.L. Peterson, (eds.), Univ. of Guelph, Guelph, Ontario, Canada, pp. 139–161.

Kartha, K.K. (1985). In: Cryopreservation of Plant Cells and Organs. K.K. Kartha (ed.), CRC Press, Boca Raton, Florida, pp. 276.

Kartha, K.K. (1987). Cryopreservation of secondary metabolite-producing plant cell cultures. In: Cell Culture and Somatic Cell Genetics of Plants, Cell Culture in Phytochemistry. F. Constbel and I.K. Vasil (eds.), Vol. 4. Academic Press, New York, pp. 217–227.

Kartha, K.K. and Gamborg, O.L. (1978). Meristem culture techniques in the production of disease-free plants and freeze-preservation of germplasm of tropical tuber crops and grain legumes. In: Diseases of Tropical Food Crops. H.H. Maraite and J.A. Meyer (eds.), Université Catholique, Louvain-la-Neuve, Belgium, pp. 267–283.

Kartha, K.K., Leung, N.L. and Gamborg, O.L. (1979). Freeze-preservation of pea meristems in liquid nitrogen and subsequent plant regeneration. Plant Sci. Lett. 15:7–15.

Kartha, K.K., Leung, N.L. and Pahl, K. (1980). Cryopreservation of strawberry meristems and mass propagation of plantlets. J. Am. Soc. Hort. Sci. 105:481–484.

Kartha, K.K., Leung, N.L. and Mroginski, L.A. (1982a). *In vitro* growth responses and plant regeneration from cryopreserved meristems of cassava (*Manihot escuenta* Crantz). Z. Pflanzenphysiol. 107:133–140.

Kartha, K.K., Leung, N.L., Gaudet-LaPrairie, P. and Consabel, F. (1982b). Cryopreservation of periwinkle *Catharanthus roseus* cells cultured *in vitro*. Plant Cell Rep. 1:35–138.

Kartha, K.K., Fowke, L.C., Leung, N.L., Caswell, K.L. and Hakman, I. (1988). Induction of somatic embryos and plantlets from cryopreserved cell cultures of white spruce (*Picea glauca*). J. Plant Physiol. 132:529–539.

Katano, M. (1986). Seasonal changes of freezing tolerance of apple shoot tips. Proc. Fac. Agric. Kvushu Tokai Univ. 5:1–5.

Katano, M., Ishihara, A. and Sakai, A. (1983). Survival of dormant apple shoot tips after immersion in liquid nitrogen. Hort. Sc. 18:707–708.

Kendall, E.J., Qureshi, J.A., Kartha, K.K., Leung, N.L., Chevrier, N., Caswell, K. and Chen, T.H.H. (1990) Regeneration of freezing-tolerant spring wheat (*Triticum aestivum* L.) plants from cryoselected callus. Plant Physiol. 94:1756–1762.

Kobayashi, S., Sakai, A. and Oiyama, I. (1990). Cryopreservation in liquid nitrogen of cultured navel orange (*Citrus sinensis* Osb.) necellar cells and subsequent plant regeneration. Plant Cell Tissue Org. Cult. 23:15–20.

Kuo, C.C. and Lineberger, R.D. (1985). Survival of *in vitro* cultured tissue of 'Jonathan' apples exposed to −196 °C. Hort. Sc, 20:764–767.

Kuriyama, A., Watanabe, K., Ueno, S. and Mitsuda, H. (1989). Inhibitory effect of ammonium ion on recovery of cryopreserved rice cells. Plant Sci. 64:231–235.

Latta, R. (1971). Preservation of suspension cultures of plant cells by freezing. Can. J. Bot. 49:1253–1254.

Langis, R. and Steponkus, P.L. (1990a). The toxicity of vitrification solutions by their osmotic potential. Cryobiology 27:654.

Langis, R. and Steponkus, P.L. (1990b). Cryopreservation of rye protoplasts by vitrification. Plant Physiol. 92:666–671.

Langis, R., Schnabel, B., Earle, E.D. and Steponkus, P.L. (1989). Cryopreservation of *Brassica campestris* suspensions by vitrification. Cryo-Letters 10:421–428.

Langis, R., Schnabel, B., Earle, E.D. and Steponkus, P.L. (1990). Cryopreservation of carnation shoot tips by vitrification. Cryobiology 27:657.

Levitt, J. (1980). Responses of Plants to Environmental Stress, Vol. 1. Chilling, Freezing and High Temperature Stress. Academic Press, New York.

Lineberger, D.R. and Steponkus, P.L. (1980). Cryoprotection by glucose, sucrose, and raffinose to chloroplast thylakoids. Plant Physiol. 65:298–304.

Ling, C.J., De, L.S. and Long, H.S. (1987). Sugarcane callus cryopreservation. In: Plant Biology, Vol. 5. Plant Cold Hardiness. P.H. Li (ed.), Alan R. Liss Inc., New York, pp. 323–337.

Lovelock, J.E. (1953). The protective action of glycerol against hemolysis of erythrocytes by freezing and thawing. Biochim. Biophys. Acta 11:28–36.

Lovelock, J.E. and Bishop, M.W.H. (1959). Prevention of freezing damage to living cells by dimethylsulfoxide. Nature 183:1394–1395.

Luyet, B.J. (1937). The vitrification of organic colloids and of protoplasms. Biodynamica 1:1–14.

Maddox, A.D., Gonsalves, F.G. and Shields, R. (1982/83). Successful cryopreservation of suspension cultures of three *Nicotiana* species at the temperature of liquid nitrogen. Plant Sci. Lett. 28:157–162.

Mannonen, L., Toivonen, L. and Kauppinen, V. (1990). The effect of long term preservation on growth and productivity of *Panax ginseng* and *Catharanthus roseus* cell cultures. Plant Cell Rep. 9:173–177.

Manzhulin, A.V., Butenko, R.G.J and Popov, A.S. (1983). Effect of pretreatments of potato apices on surviving after deep freezing. Soviet Plant Physiol. 30:1188–1194 (in Russian with English summary).

Marin, M.L. and Duran-Vila, N. (1988). Survival of somatic embryos and recovery of plants of sweet orange (*Citrus sinensis* L. Osb.). Plant Cell Tissue Org. Cult. 14:51–57.

Marin, M.L., Mafla, G., Roca, W.M. and Withers, L.A. (1990a). Conservation of cassava (*Manihot esculenta* Crantz): The role of cryopreservation. In: Abstracts, 7th Intl. Congr. Plant Tissue and Cell Culture, Amsterdam, p. 371.

Marin, M.L., Mafla, G., Roca, W.M. and Withers, L.A. (1990b). Cryopreservation of cassava zygotic embryos and whole seeds in liquid nitrogen. Cryo-Letters 11:257–264.

Mazur, P. (1969). Freezing injury in plants. Annu. Rev. Plant Physiol. 20:419–448.

Mazur, R.A. and Hartmann, J.X. (1978). Freezing of plant protoplasts. In: Plant Cell and Tissue Culture, Principles and Applications. W.R. Sharp, P.O. Larson, E.F. Paddox and V Raghavan (eds.), Ohio State Univ. Press, Clumbus, Ohio, p. 876.

Mazur, R.A. and Miller, R.H. (1976). Permeability of the human erythroyte to 1 and 2 M solutions at 0 or 20 °C. Cryobiology 13:507.

McAdams, S., Ratnasabapathi, D. and Smith, R.A. (1991). Influence of days of culture on cryoprotectant-supplemented medium and of terminal freezing temperature on the survival of cryopreserved pea shoot tips. Cryobiology 28:288–293.

Meijer, E.G.M., Van Iren, F., Schrinjnemakers, E., Van Zijderveld, M. and Schilperoort, R.A. (1990). Regeneration of rice (*Oryza sativa*) from cryopreserved cell suspension cultures. In: Abstracts, 7th Intl. Congr. Plant Tissue and Cell Culture, Amsterdam, p. 74.

Meryman, H.T., Williams, R.J. and Douglas, M.S.J. (1977). Freezing injury from 'solution effects' and its prevention by natural or artificial cryoprotectants. Cryobiolgy 14:287–302.

Mohapatra, S.S., Poole, R.J. and Dhindsa, R.S. (1987). Changes in protein patterns and translatable messenger RNA populations during cold acclimation of alfalfa. Plant. 84:1172–1176.

Mora, A., Abdelnour, A. and Villalobos, V. (1991). Cryopreservation of *Musa* zygotic embryos. In: Proc. 4th IPBNet Conf., Biotechnology for Tropical Crop Improvement in Latin America, San Jose, Costa Rica, January 14–18, 1991. 2.

Nag, K.K. and Street, H.E. (1973). Carrot embryogenesis from frozen cultured cells. Nature 245:270–272.

Nag, K.K. and Street, H.E. (1975a). Freeze-preservation of cultured plant cells, I: The pretreatment phase. Physiol. Plant. 340: 254–260.

Nag, K.K. and Street, H.E. (1975b). Freeze-preservation of cultured plant cells, II: The freezing and thawing phases. Physiol. Plant. 34:261–165.

Nash, T. (1966). Chemical constitution and physical properties of compounds able to protect living cells against damage due to freezing and thawing. In: Cryobiology. H.Y. Meryman (ed.), Academic Press, New York, pp. 197–211.

Normah, M.N., Chin, H.F. and Hor, Y.L. (1986). Desiccation and cryopreservation of embryonic axes of *Hevea brasiliensis* Muell. Arg. Pertanika 9:299–303.

Orr, W., Keller, W.A. and Singh, J. (1986). Induction of freezing tolerance in an embryogenic cell suspension culture of *Brassica napus* by abscisic acid at room temperature. J. Plant Physiol. 126:23–32.

Panis, B.J., Withers, L.A. and De Langhe, E.A.L. (1990). Cryopreservation of *Musa* suspension and regeneration of plants. Cryo-Letters 11:337–350.

Pence, V.C. (1990). *In vitro* collection, regeneration, and cryopreservation of *Brunfelsia densifolia*. In: Abstracts, 7th Intl. Congr. Plant Tissue and Cell Culture, Amsterdam, p. 377.

Pence, V.C. (1991). Cryopreservation of immature embryos of *Theobroma cacao*. Plant Cell 10:144–147.

Pence, V.C. and Dresser, B.L. (1988). Embryo cryostorage as a technique for germplasm preservation of several large-seeded tree species. In: Abstracts, Beltsville Symp. in Agric. Res. XIII. Biotic Diversity and Germplasm Preservation - Global Imperatives. May 9–11, p. 24.

Perras, M. and Sarhan, F. (1984). Energy state of spring and winter wheat during cold hardening. Soluble sugars and adenine nucleotides. Physiol. Plant. 60:129–132.

Perras, M. and Sarhan, F. (1989). Synthesis of freezing tolerance proteins in leaves, crown, and roots during cold acclimation of wheat. Plant Physiol. 89:577–585.

Polge, C., Smith, A.U. and Parkes, A.S. (1949). Revival of spermatozoa after vitrification and dehydration at low temperatures. Nature 164:666.

Pritchard, H.W., Grout, B.W.W. and Short, K.C. (1986a). Osmotic stress as a pregrowth procedure for cryopreservation, 2: Water relations and metabolic state of sycamore and soybean cell suspensions. Ann. Bot. 57:371–378.

Pritchard, H.W., Grout, B.W.W. and Short, K.C. (1986b). Osmotic stress as a pregrowth procedure for cryopreservation, 3: Cryobiology of sycamore and soybean cell suspensions. Ann. Biot. 57:379–387.

Quatrano, R.S. (1968). Freeze-preservation of cultured flax cells utilizing DMSO. Plant 43:2057–2061.

Rall, W.F. and Fahy, G.M. (1985). Ice-free cryopreservation of mouse embryos at −196 °C by vitrification. Nature 313:573–575.

Reed, B.M. (1988). Cold acclimation as a method to improve survival of cryopreserved *Rubus* meristems. Cryo-Letters 9:166–171.

Reuf, I. (1987). Untersuchungen zur kryokonservierung pflanzlicher Zellkulturen am Beispiel von *Coleus blumei* und *Berbis wilsoniae*. Dissertation zur Erlangung des Grades eines Doktors des Naturwissenschaften. Univ. Eberhard Karl. Tubingen, pp. 130.

Reuff, I., Sietz, U., Ulbrich, B. and Reinhard, E. (1988). Cryopreservation of *Coleus blumei* suspension and callus cultures. J. Plant Physiol. 133:414–418.

Robertson, A.J., Gusta, L.V., Reaney, M.J.T. and Ishikawa, M. (1987). Protein synthesis in bromegrass (*Bromus inermis* Leyss) cultured cells during the induction of frost tolerance by abscisic acid or low temperature. Plant Physiol. 84:1331–1336.

Sakai, A. (1956). Survival of plant tissues at super-low temperatures, I. Low Temp. Sci. Ser. B 14:17–23.

Sakai, A. (1958). Survival of plant tissue at super-low temperature, II. Low Temp. Sci. Ser. B 16:41–53.

Sakai, A. and Nishiyama, Y. (1978). Cryopreservation of winter vegetative buds of hardy fruit trees in liquid nitrogen. Hort. Sc. 13:225–227.

Sakai, A. and Sugawara, Y. (1973). Survival of poplar callus at super-low temperatures after cold acclimation. Plant Cell Physiol. 14:1201–1204.

Sakai, A., Yamakawa, M., Sakato, D., Harada, T. and Yakuwa, T. (1978). Development of a whole plant from an excised strawberry runner apex frozen to −196 °C. Low Temp. Sci. Ser. B. 36:31–38.

Sakai, A., Kobayashi, S. and Oiyama, I. (1990). Cryopreservation of nucellar cells of navel orange (*Citrus sinensis* Osb. var. *brasiliensis* Tanaka) by vitrification. Plant Cell Rep. 9:30–33.

Sakai, A., Kobayashi, S. and Oiyama, I. (1991a). Survival by vitrification of navel orange (*Citrus sinensis* var. *brasiliensis* Tanaka) cooled to −196 °C. J. Plant Physiol. 137:465–470.

Sakai, A., Kobayashi, S. and Oiyama, I. (1991b). Cryopreservation of nucellar cells of navel orange (*Citrus sinensis* Osb.) by a simple freezing method. Plant Sci. 74: 243–248.

Sala, F., Cella, R. and Rollo, F. (1979). Freeze-preservation of rice cells. Physiol. Plant. 45:170–176.

Santarius, K.A. (1971). The effects of freezing on thylakoid membranes in the presence of organic acids. Plant Physiol. 48:156–162.

Santarius, K.A. (1973). The protective effects of sugars on chloroplast membranes during temperature and water stress and its relationship to frost, desiccation, and heat resistance. Planta 113:105–114.

Schrijnemakers, E.W.W., McLellan, M.R. and Van Iren. (1990). Cryopreservation of cell suspensions: A cryomicroscopic study of resistant and susceptible cell lines. In: Abstracts, 7th Intl. Congr. Plant Tissue and Cell Culture, Amsterdam, p. 378.

Seibert, M. (1976). Shoot initiation from carnation shoot apices frozen to −196 °C. Science 191:1178–1179.

Seibert, M. (1977). Process for storing and recovering plant tissue. US Patent No. 4,052,817.

Seibert, M. and Wetherbee, P.M. (1977). Increased survival and differentiation of frozen herbaceous plant organ cultures through cold treatment. Plant Physiol. 59:1043–1046.

Seitz, U. and Reinhard, E. (1987). Growth and ginsenoside patterns of cryopreserved *Panax ginseng* cell cultures. J. Plant Physiol. 131:215–223.

Seitz, U., Alfermann, A.W. and Reinhard, E. (1983). Stability of biotransformation capacity in *Digitalis lanata* cell cultures. Plant Cell Rep. 2:273–276.

Shillito, R.D., Carswell, G.K., Johnson, C.M., DiMaio, J.J. and Harms, C.T. (1989). Regeneration of fertile plants from protoplasts of elite inbred maize. Bio/Technology 7:581–587.

Standke, K.H.C. (1978). Tiefgefrierung nodaler segmente von Kartoffeln mittels flussigen Stickstoff. Landbaufforschung Volkenrode 28:77–78.

Steponkus, P.L. and Lanphear, F.O. (1967). Refinement of the triphenyltetrazolium chloride method of determining cold injury. Plant Physiol. 4:1432–1436.

Sudarmonowati, E. and Henshaw, G.G. (1990). Cryopreservation of cassava somatic embryos. In: Abstracts, 7th Intl. Congr. Plant Tissue and Cell Culture, Amsterdam, p. 378.

Sugawara, Y. and Sakai, A. (1974). Survival of suspension cultured sycamore cells cooled to the temperature of liquid nitrogen. Plant Physiol. 54:772–774.

Sun, C.N. (1958). The survival of excised pea seedlings after drying and freezing in liquid nitrogen. Bot. Gaz. (Chicago) 19:234–236.

Takeuchi, M., Matsushima, H. and Sugawara, Y. (1980). Long-term freeze-preservation of protoplasts of carrot and Marchantia. Cryo-Letters 1:519–524.

Takeuchi, M., Matsushima, H. and Sugawara, Y. (1982). Totipotency and viability of protoplasts after long-term freeze-preservation. In: Plant Tissue Culture, 1982. A. Fujiwara (ed.), Maruzen, Tokyo, pp. 797–798.

Tisserat, B., Ulrich, J.M. and Finkle, B.J. (1981). Cryogenic preservation and regeneration of date palm tissue. Hort. Sc. 16:47–48.

Thorn, E.C. (1990). Cryopreservation of red clover (*Trifolium pratense*) meristems. In: Abstracts, 7th Intl. Congr. Plant Tissue and Cell Culture, Amsterdam, p. 78.

Toivonen, P.M.A. and Kartha, K.K. (1989). Cryopreservation of cotyledons of nongerminated white spruce [*Picea glauca* (Moench) Voss] embryos and subsequent plant regeneration. Plant Cell Rep. 134:766–768.

Towill, L.E. (1981a). *Solanum etuberosum:* A model for studying the cryobiology of shoot-tips in the tuber bearing *Solanum* species. Plant Sci. Lett. 20:315–124.

Towill, L.E. (1981b). Survival at low temperatures of shoot-tips from cultivars of *Solanum tuberosum* group Tuberosum. Cryo-Letters 2:373–382.

Towill, L.E. (1983). Improved survival after cryogenic exposure of shoot-tips derived from *in vitro* plantlet cultures of potato. Cryobiology 20:567–573.

Towill, L.E. (1990a). Cryopreservation of isolated mint shoot tips by vitrification. Plant Cell Rep. 9:178–180.

Towill, L.E. (1990b). Cryopreservation of shoot tips by vitrification. In: Abstracts, 7th Intl. Congr. Plant Tissue and Cell Culture, Amsterdam, p. 79.

Towill, L.E. and Mazur, P. (1976). Osmotic shrinkage as a factor in freezing injury in plant tissue cultures. Plant Physiol. 57:290.

Tyler, N.J. and Stushnoff, C. (1988a). The effects of prefreezing and controlled dehydration on cryopreservation of dormant vegetative apple buds. Can J. Plant Sci. 68:1163–1167.

Tyler, N.J. and Stushnoff, C. (1988b). Dehydration of dormant apple buds at different stages of cold acclimation to induce cryopreservability in different cultivars. Can. J. Plant Sci. 68:1169–1176.

Uemura, M. and Sakai, A. (1980). Survivai of carnation (*Dianthus caryophyllus* L.) shoot apices frozen to the temperature of liquid nitrogen. Plant Cell Physiol. 21:85–94.

Ulrich, J.M., Finkle, B.J., Moore, P.H. and Ginoza. (1979). Effect of a mixture of cryoprotectants in attaining liquid nitrogen survival of callus cultures of a tropical plant. Cryobiology 16:550–556.

Ulrich, J.M., Finkle, B.J., Mackey, B.E., Schaeffer, G.W. and Sharpe, F., Jr. (1984a). Responses of six rice callus cultures to deep-frozen temperatures. Crop Sci. 24:82–85.

Ulrich, J.M., Mickler, R.A., Finkle, B.J. and Karnosky, D.F. (1984b). Survival and regeneration of American elm callus cultures after being frozen in liquid nitrogen. Can. J. For. Res. 14:750–753.

Uragami, A., Sakai, A., Nagai, M. and Takahashi, T. (1989). Survival of cultured cells and somatic embryos of *Asparagus officinalis* cryopreserved by vitrification. Plant Cell Rep. 8:418–421.

Volger, H.G. and Heber, U.W. (1975). Cryoprotective leaf proteins. Biochim. Biophys. Acta 412:335–349.

Von Arnold, D. and Eriksson, T. (1981). *In vitro* studies of adventitious shoot formation in *Pinus contorta*. Can. J. Bot. 59:870–874.

Watanabe, K., Mitsuda, H. and Yamada, Y. (1983). Retention of metabolic and differentiation potentials of green *Lavandula vera* callus after freeze preservation. Plant Cell Physiol. 24:119–122.

Watanabe, K., Kuriyama, A., Ueno, S. and Mitsuda, H. (1990). Cryoprotectability of cultured *Lavandula vera* cells and retention of metabolic and differentiation potentials of the cells after storage in liquid nitrogen. In: Abstracts, 7th Intl. Congr. Plant Tissue and Cell Culture, Amsterdam, p. 379.

Weber, G., Roth, E.J. and Schweiger, H.-G. (1983). Storage of cell suspensions and protoplasts of *Glycine max* (L.) Merr., *Braasica napus* (L.), *Datura innoxia* (Mill) and *Daucus carota* (L.) by freezing. Z. Planzenphysiol. 10:23–29.

Wehner, T.C. (1988). Genetic considerations in germplasm collection and maintenance: A summary. Hort. Sc. 23:95–97.

Widholm, J.M. (1972). The use of fluorescein diacetate and phenosafranine for determining viability of cultured plant cells. Stain Technol. 47:189–194.

Williams, R.J. and Meryman, H.T. (1970). Freezing injury and resistance to spinach chloroplast grana. Plant Physiol. 45:752–755.

Withers, L.A. (1978). Freeze-preservation of synchronously dividing cells of *Acer pseudoplatanus*. Cryobiology 15:87–92.

Withers, L.A. (1979). Freeze-preservation of somatic embryos and clonal plantlets of carrot (*Daucus carota* L.). Plant Physiol. 63:460–467.

Withers, L.A. (1980). Preservation of germplasm. Int. Rev. Cytol. Suppl. 11B:101–136.

Withers, L.A. (1982). The develoment of cryopreservation techniques for plant cell, tissue and organ cultures. In: Plant Tissue Culture 1982. A. Fujiwara (ed.), Maruzen, Tokyo, pp. 793–794.

Withers, L.A. (1985). Cryopreservation of cultured plant cells and protoplasts. In: Cryopreservation of Plant Cells and Organs. K.K. Kartha, (ed.), CRC Press, Boca Raton, Florida, pp. 243–267.

Withers, L.A. and King, P.J. (1979). Proline: A novel cryoprotectant for the freeze-preservation of cultured cells of *Zea mays* L. Plant Physiol. 64:675–678.

Withers, L.A. and King, P.J. (1980). A simple freezing unit and cryopreservation method for plant cell suspensions. Cryo-Letters 1:213–220.

Withers, L.A. and Street, H.E. (1977). Freeze-preservation of cultured plant cells, III: The pregrowth phase. Physiol Plant. 39:171–178.

Wu, M.T. and Wallner, S.J. (1985). Effect of temperature on freezing and heat stress tolerance of cultured plant cells. Cryobiology 22:191–195.

Yamada, T., Sakai, A., Matsumura, T. and Higuchi, S. (1991). Cryopreservation of apical meristems of white clover (*Trifolium repens* L.). Plant Sci. 73:111–116.

Zandvoort, E.A. (1987). *In vitro* germplasm conservation of tropical aroids. Acta Bot. Neerl. 36:150.

Zeevaart, J.A.D. and Crelman, R.A. (1988). Metabolism and physiology of abscisic acid. Ann. Rev. Pl. Physiol. 39:439–473.

Ziebolz, B. and Forche, E. (1985). Cryopreservation of plant cells to retain special attributes. In: Advances in Agriculture and Biotechnology. A. Schaefer-Menhur, (ed.), Kluwer Academic Publishers, Dordrecht, p. 181.

10. Plant Transformation

M.A.W. Hinchee, D.R. Corbin, Ch.L. Armstrong, J.E. Fry, S.S. Sato, D.L. DeBoer, W.L. Petersen, T.A. Armstrong, D.V. Connor-Ward, J.G. Layton and R.B. Horsch

Contents

1. Introduction

In the mid 1980s, the first transgenic plants were produced via *Agrobacterium*-mediated transformation of *Nicotiana tabacum* (tobacco) protoplasts (Horsch et al., 1984; DeBlock et al., 1984). Since then, over 30 different plant species have been genetically engineered by a variety of techniques (Gasser and Fraley, 1989). The three major breakthroughs in the transformation of dicotyledonous plants were: (1) the development of shuttle vectors for harnessing the natural gene transfer capability of *Agrobacterium* (Fraley et al., 1986), (2) the methods to use these vectors for the direct transformation of regenerable explants obtained from plant organs (Horsch et al., 1985), and (3) the development of selectable markers. Genetic modification of many

Indra K. Vasil and Trevor A. Thorpe (eds.), Plant Cell and Tissue Culture, 231–270.

field crops, as well as horticultural and tree species, has been achieved with *Agrobacterium* vectors.

Unfortunately, few monocotyledonous plants are natural hosts for *Agrobacterium*, although there is some evidence that limited gene transfer is possible in cereals (Grimsley et al., 1987; Raineri et al., 1990). To date, there are no well documented reports of regeneration of transgenic plants from *Agrobacterium* infection of cereal cells or tissues. For plants which are not susceptible to *Agrobacterium*-mediated gene transfer, techniques for transfer of free DNA have been developed.

Free DNA transfer methods involve the physical or chemical introduction of DNA into plant cells (Davey et al., 1989). DNA uptake into plant protoplasts has been facilitated by electroporation (Fromm et al., 1985), chemicals such as polyethylene glycol (Krens et al., 1982), or microinjection (Neuhaus and Spangenberg, 1990). Recently, mechanical methods such as high velocity microparticle bombardment (Klein et al., 1987) and silicon carbide fibers (Kaeppler et al., 1990) have been used to facilitate DNA transport into intact plant cells. With continued progress in the development of the above technologies, it should be possible to transform any plant species and genotype in the future.

These transformation technologies have several components in common. In all cases, the usefulness of the method of introduction is dependent on the frequency of gene transfer and the targeting of transformation to cells and tissues which are highly regenerable. Optimization of the survival and growth of these target cells after gene transfer is essential for efficient transformation. Since the gene transfer process is usually very inefficient, most techniques rely on the introduction of selectable marker genes for the positive selection of transformed cells. This chapter first describes the selection tools which are available for all plant transformation techniques, and then provides an overview of the different technologies. Applications of some of the technologies are presented in order to provide more detail into some of the more widely used techniques.

2. Selectable Markers

Many antibiotic and herbicide herbicide genes can be used as selectable markers for the transformation of both dicotyledonous and monocotyledonous plants. Selectable markers are usually necessary for efficient production of transgenic cells and plants. After gene transfer, transformed cells are greatly outnumbered by nontransformed cells. A selectable marker gene allows the preferential growth of transformed cells in the presence of the corresponding selective agent. The availability of multiple selectable markers is useful in developing efficient transformation protocols for diverse species as well as for the introduction of multiple novel traits through successive transformation and regeneration cycles. Several parameters, described below, are critical to the efficiency of a selectable marker in a chosen transformation protocol.

2.1. *Factors Influencing Selection Efficiency*

Selective agents differ in their toxicity to plants, and their toxicity is influenced by the size and developmental state of the plant cells or tissues. Undifferentiated cells and cell clumps will react very differently than plant organ explants such as leaves, stems and cotyledons. Different species also have differing sensitivities to selective agents. For each transformation and regeneration protocol, it is necessary to carefully titrate for the concentration of the selective agent which permits growth and development of the transformed cells while limiting the growth of non-transformed cells. The use of kanamycin as a selective agent is illustrative of this. In many species, such as *Lycopersicon esculentum* (tomato) (Fillati et al., 1987), *Brassica napus* (oil seed rape) (Moloney et al., 1989) and *Lactuca sativa* (lettuce) (Michelmore et al., 1987), relatively low kanamycin concentrations (15–100 mg/l) provide efficient selection of transgenic plants. However, some regeneration systems are not very sensitive to kanamycin and increased concentrations are required. For example, high kanamycin (400 mg/l) selection pressure during shoot organogenesis of *Agrobacterium*-inoculated *Beta vulgaris* (sugarbeet) cotyledon explants results in very poor selection. Only 5% of the shoots produced are actually transgenic, while the rest are 'escapes' from the selection (J. Fry, unpublished results). Unfortunately, higher kanamycin concentrations cause these explants to become necrotic and overgrown with *Agrobacterium* even in the presence of high concentrations of bacteriostatic antibiotics.

Herbicides are usually more toxic to plant tissues than antibiotics. This is in part due to their specific mode-of-action in plant cells as well as their efficient uptake and translocation within plant tissues. Herbicides, and other highly toxic selective agents, may require delayed application in order to ensure that transformed cells have produced sufficient quantities of the resistance enzyme in order to confer protection. For example, Basta® herbicide selection of transformed corn protoplasts required a delay period prior to herbicide application (Donn et al., 1990). Selection with the herbicide glyphosate has also proven most efficient when herbicide application is delayed several days (J. Fry et al., unpublished results).

Selection efficiency is also influenced by regeneration response. Explants which first initiate callus prior to organogenesis or embryogenesis (Umbeck et al., 1987; Michelmore et al., 1986) can usually be selected more efficiently than explants which undergo direct organogenesis or embryogenesis (Hinchee et al., 1988; Parrott et al., 1989).

Media manipulations which allow a relatively long selection period prior to regeneration can potentially reduce the number of escape regenerates.

For example, when *Agrobacterium*-inoculated tomato leaf discs were placed immediately on shoot induction medium containing kanamycin, only escape shoots were produced. Using the same kanamycin concentrations, a high percentage of transformed tomato shoots were produced when the leaf discs are first transferred to a callus induction medium for 2 weeks prior to transfer to shoot induction medium (J. Fry, unpublished results).

2.2. *Antibiotic Markers*

2.2.1. *Aminoglycosides*

The neomycin phosphotransferase, type II (NPT II) gene has been used for the transformation of more plant species than any other selectable marker. The coding sequence was originally isolated from the a bacterial transposon, *Tn5*. NPTII provides resistance to certain aminoglycoside antibiotics such as kanamycin (see Section 4.5: Applications), paromomycin (Masson et al., 1989), and Geneticin (G418) (Michelmore et al., 1987; Kiernan et al., 1989, DeKeyser et al., 1989). Kanamycin selection has proven to be the most widely applicable, and has been used in the majority of transformation procedures for dicotyledonous plants. The NPT II coding sequence has been fused to constitutively regulated promoters, such as the nopaline synthase (NOS) and Cauliflower Mosaic Virus (CaMV) 35S regulatory sequences (Fraley et al., 1983) and cloned into plant transformation vectors by many researchers (An et al., 1985; Bevan, 1984; Klee et al., 1985). The different constitutive promoters may confer different selection efficiencies. Kanamycin selection was improved in an oil seed rape cotyledon transformation protocol when the NPT II coding sequence was driven by the CaMV 35S promoter instead of the NOS promoter (Moloney, 1989).

2.2.2. *Hygromycin*

Hygromycin selection is based on the use of an *E. coli* bacterial gene, *aph* IV, that encodes for hygromycin phosphotransferase. This enzyme inactivates hygromycin and confers a resistant phenotype on the plant cells expressing the gene (Waldron et al., 1985). Hygromycin is highly toxic to most plant species and the frequency of transformants after hygromycin selection in *Petunia hybrida* (petunia) and tobacco is lower than with kanamycin selection (R. Horsch and N. Mathis, unpublished results). However it has been used for efficient transformation of *Oryza sativa* (rice) (Hayashimoto et al., 1990), *Arabidopsis thaliana* (Lloyd et al., 1986), *Medicago sativa* (alfalfa) (D'Halluin et al., 1990), *Glycine max* (soybean) (Finer et al., 1991; Dhir et al., 1991), and *Pisum sativum* (pea) (Lulsdorf et al., 1991).

2.2.3. *Gentamicin*

Gentamicin-3-N-aceytransferases [AAC3( )]-type IV and type IV, [AAC3(3)] inactivate gentamicin by acetylation. These two enzymes, functionally similar although isolated from unrelated sources, confer broad substrate specificities to aminoglycoside compounds *in vitro*. The AAC(3) coding sequences, fused to the CaMV 35S promoter, confer resistance to gentamicin and have been used as selectable markers for the production of transformed tobacco, oil seed rape, and alfalfa plants (Hayford et al., 1988).

2.2.4. *Streptomycin and Spectinomycin*

Streptomycin and spectinomycin resistance is conferred by the *E. coli aad* A gene encoding for the enzyme aminoglycoside-3′-adenyltransferase. Chimeric *aad* A genes, driven by either the CaMV35S or the *Agrobacterium tumefaciens* 2′ promoter, allowed the direct selection of transgenic tobacco callus and shoots in the presence of either spectinomycin or streptomycin (Svab et al., 1990). The use of this marker is unique in that resistant callus and shoots were identified by their green color in presence of these antibiotics which normally cause tissue bleaching.

2.3. *Herbicide Markers*

2.3.1. *Phosphinothricin (PPT)*

DL-Phosphinothricin (PPT) is the active ingredient in the commercial herbicide Basta® (Hoechst Co.) which inhibits glutamine synthetase (GS), an enzyme critical for the assimilation of ammonia and for general nitrogen metabolism in plants (Bayer et al., 1972). A tripeptide form of this herbicide, bialaphos, is produced by *Streptomyces hygroscopicus* (Kondo et al., 1973). The BAR gene from *Streptomyces hygroscopicus* encodes for phosphinothricin acetyltransferase which provides resistance by acetylating phosphinothricin (Murakami et al., 1986). PPT or bialaphos selection has been used successfully in the transformation of *Solanum tuberosum* (potato), tobacco (DeBlock et al., 1987), *Brassica* species (DeBlock et al., 1989), *Populus* hybrids (DeBlock, 1990) *Zea mays* (maize) (Fromm et al., 1990; Gordon-Kamm et al., 1990; Donn et al., 1990) and *Triticum aestivum* (wheat) (Vasil et al., 1992).

2.3.2. *Glyphosate*

Glyphosate inhibits the enzyme 5-enolpyruvylshikamate-3-phosphate synthase (EPSP synthase) by blocking the aromatic amino acid biosynthesis in plants and bacteria (Steinrucken and Amrhein, 1980). Mutant EPSP synthase coding sequences confer glyphosate tolerance in plants when fused to strongly expressed promoters (Shah et al., 1986; Padgette et al., 1989). Selection with sub-lethal doses of glyphosate allows transformed shoots to grow normally while non-transformed shoots demonstrate an abnormal phenotype. Glyphosate selection protocols have been developed for oil seed rape, sugar-beet, tobacco, potato, tomato, and maize (J. Fry, A. Barnason, unpublished results).

2.3.3. *Sulfonylureas*

The sulfonylurea herbicides inhibit acetolactate synthase (ALS or acetohyd-roxacid synthase), the first enzyme in the biosynthetic pathway for the branched chain amino acids valine, leucine, and isoleucine (Chaleff and Mauvais, 1984) Herbicide tolerant and wildtype acetolactate synthase (ALS) genes have been isolated from tobacco (Chaleff and Ray, 1984), yeast (Yadav et al., 1986), Arabidopsis (Smith et al., 1988), and maize (M. Fromm, unpublished results). Sulfonylurea tolerant ALS coding sequences fused to strong plant promoters have been used as selectable markers for tobacco (Gabard et al., 1989; Charest et al., 1990) and maize (Fromm et al., 1990) transformation. The herbicide chlorsulfuron is most commonly used as the selective agent.

2.4. *Other Selective Agents*

2.4.1. *Methotrexate*

Methotrexate (MTX), a potent inhibitor of dihydrofolate reductase (DHFR) is extremely toxic to plant cells. The DHFR coding sequence, fused to constitutive plant promoters, has been a successful selectable marker for the transformation of *Petunia hybrida* (Eichholtz et al., 1987), tobacco (N. Mathis, unpublished results) and oil seed rape (Pua et al., 1987). Transformation efficiencies with MTX selection were lower than with kanamycin, possibly because only the cells which expressed DHFR the greatest were able to survive. Methotrexate may not be a useful selectable marker in maize. Without transformation, methotrexate selection pressure has produced resistant maize cell lines which presumably contain an amplication of an endogenous DHFR gene (Tuberosa and Phillips, 1986).

2.4.2. *Bleomycin and Phleomycin*

Bleomycin and phleomycin are mutagens which create single and double stranded breaks in eukaryotic DNA. Two resistance genes have been identified, one from the Tn5 transposon (Tn5Ble) and the other from the chromosome of *Streptoalloteichus hindustanus* (ShBle). These proteins bind irreversibly to bleomycin and phleomycin and render them inactive. The Tn5Ble gene was used to confer resistance to bleomycin in plant cells, but could not be used as a direct selectable marker (Gatignol et al., 1988; Perez et al., 1989). However, a chimeric gene, containing the ShBle coding sequence driven by the CaMV 35S promoter, produced transgenic tobacco plants with phleomycin selection (Perez et al., 1989). The phleomycin selection efficiency was similar to that obtained with a NOS/NPT II gene and kanamycin selection. Bleomycin was unsuitable as a selective agent since it inhibited plant morphogensis and transgenic plant production (Perez et al., 1989).

3. Reporter Genes

The co-incident regeneration of non-transformed plants (escapes) along with transgenic plants can occur in all transformation protocols. The early identification of genuine transformants is advantageous since it can eliminate a large number of useless plants from later analysis. There are a number of strategies for screening regenerated plants so that the transformants can be identified and selected, but most of them are laborious and time consuming. For example, DNA can be isolated from the putative transformants and subjected to Southern blot analysis, but plants must be at a relatively mature stage to obtain sufficient amounts of tissue for DNA extraction without endangering plant viability or vigor. Another strategy is to employ a recallusing assay in which tissue explants from transgenic candidates are transferred to callusing medium containing the selective agent which permits only the transgenic tissue to callus. Both Southern blotting and the recallusing assay can take several weeks to complete. Enzymatic assays and antibody assays (e.g., ELISA) can be used to identify transgenic plants by detecting selectable marker gene products (e.g., neomycin phosphotransferase), but these assays are laborious and require either radioactive substrates or appropriate antibodies.

A strategy that circumvents the above drawbacks is to include an independent reporter gene in the gene transfer cassette whose presence in regenerated plants can be easily monitored and measured. Two genes that are now widely used for this purpose are those coding for β-glucuronidase (GUS) and luciferase (LUC). The coding regions of these non-plant genes have been fused to plant promoters and polyadenylation sequences such that they confer high levels of expression in plant cells.

3.1. *β-glucuronidase*

A GUS gene isolated from *E. coli* has proven to be very useful in plant transformation studies (Jefferson et al., 1986). Its expression in plant tissues can be very easily detected in a histochemical assay in which a piece of tissue is incubated for several hours in a buffer containing the indigogenic substrate 5-bromo-4-chloro-3-indoyl-1-glucuronide (Jefferson et al., 1986). Tissue that turns blue is transformed; nontransformed tissue typically does not react. This assay needs only very small quantities of tissue and this permits its use very soon after regeneration. One potential drawback to this method is that background expression may exist from intrinsic GUS-like activity in plant tissue (Hu et al., 1990) or from expression of the GUS gene in residual *Agrobacterium* (Janssen and Gardner, 1989). Residual GUS activity from *Agrobacterium* can occur due to expression of the GUS gene from its own promoter or from cryptic bacterial promoters on the plant transformation vector. This problem can be most effectively eliminated by introducing a plant intron into the GUS coding sequence which abolishes bacterial expression while continuing to permit plant expression (Vancanneyt et al., 1990). GUS activity can also be determined by quantitative fluorimetric and spectrophotometric assays (Jefferson, 1987).

3.2. *Luciferase*

The firefly LUC gene encodes a single peptide that requires ATP and an organic substrate, called luciferin, for light emission (Ow et al., 1986). Luciferin is not phytoxic and easily infiltrates into many plant tissues so that light emission from LUC-expressing transgenic plants can be detected. Even very low levels of luciferase can be accurately measured in a luminometer. If activity is high enough, it can be directly visualized by night-scope photography (Barnes, 1990) or by direct exposure of the tissue to photographic film (Fromm et al., 1990). In cases of very high luciferase expression, the light emission can even be visualized by the naked eye (Barnes, 1990).

3.3. *Anthocyanin*

The anthocyanin regulatory genes can serve as a unique reporter system in maize and some other cereals. Transformed tissues will accumulate reddish-purple anthocyanin pigments (Ludwig et al., 1990). The anthocyanin genes are extremely sensitive markers, requiring only low levels of gene expression for visualization of individual pigmented cells. They have been used successfully as non-destructive and rapid stable transformation markers for corn (Bowen, 1992) as well as in *Arabidopsis* and *Nicotiana* transgenic plants (Lloyd et al., 1992).

4. Agrobacterium DNA Transfer

Agrobacterium mediated DNA transfer is the most common and widely used method for the transformation of dicotyledonous plants. *Agrobacterium* was first used to produce transgenic *Nicotiana plumbaginifolia* plants by transformation of protoplasts (Horsch et al., 1984). Since then, *Agrobacterium* mediated transformation of protoplasts has been applied to crop plants, including several species of *Brassica* (Thomzik and Hain, 1990; Sacristan et al., 1989.; Ohlson and Eriksson, 1988.) However, by far the greatest success with *Agrobacterium* transformation has been its ability to transform intact, regenerable plant tissues and organs (Horsch et al., 1985). This capability has made *Agrobacterium*-mediated transformation the method of choice for plant transformation, with its only limitation being host range.

4.1. *Agrobacterium in Nature*

Agrobacterium tumefaciens is a uniquely interesting and useful plant pathogenic bacterium because it has evolved the ability to transfer a discrete portion of its tumor-inducing (Ti) plasmid to plant chromosomes (for review, see Klee and Rogers, 1989; Binns and Thomashow, 1988; Ream, 1989; Zambryski, 1988; Kado, 1991). There are two critical regions of the Ti plasmid: the oncogene-containing T-DNA and the virulence (*vir*) genes which encode proteins required for T-DNA transfer. None of the genes encompassed by the T-DNA are required for DNA transfer, but transfer is dependent upon a 25 bp DNA sequence which is repeated in direct orientation at each end of the T-DNA. The expression of *vir* genes and consequent transfer of T-DNA are inducible by specific signal molecules typically exuded by wounded plant cells. The primary *vir*-inducers are phenolic compounds derived from the phenylpropanoid biosynthetic pathway, but for optimal induction at least one of a number of monosaccharides is also required (Cangelosi et al., 1990). The *vir* gene products are involved in recognizing and cleaving one strand of DNA at each of the 26 bp border sequences and transferring the intervening DNA to the plant cell, where it becomes covalently integrated into the plant genome. Several chromosomal genes in *Agrobacterium*, some of which are involved in polysaccharide production, are required for bacterial attachment and virulence.

4.2. *Agrobacterium in the Laboratory*

For any strain of *Agrobacterium* to be of practical use in transforming plants, it must first be 'disarmed' by deleting the oncogenic T-DNA. This is because the overproduction of phytohormones interferes withnormal plant development. The most frequently used strategy for disarming *Agrobacterium* strains is to remove T-DNA by an *in vivo* recombination event that replaces it with another genetic marker. The resulting strains retain the *vir* genes and are suitable hosts for plant transformation vectors. The Ti plasmids of several *Agrobacterium* strains is to remove T-DNA by an *in vivo* recombination event that replaces it with another genetic marker. The resulting strains retain the *vir* genes and are suitable hosts for plant transformation vectors. The Ti plasmids of several *Agrobacterium* strains of diverse origin have been disarmed and proven to be effective vehicles for foreign gene transfer. pLBA4404 (Hoekema et al., 1983) and pTiB6S3SE (Fraley et al., 1985) are derived from octopine-type Ti plasmids; pGV3850 (Zambryski et al., 1983), pTiT37SE (Rogers et al., 1897) and pMP90 (Koncz and Schell, 1986) are derived from nopaline-type Ti plasmids, and pEHA101 (Hood et al., 1986) is derived from an L,L-succinamophine type plasmid (pTiBo542). More recently, the entire *vir* region of a nopaline-type Ti plasmid (pTiC58) has been cloned into an autonomously replicating broad host range plasmid (Rogowsky et al., 1990). This plasmid can be transferred to any *Agrobacterium* strain which lacks a Ti plasmid, where it can serve as a helper for gene transfer (Zyprian and Kado, 1990).

Differences in virulence and effective host range of commonly used *Agrobacterium* strains have been reported. For example, nopaline strains have generally been found to be more virulent on some crop species than octopine strains (Godwin et al., 1991; McClean et al., 1991; Byrne et al., 1987; Owens and Cress, 1985), but octopine strains are strongly virulent on *Nicotiana glauca* whereas nopaline strains are not (Melchers et al., 1990). Strain A281, which contains pTiBo542, has been shown to be supervirulent on a number of plants (Hood et al., 1986; Jin et al., 1987; Pythoud et al., 1987; Hood et al., 1990). In cases where a species or cultivar is not sufficiently susceptible to transformation by existing disarmed strains, it may be fruitful to test a wide spectrum of wild type *Agrobacterium* strains in order to identify one with improved virulence properties (Stomp et al., 1990; van Wordragen et al., 1991). However, since virulence observed with wild type strains in tumor assays can be a reflection of the behavior of the transferred oncogenes as well as transformation efficiency, the corresponding disarmed strains may not necessarily be more effective hosts for transformation vectors.

4.3. *Plant Transformation Vectors*

Plant transformation vectors are designed to function in three organisms: *E. coli*, in which routine clones manipulations are easily accomplished; *A. tumefaciens*, from which genes can be transferred to plant cells; and plant cells, in which the gene(s) of interest are permanently incorporated into the plant genome and confer the desired trait(s) to the plant. All vectors must have certain basic features. There should be two directly repeated T-DNA border sequences in proper orientation that will define the boundaries of the gene transfer cassette. Although one border is sufficient to mark the initiation of transfer, vector sequences downstream of the gene transfer cassette will also be transferred unless a second border marks termination. There must be two features situated between these border repeats: a suitable plant selectable marker gene and convenient restriction enzyme sites for insertion of the foreign genes. Required features situated outside of the gene transfer cassette are plasmid replication elements and a bacterial selectable marker that function in *E. coli* and *Agrobacterium*, and an RK2 origin of plasmid transfer for shuttling the vector from *E. coli* to *Agrobacterium*. Two different styles of vectors have been developed; co-integrating and binary. Co-integrating vectors rely upon recombination and co-integration with the disarmed Ti plasmid for maintenance of the gene transfer cassette in *Agrobacterium*. Binary vectors replicate autonomously in *Agrobacterium*. The advantages and disadvantages of these two systems have been discussed in detail in a previous review (Klee and Rogers, 1989). Binary vector systems are the easiest to handle and consequently are now more widely used than cointegrating types.

In preparation for transformation, the foreign gene(s) of interest is cloned between the T-DNA borders of the basic vector using standard recombinant DNA techniques and recovered in *E. coli* by transformation and selection. Recombinants are characterized by restriction enzyme analysis of plasmid DNA isolated from the transformed *E. coli*. Once identified, the completed vector is conjugally transferred from *E. coli* to *Agrobacterium* using a triparental mating strategy (Ditta et al., 1980) that introduces a helper plasmid, pRK2013, carrying RK2 plasmid mobilization and transfer functions (Figurski and Helinski, 1979). This is most easily done as a 'smudge' mate where samples of each of the three participants (the disarmed *Agrobacterium*, *E. coli* containing pRK2013 and *E. coli* containing the vector) are mixed together on solid medium and incubated for 24 hours at 30 °C. *Agrobacterium* transconjugants carrying the vector are selected from the mate by streaking on minimal medium containing the antibiotic to which the vector confers resistance (*E. coli* strains used are auxotrophs and are counterselected under these conditions). Alternatively, the selection can be performed on rich medium if the *Agrobacterium* recipient carries an additional selectable marker. Confirmation of vector integrity can be accomplished by isolating crude plasmid DNA from the *Agrobacterium* transconjugants, re-transforming the vectors into *E. coli*, and performing restriction enzyme analyses on plasmid DNA reisolated from *E. coli*.

4.4. *Plant Transformation*

Agrobacterium carrying the shuttle vector is grown in a basic bacterial growth medium (such as *Luria broth* or AB (minimal medium) with appropriate selective antibiotics. The cells are collected by centrifugation and resuspended in the bacterial medium or a basic plant growth medium. Typically, tissue explants (e.g. hypocotyls, cotyledons, leaves, stems or roots) are incubated in the bacterial suspension for the time and final bacterial cell density that gives adequate transformation frequencies without excessive stress to the plant tissue. Inoculated explants are then incubated on an appropriate solid medium for 48–72 hours to allow for DNA transfer and integration. In some systems the explants are incubated upon a sterile filter paper disk that covers a layer of cultured tobacco cells (Horsch et al., 1985). Such 'feeder cell' layers often improve transformation frequencies, either by supplying metabolites that improve explant vigor and regenerability, or by secreting *vir* gene inducers. The specific inclusion of *vir* gene inducers during co-cultivation enhances vir gene expression relative to levels attained with feeder layers alone (D. Corbin, unpublished results) and has a positive effect on transformation frequency in some species (Matthews et al., 1990; Sheikholeslam and Weeks, 1987). At the end of the co-cultivation period, explants are transferred to regeneration medium containing an antibiotic (such as carbenicillin or cefotaxime) that will specifically inhibit bacterial growth and a selective agent that will select those cells receiving and expressing the gene transfer cassette. Transgenic cells within an explant will directly form either shoots, somatic embryos, or regeneration competent callus after one to several weeks on the selective medium. The frequency of trangenic plants produced by this process varies depending on the species, regeneration protocol, and selection regime. Transgenic plants are typically produced at frequencies ranging from 1–10% (number of transgenic plants per explant) but can be produced at frequencies as high as 60% (Burow et al., 1990).

One exciting development in *Agrobacterium* transformation is the ability to directly transform plant apical meristems. This provides the potential to produce transgenic plants directly without the tissue culture manipulations required for plant regeneration. Germinating seeds of *Arabidopsis* (Feldman et al., 1987) and soybean (Chee et al., 1989), which were inoculated with *Agrobacterium* and allowed to grow into plants, have produced transgenic seed and progeny. However, these procedures are very inefficient and require screening many plants or progeny to recover transformants. Transformation of isolated shoot tips is also possible, and transgenic plants have been produced by *Agrobacterium* inoculation of apices of petunia (Ulian et al., 1988) and *Helianthus annuus* (sunflower) (Schrammeijer et al., 1990).

In some applications it may be desirable or necessary to determine the number of T-DNAs present in the transgenic plants and to verify that the gene(s) of interest is intact. DNA rearrangements and multiple insertions of T-DNA can occur (Fry et al., 1987). In addition, truncated T-DNAs can arise in which sequences from either end of the gene transfer cassette may be absent from the integrated form (Herman et al., 1990; Gheysen et al., 1990). Thus, transgenic plants that express selectable and scoreable marker genes do not necessarily carry the gene of interest. Southern blot analysis of genomic DNA from Ro plants can be used to identify plants which contain a single full length copy of the gene transfer cassette.

4.5. *Applications*

4.5.1. *Tomato Transformation*

Transformation of several tomato genotypes (UC82B, UC204C, and VF36) with efficiencies of 5–8% (number transgenic plants per explant) has been achieved using the following protocol. Tomato seeds, surface sterilized by soaking for 20 min in 30% Clorox® and rinsing three times in sterile deionized water, are germinated at 23 °C in Plant Cons (Flow Labs) containing 4.4 g/l MS salts (Sigma), 20 g/l sucrose, Nitsch vitamins (Thomas and Pratt, 1981), at pH 5.8. Cotyledons from 7–8 day old seedlings are excised and gently cut with a #15 feather blade at both ends while immersed in 5 ml of liquid regeneration medium in a petri plate. It is critical to submerge the segments because tomato cotyledons are very sensitive to dessication. Cotyledons are precultured for 24 hours on agar solidified feeder plates composed of preculture medium containing 4.4g/l Sigma MS salts, 30 g/l sucrose, 0.2 g/ KH_2PO_4, 0.88 g/l thiamine HCl, 0.1 mg/l, kinetin, 0.2 mg/l 2,4-D, and 200 μM acetosyringone, at pH 6.0 (Fillatti et al., 1987), which is overlayed with 2 ml of *N. tabacum* ssp. *xanthi* (TXD) tobacco suspension cells and a sterile Whatman filter (Horsch et al., 1985). Cotyledons, placed 20–30 per plate, are cultured so that the best contact between the cotyledons and the feeder layer is attained. *Agrobacterium*, containing a gene transfer cassette, are grown overnight in liquid Luria Broth (LB) medium containing the appropriate antibiotics. They are subcultured by making a 1 to 10 dilution into fresh LB media and are grown for 3 to 4 hours so that they are at a log stage of growth when used for inoculation. The bacterial concentration is adjusted to 5×10^8 cells/ml by diluting in the tobacco suspension culture medium (Horsch et al., 1985). Tomato segments are soaked for 6 minutes in 8 ml of bacterial suspension. Excess bacteria are blotted away on a sterile Whatman filter, taking care to avoid dessication of the explants. The explants are returned to the original feeder plates and are co-cultured for 48 hours. They are then transferred to selection/regeneration medium containing 4.4 g/l MS salts (Sigma), 30 g/l sucrose, Nitsch's vitamins, 2 mg/l zeatin, 100 mg/l kanamycin, 500 mg/l carbenicillin, at pH 5.8 (Fillatti et al., 1987) in 10 × 15 mm petri plates (approximately 10–12 explants/plate). After 3 weeks, explants which have nodular green callus and/or shoots are transferred to fresh 100 × 25 mm plates containing selection medium. Shoots which develop are moved to rooting selection medium containing 1/2 strength MS salts, 30 g/l, sucrose, B5 vitamins, 60 mg/l kanamycin, 500 mg/ml carbenicillin, at pH 5.7. Shoots will either root, remain green or bleach. Those shoots which remain green, but fail to root, are cut again at the base and transferred to fresh rooting selection medium to encourage root development. After 3 weeks, transgenic shoots with roots are moved to soil (McCormick et al., 1986).

4.5.2. *Cotton Transformation*

Cotton has been transformed via somatic embryogenesis of callus derived from hypocotyls and cotyledons (Firoozabady et al., 1987). The formation of embryogenic calli is genotype dependent and the genotypes Coker 310, 312, 315, and 5110 are relatively responsive. Regeneration from cotton hypcotyls (Trolinder and Goodin, 1987) is very adaptable to *Agrobacterium*-mediated transformation. Transgenic plants have been produced by several groups (Umbeck et al., 1987; Slade et al., 1991; Perlak et al., 1990) using basically the same protocol described below.

Surface-sterilized seeds are germinated in the dark on semi-solid media for 7 days. Hypocotyls are cut into 4–7 mm pieces, inoculated in an overnight culture of *Agrobacterium* containing a gene transfer cassette and then thoroughly blotted on sterile filter paper. The inoculated tissue is cultured with the *Agrobacterium* for 2–3 days on callus initiation media which contains MS salts (Murashige and Skoog, 1962) plus B5 vitamins (Gamborg et al., 1968), 30 g/l glucose, 0.1 mg/l 2,4-D, 0.5 mg/l kinetin, and 2 g/l Gelrite, at pH 5.8. After co-culture, the hypocotyls are transferred to callus initiation media containing bacteriostatic antibiotics (such as 500 mg/l carbenicillin) and a selective agent (such as 50–100 mg/l kanamycin). In 4 weeks the tissue is transferred again to selective callus initiation medium. After 2–4 weeks, callus is removed from the hypocotyl section and proliferated on the same medium, with subcultures every 4 weeks. After 2–3 subcultures, embryogenic calli are removed from the callus and cultured on embryogenic maturation media (Murashige and Skoog salts plus B5 vitamins, 30 g/l glucose, 1.9 g/l KN03, and 2 g/l Gelrite at pH 5.8). Embryogenic calli are subcultured every 4 weeks. Embryos 5 mm or larger are removed and placed on embryo conversion media (Stewart and Hsu, 1977) solidified with 2g/l Gelrite. When plantlets have produced 3–4 leaves and a root system, they are transferred to 2 inch pots with Metro 350 soil-less potting mix. The plantlets are hardened offin a tray with a cover. The cover is gradually removed over a 5–6 day period. Greater than 90% of the embryogenic callus and plants produced using this protocol test positive for expression of the selectable marker gene, indicating that they are transformed.

4.5.3. *Soybean Transformation*

Soybean has been successfully transformed by *Agrobacterium* using explants such as seedling cotyledons (Hinchee et al., 1988), immature cotyledons (Parrott et al., 1989) and germinating seeds (Chee et al., 1989) The soybean cultivar Peking has been shown to be the most susceptible genotype to

Agrobacterium infection. The following describes the procedure used to regenerate transgenic plants from cotyledons using kanamycin selection.

Soybean seeds are washed in water containing Tween 20, followed by several water rinses and a 2 min soak in 70% ethanol. The seeds are sterilized in a 50% Clorox® solution for 10–13 min, followed by 5 rinses with sterile distilled water, and a soaking in an aqueous Captan suspension for 1 h before transfer to the germination medium. The germination medium consists of Gamborg's B5 basal salts (Gamborg et al., 1968) solidified with purified agar. The seeds are placed under a photoperiod of 16 hours light (cool white fluorescent light at 40 μEn s^{-1}) and 8 hours dark at 25 °C for 4–6 days. The bacterial strains are streaked on solidified LB (Luria Broth) medium containing the appropriate antibiotics 3 to 4 days prior to use. The cotyledon explants are removed from the germinated seedlings and co-cultured with an *Agrobacterium* strain, containing a gene transfer cassette which confers kanamycin resistance, for 3 days on medium containing 1/10 the major and minor salts of SH medium (Schenck and Hildebrandt, 1972) at 25 °C under a 16:8 photoperiod. After the co-culture period, the cotyledons are transferred, adaxial side down, to B5BA regeneration and selection medium (Hinchee et al. 1988). B5BA selection medium is composed of Gamborg's B6 salts, 20 mg/l sucrose, 1.15 mg/l benzyladenine (BA), and 8 g/l purified agar, at pH 5.8 prior to autoclaving. After autoclaving, 500 mg/l carbenicillin, 100 mg/l cefotaxime, and 200–300 mg/l kanamycin are added. The infected explants are transferred to B50 selection medium (identical to B5BA but without the BA) after 3–4 weeks. After two subcultures onto the B50 selection medium, putative transgenic shoots can be identified. These shoots should be transferred to a non-selective medium which is composed of B5 salts, 20 mg/l sucrose, and 8 g/l purified agar, with 500 mg/l carbenicillin and 100 mg/l cefotaxime added after autoclaving. The shoots will root on the non-selective medium. The resulting plantlets can be transferred to Metro Mix 350 in 7.5-cm inch pots and placed in the growth chamber. The plantlets should be initially covered in order to maintain high humidity. They will harden off in 1–2 weeks. Screening for transgenic plant identification is necessary due to inefficient kanamycin selection. The kanamycin resistance phenotype can be assayed by using a recallussing assay in which leaf tissue is placed on MS19 medium containing 500 mg/l carbenicillin, 100 mg/l cefotaxime, and 100 mg/l kanamycin. MS19 medium is composed of MS salts and organics with 2 mg/lBA and 0.5 mg/l NAA. Whole or cut leaflets can be placed on the medium, and if callusing occurs within 3–4 weeks, the leaflets can be scored as resistant.

5. Free DNA Transfer to Protoplasts

Agrobacterium-mediated DNA transfer has proven quite effective for a wide range of dicotyledonous plants. This is not the case for monocotyledonous plants. While convincing evidence exists for *Agrobacterium*-mediated DNA delivery into monocotyledonous cells (Mooney et al., 1991; Raineri et al., 1990; Marks et al., 1989; Dale et al., 1989; Grimsley et al., 1987; Hohn et al., 1989; Hernalsteens et al., 1984), the few reports of plant transformation (*Asparagus officininalis* – Bytebier et al., 1987; maize – Gould et al., 1991) have generally lacked adequate proof of integrative transformation (Potrykus, 1991). For the agronomically important cereal crops such as maize, wheat, and rice, free DNA delivery into protoplasts is an attractive alternative transformation method (Vasil, 1988; Potrykus, 1991; Davey et al., 1989).

Protoplast free DNA delivery methods are genotype-independent. However, production of fertile, transgenic plants with this technology is severely constrained by genetic limitations on efficient plant regeneration from protoplasts. While fertile transgenic plants are now routinely recovered after protoplast transformation of rice (Hayashimoto et al., 1990; Datta et al., 1990; Shimamoto et al., 1989) the recovery of fertile plants from protoplasts of most cereal species remains difficult. Several groups have reported regeneration of fertile plants from maize protoplasts (Prioli and Sondahl, 1989; Shillito et al., 1989; Mórocz et al., 1990; Petersen et al., 1990), but in only one case has this technology been utilized successfully for the production of fertile transgenic plants (Donn et al., 1990). Fertile plants have also recently been recovered from cultured protoplasts of *Sorghum vulgare* (sorghum) (Wei and Xu, 1990) and *Hordeum vulgare* (barley) (Jahne et al., 1991), and infertile plants have been regenerated in wheat (Chang, et al., 1991; Harris et al., 1988; Vasil et al., 1990). Continued improvements in the isolation, culture, and regeneration of monocotyledonous protoplasts, particularily from cereal crops, is needed before routine production of transgenic plants via free DNA delivery can be realized.

Direct delivery of free DNA molecules into plant protoplasts has been well-documented (Davey et al., 1989). Both physical (e.g., electroporation, microinjection) and chemical (e.g., polyethylene glycol, Polybrene) methods have been developed to facilitate DNA delivery across the plasma membrane (Krens et al., 1982; Shillito et al., 1985; Antonelli and Stadler, 1989). The choice of transformation method is dependent on a number of factors, such as the efficiency of DNA delivery, subsequent protoplast survival, and ease of use.

5.1. *Electroporation*

Electroporation causes the uptake of DNA into protoplasts by temporary permeabilization of the plasmamembrane to macromolecules. This is achieved by application of a high intensity electric field to protoplasts contained in a buffer between two electrodes. DNA diffusion occurs immediately after the electric field is applied and until the pores in the membrane reseal (Fromm et al., 1987). Optimal DNA transfer is achieved by using the appropriate electric field strength (defined as the applied voltage divided by the distance between the two electrodes). The optimum field strength is dependent on a number of parameters, including the capacitor size (which is important in determining the pulse length), buffer composition and temperature, DNA concentration, protoplast density, and protoplast size. Optimum electroporation parameters must be determined empirically for each protoplast source. Relatively harsh conditions which result in low protoplast viability (25–50% viable) are generally necessary for optimal transient gene expression and stable transformation. The optimum conditions for transient gene expression may differ from that for stable transformation.

Electroporation has been used successfully for transient and stable transformation of protoplasts from a wide range of species and tissue sources (e.g., Sheen et al., 1991; Christou et al., 1987; Fromm et al., 1985, 1986). Plating efficiencies of electroporated protoplasts placed on selection media can be as high as 0.5% (Christou et al., 1987). The highest plant transformation efficiencies have been reported for tobacco, with 0.2% of electroporated leaf mesophyll protoplasts giving rise to transgenic, regeneration competent calli (Shillito et al., 1985). Much lower plant transformation efficiencies (0.002%) have been obtained using protoplasts from rice embryogenic suspension cultures (Shimamoto et al., 1989).

5.2. *Chemical Treatments*

Polyethylene glycol (PEG) is the most widely used chemical treatment for facilitating DNA uptake into plant protoplasts. PEG-mediated transformation involves mixing freshly isolated protoplasts with DNA and immediately adding a given concentration of PEG dissolved in a buffer containing divalent cations. After incubation in the PEG/DNA mixture for up to thirty minutes, the protoplasts are rinsed and plated for further culture. Important factors for optimizing transformation frequency include but are not limited to: PEG concentration, salts and salt concentration, pH, DNA concentration, size, form (linearized, supercoiled), and culture/selection technique used. PEG-mediated transformation is generally preferred over electroporation for stable transformation of monocotyledonous protoplasts due to relatively higher survival rates after treatment (Vasil et al., 1988; Antonelli and Stadler, 1989).

PEG-mediated DNA uptake typically transforms 0.1–0.4% of the total protoplasts treated (Kruger-Lebus and Potrykus, 1987; Hayashimoto et al., 1990). However, production of transgenic plants is dependent on the regeneration competency of the resulting transformed callus. Approximately 40% of the transgenic calli derived from PEG treatment of petunia mesophyll protoplasts could be induced to regenerate fertile plants, resulting in an overall transformation efficiency of approximately 0.1% of the treated protoplasts (Kruger-Lebus and Portrykus, 1987). In rice, only 0.5% (approximately) of the transgenic calli produced from PEG-treated, embryogenic suspension protoplasts could produce fertile plants (Hayashimoto et al., 1990; Datta et al., 1990). This resulted in an overall transformation frequency of only 0.0004%. Some of the highest protoplast transformation efficiencies have been obtained using a combination of electroporation and PEG treatments. Absolute protoplast transformation efficiencies ranging from 0.7–1.0% have been achieved with tobacco (Negrutiu et al., 1987) and soybean (Dhir et al., 1991) protoplasts electroporated in the presence of PEG. Transgenic plant production in soybean was limited because only 8% of the transgenic calli regenerated plants (Dhir et al., 1991). However, since the soybean system produces approximately 600 transgenic calli per electroporation-PEG treatment, it was still possible to produce up to 40 plants per experiment.

Liposomes have also been used as carriers for the introduction of nucleic acids into plant protoplasts (Fraley et al., 1982; Deshayes et al., 1985). In early reports, PEG or other polyalcohols were included in the transformation mixture. More recently, liposome-mediated transformation has been achieved in the absence of polyalcohols by including positively charged agents such as cations in the transformation mixture (Rosenberg et al., 1988), or using the cationic liposome preparation Lipofectin(TM) (Zhu et al., 1990; Antonelli and Stadler, 1990). The polycation Polybrene® has also been used alone for both transient and stable transformation of maize protoplasts (Antonelli and Stadler, 1989, 1990). Cationic liposome and polycation-mediated DNA delivery are promising new protoplast transformation approaches, which appear to be relatively non-toxic, and may also be more suitable than other methods for the introduction of very large DNA molecules (Antonelli and Stadler, 1990). Additional studies are needed to determine the general utility of these methods relative to PEG and electroporation.

5.3. *Microinjection*

Using very fine glass needles, it is possible to physically inject a solution of DNA directly into plant cells and their nuclei. However, this method relies on visual guidance of the needle into the cell. Although precision of gene introduction is achieved, the process is laborious and requires regeneration from isolated single cells or small cell clusters since larger tissues can not be resolved clearly enough for accurate needle placement. Protoplast transformation efficiencies have ranged from 12–66% (Miki et al., 1987), with the highest efficiencies being obtained with tobacco and oil seed rape (Spangenberg and Neuhaus, 1988; Neuhaus et al., 1987a; Miki et al., 1987). Despite the potential to transform cells of any plant species, the production of transgenic plants from microinjected cells has only been reported in *Nicotiana* and *Brassica* (Neuhaus et al., 1987a; Spangenberg and Neuhaus, 1988).

Microinjection is possible through the combination of microscopy and micromanipulation equipment. The basic technique is similar to that described for animal cell microinjection (Graessman et al., 1980). In order to microinject protoplasts or other plant cells, the cells need to be immobilized. This is accomplished by the use of a holding pipette (Crossway et al., 1986, Neuhaus et al., 1987a), attachment of cells to poly-L-lysine coated coverslips (Spangenberg et al., 1986; Reich ét al., 1986 a,b), or by embedding the cells in agarose, Na-alginate, or agar (Lawrence and Davies, 1985; Aly and Owen, 1987, Toyoda et al., 1988). Glass micropipettes are prepared to have openings of approximately 0.3 μm in diameter, and are inserted into plant cell cytoplasm and nuclei with the aid of a micromanipulator device. A syringe system is used for the controlled delivery of volumes ranging from 10^{-11}–10^{-4} μl into the plant nucleus and cytoplasm (Miki et al., 1987). Most plant cells or structures are injected inside microdroplets (2–50 μl) of medium inside a chamber which is sterile, vibration-free and allows for temperature and humidity regulation. The highest reported rates for microinjection are 100–200 cells per hour (Neuhaus and Spangenberg, 1990; Miki et al., 1987).

Efficient transformation using microinjection is dependent primarily on the the plating and regeneration efficiencies of the cells after microinjection. The method of culture during and after microinjection is critical. Hanging microdroplets (Crossway et al., 1986), thin agarose or agar embedments (Lawrence and Davies, 1985; Toyoda et al., 1988), and microcultures (Spangenburg et al., 1986; Neuhaus et al., 1987a) have been used.

Other factors which influence transformation efficiency may be the site of injection and the physical structure of the DNA. Nuclear injection appears advantageous in several systems (Crossway et al., 1986; Reich et al., 1986 b,c). Lineralized DNA may also result in higher transformation efficiencies when used alone or in combination with supercoiled DNA (Miki et al., 1987; Neuhaus et al., 1987b).

5.4. *Sonication*

Mild sonication (20 kHz ultrasound) has been used to facilitate the uptake and transient expression of a chloramphenicol acetyltransferase (CAT) gene in protoplasts of sugarbeet (*Beta vulgaris* L.) and tobacco (Joersbo and Brunstedt, 1990). Transient expression was equal to or better than that previously achieved by electroporation. Protoplast plating efficiency after sonication was similar to unexposed protoplasts. However, transgenic plants have yet to be reported using this technique.

5.5. *Applications*

5.5.1. *Transient Gene Expression in Maize Cells*

Black Mexican Sweet (BMS) maize suspension culture cells are kept in an actively growing stage (i.e., subcultured at least two times per week) and protoplasts are isolated by the protocol outlined by Fromm (1987). The BMS suspension cells are digested for one hour at 50–60 rpm in petri plates containing protoplast isolation medium: 0.5% bovine serum albumin (BSA), 0.05% β-mercaptoethanol, 0.5% cellulase 'RS', 0.5% rhozyme and 0.02% Y-23 pectolyase. Forty milliters of the enzyme mix is used for every 4–7 g of tissue. The protoplasts are then spun down at 200 × g, the enzyme removed, and the protoplasts are washed with protoplast isolation media containing 7.35 g/ $CaCl \cdot 2H_2O$, 1 g/l anhydrous sodium acetate, and 45 g/l mannitol, at pH 5.8. The protoplasts are spun down at 200 × g for 5 min, the supernatant is removed, and the protoplasts are resuspended in electroporation buffer made up of 20 mg/l KH_2PO_4, 115 mg/l Na_2HPO_4, 7.5 g/l NaCl, and 36.4g/l mannitol, at pH 7.2. The protoplasts are kept on ice from this point until electroporation. Protoplast yields are determined using a hemocytometer and density adjustments to 3–4 million per ml are made prior to the electroporation. Genes of interest are engineered into high copy number plasmid vectors (e.g., pUC vectors) and large quantities of plasmid are grown in *E. coli* and purified by cesium chloride gradient centrifugation (Sambrook et al., 1989). The DNA (25–100 μg/ml) and protoplasts are mixed together and added to a cuvette. The cuvettes are kept on ice. Optimum electroporation conditions are determined by empirically testing different size capacitors to produce field strengths from 250–1250 volts/cm. Pulse lengths should be in the 5 to 25 msec range. After electroporation the samples are left on ice for 10 min and then allowed to warm up at room temperature for 10 min. The electroporated samples are added back to the original growth media plus 0.4 M mannitol. One ml of electroporated protoplasts are added to 7 ml of the growth media. Protoplast survival at this point indicates which electroporation conditions may be best. The protoplasts are allowed to grow in the dark at 26–28 °C for 24–48 hours. Protoplast survival rate is again observed before harvest. The protoplasts are pipetted from the petri plate, spun down at 200 × g, and the pellet is frozen on dry ice or assayed immediately.

5.5.2. *PEG-mediated Maize Transformation*

Suspension cultures of totipotent cells are used for the isolation of the protoplasts. To establish the suspension cultures, add approximately 1 g of highly friable embryogenic callus to 35 to 50 ml of N6 1–100–25 (Armstrong and Green, 1985) liquid medium in a 250 ml Erlenmeyer flask. The cells are cultured in the dark on a gyratory shaker at 150 rpm at 28 °C. The cells are subcultured every 7 days by transferring 1–3 ml packed cell volume into 35–50 ml fresh medium. Midway through each subculture cycle, liquid media is removed by pipetting, and replaced with fresh medium. Visual and or physical selections can be used to develop a suspension with the appropriate characteristics. Suspension cultures are selected for protoplast isolation on the basis of rapid growth rate and an abundance of small, densely cytoplasmic cells in clumps of approximately 20–400 cells. Protoplasts are isolated by first collecting suspension culture cells on a 74 μm screen 1 day post-subculture. One gram (fresh weight) of cells is placed in 20 ml of filter-sterilized enzyme solution in a 100 × 25 mm Petri dish, and then incubated for 1.5 hr in the dark on a gyratory shaker (50 rpms), at 28 °C. The enzyme solution consists of 2% cellulase (CEL, Worthington Biochemical Company), 0.1% pectolyase Y-23 (Seishin Pharmaceutical), and 0.2% BSAdissolved in 0.2 M mannitol, 80 mM calcium chloride ('protoplast wash solution' or PWS), at pH 4.0–4.5. Protoplasts are purified by filtering through a 46 μm mesh, transferring 10 ml each to 16 × 125 mm round-bottom test-tubes and centrifuging at 100 × g for 8 min. The supernatant is removed, and the protoplasts are resuspended in 10 ml of PWS. Further purification can be achieved by centrifugation of the suspended protoplasts over 2 ml of 20% sucrose at 100 × g for 10 min. After collecting the protoplasts from the PWS: sucrose interface, they are resuspended and their yield is determined by hemacytometer count. The protoplasts are pelleted by centrifugation (100 × g, 8 min) and the supernatant removed. The protoplasts are now ready for PEG mediated transformation and/or culture.

For PEG mediated transformation, a 40% PEG solution is prepared by dissolving PEG 8000 (Merck) in Krens F solution (Krens et al., 1982), adjusting the pH to 6.0 with KOH, and filter-sterilizing. The protoplasts are resuspended in PWS to obtain a protoplast density of 2.5×10^6 per ml. Protoplast aliquots of 0.5 ml are added to 16 × 125 mm round-bottom centrifuge tubes. Up to 100 μl of DNA (1 μg/ml) is added and mixed by gentle agitation, followed by 0.5 ml of 40% PEG to give a final PEG concentration of 20%. The protoplast/PEG solution is agitated gently every 5 min for 30 min, then diluted with 5 1-ml aliquots of N6 1–100–25 plus 8% mannitol at 5 min intervals. Protoplasts are collected by centrifugation at 100 × g, the supernatant removed, and the protoplasts resuspendedin culture medium for plating. The protoplasts are re-suspended in N6 1–100–25 plus 8% mannitol at a density two times the desired plating density. An equal volume of autoclaved N6 1–100–25, containing 8% mannitol and 2.4% Seaplaque agarose (FMC) and cooled to approximately 42 °C, is added to the protoplasts and the mixture is plated as a bead culture (Shillito, 1983). After solidification, 4 mls liquid plating medium containing a nurse culture (1 volume maize suspension culture: 1 volume plating medium +16% mannitol) is added over the top of the agarose embedded protoplasts. The plates are wrapped with parafilm, and incubated at 28 °C in the dark on a rotary shaker at 50 rpm (Armstrong et al., 1990).

After 7 days the liquid overlay media and 80% of the feeder cells are removed and replaced with 4 ml of N6 1–100–25 4% mannitol liquid medium. At 14 days the overlay medium is removed, and the embedded protoplasts are rinsed with 5 ml sterile water (up to 5 times) to remove all feeder cells. They are then covered with 4 ml of plating medium without additional osmotica. For transformation experiments a selective agent is added to the final liquid medium overlay and to all subsequent medium for selection of transformed colonies. About 14 days later, the agarose slab is transferred to 0.2% gelrite solidified N6 1–100–25 medium. Individual colonies are harvested after an additional 7 to 14 days for further culture. For plant regeneration, independent protoplast-derived embryogenic calli are transferred from N6 1–100–25 maintenance medium to MS (Murashige and Skoog, 1962) medium with 2% sucrose, 0.1 mg/l 2,4-D and 10^{-7} M ABA for 10–14 days, and then transferred to N6 medium with 6% sucrose (no plant growth regulators or proline) for 10 to 14 days. For both of these regeneration steps, the cultures are maintained at 28 °C in the dark. Mature somatic embryos are moved to MS 0-D medium (2% sucrose, no growth regulators) for conversion into plants (25 °C, 16 hours light: 8 hours dark photoperiod). Small rooted plants are transferred to 25 × 200 mm glass test tubes containing 15 ml MS 0-D medium solidified with 0.6% agarose (Sigma type I, Low EEO) for further development. Well-developed plants are transferred to Metro Mix 350 in 7.5-cm pots and hardened-off by gradual reduction of humidity in a growth chamber. Plants at approximately the 4-leaf stage are transplanted to 30.5-cm diameter pots and grown under greenhouse conditions.

6. Free DNA Transfer to Intact Tissues

6.1. *Acceleration of DNA Loaded Microparticles*

Microprojectile-mediated transformation is a mechanical method of introducing DNA into almost any plant species and genotype. This mechanical method may be highly advantageous when major biological barriers exist to either *Agrobacterium* or protoplast-mediated transformation. An acceleration device is used to propel microscopic metal microprojectiles (particles) carrying plasmid DNA into intact plant cells, tissues and organs (Sanford et al., 1987; Sanford, 1988). Particle-mediated plant transformation technology has been used to transform several plant species, including relatively recalcitrant species such as maize (Fromm et al., 1990), soybean (McCabe et al., 1988) and wheat (Vasil et al., 1992). DNA bombardment has also facilitated the stable transformation of higher plant chloroplasts (Ye et al., 1990; Svab et al., 1990). In addition to being tools for stable transformation, particle acceleration devices can be used to easily and rapidly test gene constructions for cell and organ-specific expression (Twell et al., 1989). Despite recent achievements, particle bombardment technology is still in its infancy. Current limitations to the technology are primarily inefficient gene transfer and reliance on plant genotypes which regenerate prolifically. Transgenic plant production in the most efficient particle-mediated transformation systems (Finer and McMullen, 1990, 1991; McCown et al., 1991) reaches what can be achieved by many *Agrobacterium*-mediated systems.

The DNA microacceleration technology has 3 basic components: (1) a particle acceleration device, (2) metal particles associated with precipitated DNA, and (3) target plant tissues which are regeneration competent and accessible to particle penetration.

6.1.1. *Particle Acceleration Devices*

A prototype of the commercially available Biolistics(TM) PDS 1000 accelerator was first used for the successful transformation of plant cells. It utilizes the explosive force of gunpowder (0.22 caliber nail gun cartridge) to accelerate a polypropylene cylindrical macroprojectile, loaded on its distal end with a drop of DNA-coated metal microprojectiles (particles), down a 0.22 caliber barrel towards the plant target (Klein et al., 1987). The polypropylene macroprojectile is stopped by a polycarbonate disk containing a small opening which allows only the particles to continue towards the tissue target. Particle velocity of approximately 2,000 ft/sec (Sanford et al., 1987), is achieved and results in penetration of both surface and underlying plant cells. Sufficient velocity for cell penetration is only achieved when the acceleration chamber is kept under partial vacuum.

Other devices have been used which are similar in basic design to the Biolistics device, but which utilize compressed air or gas to generate the motive force. Compressed air or nitrogen (approximately 150 kg/cm^2) has been used to accelerate macroprojectiles at velocities (approximately 440 m/sec) necessary to achieve DNA delivery to plant cells (Iida et al., 1990b; Oard et al., 1990; Morikawa et al., 1989).

An electric discharge particle acceleration device differs in basic design from the above devices and has proven useful for transformation of several species (Christou, 1990b; McCabe et al., 1988; Christou et al., 1988). In this device, a shock wave is generated by the discharge of 14 kV from a 2 μF capacitor through a small water drop inside of a polyvinyl chloride expansion chamber. The shock wave causes the upward acceleration of a mylar carrier sheet loaded with particles and precipitated DNA towards a 100 mesh retaining screen. When the carrier hits the retaining screen, the particles continue onward and penetrate target plant tissues inverted over the acceleration chamber.

A commercially available device, the Biolistics PDS-2000, is somewhat similar in basic design to the electric discharge device. This device uses the abrupt release of compressed helium to generate a shock wave for the acceleration of a kapton carrier sheet loaded with particles toward a stopping screen and the transformation tissue target (Ye et al., 1990). The abrupt release of helium is caused by the spontaneous rupture of a kapton plastic membrane at specific burst pressures.

Two other types of devices which accelerate DNA and/or particles without a solid carrier have also been described. One device is a modification of a vaccination gun (Miller, 1989), and the other is based on an aerosol spray mechanism (Mets, personal communication). However, only transient gene expression in wheat and the single-celled alga, *Chlamydomonas*, has been achieved with these prototypes to date.

6.1.2. *DNA/particle preparation*

The particles used in this transformation technology must be relatively dense ($19 g/cm^3$ or greater) to ensure sustained momentum through plant cell walls and plant cells. The size of the accelerated particles also needs to match the target tissue cell size. For example, relatively large cells, such as tobacco suspension culture cells (average diameter of 40 μm) can be transformed with particles which average 3–5 μm in diameter (Klein et al., 1988b). Smaller cells, such as embryogenic maize suspension culture cells (average diameter of 10 μm) are transformed best with particles which average 1 μm or less in diameter (Fromm et al., 1990). Irregularly shaped tungsten particles (GTE, DuPont) are most frequently used with the Biolistics gunpowder device due to their low cost, high efficiency, and variety of available sizes (average size 1.0–5.0 μm). Small (0.5–3 μm) gold spheres (Alfa, DuPont) have also been used with the electric discharge, compressed gas and Biolistics devices. Gold is as dense as tungsten, but unlike tungsten particles, it does not oxidize or aggregate readily.

The delivery of DNA by microscopic metal particles to plant cells is facilitated by methods to precipitate DNA onto the particles prior to bombardment. The precipitation method in widest use is one which uses a $CaCl_2$ and spermidine (free base) precipitation to associate plasmid DNA with tungsten particles (Klein et al., 1988a). The $CaCl_2$/spermidine precipitation method has been modified (relative volumes of DNA, $CaCl_2$ and spermidine during precipitation, final DNA and particle concentrations) in order to maximize gene delivery to specific plant tissues (Finer and McMullen, 1990; Gordon-Kamm et al., 1990). Ethanol precipitation of plasmid DNA onto gold particles has also been used for gene delivery and stable transformation (Iida et al., 1990a). DNA coating of particles for use in the electric discharge acceleration device involves the drying of DNA under nitrogen onto gold particles prior to resuspension in ethanol before loading on the carrier sheet (McCown et al., 1991). A precipitation method using $CaCl_2$ and NaH_2PO_4 to produce calcium phosphate crystals has also been employed for this device (D. McCabe, personal communication).

6.1.3. *Plant Target Tissues*

Cells in the target plant tissue should be meristematic, or at least regeneration competent, as well as accessible to particle penetration. Embryogenic tissues are amenable to particle transformation because they contain a large percentage of actively dividing, regenerable cells which can be present in the outer cell layers of a callus, cell cluster, or proembryonic tissue. Embryogenic cultures of corn (Fromm et al., 1990; Gordon-Kamm et al., 1990), cotton (Finer and McMullen, 1990), soybean (Finer and McMullen, 1991), *Carica payaya* (papaya) (Fitch et al., 1990) and wheat (Vasil et al., 1992) have been used successfully for particle-mediated transformation. Transformation efficiency is dependent on the regeneration response (location and efficiency of regeneration competent cells) of the target genotype. For example, success with particle mediated corn transformation has been limited to particular genotypes (A188 × B84 and A188 × B73 crosses and derivatives) which produce highly embryogenic callus and suspension cultures.

Embryogenic cells are not the only useful targets for particle-mediated transformation. Transgenic plants have also been produced from other tissues, such as tobacco leaves (Tomes et al., 1990; Klein et al., 1988c), poplar nodules (McCown et al., 1991) and cranberry stems (Serres et al., 1990) which apparently undergo shoot organogenesis from single cells. When multiple cells are involved in shoot organogenesis chimeric plants containing transgenic sectors can result. Chimeric soybean plants have been produced after particle-mediated transformation of shoot apical meristem cells undergoing multiple shoot proliferation (Christou et al., 1990a,b,c; McCabe et al., 1988). Transgenic sectors in chimeric plants may at times include germline cells and result in the production of clonal transgenic progeny.

The histochemical detection of GUS expression is commonly used to optimize bombardment conditions for specific plant tissues since the number and location of positive gene expression sites is easily visualized (Klein et al., 1988a,b). Quantitative assays, such as the fluorometric analysis of GUS activity (Hinchee, unpublished results) or analysis of chloramphenicol acetyltransferase (CAT) activity (Kartha et al., 1989), have also been used to measure gene delivery after bombardment, but these methods are more laborious and do not provide any positional information.

Transgenic plants produced from microprojectile bombardment can contain a few to many copies of the transforming gene at any one site. Restriction mapping reveals that DNA rearrangements are common (Tomes et al., 1990; Christou et al., 1989). The interspersion of multiple gene copies with nonplasmid DNA at a single insertion site is possible (Christou et al., 1989). In addition, Southern blot analysis and segregation ratios can show different fragment lengths and non-Mendelian segregation ratios in the parent and its progeny indicating that several rearrangements of the inserted gene are possible in the initial transgenic plant (Tomes et al., 1990). The possible rearrangements of the transforming DNA and the effects of different positional insertions may be a potential disadvantage of this technology. However, the rearrangements reported are not significantly different from those observed in other free DNA transformation methods (Morota and Uchimiya, 1988; Shillito and Potrykus, 1987).

6.2. *Microinjection*

Microinjection has been used to transfer DNA to diverse plant tissues such as pollen (Huang et al., 1991), tomato callus cells (Toyoda et al., 1988), and intact barley coleoptile cells (Toyoda et al., 1990). To date, transgenic plants have only been produced from microinjection of microspore-derived embryoids of oil seed rape (Neuhaus et al., 1987b). Embryoids with 8 or more cells, injected with NPT II gene constructions, were cultured in microdroplets without selection pressure. Transformation efficiencies of 27–51% were achieved based on DNA dot blot analysis of primary regenerants. Transformation was further confirmed by Southern analysis and NPT II enzyme activity assays. The primary regenerates were considered to be putative transgenic chimeras. Production of clonal, stably transformed plants was achieved via the process of secondary somatic embryogenesis (considered to be of single cell origin) from the primary embryoids. The success of this system relies on the very high regeneration capacity of oil seed rape microspores.

Success with the microinjection of microspore-derived embryos prompted attempts to produce chimeric, transgenic plants via the microinjection of zygotic proembryos. Although selectable marker genes have been introduced into proembryos of rice, wheat, barley and maize, none of the resulting plants have produced transgenic progeny (Neuhaus and Spangenburg, 1990; Potrykus, 1991).

6.3. *Alternative Transformation Technology*

6.3.1. *Laser Microbeam*

A UV laser microbeam has been used to introduce DNA into plant cells and chloroplasts. A 343 nm beam is directed through an adjustable attenuator into the optical path of an inverted microscope. The focus of the laser beam is adjusted so that it is identical with that of the objective lens, and the laser beam is essentially aimed by focusing on a specimen in the microscope. The laser can then make holes in any part of a cells which is in focus. Laser micropuncture of the cell wall and plasma membrane has allowed the movement of fluorescence-labelled plasmid DNA into single oil seed rape cells and microspores suspended in a hypertonic buffer (Weber et al., 1988, 1990). After laser introduction of a gene for hygromycin resistance, approximately 20% of the treated cells produced colonies capable of growth on hygromycin (Weber et al., 1990). However, transgenic plant regeneration from these colonies has not yet been reported. Similar technology has been used to introduce DNA into isolated plant chloroplasts (Weber et al., 1989) and into chloroplasts of intact protoplasts (Weber et al., 1990).

6.3.2. *Silicon Carbide Fibers*

Transient β-glucuronidase expression and stable transformation has been obtained in tobacco (TXD) and maize (BMS) cells after simply vortexing the cells for one minute in an Eppendorf tube containing a mixture of plasmid DNA and silicon carbide fibers (Kaeppler et al., 1990, H. Kaeppler, personal communication). The silicon fibers averaged 0.6 μm in diameter and 10–80 μm in length, and were shown by scanning electron microscopy to penetrate the plant cells after vortexing. It was estimated that 0.01% of the BMS cells treated expressed GUS transiently.

6.3.3. *Electroporation of Intact Tissues*

DNA has been electroporated into intact and organized leaf base explants of rice, corn, wheat and barley seedlings. The expression of an introduced NPT II gene was used to optimize the protocol, and it was found that the elimination of explant-released nucleases, prolonged DNA/explant incubation time and prolonged pulse lengths were critical for obtaining transient gene expression in intact tissues (DeKeyser et al., 1990). Futher evidence for gene transfer was the demonstration of the expected tissue-specific or light-regulated expression after electroporation with genes containing vascular-specific or light-inducible promoters. More recently, transgenic maize plants were obtained by direct delivery of DNA by electroporation into intact immature embryos as well as type I embryogenic callus (D'Halluin et al., 1992). Transgenic rice has also been obtained from the electroporation of zygotic embryo cells (Xu and Li, 1994).

6.3.4. *Dessication*

Topfer et al. (1989) have provided compelling evidence that naked DNA molecules can be delivered into isolated, dried embryos of several cereal and legume species during the rehydration, or imbibition, process. The approach capitalizes on the changes which occur in plant cell walls and membranes during the natural dessication process of the embryo. This method of introducing DNA into intact tissues is attractive because of its simplicity. However, only transient gene expression has been documented using this system. Recent advances in 'artificial dessication' technology for both somatic embryos and embryogenic calli (Gray et al., 1991; Kim and Janick, 1991; Shin et al., 1991) may prove useful in the application of 'DNA-uptake-by-imbibition' technology to the production of transgenic plants.

6.3.5. *DNA Transfer to Reproductive Cells*

In an effort to directly transform germline cells, DNA was injected into floral tillers of *Secale cereale* (rye) at the time of meiosis. From approximately

4,000 seeds produced following this procedure, three progeny plants were kanamycin resistant and exhibited NPT II activity (de la Pena et al., 1987). However, Mendelian heritance of the NPT II gene has not been demonstrated in subsequent generations. DNA has also been introduced into rice by applying DNA to the styles of pollinated flowers with up to 20% of the rice seeds from treated florets containing the introduced NPT II gene (Luo and Wu, 1988). Although the above reports were strongly suggestive of DNA transfer into reproductive cells, definitive proof of stable plant transformation has yet to be presented.

6.4. *Applications*

6.4.1. *Particle-Mediated Transformation of Soybean*

Embryogenic calli are initiated from cotyledons of immature, zygotic embryos obtained from soybean cultivars such as Fayette and Williams 82. The cotyledons are cultured on MS medium with B5 vitamins, 40 mg/l 2,4-D, 6% sucrose and 0.8% agar (Finer, 1988). An embryogenic suspension culture is established by placing small pieces of the embryogenic calli into 35 ml of liquid medium (maintenance medium), MS salts with 10 mM NH_4NO_3 and 30 M KNO_3 with B5 vitamins, 6% sucrose, 5 mg/l 2,4-D and 5 mM asparagine (Finer and Nagasawa, 1988) and constantly selecting for small lobed, early-staged globular embryo clumps. Every 4 weeks, 2 clumps about 4 mm in diameter, are subcultured into fresh liquid medium. Embryogenic suspensions consist entirely of clumps of proliferating globular embryos ranging in diameter from 2–8 mm. In preparation for bombardment, the entire flask of an embryogenic suspension is poured into a petri plate and the medium removed. The clumpy suspension is gently smashed with a spatula prior to bombarding two times on the shelf second from the bottom in a Biolistic™ PDS 1000 acceleration device. DNA is precipitated onto M10 (DuPont) particles using the calcium spermidine method (Klein et al., 1988a). Washed tungsten particles (12.5 µl), suspended in 50% glycerol, are gently vortexed with 2.5 µl DNA (1 µg/µl), 12.5 µl 1 M $CaCl_2$, and 5 µl 0.1 M spermidine (free base). After 10 min, 12.5 µl supernatant is removed and discarded. Just prior to bombardment, the tube containing the DNA preparation is briefly sonicated before removing 2.5 µl of the DNA coated tungsten mixture to place on the macroprojectile. After bombardment, the suspensions are grown without selection in maintenance medium for two weeks, and then on the same medium but with 50 mg/l G418 (Geneticin, Sigma) for 8 weeks, replacing the medium every 3–4 days. After 8 weeks, green embryos or clumps can be separated from the brown, presumably dead embryos and reproliferated in liquid medium without a selective agent. Further embryo development occurs after placing them on Phytagar (Gibco) solidified MS medium containing B5 vitamins and 3% maltose. After 4 weeks, the embryos are placed on the same medium, but containing 3% sucrose instead of maltose. Embryo germination is initiated after a desiccation period of 2–7 days prior to placing on the previous medium. The embryos are dessicated in empty, parafilm wrapped petri plates. Resulting plantlets are treated as in the Applications (IVE).

6.4.2. *Particle-Mediated Transformation of Maize*

Friable, embryogenic ('Type II') callus is established by explanting 1.0–1.5 mm long immature A188 X B73 (or 'Hi-II' germplasm; Maize Genetics Cooperation NewsLetter 65:92–93) embryos onto gelrite-solidified N6 1-100–25 medium containing 10 μM silver nitrate (Songstad et al., 1991). Embryogenic suspension cultures are established in N6 1–100–25 medium from Type-II callus. This step is arguably the most important and most difficult part of the procedure (see Applications VE, Green et al., 1983; Kamo et al., 1986). Log-phase suspension cells are plated under vacuum onto filter paper, taking care that no liquid layer remains on the cells. The filter is transferred to an empty Petri dish. The cells are bombarded with the Biolistics PDS 1000 device one to three times, under vacuum, with 1.0 μm diameter tungsten or gold particles coated with DNA containing an ALS selectable marker gene (Fromm et al., 1990) and other gene(s) of int rest. The filter with the bombarded cells is placed onto gelrite solidified N6 1-100–25 medium overnight for recovery. The cells are rinsed from the filter into 15 ml liquid N6 1–100–25 medium in a 100 × 25 mm Petri dish and are placed on a shaker at 60 rpm for one to several days. The cells are replated onto filters under vacuum, and then placed on N6 1–100–25 medium containing 20 ppb chlorsulfuron. Cell density is important, and a thin, uniform covering of the filter yields the best results. After two weeks, the cells are scraped from the filters and plated directly onto fresh selection medium. Cells are transferred to fresh selection medium approximately every two weeks. Chlorsulfuron-resistant calli should be visible after six to twelve weeks. Transgenic tissue is almost always friable, and generally highly embryogenic (similar to initial Type-II callus). Compact, organized tissue is generally nontransgenic escape tissue. Plant regeneration is achieved in a three-step process (see Section 5.5: Applications; Armstrong and Green, 1985).

7. Conclusions

Agrobacterium-mediated transformation has been the primary tool for the transformation of many dicotyledonous crops. It will most likely be the method of choice in plant species in which efficient T-DNA transfer is possible. This is because of the relative ease and precision of gene transfer to intact, regenerable explants. Recent reports of successful *Agrobacterium*-

mediated transformation of monocotyledonous plants, and improved understanding of the mechanism of T-DNA transfer and chromosomal insertion, indicate that *Agrobacterium* vectors have the potential to be used for the production of transgenic cereal crops in the future. Until then, recent mechanical methods for the introduction of DNA into intact plant cells show promise for the transformation of graminaceous species. Particle bombardment technology has proven itself by reproducibly facilitating the production of fertile, transgenic maize and wheat plants. Silicon carbide fiber-mediated DNA uptake is as yet unproven in transgenic plant production, but it merits further investigation because of its simplicity in achieving stable cell transformation. All the above methods are advantageous due to their ability to introduce genes into intact, readily regenerable tissues. Protoplast transformation methods, although shown to be feasible, will probably always be limited by technical difficulties in reproducibly achieving regeneration from protoplasts of most cereal genotypes. The most widely used transformation technology in the future will be one which allows for simple, precise and efficient gene transfer directly into germline cells or other plant tissues which require minimal tissue culture manipulation.

8. References

Aly, M.A. and Owens, L.D. (1987). A simple system for plant cell microinjection and culture. Plant Cell Tissue Org. Cult. 10:159–174.

An, G., Watson, B.D., Stachel, S., Gordon, M.P. and Nester, E.W. (1985). New cloning vehicles for transformation of higher plants. EMBO J. 4:277–284.

Antonelli, N.M. and Stadler, J. (1989). Chemical methods for efficient direct gene transfer to maize cells: Treatment with polyethylene glycol or Polybrene. Genet. and Breed. 4:113–122.

Antonelli, N.M. and Stadler, J. (1990). Genomic DNA can be used with cationic methods for highly efficient transformation of maize protoplasts. Theor. Appl. Genet. 80:395–401.

Armstrong, C.L. and Green, C.E. (1986). Establishment and maintenance of friable, embryogenic maize callus and the involvement of L-proline. Planta 164:207–214.

Armstrong, C.L., Petersen, W.L., Buchholz, W.G., Bowen, B.A. and Sulc, S.L. (1990). Factors affecting PEG-mediated stable transformation of maize protoplasts. Plant Cell Rep. 9:335–339.

Barnes, W.M. (1990). Variable patterns of expression of luciferase in transgenic tobacco leaves. Proc. Natl. Acad. Sci. USA 87:91839187.

Bayer, E., Guge, K.H., Hagele, K., Hogenmajer, H., Jessipow, S., Konig, W.A. and Zahner, H. (1972). Phosphinothricin and phosphinothricyl-alanyl-alanin. Helv. Chim. Acta 55:224–239.

Bevan, M. (1984). Binary *Agrobacterium* vectors for plant transformation. Nucleic Acids Res. 12:8711-8712.

Binns, A.N. and Thomashow, M.F. (1988). Cell biology of *Agrobacterium* infection and transformation of plants. Annu. Rev. Microbiol. 42:575–606.

Bowen, B. (1992). Anthocyanin genes as visual markers in transformed maize tissues. In: GUS Protocols: Using the GUS Gene as a Reporter of Gene Expression. S.R. Gallagher (ed.), Academic Press, New York, pp. 163–177.

Burow, M.D., Chem, C.A., Sen, P., Lisca, A. and Muzai, N. (1990). High frequency generation of transgenic tobacco plants after modified leaf disk co-cultivation with *Agrobacterium tumefaciens*. Plant Mol. Biol. 8:124–139.

Byrne, M.C., McDonnell, R.E., Wright, M.S. and Carnes, M.G. (1987). Strain and cultivar specificity in the *Agrobacterium* soybean interaction. Plant Cell Tissue Org. Cult. 8:3–13.

Bytebier, B., Deboeck, F., DeGreve, H., Van Montagu, M. and Hernalsteens, J-P. (1987). T-DNA organization in tumor cultures and transgenic plants of the monocotyledon *Asparagus officinalis*. Proc. Natl. Acad. Sci. USA 84:53455349.

Cangelosi, G.A., Ankenbauer, R.G. and Nester, E.W. (1990). Sugars induce the *Agrobacterium* virulence genes through a periplasmic binding protein and a transmembrane signal protein. Proc. Natl. Acad. Sci. USA 87:6708–6712.

Chaleff, R.S. and Mauvais, C.J. (1984). Acetolactate synthase is the site of action of two sulfonylurea herbicides in higher plants. Science 224:1443–1445.

Chaleff, R.S. and Ray, T.B. (1984). Herbicide resistant mutants from tobacco cell cultures. Science 223:1148–1151.

Chang, Y-F., Wang, W.C., Warfield, C.Y., Nguyen, H.T. and Wong, J.R. (1991). Plant regeneration from protoplasts isolated fromlong-term cel cultures of wheat (*Triticum aestivum* L.). Plant Cell Rep. 9:611–614.

Charest, P.J., Hattori, J., DeMoor, J., Iyer, V.N. and Miki, B. (1990). *In vitro* study of transgenic tobacco expressing *Arabidopsis* wild type and mutant acetohydroxacid synthase genes. Plant Cell Rep. 8:643–646.

Chee, P.P., Fober, K.A. and Slightom, J.L. (1989). Transformation of soybean (*Glycine max*) by infecting germinating seeds with *Agrobacterium tumefaciens*. Plant Physiol. 91:1212–1218.

Christou, P. (1990a). Morphological description of transgenic soybean chimeras created by deficiency integration and expression of foreign DNA during electric discharge particle acceleration. Annals Bot. 66:379–346.

Christou, P. (1990b). Soybean transformation by electric discharge particle acceleration. Physiologia Plantarum 794:210–212.

Christou, P., McCabe, D.E., Martinell, B.J. and Swain, W.F. (1990). Soybean genetic engineering-commercial production of transgenic plants. TIBTECH 8:145–161.

Christou, P., McCabe, D.E. and Swain, W.F. (1988). Stable transformation of soybean callus by DNA-coated gold particles. Plant Physiol. 87:671–674.

Christou, P., Murphy, J.E. and Swain, W.F. (1987). Stable transformation of soybean by electroporation and root formation from transformed callus. Proc. Natl. Acad. Sci. USA 84:3962–3966.

Christou, P., Swain, W.F., Yang and McCabe, D.E. (1989). Inheritance and expression of foreign genes in transgenic soybean plants. Proc. Natl. Acad. Sci. USA 86:7500–7504.

Crossway, A., Oakes, J.V., Irvine, J.M., Ward, B., Knauf, V.C. and Shewmaker, C. K. (1986). Integration of foreign DNA following microinjection of tobacco mesophyll protoplasts. Mol. Gen. Genet. 202:179–185.

Dale, P.J., Marks, M.S., Brown, M.M., Wooston, C.J., Gunn, H.V., Mullineaux, P.M., Lewis, D.M., Kemp, J.M., Chen, D.F., Gilmour, D.M. and Flavell, R.B. (1989). Agroinfection of wheat: Inoculation of *in vitro* grown seedlings and embryos. Science 63:237–245.

Datta, S., Peterhans, A., Dattam, K. and Potrykus, I. (1990). Genetically engineered fertile indica-rice recovered from protoplasts. Bio/Technology 8:736–740.

Davey, M.T., Rech, E.L. and Mulligan, B.J. (1989). Direct DNA transfer to plant cells. Plant Mol. Biol. 13:273–285.

DeBlock, M. (1990). Factors influencing the tissue culture and the *Agrobacterium tumefaciens*-mediated transformation of hybrid aspen and poplar clones. Plant Physiol. 93:1110–1116.

DeBlock, M., Botterman, J., Vandewiele, M., Dockx, J.,Thoen, C, Gossele, V., Mowa, N.R., Thompson, C., van Montagu, M. and Leemans, J. (1987). Engineering herbicide resistance in plants by expression of a detoxifying enzyme. EMBO J. 6:513–2518.

DeBlock, M., De Brouwer, D. and Tenning P. (1989). Transformation of *Brassica napus* and *Brassica oleracea* using *Agrobacterium tumefaciens* and the expression of the bar and neo genes in the transgenic plants. Plant Physiol. 91:694–701.

DeBlock, M., Herrera-Estrella, L., Van Montagu, M., Schell, J. and Zambryski, P. (1984). Expression of foreign genes in regenerated plants and their progeny. EMBO J. 3:1681–1689.

DeKeyser, R., Claes, B., Marichal, M., Van Montagu, M. and Caplan, A. (1989). Evaluation of selectable markers for rice transformation. Plant Physiol. 90:217–223.

DeKeyser, R.A., Claes, B., De Rycke, R.M.U., Habets, M.E., Van Montagu, M.C. and Caplan, A. (1990). Transient gene expression in intact and organized rice tissues. The Plant Cell 2:569–602.

De la Pena, A., Lorz, H. and Schell, J. (1987). Transgenic rye plants obtained by injecting DNA into young floral tillers. Nature 325:274–276.

Deshayes, A., Herrera-Estrella, L. and Caboche, M. (1985). Liposome-mediated transformation of tobacco mesophyll protoplasts by an *Escherichia coli* plasmid. EMBO J. 4:2731–2737.

D'Halluin, K., Botterman, J. and DeGreef, W. (1990). Engineering of herbicide-resistant alfalfa and evaluation under field conditions. Crop Sci. 30:866–871.

D'Halluin, K., Bonne, E., Bossut, M., de Beuckeleer, M. and Leemans, J. (1992). Transgenic maize plants by tissue electroporation. The Plant Cell 4:1495–1505.

Dhir, S.K., Dhir, S., Sturtevant, A.P. and Widholm, J.M. (1991). Regeneration of transformed shoots from electroporated soybean (*Glycine max* (L.) Merr.) protoplasts. Plant Cell Rep. 10:97–101.

Ditta, G., Stanfield, S., Corbin, D. and Helinski, D.R. (1980). Broad host range cloning system for gram negative bacteria: Construction of a gene bank of *Rhizobium meliloti*. Proc. Natl. Acad. Sci. USA 77:7347–7351.

Donn, G., Nilges, M. and Mórocz, S. (1990). Stable transformation of maize with a chimaeric, modified phosphinothricin-acetyltransferase gene from *Streptomyces virido-chromogenes*. Abstracts 7th Intl. Congress on Plant Tissue and Cell Culture. Abstract No. A2–38, p. 63.

Eichholtz, D.A., Rogers, S.G., Horsch, R.B., Kle, H.J., Hayford, M., Hoffmann, N., Bradford, S., Fink, C., Flick, J., O'Connell, K. and Fraley, R. (1987). Expression of mouse dihydrofolate reductase gene confers methotrexate resistance in transgenic petunia plants. Somatic Cell Mol. Genet. 13:67–76.

Feldmann, K.A. and Marks, M.D. (1987). *Agrobacterium*-mediated transformation of germinating seeds of *Arabidopsis thaliana*: A non-tissue culture approach. Mol. Gen. Genet. 208:1–9.

Figurski, D.H. and Helinski, D.R. (1979). Replication of an origin-containing derivative of plasmid RK2 dependent on a plasmid function provided in trans. Proc. Natl. Acad. Sci. USA 76:1648–1652.

Fillatti, J.J., Kiser, J., Rose, R. and Comai, L. (1987). Efficient transfer of a glyphosate tolerance gene into tomato using a binary *Agrobacterium tumefaciens* vector. Bio/technology 5:726–730.

Finer, J.J. and McMullen, M.D. (1990). Transformation of cotton (*Gossypium hirsutum* L.) via particle bombardment. Plant Cell Rep. 8:686–589.

Finer, J.J. and McMullen, M.D. (1991). Transformation of soybean via particle bombardment of embryogenic suspension culture tissue. *In vitro* Cell Develop. Biol. Plant 27P:175–182.

Finer, J.J. and Nagasawa, A. (1988). Development of an embryogenic suspension culture of soybean (*Glycine max* Merrill). Plant Cell Tissue Org. Cult. 15:125–136.

Finer, J.J. (1988). Apical proliferation of embryogenic tissue of soybean (*Glycine max* (L.) Merrill). Plant Cell Rep. 7:238–241.

Firoozabady, E., DeBoer, D.L., Merlo, D.J., Halk, E.L., Amerson, L.N., Rashka, K.E. and Murray, E.E. (1987). Transformation of cotton (*Gossypium hirsutum* L.) by *Agrobacterium tumefaciens* and regeneration of transgenic plants. Plant Mol. Biol. 10:105–116.

Fitch, M.M.M., Manshardt, R.M., Gonsalves, D., Slightom, J.L. and Sanford, J.C. (1990). Stable transformation of papaya via microprojectile bombardment. Plant Cell Rep. 9:189–194.

Fraley, R.T., Dellaporta, S.L. and Papahadjopoulos D. (1982). Liposome-mediated delivery of tobacco mosaic virus RNA into petunia protoplasts: A sensitive assay for monitoring liposome-mediated interactions. Proc. Natl. Acad. Sci. USA 79:1859–1863.

Fraley, R.T., Rogers, S.G., Horsch, R.B., Eichholtz, D.A., Flick, J.S., Fink, C.L., Hoffmann, N.L. and Sanders, P.R. (1985). The SEV system: A new disarmed Ti plasmid vector system for plant transformation. Bio/Technol. 3:629–635.

Fraley, R.T., Rogers, S.G., Horsch, R.B., Sanders, P.R., Flick, J.S., Adams, S.P., Bittner, M.L., Brand, L.A., Fink, C.L., Fry, J.S., Gallupi, G.R., Goldberg, S.B., Hoffmann, N.L. and Woo, S.C. (1983). Expression of bacterial genes in plant cells. Proc. Natl. Acad Sci USA 80:4803–4807.

Fromm, M., Callis, J., Taylor, L.P. and Walbot, V. (1987). Electroporation of DNA and RNA into plant protoplasts. Methods in Enzymology 153:351–366.

Fromm, M.E., Morrish, F., Armstrong, C., Williams, R., Thomas, J. and Klein, T.M. (1990). Inheritance and expression of chimeric genes in the progeny of transgenic maize plants. Bio/Technology 8:833–839.

Fromm, M., Taylor, L.P. and Walbot, V. (1985). Expression of genes transferred into monocot and dicot plant cells by electroporation. Proc. Natl. Acad. Sci. USA 82:5824–5828.

Fromm, M., Taylor, L.P. and Walbot, V. (1986). Stable transformation of maize after gene transfer by electroporation. Nature 319:791–793.

Fry, J.E., Barnason, A. and Horsch, R.B. (1987). Transformation of *Brassica napus* with *Agrobacterium tumefaciens* based vectors. Plant Cell Rep. 6:321–325.

Gabard, J.M, Charest, P.J., Iyer, V.N. and Miki, B.L. (1989). Cross-resistance to short residual sulfonylurea herbicides in transgenic tobacco plants. Plant Physiol 91:574–580.

Gamborg, O.L., Miller, R.A. and Ojina, K. (1968). Nutrient requirements of suspension cultures of soybean root cells. Exp. Cell Res. 50:151–158.

Gasser, C.S. and Fraley, R.T. (1989). Genetically engineering plants for crop improvement. Science 244:1293–1299.

Gatignol, A., Durand, H. and Tirnby, G. (1988). Bleomycin resistance conferred by a drug-binding protein. FEBS Lett. 230:171–175.

Gheysen, G., Herman, L., Breyne, P., Gielen, J., Van Montagu, M. and Depicker, A. (1990). Cloning and sequence analysis of truncated T-DNA inserts from *Nicotiana tabacum*. Gene. 94:155–163.

Graessman, A., Graessman, M. and Mueller, C. (1980). Microinjection of early SV40 DNA fragments and Tantigen. Methods Enzymol. 65:816–825.

Gray, D.J. and Purohit, A. (1991). Somatic embryogenesis and development of synthetic seed technology. Critical Reviews in Plant Sci. 10:33–61.

Green, C.E., Armstrong, C.L. and Anderson, P.C. (1983). Somatic cell genetic systems in corn. In: Advances in Gene Technology: Molecular Genetics of Plants and Animals. K. Downey, R.W. Voellmy, R. Ahmad and J. Schultz (eds.), Academic Press, New York, pp. 147–157.

Grimsley, N., Hohn, T., Davis, J.W. and Hohn, B. (1987). *Agrobacterium* mediated delivery of infectious maize streak virus into maize plants. Nature 325:177–179.

Godwin, I., Todd, G., Ford-Lloyd, B. and Newburg, H.J. (1991). The effects of acetosyringone and pH on *Agrobacterium*-mediated transformation vary according to species. Plant Cell Rep. 9:671–675.

Gordon-Kamm, W.J., Spencer, T.M., Mangano, M.L., Daines, R.J., Start, W.G., O'Brien, J.V., Chambers, S.A., Adams, W.R. Jr., Willetts, N.G., Rice, T.R., Mackey, C.J, Krueger, R.W., Kausch, A.P. and Lemaux, P.G. (1990). Transformation of maize cells and regeneration of fertile transgenic plants. The Plant Cell 2:603–618.

Gould, J., Devey, M., Hasegawa, O., Ulian, E.C., Petersen, G. and Smith, R.H. (1991). Transformation of *Zea mays* L. using *Agrobacterium tumefaciens* and the shoot apex. Plant Physiol. 95:426–434.

Harris, R., Wright, M., Byrne, M., Varnum, J., Brightwell, B. and Schuber, K. (1988). Callus and plantlet regeneration from protoplast derived from suspension cultures of wheat (*Triticum asestivum* L.). Plant Cell Rep. 7:337–340.

Hayashimoto, A., Li, Z. and Murai, N. (1990). A polyethylene glycol-mediated protoplast transformation system for production of fertile transgenic rice plants. Plant Physiol. 93:857–863.

Hayford, M., Medford, J., Hoffmann, N., Rogers, S. and Klee, H. (1988). Development of a plant transformation selection system based on expression of genes encoding gentamicin acetyltransferases. Plant Physiol. 86:12161222.

Herman, L., Jacobs, A., Van Montagu, M. and Depicker, A. (1990). Plant chromosome/marker gene fusion assay for study of normal and truncated T-DNA integration events. Mol. Gen. Genet. 224:248–256.

Hernalsteens, J.-P., Thia-toong, L., Schell, J. and Van Monatgu, M. (1984). An *Agrobacterium*-transformed cell culture from the monocot *Asparagus officinalis*. EMBO J. 3: 30394041.

Hinchee, M.A.W., Connor-Ward, D.V., Newell, C.A., McDonnell, R.E., Sato, S.J., Gasser, C.S., Fischhoff, D.A., Re, D.B., Fraley, R.T. and Horsch, R.B. (1988). Production of transgenic soybean plants using *Agrobacterium*-mediated DNA transfer. Bio/Technology 6:915–922.

Hoekema, A., Hirsch, P.R., Hooykaas, P.J.J. and Schilperoort, R.A. (1983). A binary plant vector strategy based on separation of the *vir-* and T-region of *Agrobacterium tumefaciens* Ti-plasmid. Nature 303:179–180.

Hohn, B., Koukolikova-Nicola, Z., Bakkeren, G. and Grimsley, N. (1989). *Agrobacterium*-mediated gene transfer to monocots and dicots. Genome 31, 987–992.

Hood, E.E., Clapham, D.H., Ekberg, I. and Johannson, T. (1990). T-DNA presence and opine production in tumors of *Picea abies* (L.) Karst induced by *Agrobacterium tumefaciens* A281. Plant Mol. Biol. 14:111–117.

Hood, E.E., Helmer, G.L., Fraley, R.T. and Chilton, M-D. (1986). The hypervirulence of *Agrobacterium tumefaciens* A281 is encoded in a region of pTiBo542 outside of T-DNA. J. Bacteriol. 168:1291–1301.

Horsch, R.B., Fraley, R.T., Rogers, S.G., Sanders, P.R., Lloyd, A. and Hoffmann, N. (1984). Inheritance of functional foreign genes in plants. Science 223, 496–498.

Horsch, R.B., Fry, J., Hoffmann, N., Wallroth, M., Eichholtz, D., Rogers, S. and Fraley, R. (1985). A simple and general method for transferring genes into plants. Science 227:1229–1231.

Hu, C.-Y., Chee, P.P., Chesney, R.H., Zhou, J.H., Miller, P.D. and O'Brien, W.T. (1990). Intrinsic GUS-like activities in seed plants. Plant Cell Rep. 9:1–6.

Huang, B., Bird, S., MacLellan, L., Erickson, L., Swanson, E., Kemble, R., Simmonds, D., Miki, B. and Keller, W. (1991). Haploid transformation in *Brassica*. J. Cell. Biochem. Supp. 15A:98.

Iida, A., Morikawa, H. and Yamada, Y. (1990a). Stable transformation of cultured tobacco cells by DNA-coated gold particles accelerated by gas-pressure-driven particle gun. Appl. Microbiol. Biotechnol. 33:560–653.

Iida, A., Seki, M., Kamada, M., Yamada, Y. and Morikawa, H. (1990b). Gene delivery into cultured plant cells by DNA-coated gold particles accelerated by a pneumatic particle gun.

Jahne, A., Lazzeri, P. and Lorz, H. (1991). Regeneration of fertile plants from protoplasts derived from embryogenic cell suspensions of Barley (*Hordeum vulgare* L.). Plant Cell Rep. 10:1–6.

Janssen, B.-J. and Gardner, R.C. (1989). Localized transient expression of GUS in leaf discs following cocultivation with *Agrobacterium*. Plant Mol. Biol. 14:61–72.

Jefferson, R.A. (1987). Assaying chimeric genes in plants: The GUS gene fusion system. Plant Mol. Biol. Rep. 5:387–405.

Jefferson, R.A., Burgess, S.M. and Hirsh, D. (1986). β-Glucuronidase from *Escherichia coli* as a gene-fusion marker. Proc. Natl. Acad. Sci. 83:8447–8451.

Joersbo, M. and Brunstedt, J. (1990). Direct gene transfer to plant protoplasts by mild sonication. Plant Cell Rep. 9:207–210.

Jin, S., Komari, T., Gordon, M.P. and Nester, E.W. (1987). Genes responsible for the supervirulence phenotype of *Agrobacterium tumefaciens* A281. J. Bacteriol. 169:441–74425.

Kado, C.I. (1991). Molecular mechanisms of crown gall tumorigenesis. Critical Rev. Plant Sci. 10:1–32.

Kaeppler, H.F., Weining, G., Somers, D.A., Rines, H.W. and Cockbum, A.F. (1990). Silicon carbide fiber-mediated DNA delivery into plant cells. Plant Cell Rep. 9:415–418.

Kamo, K.K. and Hodges, T.K. (1986). Establishment and characterization of embryogenic maize callus and cell suspension cultures. Plant Sci. 45:111–117.

Kartha, K.K., Chibbar, R.N., Georges, F., Leung, N., Caswell, K., Kendall, E. and Qureshi, J. (1989). Transient expression of chloramphenicol acetyltransferase (CAT) gene in barley cell cultures and immature embryos through microprojectile bombardment. Plant Cell Rep. 8:429–432.

Kiernan, J.M., Goldber, K.-B., Young, M.J., Schoelz, J.E. and Shepard, R.J. (1989). Transformation and regeneration of *Nicotiana edwarsonii*. Plant Sci. 64:67–78.

Kim, Y. and Janick, J. (1991). Abscisic acid and proline improve desiccation tolerance and increase fatty acid content of celery somatic embryos. Plant Cell Tissue Org. Cult. 24:83–89.

Klee, H.J. and Rogers, S.G. (1989). Plant gene vectors and genetic transformation: Plant transformation systems based on the use of *Agrobacterium tumefaciens*. In: Cell Culture and Somatic Cell Genetics of Plants. Vol. 6. Academic Press, pp. 1–23.

Klee, H.J., Yanofsky, M. and Nester, E.W. (1985). Vectors for transformation of higher plants. Bio/Technology 3:637–42.

Klee, H.J. and Rogers, S.G. (1989). Plant gene vectors and genetic transformation: Plant transformation systems based on the use of *Agrobacterium tumefaciens*. In: Cell Culture and Somatic Cell Genetics of Plants, Vol. 6. Academic Press, pp. 1–23.

Klein, T.M., Fromm, M.E., Weissinger, A., Tomes, D., Schaaf, S., Sletten, M. and Sanford, J.C. (1988a). Transfer of foreign genes into intact maize cells using high-velocity microprojectiles. Proc. Natl. Acad. Sci. USA 85:4305–4309.

Klein, T.M., Gradziel, T., Fromm., M.E. and Sanford, J.C. (1988b). Factors influencing gene delivery into *Zea mays* cells by high velocity microprojectiles. Bio/Technology 6, 559–563.

Klein, T.M., Harper, E.C., Svab, Z., Sanford, J.C., Fromm, M.E. and Maliga, P. (1988c). Stable genetic transformation of intact *Nicotiana* cells by the particle bombardment process. Proc. Natl. Acad. Sci. USA 85:8502–8605.

Klein, T.M., Wolf, E.D., Wu, R. and Sanford, J.C. (1987). High-velocity microprojectiles for delivering nucleic acids into living cells. Nature 327:70–73.

Kondo., Y., Samura, T., Ogawa, Y., Tsuruoka, T., Watanaba, A., Totukaya, K., Suzuki, T., Moriyama, C., Yoshima, J. and Inouye, S. (1973). Studies of a new antibiotic, SF1293, I. Isolation and physio-chemical and biological characterization of SF-1295 substances. Sci. Rep. Meij. Seika 13:34–41.

Koncz, C. and Schell, J. (1986). The promoter of TL-DNA gene controls the tissue-specific expression of chimaeric genes carried by a novel type of *Agrobacterium* binary vector. Mol. Gen. Genet. 204:383–396.

Krens, F.A., Molendijk, L., Wullems, G.J. and Schilperoort, R.A. (1982). *In vitro* transformation of plant protoplasts with Ti-plasmid DNA. Nature 296:72–74.

Kruger-Lebus, S. and Potrykus, I. (1987). A simple and efficient method for direct gene transfer to *Petunia hybrida* without electroporation. Plant Mol. Biol. Rep. 5:289–294.

Lawrence, W.A. and Davies, D.R. (1985). A method for the microinjection and culture of protoplasts at very low densities. Plant Cell Rep. 4: 33–35.

Lloyd, A., Barnason, A., Rogers, S.G., Byrne, M., Fraley, R.T. and Horsch, R.B. (1986). Transformation of *Arabidopsis thaliana* with *Agrobacterium tumefaciens*. Science 234:464–66.

Lloyd, A.M., Walbot, V. and Davis, R.W. (1992). *Arabidopsis* and *Nicotiana* anthocyanin production activated by maize regulators *R* and *Cl*. Science 258:1773–1775.

Ludwig, S.R., Bowen, B., Beach, L. and Wessler, S.R. (1990). A regulatory gene as a novel visible marker for maize transformation. Science 247:449–450.

Lulsdorf, M.M., Rempel, H., Jackson, J.A., Baliski, D.S. and Hobbs, S.L.A. (1991). Optimizing the production of transformed pea (*Pisum sativum* L.) callus using disarmed *Agrobacterrum tumefaciens* strains. Plant Cell Rep. 9:479–483.

Luo, Z.-X. and Wu, R. (1988). A simple method for the transformation of rice via the pollen-tube pathway. Plant Mol. Biol. Rep. 6:165–174.

Lyznik, L.A., Peng, J.Y. and Hodges, T.K. (1991). Simplified procedure for transient transformation of plant protoplasts using polyethylene glycol treatment. Biotechniques 10:166.

Masson, J., Bellini, C., Charpentier, A. and Pelletier, G. (1989). Selection of somatic hybrids between diploid clones of potato (*Solanum tuberasum* L.) transformed by direct gene transfer. Theor. Appl. Genet. 78:153–159.

Marks, M.S., Kemp, J.M, Wooston, C.J. and Dale, P.J. (1989). Agroinfection of wheat: A comparison of *Agrobacterium* strains. Plant Sci. 63:247–256.

Matthews, H., Bharathan, N., Litz, R.E., Narayanan, K.R., Rao, P.S. and Bhatia, C.R. (1990). The promotion of *Agrobacterium* mediated transformation in *Atropa belladona* L. by acetosyringone. J. Plant Physiol. 136:404–409.

McCabe, D.E., Swain, W.F., Martinell, B.J and Christou, P. (1988). Stable transformation of soybean (*Glycine max*) by particle acceleration. Bio/Technology 6:923–926.

McClean, P., Chee, P., Held, B., Simental, J., Drong, R.F. and Slightom, J. (1991). Susceptibility of dry bean (*Phaseolus vulgaris* L.) to *Agrobacterium* infection: transformation of catyledonary and hypocotyl tissues. Plant, Cell, Tissue, Organ Culture 24:131–138.

McCormick, S., Niedermeyer, J., Fry, J., Barnason A., Horsch, R. and Fraley, R. (1986). Leaf disc transformation of cultivated tomato (*L. esculentum*) using *Agrobacterium tumefaciens*. Plant Cell Rep. 5:81–84.

Melchers, L.S., Maroney, M.J., den Dulk-Ras, A., Thompson, D.V., van Vuuren, H.A.J., Schilperoort, R.A. and Hooykaas, P.J.J. (1990). Octopine and nopaline strains of *Agrobacterium tumefaciens* differ in virulence; molecular characterization of the *vir*F locus. Plant Mol. Biol. 14:249–259.

McCown, B.H., McCabe, D.E., Russell, D.R., Robinson, D.J., Barton, K.A. and Raffa, K.E. (1991). Stable transformation of *Populus* and incorporation of pest resistance by electric discharge particle acceleration. Plant Cell Rep. 9:590–594.

Michelmore, R., Marsh, E., Seely, S. and Landry, R. (1987). Transformation of lettuce (*Lactuca sativa*) mediated by *Agrobacterium tumefaciens*. Plant Cell Rep. 6:439–442.

Miki, B.L.A., Reich, T.J. and Iyer, V.N. (1987). Microinjection: an experimental tool for studying and modifying plant cells. In: Plant DNA Infectious Agent. T. Hohn and J. Schell (eds.), Springer Verlag, New York, pp. 349–365.

Miller, P.D. and Lauritis, J.A. (1989). DNA delivery by microparticles accelerated by a vaccination gun. J. Cell Biochem. Suppl. 13D: 264.

Mollenhauser, C., Smart, C. and Amrhein, N. (1987). Glyphosate toxicity in the shoot apical region of tomato plants. Pest. Biochem. Physiol. 29:55–65.

Moloney, M.M., Walker, J.M. and Sharma, K.K. (1989). High efficiency transformation of *Brassica napus* using *Agrobacterium vectors*. Plant Cell Rep. 8:238–242.

Mooney, P.A., Goodwin, P.B., Dennis, E.S. and Llewellyn, D.J. (1991). *Agrobacterium tumefaciens*-gene transfer into wheat tissues. Plant Cell Tissue Org. Cult. 25:209–218.

Morikawa, H., Iida, A. and Yamada, Y. (1989). Transient expression of foreign genes in plant cells and tissues obtained by a simple biolistics device (particle gun). App. Micr. and Biotech. 31:320–322.

Mórocz, S., Donn, G., Nemeth, J. and Dudits, D. (1990). An improved system to obtain fertile regenerants via maize protoplasts isolated from a highly embryogenic suspension culture. Theor. Appl. Gene. 80:721–726.

Morota, H. and Uchimiya, H. (1988). Inheritance and structure of foreign DNA in progenies of transgenic tobacco obtained by direct gene transfer. Theor. Appl. Genet. 76:161–164.

Murashige, T. and Skoog, F. (1962). A revised medium for rapid growth and bioassays with tobacco cultures. Physiol. Plant. 15:473–497.

Murakami, T., Anzai, H., Imai, S., Sato, A., Nagaoka, K. and Thompson C.J. (1986). The bialaphos biosynthetic genes of *Streptomyces hygroscopicus*: molecular cloning and characterization of the gene cluster. Mol. Gen. Genet. 205:42–50.

Negrutiu, I., Shillito, R., Potrykus, I., Biansini, G. and Sala, F. (1987). Hybrid genes in the analysis of transformation conditions. Plant. Mol. Biol. 8:363–373.

Neuhaus, G. and Spagenberg, G. (1990). Plant transformation by microinjection techniques. Physiol. Plant. 79:213–217.

Neuhaus, G., Kranz, E., Spangenberg, G. and Schweiger, H.G. (1987a). Transformation of *Nicotiana tabacum* and *Brassica napus* by microinjection into protoplasts. Eur. J. Cell Biol. 43:39.

Neuhaus, G. and Spangenberg, G. (1990). Plant transformation by microinjection techniques. Physiol. Plant. 79:213–217.

Neuhaus, G., Spangenberg, G., Mittelsten Scheid, O. and Schweiger, H.-G. (1987b).Transgenic rapeseed plants obtained by the microinjection of DNA intro microspore-derived embryoids. Theor. Appl. Genet. 75:30–36.

Oard, J.H., Paige, D.F., Simmonds, J.A. and Gradziel, T.M. (1990). Transient gene expression in maize, rice and wheat cells using an airgun apparatus. Plant Physiol. 92: 334–339.

Ohlson, M. and Ericksson, T. (1988). Transformation of *Brassica campestris* protoplasts with *Agrobacterium tumefaciens*. Hereditas 108:173–177.

Ow, D.W., Wood, K.V., DeLuca, M., de Wet, J.R., Helinski, D.R. and Howell, S.H. (1986). Transient and stable expression of the firefly luciferase gene in plant cells and transgenic plants. Science 234:856–859.

Owens, L.D. and Cress, D.E. (1985). Genotypic variability of soybean response to *Agrobacterium* strains harboring the Ti or Ti plasmids. Plant Physiol. 77:87–94.

Padgette, S.R., della-Cioppa, G., Shah, D.M., Fraley, R.T. and Kishore, G.M. (1989). Selective herbicide tolerance through protein engineering. In: Cell Culture and Somatic Cell Genetics, Vol. 6. J. Schell and I.K. Vasil (eds.), Academic Press, New York, pp. 441–475.

Parrott, W.A., Hoffmann, L.M., Hildebrand, D.F., Williams, E.G. and Collins, G.B. (1989). Recovery of primary transformants of soybean. Plant Cell Rep. 7:615–617.

Perez, P., Tiraby, G., Kallerhoff, J. and Perrat, J. (1989). Phleomycin resistance as a dominant selectable marker for plant cell transformation. Plant Mol. Biol. 13:365–373.

Perlak, F.J., Deaton, R.W., Armstrong, T.A., Fuchs, R.L., Sims, S.R., Greenplate, J.T.L. and Fischhoff, D.A. (1990). Insect resistant cotton plants. Bio/Technology 8:939–943.

Petersen, W.L., Sulc, S. and Armstrong, C.L. (1990). Embryogenic callus and fertile plants from protoplasts of maize suspension cultures. Agronomy Abstracts 82:201.

Planckaert, F. and Walbot, V. (1989). Transient gene expression after electroporation of protoplasts derived from embryogenic maize callus. Plant Cell Rep. 8:144–147.

Potrykus, I. (1991). Gene transfer to plants – assessment of published approaches and results. Ann. Rev. Plant Physiol. Plant. Mol Biol. 42:205–225.

Potrykus, I., Datta, S.K., Neuhaus, G. and Spangenberg, G. (1989). Approach to cereal transformation via microinjection into microspore-derived and into zygotic proembryos. J. Cell. Biochem. Suppl. 13D:228.

Prioli, L. and Sondahl, M. (1989). Plant regeneration and recovery of fertile plants from protoplasts of maize (*Zea mays* L.). Bio/Technology 7:589–594

Pua, E., Mehra-Palta, S., Nagy, F. and Chua, N.-H. (1987). Transgenic plants of *Brassica napus* L. Bio/Technology 5:815–817.

Pythoud, F., Sinkar, V.P., Nester, E.W. and Gordon, M.P. (1987). Increased virulence of *Agrobacterium* rhizogenes conferred by the *vir* region of pTiBo542: application to genetic engineering of poplar. Bio/Technology 5:1323–1327.

Raineri, D.M., Bottino, P., Gordon, M.P. and Nester, E.W. (1990). *Agrobacterium*-mediated transformation of rice (*Oryza sativa* L.). Bio/Technology 8:33–38.

Ream, W. (1989). *Agrobacterium tumefaciens* and interkingdom genetic exchange. Ann. Rev. Phytopathol. 27:583–618.

Reich, T.J., Iyer, V.N., Haffner, M., Holbrook, L.A. and Miki, B.L. (1986a). The use of fluorescent dyes in the microinjection of alfalfa protoplasts. Can. J. Bot. 64:1259–1267.

Reich, T.J., Iyer, V.N. and Miki, B.L. (1986b). Efficient transformation of alfalfa protoplasts by the intranuclear microinjection of Ti plasmids. Bio/Technology 4:1001–1004.

Reich, T.J., Iyer, V.N., Scobie, B. and Miki, B.L. (1986c). A detailed procedure for the intranuclear microinjection of plant protoplasts. Can. J. Bot. 64:1259–1267.

Riggs, C.D. and Bates, G.W. (1986). Stable transformation of tobacco by electroporation: Evidence for plasmid concatenation. Proc. Natl. Acad. Sci. USA 83:5602–5606.

Rogers, S.G., Klee, H.J., Horsch, R.B. and Fraley, R.T. (1987). Improved vectors for plant transformation: expression cassette vectors and new selectable markers. Meth. Enzymol. 153:253–277.

Rogowsky, P.M., Powell, B.S., Shirasu, K., Lin, T.-S., Morel, P., Zyprian, E.M., Steck, T.R. and Kado, C.I. (1990). Molecular characterization of the *vir* regulon of *Agrobacterium tumefaciens*: Complete nucleotide sequence and gene organization of the 28.63-kbp regulon cloned as a single unit. Plasmid 23:85–106.

Rosenberg, N., Gad, A.E., Altman, A., Navot, N. and Czosnek, H. (1988). Liposome-mediated introduction of the chloramphenicol acetyl transferase (CAT) gene and its expression in tobacco protoplasts. Plant Mol. Biol. 10:185–191.

Sacristan, M.D., Gerdemann, Knörck, M. and Schneider, O. (1989). Incorporation of hygromycin resistance in *Brassica nigra* and its transfer to *B. napus* through asymmetric protoplast fusion. Theor. Appl. Genet. 78:194–200.

Sambrook, J., Fritsch, E.F. and Maniatis, T. (1989). Molecular Cloning: A Laboratory Manual. Cold Spring Harbor Laboratory Press, New York.

Sanford, J.C. (1998). The Biolistic process. TIBTECH 6:299–302.

Sanford, J.C., Klein, T.M., Wolf, E.D. and Allen, N. (1987). Delivery of substances into cells and tissues using a particle bombardment process. Part. Sci. and Tech. 5:27–37.

Schenk, R.V. and Hildebrandt, A.C. (1972). Medium and techniques for induction and growth of monocotyledonous and dicotyledonous plant cell cultures. Can. J. Bot. 50:199–204.

Schrammeijer, B., Sijmons, P.C., van den Elzen, P.J.M., and Hoekma, A. (1990). Meristem transformation of sunflower via *Agrobacterium*. Plant. Cell. Rep. 9:55–60.

Serres, R., McCown, B., McCabe, D., Stang, E., Russell, D. and Martinell, B. (1990). Stable genetic transformation of cranberry using electric discharge particle acceleration. Hort. Sc. 25:1130.

Shah, D.M., Horsch, R.B., Klee, H.J., Kishore, G.M., Winter, J.A., Tumer, N.E., Hironaka, C.A., Sanders, P.R., Gasser, C.S., Aykent, S., Siegel, N.R., Rogers, S.G. and Fraley, R.T. (1986). Engineering herbicide tolerance in transgenic plants. Science 233:478–481.

Sheen, J. (1991). Molecular mechanisms underlying the differential expression of maize pyruvate, orthophosphate dikinase genes. The Plant Cell 3:225–245.

Sheikholeslam, S.N. and Weeks, D.P. (1987). Acetosyringone promotes high efficiency transformation of *Arabidopsis thaliana* explants by *Agrobacterium tumefaciens*. Plant Mol. Biol. 8:291–298.

Shillito, R.D., Carswell, G.K., Johnson, C.M., DiMaio, J.J. and Harms, C.T. (1989). Regeneration of fertile plants from protoplasts of elite inbred maize. Bio/Technology 7:581–587.

Shillito, T.D., Saul, M.W., Paszkowski, J., Muller, M. and Potrykus, I. (1985). High efficiency direct gene transfer to plants. Bio/Technology 3:1099–1103.

Shillito, R.D., Paszkowski, J.I and Potrykus, I. (1983). Agarose plating and bead type culture technique enable and stimulate development of protoplast-derived colonies in a number of plant species. Plant Cell Rep. 2:244–247.

Shillito, R.D. and Potrykus, I. (1987). Direct gene transfer to protoplasts of dicotyledonous and monocotyledonous plants by a number of methods, including electroporat ion. Meth. Enzymol. 5:313–336.

Shimamoto, K., Terada, R., Izawa, T. and Fujimoto, H. (1989). Fertile transgenic rice plants regenerated from transformed protoplasts. Nature 338:274–276.

Shin, D.H., Virigool, S., Shinozaki, K.Y. and Oono, K. (1991). Survival mechanism of dried calli and regeneration of plants in rice. Jap. J. Genet. 66:13–25.

Slade, D. and McCall, C. (1991). Examination of transformation efficiencies in cotton (*Gossypium hirsutum* L.). *In vitro*. 27 part 11:151A.

Smith, J.K., Mauvais, C.J., Knowlton, S. and Mazur, B.J. (1988). Molecular biology of resistance to sulfonylurea herbicides. Proc. ACS Symp. Biotechnol. Corp. Protect., Washington DC, pp. 25–36.

Songstad, D.D., Armstrong, C.L. and Petersen, W.L. (1991). $AgNO_3$ increases type II callus production from immature embryos of maize inbred B73 and its derivatives. Plant Cell Rep. 9:699–702.

Spagenberg, G. and Neuhaus, G. (1988). Gene transfer by individual manipulation of plant cells. In Progress in Plant Protoplast Research. K.J. Puite, J.J.M. Dons, H.J. Huizing, A.J. Kool, M. Koornneff and F.A. Krens (eds.), Academic Publishers, Boston, pp. 151–157.

Spagenberg, G., Neuhaus, G. and Schweiger, H.G. (1986). Expression of foreign genes in a higher plant cells after electrofusion-mediated cell reconstitution of a microinject karyoplast and a cytoplast. Eur. J. Cell. Biol. 42:236–238.

Steinrucken, H.C. and Amrhein, N. (1980). The herbicide glyphosate is a potent inhibitor of 5-enolpyruvly-shikimic acid-3-phophate synthase. Biochem. Biophys. Res Commun. 94:1207–1212.

Stewart, J. and Hsu, C.L. (1977). *In-ovulo* embryo culture and seedling development of cotton. Planta 137:113–117.

Stomp, A.M., Loopstra, C., Chilton, W.S., Sederoff, R.R. and Moore, L.W. (1990). Extended host range of *Agrobacterium tumefaciens* in the Genus *Pinus*. Plant Physiol. 92:1226–1232.

Svab, Z., Hajdukiewicz, P. and Maliga, P. (1990). Stable transformation of plastids in higher plants. Proc. Nat Acad. Sci. USA 82:8626–8531.

Svab, Z., Harper, F.C., Jones, J.D.G. and Maliga, P. (1990). Aminoglycoside-3-adenyltransferase confers resistance to spectitnomycin and streptomyein in *Nicotiana tabacum*. Plant Mol. Biol. 14:197–205.

Thomas, B.R. and Pratt, D. (1981). Efficient hybridization between *Lycopsercion esculentum* and *L. peruvanum* via *embryo callus*. Theor. Appl. Genet. 59:215–219.

Tomes, D.T., Weissinger, A.K., Ross, M., Higgins, R., Drummond, B.J., Schaaf, S., Malone-Schoneberg, J., Staebell, M., Flynn, P., Anderson, J. and Howard, J. (1990). Transgenic tobacco plants and their progeny derived by microprojectile bombardment of tobacco leaves. Plant Mol. Biol. 14:261–268.

Thomzik, J.E. and Hain, R. (1990). Transgenic *Brassica napus* plants obtained by cocultivation with *Agrobacterium tumefaciens*. Plant Cell Rep. 9:223–236.

Topfer, R., Gronenborn, B., Schell, J. and Steinbiss, H. (1989). Uptake and transient expression of chimeric genes in seed-derived embryos. The Plant Cell 1:133–139.

Toyoda, H., Matsuda, Y., Utsumi, R. and Ouchi, S. (1988). Intranuclear microinjection for transformation of tomato callus cells. Plant Cell Rep. 7:293–296.

Toyoda, H., Yamaga, T., Matsuda, Y. and Ouchi, S. (1990). Transient expression of the β-glucuronidase gene introduced into barley coleoptile cells by microinjection. Plant Cell Rep. 9:299–302.

Trolinder, N.L. and Goodin, J.R. (1987). Somatic embryogenesis and plant regeneration in cotton (*Gossypium hirsutum*). Plant Cell Rep. 6:231–234.

Tuberosa, R. and Phillips, R.L. (1986). Isolation of methothrexate-tolerant cell lines of corn. Maydica 31:215–225.

Twell, D., Klein, T.M, Fromm, M.E. and McCormick, S. (1989). Transient expression of chimeric genes delivered into pollen by microprojectile bombardment. Plant Physiol. 91:1270–1274.

Ulian, E., Smith, R., Gould, J. and McKnight, T. (1988). Transformation of plants via the shoot apex. *In vitro* Cell Dev. Biol. 24:951–954.

Umbeck, P., Johnson, G., Barton, K. and Swain, W. (1987). Genetically transformed cotton (*Gossypium hirsutum* L.). Bio/Technology 5:263–266.

Vancanneyt, G., Schmidt, R., O'Connor-Sanchez, A., Willmitzer, L. and Rocha-Sosa, M. (1990). Construction of an intron-containing marker gene: Splicing of the intron in transgenic plants and its use in monitoring early events in *Agrobacterium* mediated plant transformation. Mol. Gen. Genet. 220:245–250.

van Wordragen, M.F., de Jong, J., Huitema, H.B.M. and Dons, J.M. (1991). Genetic transformation of *Chrysanthemum* using wild type *Agrobacterium* strains: Strain and cultivar specificity. Plant Cell Reports 9:505–508.

Vasil, I.K. (1988). Progress in the regeneration and genetic manipulation of cereal crops. Bio/Technology 6:397–402.

Vasil, V., Castillo, A.M., Fromm, M.E. and Vasil, I.K. (1992). Herbicide resistant fertile transgenic wheat plants obtained by microprojectile bombardment of regenerable callus. Bio/Technology 10:667–674.

Vasil, V., Hauptmann, R.M., Morrish, F.M. and Vasil, I.K. (1988). Comparative analysis of free DNA delivery and expression into protoplasts of *Panicum maxium* Jacq. (Guinea grass) by electroporation and polyethylene glycol. Plant Cell Rep. 7:499–503.

Vasil, V., Redway, F. and Vasil, I.K. (1990). Regeneration of plants from embryogenic suspension culture protoplasts of wheat (*Triticum aestivum* L.). Bio/Technology 8:429–434.

Waldron, C., Murphy, E.B., Roberts, J.L., Gustafson, G.D., Armour, S.L. and Malcom, S.K. (1985). Resistance to hygromycin-B. Plant Mol Biol. 5:103–108.

Wang, Y., Klein, T.M., Fromm, M., Cao, J., Sanford, J.C. and Wu, R. (1988). Transient expression of foreign genes in rice, wheat, and soybean cells following particle bombardment. Plant Mol. Biol. 11:433–439.

Weber, G. (1988). Microperforation of plant tissue with a UV-laser microbeam and injection of DNA into cells. Naturwissenschaften 75:35–36.

Weber, G., Monajembashi, S., Greulich, K.-O. and Wolfrum, J. (1989). Uptake of DNA in chloroplasts of *Brassica napus* (L.) facilitated by a UV-laser microbeam. Eur. J. Cell Biol. 49:73–79.

Weber, G., Monajembashi, S., Wolfrum, J. and Gruelich, K.O. (1990). Genetic changes in breeding higher plant cells by a laser microbeam. Physiol. Plant. 79:190–193.

Wei, Z. and Xu, Z. (1990). Regeneration of fertile plants from embryogenic suspension culture protoplasts of *Sorghum vulgare*. Plant Cell Rep. 9:51–53.

Weissinger, A., Tomes, D., Maddock, S., Fromm, M. and Sanford, J. (1988). Maize transformation via microprojectile bombardment. In: Genetic Improvements of Agriculturally Important Crops. Fraley, R.T., Frey, N.M. and Schell, J. (eds.), Cold Spring Harbour Laboratory Press, New York, pp. 21–25.

Xu, X.P. and Li, B.J. (1994). Fertile transgenic indica-rice plants obtained by electroporation of the seed embryo cells. Plant. Cell. Rep. 13:237–242.

Yadav, N., McDevitt, R.E., Benard, S. and Falco, S.C. (1986). Single amino acid substitutions in the enzyme acetolactate synthase confer resistance to the herbicide sulfonmeturon methyl. Proc. Natl. Acad. Sci. USA 83: 4418–4422.

Ye, G., Daniell, H. and Sanford, J.C. (1990). Optimization of delivery of foreign DNA into higher-plant chloroplasts. Plant Mol. Biol. 15: 809–819.

Zambryski, P. (1988). Basic processes underlying *Agrobacterium*-mediated DNA transfer to plant cells. Annu. Rev. Genet. 22:1–30.

Zambryski, P., Joos, H., Genetello, C., Leemans, J., Van Montagu, M. and Schell, J. (1983). Ti plasmid vector for the introduction of DNA into plant cells without alteration of their normal regeneration capacity. EMBO J. 2:2143–2150.

Zhu, Z., Hughes, K.W., Huang, L. and Li, X. (1990). Cationic liposome-mediated transformation of rice protoplasts. Focus 12, 41–44 (BRL Technical Services publication).

Zyprian, E. and Kado, C.I. (1990). *Agrobacterium*-mediated plant transformation by novel-T vectors in conjunction with a high-copy *vir* region helper plasmid. Plant Mol. Biol. 15:245–256.

11. Cell Culture for Production of Secondary Metabolites

F. Constabel and R.T. Tyler

Contents

1. Introduction

Phytochemicals derived from secondary metabolism have long been processed for pharmaceuticals, food additives, flavors and fragrances, and for products like latex and tannins. Traditionally, phytochemicals have been obtained by extraction from plants growing in the wild or in plantations. Wild plants, however, have become the subject of environmental concerns and may not be open to harvesting as before. Plantation crops are subject to biological and climatic adversities, as well as economic and political instabilities. Alternatives to collecting wild plants and growing plantations are the result of new technology, in particular, the development of plant tissue and cell culture. The *in vitro* mode of production, however, has yet to prove its profitability and may in future be surpassed by medicinal and industrial plants improved by plant breeding and genetic engineering. Still, the demonstration of secondary metabolites in plant cell cultures and strategies to exploit their potential is valid (Constabel, 1990).

The observation that some callus and cell suspension cultures turn red with age, show lignified tracheids, or emit a fragrance is an indication of the potential of such cultures to produce secondary metabolites. The conditions

Indra K. Vasil and Trevor A. Thorpe (eds.), Plant Cell and Tissue Culture, 271–289.

Table 1. Species of cells grown as suspension with capacity for high accumulation of secondary metabolites

Plant species	Product	References
Morinda citrifolia	Anthraquinones	Zenk et al., 1975
Coleus blumei	Rosmarinic acid	Razzaque and Ellis, 1977
Daucus carota	Anthocyanin	Noé et al., 1980
Lithospermum erythrorhizon	Shikonin	Fujita et al., 1981
Coffea arabica	Caffeine	Waller et al., 1983
Trypterigium wilfordii	Tripdiolide	Kutney et al., 1983
Coptis japonica	Berberine	Sato and Yamada, 1984
Thalictrum minus	Berberine	Nakagawa et al., 1984
Papaver somniferum	Sanguinarine	Eilert et al., 1985
Catharanthus roseus	Catharanthine	Smith et al., 1987
Thalictrum rugosum	Berberine	Piehl et al., 1988
Vitis vinifera	Anthocyanin	Do and Cormier, 1990
Panax ginseng	Ginsengosides	Strogov et al., 1990
Catharanthus roseus	Serpentine	Scragg et al., 1990

which favor such production and a description of compounds found in cultures have been discussed in detail (Constabel and Vasil, 1987, 1988). Given suitable culture conditions and adequate analytical methods, cell cultures will display compounds as known for the source plants. Non-occurrence of compounds expected may motivate variation of culture conditions or genetic manipulation of cell cultures. Greater diversity of compounds in cell cultures than in source plants has been detected on occasion. For example, anthraquinones have been found in addition to known alkaloids in *Cinchona* spp. cultures (Mulder-Krieger et al., 1982; Koblitz, 1988).

Generally, cell cultures differ from source material in the relative amount of individual compounds (Zenk et al., 1977). Differences may even be detected among cell lines of the same source (Constabel et al., 1981). Availability of precursors, physiological conditions of the cells, and genetic make-up as affected by somaclonal variation, all contribute to diversity in the quantity of products synthesized and accumulated.

Occurrence of high concentrations of valuable compounds in cultured cells, as found with various cell strains (Table 1) would suggest industrial exploitation of plant cell cultures. A strategy for profitability of such a process requires chemicals to accumulate in cultured cells in concentrations higher than in traditional material, scale-up and development of bioreactor processes, cost-effective extraction of compounds of interest and, importantly, certification that compounds extracted from cell cultures are admissible for prescription drugs and food stuffs in accordance with regulations which govern the industry. The latter point is often ill researched and makes for development which is science rather than market driven.

Cell cultures have been employed not only as source material for extraction of secondary metabolites but also for biotransformations. The most advanced systems involve the hydroxylation of β-methyldigitoxin to β-methyldigoxin and biotransformation of digitoxin to deacetyllanatoside C with cells derived from *Digitalis lanata* (Reinhard et al., 1989; Kreis and Reinhard, 1990).

2. Cell Material

Plant cells cultured *in vitro* show several characteristics which must be considered before attempting to employ such material for scale-up of metabolite production: cell size and rigidity which may invite shear stress, slow growth, tendency to aggregate and form clumps, competence or level of differentiation/specialization for product accumulation, and intravacuolar storage of products (cf. Taticek et al., 1991).

2.1. *Shear Stress*

Plant cells measure 10–100 μm in diameter (30–50 μm on average) and are encased in a fairly rigid cellulose-based wall. Suspended in 50 ml liquid medium in small flasks, for example, the cells will tolerate agitation by reciprocal or gyratory shakers (100–150 rpm) over any length of time without damage to their structure and viability. When suspended as heavy biomass in large volume bioreactors, however, agitation and aeration must be controlled so as not to effect destructive shear stress. Kurz' (1971) chemostat with aeration and agitation by air bubbles (Fig. 1) and the advanced airlift bioreactors of today have become satisfactory solutions to reduce shear stress. An alternative approach may be selection of shear tolerant cell lines, which would permit large scale culture in conventional stirred tank reactors and much higher biomass loadings than is likely possible in any air-driven design (Fowler, 1987).

2.2. *Growth Kinetics*

Plant cells grown in optimal nutrient medium may divide every 22 hours. In general, however, divisions occur at much longer intervals, i.e. 30 to 70 hours (Street, 1973). Also, some cells may cease to divide early in the culture period and elongate or specialize to accumulate secondary metabolites, i.e. pigments, alkaloids, phenolics. The consequences of relatively slow growth are two-fold: cell suspensions require long periods of culture before substantial amounts of biomass and products have accumulated and the suspensions need to be carefully protected from contamination. Bioreactors up to 15,500 l have, however, successfully been run (Noguchi et al., 1977) and over several months.

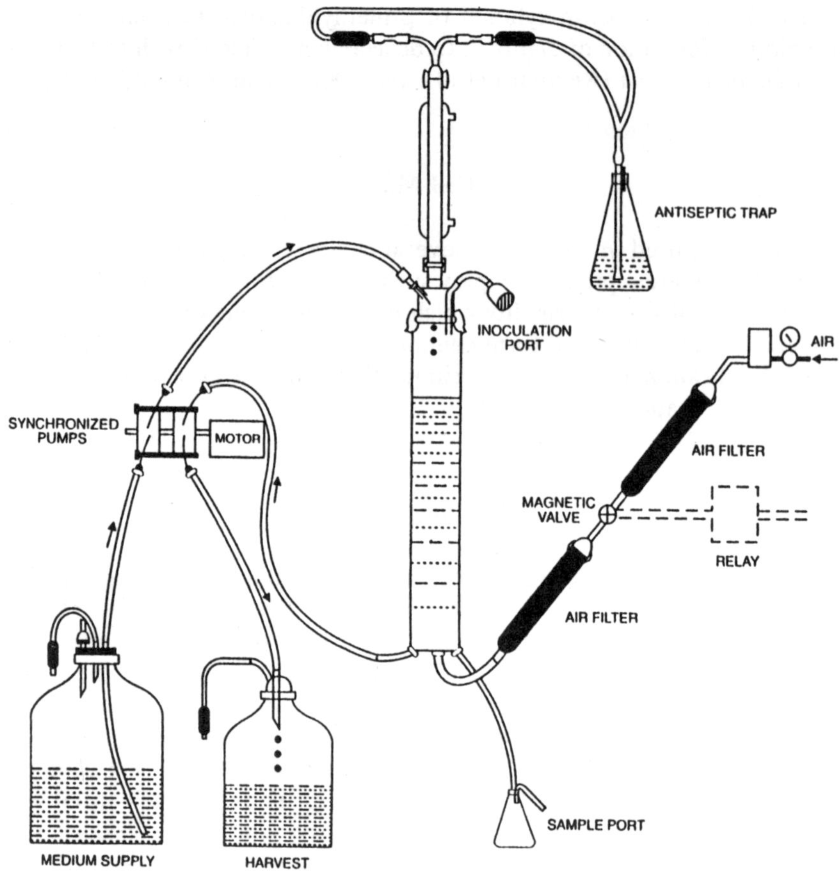

Figure 1. Chemostat after Kurz (1971).

2.3. *Aggregation*

As mentioned in Section 1, cell suspensions generally do not consist of isolated single cells, but of aggregates ranging from a few to several hundred cells. All efforts to achieve single cell suspensions have failed so far, or are impractical when aiming at metabolite production. Clumping is the result of non-dehiscence of daughter cells subsequent to a cell division. Such behavior may persist until mechanical forces or rounding off of maturing cells leads to dissociation. A problem which results from clumping is the occurrence of gradients of nutrients and differentiation of cells. Since such gradients impede control over growth and product formation in bioreactors, clumping has to be reduced as much as possible (cf. clumping as a process of self-immobilization as used for solasodine production with *Solanum aviculare*, Tsoulpha and Doran, 1991). Aggregation becomes less of a problem when over time and due to selection of fast growing cells, suspensions become more homogenous, rather fine, and most suitable for manipulation in bioreactors. Aggregation remains a serious impediment to the application in cell culture of continuous culture techniques developed for microbial systems (Rokem, 1987).

One particular problem is a tendency of plant cells to adhere to glass or plastic surfaces at the air-liquid interface of a culture. Magnetized scrubbers inside vessels to manually remove cell aggregates from glass walls may help. Bubble action in airlift fermenters also counters aggregation and adhesion of cells. Still, development of a meringue-like clump of cells and foam at the air-liquid interface may be problematic and require the use of anti-foam.

2.4. *Competence*

Growth by cell division and product formation often exclude each other. While some biosynthesis of secondary metabolites may occur during the log-phase of the culture period, product accumulation will generally become significant only during deceleration of growth and during the idiophase, when cells have ceased dividing and have specialized in product accumulation. For example, a red inoculum of *Haplopappus gracilis* cells can be seen to become pale subsequent to subculturing and turning red a few days later. Frequent subcultures can cause temporary loss of all pigmentation. Given proper culture conditions, *Haplopappus* cells have been observed to restore their ability to accumulate pigments, even after 10 years of growing as a pale suspension (F. Constabel, unpubl. results).

The incompatibility of cell proliferation and metabolite accumulation in cells has remained of considerable concern regarding industrial exploitation of cell cultures. It has led to the concept of two-stage cultures, growth of biomass representing stage 1, and stage 2 the production of metabolites. The concept was subsequently realized in single and double bioreactors, the latter operated in tandem. The change from growth to idiophase is gradual. Termination of cell division simply results from starvation, in particular from lack of sufficient carbohydrates. It may be induced by transfer of cells to hormone-free media. It may be provoked also by biotic stress as effected by fungal mycelia or fractions thereof (Eilert, 1987), or abiotic stress as caused by heavy metal ions, vanadyl sulphate for instance (Smith, 1987), organics like arachidonic acid, or chitosan (Eilert, 1987), or changes in temperature (Lockwood, 1984). Even cell density has been found to exert stress which induced product formation (Fujita et al., 1986).

Maximum accumulation of secondary metabolites is often realized only after extensive optimization of the production medium. An often overlooked but perhaps important consideration is the composition of the culture atmosphere, as the levels of dissolved oxygen or carbon dioxide in the medium may influence culture growth or productivity. This may be of particular importance in air-driven bioreactors, particularly at high ventilation rates (Fowler, 1987).

The use of cloning techniques for selection of superior cell lines was discussed by Dougall (1987); the examples presented confirm the utility of this approach in improving metabolite yields from cell cultures, although the long term stability of selected high yielding strains remains a concern. Production of some metabolites may depend upon attaining a higher level of cellular differentiation in culture, such as the selection/development of cultured cells containing functional chloroplasts (Wink and Hartmann, 1980).

2.5. *Excretion*

Compartments for product accumulation are, for example, plastids in the case of carotenoids, cell walls for phenolics and lignin, and vacuoles for the majority of products of interest. As a result, cell material has to be destroyed during product recovery. Since this is a costly proposition, interest has shifted to cultures which show natural excretion or inducible product release. The latter has been observed to follow treatment of cells with stress agents like fungal mycelia (Eilert et al., 1985, 1986).

Natural excretion of products from cells would be a most desirable process, since this would permit extraction procedures known for micro-organisms including continuous adsorption of products from the medium without loss of biomass (cf., Beiderbeck and Knoop, 1987). The formation of berberine by cells of *Thalictrum minus* is accompanied by excretion, a rare example of such behavior, so far (Nakagawa et al., 1984). The tolerance of berberine producing cells (*Coptis japonica*, *Thalictrum flavum*, *T. minus*) to extracellular berberine is remarkable. With non-berberine producing cells, small amounts of berberine will cause growth inhibition (Sato et al., 1990). Shikonin particles are deposited between cell membrane and cell wall (Tsukuda and Tabata, 1984). Recently, hairy root cultures of *Lithospermum erythrorhizon* were found to not only accumulate sizeable amounts of shikonin, but to release 85–90% of it into the medium. The shikonin could now be trapped in resin columns and permit a continuous culture over several months (Shimomura et al., 1991). Due to the importance of this phenomenon, it is conceivable that genetic manipulation of cells for product excretion may become imperative; success may pave the way for ultimate bioreactor design.

3. Suspension Cell Culture

3.1. *One-cycle Batch Culture*

The simplest system for production of secondary metabolites by plant cell cultures requires cell material with proven capacity for product accumulation and a culture vessel to hold several litres of suspension. A first experiment could be performed with Kurz' (1971) lab-made chemostat (Fig. 1) or a commercially available generic bench-top fermenter of several litres capacity, with an impeller for agitation, an air sparger for aeration, and separate ports for addition of fresh medium and sampling of suspension. For example, with *Catharanthus roseus* or *Papaver somniferum* cell suspensions, 4.8 l of B5 medium were inoculated with 0.2 l of sedimented cells, agitated by a marine impeller at about 200 rpm, and aerated by a disk sparger at 150 ml/min. Over culture periods of 10–14 days, foaming, aggregation and adhesion to surfaces did not become a problem.

Use of an airlift fermenter permits agitation and aeration in one process and counters the tendency of cells to aggregate as well as reducing mechanical injury to cells as effected by impellers. Figure 2 shows the dimensions of a 20-l airlift fermenter as suggested by Alfermann et al. (1977). Air under pressure escapes through the pores of a sparger ring which is situated within the inner glass tube near the bottom of the fermenter. The mixture of air, medium and cells rises and after passing the upper end of the inner glass tube, the medium and cells, now largely free of air bubbles and, therefore, of higher density, return to the bottom of the fermenter via the peripheral space. The result is a continuous circulation of the cell suspension. For operation of the airlift fermenter as well as kinetics of growth and product formation, see Seitz et al. (1985).

The introduction of a two-stage system in order to improve product output can be achieved with one culture vessel and would require an exchange of growth medium for production medium as suggested by Zenk et al. (1977). The composition of the two media differs substantially in the concentration of mineral salts, i.e. nitrate, ammonium, and phosphate, carbohydrates and hormones. The production medium may be reduced to a carbohydrate solution. The media formulations by Payne et al. (1988) can be suggested as an example: The medium for maintenance and growth of *Catharanthus roseus* cells was a Murashige and Skoog (1962) salts and vitamins solution with 30 g/l glucose and 1 μM 2,4-D. The production medium was prepared by making a 10-fold dilution of a Murashige and Skoog salts and vitamins solution in which both the nitrogen and phosphorous salts were deleted. Glucose (30 g/l) was added, but no growth regulators were used in the production medium. Or, with *Lithospermum erythrorhizon* cells, medium after Linsmaier and Skoog (1965) was used for growth promotion, and White's medium (1963) for metabolite production (Fujita et al., 1981).

A similar procedure is required to achieve biotransformations. Cell suspensions derived from high-yielding callus strains of *Digitalis lanata* were first cultured in growth medium modified after Murashige and Skoog (1962). After substantial proliferation of the cells, the medium was exchanged for production medium consisting of pure 8% glucose solution, followed by addition of digitoxin as substrate and harvest of cells with high levels of deacetyllanatoside C. The process was performed in two 20-l air-lift bioreac-

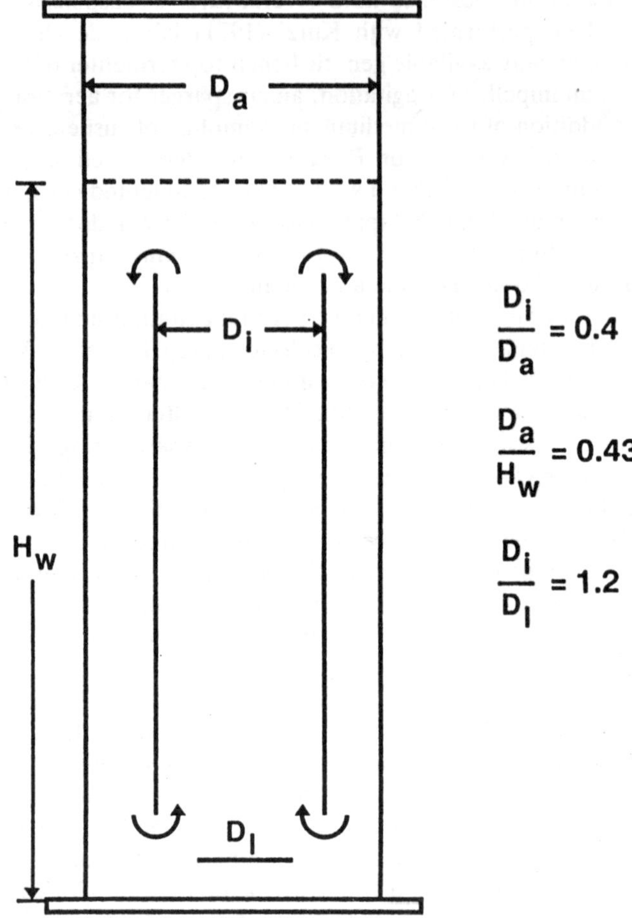

Figure 2. 20-l airlift fermentor after Alfermann et al. (1977).

tors and designed to accommodate 6 cycles of proliferation and production in a semi-continuous fashion (Kreis and Reinhard, 1990).

The problem with two-stage systems is media exchange without loss or damage of the cells cultured. A general method would be to allow cells to settle and to siphon off the spent growth medium, followed by several rinses of the cell mass with production medium prior to culture in this medium. Payne et al. (1988) suggest a more elegant method by using a coarse cotton filter and coupling filtration to aeration. With this filtration system, media can be rapidly and completely exchanged and the filter can be quickly and effectively backwashed. By coupling filtration to aeration, a two-stage system can be employed while requiring only one single bioreactor.

Recently the two-stage system has given way to one-stage systems with

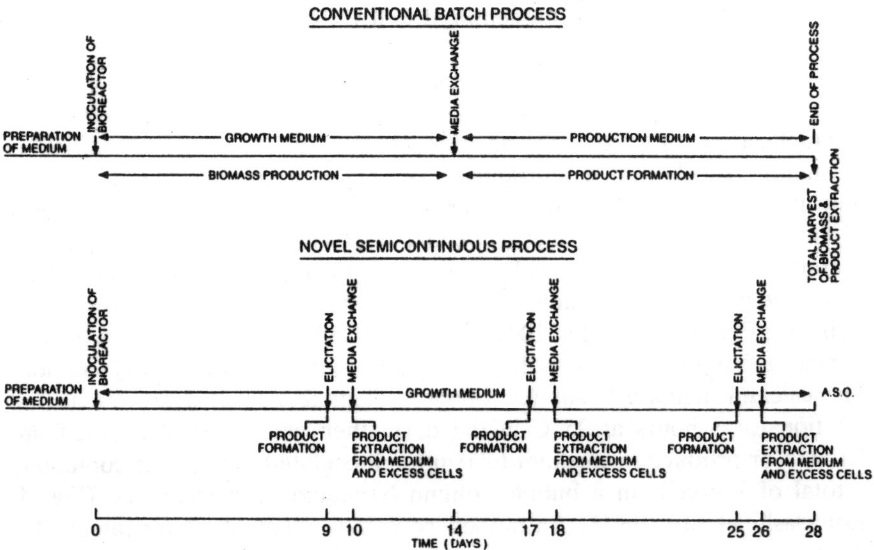

Figure 3. Repeated elicitor treatment for semi-continuous metabolite production after Constabel et al. (1990).

control of the idiophase by addition to the culture of elicitors. These may be heavy metal salts like vanadyl sulphate (Smith et al., 1987) or organics like arachidonic acid or complex substances obtained with fungal mycelia and fractions thereof (Eilert, 1987). In fact, elicitation of cell cultures can improve the efficiency of product accumulation in several ways: the time required to obtain a product may be reduced from weeks to a few days (1), the exchange of growth for production medium becomes superfluous (2), metabolites may now-be released into the culture medium and recovered from there (3), and reelicitation may be possible, allowing a semi-continuous system (4).

Treatment of cultured cells with elicitors not only induces cell specialization and product formation, but may lead to leakiness of cultured cells also. The latter leads to product accumulation in the surrounding medium and would permit non-destructive extraction of at least a portion of the product (Eilert et al., 1985, 1986). This observation led to the concept of a semi-continuous culture protocol consisting of alternate periods of growth and elicitor treatment with media drainage and product extraction, followed by resuspension in fresh medium and regrowth of the culture (Fig. 3). The comparatively lower amount of products may be compensated by the retainment of the entire biomass. Depending on the tolerance towards the elicitor, employed cells may be submitted to 3 cycles of semi-continuous culture (Tyler et al., 1989).

4. Immobilized Cell Culture

4 1. *Entrapment of Cells*

Immobilization of plant cells as pioneered by Brodelius et al. (1979) is technology which enables entrapment of cells in a gel of calcium alginate, polyvinyl alcohol resin or other polymer, or restrains cells on a fixed support of foam, fabric or hollow fibres. Yeoman (1987) has presented techniques, characteristics, properties, and commercial potential of immobilized plant cells. The procedure to follow in the case of entrapment in gels is as follows: Cells (8 g FW) suspended in 100 ml alginate (1.5%) are dripped from nozzles (1 mm in diameter) or the orifice of a pipette into a 50 μm $CaCl_2$ solution. The calcium alginate beads (ca 2 mm in diameter) formed are left in the solution for 3 hours at 25 °C in the dark, then washed in 30 ml medium. Growth or production medium (300 ml) is inoculated with beads containing a total of 7 g cells in a bubble column bioreactor and incubated (Fig. 4, Kobayashi et al., 1991). Entrapped cells have been used repeatedly for product formation.

Using a different matrix, cultured cells of *Lavandula vera* have been entrapped in synthetic resin pre-polymers which are derivates of polyvinyl alcohol (degree of polymerization 1,700 to 1,800) with photosensitive functional sites which initiate dimerization by irradiation with visible light. Entrapped cells grew well inside gels and synthesized blue pigment in the presence of L-cysteine as an inducer. The entrapped cells were superior to calcium alginate entrapped cells in both cell growth and metabolite production. Since blue pigment was released into the medium, cells could be reused for metabolite production several times (Nakajima et al., 1986).

4.2. *Surface Immobilization*

The tendency of cells suspended in liquid medium to adhere to glass surfaces in culture vessels can be exploited for immobilization. Lindsey and Yeoman (1984) found that after introduction of particles of polyurethane foam into cell suspensions of *Capsicum frutescens* and *Daucus carota*, cells settled preferentially in the pores of the foam. The resulting aggregates were stable and could be used in bioreactor systems. Archambault et al. (1990) proposed a system to be operated in an airlift fermenter. Here, the surface consisted of sheets of non-woven short fibre polyester material (#7607, Texel Inc., Ste. Marie de Beauce, Quebec, Canada), formed into a square spiral configuration on a supporting structure made of welded stainless steel rods. The structure was placed into a bioreactor (Fig. 5). *Catharanthus roseus* inoculum (2.5% v/v) attached readily and uniformly to the immobilizing structure. All cells were retained within the first 24 hours of culture. In B_5 medium (Gamborg et al., 1968) a constant carbohydrate consumption of ca. 2 g/l/day

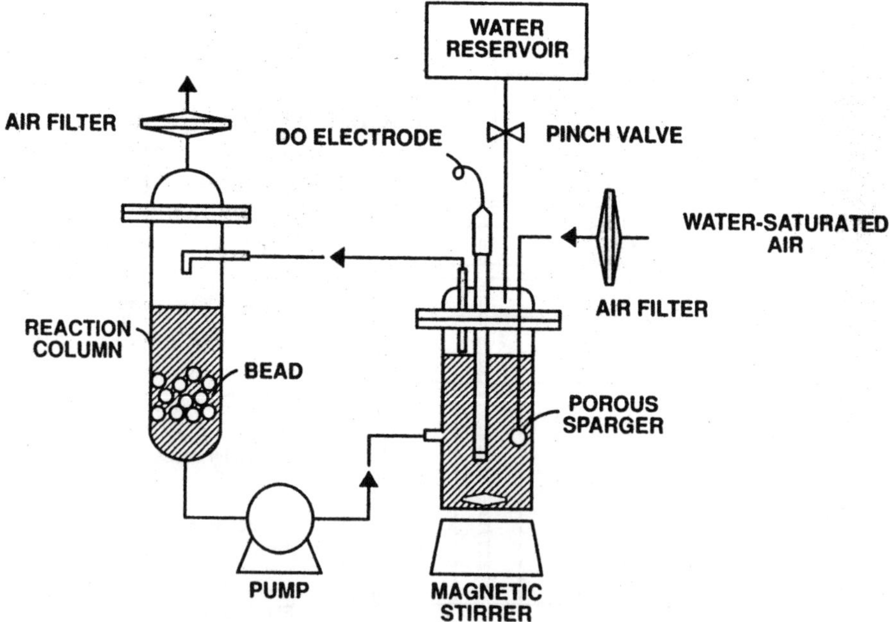

Figure 4. Bubble column reactor system used for berberine production by immobilized cells of *Tralictrum minus* after Shimomura et al. (1991).

was observed during the growth phase. The stationary or idiophase was characterized by a much lower sugar consumption rate of ca. 0.5 g/l/day.

Adjustments will be required if high biomass concentration and product formation are to be attained with other cell cultures. For example, with *Papaver somniferum* cells the medium had to be supplemented with an elicitor (*Botrytis* spp. mycelium homogenate) 7.5 days after inoculation in order to stimulate metabolite production (sanguinarine and dihydrosanguinarine). After 6.5 days of culture, immobilized cells in a 2-l culture had released into the medium 6 mg of sanguinarine (total, i.e. cells and medium = 690 mg, Kurz et al., 1990).

Immobilization creates for cells a situation which is to imitate membership in a tissue of a whole plant. As such, cells are expected to cease to grow and accumulate metabolites. The technical advantage of this system clearly is superior metabolite production compared to cell suspensions and media exchange without loss of cells. Some recent systems are presented in Table 2. Development of cell immobilization conceivably leads to continuous culture and extreme reduction of upstream costs due to saving of biomass, were not growth control and product release major stumbling blocks.

Immobilized cells would also be the perfect catalyst for biotransformation. Corchete and Yeoman (1989) reported the conversion of (-)-codeinone to

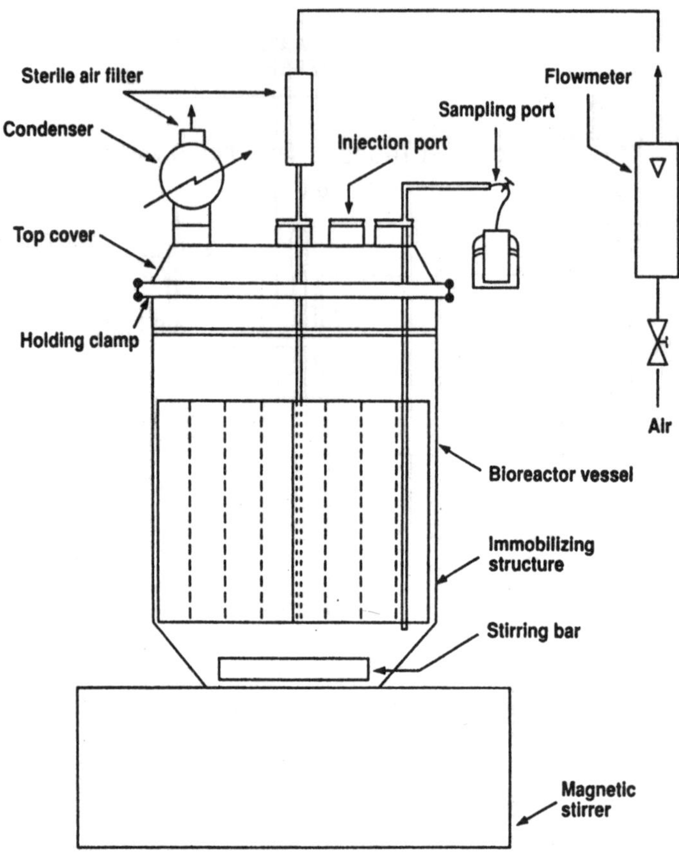

Figure 5. Bioreactor for culture of surface immobilized cells after Archambault et al. (1990).

(-)-codeine by 10 g FW of *Papaver somniferum* cells retained in 8–10 days in four reticulate polyurethane foam pieces, 1 cm^3, 20 pores/25 mm. Similarly, theobromine was completely converted to caffeine by cells of *Coffea arabica* (Furuya et al., 1991).

5. Hairy Root Culture

The advent of hairy root cultures may bridge problems which result from deficiencies in product formation with cell cultures and also could be admissible substitutes for traditional plants, since hairy roots are cultured in hormone-free medium, much like hydroponics. The technology of culturing hairy roots is advancing fast and may lead to success where cell cultures have failed. The problem with cells suspended in liquid medium having to undergo phases of growth and specialization, with intervention required to accommo-

Table 2. Recent systems for production of secondary metabolites by *in vitro* culture of immobilized plant cells

Plant species	Product	Immobilized Substratum	References
Catharanthus roseus	Ajmalicine	Hollow fibres	Asada and Shuler, 1989
Catharanthus roseus	Alkaloids	Hollow fibres	Ziyad Mohamed and Scragg, 1990
Catharanthus roseus	Alkaloids	Fibreglass mats, coated	Facchini and DiCosmo, 1990
Lithospermum erythrorhizon	Shikonin	Polyurethane foam	Park et al., 1990
Lithospermum erythrorhizon	Shikonin	Calcium alginate	Kim and Chang, 1990
Capsicum annuum	Capsaicin	Calcium alginate	Johnson et al., 1990
Coffea arabica	Caffeine	Polyurethane foam	Furuya et al., 1990
Daucus carota	Anthocyanins	Cell aggregates	Prenosil and Hegglin, 1990
Thalictrum rugosum	Protoberberine	Fibreglass mats, coated	Facchini and DiCosmo, 1990
Solanum aviculare	Solasodine	Cell aggregates, hormone induced	Tsoulpha and Doran, 1991
Thalictrum minus	Berberine	Calcium alginate	Kobayashi et al., 1991

date the latter, can be overcome by shifting to a less demanding procedure, namely culturing fast growing roots, so-called hairy roots.

Hairy roots are obtained in a rather straight forward manner. They are induced by direct infection of axenic shoot cultures with virulent strains of *Agrobacterium rhizogenes*. The infection may be achieved with aseptic plants, detached leaves or other organs, leaf disks, or protoplasts. Scratching the midrib of a leaf with a needle containing a bacterial suspension (ca. 10^{10} viable bacteria/ml) allows inoculation with small droplets (5–10 μl) containing *A. rhizogenes*. Alternatively, infection may be induced by co-cultivation of explants with bacteria. After 24–48 h the bacteria are eliminated on medium containing 1 g/l carbenicillin. The hairy roots are sub-cultured to solid medium and later to liquid root culture medium. Cultures are maintained on a rotary shaker at 100 rpm at 25 °C in the dark. Transformation is confirmed by the detection of opines using paper electrophoresis according to Petit et al. (1983).

Hairy roots can be cultured as batches, harvested and extracted for metabolites. By way of example, a 2-l air-lift fermenter is connected to an XAD-2 column (25 g) via a peristaltic pump. Hairy roots (200 mg) of *Lithospermum erythrorhizon* obtained from cultures in Murashige and Skoog (1962) medium are subcultured in 2-l of root culture medium with 3% sucrose at pH 5.7 at 25 °C in the dark. Medium and XAD-2 column are exchanged every 4–6 weeks. The XAD-2 is washed with water and extracted with $CHCl_3$ for

Table 3. Hairy root cultures with capacity for accumulation of secondary metabolites

Plant	Product	References
Datura stramonium	Hyoscyamine	Rhodes et al., 1988
Chaenactis douglasii	Thiarubrines	Constabel et al., 1988
Catharanthus roseus	Alkaloids	Parr et al., 1989
		Toivonen et al., 1989
Datura candida	Alkaloids	Christen et al., 1990
Lobelia inflata	Lobeline	Yonemitsu et al., 1990
Tagetes patula	Nematocides	Kyo et al., 1990
Lippia dulcis	Hernandulcin	Sauerwein et al., 1991
Lithospermum erythrorhizon	Shikonin	Shimomura et al., 1991
Echinacea purpurea	Rhamnogalactans	Trypsteen et al., 1991

recovery of shikonin (Shimomura et al., 1991). Employed in a continuous mode over extended periods of time, major problems would, again, be product release and control of growth. While the latter remains unresolved, product release has been observed in a number of cases and may be subject to improvement. *Nicotiana rustica* and *Lithospermum eryhtrorhizon* cultures, for instance, have shown 30–50% of product release (cf., Hamill et al., 1987). With *Datura stramonium*, release of hyoscyamine from hairy roots was shown to be clone dependent. The concentration of alkaloid at day 35 varied from 0.3 to 1.0 mg/g FW of hyoscyamine and for most cultures less than 10% of the alkaloid produced was found in the medium. In several clones, however, up to 45% of hyoscyamine was detected in the medium during a first cycle of culture, a lesser percentage in subsequent cycles (Rhodes et al., 1988).

Hairy root cultures are potentially applicable to the production of all root-derived metabolites from dicotyledonous plants. Table 3 lists some of the more recent cultures employed and products obtained.

6. Outlook

Cell culture enables the production of secondary metabolites. This observation has attracted both academic and industrial interest. With emphasis shifting to genetic manipulation of secondary metabolism in order to better understand the synthesis, accumulation and compartmentalization of products in plant cells, cell cultures may or may not continue to retain a prominent place in laboratories. Plants may be preferred as experimental and as source material, since they offer greater genetic stability.

Cell cultures as sources of biological commodities would appear to present a different prospect. But even here, improvement by genetic engineering may favor plants and plantations over *in vitro* cultures. The transformation of *Atropa belladonna* plants to produce scopolamine (Y. Yamada, unpubl.

results) would present an example of experiments to come. Two conditions, however, will continue to promote *in vitro* culture of plant cells for product formation. One condition arises from the fact that some plants of industrial (pharmaceutical) interest are rare and possibly endangered by extinction (*Coleus forskohlii*, Krombholz et al., 1990; Sharma et al., 1991), others are growing too slowly to match a sudden demand (*Taxus brefolia* Nutt.). Attractive, also, are cell cultures which successfully compete with plants by producing a single compound of interest, rather than a gemisch of substances, which should reduce downstream costs (*Papaver somniferum* – sanguinarine, Eilert et al., 1985).

The goals of achieving cost effective production of secondary metabolites would be aided by the stabilization of cells in the idiophase and the control of excretion of products into the medium. These two conditions, if met, would promise new horizons for plant cell culture.

7. References

Alfermann, A.W., Borg, H.M., Döller, P.D., Hagedorn, W., Heins, M., Wahl, J. and Reinhard, E. (1977). Biotransformation of cardiac glycosides by plant cell cultures. In: Plant Tissue Culture and its Bio-technological Application. W. Barz, E. Reinhard and M.H. Zenk (eds.), Springer Verlag, Berlin, pp. 125–141.

Archambault, J., Volesky, B. and Kurz, W.G.W. (1990). Development of bioreactors for the culture of surface immobilized plant cells. Biotechnol. Bioeng. 35:702–711.

Asada, M. and Shuler, M.L. (1989). Stimulation of ajmalicine production and excretion from *Catharanthus roseus*: effects of adsorption *in situ* elicitors and alginate immobilization. Appl. Microbiol. Biotechnol. 30:475–481.

Beiderbeck, R. and Knoop, B. (1987). Two phase culture. In: Cell Culture and Somatic Cell Genetics of Plants, Vol. 4. F. Constabel and I.K. Vasil (eds.), Academic Press, New York, pp. 255–266.

Brodelius, P., Deus, B., Mosbach, K. and Zenk, M.H. (1979). Immobilized plant cells for the production and transformation of natural products. FEBS Lett. 103:93–97.

Christen, P., Roberts, M.F., Phillipson, J.D. and Evans, W.C. (1990). Alkaloids of hairy root cultures of a *Datura candida* hybrid. Plant Cell Rep. 9:101–104.

Constabel, C.P. and Towers, G.H.N. (1988). Thiarubrine accumulation in hairy root cultures of *Chaenactis douglasii*. J. Plant Physiol. 133:67–70.

Constabel, F. (1990). Medicinal Plant Biotechnology. Planta Med. 56:421–425.

Constabel, F., Rambold, S., Chatson, K.B., Kurz, W.G.W. and Kutney, J.P. (1981). Alkaloid production in *Catharanthus roseus* (L.) G. Don. VI. Variation in alkaloid spectrum of cell lines derived from one single leaf. Plant Cell Rep. 1:3–5.

Constabel, F. and Vasil, I.K. (eds.) (1987). Cell Culture in Phytochemistry. Cell Culture and Somatic Cell Genetics of Plants, Vol. 4. Academic Press, New York.

Constabel, F. and Vasil, I.K. (eds.) (1988). Phytochemicals in Plant Cell Cultures. Cell Culture and Somatic Cell Genetics of Plants, Vol. 5. Academic Press, New York.

Constabel, F., Kurz, W.G.W. and Eilert, U.F.K. (1990). Repeated elicitor treatment, a method for semicontinuous metabolite production by plant cells cultured *in vitro*. Can. Patent #496, 984.

Corchete, P. and Yeoman, M.M. (1989). Biotransformation of (–)-codeinone to (–)-codeine by *Papaver somniferum* cells immobilized in reticulate polyurethane foam. Plant Cell Rep. 8:128–131.

Do, C.G. and Cormier, F. (1990). Effects of low nitrate and high sugar concentrations on anthocyanin content and composition of grape (*Vitis vinifera* L.) cell suspensions. Plant Cell Rep. 9:500–503.

Dougall, D.K. (1987). Cell cloning and the selection of high yielding strains. In: Cell Culture and Somatic Cell Genetics of Plants, Vol. 4. F. Constabel, and I.K. Vasil (eds.), Academic Press, New York, pp. 117–124.

Eilert, U. (1987). Elicitation: Methodology and aspects of application. In: Cell Culture and Somatic Cell Genetics of Plants, Vol. 4. F. Constabel and I.K. Vasil (eds.), Academic Press, New York, pp. 153–196.

Eilert, U., Kurz, W.G.W. and Constabel, F. (1985). Stimulation of sanguinarine accumulation in *Papaver somniferum* cell cultures by fungal elicitors. J. Plant Physiol. 119:65–76.

Eilert, U., Constabel, F. and Kurz, W.G.W. (1986). Elicitor stimulation of monoterpene indole alkaloid formation in suspension cultures of *Catharanthus roseus*. J. Plant Physiol. 126:11–22.

Facchini, P.J. and DiCosmo, F. (1990). Immobilization of cultured *Catharanthus roseus* cells using a fibreglass substratum. Appl. Microbiol. Biotechnol. 33:36–42.

Facchini, P.J. and DiCosmo, F. (1991). Plant cell bioreactor for the production of protoberberine alkaloids from immobilized *Thalictrum rugosum* cultures. Biotech. Bioeng. 37:397–403.

Fowler, M.W. (1987). Process systems and approaches for large scale plant cell culture. In: Plant Tissue and Cell Culture. C.E. Green, D.A. Somers, W.P. Hackett and D.D. Biesboer (eds.), Alan R. Liss, New York, pp. 459–471.

Fujita, Y., Hara, Y., Suga, C.L. and Morimoto, T. (1981). Production of shikonin derivatives by cell suspension cultures of *Lithospermum erythrorhizon* II. A new medium for the production of shikonin derivatives. Plant Cell Rep. 1:61–63.

Fujita, Y. and Tabata, M. (1986). Secondary metabolites from plant cells. In: Plant Tissue and Cell Culture. C.E. Green, D.A. Somers, W.P. Hackett and D.D. Biesboer (eds.), Alan R. Liss, New York, pp. 169–185.

Furuya, T., Koge, K. and Orihara, Y. (1990). Long term culture and caffeine production of immobilized coffee *Coffea arabica* L. cells in polyurethane foam. Plant Cell Rep. 9:125–128.

Furuya, T., Orihara, Y. and Koge, K. (1991). Biotransformation of theobromine to caffeine in suspension and polyurethane foam immobilized coffee (*Coffea arabica* L.) cells. Plant Cell Rep. 9:659–662.

Gamborg, O.L., Miller, R.A. and Ojima, K. (1968). Nutrient requirements of suspension cultures of soybean root cells. Exp. Cell Res. 50:151–158.

Hamill, J.D., Parr, A.J., Rhodes, M.J.C., Robins, R.J. and Walton, N.J. (1987). New routes to plant secondary products. Biotechn. 5:800–804.

Johnson, T.S., Ravishankar, G.A. and Venkataraman, L.V. (1990). *In vitro* capsaicin production by immobilized cells and placental tissues of *Capsicum annuum* L. grown in liquid medium. Plant Sci. 70:223–230.

Kim, D.J. and Chang, H.N. (1990). Enhanced shikonin production from *Lithospermum erythrorhizon* by *in situ* extraction and calcium alginate immobilization. Biotechnol. Bioeng. 36:460–466.

Kobayashi, Y., Fukui, H. and Tabata, M. (1991). Effect of carbon dioxide and ethylene on berberine production and cell browning in *Thalictrum minus* cell cultures. Plant Cell Rep. 9:496–499.

Koblitz, H. (1988). Anthraquinones. In: Cell Culture and Somatic Cell Genetics of Plants Vol. 5. F. Constabel and I.K. Vasil (eds.), Academic Press, New York, pp. 113–139.

Kreis, W. and Reinhard, E. (1990). Two-stage cultivation of *Digitalis lanata* cells: semicontinuous production of deacetyllanatoside C in 20 litre airlift bioreactor. J. Biotech. 16:123–136.

Krombholz, R., Mersinger, R. and Reinhard, E. (1990). Root cultures of *Coleus forskohlii*: A system for continuous production of forskolin. Planta Med. 56:629–630.

Kurz, W.G.W. (1971). A chemostat for growing higher plant cells in single cell suspension cultures. Exp. Cell Res. 64:477–479.

Kurz, W.G.W., Parva, N.L. and Tyler, R.T. 1990. Biosynthesis of sanguinarine by elicitation of surface immobilized cells of *Papaver somniferuml*. Current plant science and biotechnology in agriculture, Vol. 9, Progress in Plant Cellular and Molecular Biology, Proceedings of the 7th Intl. Congress on Plant Tissue and Cell Culture, Amsterdam, H.J.J. Nijkamp, L.H.W. Van Der Plas. and J. Van Aartrijk, (eds.), Kluwer Academic Publishers, Dordrecht, pp. 682–688.

Kutney, P.J., Choi, L.S.L., Duffin, R., Hewitt, G., Kawamura, N., Kurihara, T., Salisbury, P., Sindelar, R., Stuart, K.L., Townsley, P.R., Chalmen, W.T., Wellster, E. and Jacoli, G.G. (1983). Cultivation of *Tripterygium wilfordii* tissue cultures for the production of the cytotoxic diterpene tripdiolide. Planta Med. 48:158–163.

Kyo, M., Miyauchi, Y., Fujimoto, T. Mayama, S. (1990). Production of nematocidal compounds by hairy root cultures of *Tagetes patula* L. Plant Cell Rep. 9:393–397.

Lindsey, K. and Yeoman, M.M. (1984). The viability and biosynthetic activity of cells of *Capsicum frutescens* Mill. cv. annuum immobilized in reticulate polyurethane foam. J. Exp. Bot. 35:1684–1696.

Linsmaier, E.M. and Skoog, F. (1965). Organic growth factor requirements of tobacco tissue cultures. Physiol. Plant. 18:100–127.

Lockwood, B. (1984). Alkaloids of cell suspensions derived from four Papaver species and the effect of temperature stress. Z. Pflanzenphysiol. 114:361–363.

Mulder-Krieger, T., Verpoorte, R., de Water, A., van Gessel, M., Oeveren, B.C.J.A. and Baerheim-Svendsen, A. (1982). Identification of the alkaloids and anthraquinones in *Cinchona ledgeriana* callus cultures. Planta Med. 46:19–24.

Murashige, T. and Skoog, F. (1962). A revised medium for rapid growth and bioassays with tobacco tissue cultures. Physiol. Plant. 15:473–497.

Nakagawa, K., Konagai, A., Fukui, H. and Tabata, M. (1984). Release and crystallization of berberine in the liquid medium of *Thalictrum minus* cell suspension cultures. Plant Cell Rep. 3:254–257.

Nakajima, H., Sonomoto, K., Morikawa, H., Sato, F., Ichimura, K., Yamada, Y. and Tanaka, A. (1986). Entrapment of *Lavandula vera* cells with synthetic resin prepolymers and its application to pigment production. Appl. Microbiol. Biotechnol. 24:266–270.

Noé, W., Langebartels, C. and Seitz, H.U. (1980). Anthocyanin accumulation and PAL-activity in a suspension culture of *Daucus carota* L. Inhibition by L-α-aminooxy-β-phenylpropionic acid and *t*-cinnamic acid. Planta 149:283–287.

Noguchi, M., Matsumoto, T., Hirata, Y., Yamamoto, K., Kato, A., Katsuyama, A., Azechi, A. and Kato, K. (1977). Improvements of growth rates of plant cell cultures. In: Plant Tissue Culture and its Bio-technological Application. W. Barz, E. Reinhard and M.H. Zenk (eds.), Springer Verlag, Berlin, pp. 85–94.

Park, Y.H., Seo, W.T. and Liu, J.R. (1990). Enhanced production of shikonin by *Lithospermum erythrorhizon* cells immobilized in polyurethane foam matrices. J. Ferment. Bioeng. 70:317–321.

Parr, A.J., Peerless, A.C.J., Hamill, J.D., Walton, N.J., Robins, R.J. and Rhodes, M.J.C. (1989). Alkaloid production by transformed root cultures of *Catharanthus roseus*. Plant Cell Rep. 7:309–311.

Payne, G.F., Payne, N.N. and Shuler, M.L. (1988). Bioreactor considerations for secondary metabolite production from plant cell tissue culture: Indole alkaloids from *Catharanthus roseus*. Biotechn. Bioeng. 31:905–912.

Petit, A., David, C., Dahl, G.A., Ellis, J.G., Guyon, P., Casse-Delbart, F. and Tempé, J. (1983). Further extension of the opine concept:Plasmids in *Agrobacterium tumefaciens* cooperate for opine degradation. Mol. Gen. Genet. 190:204–214.

Piehl, G.W., Berlin, J., Mollenschott, C. and Lehmann, J. (1988). Growth and alkaloid production of a cell suspension culture of *Thalictrum rugosum* in shake flasks and membrane stirrer reactors with bubble free aeration. Appl. Microbiol. Biotechnol. 29:456–461.

Prenosil, J.E. and Hegglin, M. (1990). Self-immobilized plant cell aggregates in a bioreactor system with low shear stress. Annals of the New York Academy of Sciences. 613:234–247.

Razzaque, A. and Ellis, B.E. (1977). Rosmarinic acid production in *Coleus blumei*. Planta 137:287–292.

Reinhard, E., Kreis, W., Barthlen, W. and Helmbold U. (1989). Semicontinuous cultivation of *Digitalis lanata* cells. Production of β-methyldigoxin in a 300 l airlift bioreactor. Biotechn. Bioeng. 34:502–508.

Rhodes, M.J.C., Robins, R.J., Aird, E.L.H., Payne, J., Pan, A J. and Walton, N.J. (1988). Regulation of secondary metabolism in transformed root cultures. In: Primary and Secondary Metabolism of Plant Cell Cultures II. W.G.W. Kurz (ed.), Springer Verlag, Berlin, pp. 58–72.

Rokem, J.S. (1987). Continuous culture of plant cells. In: Cell Culture and Somatic Cell Genetics of Plants, Vol. 4. F. Constabel and I.K. Vasil (eds.), Academic Press, New York, pp. 267–286.

Sato, H., Kobayashi, Y., Fukui, H. and Tabata, M. (1990). Specific differences in tolerance to exogenous berberine among plant cell cultures. Plant Cell Rep. 9:133–136.

Sato, F. and Yamada, Y. (1984). High berberine producing cultures of *Coptis japonica* cells. Phytochem. 23:281–285.

Sauerwein, M., Yamazaki, T. and Shimomura, K. (1991). Hernandulcin in hairy root cultures of *Lippia dulcis*. Plant Cell Rep. 9:579–582.

Scragg, A.H., Ashton, S., York, A., Bond, P., Stepan-Sarkissian, G. and Grey, D. (1990). Growth of *Catharanthus roseus* suspensions for maximum biomass and alkaloid accumulation. Enz. Microb. Technol. 12:292–298.

Seitz, H.U., Seitz, U. and Alfermann, W. (1985). Pflanzliche Gewebehultur. Gustav Fischer Verlag, Stuttgart, pp. 76–84.

Sharma, N., Changel, K.P.S. and Srivastava, V.K. (1991). *In vitro* propagation of *Coleus forskohlii* Brig. a threatened medicinal plant. Plant Cell Rep. 10:67–70.

Shimomura, K., Sudo, H., Saga, H. and Kamada, H. (1991). Shikonin production and secretion by hairy root cultures of *Lithospermum erythrorhizon*. Plant Cell Rep. 10:282–285.

Smith, J., Smart, N.J., Misawa, M., Kurz, W.G.W., Tavelli, S.G. and DiCosmo, F. (1987). Stimulation of indole alkaloid production in *Catharanthus roseus* (L.) G. Don by vanadyl sulfate. Plant Cell Rep. 6:142–145.

Street, H.E. (ed.) (1973). Plant Tissue and Cell Culture. Blackwell Scientific Publ., Oxford.

Strogov, S.E., Zaitseva, G.V., Balonsova, I.M., Shamkow, N.V. and Simonova, G.M. (1990). Large-scale cultivation of ginseng cells in a suspension I. Scaling-up of a pilot plant. Biotechnologia 1:43–45.

Taticek, R.A., Moo-Young, M. and Legge, R.L. (1991). The scale-up of plant cell culture: Engineering considerations. Plant Cell Tissue Org. Cult. 24:139–158.

Toivonen, L., Balsevich, J. and Kurz, W.G.W. (1989). Indole alkaloid production by hairy root cultures of *Catharanthus roseus*. Plant Cell Tissue Org. Cult. 18:79–94.

Trypsteen, M., van Lijsebettens, M., van Severen, R. and van Montagu, M. (1991). *Agrobacterium rhizogenes*-mediated transformation of *Echinacea purpurea*. Plant Cell Rep. 10:85–89.

Tsoulpha, P. and Doran P.M. (1991). Solasodine production from self-immobilized *SOLANUM AVICULARE* cells. J. Biotechnol. 19(1):99–110.

Tsukada, M. and Tabata, M. (1984). Intracellular localization and secretion of naphthoquinone pigments in cell cultures of *Lithospermum crythrorhizon*. Planta Med. 50:338–340.

Tyler, R.T., Eilert, U., Rijnders, C.O.M., Roewer, I.A. and Kurz, W.G.W. (1989). Semi-continuous production of sanguinarine and dihydrosanguinarine by *Papaver somniferum* L. cell suspension cultures treated with fungal homogenate. Plant Cell Rep. 7:410–413.

Waller, G.R. MacVean, C.D. and Suzuki, T. (1983). High production of caffeine and related enzyme activities in callus cultures of *Coffea arabica* L. Plant Cell Rep. 2:109–112.

White, P.R. (1963). The Cultivation of Animal and Plant Cells. 2nd ed., Ronald Press, New York.

Wink, M. and Hartmann, T. (1980). Production of quinolizidine alkaloids by photomixotrophic cell suspension cultures:Biochemical and biogenetic aspects. Planta Med. 40:149–155.

Yeoman, M.M. (1987). Techniques, characteristics, properties, and commercial potential of immobilized plant cells. In: Cell Culture and Somatic Cell Genetics of Plants, Vol. 4. F. Constabel. and I.K. Vasil, I.K. (eds.), Academic Press, New York, pp. 197–216.

Yonemitsu, H., Shimomura, K., Satake, M., Mochida, S., Tanaka M., Endo, T. and Kaji A. (1990). Lobeline production by hairy root culture of *Lobelia inflata* L. Plant Cell Rep. 9:307–310.

Zenk, M.H., El-Shagi, H. and Schulte, U. (1975). Anthraquinone production by cell suspension cultures of *Morinda citrifolia*. Planta Med. 1975, suppl. 79–81.

Zenk, M.H., El-Shagi, H., Aren, H., Stöckigt, J., Weiler, E.W. and Deus B. (1977). Formation of the indole alkaloids serpentine and ajmalicine in cell suspension cultures of *Catharanthus roseus*. In: Cell Culture and its Bio-Technological Application. W. Barz, E. Reinhard and M.H. Zenk (eds.), Springer Verlag, Berlin, pp. 26–43.

Ziyad Mohamed, M.T. and Scragg, A.H. (1990). A comparison of the growth and alkaloid production by suspension and immobilized *Catharanthus roseus* cells in 6L and 7L bioreactors. In: Physiology of Immobilized Cell. J.A.M. De Bont (ed.), Elsevier, Amsterdam, pp. 335–342.

PART B

In vitro Cultures

12. *In vitro* Culture of Cereals and Grasses

Indra K. Vasil and Vimla Vasil

Contents

1. Introduction

The origins of the grass family Gramineae (*Poaceae*) remain obscure and debatable. However, the grasses are known to have evolved early and become prominent about 70 million years ago at the end of the Cretaceous. Since then, grasses have become a highly diversified group of plants that are distributed worldwide in virtually all climatic zones. One of the largest families of angiosperms, the Gramineae comprise nearly 785 genera, and more than 10,000 species, ranging from some of the smallest angiosperms to the giant bamboos. By virtue of including such major food crops as the cereals and sugarcane for human beings, and forage grasses for animals, it is clearly the most important family of angiosperms. The primitive forms of several cereal species were amongst the handful of plants that were first domesticated by early man. The choices made by the Neolithic man, nearly 10,000 years ago, have not only proven to be remarkably durable but have also played a major role in sustaining the development of the human civilization. The major cereal species, such as wheat, rice and maize, have long and venerable

Indra K. Vasil and Trevor A. Thorpe (eds.), Plant Cell and Tissue Culture, 293–312.

histories. Even today they form the centerpiece of world agriculture by providing more than half of all food for mankind. It is also noteworthy that grasses in general are excellent and vigorous colonizers, especially of disturbed environments. With the ever increasing human population and industrialization, and the severe resultant strains on the environment, the grasses are destined to become even more important for mankind in the future.

The yield as well as the quality of cereal grains (Borlaug and Dowswell, 1988; Vasil, 1990) as well as forage grasses, have been gradually but substantially improved during the past several decades by breeding and selection. Realistically, such increases can not be sustained indefinitely. More importantly, plant improvement by the traditional breeding methods is severely hampered by the rather limited gene pool owing to sexual incompatibility even between related species, and the time scale of most breeding programs. With a doubling of the world population expected within the next 40 years, there is an urgent need to increase food production rapidly to meet the needs of the expanded humanity. It is in this context that the powerful techniques of cell culture and molecular biology of cereal and grass species have attracted much attention and give some hope. They offer the opportunity for marker assisted breeding by providing novel means for rapid identification and characterization of agronomically important genes (Phillips and Vasil, 1994) to complement and supplement the very successful methods of traditional breeding and selection. In addition, they make it possible to introduce such genes as well as useful genes obtained from other living organisms directly into these crops of immense importance to man by genetic transformation (Vasil, 1993).

The application of these technologies to the improvement of plants is dependent on the development of efficient systems for the regeneration of plants from cultured cells and tissues, preferably from single cells, and the delivery, stable integration and expression of alien genes into plants of interest. In this chapter we provide a summary of the progress made since about 1980 in cell culture, plant regeneration, and cellular and molecular genetic improvement of cereal and grass species. No effort is made to provide an exhaustive listing of the literature on the subject, but rather to highlight the significant developments. Information on anther and microspore culture, the subject of Chapter 7, is not included. A detailed account of the methods used for genetic transformation is provided in Chapter 10.

2. Early History of Tissue Culture of the Gramineae

The difficulties encountered by the early workers to obtain reproducible plant regeneration from tissue cultures of gramineous species discouraged many others, and were largely responsible for the low level of activity in this area of research. Slowly, a dogma developed that monocotyledonous species in general, but the Gramineae in particular, were not amenable to sustained growth and/or regeneration *in vitro* as they lacked, in comparison to their dicotyledonous counterparts, the natural ability for secondary growth by cambium or cambium-like tissues. Most of these problems could be attributed to the use of mature and differentiated organ or tissue explants that were used to initiate cultures. The resulting cultures were often masses of proliferating root meristems which sloughed off terminally differentiated cells (Cure and Mott, 1978). Some of the earliest successful reports include the culture of young hybrid embryos to recover plants (Brink et al., 1944), and the establishment of long term non-regenerable cultures from the immature endosperm of *Zea mays* (LaRue, 1949; Straus and LaRue, 1954) and *Lolium perenne* (Norstog, 1956). Plants were first regenerated from tissue cultures of rice (Nishi et al., 1968) and sugarcane (Heinz and Mee, 1969). During the next decade plants were regenerated from a limited number of genotypes in several other species. However, regeneration was found to be transient and sporadic in nature, the number of plants recovered was rather small, and the competence for regeneration was lost within a short time. There was still no clear understanding of the various factors which determined plant regeneration *in vitro*.

3. Plant Regeneration from Callus Cultures

Beginning in 1980, significant progress has been made in the establishment of regenerable tissue cultures of gramineous species. Plants have been regenerated from scores of varieties in almost all of the important cereal and grass species, and several species of bamboos (Bright and Jones, 1985; Vasil and Vasil, 1986a; Morrish et al., 1987). This was made possible by defining the physiological and developmental states of the explants placed in culture, the use of defined high salt nutrient solutions containing relatively high concentrations of strong auxins, and by the deliberate selection and maintenance of highly regenerable sectors of the callus tissues.

3.1. *Explants*

Perhaps the most dramatic difference between the earlier studies, and those conducted since about 1980, has been in the choice of explants. Instead of the differentiated and mature tissues, immature developing embryos, and tissue segments obtained from young inflorescences and bases of young leaves were used to initiate cultures (Dale, 1980; Vasil and Vasil, 1980, 1981a; Wernicke and Brettell, 1980; Haydu and Vasil, 1981; Lu and Vasil, 1981a). Such explants were largely made up of meristematic and undifferentiated cells which were not yet committed to any specific developmental pathways. The cultures derived from such cells were highly regenerable. The physiological condition and developmental stage of the explants were found to be critical for obtaining a desirable response. It was further shown that explants cultured only during a well defined but narrow 'competence window' of development were responsive. The explants exhibited clear gradients of morphogenetic competence. Thus, older or younger explants were relatively or completely ineffective. The gradients of morphogenetic competence corresponded to the gradients of endogenous levels of certain plant growth regulators in the explants (Rajasekaran et al., 1987a; Carnes and Wright, 1988; Wenck et al., 1988). Perturbation of the endogenous levels of key plant growth regulators resulted in the abolition of regenerative ability in otherwise competent explants cultured under permissive conditions (Rajasekaran et al., 1987b).

Regenerable cultures have been established by the careful choice and selection of explants obtained from donor plants grown under optimal environmental conditions and cultured on a suitable nutrient medium. Nevertheless, some agronomically elite genotypes of major cereal crops are still difficult to manipulate *in vitro*. Studies with some of these reclacitrant genotypes suggest that the response *in vitro* may be under genetic control (Nesticky et al., 1983; Tomes and Smith, 1985; Morocz et al., 1990).

3.2. *Nutrient Media*

Nutrient requirements of grass cells *in vitro* are rather simple. In most instances excellent growth is obtained on the defined medium of Murashige and Skoog (1962). In addition, the N6 (Chu et al., 1975) or the B5 (Gamborg et al., 1968) media have proven superior for some species. Doubling of the inorganic components of the medium and higher amounts of sucrose (3–12%) are beneficial for callus growth and plant regeneration (Lu et al., 1983; Ozias-Akins and Vasil, 1983; Redway et al., 1990a). Often the only plant growth regulator added to the medium to induce active cell proliferation is the auxin 2,4-dichlorophenoxyacetic acid (2,4-D), with strong herbicidal activity, in a range of 1–5 mg/l. In species such as orchard grass, best results are obtained with dicamba (Conger et al., 1983). In still others, picloram has been used with equal success. A number of other media additives have been reported to induce better callus growth and/or regeneration. These include, cytokinins (e.g. benzyl adenine), proline, silver nitrate, abscisic acid, etc. Complex organic additives such as coconut milk, etc., are neither required nor recommended.

3.3. *Plant Regeneration*

DNA synthesis and cell divisions in the cultured explants begin within three days of culture, and within a week visible signs of cell proliferation become evident. Cell divisions are most often initiated at well defined and predictable loci close to pro-cambial/vascular tissues. This may be related to the presumed high levels of plant growth regulators and nutrients in such tissues. In cultured immature embryos, cell proliferation is localized in the epithelial and subepithelial cells of the scutellum at the coleorhizal end (Vasil and Vasil, 1981a, 1982b; Ozias-Akins and Vasil, 1982; Vasil et al., 1985; Jones and Rost, 1989; Fransz and Schel, 1991). Similarly, in leaf and inflorescence explants, cell divisions are restricted to the vicinity of the vascular tissues (Haydu and Vasil, 1981; Ho and Vasil, 1983a). In immature embryos of rice active cell divisions take place in the scutellar epithelium (Jones and Rost, 1989).

The presence of mature and more differentiated tissues in the explants is not conducive to the formation of regenerable cultures. For example, it is not possible to obtain regenerable cultures from mature embryos or stem tissues in most grass species. However, shoot meristems and young leaf primordia excised from mature embryos of *Pennisetum americanum* (Botti and Vasil, 1983) or thin cell layers obtained from stem internodes of *P. purpureum* (Franco and Vasil, unpublished results) form embryogenic cultures from which green plants can be readily regenerated.

Prior to 1980, plant regeneration in tissue cultures of grass species was reported to take place by the *de novo* formation of shoot meristems or the multiplication of pre-existing axillary buds (Yamada, 1977; Green, 1978). Formation of somatic embryos was first demonstrated in 1980 (Dale, 1980; Vasil and Vasil, 1980; Wernicke and Brettell, 1980). Since then somatic embryogenesis has been described in almost all the species of grasses that have been regenerated in culture, and is considered to be the predominant mode of plant regeneration (Vasil, 1987). In a few species organogenesis may be the primary method of regeneration, while in others both somatic embryogenesis and organogenesis may occur simultaneously or successively. Somatic embryos arise either directly from single cells (Vasil and Vasil, 1982b; Conger et al., 1983; Ho and Vasil, 1983a; Jones and Rost, 1989) or indirectly from proembryogenic masses which themselves are derived from single cells. They are physiologically and morphologically very similar to their zygotic counterparts, and can be encapsulated to produce artificial seeds (Datta and Potrykus, 1989; Mukunthakumar and Mathur, 1992). On the other hand, the development of shoot meristems by organogenesis is generally multicellular in nature.

The typical embryogenic callus formed in tissue cultures of grass species has several well defined characteristics. It is compact, highly organized, slow growing, and pale white to light yellow in color. It is comprised almost entirely of tightly packed, small, isodiametric, thin-walled, highly basophilic and richly cytoplasmic meristematic cells which lack any large vacuoles but contain many amyloplasts and lipid droplets. This type of callus, which is most prevalent and regenerable, has been described as Type I callus.

In some species, such as maize, Type I callus can neither be maintained for more than a few months, nor is it suitable for the establishment of cell suspension cultures. In such species another type of callus, termed the Type II callus, which is soft, friable, fast growing, and highly regenerative, is formed at very low frequencies (Vasil et al., 1984; Armstrong and Green, 1985; Vasil and Vasil, 1986b). Type II callus can be maintained in culture for long periods of time and readily forms cell suspension cultures. The typical embryogenic callus of wheat (Type A) too can not be maintained for long periods in culture, nor does it go into suspension (Redway et al., 1990a). Here too, a new type of callus, described as Type C, occurs at low frequencies (Redway et al., 1990b). It differs from the Type I callus of maize in neither being very soft nor friable, but like the Type II callus of maize it maintains its competence for regeneration for long periods of time and forms suspension cultures.

Formation of somatic embryos from single cells begins with internal segmenting divisions, thickening of outer cell walls, and early severance of all plasmodesmatal connections. Species specific patterns of cell division reminiscent of zygotic embryos are generally not followed, although very soon globular embryos appear, with all the characteristics of zygotic embryos. This is followed by the development of the scutellum, the coleoptile, differentiation of shoot and root meristems, and coleorhiza. Many embryos often develop a suspensor like structure, which is loosely attached to the underlying callus tissue. In some instances where somatic embryos appear to be 'multicellular' in origin, they arise directly from folding of the proliferating scutellar tissue. In such cases the scutellum may be considered to represent the proembryonic mass, derived from the zygote, a single cell (Vasil et al., 1985).

Addition of abscisic acid, cytokinins and higher concentrations of sucrose to the nutrient media help in the maturation of the somatic embryos and prevent their precocious germination (Vasil and Vasil, 1986a). At this time, inorganic components of the medium can be reduced to half the usual concentration, and 2,4-D reduced to about 10% of its normal concentration. The presence of a shoot-root axis in the embryo results in the direct formation of rooted plantlets, which are easily transferred to soil and grown to maturity. Older cultures, in which the ratio of embryogenic cells to non-embryogenic cells is sharply reduced, either form plants by organogenesis or lose their capacity for plant regeneration. The apparent rapid metabolism of 2,4-D in actively growing embryogenic cultures causes cell enlargement, vacuolation, loss of starch grains, and the thickening of cell walls (Karlsson and Vasil, 1986a, b). This leads to the differentiation and senescence of embryogenic cells into non-embryogenic cells resulting in the loss of regenerative ability. Rapid subculture (every 10–14 days) of selected portions of embryogenic calli to fresh media can often alleviate this problem by maintaining optimal levels of 2,4-D in fresh media (Chandler and Vasil, 1984).

Unlike the prevalence of albino plants produced in anther and microspore cultures of grasses, most plants obtained from the culture of somatic tissue explants are green. However, preferential growth and selection of albino sectors in older cultures can significantly increase the incidence of albinos.

4. Plant Regeneration from Cell Suspension Cultures

Most of the attempts to establish cell suspension cultures of grass species were dictated by their possible use as sources of protoplasts. The normally non-friable nature of grass callus tissues, and their inability to be maintained in culture for long periods of time, made it particularly difficult to establish suspension cultures from them. Nevertheless, a few finely dispersed and fast growing cell lines were successfully established in the 1970's (Maretzki and Nickell, 1973; King et al., 1978; Vasil and Vasil, 1979). Although none of these were regenerable in nature, they did provide valuable information about the important parameters to consider for the establishment of totipotent cell suspension cultures.

4.1. *Establishment and Characterization of Cell Cultures*

Regenerable embryogenic suspension cultures of grasses were established in the early 1980's in several species (Vasil and Vasil, 1980, 1981b; Lu and Vasil, 1981b; Green et al., 1983; Ho and Vasil, 1983b; Vasil et al., 1983). These finely dispersed and fast growing suspensions were comprised largely of small groups of the characteristic embryogenic cells described in Section 3.3. The cultures have cell densities of 10–20 million cells/ml, with cell doubling times of 27–32 hr (Vasil and Vasil, 1982a; Karlsson and Vasil, 1986a, b). The cultures must be sub-cultured every 3–7 days in order to maintain high levels of 2,4-D in the medium and their embryogenic nature. The formation of somatic embryos can be induced by transfer to low 2,4-D media or by prolonged culture on the maintenance medium during which the amounts of 2,4-D in the medium are apparently reduced by metabolism. Development of somatic embryos in the liquid medium normally does not proceed beyond the globular stage. It is best, therefore, to transfer the embryogenic suspensions directly to the surface of low 2,4-D agar media where complete embryo development and germination into plantlets can take place.

Most of the regenerable embryogenic suspension cultures of grasses described in the literature were initiated from Type I callus cultures (Table 1). However, as stated earlier, such primary callus tissues of maize and wheat have not proven suitable for this purpose. In both species, suspension cultures could be established only from a different callus phenotype (Type II in maize, and Type C in wheat) which occurs at rather low frequencies (Vasil et al., 1984; Armstrong and Green, 1985; Vasil and Vasil, 1986b; Redway et al., 1990a).

The establishment as well as the maintenance of embryogenic suspension cultures are extremely difficult, labor intensive and time consuming. In addition, the quality of the cell lines and their ability to regenerate plants deteriorates with time. There is also the added possibility of mutational

Table 1. Regeneration of plants from protoplasts isolated from embryogenic suspension cultures.

Pennisetum americanum	Vasil and Vasil (1980)
Panicum maximum	Lu et al. (1981)
Pennisetum purpureum	Vasil et al. (1983); Wan and Vasil (1994)
Saccharum sp.	Srinivasan and Vasil (1986); Chen et al. (1988)
Oryza sativa	Yamada et al. (1986); Kyozuka et al. (1987) Datta et al. (1990)
Polypogon fugax	Chen and Xia (1987)
Dactylis glomerata	Horn et al. (1988a)
Festuca arundinacea	Dalton (1988)
Lolium perenne	Dalton (1988); Creemers-Molenaar et al. (1989)
Lolium multiflorum	Dalton (1988); Wang et al. (1993a)
Poa pratensis	van der Valk et al. (1988)
Zea mays	Rhodes et al. (1988a); Prioli and Sondahl (1989) Shillito et al. (1989); Morocz et al. (1990)
Triticum aestivum	Vasil et al. (1990); Ahmed and Sagi (1993)
Setaria italica	Dong and Xia (1990)
Agrostis alba	Asano and Sugiura (1990)
Sorghum vulgare	Wei and Xu (1990)
Hordeum vulgare	Jahne et al. (1991)
Oryza rufipogon	Baset et al. (1991)
Agrostis palustris	Terakawa et al. (1992)
Paspalum dilatatum	Akashi and Adachi (1992)
Festuca pratensis	Wang et al. (1993b)
Oryza granulata	Baset et al. (1993)
Triticum durum	Yang et al. (1993)

changes, and loss owing to infection or equipment failure. Fortunately, each of these problems has been overcome by the development of highly efficient procedures to cryopreserve cell lines (Shillito et al., 1989; Gnanapragasam and Vasil, 1990; Meijer et al., 1991; Fretz et al., 1992; Lu and Sun, 1992), and callus tissues or immature embryos (Chen et al., 1985; Gnanapragasam and Vasil, 1992; Lu and Sun, 1992). The cryopreserved lines can be periodically thawed to reestablish cultures that are similar to the original lines, or to regenerate plants.

5. Plant Regeneration from Protoplasts

Ever since the first report of plant regeneration from mesophyll protoplasts of tobacco (Takebe et al., 1971), mesophyll protoplasts of dicotyledonous species have been widely used for the regeneration of plants. In contrast, extensive and intensive efforts to induce sustained cell divisions in mesophyll protoplasts of gramineous species have not been successful (Potrykus, 1980; Vasil, 1983; Vasil and Vasil, 1992). The recent reports of cell colonies and plants obtained from mesophyll protoplasts of oat (Hahne et al., 1989) and rice (Gupta and Pattanayak, 1993), respectively, need further confirmation.

During the 1970's, a number of investigators demonstrated that protoplasts isolated from non-regenerable cell suspensions of grass species could be induced to undergo sustained cell divisions in culture to form protocallus (Maretzki and Nickell, 1973; Nemet and Dudits, 1977; Vasil and Vasil, 1979; Potrykus et al., 1979). These encouraging results soon led to the establishment and use of regenerable embryogenic suspension cultures as a source of totipotent protoplasts (Vasil and Vasil, 1980; Lu et al., 1981; Vasil et al., 1983). In these early studies, protoplast plating efficiencies were rather low and the few plantlets obtained could not be grown to maturity in soil. During the next ten years this system was considerably improved by obtaining more dispersed, finely grained and fast growing embryogenic suspension cultures, and by the use of better cell wall hydrolyzing enzymes, and feeder layers or agarose embedding of protoplasts. Currently it is the only reliable method for the regeneration of plants from protoplasts of grass species, including such important crops as wheat, rice, maize, barley, sugarcane, etc. (Table 1). In most instances protoplast derived plants are green, fertile and can be grown to maturity. Problems of albinism and aberrant plants increase with the age of the cell lines and the gradual loss of embryogenic potential, but can be overcome either by establishing new lines or better still by making use of superior lines which have been previously tested and cryopreserved.

6. Fidelity of Plants Regenerated *in vitro*

Most plant cell cultures are cytologically quite heterogeneous owing to the polysomatic nature of the initial explants and the genetic changes that often take place *in vitro*, although it is difficult to distinguish between the two types of variation (D'Amato, 1985; Vasil, 1987, see also Chapter 6). Much of the observed variation is epigenetic in nature, is lost due to diplontic selection, and is neither heritable nor novel. Indeed, most often culture induced variation is quite similar to what is generated during sexual or mutation breeding. In addition, it is often of no agronomic interest. Nevertheless, it has been proposed that the tissue culture process produces novel variation that can be useful in the production of new breeding lines for crop improvement (Larkin and Scowcroft, 1981). In order to be useful, such variability must not only be novel, but also be stable and heritable. Unfortunately, in spite of extensive attempts, there are as yet no examples of any major new commercial varieties that have been developed as a result of random *de novo* variation arising in tissue culture of any important grass species (Maretzki, 1987; Vasil, 1988). Indeed, major efforts undertaken to exploit somaclonal variation in cereals were abandoned because it did not prove to be of any benefit to the breeding program (Baillie et al., 1992; Qureshi et al., 1992).

The presence of variant cells in embryogenic tissue cultures of grass species is not uncommon. It is significant, however, that plants derived from somatic embryos formed in such mixoploid cultures are generally devoid of any significant variation. This observation led Swedlund and Vasil (1985) to propose that there is stringent selection against cytologically aberrant cells during the formation of somatic embryos (Conversely, plants derived through organogenesis show genetic variability probably because the threshold of tolerance for genetic anomalies is higher for shoot morphogenesis). Additional evidence in support of this theory has been provided by a number of studies (Hanna et al., 1984; Rajasekaran et al., 1986; Cavallini et al., 1987; Kobayashi, 1987; Cavallini and Natali, 1989; Morrish et al., 1990; Gmitter et al., 1991), including the recent detailed biochemical and molecular characterization of plants regenerated via somatic embryogenesis (Breiman et al., 1989; Shimron-Abarbanell and Breiman, 1991; Shenoy and Vasil, 1992; Chowdhury and Vasil, 1993; Isabel et al., 1993; Valles et al., 1993; Chowdhury et al., 1994). This unique characteristic of plants derived from somatic embryos makes them suitable not only for the purposes of clonal propagation but also for genetic transformation.

7. Somatic Hybridization

Fusion of protoplasts has been used to hybridize normally incompatible species and to obtain novel combinations of cytoplasmic genomes in several dicotyledonous species. Such experiments were not carried out with grass species until suitable protoplast systems became available in 1986. In the early work, somatic hybrid cell lines were obtained from the fusion of protoplasts isolated from cell suspension cultures of *Pennisetum americanum* (+) *Saccharum officinarum* (Tabaeizadeh et al., 1986), *Panicum maximum* (+) *Pennisetum americanum* (Ozias-Akins et al., 1986), and *Triticum monococcum* (+) *Pennisetum americanum* (Vasil et al., 1988). No hybrid plants were obtained as the donor cell lines were themselves non-regenerable. Much of the later work was limited to the production of somatic hybrids of rice and several wild species of *Oryza* (Hayashi et al., 1988) or *Echinochloa oryzicola* (Terada et al., 1987), and the transfer of cytoplasmic male sterility through cybrid formation (Akagi et al., 1989; Kyozuka et al., 1989; Yang et al., 1989). More recently, fertile intergeneric somatic hybrids of *Festuca arundinacea* (+) *Lolium multiflorum* have been obtained (Takamizo et al., 1991).

The regeneration of fertile maize plants from zygotic embryos obtained following electrofusion of sperm and egg cell protoplasts (Kranz and Lorz, 1993), provides yet another method for overcoming sexual incompatibilities and recovery of hybrid plants. In addition, the procedure provides an excellent opportunity to study the process of fertilization, and perhaps also an additional avenue for genetic transformation.

Based on the experience with dicotyledonous species, it is doubtful that many useful fertile somatic hybrids will be obtained by protoplast fusion. In most instances the transfer of only a few useful genes, and not the mixing

Table 2. Transgenic grasses obtained by direct delivery of DNA into protoplasts (P), or by high velocity microprojectile bombardment (B) or electroporation (E) into intact cells.

Species	Method of transformation	References
Dactylis glomerata	P	Horn et al. (1988b)
Zea mays	P	Rhodes et al. (1988b); Omirulleh et al. (1993); Golovkin et al. (1993)
	B	*Gordon-Kamm et al. (1990); *Fromm et al. (1990); **Koziel et al. (1993); ***Murry et al. (1993)
	E	D'Halluin et al. (1992)
Oryza sativa	P	Shimamoto et al. (1989); Datta et al. (1990); ***Hayakawa et al. (1992); **Fujimoto et al. (1993)
	B	*Christou et al. (1991); Li et al. (1993)
Avena sativa	B	*Somers et al. (1992)
Festuca arundinacea	P	*Wang et al. (1992); Ha et al. (1992)
Saccharum spp. hybrid	B	Bower and Birch (1992)
Triticum aestivum	B	*Vasil et al. (1992, 1993); *Weeks et al. (1993)
Hordeum vulgare	B	*Wan and Lemaux (1994)
Agrostis palustris	B	Zhong et al. (1993)

*Herbicide resistant; **Insect resistant; ***Virus resistant.

of entire genomes, is the desired objective. Therefore, assymetric hybridization – though more difficult and random at present – may be a more attractive and useful alternative. Perhaps, the best use of protoplast fusion will be in the production of cytoplasmic male sterile lines.

8. Genetic Transformation

The soil borne bacteria *Agrobacterium tumefaciens* and *A. rhizogenes* are widely used for the genetic transformation of dicotyledonous species. They are able to naturally transfer and integrate a part of their DNA (the transfer or T-DNA) into the nuclear genome of host plant cells. With a few rare exceptions, the host range of *Agrobacterium* does not extend to monocotyledonous, especially grass, species. Therefore, entirely new strategies have had to be developed for the transformation of grass species. Although DNA can be delivered efficiently into plant cells by a variety of methods, stable integration of the transferred DNA, selection of the rare transformed cells and regeneration of transgenic plants from them remain the principal problems. The following three methods of direct DNA delivery have proven to be most suitable for grass species (Table 2):

1. DNA delivery by osmotic (polyethylene glycol treatment) or electric (electroporation) shock into regenerable protoplasts isolated from embryogenic suspension cultures.
2. DNA delivery by bombardment with high velocity microprojectiles into intact cells, tissues or organs.
3. DNA delivery by electroporation into intact cells, tissues or organs.

Rapid progress made during the past few years in improving the regeneration of plants from cultured cells, in developing efficient methods for the direct delivery of DNA, and the choice of suitable selectable marker genes have together made it possible to produce transgenic plants in some forage grass species, but more importantly in wheat, rice and maize, the three most important cereal crops.

Direct delivery of DNA into regenerable protoplasts appears to be the method of choice for forage grass species such as *Dactylis* (Horn et al., 1988b) and *Festuca* (Wang et al., 1992). It has also been used successfully in maize (Rhodes et al., 1988b; Omirulleh et al., 1993; Golovkin et al., 1993), and remains the most widely used method for rice transformation (Shimamoto et al., 1989; Datta et al., 1990). On the other hand, high velocity microprojectile bombardment has proven to be suitable for several species, including wheat (Vasil et al., 1992, 1993; Weeks et al., 1993), maize (Gordon-Kamm et al., 1990; Fromm et al., 1990), rice (Christou et al., 1991; Li et al. 1993), sugarcane (Bower and Birch, 1992), oats (Somers et al., 1992), and barley (Wan and Lemaux, 1994).

In many instances improved crops have been produced by the introduction of agronomically useful genes which provide resistance to herbicides (wheat: Vasil et al., 1992, 1993; maize: Gordan-Kamm et al., 1990; Fromm et al., 1990; rice: Datta et al., 1992; oats: Somers et al., 1992), viruses (rice: Hayakawa et al., 1992; maize: Murry et al., 1993; rice: Fujimoto et al., 1993), and insects (maize: Koziel et al., 1993). It has also been reported that herbicide (bialaphos) treatment of transgenic rice plants containing the herbicide resistant *bar* gene protects them from infection by the fungal pathogen *Rhizoctonia solani* (Uchimiya et al., 1993).

9. The Future

Since 1980, remarkable progress has been made in developing highly efficient systems for the regeneration of fertile plants from tissue, cell and protoplast cultures of all of the important cereal and grass species. By forging a productive partnership between these technologies and the powerful new methods of molecular biology, the molecular genetic improvement of important food crops has become a reality. Based on the results of field trials, it is expected that transgenic cereals will be commercially grown before the end of the century. Further progress in this field will be limited more by the availability of well defined and characterized agronomically important genes, rather than by the lack of efficient protocols for regeneration and/or transformation.

10. References

Ahmed, K.Z. and Sagi, F. (1993). Culture of and fertile plant regeneration from regenerable embryogenic cell-derived protoplasts of wheat (*Triticum aestivum* L.). Plant Cell Rep. 12:175–179.

Akagi, H., Sakamoto, M., Negishi, T. and Fujimura, T. (1989). Construction of rice hybrid plants. Molec. Gen. Genet. 215:501–506.

Akashi, R. and Adachi, T. (1992). Plant regeneration from suspension culture-derived protoplasts of apomictic dallisgrass (*Paspalum dilatatum* Poir.). Plant Sci. 82: 219–225.

Armstrong, C.L. and Green, C.E. (1985). Establishment and maintenance of friable embryogenic maize callus and the involvement of L-proline. Planta 164: 207–214.

Asano, Y. and Sugiura, K. (1990). Plant regeneration from suspension culture-derived protoplasts of *Agrostis alba* L. (Redtop). Plant Sci. 72: 267–273.

Baillie, A.M.R., Rossnagel, B.G. and Kartha, K.K. (1992). Field evaluation of barley (*Hordeum vulgare* L.) genotypes derived from tissue culture. Canad. J. Plant Sci. 72: 725–733.

Baset, A., Cocking, E.C. and Finch, R.P. (1993). Regeneration of fertile plants from protoplasts of the wild rice species *Oryza granulata*. J. Plant Physiol. 141:245–247.

Baset, A., Finch, R.P. and Cocking, E.C. (1991). Plant regeneration from protoplasts of wild rice (*Oryza rufipogon* Griff.). Plant Cell Rep. 10:200–203.

Borlaug, N.E. and Dowswell, C.R. (1988). World revolution in agriculture. In: Book of the Year 1988. Encyclopedia Britannica, Chicago, pp. 5–14.

Botti, C. and Vasil, I.K. (1983). Plant regeneration by somatic embryogenesis from parts of cultured mature embryos of *Pennisetum americanum* (L.) K. Schum. Z. Pflanzenphysiol. 111:319–325.

Bower, R. and Birch, R.G. (1992). Transgenic sugarcane plants via microprojectile bombardment. The Plant J. 2:409–416.

Breiman, A., Felsenburg, T. and Galun, E. (1989) Is *Nor* region variability in wheat invariably caused by tissue culture? Theor. Appl. Genet. 77:809–814.

Bright, S.W. and Jones, M.G.K. (eds). (1985). Cereal Tissue and Cell Culture. Nijhoff/W. Junk Publishers, Dordrecht.

Brink, R.A., Cooper, D.C. and Ausherman, L.E. (1944). A hybrid between *Hordeum jubatum* and *Secale cereale* reared from an artificially cultivated embryo. J. Hered. 35:67–75.

Carnes, M.G. and Wright, M.S. (1988). Endogenous hormone levels of immature corn kernels of A188, Missouri-17, and DeKalb XL-12. Plant Sci. 57:195–203.

Cavallini, A., Lupi, M.C., Cremonini, R. and Bennici, A. (1987). *In vitro* culture of *Bellevalia romana* (L.) Rchb. III. Cytological study of somatic embryos. Protoplasma 139:66–70.

Cavallini, A. and Natali, L. (1989). Cytological analyses of *in vitro* somatic embryogenesis in *Brimura amethystina* Salisb. (*Liliaceae*). Plant Sci. 62:255–261.

Chandler, S.F. and Vasil, I.K. (1984). Optimization of plant regeneration from long term embryogenic callus cultures of *Pennisetum americanum* Schum. (Napier grass). J. Plant Physiol. 117:147–156.

Chen, D. and Xia, Z. (1987). Mature plant regeneration from cultured protoplasts of *Polypogon fugax* Nees ex Steud. Sci. Sinica B 30:698–704.

Chen, T.H.H., Kartha, K.K. and Gusta, L.V. (1985). Cryopreservation of wheat suspension culture and regenerable callus. Plant Cell Tissue Org. Cult. 4:101–109.

Chen, W.H., Davey, M.R., Power, J.B. and Cocking, E.C. (1988). Sugarcane protoplasts: factors affecting division and plant regeneration. Plant Cell Rep. 7: 344–347.

Chowdhury, M.K.U. and Vasil, I.K. (1993). Molecular analysis of plants regenerated from embryogenic cultures of hybrid sugarcane cultivars (*Saccharum* spp.). Theor. Appl. Genet. 86:181–188.

Chowdhury, M.K.U., Vasil, V. and Vasil, I.K. (1994). Molecular analysis of plants regenerated from embryogenic cultures of wheat (*Triticum aestivum* L.). Theor. Appl. Genet. 87:821–828.

Christou, P., Ford, T.L. and Kofron, M. (1991). Production of transgenic rice (*Oryza sativa* L.) plants from agronomically important Indica and Japonica varieties via electric discharge particle acceleration of exogenous DNA into immature zygotic embryos. Bio/Technology 9:957–962.

Chu, C.C., Wang, C.C., Sun, C.S., Hsu, C., Yin, K.C. and Chu, C.Y. (1975). Establishment of an efficient medium for anther culture of rice through comparative experiments on the nitrogen sources. Sci. Sin. 18:659–668.

Conger, B.V., Henning, G.E., Gray, D.J. and McDaniel, J.K. (1983). Direct embryogenesis from mesophyll cells of orchardgrass. Science 221:850–851.

Creemers-Molenaar, J., van der Valk, P., Loeffen, J.P.M. and Zaal, M.A.C.M. (1989). Plant regeneration from suspension cultures and protoplasts of *Lolium perenne* L. Plant Sci. 63:167–176.

Cure, W.W. and Mott, R.L. (1978). A comparative anatomical study of organogenesis in cultured tissues of maize, wheat and oats. Physiol. Plant. 42:91–96.

Dale, P.J. (1980). Embryoids from cultured immature embryos of *Lolium multiflorum*. Z. Pflanzenphysiol. 100:73–77.

Dalton, S.J. (1988). Plant regeneration from cell suspension protoplasts of *Festuca arundinacea* Schreb. (tall fescue) and *Lolium perenne* L. (perennial ryegrass). J. Plant Physiol. 132:170–175.

D'Amato, F. (1985). Cytogenetics of plant cell and tissue cultures and their regenerants. CRC Crit. Rev. Plant Sci. 3:73–112.

Datta, S.K., Datta, K., Soltanifar, N., Donn, G. and Potrykus, I. (1992). Herbicide-resistant Indica rice plants from IRRI breeding line IR72 after PEG-mediated transformation of protoplasts. Plant Molec. Biol. 20:619–629.

Datta, S.K., Peterhans, A., Datta, K. and Potrykus, I. (1990). Genetically engineered fertile Indica rice recovered from protoplasts. Bio/Technology 8:736–740.

Datta, S.K. and Potrykus, I. (1989). Artificial seeds in barley: encapsulation of microspore-derived embryos. Theor. Appl. Genet. 77:820–824.

D'Halluin, K., Bonne, E., Bossut, M., de Beuckeleer, M. and Leemans, J. (1992). Transgenic maize plants by tissue electroporation. The Plant Cell 4:1495–1505.

Dong, J. and Xia, Z. (1990). Protoplast culture and plant regeneration of foxtail millet. Chinese Sci. Bull. 35:1560–1564.

Fransz, P.F. and Schel, J.H.N. (1991. An ultrastructural study on the early development of *Zea mays* somatic embryos. Can. J. Bot. 69:858–865.

Fretz, A., Jahne, A. and Lorz, H. (1992). Cryopreservation of embryogenic suspension cultures of barley (*Hordeum vulgare* L.). Bot. Acta 105:140–145.

Fromm, M.E., Morrish, F., Armstrong, C., Williams, R., Thomas, J. and Klein, T.M. (1990). Inheritance and expression of chimeric genes in the progeny of transgenic maize plants. Bio/Technology 8:833–839.

Fujimoto, H., Itoh, K., Yamamoto, M., Kyozuka, J. and Shimamoto, K. (1993). Insect resistant rice generated by introduction of a modified δ-endotoxin gene of *Bacillus thuringiensis*. Bio/Technology 11:1151–1155.

Gamborg, O.L., Miller, R.A. and Ojima, K. (1968). Nutrient requirements of suspension cultures of soybean roots. Exp. Cell Res. 50:151–158.

Gmitter, F.G. Jr., Ling, X., Cai, C. and Grosser, J.W. (1991). Colchicine induced polyploidy in *Citrus* embryogenic cultures, somatic embryos and regenerated plantlets. Plant Sci. 74:135–141.

Gnanapragasam, S. and Vasil, I.K. (1990). Plant regeneration from a cryopreserved embryogenic cell suspension of a commercial sugarcane hybrid (*Saccharum* sp.). Plant Cell Rep. 9:419–423.

Gnanapragasam, S. and Vasil, I.K. (1992). Cryopreservation of immature embryos, embryogenic callus and cell suspension cultures of gramineous species. Plant Sci. 83:205–215.

Golovkin, M.V., Abraham, M., Morocz, S., Bottka, S., Feher, A. and Dudits, D. (1993). Production of transgenic maize plants by direct uptake of DNA into embryogenic protoplasts. Plant Sci. 90:41–52.

Gordon-Kamm, W.J., Spencer, T.M., Mangano, M.L., Adams, T.R., Daines, R.J., Start, W.G., O'Brien, J.V., Chambers, S.A., Adams Jr. W.R., Willets, N.G., Rice, T.B., Mackey, C.J., Krueger, R.W., Kausch, A.P. and Lemaux, P.G. (1990). Transformation of maize cells and regeneration of fertile transgenic plants. The Plant Cell 2:603–618.

Green, C.E. (1978). *In vitro* plant regeneration in cereals and grasses. In: Frontiers of Plant Tissue Culture 1978. T.A. Thorpe (ed.), University of Calgary, Calgary, pp. 411–418.

Green, C.E., Armstrong, C.L. and Anderson, P.C. (1983). Somatic cell genetic systems in corn. In: Advances in Gene Technology: Molecular Genetics of Plants and Animals. K. Downey, R.W. Voellmy, J. Schultz and F. Ahmad (eds.), Academic Press, New York, pp. 147–157.

Gupta, H.S. and Pattanayak, A. (1993). Plant regeneration from mesophyll protoplasts of rice (*Oryza sativa* L.). Bio/Technology 11:90–94.

Ha, S., Wu, F. and Thorne, T.K. (1992). Transgenic turf-type tall fescue (*Festuca arundinacea* Schreb.) plants regenerated from protoplasts. Plant Cell Rep. 11:601–604.

Hahne, B., Fleck, J. and Hahne, G. (1989). Colony formation from mesophyll protoplasts of a cereal, oat. Proc. Natl. Acad. Sci. USA 86:6157–6160.

Hanna, W.W., Lu, C. and Vasil, I.K. (1984). Uniformity of plants regenerated from somatic embryos of *Panicum maximum* Jacq. (Guinea grass). Theor. Appl. Genet. 67:155–159.

Hayakawa, T., Zhu, Y., Itoh, K., Kimura, Y., Izawa, T., Shimamoto, K. and Toriyama, S. (1992). Genetically engineered rice resistant to rice stripe virus, an insect-transmitted virus. Proc. Natl. Acad. Sci. USA 89:9865–9869.

Hayashi, Y., Kyozuma, J. and Shimamoto, K. (1988). Hybrids of rice (*Oryza sativa* L.) and wild *Oryza* species obtained by cell fusion. Molec. Gen. Genet. 214:6–10.

Haydu, Z. and Vasil, I.K. (1981). Somatic embryogenesis and plant regeneration from leaf tissues and anthers of *Pennisetum purpureum*. Theor. Appl. Genet. 59:269–273.

Heinz, D.J. and Mee, G.W.P. (1969). Plant differentiation from callus tissue of *Saccharum* species. Crop Sci. 9:346–348.

Ho, W. and Vasil, I.K. (1983a). Somatic embryogenesis in sugarcane (*Saccharum officinarum* L.), I: The morphology and physiology of callus formation and the ontogeny of somatic embryos. Protoplasma 118:169–180.

Ho, W. and I.K. Vasil, I.K. (1983b). Somatic embryogenesis in sugarcane (*Saccharum officinarum* L.), II: The growth of and plant regeneration from embryogenic cell suspension cultures. Ann. Bot. 51:719–726.

Horn, M.E., Conger, B.V. and Harms, C.T. (1988a). Plant regeneration from protoplasts of embryogenic suspension cultures of orchardgrass (*Dactylis glomerata* L.). Plant Cell Rep. 7:371–374.

Horn, M.E., Shillito, R.D., Conger, B.V. and Harms, C.T. (1988b). Transgenic plants of orchardgrass (*Dactylis glomerata* L.) from protoplasts. Plant Cell Rep. 7:469–472.

Isabel, N., Tremblay, L., Michaud, M., Tremblay, F.M. and Bousquet, J. (1993). RAPDs as an aid to evaluate the genetic integrity of somatic embryogenesis-derived populations of *Picea mariana* (Mill.) B.S.P. Theor. Appl. Genet. 86:81–87.

Jahne, A., Lazzeri, P.A. and Lorz, H. (1991). Regeneration of fertile plants from protoplasts derived from embryogenic cell suspensions of barley (*Hordeum vulgare* L.). Plant Cell Rep. 10:1–6.

Jones, T.J. and Rost, T.L. (1989). The developmental anatomy and ultrastructure of somatic embryos from rice (*Oryza sativa* L.) scutellum epithelial cells. Bot. Gaz. 150:41–49.

Karlsson, S.B. and Vasil, I.K. (1986a). Morphology and ultrastructure of embryogenic cell suspension cultures of *Panicum maximum* Jacq. (Guinea grass) and *Pennisetum purpureum* Schum. (Napier grass). Am. J. Bot. 73:894–901.

Karlsson, S.B. and Vasil, I.K. (1986b). Growth, cytology and flow cytometry of embryogenic cell suspension cultures of *Panicum maximum* Jacq. (Guinea grass) and *Pennisetum purpureum* Schum. (Napier grass). J. Plant Physiol. 123:211–227.

King, P.J., Potrykus, I. and Thomas, E. (1978). *In vitro* genetics of cereals: Problems and perspectives. Physiol. Veg. 16:381–399.

Kobayashi, S. (1987). Uniformity of plants regenerated from orange (*Citrus sinensis* Osb.) protoplasts. Theor. Appl. Genet. 74:10–14.

Koziel, M.G., Beland, G.L., Bowman, C., Carozzi, N.B., Crenshaw, R., Crossland, L., Dawson, J., Desai, N., Hill, M., Kadwell, S., Launis, K., Lewis, K., Maddox, D., McPherson, K., Meghji, M.R., Merlin, E., Rhodes, R., Warren, G.W., Wright, M. and Evola, S.V. (1993). Field performance of elite transgenic maize plants expressing an insecticidal protein derived from *Bacillus thuringiensis*. Bio/Technology 11:194–200.

Kranz, E. and Lorz, H. (1993). *In vitro* fertilization with isolated, single gametes results in zygotic embryogenesis and fertile plants of maize. The Plant Cell 5:739–746.

Kyozuka, J., Hayashi, Y. and Shimamoto, K. (1987). High frequency plant regeneration from rice protoplasts by novel nurse culture methods. Molec. Gen. Genet. 206:408–413.

Kyozuka, J., Taneda, K. and Shimamoto, K. (1989). Production of cytoplasmic male sterile rice (*Oryza sativa* L.) by cell fusion. Bio/Technology 7:1171–1174.

Larkin, P.J. and Scowcroft, W.R. (1981). Somaclonal variation – a novel source of variability from cell cultures for plant improvement. Theor. Appl. Genet. 60:197–214.

LaRue, C.D. (1949). Cultures of endosperm of maize. Am. J. Bot. 36:798.

Li, L., Qu, R., de Kochko, A., Fauquet, C. and Beachy, R.N. (1993). An improved rice transformation system using the biolistic method. Plant Cell Rep. 12:250–255.

Lu, C. and Vasil, I.K. (1981a). Somatic embryogenesis and plant regeneration from leaf tissues of *Panicum maximum* Jacq. Theor. Appl. Genet. 59:275–280.

Lu, C. and Vasil, I.K. (1981b). Somatic embryogenesis and plant regeneration from freely suspended cells and cell groups of *Panicum maximum in vitro*. Ann. Bot. 47:543–548.

Lu, C., Vasil, V. and Vasil, I.K. (1981). Isolation and culture of protoplasts of *Panicum maximum* Jacq. (Guinea grass): somatic embryogenesis and plantlet formation. Z. Pflanzenphysiol. 104:311–318.

Lu, C., Vasil, V. and Vasil, I.K. (1983). Improved efficiency of somatic embryogenesis and plant regeneration in tissue cultures of maize (*Zea mays* L.). Theor. Appl. Genet. 62:285–290.

Lu, T.G. and Sun, C.S. (1992). Cryopreservation of millet (*Setaria italica* L.). J. Plant Physiol. 139:295–298.

Maretzki, A. (1987). Tissue culture: its prospects and problems. In: Sugarcane Improvement Through Breeding. D.J. Heinz (ed.), Elsevier, Amsterdam, pp. 343–384.

Maretzki, A. and Nickell, L.G. (1973). Formation of protoplasts from sugarcane cell suspensions and the regeneration of cell cultures from protoplasts. Colloq. Int. CNRS 212:51–53.

Meijer, E.G.M., van Iren, F., Schrijnemakers, E., Hensgens, L.A.M., van Zijderveld, M. and Schilperoort, R.A. (1991). Retention of capacity to produce plants from protoplasts in cryopreserved cell lines of rice (*Oryza sativa* L.). Plant Cell Rep. 10:171–174.

Morocz, S., Donn, G., Nemeth, J. and Dudits, D. (1990). An improved system to obtain fertile regenerants via maize protoplasts isolated from a highly embryogenic suspension culture. Theor. Appl. Genet. 80:721–726.

Morrish, F.M., Vasil, V. and. Vasil, I.K (1987). Developmental morphogenesis and genetic manipulation in tissue and cell cultures of the Gramineae. Adv. Genet. 24:431–499.

Morrish, F.M., Hanna, W.W. and Vasil, I.K. (1990). The expression and perpetuation of inherent somatic variation in regenerants from embryogenic cultures of *Pennisetum glaucum* (L.) R.Br. (pearl millet). Theor. Appl. Genet. 80:409–416.

Mukunthakumar, S. and Mathur, J. (1992). Artificial seed production in the male bamboo *Dendrocalamus strictus* L. Plant Sci. 87:109–113.

Murashige, T. and Skoog, F. (1962). A revised medium for rapid growth and bioassay with tobacco tissue culture. Physiol. Plant. 15:473–497.

Murry, L.E., Elliott, L.G., Capitant, S.A., West, J.A., Hanson, K.K., Scarafia, L., Johnston, S., DeLuca-Flaherty, C., Nichols, S., Cunanan, D., Dietrich, P.S., Mettler, I.J., Dewald, S., Warnick, D.A., Rhodes, C., Sinibaldi, R.M. and Brunke, K.J. (1993). Transgenic corn plants expressing MDMV strain B coat protein are resistant to mixed infections of maize dwarf mosaic virus and maize chlorotic mottle virus. Bio/Technology 11:559–564.

Nemet, G. and Dudits, D. (1977). Potential of protoplast, cell and tissue culture in cereal research. In: Use of Tissue Cultures in Plant Breeding. F.J. Novak (ed.), Czechoslovak Academy of Sciences, Prague, pp. 145–163.

Nesticky, M., Novak, F.J., Piovarci, A. and Dolezelova, M. (1983). Genetic analysis of callus growth of maize (*Zea mays* L.) *in vitro*. Z. Pflanzenphysiol. 91:322–328.

Nishi, T., Yamada, Y. and Takahashi, E. (1968). Organ differentiation and plant restoration in rice callus. Nature 219:508–509.

Norstog, K. (1956). Growth of rye grass endosperm *in vitro*. Bot. Gaz. 117:253–259.

Omirulleh, S., Abraham, M., Golovkin, M., Stefanov, I., Karabaev, M.K., Mustardy, L., Morocz, S. and Dudits, D. (1993). Activity of a chimeric promoter with the doubled CaMV 35S enhancer element in protoplast-derived cells and transgenic plants in maize. Plant Mol. Biol. 21:415–428.

Ozias-Akins, P. and Vasil, I.K. (1982). Plant regeneration from cultured immature embryos and inflorescences of *Triticum aestivum* L. (wheat): Evidence for somatic embryogenesis. Protoplasma 110:95–105.

Ozias-Akins, P. and Vasil, I.K. (1983). Improved efficiency and normalization of somatic embryogenesis in *Triticum aestivum* (wheat). Protoplasma 117:40–44.

Ozias-Akins, P., Ferl, R. and Vasil, I.K. (1986). Somatic hybridization in the Gramineae: *Pennisetum americanum* (L.) K. Schum. (pearl millet) + *Panicum maximum* Jacq. (Guinea grass). Molec. Gen. Genet. 203:365–370.

Phillips, R.L. and Vasil, I.K. (eds.) (1994). Advances in Cellular and Molecular Biology of Plants, Vol. 1. DNA-Based Markers in Plants. Kluwer Academic Publishers, Dordrecht.

Potrykus, I. (1980). The old problem of protoplast culture: cereals. In: Advances in Protoplast Research. L. Ferenczy, G.L. Farkas and G. Lazar (eds.), Akademiai Kiado, Budapest, pp. 243–254.

Potrykus, I., Harms, C.T. and. Lorz, H (1979). Callus formation from cell culture protoplasts of corn (*Zea mays* L.). Theor. Appl. Genet. 54:209–214.

Prioli, L.M. and Sondahl, M.R. (1989). Plant regeneration and recovery of fertile plants from protoplasts of maize (*Zea mays* L.). Bio/Technology 7:589–594.

Qureshi, J.A., Hucl, P. and Kartha, K.K. (1992). Is somaclonal variation a reliable tool for spring wheat improvement? Euphytica 60:221–228.

Rajasekaran, K., Hein, M.B., Davis, G.C., Carnes, M.G. and Vasil, I.K. (1987a). Endogenous plant growth regulators in leaves and tissue cultures of napier grass (*Pennisetum purpureum* Schum.). J. Plant Physiol. 130:13–25.

Rajasekaran, K., Hein, M.B. and Vasil, I.K. (1987b). Endogenous abscisic acid and indole-3-acetic acid and somatic embryogenesis in cultured leaf explants of *Pennisetum purpureum* Schum.: effects *in vivo* and *in vitro* of glyphosate, fluridone and paclobutrazol. Plant Physiol. 84:47–51.

Rajasekaran, K.,Schank, S.C. and Vasil, I.K. (1986). Characterization of biomass production and phenotypes of plants regenerated from embryogenic callus cultures of *Pennisetum americanum* × *P. purpureum* (hybrid Napiergrass). Theor. Appl. Genet. 73:4–10.

Redway, F.A., Vasil, V., Lu, D. and Vasil, I.K. (1990a). Identification of callus types for long-term maintenance and regeneration from commercial cultivars of wheat (*Triticum aestivum* L.). Theor. Appl. Genet. 79:609–617.

Redway, F.A., Vasil, V. and Vasil, I.K. (1990b). Characterization and regeneration of wheat (*Triticum aestivum* L.) embryogenic cell suspension cultures. Plant Cell Rep. 8:714–717.

Rhodes, C.A., Lowe, K.S. and Ruby, K.L. (1988a). Plant regeneration from protoplasts isolated from embryonic maize cell cultures. Bio/Technology 6:56–60.

Rhodes, C.A., Pierce, D.A., Mettler, I.J., Mascarenhas, D. and Detmer, J.J. (1988b). Genetically transformed maize plants from protoplasts. Science 240:204–207.

Shenoy, V.B. and Vasil, I.K. (1992). Biochemical and molecular analysis of plants derived from embryogenic tissue cultures of Napier grass (*Pennisetum purpureum* Schum.). Theor. Appl. Genet. 83:947–955.

Shillito, R.D., Carswell, G.K., Johnson, C.M., DiMaio, J.J. and Harms, C.T. (1989). Regeneration of fertile plants from protoplasts of elite inbred maize. Bio/Technology 7:581–587.

Shimamoto, K., Terada, R., Izawa, T. and Fujimoto, H. (1989). Transgenic rice plants regenerated from transformed protoplasts. Nature 338:274–276.

Shimron-Abarbanell, D. and Breiman, A. (1991). Comprehensive molecular characterization of tissue-culture-derived *Hordeum marinum* plants. Theor. Appl. Genet. 83:71–80.

Somers, D.A., Rines, H.W., Gu, W., Kaeppler, H.F. and Bushnell, W.R. (1992). Fertile, transgenic oat plants. Bio/Technology 10:1589–1594.

Srinivasan, C. and Vasil, I.K. (1986). Plant regeneration from protoplasts of sugarcane (*Saccharum officinarum* L.). J. Plant Physiol. 126:41–48.

Straus, J. and LaRue, C.D. (1954). Maize endosperm tissue grown *in vitro*, I: Culture requirements. Am. J. Bot. 41:687–694.

Swedlund, B. and Vasil, I.K. (1985). Cytogenetic characterization of embryogenic callus and regenerated plants of *Pennisetum americanum* (L.) K. Schum. Theor. Appl. Genet. 69:575–581.

Tabaeizadeh, Z., Ferl, R.J. and Vasil, I.K. (1986). Somatic hybridization in the Gramineae: *Saccharum officinarum* L. (sugarcane) + *Pennisetum americanum* (L.) K. Schum. (pearl millet). Proc. Natl. Acad. Sci. USA 83:5616–5619.

Takebe, I., Labib, G. and Melchers, G. (1971). Regeneration of whole plants from isolated mesophyll protoplasts of tobacco. Naturwiss. 58:318–320.

Takamizo, T., Spangenberg, G., Suginobu, K. and Potrykus, I. (1991). Intergeneric somatic hybridization in Gramineae: Somatic hybrid plants between tall fescue (*Festuca arundinacea* Schreb.) and Italian rye grass (*Lolium multiflorum* Lam.). Mol. Gen. Genet. 231:1–6.

Terada, R., Kyozuka, J., Nishibayashi, S. and Shimamoto, K. (1987). Plantlet regeneration from somatic hybrids of rice (*Oryza sativa* L.) and barnyard grass (*Echinochloa oryzicola* Vasing). Mol. Gen. Genet. 210:39–43.

Terakawa, T., Sato, T. and Koike, M. (1992). Plant regeneration from protoplasts isolated from embryogenic suspension cultures of creeping bentgrass (*Agrostis palustris* Huds.). Plant Cell Rep. 11:457–46 1.

Tomes, D.T. and Smith, D.S. (1985). The effect of parental genotype on initiation of embryogenic callus from elite maize (*Zea mays* L.) germplasm. Theor. Appl. Genet. 70:505–509.

Uchimiya, H., Iwata, M., Nojiri, C., Samarajeewa, P.K.,Takamatsu, S., Ooba, S., Anzai, H., Christensen, A.H., Quail, P.H. and Toki, S. (1993). Bialaphos treatment of transgenic rice plants expressing a bar gene prevents infection by the sheath blight pathogen (*Rhizoctonia solani*). Bio/Technology 11:835–836.

Valles, M.P., Wang, Z.Y., Montavon, P., Potrykus, I. and Spangenberg, G. (1993). Analysis of genetic stability of plants regenerated from suspension cultures and protoplasts of meadow fescue (*Festuca pratensis* Huds.). Plant Cell Rep. 12:101–106.

Van der Valk, P., Zaal, M.A.C.M. and Creemers-Molenaar, J. (1988). Regeneration of albino plantlets from suspension culture derived protoplasts of *Poa pratensis* L. (Kentucky bluegrass). Euphytica S:169–176.

Vasil, I.K. (1983). Isolation and culture of protoplasts of grasses. Int. Rev. Cytol. Suppl. 16:79–88.

Vasil, I.K. (1987). Developing cell and tissue culture systems for the improvement of cereal and grass crops. J. Plant Physiol. 128:193–218.

Vasil, I.K. (1988). Progress in the regeneration and genetic manipulation of cereal crops. Bio/Technology 6:397–402.

Vasil, I.K. (1990). The realities and challenges of plant biotechnology. Bio/Technology 8:296–301.

Vasil, I.K. (1993). Cell culture, molecular biology and the improvement of grasses. IAPTC Newslet 72:1–10.

Vasil, I.K. and Vasil, V. (1986a). Regeneration in cereal and other grass species. In: Cell Culture and Somatic Cell Genetics of Plants, Vol. 3. Plant Regeneration and Genetic Variability. I.K. Vasil (ed.), Academic Press, Orlando, pp. 121–150.

Vasil, I.K. and Vasil, V. (1992). Advances in cereal protoplast research. Physiol. Plant. 85:279–283.

Vasil, V. and Vasil, I.K. (1979). Isolation and culture of cereal protoplasts. I. Callus formation from pearl millet (*Pennisetum americanum*) protoplasts. Z. Pflanzenphysiol. 92:379–383.

Vasil, V. and Vasil, I.K. (1980). Isolation and culture of cereal protoplasts. II. Embryogenesis and plantlet formation from protoplasts of *Pennisetum americanum*. Theor. Appl. Genet. 56:97–99.

Vasil, V. and Vasil, I.K. (1981a). Somatic embryogenesis and plant regeneration from tissue cultures of *Pennisetum americanum* and *P. americanum* × *P. purpureum* hybrid. Am. J. Bot. 68:864–872.

Vasil, V. and Vasil, I.K. (1981b). Somatic embryogenesis and plant regeneration from suspension cultures of pearl millet (*Pennisetum americanum*). Ann. Bot. 47:669–678.

Vasil, V. and Vasil, I.K. (1982a). Characterization of an embryogenic cell suspension culture derived from inflorescences of *Pennisetum americanum* (pearl millet; Gramineae). Am. J. Bot. 69:1441–1449.

Vasil, V. and Vasil, I.K. (1982b). The ontogeny of somatic embryos of *Pennisetum americanum* (L.) K. Schum.: in cultured immature embryos. Bot. Gaz. 143:454–465.

Vasil, V. and Vasil, I.K. (1986b). Plant regeneration from friable embryogenic callus and cell suspension cultures of *Zea mays* L. J. Plant Physiol. 123:211–227.

Vasil, V., Wang, D. and Vasil, I.K. (1983). Plant regeneration from protoplasts of *Pennisetum purpureum* Schum. (Napier grass). Z. Pflanzenphysiol. 111:319–325.

Vasil, V., Vasil, I.K. and Lu, C. (1984). Somatic embryogenesis in long-term callus cultures of *Zea mays*. Am. J. Bot. 71:158–161.

Vasil, V., Lu, C. and Vasil, I.K. (1985). Histology of somatic embryogenesis in cultured embryos of maize (*Zea mays* L.). Protoplasma 127:1–8.

Vasil, V., Ferl, R.J. and Vasil, I.K. (1988). Somatic hybridization in the Gramineae: *Triticum monococcum* L. (Einkorn) + *Pennisetum americanum* (L.) K. Schum. (pearl millet). J. Plant Physiol. 132:160–163.

Vasil, V., Redway, F.A. and Vasil, I.K. (1990). Regeneration of plants from embryogenic suspension culture protoplasts of wheat (*Triticum aestivum* L.). Bio/Technology 8:429–433.

Vasil, V., Castillo, A.M., Fromm, M.E. and Vasil, I.K. (1992) Herbicide resistant fertile transgenic wheat plants obtained by microprojectile bombardment of regenerable embryogenic callus. Bio/Technology 10:667–674.

Vasil, V., Srivastava, V., Castillo, A.M., Fromm, M.E. and Vasil, I.K. (1993). Rapid production of trangenic wheat plants by direct bombardment of cultured immature embryos. Bio/Technology 11:1553–1558.

Wan, C. and Vasil, I.K. (1994). Plant regeneration from protoplasts, embryogenic suspensions, and cryopreserved cells of Napier grass (*Pennisetum purpureum* Schum.) (in prep.).

Wan, Y. and Lemaux, P.G. (1994). Generation of large numbers of independently transformed fertile barley plants. Plant Physiol 104:37–48.

Wang Z., Nagel, J., Potrykus, I. and Spangenberg, G., (1993a). Plants from cell suspension derived protoplasts of *Lolium* species. Plant Sci. 94: 179–193.

Wang, Z., Takamizo, T., Iglesias, V.A., Osusky, M., Nagel, J., Potrykus, I. and Spangenberg, G. (1992). Transgenic plants of tall fescue (*Festuca arundinacea* Schreb.) obtained by direct gene transfer to protoplasts. Bio/Technology 10:691–696.

Wang, Z., Valles, M.P., Montavon, P., Potrykus, I. and Spangenberg, G. (1993b). Fertile plant regeneration from protoplasts of meadow fescue (*Festuca pratensis* Huds.). Plant Cell Rep. 12:95–100.

Weeks, J.T., Anderson, O.D. and Blechl, A.E. (1993). Rapid production of multiple independent lines of fertile transgenic wheat. Plant Physiol. 102:1077–1084.

Wei, Z. and Xu, Z. (1990). Regeneration of fertile plants from embryogenic suspension culture protoplasts of *Sorghum vulgare*. Plant Cell Rep. 9:51–53.

Wenck, A.R., Conger, B.V., Trigiano, R.N. and Sams, C.E. (1988). Inhibition of somatic embryogenesis in orchard grass by endogenous cytokinins. Plant Physiol. 88:990–992.

Wernicke, W. and Brettell, R. (1980). Somatic embryogenesis from *Sorghum bicolor* leaves. Nature 287:138–139.

Yamada, Y. (1977). Tissue culture studies on cereals. In: Applied and Fundamental Aspects of Plant Cell, Tissue and Organ Culture. J. Reinert and Y.P.S. Bajaj (eds.), Springer-Verlag, Berlin, pp. 144–159.

Yamada, Y., Yang, Z.Q. and Tang, D.T. (1986). Plant regeneration from protoplast derived callus of rice (*Oryza sativa* L.). Plant Cell Rep. 4:85–88.

Yang, Y.M., He, D.D. and Scott, K.J. (1993). Plant regeneration from protoplasts of durum wheat (*Triticum durum* Desf. cv. D6962). Plant Cell Rep. 12:320–323.

Yang, Z., Shikanai, T., Mori, K. and Yamada, Y. (1989). Plant regeneration from cytoplasmic hybrids of rice (*Oryza sativa* L.). Theor. Appl. Genet. 77:305–310.

Zhong, H., Bolyard, M.G., Srinivasan, C. and Sticklen, M.B. (1993). Transgenic plants of turfgrass (*Agrostis palustris* Huds.) from microprojectile bombardment of embryogenic callus. Plant Cell Rep. 13:1–6.

13. *In vitro* Culture of Legumes

M.R. Davey, V. Kumar and N. Hammatt

Contents

1. Introduction

The Leguminosae (Syn. Fabaceae, Papilionaceae, Caesalpinaceae, Mimosaceae) is one of the largest and most widespread families of flowering plants, with representatives in most habitats. The plants range from small herbs to large trees. The economic importance of this family derives from the large number of legumes exploited for human consumption, being used in several forms as immature seed, dry grain, roasted grain, in flour production, as condiments and as fermented products. The high level of lysine makes grain legumes a good supplement for cereals, which are deficient in this amino acid. In addition to their protein content, the seeds of many grain legumes are also rich in oils (e.g., seeds of *Glycine max* contain 40% protein and 20% oil). Owing to their high protein content, legumes are important in the production of livestock and fish food, while forage legumes such as species of *Medicago*, *Lotus* and *Trifolium* are used as fodder, hay and silage. Legumes are an excellent source of vitamins, thiamine and niacin, as well as minerals, calcium and iron. They also contain about 60% carbohydrates,

Indra K. Vasil and Trevor A. Thorpe (eds.), Plant Cell and Tissue Culture, 313–329.

mainly as starch. Many legumes (e.g, species of *Acacia*, *Albizia*, *Dalbergia*, *Leucaena* and *Prosopsis*) are used as sources of timber and fuel, particularly in tropical regions, while others are grown as ornamentals (e.g., *Lathryus odoratus* and species of *Clianthus*, *Laburnum* and *Lupinus*). The oils of grain legumes are employed extensively for industrial purposes. A characteristic feature of the majority of legumes is their ability to develop root nodules in response to infection by soil bacteria of the genus *Rhizobium*. The atmospheric nitrogen fixed within such structures enables the host plants to grow on nitrate-deficient soils and, ultimately, to improve soil fertility.

The ability to culture plant cells and tissues has provided the opportunity to micropropagate specific, and often elite, plant clones by the stimulation of axillary buds and, latterly, through the production of somatic embryos. Furthermore, manipulation of culture conditions has facilitated the adventitious regeneration of plants from whole organs, explants, callus and isolated protoplasts. Advances achieved during the 1970s and 1980s in the application of *in vitro* techniques to the genetic manipulation of members of plant families, such as the Solanaceae, stimulated interest in extending such procedures to other economically important crop plants, including legumes. This article summarises advances made in the culture of legumes, and suggests ways in which some of these techniques may be applied to legume improvement.

2. Tissue Culture of Legumes

2.1. *Culture of Excised Meristems and Micropropagation*

Many herbaceous and woody plants can be propagated rapidly *in vitro* through the stimulation of axillary branching, followed by rooting of the resulting microcuttings. The micropropagation of legumes was first reported by Kartha and Gamborg (1978) who cultured excised meristems of pea (0.2–0.3 mm in diameter) on a B5-based medium supplemented with the growth regulators 6-benzylaminopurine (BAP) and α-naphthaleneacetic acid (NAA). Such micropropagated shoots also provided a source of axillary meristems for further multiplication. Several herbaceous legumes have now been micropropagated, such as *Trifolium* species (Williams et al., 1990), but this approach has only just begun to be investigated for leguminous trees such as *Acacia* (Mittal et al., 1989).

2.2. *Embryo Culture*

The culture of excised embryos was one of the earliest techniques devised for legumes. Indeed, methods have been developed to culture embryos after their excision from the pods, at all stages of development from globular

(Hammatt and Davey, 1987) to cotyledonary structures, into whole plants (Bajaj, 1990a). Although a range of culture media has been employed with various salt and vitamin combinations, media based on the formulation of Murashige and Skoog (1962) have been used most extensively. Frequently, media for embryo culture have been supplemented with endosperm obtained from legumes and non-leguminous species. An important aspect of embryo culture is the staging of the seeds within the pods to facilitate isolation of embryos of the required size, and guidelines for this procedure have been documented in soybean (Ranch et al,, 1986). In some species, procedures involving the culture of ovules prior to embryo excision have been important for successful culture of the embryos, while the culture of intact pods following pollination has also been a useful technique in legumes such as *Medicago* species.

2.3. *Culture of Cells and Isolated Protoplasts*

Pioneering work on legume cell culture was performed by Gamborg et al, (1968), who initiated callus from root explants of *Glycine max* using their B5 formulated medium. These tissues were used to establish cell suspensions from which protoplasts were isolated. Although the same research group observed, for the first time, division in isolated plant protoplasts (Kao et al., 1970), they were unable to regenerate plants from protoplast-derived cells or from the original cell suspensions. Nevertheless, their work stimulated interest in the isolation of protoplasts from cultured cells of most of the economically important groups of legumes. Soybean cell suspension cultures are still often used as a model system for investigating a range of plant physiological processes. In many reports, axenic shoot cultures and seedlings have been the material of choice for protoplast isolation. The use of axenic shoots has enabled genotypes, with known plant regeneration capability in culture, to be multiplied as a source of protoplasts. The advantages of using seedlings are that they do not require regular subculture, and they are available for use shortly after seed sowing. Leaflets of axenic shoots have released workable numbers of mesophyll protoplasts, while in the case of seedlings, protoplasts have been isolated from cotyledons and hypocotyls of both green and dark-grown material, and from roots. Extensive studies with *Medicago* and *Glycine* species have established the protocols for seedling protoplast isolation and culture (Gilmour et al., 1989; Hammatt and Davey, 1990), and such procedures can be applied readily to other genera such as *Stylosanthes* (Vieira et al., 1990).

2.4. *Plant Regeneration*

Adventitious regeneration occurs following the induction of shoot meristems or somatic embryos *de novo* in organ explants, explant-derived callus, isolated cells, microspores and from protoplast-derived tissues. As recently as ten years ago, little progress had been made in the development of techniques to induce adventitious regeneration in tissues of legumes, which led to these plants being considered as recalcitrant. However, evidence has since shown that there is little to distinguish the conditions required for adventitious regeneration in legume cultures from those required for non-leguminous plants. The lack of early progress with plant regeneration in legumes probably resulted from the species initially examined, since most attention focussed on grain legumes. Indeed, *G. max* and *Vigna* species (Bhadra et al., 1989) still remain relatively difficult to manipulate *in vitro*, in contrast to small seeded legumes, particularly those used as forage crops. A further major difficulty that has hindered the development of regeneration techniques for legumes, is that, in common with many non-legumes, the induction of shoot regeneration or somatic embryogenesis appears to be genotype specific. Nevertheless, considerable progress has been achieved in the development of plant regeneration procedures in recent years and has encompassed several of the most economically important legumes, with adventitious shoots being obtained in a wide range of forage, grain and timber species (see Bajaj, 1990b). Significantly, this developmental pathway has led to the first report of plant regeneration from protoplasts of six cultivars of *G. max* and its conspecific progenitor *G. soja* (Wei and Xu, 1988). It remains to be seen whether this procedure is generally applicable to Western cultivars of soybean.

An alternative means to obtain adventitious regeneration is through somatic embryogenesis, and this process is now well established in several cases (Hammatt et al., 1986). Indeed, it has been reported, commonly for legumes, that the formation of both adventitious shoots and somatic embryos may occur in the same culture, although often it has been unclear whether structures reported as somatic embryos have been truly embryonic in nature. Since Saunders and Bingham (1972) first observed somatic embryogenesis in alfalfa, this has been the primary route to regenerate plants from both complex explants, including those from seedlings, and protoplast-derived tissues of *Medicago* species (see Hammatt et al., 1986; Gilmour et al., 1989b). Most methods published on somatic embryo formation in soybean, such as those reported by Hammatt and Davey (1987) and Ranch et al. (1986), have substantiated the original observations of Christianson et al. (1983) that high levels of 2,4-dichlorophenoxyacetic acid (2,4-D) induce the formation of somatic embryos on explants of cotyledons excised from zygotic embryos. The approach adopted in soybean is similar to that used for other difficult groups of plants, in particular, cereals, trees and other grain legumes, where plant regeneration has also been facilitated by the use of embryogenic tissues.

Unfortunately, for soybean and several other pulses, the ability to obtain somatic embryogenesis in such primary organs has not yet led to methods to achieve this process in secondary tissues.

Regenerated plants of more outbreeding, heterozygous species, obtained from embryo or seedling sources via somatic embryogenesis or shoot induction, are likely to differ more from the seed donor than those from more inbreeding, homozygous plants. This may limit the use of explants of embryos and seedlings in genetic manipulations *in vitro* for cultivar improvement. However, for cases in which regenerants differ significantly from the parental phenotype, conventional plant breeding techniques could be useful in eliminating any undesirable characters.

2.5. *Anther Culture*

In many *Solanaceous* crops, *Brassicas* and cereals, culture conditions have been optimised to facilitate the regeneration of large populations of morphologically normal, haploid or dihaploid plants from cultured anthers or microspores. Although haploid plants have been obtained in legumes such as *Medicago* species, *Trifolium alexandrium* and *Albizia lebbeck* (see Hammatt et al., 1986), the few reports on anther culture have described primarily the production of haploid, aneuploid or diploid callus. On rare occasions in which plants have been obtained, they were often eudiploids or aneudiploids. Extensive empirical research is required still to develop efficient and reliable techniques that can be used to regenerate large numbers of haploid plants from cultured anthers of legumes.

3. Genetic Manipulation of Legumes

The ability to regenerate plants from cultured explants, tissues and protoplasts affords the opportunity to genetically manipulate legumes through somatic cell techniques.

3.1. *Somaclonal Variation*

The simple procedure of culturing plant cells frequently exposes genetic variation and, as such, can be considered as the most simple form of genetic manipulation. Although the precise basis of such variation is still not known, it is rarely associated with gross cytological changes and is more probably related to more subtle modifications in nucleic acid composition. One of the most comprehensive studies of somaclonal variation in legumes has been that reported by Barwale and Widholm (1990), involving the regeneration of plants by somatic embryogenesis from cotyledons of zygotic embryos of soybean. Plants were identified with chlorophyll deficiency, wrinkled leaves, abnormal leaflet numbers, dwarfism and multiple shoots. Some individuals exhibited complete or partial sterility. The culture of isolated protoplasts also exposes variation, with plants derived from leaf mesophyll protoplasts of alfalfa exhibiting differences in morphology together with photosynthetic and nitrogen fixing capabilities (Golds, 1989). Another detailed investigation of somaclonal variation in forage legumes is that reported in *Lotus corniculatus* cv Leo (Webb and Watson, 1991), in which plants were regenerated from leaf explants and from protoplasts isolated from leaves and cotyledons. All three populations of plants differed from control, seed-grown plants, with plants derived from leaf explants differing in fewer characteristics than those from protoplasts. Generally, plants from cultured material were shorter than seed grown controls, with smaller leaves and shorter petioles. Plants showed variation in their ability to survive transfer from *in vitro* conditions to soil. Although most regenerants had the normal tetraploid chromosome number ($2n = 4x = 24$), plants with chromosome numbers near to 24, near to 48 and with more than one chromosome number (mixoploids/chimaeras), were identified in the three populations. Of interest was the fact that regenerated plants with both normal and abnormal numbers of chromosomes produced progeny which were cytologically normal and which showed only minor morphological changes compared to control plants. This indicated that much of the variation was expressed only in the initial regenerants and was not transmitted to their seed progeny.

3.2. *Transformation*

The progress achieved in the transformation of legumes has been reviewed recently (Kumar and Davey, 1991) and only the most relevant aspects of this subject will be summarised in the present discussion. As in most crop plants, *Agrobacterium*-mediated transformation has been the principal route for gene insertion into legumes. In studies of the interaction of a range of oncogenic strains of *A. tumefaciens* with varieties of *M. sativa*, *Glycine* species (primarily *G. max*) and *Pisum sativum*, a plant genotype-bacterial strain specificity was apparent (Mariotti et al., 1984; Lianzheng et al., 1984; Owens and Cress, 1985; Byrne et al., 1987; Delzer et al., 1990; Hobbs et al, 1989; Bhatt et al, 1991). The host tissue also influences infection by *Agrobacterium*, with hypocotyls of seedlings of *G. clandestina* and *G. canescens* being more responsive than cotyledons to infection (Rech et al., 1988). Although tumours induced by oncogenic strains of *A. tumefaciens* generally fail to produce shoots, transformed roots incited by *A. rhizogenes* may regenerate shoots spontaneously, as in *Lotus corniculatus* (Petit et al., 1987), or produce shoots following a callus induction phase and transfer to a suitable shoot-inducing medium, as in *G. canescens* (Rech et al., 1988).

The neomycin phosphotransferase (*npt*II) gene has been employed extensively as a genetic marker in establishing transformation systems using engineered vectors. In the case of agrobacteria carrying co-integrate plasmids, the strain R1601 of *A. rhizogenes* has been used to transform *G. canescens* (Rech et al., 1989). Indeed, this strain is particularly useful in effecting gene transfer into legumes and other dicotyledons because of its 'supervirulent' phenotype conferred by pTVK291 *in trans*. As in other dicotyledons, dis-armed vectors have been employed increasingly to transform legumes in both co-integrate and binary systems. Examples of the use of dis-armed co-integrate vectors include the production of kanamycin resistant transgenic plants of *Vigna unguiculata* (Garcia et al., 1986) and *G. max* (Hinchee et al., 1988) following inoculation of leaf discs and cotyledon explants respectively. *Pisum sativum* has also been transformed to hygromycin resistance by a dis-armed co-integrate vector (Puonti-Kaerlas et al., 1990). In the case of binary vectors, those carried in *A. tumefaciens* have been used to transform *M. sativa* (Shahin et al., 1986), *T. repens* (White and Greenwood, 1987) and *Stylosanthes humilis* (Manners, 1987). Additionally, significant progress has been made by infecting plumules, cotyledonary nodes and surrounding regions of axenic seedlings of soybean with *A. tumefaciens* carrying binary vectors. Using this approach, Chee et al. (1990) produced transgenic plants and seed-derived progeny, overcoming the complications of regenerating plants from tissue-cultured material. *S. humilis*, together with *M. sativa*, has also been transformed by strains of *A. rhizogenes* carrying dis-armed binary vectors (Manners and Way, 1989; Sukhapinda et al., 1987). As in other dicotyledons, phenolic compounds such as syringaldehyde and acetosyringone, stimulate transformation of some legumes by agrobacteria.

Whilst explants have been the targets in most transformation experiments, there are reports of the co-cultivation of protoplast-derived cells with *Agrobacterium* strains carrying co-integrate vectors, as in the case of soybean (Baldes et al., 1987) and *Vigna aconitifolia* (Eapen et al., 1987). Additionally, isolated protoplasts have been used as recipients of foreign genes in transformation experiments involving the direct uptake of DNA. Thus, kanamycin resistant tissues of *V. aconitifolia* (Köhler et al., 1987) and *Psophocarpus tetragonolobus* (Gill, 1990) resulted from PEG-induced uptake of plasmid into protoplasts. Electroporation has been used to introduce plasmids into protoplasts of *G. max* cv Clark 63, optimisation of the DNA uptake conditions being established using the chloramphenicol acetyltransferase (*cat*) and β-glucuronidase (*gus*) genes as readily assayable reporter systems in transient gene expression experiments (Dhir et al., 1991). Subsequently, transgenic plants were regenerated from transformed tissues following electroporation-induced plasmid uptake into protoplasts (Widholm et al., 1991). Electroporation has also resulted in the production of kanamycin resistant plants of *Medicago borealis* (Kuchuk et al., 1990).

Other techniques for DNA delivery have been evaluated in order to circumvent any host range restrictions of *Agrobacterium* and the need to select transformed cells. Intranuclear micro-injection of DNA into protoplasts of alfalfa that had partially resynthesised their cell walls, resulted in 15–26% of the injected protoplasts being transformed (Reich et al., 1986). Currently, the acceleration of DNA coated tungsten or gold particles into suitable target tissues is a research area which is being vigorously pursued and which has already met with success in legumes. CAT expression has been demonstrated in soybean suspension cultured cells (Wang et al., 1988), while protoplasts isolated from soybean embryos bombarded with DNA carrying the *npt*II gene produced kanamycin resistant tissues (Christou et al., 1988). Meristems of immature soybean seedlings have also been the targets for particle bombardment, with plants regenerated from such meristems expressing the *npt*II and *gus* genes which were transmitted to their seed progeny (McCabe et al., 1988).

3.3. *Somatic Hybridization*

Although there are fewer reports of somatic hybridization than transformation, protoplast fusion has given rise to somatic hybrid tissues and somatic hybrid plants in legumes. Almost a decade has elapsed since the first somatic hybrid was reported between a male sterile line of *M. sativa* and a male fertile line of *M. falcata* following PEG-induced fusion of isolated protoplasts (Téoulé 1983). Subsequently, somatic hybrid tissues were reported involving *M. sativa* with *M. borealis*, *M. falcata* and *M. quasifalcata* after chemically-induced fusion or electrofusion of protoplasts. Micro-manipulation was used to select heterokaryons or albino complementation to select hybrid cells (Gilmour et al., 1987, 1989a). However, these somatic hybrid tissues failed to produce plants. Indeed, it is only recently that somatic hybrid plants have been produced again between *M. sativa* and *M. falcata* (Mendis et al., 1991). The parental material employed in this case could be crossed sexually, but only with extreme difficulty using embryo rescue. In the somatic hybridization experiments, heterokaryons were selected by micro-manipulation and cultured in the presence of nurse protoplasts from an embryogenic cell suspension of *M. sativa*.

Somatic hybrid plants have been reported in the genus *Lotus* between the sexually incompatible species *L. corniculatus* and *L. conimbricensis*. The selection system in this case involved the fusion of iodoacetate inactivated protoplasts from hypocotyls of etiolated seedlings of *L. corniculatus* with cell suspension protoplasts of the other partner (Wright et al., 1987). Selection systems have been devised which involve the transformation of one partner prior to protoplast isolation and fusion. For example, transformed roots were induced on seedlings of *L. tenuis* by *A. rhizogenes* carrying a binary vector conferring kanamycin resistance on plant cells, and callus produced from these roots was used to initiate cell suspensions (Aziz et al., 1990). Protoplasts isolated from the suspension cultures were treated with a sub-lethal concentration of sodium iodoacetate prior to their electrofusion with protoplasts from cotyledons of green seedlings of *L. corniculatus*. Putative somatic hybrid colonies, expressing the *npt*II gene, were selected on medium containing kanamycin sulphate. These colonies regenerated plants, the hybridity of which was confirmed by plant morphology, esterase banding patterns, the presence of condensed tannins in leaves and stems and their chromosome complements.

4. Applications and Future Prospects

4.1. *Micropropagation*

Leguminous trees are of considerable economic importance, particularly in tropical regions, as sources of construction materials and fuel. Thus, clonal plantations of superior genotypes with above-average, long-term forest performance, could contribute to significant improvements in timber yield. Micropropagation techniques, either through axillary branching or somatic embryogenesis, are beginning to be developed for leguminous trees and, eventually, these techniques could provide a source of clonal planting material.

Experience with non-leguminous species suggests that repeated subculture of axillary shoot cultures of mature trees can result in propagules with desirable juvenile characters, including enhanced wood accumulation and improved rooting of conventional cuttings (Jones, 1991). These observations have led to the speculation that micropropagation may lead to the rejuvenation of mature trees. It is possible, therefore, that micropropagation of leguminous trees may result in propagules that also possess advantageous, juvenile features.

4.2. *Embryo Rescue*

Wild relatives of crop plants comprise an important germplasm resource for plant improvement. Access to such tertiary and quarternary gene pools is usually prevented by barriers of sexual incompatibility, which can be expressed both before, or after, fertilisation. One important mechanism post-fertilisation derives from incompatibility between the hybrid embryo and endosperm, resulting in embryo abortion. Hybrid embryos removed from the pods can often be cultured through to plants *in vitro*. Such an approach has been adopted for several legumes, resulting in hybrids from a number of genera, including *Arachis* and *Phaseolus* (Bajaj, 1990a), *Glycine* (Grant, 1990), *Vigna* (Bhadra et al., 1989) and *Trifolium* (Williams et al., 1990). Additionally, adventitious shoots have been induced on hybrid embryos such as those of *G. argyrea* × *G. clandestina* (Grant, 1990). In some combinations, an ovule-embryo culture method has been preferable, as in the case of diploid

M. sativa × diploid *M. rupestris*, where the ovules were cultured for 6–9 days prior to embryo removal and transfer to fresh medium (McCoy, 1985). The excision of intact pods about 3 days after pollination and their subsequent culture has been used to produce viable seeds in annual and perennial *Medicago* species. This technique was most successful in the annual diploids *M. blancheana* and *M. disciformis*, where over 80% of the cultured pods produced seeds.

4.3. *Artificial Seeds*

The highly heterozygous state of many leguminous crops derives partly from the outbreeding nature of these species. Normally, cultivars consisting of single genotypes of above average field performance can be developed only through vegetative propagation. The maintenance of male sterile lines, which is expensive since currently it can only be achieved using maintainers and restorers, could also benefit from vegetative propagation. However, the use of conventional cuttings or micropropagation by the stimulation of axillary branching for seed legumes are both unfeasible economically. Somatic embryogenesis may offer an alternative, economically viable technique to clonally propagate a range of crops. This is based on observations that somatic embryos frequently produce further generations of secondary somatic embryos and that this mode of development can be obtained for legumes in agitated liquid medium. This provides a convenient basis on which to investigate the automation of the propagation process. Furthermore, somatic embryos encapsulated in suitable matrices to produce 'artificial seeds', will germinate in soil. The potentially low cost of storing such artificial seeds, combined with direct sowing into the field, should lead to further reductions in propagation costs.

In legumes, most success with artificial seeds has been achieved with alfalfa, using highly embryogenic cultivars as model systems (Redenbaugh and Walker, 1990). Indeed, several detailed investigations have commenced into factors that affect the conversion of somatic embryos of legumes into vigorous plantlets. Thus, Fujii et al. (1990) found that maturation of alfalfa embryos in the presence of abscisic acid (ABA) stimulated germination of undesiccated embryos in soil, while desiccation of soybean embryos enhanced their germination *in vitro* (Hammatt and Davey, 1987; Buccheim et al., 1989) Studies with alfalfa have shown that treatment during embryo maturation, with ABA, heat shock and high osmoticum, resulted in somatic embryos that were more tolerant of the process of desiccation, which is a prerequisite to the formation of dry artificial seeds (Senaratna et al., 1989a, b; Anandarajah and McKersie, 1990a, b). The beneficial effects of these factors were realised only if they were applied at specific stages of embryo development. As many as 68% of desiccated somatic embryos of alfalfa have germinated (Anandarajah and McKersie, 1990b), but the mode of action by which these factors stimulate conversion of somatic embryos into plantlets are, as yet, unclear. However, it is likely that survival of the desiccation process and subsequent vigorous growth are dependent, at least in part, upon the accumulation of sufficient storage reserves during the maturation period. Thus, Xu et al. (1990) observed the accumulation of starch grains only in somatic embryos matured with ABA. Using excised zygotic embryos of alfalfa, Xu et al. (1990) demonstrated that treatment with high osmoticum, but not ABA, resulted in the synthesis of proteins similar to those produced during embryogenesis *in situ*.

4.4. *Somaclonal Variation, Transformation and Somatic Hybridization*

Although many somaclonal variants exhibit characteristics which are detrimental compared to those of parental plants, individuals have been produced which express superior traits. An example in legumes is provided by alfalfa, where a protoplast-derived plant of the variety Europe was resistant to *Verticillium* wilt (Latunde-Dada and Lucas, 1983). Such characteristics are of agronomic importance, provided they are maintained through seed generations.

Many possibilities exist for the exploitation of transformed tissues and transgenic leguminous plants in terms of basic studies of gene expression. Thus, Facciotti et al. (1985) transformed *G. max* cv Forrest with the *npt*II gene linked to the 5′ portion of a soybean small subunit carboxylase gene. Light-inducible expression of the rubisco-*npt*II gene in transformed cells was demonstrated by the increase in levels of kanamycin resistance. Gene transfer can also be used for objectives specific to legumes. It may assist in understanding the regulatory mechanisms associated with seed storage proteins such as vicillin and legumin, which are differentially regulated during seed development (Boulter et al., 1987). The lack of sulphur-containing amino-acids in the seed storage proteins of pulses could be rectified by manipulating the regulation of expression of less desirable protcins. Alternatively, genes might be inserted encoding for foreign proteins, such as that found in Brazil nuts which is rich in methionine, linked to suitable promoters (Altenbach et al., 1987). Interestingly, Weir et al. (1988) transformed *Lotononis bainesii* with DNA coding for essential amino-acids such as isoleucine, lysine, methionine, threonine and tryptophan. In Australia, the pea albumin gene has been transferred into alfalfa to develop a variety with high levels of albumin protein rich in sulphur amino acids (Ford, 1988). Plants are being assessed to determine whether sheep grazing on this genetically engineered plant variety exhibit increased wool production. The recent ability to transform *Pisum sativum* (Puonti-Kaerlas et al., 1990), should permit studies of the expression of seed storage protein genes in this crop.

The nutritional value of forage legumes such as *M. sativa*, *Trifolium repens* and *T. pratense* is often reduced by bloating which they induce in ruminant animals. In the longer term, it should be feasible to clone the genes for tannin biosynthesis from forage legumes such as *Lotus corniculatus* and *Onobrychis viciifolia* and to transfer such genes into alfalfa and clover, thereby reducing or eliminating bloat in ruminants grazing on these latter plants.

Genetic transformation also enables studies to be performed of the genes involved in symbiotic nitrogen fixation. Stougaard et al. (1986) demonstrated nodule-specific expression of the chimaeric soybean leghaemoglobin-*cat* gene in nodules which developed on the roots of transgenic plants of *L. corniculatus*. Expression of the chimaeric gene occurred at the correct stage of nodule development. There have been several reports of the effects of Ri plasmid T-DNA genes on nodulation of forage legumes. Some reports confirmed that root nodules which developed on plants of *L. corniculatus* transformed by wild type Ri plasmids had nitrogenase levels comparable to those of nodules induced on non-transformed seedlings (Petit et al., 1987), but Ri transformed plants of *M. sativa* and *O. viciifolia* failed to nodulate (Golds et al., 1991). The presence of Ri T-DNA in transformed roots induced on nontransformed seedlings of *T. repens*, *Macroptilum atropurpureum* and *M. sativa* was also thought to inhibit nodulation by *Rhizobium* (Beach and Gresshoff, 1988). The use of strains of *A. rhizogenes* deleted of the pRi *rol*A, *rol*B, *rol*C or *rol*D genes may confirm whether or not this is the case. One of the wider applications of transformation, as in most crop plants, is in the insertion into legumes of insect resistance and tolerance to herbicides. Indeed, transgenic soybean plants resistant to the herbicide glyphosate have been produced following transformation of seedling cotyledon explants by *A. tumefaciens* carrying pTiT37SE::pMON894 (Hinchee et al., 1988).

Somatic hybridisation can be employed to overcome sexual incompatibility barriers and also to produce novel nuclear-cytoplasmic combinations. Most progress in the application of this technology to legumes has been achieved in forage crops in which plants can be regenerated easily from protoplasts. In *Lotus*, experiments on the somatic hybridization of *L. corniculatus* with *L. conimbricensis* were performed to transfer pod indehiscence from the latter species into *L. corniculatus* to reduce seed loss during harvesting (Wright et al., 1987). Somatic hybridization of *L. corniculatus* with *L. tenuis* was designed to combine the 'bloat safe' tannin synthesis of *L. corniculatus* with rapid recovery from defoliation during grazing of *L. tenuis* (Aziz et al., 1990). Unfortunately, the somatic hybrid plants produced in *Lotus* were infertile and the nature of this infertility still needs to be characterised. In contrast, some of the somatic hybrids between *M. sativa* and *M. falcata* were fertile (Mendis et al., 1991), enabling such plants to be incorporated into conventional breeding programmes. Somatic hybridization also has the potential for improving grain legumes such as soybean and *Vigna* pulses (Bhadra et al., 1990), but application of this technology to grain legumes still depends upon the development of reproducible protoplast-to-plant systems. In the case of soybean, heterokaryons between *G. max* cv HP-20–20 and perennial wild *Glycine* G1171 produced shoots (Hammatt et al., 1992), which were confirmed to be somatic hybrid in nature from analysis of aspartate aminotransferase. However, these shoots failed to develop into plants, despite the fact that protoplast-derived shoots of the wild partner readily develop into plants which flower and set seed.

In conclusion, the issue fundamental to all aspects of the genetic manipulation of legumes and, indeed, many other crop plants, is the inability to regenerate plants reproducibly from cultured explants and protoplast-derived tissues. Until our understanding of the physiological and genetic control of the expression of totipotency in plants improves, the development of regeneration procedures for a range of important legume cultivars will remain problematical.

5. References

Altenbach, S.B., Pearson, K.W., Leung, F.W. and Sun, S.M. (1987). Cloning and sequence analysis of a cDNA encoding Brazil nut protein exceptionally rich in methionine. Plant Mol. Biol. 8:239.

Anandarajah, K. and McKersie, B.D. (1990a). Enhanced vigor of dry somatic embryos of *Medicago sativa* L. with increased sucrose. Plant Sci. 71:261–266.

Anandarajah, K. and McKersie, B.D. (1990b). Manipulating the desiccation tolerance and vigor of dry somatic embryos of *Medicago sativa* L. with sucrose, heat shock and abscisic acid. Plant Cell. Rep. 9:451–455.

Aziz, M.A., Chand, P.K., Power, J.B. and Davey, M.R. (1990). Somatic hybrids between the forage legumes *Lotus corniculatus* L. and *L. tenuis* Waldst et Kit. J. Exp. Bot. 41:471–479.

Baldes, R., Moos, M. and Geider, K. (1987). Transformation of soybean protoplasts from permanent suspension cultures by cultivation with cells of *Agrobacterium tumefaciens*. Plant Mol. Biol. 9:135–146.

Bajaj, Y.P.S. (1990a). Wide hybridization in legumes and oilseed crops through embryo, ovule and ovary culture. In: Biotechnology in Agriculture and Forestry, Vol 10. Legumes and Oilseed Crops I. Y.P.S. Bajaj (ed.), Springer-Verlag, Heidelberg, pp. 3–37.

Bajaj, Y.P.S. (1990b), Biotechnology in Agriculture and Forestry, Vol 10. Legumes and Oilseed Crops I. Springer-Verlag, Heidelberg.

Barwale, U.B. and Widholm, J.M. (1990). Soybean: Plant regeneration and somaclonal variation. In: Biotechnology in Agriculture and Forestry, Vol. 10. Legumes and Oilseed Crops I. Y.P.S. Bajaj (ed.), Springer-Verlag, Heidelberg, pp. 114–133.

Beach, K.H. and Gresshoff, P.M. (1988). Characterization and culture of *Agrobacterium rhizogenes* transformed roots of forage legumes. Plant Sci. 57:73–81.

Bhadra, S.K. Hammatt, N. and Davey, M.R. (1989). Prospects for the use of *in vitro* techniques in the improvement of *Vigna* pulses. SABRAO J. 21:75–91.

Bhatt, D,., Parrott, W.A., Collins, G.B. and Hildebrandt, D.F. (1991). *Agrobacterium* induced gall formation in lipoxygenase mutant isolines of soybeans. Plant Cell Rep. 9:651–654.

Boulter, D., Evans, I.M., Ellis, J.R., Shirrat, A., Gatehouse, J.A. and Croy, R.R.D. (1987). Differential gene expression in the development of *Pisum sativum*. Plant Physiol. Biochem. 25:283–289.

Buccheim, J.A., Colburn, S.M. and Ranch, J.P. (1989). Maturation of soybean somatic embryos and the transition to plantlet growth. Plant Physiol. 89:768–775.

Byrne, M.C., McDonnell, R.E., Wright, M.S. and Carnes, M.G. (1987). Strain and cultivar specificity in *Agrobacterium*-soybean interaction. Plant Cell Tissue Org. Cult. 8:3–15.

Chee, P.P., Fober, K.A. and Slightom, J.L. (1990). Transformation of soybean (*Glycine max*) by infecting germinating seeds with *Agrobacterium tumefaciens*. Plant Physiol. 91:1212–1218.

Christianson, M.L., Warnick, D.A. and Carlson, P.S. (1983). A morphogenetically competent soybean suspension culture. Science 222:632–634.

Christou, P., McCabe, D.E. and Swain, W.F. (1988). Stable transformation of soybean callus by DNA coated gold particles. Plant Physiol. 87:671–674.

Delzer, B.W., Somers, D.A. and Orf, J.H. (1990). *Agrobacterium tumefaciens* susceptibility and plant regeneration of 10 soybean genotypes in maturity groups OO to II. Crop Sci. 30:320–322.

Dhir, S.K., Dhir, S., Hepburn, A. and Widholm, J.M. (1991). Factors affecting transient gene expression in electroporated *Glycine max* protoplasts. Plant Cell Rep. 10:106–110.

Eapen, S., Köhler, F., Gerdemann, M. and Schieder, O. (1987). Cultivar dependence of transformation rates in moth bean after co-cultivation of protoplasts with *A. tumefaciens*. Theor. Appl. Genet. 75:207–210.

Facciotti, D., O'Neal, J.K., Lee, S. and Shewmaker, C.K. (1985). Light-inducible expression of a chimeric gene in soybean tissue transformed with *Agrobacterium*. Bio/Technology 3:214–216.

Ford, J. (1988). Sheep will grow woolier on a bioengineered diet. New Sci. 117:24.

Fujii, J.A., Slade, D., Olsen, R., Ruzin, S.E. and Redenbaugh, K. (1990). Alfalfa somatic embryo maturation and conversion to plants. Plant Sci. 72:931–100.

Gamborg, O.L., Miller, R.A. and Ojima, K. (1968). Nutrient requirements for suspension cultures of soybean root cells. Exptl. Cell Res. 50:151158.

Garcia, J.A., Hille, J. and Goldback, R. (1986). Transformation of cowpea *Vigna unguiculata* cells with an antibiotic resistance gene using a Ti plasmid derived vector. Plant Sci. 44:37–46.

Gill, R. (1990). Direct gene transfer in *Psophocarpus tetragonolobus: resistance to kanamycin*. Ann. Bot. 66:31–39.

Gilmour, D.M., Davey, M.R. and Cocking, E.C. (1987). Isolation and culture of heterokaryons following fusion of protoplasts from sexually compatible and sexually incompatible *Medicago* species. Plant Sci. 53:263–270.

Gilmour, D.M., Davey, M.R. and Cocking, E.C. (1989a). Production of somatic hybrid tissues following chemical and electrical fusion of protoplasts from albino suspensions of *Medicago sativa* and *M. borealis*. Plant Cell Rep. 8:29–32.

Gilmour, D.M., Golds, T.J. and Davey, M.R. (1989b). *Medicago* protoplasts: Fusion, culture and plant regeneration. In: Biotechnology in Agriculture and Forestry, Vol. 8. Plant Protoplasts and Genetic Engineering I. Y.P.S. Bajaj (ed.), Springer-Verlag, Heidelberg, pp. 370–388.

Golds, T.J. (1989). Somaclonal Variation and Transformation in Forage Legumes. Ph.D. Thesis, University of Nottingham.

Golds, T.J., Lee, J.Y., Husnain, T., Ghose, T.K. and Davey, M.R. (1991). *Agrobacterium rhizogenes* mediated transformation of the forage legumes *Medicago sativa* and *Onobrychis viciifolia*. J. Exp. Bot. 42:1147–1157.

Grant, J.E. (1990). Soybean: Wide hybridization through embryo culture. In: Biotechnology in Agriculture and Forestry, Vol. 10. Legumes and Oilseed Crops I. Y.P.S. Bajaj (ed.), Springer-Verlag, Heidelberg, pp. 134–148.

Hammatt, N. and Davey, M.R. (1987). Somatic embryogenesis and plant regeneration from cultured zygotic embryos of soybean (*Glycine max* L. Merr.). J. Plant Physiol. 128:219–226.

Hammatt, N. and Davey, M.R. (1990). Soybean: isolation, culture, and fusion of protoplasts. In: Biotechnology in Agriculture and Forestry, Vol.10. Legumes and Oilseed Crops I. Y.P.S. Bajaj (ed.), Springer-Verlag, Heidelberg, pp. 149–169.

Hammatt, N., Ghose, T.K. and Davey, M.R. (1986). Regeneration in legumes. In: Cell Culture and Somatic Cell Genetics of Plants. Vol. 3. Plant Regeneration and Genetic Variability. I.K. Vasil (ed.), Academic Press, London, pp. 67–95.

Hammatt, N., Lister, A., Jones, B., Cooking, E.C. and Davey, M.R. (1992). Shoot formation from somatic hybrid callus between soybean and a perennial relative. Plant Sci. 85:215–222.

Hinchee, M.A.W., Conner-Ward, D.V., Newell, C.A., McDonnell, R.E., Sato, S.J., Gasser, C.S,. Fischhoff, D.A., Re, D.B., Fraley, R.T. and Horsch. R.B. (1988). Production of transgenic soybean plants using Agrobacterium mediated DNA transfer. Bio/Technology 6:915–922.

Hobbs, S.L.A., Jackson, J.A. and Mahon, J.D. (1989). Specificity of strain and genotype in the susceptibility of pea to Agrobacterium tumefaciens. Plant Cell Rep. 8:274–277.

Jones, O.P. (1991). The role of biotechnology in the multiplication and improvement of woody plants. Acta Horticulturae 289:35–44.

Kao, K.N. and Michayluk, M.R. (1975). Nutrient requirements for growth of Vicia hajastana cells and protoplasts at very low population density in liquid medium. Planta 126:105–110.

Kartha, K.K. and Gamborg, O.L. (1978). Meristem culture techniques in the production of disease-free plants and freeze-preservation of germplasm of tropical tuber crops and grain legumes. In: Diseases of Tropical Food Crops. H. Maraite and J.A. Meyer (eds.), Univ. Catholique de Leuvain, Belgium, pp. 267–283.

Kao, K.N., Keller, W.A. and Miller, R.A. (1970). Cell divisions in newly formed protoplasts of soybean. Exptl. Cell Res. 62:338–340.

Köhler, F., Golz, C., Eapen, S., Köhn, H. and Schieder; O. (1987). Stable transformation of moth bean Vigna aconitifolia via direct gene transfer. Plant Cell Rep. 6:313–317.

Kuchuk, N., Komarnitski, I., Shakhovsky, A. and Gleba, Y. (1990). Genetic transformation of Medicago species by *Agrobacterium tumefaciens* and electroporation of protoplasts. Plant Cell Rep. 8:660–663.

Kumar, V. and Davey, M.R. (1991). Genetic improvement of legumes using somatic cell and molecular techniques. Euphytica 55:157–169.

Latunde-Dada, A.O. and Lucas, J.A. (1983). Somaclonal variation and reaction to *Verticillium* wilt in *Medicago sativa* L. plants regenerated from protoplasts. Plant Sci. Lett. 32:205–211.

Lianzheng, W., Guangchu, Y., Jiaoten, L., Bojun, L., Jian, W., Zhenchum, Y., Xiulan, L., Qiquan, S., Xingcum, J. and Zequi, Z. (1984). A study of tumor formation of soybean and gene transfer. Scientia Sinica 27:391–397.

Manners, J.M. (1987). Transformation of *Stylosanthes* species using *Agrobacterium tumefaciens*. Plant Cell Rep. 6:204–207.

Manners, J.M. and Way, H. (1989). Efficient transformation with regeneration of the tropical pasture legume *Stylosanthes humilis* using *Agrobacterium rhizogenes* and a Ti plasmid-binary vector system. Plant Cell Rep. 8:341–345.

Mariotti, D.M., Davey, M.R., Draper, J., Freeman, J.P. and Cocking, E.C. (1984). Crown gall tumorigenesis in the forage legume *Medicago sativa* L. Plant Cell Physiol. 25:473–482.

McCabe, D.E., Swain, W.F., Martinell, B.J., Christou, P. (1988). Stable transformation of soybean (*Glycine max*) by particle acceleration. Bio/Technology 6:923–926.

McCoy, T.J. (1985). Interspecific hybridization of *Medicago sativa* L. and *M. rupestris* M.B. using ovule-embryo culture. Can. J. Genet. Cytol. 27:238–245.

Mendis, M.H., Power, J.B. and Davey, M.R. (1991), Somatic hybrids of the forage legumes *Medicago sativa* L. and *M. falcata* L. J. Exp. Bot. 42:1565–1573.

Mittal, A., Agarwal, R. and Gupta, S.C. (1989). *In vitro* development of plantlets from axillary buds of *Acacia auriculiformis* – a leguminous tree. Plant Cell Tissue Org. Cult. 19:65–70.

Murashige, T. and Skoog, F. (1962). A revised medium for rapid growth and bioassays with tobacco tissue cultures. Physiol. Plant. 15:473–497.

Owens, L.D. and Cress, D.E. (1985). Genotypic variability of soybean response to *Agrobacterium* strains harbouring the Ti or Ri plasmids. Plant Physiol. 77:87–94.

Petit, A., Stougaard, J., Kühle, A., Marcker, K.A. and Tempé, J. (1987). Transformation and regeneration of the legume *Lotus corniculatus*: A system for molecular studies of symbiotic nitrogen fixation. Molec. Gen. Genet. 207:245–250.

Puonti-Kaerlas, J., Eriksson, T. and Engstrom, P. (1990). Production of transgenic pea (*Pisum sativum*) plants by *Agrobacterium tumefaciens*-mediated gene transfer. Theor. Appl. Genet. 80:246–252.

Ranch, J.P., Ogelsby, L. and Zielinski, A.C. (1986). Plant regeneration from tissue cultures of soybean by somatic embryogenesis. In: Cell Culture and Somatic Cell Genetics of Plants, Vol. 3. Plant Regeneration and Genetic Variability. I.K. Vasil (ed.), Academic Press, London, pp. 97–119.

Rech, E.L., Golds, T.J., Hammatt, N., Mulligan, B.J. and Davey, M.R. (1988). *Agrobacterium rhizogenes* mediated transformation of the wild soybeans *Glycine canescens* and *G. clandestina*: Production of transgenic plants of *G. canescens*. J. Exp. Bot. 39:1275–1285.

Rech, E.L., Golds, T.J., Vainstein, M.H., Jones, B., Hammatt, N., Mulligan, B.J. and Davey, M.R. (1989). Expression of a chimaeric kanamycin resistance gene introduced into the wild soybean *Glycine canescens* using a co-integrate Ri plasmid vector. Plant Cell Rep. 8:33–36.

Redenbaugh, K. and Walker, K. (1990). Role of artificial seeds in alfalfa breeding. In: Developments in Crop Science 19. Plant Tissue Culture: Applications and Limitations. S.S. Bhojwani (ed.), Elsevier, Amsterdam, pp. 102–135.

Reich, T.J., Iyer, V.N. and Miki, B.L. (1986). Efficient transformation of alfalfa protoplasts by the intranuclear microinjection of Ti plasmids. Bio/Technology 4:1001–1004.

Saunders, J.W. and Bingham, E.T. (1972). Production of alfalfa plants from callus tissue. Crop Sci. 12:804–808.

Senaratna, T., McKersie, B.D. and Bowley, S.R. (1989a). Desiccation tolerance of alfalfa (*Medicago sativa* L.) somatic embryos. Influence of abscisic acid, stress pretreatments and drying rates. Plant Sci. 65:253–259.

Senaratna, T., McKersie, B.D. and Bowley, S.R. (1989b). Artificial seeds of alfalfa: induction of desiccation tolerance in somatic embryos of alfalfa. *In vitro* Cell Dev. Biol. 26:85–90.

Shahin, E.A., Spielmann, A., Sukhapinda, K., Simpson, R.B. and Yashar, M. (1986). Transformation of cultivated alfalfa using dis-armed *Agrobacterium tumefaciens*. Crop Sci. 26:1235–1239.

Stougaard, J., Marcker, K.A., Otten, L. and Schell, J. (1986). Nodule-specific expression of a chimaeric soybean leghaemoglobin gene in transgenic *Lotus corniculatus*. Nature 321:667–674.

Sukhapinda, K., Spivey, R. and Shahin, E.A. (1987). Ri-plasmid as a helper for introducing vector DNA into alfalfa plants. Plant Molec. Biol. 8:209–216.

Téoulé, E. (1983). Hybridization somatique entre *Medicago sativa* L. et *Medicago falcata* L. C. R. Acad. Sci., Paris, Serie III 297:13–16.

Uchimiya, H. and Murashige, T. (1974). Evaluation of parameters in the isolation of viable protoplasts from cultured tobacco cells. Plant Physiol. 54:936–944.

Vieira, M.L.C., Jones, B., Cocking, E.C. and Davey, M.R. (1990). Plant regeneration from protoplasts isolated from seedling cotyledons of *Stylosanthes guianensis*, *S. macrocephala* and *S. scabra*. Plant Cell Rep. 9:289–292.

Wang, Y-C., Cao, J., Sanford, J.C. and Wu, R. (1988).Transformation of rice, wheat and soybean by the microprojectile method. J. Cell Biochem. Suppl. 12 part C:189.

Webb, K.J. and Watson, E.J. (1991). *Lotus corniculatus* L.: Morphological and cytological analysis of regenerants from three sources of tissue and selected progeny. Plant Cell Tissue Org. Cult. 25:27–33.

Wei, Z.M. and Xu, Z.-H. (1988). Plant regeneration from protoplasts of soybean (*Glycine max* L.). Plant Cell Rep. 7:348–351.

White, D.W.R. and Greenwood, D. (1987). Transformation of the forage legume *Trifolium repens* L. using binary *Agrobacterium* vectors. Plant Mol. Biol. 8:461–469.

Widholm, J., Dhir, S. and Dhir, S. (1991). Production of transformed soybean plants by electroporation of protoplasts. Physiol. Plant. 85:357–361.

Williams, T.G., Collins, G.B. and Myers, J.R. (1990). Clovers (*Trifolium* spp.) In: Biotechnology in Agriculture and Forestry. Vol.10. Legumes and Oilseed Crops. Y.P.S. Bajaj (ed.), Springer-Verlag, Heidelberg, pp. 243–287.

Wright, R.L., Somers, D.A. and McGraw, R.L. (1987). Somatic hybridization between birdsfoot trefoil (*Lotus corniculatus* L.) and L. conimbricensis Willd. Theor. Appl. Genet. 75:151–156.

Xu, N., Coulter, K.M. and Bewley, J.D. (1990). Abscisic acid and osmoticum prevent germination of developing alfalfa embryos, but only osmoticum maintains the synthesis of developmental proteins. Planta 182:382–390.

14. *In vitro* Culture of Vegetable Crops

John F. Reynolds

Contents

1. Introduction

Plant cell regeneration potential can be grouped into two general categories: organogenesis and embryogenesis. Although there are several types of organogenesis, adventitious shoot formation or expression of axillary buds are most useful. These adventitious shoots can lead to plants and seed, continuing any genetic manipulations from *in vitro* techniques. Embryogenesis is the formation of somatic embryos directly from initial explant tissues or through a callus intermediate. Somatic embryogenesis is useful for high frequency propagation or the production of synthetic clonal seed. All vegetable species discussed in this chapter regenerate via one or both of these routes.

Regeneration by adventitious shoot organogenesis has been studied in detail in tomato and *Brassica*. Both have a high potential for shoot formation directly from the initial explant. *Brassica* has a good potential for shoot formation from callus while tomato does not. Organogenesis is most influenced by genotype in both species. Adventitious shoot formation is also

Indra K. Vasil and Trevor A. Thorpe (eds.), Plant Cell and Tissue Culture, 331–362.

possible in eggplant, melon, watermelon and pepper. However, extensive studies have not been done on these species.

Embryogenesis has been studied in detail using the carrot model system. Empirical as well as basic biochemistry and gene activity associated with embryogenesis has been advanced using the carrot system. Recently, eggplant has also been shown to have good embryogenic potential to the point of use in more basic studies. Embryogenesis via microspores is also well studied in *Brassica*.

Other species observed to form somatic embryos include onion, pea, melon, cucumber, and squash. The cucurbit vegetables typically have the problem of aberrant somatic embryogenesis, therefore, special procedures need to be developed to produce normal plants.

The most recent application in vegetable regeneration potential has been in the area of transformation. It becomes obvious that the vegetable species having the highest and most studied regeneration capabilities are usually those that have most success in transformation.

The following is a continuation of the chapter by Reynolds (1986) and include regeneration studies on ten species and two genera of vegetables. In addition, transformation studies are also addressed.

2. Brassica

The genus *Brassica* has been one of the more easily manipulated genera in cell and tissue culture studies. There are a number of reports on regeneration, anther culture as well as protoplast culture and fusion. This section will cover several species and varieties of vegetable *Brassica* including *B. oleracea* varieties *capitata*, *italica*, *botrytis*, *gemmifera*, *gongurlodes*, *acephala*, *B. alboglabra*, and *B. campestris* ssp. *chinensis*.

A study on callus production and callus regeneration of many *Brassica* species was conducted by Murata and Orton (1987). This study was similar to Dietert et al. (1982) in that they found differences in shoot forming ability and callus initiation between species and varieties within species. They showed that within the vegetable species tested *B. oleracea*, variety *acephala* (curly kale) had the highest shoot forming potential from callus. When grouping all eleven *B. oleracea* varieties and comparing them to other species, *B. oleracea* was second to *B. napus*. Best callus formation in this study was induced on MS medium with 1.0 mg/l 2,4-D and 0.1 mg/l kinetin or BA. Best shoot formation from callus was observed with 4.0 mg/l kinetin. A reduction in the sucrose level from 3 to 2% was critical for success in shoot formation. They also presented a scenario of gene control on regeneration. The *Brassica* species tested are all interfertile and are made up of one or more of genomes A, B, or C. B. *napus* and *B. oleracea* have the highest regeneration potential and contain the C genome. They suggested that the genetic mechanism for regeneration was most prominent on the C genome, however, *B. carinata* also has C as well as B but has a low potential for regeneration.

In a study by Wong and Loh (1988) *B. alboglabra* Bailey shoots could be regenerated directly from root explants of four or seven day old seedlings in MS medium with 4.5×10^{-7} M BA. Any addition of auxin to the medium inhibited shoot formation. A significant effect on shoot formation was found in the type of gelling agent. Gelrite (Kelco, San Diego, USA) was superior to agar or liquid medium. Shoots could easily root in basal MS medium and could successfully be established in soil.

Somatic embryogenesis in *B. oleracea* var. *botrytis* was reported by Singh and Chandra (1985). They used 5–6 month old callus or suspensions. Shoot formation was on MS medium with 5.0 mg/l BA. Embryo formation in suspension culture was observed in MS medium with 0.5 mg/l kinetin. Embryos formed from suspension culture cell clumps when IAA was removed from the medium. Embryos were assumed to form from outer layers of cell clumps and were thought to be of single cell origin. Development of the embryos to plants was best achieved in MS medium with IAA (0.5 mg/l), kinetin (0.2 mg/l) and GA (0.2 mg/l).

In a study by Chi and Pua (1989), the difficult to regenerate *B. campestris* ssp. *chinensis* was examined. There were several critical manipulations that resulted in success. The best regeneration medium contained BA (4.4–17.6 μM) and NAA (2.7–5.4 μM) in a MS basal formulation. Explant source was critical, with best results from three day old seedling cotyledons. Five day old cotyledons, leaves, stems or hypocotyls were not as good. The most critical factor for regeneration was the presence of ethylene inhibitors. $AgNO_3$ or Ag_2SO_4 at 7.5–6.0 μM resulted in 70–85% of the explants producing shoots. The inhibitor aminoethoxy vinylglycine (AVG) in medium with 8.8 μM BA and 5.4 μM NAA also induced shoot formation at high frequency. Two other inhibitors, aminooxyacetic acid (AOA) and 2,4-dinitrophenol (DNP) did not give similar results. This study suggests that genetic factors thought to control regeneration can be overcome by close empirical study of factors influencing regeneration.

The ease of manipulation of *Brassica* species is illustrated by the large number of successful protoplast regeneration reports. Fu et al. (1985), using the first true leaves of 2–4 week old greenhouse seedlings, were able to isolate large quantities of protoplasts with a 95% survival rate. Cell division was first observed at 4–5 days with a rapid acceleration of division at 7–8 days. Mannitol at 0.35 M was found critical for cell division. Callus growth and maintenance was possible in medium with 0.1–0.5 mg/l 2,4-D and 3–4 mg/l kinetin. Differentiation of callus was possible on medium with 3–4 mg/l kinetin and GA_3 at 0.1–0.5 mg/l. Plant regeneration was also reported to be via somatic embryogenesis but at a much lower frequency.

Success in protoplast regeneration from Chinese kale, *B. alboglabra* Bailey, was reported by Pua (1987). Tissue selection was critical for success. Best response was observed from stem tissue of 10–14 day old *in vitro* germinated seed, precultured on MS medium with 1.0 mg/l NAA and 2 mg/l BA. Similar results could be obtained using untreated hypocotyl tissue. The best cell division was observed in medium with 0.2 mg/l 2,4-D and 0.2 mg/l BA. Regeneration of callus was possible on medium with only BA at 0.5 mg/l. The stability of the regenerated plants, determined by chromosome number, was 95% normal.

Success in regenerating protoplasts of cabbage, *B. oleracea* var. *capitata* was reported by Nishio et al. (1987). Many media combinations were tested for callus and shoot formation from established colonies. Callus cultured in 0.5 mg/l 2,4-D, 5 mg/l kinetin regenerated best when transferred to MS medium with 1 mg/l BA and 0.035 mg/l GA_3. Lillo and Olsen (1989) also regenerated protoplasts of cabbage as well as curly kale. Tissue source was hypocotyls from 5–7 day old *in vitro* germinated seed. Osmotic level was critical for callus formation. Sucrose at 60 mM was best for callus formation but inhibited shoot formation. Nitrate and ammonium were necessary for callus and shoot formation and could be added over a range of 23–100 mM for nitrate and 8 or 21 mM ammonium. Organic nitrogen compounds were investigated with 0.5 mM putrescine giving more callus and shoot formation. Photoperiod, intensity or quality of light was not found to be critical for shoot production.

Several varieties of *B. oleracea* including cabbage, cauliflower, and brussels sprouts were reported to regenerate from protoplasts by Bauer (1988). Best explants were two week old leaves preincubated in protoplast culture medium at 26 °C in darkness for 16 hours. Callus formation was most effective in 0.5 mg/l 2,4-D. Highest regeneration rate was on medium with 1 mg/l BA and 0.04 mg/l GA_3. Cauliflower callus regenerated best, followed by cabbage with brussels sprouts being reported to regenerate for the first time.

Most all discussion of protoplast culture in *Brassica* concerns the production of phenolic compounds in the culture medium. In general, to circumvent the problem rapid division of protoplasts to callus is necessary. Low light conditions and frequent dilutions of cultures also reduce browning.

The ultimate use of protoplast isolation procedures is the fusion of protoplasts for the purpose of moving genetic elements between or among genera or species. The importance of cytoplasimic male sterility (CMS) in *Brassica* breeding strategies has been known for decades. The incorporation of the radish CMS in vegetable species by protoplast fusion was reported by Kameya et al. (1989). Two of their plants developed and were shown to be somatic hybrids with the nucleus of cabbage and the cytoplasm of radish. Seed from these plants flowered and were all sterile. The possible drawback of this procedure is that the cytoplasm probably contains radish chloroplast which would likely cause chlorosis and poor flower development. To make this strategy work seed from a number of somatic hybrids need to be germinated at 8 °C to test for chlorosis. The fusions should also be done without a cytoplasmic inhibitor in the fertile donor so that *Brassica* chloroplasts can be made available in the fusion pool of genomes.

Jourdan et al. (1990) did some preliminary work on cauliflower (*B. oleracea* ssp. *botrytis*). They examined 1000 protoplast derived plants from CMS and fertile lines of cauliflower. They found that most all of the plants had normal morphology and no change in their cytoplasm as tested by mt DNA probes. This suggests that fusion procedures can be used for production of normal plants.

Fusion between *B. oleracea* and chinese cabbage *B. campestris* was reported by Taguchi and Kameya (1986). The regenerated plants from two callus colonies were suggested to be fusions as indicated by morphology, chromosome marker, and isozyme patterns. This was the first report of fusions between two heading type vegetables.

Another indicator of *Brassica* tissue culture manipulability is the large number of successful anther culture studies. Cauliflower, *B. oleracea* var. *botrytis*, has been studied by Phippen and Ockendon (1990). They investigated eleven F_1 hybrids with maturity dates representing summer, autumn and winter types. Somatic embryos could be produced in all cultivars tested with the autumn and winter types responding better than the summer varieties. The largest response variation was seen between culture vessels. The second largest variation was plant to plant within a genotype followed by a genotype flower bud size interaction. Best medium was a B5 agar with 0.3 mg/l 2,4-D.

Success in cabbage, *B. oleracea* var. *capitata*, anther culture was demonstrated by Lillo and Hansen (1987) on liquid culture medium with 1 mg/l 2, 4–D. Best anther response was from donor plants grown at 1 °C in light and 12 °C dark or at a constant 12 °C. Embryos usually did not develop into plants. Whole plants were produced by adventitious shoot formation from embryos on medium with 1.0 mg/l BA and 0.1 mg/l GA_3. They noted that 72% of the plants were diploid, 16% tetraploid, 5% haploid, 4% triploid and 3% had no fate reported.

Chiang et al. (1985) also had success in cabbage anther culture. Their objective was to advance breeding material selected in field screens for resistance to clubroot, *Plasmodiophora brassicae* Wor. Their anther derived plants regenerated directly from embryos and were 36% diploid, 55% tetraploid, 8% haploid, and 0.3% aneuploid. No tetraploids were reported. The similarities in haploid plant production in the two previous studies indicate no effect of the two step regeneration step versus direct embryogenesis.

Rouland et al. (1990) also studied cabbage anther culture. They noted success in forming embryos from 11 varieties of head cabbage. They also noted, as most other workers have reported, differences in genotype response and effects of environment on donor plant response. Plants grown in field conditions were always better than greenhouse grown plants. Carbohydrate type and concentration in culture medium were also addressed. Sucrose and maltose were examined with sucrose at 13% giving the better response.

One of the more difficult *Brassica* species to culture has been *B. campestris*. Sato et al. (1989a, b) successfully obtained embryos from anthers of chinese cabbage *B. campestris* ssp. *pekinensis*. Nine cultivars were tested with all responding positively, however, the Kenshin group showed the best response. A specific variety, Formosa 45 days, was the best of all tested. It was important to treat anthers at 35 °C for one day for highest embryogenic response. One of the major problems in these studies was the conversion of embryos to plants. Only 2–5% of the embryos were capable of forming plants.

Considerable effort has been made in developing successful culture strategies in brussels sprouts (*B. oleracea* var. *gemmifera*). Ockendon (1985) noted that genotype had a large influence in anther response. Of the seven genotypes tested, one was very responsive, two were moderate and four were poor. Further genetic analysis showed that about half of the variation in anther response was due to genetic background (Ockendon and Sutherland, 1987).

Increased regeneration in poorly responding lines of brussels sprouts was accomplished by the addition of the ethylene inhibitor, silver nitrate ($AgNO_3$), to the medium (Biddington et al., 1988). A range of 1–10 mg/l was found to give best results. Biddington and Robinson (1990) also examined temperature effects on anther response. Temperatures of 20, 32, 35, and 38 °C were used to treat anthers for 24 hours. Maximum response was at either 32 or 35 °C.

In summary, Ockendon (1986) emphasized some of the more important factors to consider in anther culture of *Brassica*. The two most important factors are the condition and genotype of the donor plant. Optimal bud stage is also important followed by medium composition. A large study by Arnison and Keller (1990) examined 95 varieties covering 7 ssp. of *B. oleracea*. They noted a response from most lines, with cabbage being best. All of the varieties were grown under less than optimal field conditions. Despite poor conditions, the study did show that controlled environmental conditions were not required to achieve success in producing inbred lines from a variety of vegetable *Brassica*.

The extensive studies on vegetable *Brassica* regeneration, protoplast culture and anther culture should be useful for developing transformation procedures. However, there are few reports. David and Tempé (1988) inoculated hypocotyls of sterile seedlings with strains of *Agrobacterium rhizogenes*. Roots produced at the wound site were cultured on solid medium and eventually formed buds and plants. The plants were analyzed and found to be transformed by the presence of opines in the plants and a positive Southern blot analysis. Plants from the atropine-type Ri strain produced normal plants from the root culture while the mannopine type produced typical hairy root morphology of wrinkled leaves, plagiotropic roots, and short internodes. The plants that regenerated from the transformed roots took 1.6 to 6 times longer to regenerate than untransformed roots.

3. Artichoke

Artichoke is one of the less cultivated species discussed in this review. The lack of literature also reflects the lack of economic importance of the species. A study by Onisei et al. (1988) was conducted for the purpose of propagating *in vitro* valuable field material and to eventually use cell culture as a source of compounds for medicinal use. Seedling and mature field grown plants were used as explant source. The only explants that resulted in plants were seedling mesocotyls. Best medium for callus and embryo formation was either BA or kinetin at 1–2 mg/l with 2,4-D at 1.0 mg/l. They speculated that shoots originated via embryogenesis due to the ease of separating the shoots. Plants were difficult to establish in soil due to leaf vitrification and poor rooting. Plants were established by in vivo rooting under low light conditions.

4. Bean

Bean (*Phaseolus vulgaris*) is another minor vegetable in terms of world market and is also studied infrequently using tissue culture systems.

In studies by Breuer et al. (1986) and Jacobsen and Kysely (1986), a procedure was outlined for obtaining somatic embryogenesis in bean as well as pea and other legumes. They stressed the importance of screening a number of genotypes for a positive response. Best explant source in bean was leaf segments or immature embryos cultured on MS, B5 or RBI medium with 0.2 mg/l BA and 0.01 mg/l picloram. After initial callus growth tissue was transferred to K1S1 or B5 medium with 1.0 mg/l picloram. Embryos would form in liquid medium without growth regulators. No plants were reported to form from any of the somatic embryos.

Organogenesis was reported in green bean cv. Park Red Kidney by Franklinet al. (1991). Cotyledons with a small portion of the embryonic node axis were grown on MS medium minus inorganic nitrogen plus 10 mM L-glutamine, MS vitamins, 15 μM BA, 2 μm GA, and 1% sucrose. Shoots rose from a meristematic ring at the base of the axillary bud at the cotyledonary node. Plants could be rooted in MS medium with 2 μm NAA, 0.4 μM GA_3 and 3% sucrose. They also noted a genotype effect on shoot regeneration. Similar but reduced regeneration was described by McClean and Grafton (1989), however, the axillary buds were removed. In a later study by McClean et al. (1991), transformation was attempted using cotyledonary node explants without success. Lack of success may have been caused by removal of the axillary bud in the initial explant source.

5. Eggplant

Eggplant (*Solanum melongena*) is one of the more manipulable species in tissue culture systems. The outstanding ability to respond in culture was utilized to study effects of polyamines on somatic embryogenesis. As mentioned in the section on carrot tissue culture, polyamines are known to effect embryogenesis. Similar observations were made by Fobert and Webb (1988). They noted that embryogenesis from cotyledon explants on 2.5–5.0 mg/l NAA was associated with increased levels of spermidine. Explants that were competent had higher levels of putrescine and spermidine than non-competent explants. Inhibitors of polyamine biosynthesis reduced embryogenesis while putrescine restored it.

A number of studies have been conducted in the area of protoplast culture. Gleddie et al. (1986) showed that yield and cell division of protoplasts were best when suspension cultures initiated from embryogenic callus cultures were maintained in 2 mg/l 2,4-D. Suspensions maintained in 10 mg/l NAA were not as good. Conversion of embryos to plants, however, was low with only 25% responding. Shoots could also be regenerated from protoplasts of eggplant via organogenesis. Nishio et al. (1987) regenerated protoplasts from hypocotyls of *in vitro* germinated seed. Protoplast callus regenerated best using 1–3 mg/l zeatin. Protoplast from leaf tissue also could regenerate through organogenesis (Saxena et al., 1987). In a tissue comparison study, Sihachakr and Ducreux (1987) noted that highest percentage of dividing protoplasts was obtained from petiole material of *in vitro* grown plants. Leaf lamina or stems were not as good. Regeneration of the protoplast callus in this study was through an embryogenic type of structure on medium with zeatin.

It is not surprising that a number of papers have been published on transformation of eggplant. The first study was conducted by Guri and Sink (1988). They used leaves from 2 week old in vitro grown seedlings as explant source. Sliced leaves were co-cultured for 48 hours with *Agrobacterium* followed by a sterile water rinse and culture on an adventitious shoot inducing medium containing 2 mg/l zeatin and 0.1 mg/l kanamycin or 0.025 mg/l hygromycin. The successful transformation used a cointegrated plasmid, pMON200. Two binary plasmids, 715 and pCIB 10, were not successful. A total of 28 kanamycin resistant plants were regenerated, three of which were analyzed in detail. The three contained different copy numbers of the chimeric NOS-NPT II genes. Seed from two plants showed segregation ratios expected for a single dominant trait. Filippone and Lurquin (1989) were able to obtain transformation of eggplant cotyledons and leaves using the binary containing hypervirulent strain A281. Transformation was best using cotyledon explants. Regeneration was through organogenesis and not embryogenesis and took six months to achieve. They speculated that the *onc* genes caused the delay and the type of regeneration observed in this tissue. Transformation with a disarmed binary vector was reported by Rotino and

Gleddie (1990). They used a C58 strain GV3101 with binary system pBCATl and pGV3850. Leaves from *in vitro* plants were used as explants. Regeneration from transformed callus was through organogenesis on a 1 mg/l zeatin medium. A total of 120 shoots were regenerated. Five were self pollinated and found to segregate near a 3:1 ratio. CAT and NPT II activity as well as Southern blot analysis confirmed transformation.

It is interesting that regeneration in all three of the transformation reports was through an organogenic system. Eggplant has a prolific capability to undergo somatic embryogenic but was not used or achieved in these studies.

6. Garlic

Little recent work has been reported using tissue culture systems in garlic (*Allium sativum* L.). Malpathak and David (1986) determined that alliin content of garlic cultures was higher when cultures were induced to differentiate. Maggioni et al. (1989) described a method of regenerating leaf derived callus on an MS medium with 2 mg/l IAA and 4.5 mg/l kinetin. It was noted that callus cultures increased in ploidy over time, however, most regenerated plants were diploid. The purpose of this study was to generate somaclonal variability for use in this vegetatively propagated crop.

7. Onion

A procedure was developed that produced somatic embryos in field grown hybrids of *Allium fistulosum* × *A. cepa* (Lu et al., 1989). The project's goal was to induce chromosomal rearrangements and increase the low fertility hybrid. The increased fertility may make it possible to introduce genes from *A. fistulosum*. Callus was initiated from basal plate tissue or immature inflorescence on medium with 0.75 mg/l picloram and 2.0 mg/l BA. Embryos could be initiated when vitamins, glycine and 2.5 mg/l proline were added to the medium. Embryo development continued when picloram was lowered to 0.03 mg/l and 2iP was added at 0.32 mg/l. Plants were rooted on 1/2 strength BDS medium and climatized in plastic corsage boxes.

Two reports were recently published on production of haploid onions by ovule cultures (Campion and Alloni, 1990; Keller, 1990). In the first study flowers were precultured on MS medium with adenine sulfate at 0.3 mg/l. After 15–20 days ovules were cultured on MS medium with NAA at 1.0 mg/l and 2iP at 2.0 mg/l. Only one embryo was produced from each competent ovule with no callus intermediates. Embryo induction was low at 0.28%. The latter study produced ovule callus on medium with 2 mg/l IBA, 12.5 mg/l BA, and 3.5 mg/l GA_3. All plants regenerated from the callus were haploid.

A recent report by Dommisse et al. (1990) showed that onion was capable of being transformed by *Agrobacterium*. Twenty-five strains of *Agrobacter-*

Table 1. Transformation protocols for five vegetable species

Species	Explant	Pre-culture	*A. tumefaciens* strain	Cocultivation	Selection	Rooting shoots	References
Eggplant *Solanum melongena* L.	Younger leaves from 4 weeks old *in vitro* plants cut 10 × 15 mm	2 day on MS mod 2 mg/l glycine 1 mg/l zeatin 2% sucrose 0.2% gelrite	pGV3850 (Zambryski et al., 1983). Explants dipped in late log stage culture	2 days on MS mod 2 mg/l glycine 1 mg/l zeatin 2% sucrose 0.2% gelrite	MS mod 2 mg/l glycine 1 mg/l zeatin 2% sucrose 0.2% gelrite 500 mg/l cefotaxime 100 mg/l kanamycin Subculture every 3 weeks	1/2 B5 1% sucrose 0.3% gelrite 100 mg/l kanamycin	Rotino and Gleddie (1990)
Tomato *Lycopersicon esculentum* cv. UC82B	Cotyledons from 7–8 day old *in vitro* seedlings mid section cut 0.5 × 0.25 cm	1 day filter paper over a 2 day old tobacco feeder plate growing on MS salts 0.2 mg/l 2,4-D 0.1 mg/l kinetin 0.9 mg/l thiamine HCL 200 mg/l potassium acid phosphate Bactoagar 0.8%	LBA4404/587/85 Explants in a 30 min treatment at a concentration of 5×10^8 cfu/ml	2 days on filter paper over tobacco feeder growing on MS salts 0.2 mg/l 2,4-D 0.1 mg/l kinetin 0.9 mg/l thiamine HCL 200 mg/l potassium acid phosphate Bactoagar 0.8%	MS salts 2 mg/l zeatin Nitsch vitamins 100 mg/l myo-inositol 3% sucrose 0.8% Bactoagar 500 mg/l carbenicillin 50 mg/l kanamycin	MS salts 3% sucrose Nitsch vitamins 0.8% Bactoagar 50 mg/l kanamycin 500 mg/l carbenicillin	Fillatti et al. (1987)
Melon *Cucumis melo* L.	Cotyledons from 4–5 day *in vitro* seedlings cut on all sides	None	LBA4404 pCIB10 Rothstein et al., 1987 Explants in 10–30 min treatment at concentration of 10^{-7}–10^{-8}/ml	3 days on MS salts 5 µm IAA 5 µm BA 1 µm ABA 3% sucrose	MS salts 5 µm IAA 5 µm BA 1 µm ABA 3% sucrose 75 mg/l kanamycin 100 mg/l cefotaxime	MS salts 3% sucrose 100 mg/l cefotaxime 50 mg/l kanamycin	Fang and Grumet, 1990

Table 1. Continued

Species	Explant	Pre-culture	*A. tumefaciens* strain	Cocultivation	Selection	Rooting shoots	References
Carrot *Daucus carota* Danvers 126	10–15 mm hypocotyls from 7 day old *in vitro* seedlings	2 days on B5 medium 9 µm 2,4-D 8 g/l Bactoagar Dark culture	LBA4404 5 min dip in overnight culture	2–3 days on B5 medium 9 µm 2,4-D 5 g/l Bactoagar	B5 medium 9 µm 2,4-D 8 g/l Bactoagar 100 µg/ml kanamycin 400 µg/ml carbenicillin subculture every 2, 6 weeks Initiate suspensions in similar medium minus agar. Regenerate embryos by auxin removal	Torpedo stage embryos to 1/2 MS macro salts 1X micro salts 2 g/l sucrose 8 g/l glucose 0.4 mg/l thiamine 8 g/l Bactoagar 400 µg/ml carbenicillin 100 µg/ml kanamycin	Thomas et al. (1989)
Cucumber *Cucumis sativus* L. cv. Poinsett 76	5 mm^2 coytyledon sections from 3–5 day old *in vitro* seedlings	None	C582707 + pGA482 (An, 1986) Explants dipped in bacteria at 2 × 10^8 cells/ml	4 days on MS 20 mg/l 2,4-D 0.5 mg/l kinetin	MS salts 2.0 mg/l 2,4-D 0.5 mg/l kinetin 500 mg/l carbenicillin 100 mg/l kanamycin Culture for 5 weeks in dark Callus Regeneration on MS salts 1.0 mg/l NAA 0.5 mg/l kinetin 100 mg/l kanamycin 500 mg/l carbenicillin	MS salts 50 mg/l kanamycin	Chee (1990)

ium were used to inoculate onion bulbs by stabbing with needles dipped in *Agrobacterium*. Eleven *A. tumefaciens*, six *A. rhizogenes* and one *A. rubi* strain caused tumor growth on the onion bulbs. Tumors were analyzed for opine production and were found positive. Tumor free inoculation areas did not show opine production. The use of acetosyringone increased the rate of tumor formation but not the range of strain response.

8. Pea

Over the past few years advances in tissue culture of pea, *Pisum sativum* L., have been substantial. Propagation of apical or axillary shoot explants by axillary expression or adventitious shoot formation was reported by Griga et al. (1986). Shoot explants from 4 to 6 day old seedlings were cultured on MS medium plus 20 μM BA and 0.1 μM NAA. Shoot cultures maintained their regenerative ability over a 20 μmonth period. Plants analyzed were diploid. Jackson and Hobbs (1990) obtained similar results from cotyledonary node explants. Adventitious shoot formation from immature embryo explants was reported by Natali and Cavallini (1987). Embryos 3–4 mm long formed callus on MS medium with BA (3–0.05 mg/l) and NAA (2–3 mg/l). Transfer of the callus to similar medium resulted in shoot formation. Embryonic axis produced the most regenerable callus. Plants produced from the callus were 50% diploid. They noted that 2,4-D based medium was not useful in producing organized structures.

Somatic embryogenesis was possible on 2,4-D or picloram based medium using immature embryos as explants (Kysely et al., 1987). Best embryo development was from picloram based medium with further embryo development on 1.0 mgA BA followed by 0.05 mg/l NAA and 0.017 mg/l BA, kinetin and zeatin. Two-thirds of the plants tested were diploid. In a subsequent study, Kysely and Jacobsen (1990) noted that morphogenic callus was produced only from the embryonic axis. Embryo explant size of 3–9 mm on medium with 0.2–0.4 μM picloram gave best embryo formation.

Protoplast culture and regeneration of pea has made rapid progress. Isolation and division of protoplasts and subsequent callus formation were reported by Xu et al. (1985). They obtained high division rates (91%) when protoplasts were cultured in agarose beads two days after isolation.

Puonti-Kaerlas and Eriksson (1988) noted that protoplast derived callus regenerated shoots at a low frequency after 6–9 months in culture but formed no roots. Similar results were obtained by Lehminger-Mertens and Jacobsen (1989a) using bead culture. Shoots but no roots were observed after three months culture. Embryonic axes precultured without the cotyledons were found to be the best explant source. In a following study, Lehminger-Mertens and Jacobsen (1989b) were able to produce somatic embryos and plants from embryogenic axis derived protoplasts. Protoplast callus placed on picloram or 2,4-D at 4–15 mM produced embryos. Embryo maturation was best on

MS medium with casein hydrolysate and 0–5% mannitol. Germination was possible on 1.5–2.9 μM GA_3. Embryos developed into morphologically normal plants.

Further evidence for rapid advancement of pea tissue culture is the number of studies on transformation. Puonti-Kaerlas et al. (1989) examined the susceptibility of five pea cultivars to seven strains of *Agrobacterium tumefaciens*. Greenhouse grown plants, shoot cultures and embryo axes were inoculated and monitored for tumor formation. No cultivar differences were seen with wild strain A207 being most virulent. Strains with kanamycin selectable markers were used to transform *in vitro* seedling explants and protoplast mini calli. Callus capable of growing on 75 mg/l kanamycin was obtained from all tissue sources. Opine production and DNA analysis showed most of the callus was transformed but no shoot production was reported. Hussey et al. (1989) described tumor formation on shoot tips of *in vitro* cultures. Shoot apices and stem sections were inoculated with needles carrying *A. tumefaciens* wildtype C-58, ACH-5 or *A rhizogenes* 9402. Best reaction came from youngest stem pieces or apex tissue using strain 9402. No plants were reported. Shoot formation was described from transformed callus tissue of pea nodal explants (DeKathen and Jacobsen, 1990). Binary or cointegrate vectors GV2260 (p35S GUS INT) or GV385 APT worked equally well. Callus growth was on 0.1 mg/l BA and 0.1 mg/l picloram. Shoot formation was obtained on 1 mg/l NAA for one week followed by 4.5 mg/l BA and 0.02 mg/l NAA. Shoots were successfully transferred to the greenhouse but did not set seed. Plants but no seed were also reported from transformed pea cultures by Puonti-Kaerlas et al. (1990). Best response was from shoot cultures or epicotyls cocultured with *A. tumefaciens* strains containing a hygromycin resistant gene GV3101 (pGV2260:pGV1503). Best media sequence was 0.5 mg/l BA and 2,4-D for callus (1–2 months) followed by 1 mg/l BA and 0.5 mg/l NAA (1–2 months) with a single passage on 0.5 mg/l picloram. Final shoot development was on 5 mg/l BA, kinetin and 1 mg/l ABA. Plants were positive for transformation by DNA analysis.

9. Pepper

There are few papers dealing with successful pepper (*Capsicum annum*) regeneration systems that could lead to transformation. Phillips and Hubstenberger (1985) described shoot organogenesis from sweet and chili pepper types. Shoot organogenesis was best and longest from meristematic tissue of both varieties using MS media with glucose (3%) and IAA at 0.05–4 mg/l and 10–50 mg/l BA. Continuous light at 28 °C prolonged shoot formation for eight months. Whole plants were not described in this report. Sripichitt et al. (1987) regenerated red pepper from cotyledon explants of 12 day old seedlings. Best medium contained only BA at 3.0 mg/l. Whole plants were produced by rooting shoots on 0.5 mg/l IAA or NAA. This method was eventually used by these workers in attempts to isolate mutants by gamma radiation (Sripichitt et al., 1988). Cotyledon explants of the pepper cultivar Mathania gave best adventitious shoot formation in medium with BA at 5.0 mg/l (Agrawal et al. 1989). Plants were rooted in IBA or NAA at 0.1 mg/l and were found to be diploid by chromosome analysis. Histology showed that shoots formed directly from cotyledon tissue without a callus intermediate. Genotype response of 16 pepper varieties was tested by Ochoa-Alejo and Ireta-Moreno (1990). Approximately one third of the varieties were good adventitious shoot responders from one month old hypocotyl explants.

An anther culture method for a wide variety of pepper cultivars was developed by Morrison et al. (1986). Embryos and plants were produced from anthers pretreated for 100 hours at 40 °C then floated on liquid medium overlaying an agar medium with charcoal.

10. Tomato

The *in vitro* culture of tomato tissues and organs is highly advanced. This is illustrated by the recent lack of basic studies on tomato regeneration and the abundance of studies on transformation and protoplast culture.

A study by Branca et al. (1990) showed that the synthetic auxin BOA (1,2-benzisoxazole-3-acetic acid) and BLA (1,2-benzisothiazole-3-acetic acid) produced more shoots on cotyledon explants than IAA when used with 0.7 μm zeatin riboside.

Environmental factors for protoplast isolation were studied by Niedz and Sink (1988). They studied cold treatments of 12 and 24 hours and dark treatments of 0, 24, or 48 hours on donor seedlings grown *in vitro* or *in vivo*. Best results in terms of yield, viability, callus production and regeneration were from both tissue sources grown 48 hours in dark, 12 hours in cold. Gleddie et al. (1989) isolated protoplasts from stem cortex tissue of a tomato hybrid (*L. esculentum* × *S. lycopersicoides*). Minicalli were produced in 30 days with 60% regeneration on medium with zeatin or kinetin (2 mg/l) plus IAA (0.1 mg/l). Two tomato lines did not respond to the procedure. The advantage of the procedure was that whole plants did not require pretreatment and tissue manipulation was easier and quicker. Protoplast fusion between *L. esculentum Mill.* and *L. peruvianum Mill.* was demonstrated by San et al. (1990). PEG and electrofusion techniques were used resulting in 19 and 24 hybrids from the two methods respectively. Analysis of plant morphology and Rubisco small subunit showed the nuclear hybridity of the plants. Chloroplast DNA restriction patterns and CFl-ATPase markers showed only one parental type chloroplast in all hybrids with no preferential selection from either parent. Mitochondrial restriction analysis showed recombination of mtDNAs from both parents. A similar study by Ratushnyak et al. (1991) used gamma treated *L. peruvianum* var. *dentatum* Dun. protoplasts. Efficient selection of hybrids was possible due to the lack of division in the gamma treated protoplasts and the white colony marker from the albino tomato mutant (plastome mutant albina Pl-alb). Hybrid analysis showed that two types of plants were produced. Hybrid nucleus with a wildtype plastome and a cultivated tomato mitochondria or a cultivated tomato nucleus with a wildtype plastome. These results differ from the previous study probably due to the use of gamma treatment of one parent.

Derks et al. (1990) also described the regeneration of a cytoplasmic albino mutant of the cv Large Red Cherry. This could be used for selection of cytoplasmic fusions at the cell level as described in the previous study.

A large number of studies on tomato transformation have been presented over the past five years. Fillatti et al. (1987) described a particularly useful procedure. Middle cotyledon sections (0.5 × 0.25 cm) from 7–8 day old *in vitro* grown tomato variety UC82 were used as explant source. Tissue was incubated on a tobacco feeder plate for 24 hours. Sections were dipped in binary *Agrobacterium* LBA4404/587/85 (5×10^8 cfu/ml) with a NPT II gene and an *aroA* gene. After 30 min sections were blotted dry and placed back on the feeder for 48 hours. Sections were moved to a regeneration medium containing MS salts, 100 mg/l inositol, Nitsch vitamins, 2 mg/l zeatin and 0.8% agar. Carbenicillin (500 mg/l) was used to inhibit *Agrobacterium* and 100 mg/l kanamycin was used for selection. Shoots could be rooted on MS medium with Nitsch vitamins and 0.8% Bactoagar. They describe the production of over 100 transgenic plants that have the NPT II gene and express the *aroA* protein, which conveys tolerance to glyphosate.

A study by Necasek et al. (1988) used armed strains of *Agrobacterium* T37 to produce tumors on tomato. Regeneration was possible from the tumors with the resulting plants producing opines. The plants were male sterile which was one of the study's goals, however they were also reduced in female fertility.

Additional methods of transformation have also been reported for tomato. Toyoda et al. (1988) used microinjection to transform friable callus tissue obtained from tomato leaf explants. Nuclei were microinjected with a DNA fragment from plasmid pE2KX containing NPT II. Colonies formed on kanamycin medium and were shown to contain the NPT II gene. No plants were regenerated from the callus tissue. Transformation of tomato protoplasts was reported by Bellini et al. (1989). *L. peruvianum* was also transformed in this study using plasmid pABD1 containing NPT II. Plants were regenerated only from *L. peruvianum* and were shown to contain the foreign gene which could be transmitted to subsequent generations.

A general technique of callus transformation was reported by Komari (1989) using the binary plasmid pTOK119 in several *Agrobacterium* strains. Callus was suspended in liquid medium with the *Agrobacterium* and cocultured for 48 hours. Bacteria were then washed off and the callus was cultured on selection medium with 30 mg/l kanamycin. Strain A281 was best for transformation, producing callus tissue containing the NPT II gene. No plants were reported. Strain A281 was also found best for transformation when compared to strains A66 or A6 (Davis et al., 1991). They also noted cultivar differences with Oh7870 and Roma transforming better than UCD82b. The first true leaf was also as good as cotyledon tissue for transformation. However, shoots formed only from cotyledons of all three cultivars. Shoots were not analyzed for transformation.

11. Carrot

Carrot is by far the most atypical of the vegetable species discussed in this chapter. It is the one vegetable species that is being used as a model system to study plant cell differentiation. Carrot is well known for its potential of forming somatic embryos. This ability is exploited to study effects of exogenous compounds on embryogenesis as well as associated biochemistry.

In the area of exogenous compound effects, Litvay et al. (1985) noted that a medium developed for loblolly pine was superior to the standard wild carrot medium for growth and embryogenesis. When levels of CA^{2+} and Mg^{2+} in the carrot medium were increased to equal that of the pine medium, growth was inhibited. They suggested counteracting factors in the pine medium enabled the carrot to grow at higher levels of CA^{2+} and Mg^{2+}.

Roustan et al. (1989) showed that the addition of cobalt ($CoCl_2$) or nickel ($NiCl_2$) to carrot cell cultures increased the production of somatic embryos. This was correlated to the reduction in the amount of ethylene in the cultures. The addition of Ethephon, an ethylene producing compound, counteracted the effect of the metals. This suggested that in part the effect of the metal additions on improving embryogenesis was due to a reduction in ethylene action. Further studies by Roustan et al. (1990) showed that silver, $AgNo_3$, also increased embryogenesis. It was shown that $AgNO_3$ stimulated arginine decarboxylase, an important enzyme in polyamine biosynthesis. Ethylene, supplied as Ethephon, decreased its activity. Polyamine synthesis was known to be associated with embryogenesis and it was suggested that the positive effect of silver was due to its counteracting effect on ethylene.

Other element additions to medium were also found to increase embryogenesis. Jansen et al. (1990) showed that increasing calcium from 10^{-3} M to 10^{-2} M increased embryogenesis twofold. They also reported no changes in extracellular proteins with the different calcium levels.

The induction of somatic embryos can be induced by mechanisms other than auxin treatments. Kamada et al. (1989) showed that carrot apical meristems formed embryos when treated with 0.7 mM sucrose or 0.25–1.0 mM cadmium ion on a growth regulator free MS medium. Embryos formed on the surface of the explants when cultured on MS medium with 0.1 M sucrose. When seed were treated with a 6% sodium hypochlorite solution for one hour, embryos formed on abnormal seedlings. These embryos were fractionated and encapsulated in calcium alginate for the purpose of synthetic seed production. When the synthetic seed produced from the above three methods were compared to 2,4-D induced embryos, conversion of synthetic seed to plants was always higher on the non 2,4-D produced seed.

A compliment to the production of cloned seeds would be the storage of its progenitor callus. In a short communication by Kaimori and Takahashi (1989) a method was described for storing callus by desiccation. Callus was dried to about 3% water content using silica gel, and stored for four days. When dried callus was cultured, 90% of the calli recovered, grew and formed somatic embryos. This may be an easy alternative to cryopreservation.

The carrot cell culture system has been shown to be well adapted for use in the study of protoplast fusion as well as the transfer of genetic traits. Kanchanapoom and Boss (1986) selected lines of wild carrot suspensions that were fusion permissive. They were able to fuse protoplasts by the addition of calcium or osmotic stress in the absence of calcium. They are using the system to study the basic mechanism behind membrane fusion.

The transfer of genetic traits in fusion between two species was demonstrated by Dudits et al. (1987). They fused tobacco mesophyll protoplasts with methotrexate (MTX) and 5 methyltryptophan resistant lines of carrot. The carrot protoplasts were irradiated prior to fusion. Plants were regenerated from the fusion and some had morphology normal enough to allow seed set. The plants also had isozyme patterns and immunological patterns to demonstrate the enzyme for MTX came from carrot. Progeny from the somatic hybrids also showed segregation for the MTX phenotype.

The transfer of cytoplasmic genes was demonstrated by Ichikawa et al. (1987), Tanno-Suenaga et al. (1988) and Tanno-Snenaga and Imamura (1991). In their series of reports they first demonstrated the donor recipient method of cell fusion was possible in carrot and related carrot species. *D. capillifolius* was X-irradiated and fused to a 15 mM iodocetamide treated line of *D. carota*. Cybrid plants had unique mtDNA restriction patterns not present in either parent. Further work fused an x-rayed cytolasmic male sterile (CMS) carrot line to a fertile line. A total of 73 plants were regenerated with 26 analyzed in detail. Eleven plants had mtDNA patterns close to the CMS pattern with ten showing sterility. Another study fused an X-rayed CMS line to a genic brown anther male sterile. Seventeen cybrids were analyzed by hybridization with 12 heterologous mitochondrial gene probes. Six of the cybrids had patterns identical to the CMS parent with the remaining 11 showing novel patterns different from the parents. It should be mentioned that the nuclear background was a genic brown anther male sterile, however, all but one of the cybrids were petaloid CMS. This suggests an override of the brown anther phenotype with the petaloid phenotype. In the one brown anther type there must have been a loss of function in the CMS gene allowing the brown anther type to develop.

The carrot somatic embryogenic system has been used extensively to study the basic mechanisms behind embryogenesis in plants. Most approaches fall into the categories of genetic, biochemical and molecular.

In the areas of genetic investigations there has been an effort to develop temperature sensitive carrot cell lines which have a variety of morphological stage blocks initiated at a restrictive temperature of 33 °C (Schiavo et al., 1988; Schnall et al., 1988). These mutants can be used to compare gene expression at a variety of stages during embryogenesis.

The biochemical approaches to study embryogenesis are wide ranging. Timmers et al. (1989) used the fluorescent marker chlortetracycline (CTC) and fluphenazine fluorescence to study membrane-bound calcium and activated calmodulin. They observed that calcium was evenly distributed over the proembryonic mass. Calmodulin was more specific in its concentration. In developing embryos, calmodulin was located in the basal region of the embryos. This suggests that a polarity for embryogenesis is initiated in the proembryonic mass long before a recognizable embryo is initiated.

A study of lipid biochemistry was done by Dutta and Appelqvist (1989) during the development of carrot somatic embryos. They noted that the carrot family specific fatty acid petroselinic acid was weakly expressed in developing somatic embryos. This implies that although the morphological appearance of the somatic embryo resembles that of the zygotic, the gene expression at least with respect to lipid synthesis is not the same.

A number of studies utilized the presence of isozymes as markers for the identity of competent carrot cell lines. Chibbar et al. (1988) noted two esterase isozymes that differed in their presence in embryogenic and nonembryogenic carrot lines. Joersbo et al. (1989) noted similar results in an isoperoxidase enzyme.

Other proteins have been studied during embryogenesis. Cyr et al. (1987) studied tubulin and microtubule formation. They noted that during early embryogenesis there was no correlation with tubulin levels and production of microtubules. Later stages of development correlated well. Stacey et al. (1990) showed an arabinogalactan-protein epitope, recognized by a monoclonal antibody, to be associated with somatic embryogenesis. The protein could be a marker for development of tissue patterns and could be involved in positional distinctions between cells.

Another interesting study on the role of proteins in somatic embryogenesis was conducted by de Vries et al. (1988). They showed that a series of high molecular weight proteins appeared during somatic embryogenesis and were reduced or absent in the presence of 2,4-D. Cells grown in the presence of 2,4-D produced a different set of proteins. Non-competent cell lines did not produce the full compliment of proteins or had aberrant forms. Replacement of missing or aberrant proteins with normal proteins replaced embryogenic competence. This is a clear first step in identifying genes associated with embryogenesis.

Nucleic acid biochemistry has also been studied at length during the process of carrot somatic embryogenesis. Nomara and Komamine (1986a, b) showed the polarity of DNA synthesis of single isolated carrot cells was evident as early as the four-cell stage.

The methylation pattern of two carrot DNA areas, were shown to change during embryo development (Vergara et al., 1990). One of the two areas also changed transcriptionally during development. They speculated that this may be a method to identify developmentally regulated genes specifically associated with embryogenesis.

Transformation of carrot cells has been achieved by many investigators. Langridge et al. (1985) transformed carrot protoplasts with naked DNA from the Tiplasmid of the wild type pTiC68 *Agrobacterium tumefaciens*. Transformed cells were detected by hormone independent growth and regeneration of protoplasts. Nopaline synthesis in teratomas originating from embryos, and Southern hybridization with a DNA fragment of the ptiC58 T-DNA also proved transformation. The incorporation of DNA into carrot protoplast was successful by the use of PEG (Ballas et al., 1987). However, the expression was only transient as monitored by CAT enzyme activity from the pCaMVCAT plasmid introduced into the cells. Transformation of carrot cell suspensions was claimed by Zhi et al. (1987) using the C58 *Agrobacterium tumefaciens* strain. No opine synthesis was detected in the tissues or plants from the transformed cells and no molecular analysis was done on the tissue. Work by Scott and Draper (1987) showed transformation of suspension cells by a non-oncogenic Ti-plasmid vector (pGV3850 1103) carrying a kanamycin resistance gene. Proembryonic cell cultures were incubated with *Agrobacterium* and were shown to be transformed by the presence of NPT II, nopaline and the presence of T-DNA in extracted DNA. Plants regenerated from the callus tissues retained the same DNA organization and expression as seen in the parent callus. Transformation of callus tissues was accomplished by Wurtele and Bulka (1989). They used a number of C58 base vectors including a oncogenic and non oncogenic helper plasmid combined with a binary vector plasmid. Transformation was confirmed by Southern blot analysis. The number of integrations per transformation event were from 2 to 12 and maintained fidelity in plants regenerated from the callus.

A fourth method for carrot transformation was described by Thomas et al. (1989). They used hypocotyl sections of carrot seedlings. Hypocotyls were pretreated for two days prior to a 2–3 day incubation with *Agrobacterium* strains containing GUS genes. Hypocotyls were then transferred to kanamycin selection medium. After six weeks the callus was introduced into suspension and was able to undergo embryogenesis. Tissue was shown to be transformed by the presence of NPT II, GUS activity and by Southern analysis. It was mentioned that a genotype effect was observed in the ability to transform cotyledons.

12. Cucurbitaecae

12.1. *Melon*

A number of reports have described different types of regeneration as well as transformation of *Cucumis melo* by *Agrobacterium tumefaciens*.

A report by Kathal et al. (1988) described adventitious shoot formation from leaf explants. Seed were germinated *in vitro* and leaves were excised from 14, 21, 28, or 36 day old seedlings. Three size classes of leaves, 0.3–0.5, 0.6–0.9 and 1.0–1.2 cm, were excised from each age. Best medium for regeneration was BA and 2iP at 1 mM each. The most shoots were formated on 0.3–0.5 cm, 14 day old leaves. A study by Leshem (1989) showed that specific regions of melon cotyledons have higher regenerative ability. Shoots or roots could be induced on medium with BA 0.2 mg/l or IAA at 2.0 mg/l respectively. Organs were produced only at the basal part of the cotyledon or on the most basal end of transversely cut cotyledons. He speculated that the transport of endogenous growth regulators, specifically IAA, may cause localized shoot formation. The inhibition of shoot formation by the auxin transport inhibitor TIBA caused the speculation.

Regeneration from leaf and cotyledon explants was reported by Dirks and van Buggenum (1989). Fully expanded leaves and cotyledon or cotyledons from fresh seed were used as explant source. Best medium was MS with 1.0 mg/l BA. Regeneration was best from fresh cotyledons and leaf explants. Mature cotyledons gave a poor response. A genotype response was observed only from fresh cotyledon explants. Neidz et al. (1989) further investigated shoot regeneration from cotyledon explants. They noted best shoot formation on MS medium with 5 μM BA and IAA, although IAA may not be necessary. Four day old cotyledons were better than 7 or 18 day. A low light intensity, 5–30 μmol $M^{-2}S^{-1}$, was better than 60–3000 μmol $M^{-2}S^{-1}$. Temperature of 25–29 °C was better than 21°C. The use of ABA at 1–30 μM had a significant positive effect on shoot formation while GA_3 or thidiazuron did not. They also noted a genotype effect with ‘Hale's Best Jumbo’ being the best variety.

Punja et al. (1990) briefly mentioned the formation of shoots from cotyledon leaf and petiole explants of melon on media with 5.0 μM NAA, 2.5 mM BA, or 5.0 μM zeatin.

Somatic embryos were reported by Branchard and Chateau (1988) on medium with 4.52 mM 2,4-D and 0.44 mM BA. Embryos developed from cotyledon callus and could successfully be transferred to soil.

Regeneration of established callus tissue from melon cotyledons was reported by Moreno et al. (1985). Morphogenetic callus could be excised and maintained on medium with 6 mg/l kinetin and 0.5–3.0 mg/l IAA. Regenerative capacity could only be maintained if organized areas were subcultured every 2–3 weeks. Somatic embryos were also reported on cotyledon and hypocotyl derived callus on medium with 3–6 mg/l IAA. The embryos did not develop into plants.

Organogenic callus cultures were established by Kathal et al. (1986) from hypocotyls of seven day old seedlings. Callus was established on medium with 1 mg/l IAA and 0.5 mg/l kinetin. Established callus would regenerate shoots on a complex medium with BAP and 2iP 0.5 mg/l, ABA 0.03–0.13 mg/l, GA_3 0.04–0.17 mg/l and zeatin 0.01–0.11 mg/l.

Transformation of melon has been reported by a number of workers. Ben

Tahar and De Both (1988) were able to produce transformed callus and a few shoots containing the NPI II and GUS genes. Explants were cotyledons and induction medium continued 1.5 mg/l IAA and 6 mg/l kinetin. In a subsequent study, Ben Tahar et al. (1989) regenerated transformed plants. Two day old cotyledon explants were found to regenerate best on medium with 5 μM BA and 5 μM IAA. Regeneration was equal from all transversely sectioned cotyledons. Coculture was 2 days with *A. tumefaciens* containing NPT II and GUS genes in MS medium without growth regulators or antibiotics. Induction followed on the previously described medium with 50 mg/l kanamycin and 400 mg/l cefotaxime. Regenerated shoots were transferred to 3 μM BA with 50 mgA kanamycin and 400 mg/l cefotaxime. Rooting was in medium with 0.4% Gelrite.

The most recent transformation paper was by Fang and Grumet (1990). Cotylydons from 4–5 day old seedlings were cut on all edges for maximum wounding and dipped in a fresh actively growing culture of *A. tumefaciens* LBA4404 pCIB10 containing NPI II. Explants were cultured on medium by Neidz et al. (1989) for 1 to 7 days. After cocultivation, explants were washed of bacteria and incubated on similar medium with 75 mg/l kanamycin and 100 mg/l cefotaxime. Shoots were rooted in MS medium with 50 mg/l kanamycin. They noted important factors for most efficient transformation. Titer of bacteria was not important but good quality cultures gave best results. Explant incubation time in the bacteria did not make a difference up to 60 min, after which shoot response declined. Cocultivation time of three days was found best for shoot production. Tobacco nurse cultures did not seem to influence transformation frequencies.

12.2. *Cucumber*

In vitro manipulation of cucumber (*Cucumis sativus* L.) is well advanced. Reports are available showing regeneration from callus, suspension cultures and protoplasts. There are also several reports on transformation of cucumber plants.

Malepszy et al. (1985) gave a short general summary of optimal conditions for cucumber regeneration. They observed the initiation of a gel-like callus and embryoid callus from leaf explants cultured on medium with 2,4-D and BA. Initial cultures were kept in the dark for four weeks then transferred to light. Plants were formed when callus was transferred to hormone free medium. About 30–50% of the regenerated plants were abnormal.

A number of cucumber lines were tested for their regeneration ability by Trulson and Shahin (1986). Cotyledon or root tip explants were taken from five day old seedlings and placed on an MS medium with 1.0 mg/l 2,4-D, 1.0 mg/l NAA and 0.5 mg/l BA. Embryoid structures developed in 2–3 weeks from cotyledons and 3–4 weeks from roots. Leaf structures developed when embryos were transformed to MS medium with 10 mg/l NAA and

0.5 mg/l BA. Embryos fully developed in hormone free medium, sometimes requiring repeated subculture. Plants arising from the embryos looked normal. All 14 lines of cucumber formed somatic embryos and all but one formed plants.

The problems of developing normal appearing somatic embryos was addressed by Ziv and Gadasi (1986). Hypocotyls from eight day old seedlings were cultured in liquid shake flasks containing 5 μM 2,4-D, 1 μM BA and 0.1 mg/l casein hydrolysate. After 21–24 days callus formed and was transferred to medium without growth regulators for 25–30 days. Normal embryos were obtained by adding sucrose (0.1–0.3 m) mannitol (0.1 m), casein hydrolysate (0.01%) activated charcoal (0.5%) zeatin (1.0 μM), ABA (0.4 μM) or charcoal plus ABA together in the liquid medium. Most nodular callus was formed when ABA and charcoal were in the medium. Largest number of plantlets from embryos were developed when embryogenic callus was cultured on a two layer culture with charcoal in the bottom layer and ABA in the top. All plants had vitreous leaves. Plants were hardened prior to soil transfer by culturing for 6–8 days in 1/2 strength MS medium on filter paper bridges in a desiccation chamber containing anhydrous $CaSO_4$. The problem of abnormal embryo development was also addressed by Chee and Tricoli (1988). Embryogenic suspension cultures initiated from leaf callus on 5 μM 2,4,5-T and 4 μM BA were washed with MS medium with 0.5% charcoal with two additional washes in MS basal medium. Normal embryos and plants could be obtained by culture of washed cells in shaker flasks with minimal medium for 10 days followed by transfer to conversion medium (MS minimal). Aberrant embryos could also be rescued by excision of the shoot tip to a MS minimal medium.

Regeneration by organogenesis from callus tissue was reported by Kim et al. (1989). Ten cucumber cultivars were tested for callus formation and were all found to form callus equally on MS medium with 0.5 μM 2,4-D and 5 μM BA. Shoot induction was best on medium with 0.5 μM NAA and 5 μM BA. Shoot formation came from green islands of tissue after at least three monthly subcultures. Only three cultivars regenerated shoots and plants with all plants having tetraploid (4n = 28) chromosome numbers.

The genetics of regeneration was studied by Nadolska-Orczyk and Malepszy (1989). Using high, moderate and low regenerating cucumber lines, they determined that the two highest regenerating lines contained two pairs of dominant genes for regeneration.

Protoplast culture of cucumber has been reported by a number of researchers. Success to date has been regeneration of plants from various explants and genotypes, however, no work has been reported on fusion or transformation using protoplast systems. Orczyk and Malepszy (1985) isolated protoplasts from leaves of 2–4 week old *in vitro* plants. Best isolation was achieved using 0.1 M glycine in the enzyme isolation procedure. A plating efficiency of 21% was reported using a medium with 5 mg/l NAA and 3 mg/l 2iP in an agarose bead culture method, at a protoplast plating density of 1.1×10^5 per ml. The bead cultures needed to be started 7–14 days after protoplasts were initiated. Normal plants were regenerated from the protoplast cultures.

Protoplasts were isolated from cotyledons of cucumber by Jia et al. (1986). A division frequency of 46% was claimed in liquid medium with 0.5 mg/l 2,4-D and 1 mg/l kinetin. Callus colonies were formed on all 11 tested genotypes but subsequent regeneration was possible in only one. No mention of plant morphology was made. A later study by Colijn-Hooymans (1988) isolated protoplasts from both leaves and cotyledons of cucumber. They noted a 50% plating efficiency only when a agarose bead culture procedure was used. Liquid culture was inferior. Age of tissue was important for maximum protoplast isolation. The first fully developed leaves from 3 week old seedlings and cotyledons from 10-day old plants were best. Embryogenic callus was formed on 3 of the 4 genotypes tested using medium with 4 μM BA and 2,4-D. Removing the callus to hormone free medium resulted in development of embryo like structures. Both cotyledon and leaves produced the structures, however, normal plants could only be recovered from leaf derived protoplasts. Flow cytometric analysis of cotyledons and leaves show a much higher percentage of 4C and 8C DNA in cotyledon nuclei. This suggests a note of caution when using cotyledons as explant source. Trulson and Shahin (1986) used cotyledons as explant source and reported that 5–10% of protoplast derived minicalli formed embryo-like structures. However, less than 1% of the structures developed into mature embryos and only six plants were transferred to the greenhouse. They appeared normal but no cytogenetics were reported. The low level of embryo conversion to plants may be explained by ploidy problems in the original cotyledon explant tissue.

Transformation of cucumber has been reported using either *Agrobacterium rhizogenes* or *A. tumefaciens*. Trulson et al. (1986) used hypocotyl sections of cucumber to inoculate with *A. rhizogenes* strain A4 containing the resident R_1 plasmid and a vector derived from *A. tumefaciens* (pARC8) containing a NPT II selectable marker. Roots produced on the inoculated hypocotyls regenerated at a low frequency on an embryogenic medium containing MS salts, 5 μM 2,4-D, 5 μM NAA, 2 μM BA, and 25 mg/l kanamycin for selection. Embryos were matured on 5 μM NAA and 2 μM BA. Shoots were produced by transfer to hormone free medium. Plants were analyzed for the presence of the vector or Ri-plasmid. Three out of five plants had evidence of both DNAs and 2 had only vector DNA. NPT II negative plants regenerated on 25 mg/l kanamycin selection, however, 50 mg/l inhibited regeneration and caused abnormal plant formation. Mature plants were not normal with small stature and abscision of male flowers. This may be due to the presence of T-DNA from *A. rhizogenes*.

Transformation of cucumber cotyledons by *A. tumefaciens* was reported by Chee (1990). Cotyledons from 3–5 day old *in vitro* grown seedlings were inoculated with a C582707 strain of *A. tumefaciens* containing a binary plasmid pGA482 with NPT II as a selectable marker. After 4 days in the dark at 26 °C on medium with 2 mg/l 2,4-D and 0.5 mg/l kinetin, tissue was transferred to similar medium with 100 mg/l kanamycin and 500 mg/l carbenicillin. After 5 weeks embryogenic callus was placed on MS medium with 10 mg/l NAA and 0.5 mg/l kinetin. This caused maturation of embryos which were converted to plants by culture on hormone free medium. Using this procedure over 100 kanamycin resistant R_0 plants were produced all of which were claimed normal and had NPT II enzyme activity. Southern analysis showed the integration of the NPT II gene and heritability in the R_1 generation.

12.3. *Squash*

The regeneration of squash (*Cucurbita pepo* L.) was reported by Chee (1991). Shoot apices from 7 day old *in vitro* dark germinated seed were cut in half longitudinally and cultured in MS medium with 1.2 mg/l 2,4,5-T, 0.8 mg/l BA and 0.1 mg/l kinetin. Embryo conversion medium contained 0.05 mg/l NAA + 0.05 mg/l kinetin.

12.4. *Pumpkin*

Embryogenesis from 15 year old or newly established pumpkin (*Cucurbita pepo* L.) callus was described by Juretic' and Jelaska (1991). Callus lines that were induced and maintained on MS medium with IBA converted more embryos to plants. Best conversion medium was IAA at 11.3 μm or MS with 2% charcoal.

12.5. *Watermelon*

Shoot regeneration from 7–8 day old *in vitro* cotyledon or hypocotyl explants of watermelon (*Citrullus vulgaris* Schrad.) cv Melitopolski was reported by Srivastava et al. (1989). Best regeneration was from cotyledon explants with 4.5 μm BA. Root formation from induced shoots was accomplished on 0.5 mg/l NAA.

13. Progress in Vegetable Tissue Culture

There has been much progress in vegetable tissue culture systems. Regeneration is possible in most vegetables with best recent progress being made in peas. Species still problematic are beans and pepper. Studies on regeneration have shown the importance of ethylene in tissue culture. Recent studies showed that ethylene inhibitors such as $AgNO_3$, $AgSO_4$, ADA, DNP,

$CoCl_2$, and $NiCl_2$ increase carrot somatic embryogenesis and production of haploids in *Brassica*.

Basic knowledge in regeneration is creeping forward using the carrot embryogenic model system. Protein compliments have been shown to be associated with cell competence and may be useful in fishing out genes regulating embryogenesis. Gene expression has also been studied with specific clones being identified which are transcribed during the embryogenic process. (Aleith and Rehter, 1990).

The production of haploid onion by ovule culture is new and anther culture is becoming more routine and widespread throughout the *Brassica* species.

Protoplast culture has come to practical fruition by the success in moving CMS into fertile lines of carrots and *Brassica*. This will rapidly produce seed parents in carrot and non chlorotic *Brassica* CMS lines. Chlorosis is eliminated by combining the *Brassica* nuclear genome with their own chloroplasts plus sterility from radish.

The most significant progress in vegetable tissue culture has been the transformation of most species. Model systems are available for transformation in tomato and carrot. Transformation is frequently reported in eggplant, melon, and cucumber. Successful transformation has also been reported in pea and *Brassica*, however no seed was reported in pea. Species that are still not reported as transformed are pepper, bean and artichoke. Onion has also been shown to be .susceptible to some strains of *Agrobacterium* and may possibly be transformed in the near future.

14. Future Prognosis

The future of plant biotechnology and the role of plant tissue culture in its development has never been brighter. The direction of this research is toward introduction of foreign genes. The introduction of these genes requires a practical transformation protocol resulting in significant numbers of plants for testing. Established transformation protocols for five vegetable species are presented in Table 1. Our challenge for the next five years will be developing the pea, pepper, bean and *Brassica* transformation systems and to determine if onion can in reality be transformed using *Agrobacterium*. The more difficult species may employ microprojectile technology as an alternative to *Agrobacterium*.

Genes to be introduced into vegetable species will include virus coat protein genes to protect the cucurbits, tomato, pepper, and fungal resistant genes such as chitinase to protect almost all vegetable species. Anther specific genes producing male sterility would aid in the expensive hand-pollinated hybrid production as in melon, tomato and pepper. Sterile lines could also be produced in *Brassica* to circumvent self-incompatible systems. Herbicide and insect resistant genes such as BT could be targets for all vegetable species.

15. References

Agrawal, S., Chandra, N. and Kothari, S.L. (1989). Plant regeneration in tissue cultures of pepper (*Capsicum annuum* L. cv. *Mathania*). Plant Cell Tissue Org. Cult. 16:47–55.

Aleith, F. and Richter, G. (1990). Gene expression during induction of somatic embryogenesis in carrot cell suspensions. Planta 183:17–24.

An, G. (1986). Development of plant promoter expression vectors and their use for analysis of differential activity of nopaline synthase promoter in transformed tobacco cells. Plant Physiol. 81:86–91.

Arnison, P.G. and Keller, W.A. (1990). A survey of the anther culture response of *Brassica oleracea* L. cultivars grown under field conditions. Plant Breeding 104:125–133.

Ballas, N., Zakai, N. and Loyter, A. (1987). Transient expression of the plasmid pCaMVCAT in plant protoplasts following transformation with polyethyleneglycol. Experimental Cell Research 170:228–234.

Bauer, R. (1988). Protoplast manipulation in the genus *Brassica*. I. Plant regeneration from leaf and callus protoplasts of difference *Brassica oleracea* varieties. Biol. Zent. Bl. 107:533–539.

Bauer, R., Kysely, W. and Jacobsen, H-J (1986). Recent results on somatic embryogenesis in pea and bean. Genetic Manipulation in Plant Breeding. International Symposium 279–282.

Bellini, C, Chupeau, M-C., Guerche, P., Vastra G. and Chupeau, Y. (1989). Transformation of *Lycopersicon peruvianum* and *Lycopersicon esculentum* mesophyll protoplasts by electroporation. Plant Sci. 65:63–75.

Ben Tahar, S., DeBoth, M. and Thion, L., (1989). Regeneration and Transformation of Muskmelon. Cucurbitaceae 89: Evaluation and Enhancement of Cucurbit Germplasm, Charleston, South Carolina, Edited by C.E. Thomas, USDA, pp. 21–27.

Ben Tahar, S. and De Both, J.T.J. (1988). Introduction of Foreign Genes into Melo (*Cucumis Melo* L.) using *Agrobacterium tumefaciens*. Proc. *Cucurbitaceae* France.

Biddington, N.L. and Robinson, H.T. (1990). Variations in response to high temperature treatments in anther culture of brussels sprouts. Plant Cell Tissue Org. Cult. 22:48–54.

Biddington, N.L., Sutherland, R.A. and Robinson, H.T. (1988). Silver nitrate increases embryo production in anther culture of brussels sprouts. Annals of Botany 62:181–185.

Branca, C., Torelli, A. and Bassi, M. (1990). Effects of benzisoxazole and benzisothiazole on tomato plant regeneration *in vitro*. Plant Cell Tissue Org. Cult. 21:17–19.

Branchard, M. and Chateau, M. (1988). Obtention de plantes de melon (*Cucumus melo* L.) par embryogenese somatique. C.R. Acad. Sci. Paris 307:777–780.

Breuer, R., Kysely, W. and Jacoben, H.J. (1986). Recent results on somatic embryogenesis in pea and bean. Genetic Manipulation in Plant Breeding 279–281.

BV, N.Z. and van Buggenum, M. (1989). *In vitro* plant regeneration from leaf and cotyledon explants of *Cucumis melo* L. Plant Cell Rep. 7:626–627.

Campion, B. and Alloni, C. (1990). Induction of haploid plants in onion (*Allium cepa* L.) by *in vitro* culture of unpollinated ovules. Plant Cell Tissue Org. Cult. 20:1–6.

Chee, P.P. (1991). Somatic embryogenesis and plant regeneration of squash *Cucurbita pepo* L. cv. YC 60. Plant Cell Rep. 9:620–622.

Chee, P.P. (1990). Transformation of *Cucumis satiuus* tissue by *Agrobacterium tumefaciens* and the regeneration of transformed plants. Plant Cell Rep. 9:245–248.

Chee, P.P. and Tricoli, D.M. (1988). Somatic embryogenesis and plant regeneration from cell suspension cultures of *Cucumis sativus* L. Plant Cell Rep. 274–277.

Chi, Gek-Lan and Pau, Eng-Chong (1989). Ethylene inhibitors enhanced de novo shoot regeneration from cotyledons of *Brassica campestris* ssp. *chinensis* (Chinese cabbage) *in vitro*. Plant Sci. 64:243–250.

Chiang, M.S., Frechette, S., Kuo, C.G., Chong, C. and Delafield, S.J. (1985). Embryogenesis and haploid plant production from anther culture of cabbage (*Brassica oleracea* var. *capitata* L.). Can. J. Plant Sci. 65:1033–1037.

Chibbar, R.N., Shyluk, J., Georges, F., Mallard, C.S. and Constabel, F. (1988). Esterase isozymes as markers of somatic embryogenesis in cultured carrot cells. J. Plant Physiol. 133:367–370.

Colijn-Hooysmans, C.M., Bouwer, R., Orczyk, E. and Dons J.J.M. (1988). Plant regeneration from cucumber (*Cucumis sativus*) protoplasts. Plant Sci. 57:63–71.

Cyr, R.J., Bustos, M.M., Guiltinan, M.J. and Fosket, D.E. (1987). Developmental modulation of tubulin protein and mRNA levels during somatic embryogenesis in cultured carrot cells. Planta 171:365–376.

David, C. and Tempe, J. (1988). Genetic transformation of cauliflower (*Brassica oleracea* L. var. *botrytis*) by *Agrobacterium rhizogenes*. Plant Cell Rep. 7:88–91.

Davis, M.E., Lineberger, R.D. and Miller A.R. (1991). Effects of tomato cultivar, leaf age, and bacterial strain on transformation by *Agrobacterium tumefaciens*. Plant Cell Tissue Org. Cult. 24:115–121.

de Vries, S.C., Booij, H., Janssens, R., Vogels, R., Saris, L., LoSchiavo, F., Terzi, M. and Kammen, A.V. (1988). Carrot somatic embryogenesis depends on the phytohormone-controlled presence of correctly glycosylated extracellular proteins. Genes & Development 2:462–476.

DeKathen, A. and Jacobsen, H-J (1990). *Agrobacterium tumefaciens*-mediated transformation of *Pisum sativum* L. using binary and cointegrate vectors. Plant Cell Rep. 9:276–279.

Derks, F.H.M., Zelcer, A. and Colijn-Hooymans, C.M. (1990). Plant regeneration of mesophyll protoplasts from cytoplasmic albino mutant of *Lycopersicon esculentum* cv. large red cherry. Plant Cell Tissue Org. Cult. 21:211–216.

Dietert, M.F., Barron, S.A. and Yoder, O.C. (1982). Effects of genotypes on *in vitro* culture in the genus *Brassica*. Plant Sci. Lett. 26:233–240.

Dirks, R. and van Buggenum, M. (1989). *In vitro* plant regeneration from leaf and cotyledon explants of *Cucumis melo* L. Plant Cell Rep. 7:626–627.

Dommisse, E.M., Leung, D.W.M., Shaw, M.L. and Conner, A.J. (1990). Onion is a monocotyledonous host for *Agrobacterium*. Plant Sci. 69:249–257.

Dudits, D., Maroy, E., Praznovsky, T., Olah, Z., Gyorgyey, J. and Cella, R. (1987). Transfer of Resistance Traits From Carrot into Tobacco by Asymmetric Somatic Hybridization: Regeneration of Fertile Plants. Proc. Natl. Acad. Sci. 84:8434–8438.

Dutta, P.C. and Appelqvist, L.A. (1989). The effects of different cultural conditions on the accumulation of deport lipids notably petroselinic acid during somatic embryogenesis in *Daucus carota* L. Plant Sci. 64:167–177.

Fang, G. and Grumet, R., (1990). *Agrobacterium tumefaciens* mediated transformation and regeneration of muskmelon plants. Plant Cell Rep. 9:160–164.

Filippone, E. and Lurquin, P.F. (1989). Stable transformation of eggplant (*Solanum melongena* L.) by cocultivation of tissues with *Agrobacterium tumefaciens* carrying a binary plasmid vector. Plant Cell Rep. 8:370–373.

Fillatti, J.J., Kiser, J., Rose, R. and Comai, L. (1987). Efficient transfer of a glyphosate tolerance gene into tomato using a binary *Agrobacterium tumefaciens* vector. Bio/Technolog. 5:726–730.

Fobert, P.R. and Webb, D.T. (1987). Effects of polyamines, polyamine precursors, and polyamine biosynthetic inhibitors on somatic embryogenesis from eggplant (*Solanum melongena*) Cotyledons. Can. J. Bot. 66:1734–1742.

Franklin, C.I., Trieu, T.N., Gonzales, R.A. and Dixon, R.A. (1991). Plant regeneration from seedling explants of green bean (*Phaseolus vulgaris* L.) via organogenesis. Plant Cell Tissue Org. Cult. 24:199–206.

Fu, You-ying, Jia, Shi-rong and Lin, Yun, (1985). Plant regeneration from mesophyll protoplast culture of cabbage (*Brassica oleracea* var 'capitata'). Theor. Appl. Genet. 71:495–499.

Gleddie, S., Keller, W.A. and Poysa, V. (1989). Plant regeneration from stem cortex protopasts of a tomato hybrid. Plant Cell Rep. 8:21–24.

Gleddie, S., Keller, W.A. and Setteffield, G. (1986). Somatic embryogenesis and plant regeneration from cell suspension derived protoplasts of *Solanum melongena* (eggplant). Can. J. Bot. 64:355–361.

Griga, M., Tejklova, E., Novak, F.J. and Kubalakova, M. (1986). *In vitro* clonal propagation of *Pisum sativum* L. Plant Cell Tissue Org. Cult. 6:95–104.

Guri, A. and Sink, K.C. (1988). *Agrobacterium* transformation of eggplant. J. Plant Physiol. 133:52–55.

Horn, W., Jensen, C.J., Odenbach, W. and Schieder, O. (1986). Genetic Manipulation in Plant Breeding. Proceedings International Symposium. EUCARPIA, September 8–13, 1985, Berlin (West), Germany. Horn, W. Jensen, Odenbach, W. and Schieder, O. (eds.).

Hussey, G., Johnson, R.D. and Warren, S. (1989). Transformation of meristematic cells in the shoot apex of cultured pea shoots by *Agrobacterium tumefaciens* and *A. rhizogenes*. Protoplasm 148:101–105.

Ichikawa, H., Tanno-Suenaga, L. and Imamura, J. (1987). Selection of *Daucus* cybrids based on metabolic complementation between X-irradiated D. *capillifolius* and iodoacetamide-treated D. *carota* by somatic cell fusion. Theor. Appl. Genet. 74:746–752.

Jackson, J.A. and Hobbs, S.L.A. (1990). Rapid multiple shoot production from cotyledonary node explants of pea (*Pisum sativum* L.). *In vitro* Cell. Dev. Biol. 26:835–838.

Jacobsen, J. and Kysely, W. (1986). Induction of *in vitro*-regeneration via somatic embryogenesis in pea (*Pisum sativum*) and bean (*Phaseolus vulgaris*). Genetic Manipulation in Plant Breeding 445–448.

Jansen, M.A., Booij, H., Schel, J.H.N. and de Vries, S.C. (1990). Calcium increases the yield of somatic embryos in carrot embryogenic suspension cultures. Plant Cell Rep. 9:221–223.

Jia, S-R., Fu., Y-Y. and Lin, Y. (1986). Embryogenesis and plant regeneration from cotyledon protoplast culture of cucumber (*Cucumis sativus* L.). J. Plant Physiol. 124:393–398.

Joersbo, M., Andersen, J.M., Okkels, F.T. and Rajagopal, R. (1989). Isoperoxidases as markers of somatic embryogenesis in carrot cell suspension cultures. Physiologia Plantarum 76:10–16.

Jourdan, P.S., Earle, E.D. and Mutschler, M.A. (1990). Improved protoplast culture and stability of cytoplasmic traits in plants regenerated from leaf protoplasts of cauliflower (*Brassica oleracea* ssp. *botrytis*). Plant Cell Tissue Org. Cult. 21:227–236, 1990.

Jureti', B. and Jelaska, S. (1991). Plant development in long-term embryogenic callus lines of *Cucurbita pepo*. Plant Cell Rep. 9:623–626.

Kaimori, N., and Takahashi, N. (1989). Plant regeneration frorn dried callus of carrot (*Daucus carota* L.). Japan. J. Breed. 39:379–382.

Kamada, H., Kobayashi, K., Kiyosue, T. and Harada, H. (1989). Stress induced somatic embryogenesis in carrot and its application to synthetic seed production. *In vitro* Cell. Dev. Biol. 25:1163–1166.

Kameya, T., Kanzaki, H., Toki, S. and Abe, T. (1989). Transfer of radish (*Raphanus sativus* L.) chloroplasts into cabbage (*Brassica oleracea* L.) by protoplast fusion. Japan. J. Genet. 64:27–34.

Kanchanapoom, K. and Boss, W.F. (1986). Osmoregulation of fusogenic protoplast fusion. Biochimica et Biouhysica Acta 861:429–439.

Kathal, R., Bhatnagar, S.P. and Bhojawani, S.S. (1986). Regeneration of shoots from hypocotyl callus of *Cucumis melo* cv. Pusa Sharbati. J. Plant Phvsiol 126:59–62.

Kathal, R., Bhatnagar, S.P. and Bhojawani S.S. (1988). Regeneration of plants from leaf explants of Cucumis melo cv. Pusa Sharbati. Plant Cell Rep. 7:449–451.

Keller, J. (1990). Culture of unpollinated ovules, ovaries, and flower buds in some species of the genus *Allium* and haploid induction via gynogenesis in onion (*Allium cepa* L.). Euphytica 47:241–247.

Kim, S-G., Chang, J-R., Cha, H.C. and Lee K-W. (1988). Callus growth and plant regeneration in diverse cultivars of cucumber (*Cucumis sativus* L.). Plant Cell Tissue Org. Cult. 12:67–74.

Komari, T. (1989). Transformation of callus cultures of nine plant species mediated by *Agrobacterium*. Plant Sci. 60:223–229.

Kysely, W. and Jacobsen, H-J. (1990). Somatic embryogenesis from pea embryos and shoot apices. Plant Cell Tissue Org. Cult. 20:7–14.

Kysely, W., Myers, J.R., Lazzeri, P.A., Collins, G.B. and Jacobsen, H.J. (1987). Plant regeneration via somatic embryogenesis in pea (*Pisum satiuum* L.). Plant Cell Rep. 6:305–308.

Langridge, W.H.R., Bao, Li, B.J. and Szalay, A.A. (1985). Electric field mediated stable transformation of carrot protoplasts with naked DNA. Plant Cell Rep. 4:355–359.

Lehminger-Mertens, R. and Jacobsen, H-J. (1989a). Protoplast regeneration and organogenesis from pea protoplasts. *In vitro* Cellular & Developmental Biology 25:571–574.

Lehminger-Mertens, R. and Jacobsen, H-J. (1989b). Plant regeneration from pea protoplasts via somatic embryogenesis. Plant Cell Rep. 8:379–382.

Leshem, B. (1989). Polarity and responsive regions for regeneration in the cultured melon cotyledon. J. Plant Physiol. 135:237–239.

Lillo, C. and Hansen M. (1987). Anther culture of cabbage. Influence of growth temperature of donor plants and media composition on embryo yield and plant regeneration. Norwegian J. Agricul. Sci. 1:105–109.

Lillo, C. and Olsen, J.E. (1989). Growth and shoot formation in protoplast-derived calli of *Brassica oleracea* ssp. *acephala* and *ssp. capitata*. Plant Cell Tissue Org. Cult. 17:91–100.

Litvay, J.D., Verma, D.C. and Johnson, M.A. (1985). Influence of a loblolly pine (*Pinus taeda* L.). Culture medium and its components on growth and somatic embryogenesis ofthe wild carrot (*Daucus carota* L.). Plant Cell Rep. 4:325–328.

Lu, C.C., Currah, L. and Peffley, E.B. (1989). Somatic embryogenesis and plant regeneration in diploid *Allium fistulosum* X/*A. cepa* F_1 hybrid onions. Plant Cell Rep. 7:696–700.

Maggioni, L., Corti, C. and Fogher, C. (1989). Callus induction, ploidy level and plant regeneration in *in vitro* garlic (*Allium sativum* L.) cultures. J. Genet. and Breed. 43:251–254.

Malpathak, N.P. and David, S.B. (1986). Flavor formation in tissue cultures of garlic (*Allium sativum* L.). Plant Cell Rep. 5:446–447.

McClean, P., Chee, P., Held, B., Simental, J., Drong; R.F. and Slightom J. (1991). Susceptibility of dry bean (*Phaseolus vulgaris* L.) to *Agrobacterium* infection: transfornation of cotyledonary and hypocotyl tissues. Plant Cell Tissue Org. Cult. 24:131–138.

McClean, P. and Grafton, K.F. (1989). Regeneration of dry bean (*Phaseolus Vulgaris* L.) via organogenesis. Plant Sci. 60:117–122.

Moreno, V., Garcia-Sogo, M., Granell, I., Garcia-Sogo, B. and Roig, L.A. (1985). Plant regeneration From calli of melon (*Cucumis melo* L., cv 'Amarillo Oro'). Plant Cell Tissue Org. Cult. 5:139–146.

Morrison, R.A., Koning, R.E. and Evans, D.A. (1986). Anther culture of an interspecific hybrid of Capsicum. J. Plant. Physiol. 126:1–9.

Murata, M. and Orton, T.J. (1987). Callus initiation and regeneration capacities in *Brassica species*. Plant Cell Tissue and Org. Cult. 11:111–123.

Nadolska-Orczyk, W. and Malepszy S. (1985). *In vitro* culture of *Cucumis sativus* L. V. stabilizing effect of glycine on leaf protoplasts. Plant Cell Rep. 4:269–273.

Nadolska-Orczyk, A. and Malepszy, S. (1989). *In vitro* culture of *Cucumis sativus* L. 7. Genes controlling plant regeneration. Theor. Appl. Genet. 78:836–840.

Natali, L. and Cavallini, A. (1987). Regeneration of pea (*Pisum sativum* L.) plantlets by *in vitro* culture of immature embryos. Plant Breeding 99:172–176.

Necasek, J., Dusbabkova, J. and Pekarkova-Tronickova, E. (1988). Whole plants regeneration from crown galls of *Lycopersicon esculentum*. Biologia Plantarum (PRAHA) 30:1–8.

Niedz, R.P. and Sink, K.C. (1988). Multifactor analysis of environmental preconditioning of tomato seedlings on protoplast culture and development. J. Plant Physiol. 133:385–391.

Niedz, R.P., Smith, S.S., Dunbar, K.B., Stephens, C.T. and Murakishi, H.H. (1989). Factors influencing shoot regeneration from cotyledonary explants of *Cucumis melo*. Plant Cell Tissue Org. Cult. 18:313–319.

Nishio, T., Sato, T. and Takayanagi, K. (1987). Efficient plant regeneration from hypocotyl protoplasts in eggplant (*Solanum melongena* L. and *Solanum insanum* L.). Japan. J. Breed. 37:389–396.

Nishio, T., Yamagishi, H.J. and Takayanagi, K. (1987). Shoot regeneration capacity destined at an early stage of protoplast culture in cabbage. Japan. J. Breed. 37:22–28.

Nomura, K. and Komamine, A. (1986a). Molecular mechanisms of somatic embryogenesis. Oxford Surveys of Plant Molecular & Cell Biology 3:451–466.

Nomura, K. and Komamine, A. (1986b). Polarized DNA synthesis and cell division in cell clusters during somatic embryogenesis from single carrot cells. New Phytol. 104:25–32.

Ochoa-Alejo, N. and Ireta-Moreno, L. (1990). Cultivar differences in shoot-forming capacity of hypocotyl tissues of chili pepper (*Capsicum annuum* L.) cultured *in vitro*. Sci. Hort. 42:21–28.

Ockendon, D.J. (1986). Utilization of Anther Culture in Breeding Brussels Sprouts. Genetic Manipulation in Plant Breeding. Proceedings International Symposium. Horn, W., Jensen, C.J., Odenbach, W. and Schieder, O. (eds.), Organized by EUCARPIA (September 8–13, 1985), Berlin (West), Germany, pp. 265–272.

Ockendon, D.J. and Sutherland, R.A. (1987). Genetic and non-genetic factors affecting anther culture of brussels sprouts (*Brassica oleracea* var. *gemmifera*). Theor. Appl. Genet. 74:566–570.

Ockendon, D.J. (1985). Anther culture in brussels sprouts (*Brassica oleracea* var. *gemmifera*). II. Effect of genotype on embryo yields. Ann. Appl. Biol. 107:101–104.

Onisei, T., Ecaterina T., Tòth and Amariei, D., (1988). Callus induction and plant regeneration of *Cynara scolymus* L. Rev. Roum. Biol. Veget., Tome 33:115–119.

Phippen, C. and Ockendon, D.J. (1990). Genotype, plant, bud size and media factors affecting anther culture of cauliflowers (*Brassica oleracea* var. *botrytis*). Theor. Appl. Genet. 79:33–38.

Phillips, G.C. and Hubstenberger, J.F. (1987). Plant regeneration *in vitro* of selected *Allium* species and interspecific hybrids. Hort. Sci. 22:124–125.

Phillips, G.C. and Hubstenberger, J.F. (1985). Organogenesis in pepper tissue cultures. Plant Cell Tissue Org. Cult. 4:261–269.

Pua, E.C. (1987). Plant regeneration from stem-derived protoplasts or *Brassica* Alboglabra Bailey. Plant Sci. 50:153–160.

Punja, Z.K., Abbas, N., Sarmento, G.G. and Tang, F.A. (1990). Regeneration of *Cucumis sativus* var. Hard*wickii* C. *melo*, and C. *metuliferus* from explants through somatic embryogenesis and organogenesis. Plant Cell Tissue Org. Cult. 21:93–102.

Puonti-Kaerlas, J., Eriksson and Engstrom, P. (1990). Production of transgenic pea (*Pisum sativum* L.) plants by *Agrobacterium tumefaciens*-mediated gene transfer. Theor. Appl. Genet. 80:246–252.

Puonti-Kaerlas, J., Stabel, P. and Eriksson, T. (1989). Transformation of pea (*Pisum sativum* L.) by *Agrobacterium tumefaciens*. Plant Cell Rep. 8:321–324.

Puonti-Kaerlas, J. and Eriksson, T. (1988). Improved protoplast culture and regeneration of shoots in pea (*Pisum sativum* L.). Plant Cell Rep. 7:242–245.

Ratushnyak, Y.I., Latypov, S.A., Samoylov, A.M, Piven, N.M. and Gleba, Y.Y. (1991). Introgressive hybridization of tomatoes by 'gamma-fusion' of *Lycopersicon esculentum* Mill. and *Lycopersicon peruvianum* var. *Dentatum* dun. protoplasts. Plant Sci. 73:65–78.

Reyes, F., Martinez, M.J. and Vazquez, C. (1988). Maceration and protoplast production from tomato tissues with the enzymic complex from autolyzed cultures of *Alternaria alternata*. Microbiological 11:63–68.

Reynolds, J.F. (1986). Regeneration in Vegetable Species. In: Cell Culture and Somatic Cell Genetics of Plants, Vol. 3. Vasil, I.K. (ed.). Academic Press, Inc. Orlando, pp. 151–178.

Rotino, G.L. and Gleddie, S. (1990). Transformation of eggplant (*Solanum melongena* L.) using a binary *Agrobacterium tumefaciens* vector. Plant Cell Rep. 9:26–29.

Rothstein, S.J., Lahners, K.N., Lotstein, R.J., Carozzi, N.B., Jayne, S.M. and Rice, D.A. (1987). Promoter cassettes, antibiotic-resistance genes, and vectors for plant transformation. Gene 53:153–161.

Roulund, N., Hansted, L., Andersen S.B. and Farestveit, B. (1990). Effect of genotype, environment and carbohydrate on anther culture response in head cabbage (*Brassica oleracea* L. var. *capitata* (L.) Alef.). Euphytica 49:237–242.

Roustan, J.P., Latche, A. and Fallot, J. (1989). Stimulation of *Daucus carota* somatic embryogenesis by inhibitors of ethylene synthesis: Cobalt and nickel. Plant Cell Rep. 8:182–185.

Roustan, J.P., Latche, A. and Fallot, J. (1990). Control of carrot somatic embryogenesis by $AgNo_3$, an inhibitor of ethylene action: Effect on arginine decarboxylase activity. Plant Sci. 67:89–96.

San, L.H., Vedel, F., Sihachakr, D. and Remy, R. (1990). Morphological and molecular characterization of fertile tetraploid somatic hybrids produced by protoplast electrofusion with PEG-induced fusion between *Lycopersicon esculentum* Mill. and *Lycopersicon peruvianum* Mill. Mol. Gen. Genet. 221:17–26.

Sato, T., Nishio, T. and Hirai, M. (1989). Plant regeneration from isolated microspore cultures of Chinese cabbage (*Brassica campestris* ssp. *pekinensis*). Plant Cell Rep. 8:486–488.

Sato, T., Nishio, T. and Hirai, M. (1989). Varietal differences in embryogenic ability in anther culture of Chinese cabbage (*Brassica campestris* ssp. *pekinesis*). Japan. J. Breed. 39:149–157.

Saunders, J.W., Hosfield, G.L. and Levi, A. (1986). Morphogenetic effects of 2,4-Dichlorophenoxyacetic acid on pinto bean (*Phaseolus vulgaris* L.) leaf explants *in vitro*. Plant Cell Rep. 6:46–49.

Saxena, P.K., Gill, R. and Rashid A. (1987). Optimal conditions for plant regeneration from mesophyll protoplasts of eggplant (*Solanum melongena* L.). Sci. Hort. 31:186–194.

Schiavo, F.L., Giuliano, G. and Sung, Z.R. (1988). Characterization of a temperature-sensitive carrot cell mutant impaired in somatic embryogeneis. Plant Sci. 54:157–164.

Schnall, J.A., Cooke, T.J. and Cress, D.E. (1988). Genetic analysis of somatic embryogenesis in carrot cell culture: Initial characterization of six classes of temperature-sensitive variants. Developmental Genetics 9:49–67.

Scott, R.J. and Draper, J. (1987). Transformation of carrot tissues derived from proembryogenic suspension cells: A useful model system for gene expression studies in plants. Plant Mol. Biol. 8:265–274.

Sihachakr, D. and Ducreux G. (1987). Cultural behavior of protoplasts from different organs of eggplant (*Solanum melongena* L.), and plant regeneration. Plant Cell Tissue Org. Cult. 11:179–188.

Singh, S. and Chandra, N. (1983). Morphogenesis and plantlet formation in callus and suspension cultures of cauliflowers (*Brassica oleracea* var. *botrytis*). Beitr. Biol. Pflanzen 60:191–198.

Sripichitt, P., Nawata, E. and Shigenaga, S. (1988). The effects of exposure dose and dose rate of gamma radiation on *in vitro* shoot-forming capacity of cotyledon explants in red pepper (*Capsicum annuum* 1. cv. Yatsufusa). Japan. J. Breed. 38:27–34.

Sripichitt, P., Nawata, E. and Shigenaga, S., (1987). *In vitro* shoot-forming capacity of cotyledon explants in red pepper (*Capsicum annuum* L. cv. Yatsufusa). Japan. J. Breed. 37:133–142.

Srivastava, D.R., Andrianov, V.M. and Piruzian, E.S. (1989). Tissue culture and plant regeneration of watermelon (*Citrullus vulgaris* Schrad. p. Melitopolski). Plant Cell Rep. 8:300–302.

Stacey, N.J., Roberts, K. and Knox, J.P. (1990). Patterns of expression of the JIM4 arabinogalactan-protein epitope in cell cultures and during somatic embryogenesis in *Daucus carota* L. Plants 180:285–292.

Taguchi, T. and Kameya, T. (1986). Production of somatic hybrid plants between cabbage and Chinese cabbage through protoplast fusion. Japan. J. Breed. 36:185–189.

Tanno-Suenaga, L. and Imamura, J. (1991). DNA hybridization analysis of mitochondrial genomes of carrot cybrids produced by donor-recipient protoplast fusion. Plant Sci. 73:79–86.

Tanno-Suenaga, L., Ichikawa, H. and Imamura, J. (1988). Transfer ofthe CMS trait in *Daucus carota* L. by donor-recipient protoplast fusion. Theor. Appl. Genet. 76:865–860.

Thomas, J.C., Guiltinan, M.J., Bustos, S., Thomas, T. and Nessler, C. (1989). Carrot (*Daucus carota*) hypocotyl transformation using *Agrobacterium tumefaciens*. Plant Cell Rep. 8:354–357.

Timmers, A.C.J., de Vries, S.C. and Schel, J.H.N. (1989). Distribution of membrane-bound calcium and activated calmodulin during somatic embryogenesis of carrot (*Daucus carota* L.). Protoplasma 153:24–29.

Toyoda, H., Matsuda, Y., Utsumi, R. and Ouchi, S. (1988). Intranuclear microinjection for transformation of tomato callus cells. Plant Cell Rep. 7:293–296.

Trulson, A.J. and Shahin, E.A. (1986). *In vitro* plant regeneration in the genus *Cucumis*. Plant Sci. 47:35–43.

Trulson, A.J., Simpson, R.B. and Shahin, E.A. (1986). Transformation of cucumber (*Cucumis sativus* L.) plants with *Agrobacterium rhizogenes*. Theo. Appl. Genet. 73:11–15.

Vergara, R., Verde, F., Pitto, L., Schiavo, F.L. and Terzi, M. (1990). Reversible variations in the methylation pattern of carrot DNA during somatic embryogenesis. Plant Cell Rep. 8:697–700.

Wong, K.W. and Loh, C.S., (1988). *In vitro* regeneration of plantlets from root segments of *Brassica alboglabra* Bailey. Sci. Hort. 35:7–14.

Wurtele, E.S. and Bulka, K. (1989). A simple, efficient method for the *Agrobacterium*-mediated transformation of carrot callus cells. Plant Sci. 61:253–262.

Xu, Z., Davey, M.R. and Cocking, E.C. (1985). Root protoplast isolation and culture in higher plants. Scientia Sinica 4:386–393.

Zambryski, P., Joos, H., Genetello, C., Leemans, J., Van-Montagu, M. and Schell, J. (1983). TI plasmid vector for the introduction of DNA into plant cells without alteration of their normal regeneration capacity. EMBO 2:2143–2150.

Zhi, L., Mengkang, F. and Cheng, C. (Tsui C.) (1987). *In vitro* transformation of suspension cultured cells crom *Daucus carota* by *Agrobacterium tumefaciens* strain C58. Scientia Sinica 11:1144–1151.

Ziv, M. and Gadasi, G. (1986). Enhanced embryogenesis and plant regeneration from cucumber (*Cucumis sativus* L.) callus by activated charcoal in solid/liquid doublelayer cultures. Plant Sci. 47:115–122.

15. *In vitro* Culture of Potato

M.G.K. Jones

Contents

1. Introduction

The cultivated potato (*Solanum tuberosum*) is a crop of considerable importance, ranking fourth in world production after wheat, rice and maize. It is thus the most important dicotyledonous crop species, and the most important tuber crop. There are seven cultivated species of potato, which range in chromosome numbers from diploid (24 chromosomes) to triploid, tetraploid and pentaploid. *S. tuberosum* subspecies *tuberosum* is normally regarded as the cultivated potato, and is the most widely grown; cultivation of subspecies *andigena* is restricted to the Andes region of S. America. In addition to the seven cultivated species, there are documented a further 228 wild species. Thus there is a wide gene pool of genetic diversity (Hawkes, 1990).

Potatoes evolved in South America, and wild species are found from south-west USA, through central and S. America, in habitats ranging from mountains to semi-desert conditions. Although grown for food in South America for many millennia, it is generally accepted that the potato was brought to Europe, to Spain in about 1570, and to England in about 1590. From these introductions, it rapidly spread as a food crop throughout Eur-

Indra K. Vasil and Trevor A. Thorpe (eds.), Plant Cell and Tissue Culture, 363–378.

ope, to the United States, India, China and elsewhere, and is now grown in 79% of the countries of the world. The spread of the potato crop presents a fascinating study, as documented by Salaman (1949/1985), in the 'History and Social Influence of the Potato'. Perhaps the best known historical aspect was the Irish Potato Famine of 1845–47, when the entire Irish potato crop was decimated by the late blight fungus (*Phytophthora infestans*).

One consequence of the transport of tubers and associated soil, was that pests and diseases that had evolved with the potato in S. America, also spread to potato growing regions world wide. As an example, the dissemination of potato cyst-nematodes (*Globodera* spp.) provides a classic epidemiological study (Jones and Jones, 1984).

The cultivated tetraploid potato, which is normally propagated vegetatively, is highly heterozygous, and from a genetical point of view is difficult to breed. This is compounded by partial or complete sterility of some varieties. It is thus a prime candidate for *in vitro* manipulation and, as we shall see, it is fortunate that in general potato is amongst the most amenable crop plants in its response to *in vitro* culture.

2. *In vitro* Approaches for Potato

Potato has been subjected to a wide range of *in vitro* manipulations. These can be divided into two types: those in which genetic stability should be maintained, and those in which the aim is to achieve some kind of genetic modification.

2.1. *In vitro Approaches Requiring Genetic Stability*

The following *in vitro* approaches require genetic stability:

- virus elimination
- micropropagation
- germplasm storage/transport
- production of microtubers/minitubers
- ploidy manipulations.

Because potatoes are normally propagated vegetatively, they can pick up a burden of virus infections, which are frequently latent, but nevertheless significantly depress yield. All material kept in culture, and also stock plants used to produce seed tubers commercially, should ideally be free from such viruses. This can be achieved by culture of very small meristems (apical dome and one or two leaf primordia, size 0.75–1.5 mm), which contain few or no virus particles. Meristem culture can be combined with heat or chemical treatments, after which whole plants are regenerated (Wang and Hu, 1985; see also Chapter 3, pp. 37–40). These must be checked for presence or absence of viruses by standard techniques.

Using virus-free stock, potato is routinely propagated *in vitro* by serial subculture of stem nodes. From each axillary bud a new shoot grows, producing about 6 further nodes in about 4 weeks. Micropropagation is used to bulk up new varieties or breeding lines, for germplasm storage and transport, and for production of microtubers and minitubers. Molecular analysis of potatoes propagated and stored *in vitro* has indicated that micropropagation gives genetically stable plants (Potter and Jones, 1991). Most germplasm collections, such as that at the International Potato Centre (IPC) have been transferred to *in vitro* culture. The advantages over annual field growth are in savings in manpower, space, freedom from pests and diseases, the vagaries of the weather and so on. Storage of shoot cultures at reduced temperature, or in the presence of mannitol, slows growth so that only an annual subculture may be required. International transport of potato germplasm is greatly facilitated using certified virus-free shoot cultures *in vitro*, or microtubers (see below), allowing savings of transport costs and in meeting International Phytosanitary requirements.

There is considerable interest in the direct use of 'microtubers' produced *in vitro* (Gardner and Blake, 1989), and in 'minitubers'. Microtubers are produced in place of axillary buds when potato shoot cultures are grown in the presence of high levels of sucrose, whereas minitubers are produced on stolons of micropropagated plants transferred to soil and grown at high density in greenhouses. The use of microtubers and minitubers should reduce the time taken to produce seed tubers, reduce the number of field generations required, and so result in higher quality seed tubers. They can be directly sown into the soil using modified maize drills, and their small sizes ease storage and transport requirements.

Commercial potato is tetraploid (2n = 4x = 48 chromosomes) and heterozygous, and there are advantages for breeding in simplifying the genetics by reducing the ploidy level. This can be achieved either by crossing with *S. phureja* (in which *S. phureja* chromosomes are eliminated), or by anther culture to regenerate plants from microspores. Anther culture is genotype dependent, and crossing with *S. phureja* is perhaps the easiest approach to obtain the dihaploid (24 chromosomes) level. Anther culture of dihaploids has been used to extract monohaploids (12 chromosomes) or spontaneously doubled monohaploids (24 chromosomes, homozygous). Doubling up of the ploidy frequently occurs after culture, and tetraploids can be obtained from monohaploids or dihaploids simply by regenerating plants from leaf discs or other explants (Evans et al., 1985; Wheeler et al., 1985).

2.2. *In vitro Approaches to Genetic Manipulation of Potato*

The following approaches can lead to genetic changes resulting from *in vitro* culture itself, or from manipulations *in vitro*:

– regeneration of plants via a callus phase (somaclonal variation)

– cell suspension culture
– regeneration of plants from protoplasts (Section 2.3)
– somatic hybridisation (Section 2.4)
– cybrid production (Section 2.4)
– direct gene transfer into protoplasts (Section 3)
– *Agrobacterium* transformation (Section 4).

Potato plants can be regenerated from explants (leaves, stems, rachis, petioles, tubers) following a short callus phase, and also from protoplasts (Section 2.3). The response is genotype-dependent, with cultivars such as Desiree being highly responsive. Interest in plants regenerated either directly from explants, or from protoplasts, was initially highlighted by the work of Shepard and co-workers (Shepard et al., 1980), who recorded in detail that plants regenerated in this way were not all identical. This variation, subsequently termed 'somaclonal variation' (Larkin and Scowcroft, 1981), was manifested in potato by potentially useful changes in agronomic characters. (This aspect is covered in more detail in the chapter 6 by Karp in this volume, and includes discussion of somaclonal variation of potato: only most relevant aspects are discussed here). Results reported included variants with resistance to early blight and late blight (caused by *Alternaria solani* and *Phytophthora infestans*, respectively, Shepard et al., 1980; Secor and Shepard, 1981); to common scab (*Streptomyces scabies*), and changes in tuber colour and flower colour (Evans et al., 1986; Wheeler et al., 1985). Further studies of the mechanisms that underlie somaclonal variation in potato have shown a number of contributory events. These include: changes in chromosome number and structure, deletions, amplifications, mutations and changes in methylation patterns (Karp et al., 1982; Fish and Karp, 1986; Landsmann and Uhrig, 1985; Potter and Jones, 1991). Plants regenerated from tetraploid potato protoplasts are more variable than those regenerated directly from explants, and this is a result of more extensive chromosomal changes. The percentage of aneuploids in protoplast-derived plants can be 30–50% (with aneuploids around the tetraploid level of 48 chromosomes, and with higher chromosome numbers apparently derived from instability of tissues at the octaploid level), whereas leaf disc-regenerants tend to be 5–10% aneuploid. When starting tissues with lower ploidy levels are used, there is a tendency for doubling of chromosomes to occur, with reduced incidence of aneuploidy. Two exceptions should be noted. Potato cell suspension cultures rapidly accumulate cytological changes, and are thus not usually a good source of protoplasts for genetic manipulation studies, and plants regenerated from tuber protoplasts are only 11% tetraploid, with 36% octaploid and 53% aneuploid (75–97 chromosomes) (Jones et al., 1989a).

Unfortunately, improvement of potato via somaclonal variation has not had the direct impact that was originally forecast, at least for the introduction of novel traits not present in the original plants. Nevertheless, there are other potentially useful applications of somaclonal variation. A good example is the re-assortment of genes already present in existing material, in order

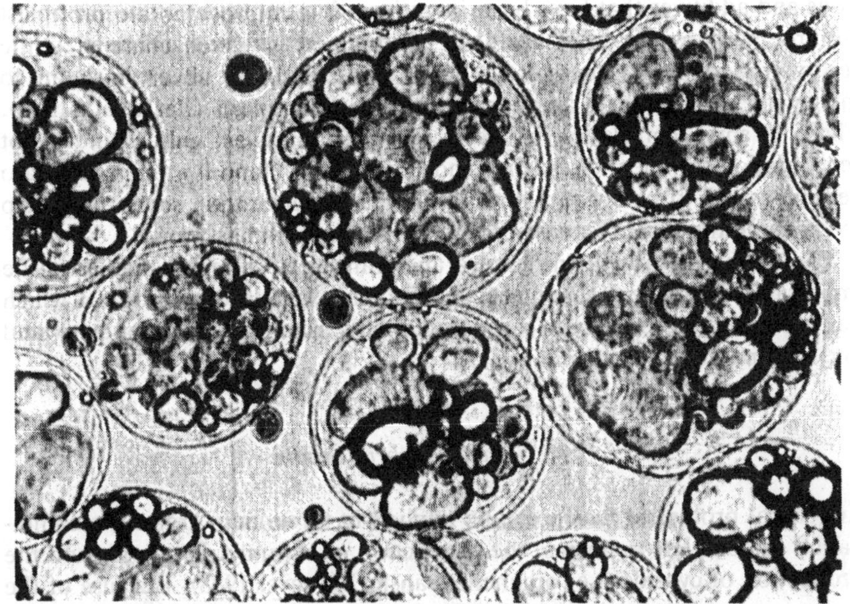

Figure 1. Potato tuber protoplasts; note the prominent starch grains.

to accelerate introgression of desirable traits from wild species into useful germplasm. Lentini et al. (1990) have used this approach to overcome restricted recombination between interspecific sexual hybrids of potato and the insect resistant *S. berthaultii*. By regenerating plants via callus, two variants with stable improvements in yield and insect resistance were obtained. These characters were heritable in backcrosses to potato and are currently being incorporated into breeding lines.

2.3. *Potato Protoplasts*: *Isolation*, *Culture*, *Regeneration*

Protoplasts can be isolated and plants regenerated from potato and many related *Solanum* species, from different tissue sources (leaf, stem, roots, tubers – Fig. 1, cell suspension cultures), from different genotypes and ploidy levels (e.g. Haberlach et al., 1985; Foulger and Jones, 1986; Jones et al., 1989). The ease of culture *in vitro* is genotype-dependent, in some cases relatively standard media can be employed, but in other cases more groundwork must be carried out to modify such media to suit a particular genotype (e.g. Foulger and Jones, 1986). In addition to culture media used, other important factors for good potato protoplasts are: the physiological and growth conditions of source tissues, the wall-degrading enzymes used, the osmolarity of the solutions, and protoplast density in culture. Amongst variations and additives that have been investigated to improve potato protoplast culture, perhaps the most useful is addition of activated charcoal to the reservoir medium (Carlberg et al., 1983) and addition of silver thiosulfate to media for shoot cultures and perhaps also to protoplasts (Perl et al., 1988). There is some correlation between ease of protoplast culture and shoot regeneration from leaf discs (Fish and Jones, 1988), and if a range of potato genotypes are to be studied, the simple leaf regeneration screen may help in choosing the most responsive genotypes for protoplast work.

The timing of subsequent culture and transfer to different media can be important for efficient shoot production. An example of a protocol which has been used successfully to regenerate plants from protoplasts of cultivated and wild potatoes is given in Fig. 2.

2.4. *Somatic Hybridization*

In general protoplast fusion can be applied in three main areas: the combination of complete genomes, the production of asymmetric hybrids and the production of cybrids (i.e. organelle transfer) (Jones, 1988). Because of the wide range of potentially useful germplasm in wild *Solanum* species (Table 1) many of which are sexually incompatible with potato, the generation of somatic hybrids is an attractive strategy for introduction of useful traits into potato germplasm. The transfer of a useful character by somatic hybridization does not require detailed knowledge of its molecular basis, and polygenic characters may also be transferred. Potato and other *Solanum* protoplasts are amenable both to chemical and electrical fusion (Austin et al., 1985; Fish et al., 1988a), and protoplasts of *S. brevidens* have been used as a model *Solanum* species to elucidate electrical parameters and principles of electrofusion (Tempelaar and Jones, 1985a, b; Tempelaar et al., 1987). With high fusion frequencies there is not necessarily any need for selection systems immediately after fusion. For example, 2.6% of shoots regenerated after chemical fusion and 12.6% after electrofusion were hybrids between potato and *S. brevidens* (Fish et al., 1987, 1988a).

The most widely studied somatic hybrids of potato are those between potato and *S. brevidens* (e.g., Barsby et al., 1984; Austin et al., 1985, 1986, 1988; Fish et al., 1987, 1988a, b; Pehu et al., 1990a; Gibson et al., 1990; Valkonnen et al., 1991). *S. brevidens* was originally identified as resistant to PLRV (Jones, 1979), but in addition, it has since been shown to confer resistance to the other two major virus diseases of potato, PVX and PVY (Gibson et al., 1988, 1990). Perhaps more surprisingly, since *S. brevidens* is non-tuber bearing, is the discovery that somatic hybrids between potato and *S. brevidens* possess strong resistance to bacterial tuber rot caused by *Erwinia* (Helgeson et al., 1986; Austin et al., 1988). Hybrids between potato and *S. brevidens* can be female fertile, although frequently they exhibit reduced or no male fertility (Ehlenfeldt and Helgeson, 1987). Analysis of a population

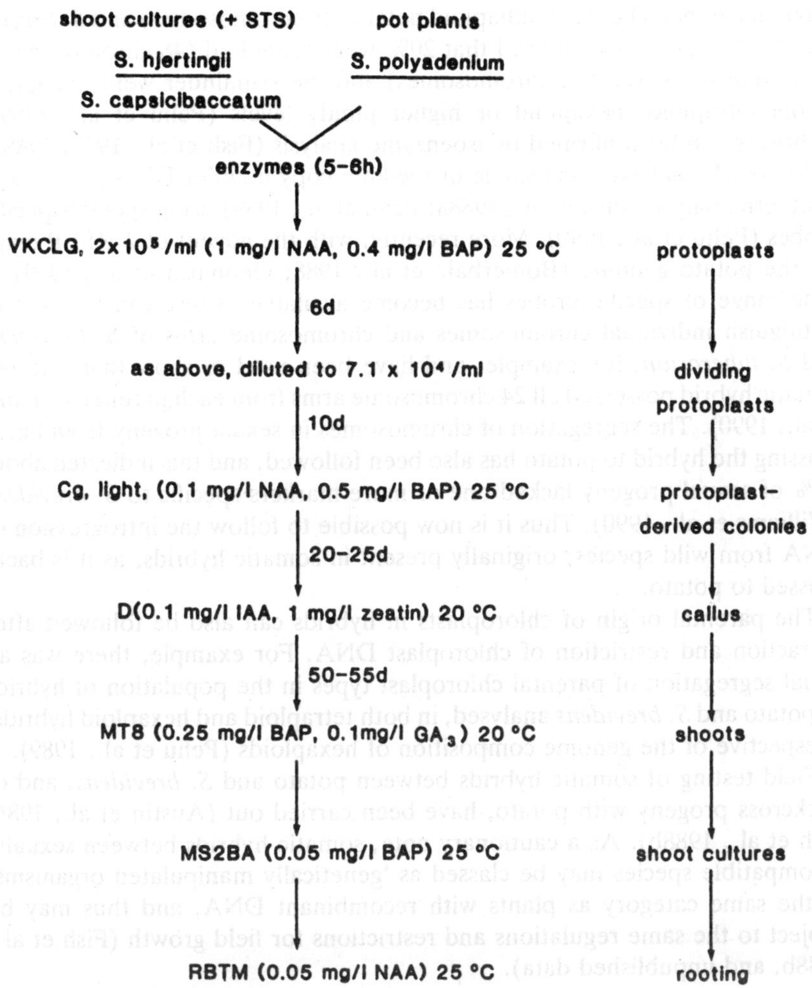

Figure 2. Regeneration protocol for *Solanum* protoplasts (Xu et al., 1991).

Table 1. Some useful agronomic characters present in wild *Solanum* spp. (*sexually incompatible with *S. tuberosum*)

Solanum species	Useful characters
**S. brevidens*	Resistance to PLRV, PVX, PVY and *Erwinia*
**S. hjertingii*	Non-browning tuber flesh when cut
S. bulbocastanum	Resistance to late blight
**S. polyadenium*	Insect resistance
**S. berthaultii*	Insect resistance
**S. capsicibaccatum*	Nematode and aphid resistance

of 60 such hybrids between dihaploid potato (24 chromosomes) and *S. brevidens* (24 chromosomes) showed that 20% were tetraploid (48 chromosomes), 14% were hexaploid (72 chromosomes) and the remainder were aneuploid at the tetraploid, hexaploid or higher ploidy levels (Pehu et al., 1989). Hybridity can be confirmed by isoenzyme analysis (Fish et al., 1987, 1988a) or by RFLP analysis using single or medium copy number DNA probes and Southern analysis (Fish et al., 1988a; Pehu et al., 1989) using species-specific probes (Pehu et al., 1990). More recently, with the advent of 2 RFLP maps for the potato genome (Bonierbale et al., 1988; Gebhardt et al., 1989), a wide range of specific probes has become available. These can be used to distinguish individual chromosomes and chromosome arms of *S. brevidens* and *S. tuberosum*, for example, and have been used to show that a fertile somatic hybrid possessed all 24 chromosome arms from each parent (Williams et al., 1990). The segregation of chromosomes in sexual progeny from backcrossing the hybrid to potato has also been followed, and this indicated about 10% of sexual progeny lacked one or more markers specific to *S. brevidens* (Williams et al., 1990). Thus it is now possible to follow the introgression of DNA from wild species, originally present in somatic hybrids, as it is backcrossed to potato.

The parental origin of chloroplasts in hybrids can also be followed after extraction and restriction of chloroplast DNA. For example, there was an equal segregation of parental chloroplast types in the population of hybrids of potato and *S. brevidens* analysed, in both tetraploid and hexaploid hybrids, irrespective of the genome composition of hexaploids (Pehu et al., 1989).

Field testing of somatic hybrids between potato and *S. brevidens*, and of backcross progeny with potato, have been carried out (Austin et al., 1986; Fish et al., 1988b). As a cautionary note, somatic hybrids between sexually incompatible species may be classed as 'genetically manipulated organisms' in the same category as plants with recombinant DNA, and thus may be subject to the same regulations and restrictions for field growth (Fish et al., 1988b, and unpublished data).

2.4.1. *Asymmetric Hybridization*

Production of somatic hybrids between potato and wild species introduces undesirable germplasm together with useful characters. There are two approaches to remove unwanted genes: backcrossing to the potato parent with selection for the desired characters, and production of asymmetric hybrids in which only chromosomal or subchromosomal fragments with the desired genes are introduced from donor to recipient protoplasts. Here again the use of widely dispersed species-specific probes can aid the identification of putative asymmetric hybrid material. A simple 'squash blot' approach, in which callus is squashed onto a membrane and subsequently probed with the donor species-specific probe, can be used to screen large numbers of calli to identify asymmetric hybrids (Pehu et al., 1990b). This approach is currently being applied to transfer virus resistances from *S. brevidens* to potato.

2.4.2. *Cybrids*

Cybrids that contain the nuclear genome from one parent and the organelles (chloroplasts, mitochondria) from another, can be produced spontaneously by elimination of one nuclear genome following protoplast fusion, or more reliably, by strongly irradiating the organelle donor (X- or γ-rays) before fusion to the recipient. The latter approach has been applied by Perl et al., 1990) to synthesise interspecific *Solanum* cybrids. Chloroplasts and mitochondria from *S. chacoense*, *S. brevidens*, *S. etuberosum* and *S. berthaultii* were transferred to a *S. tuberosum* background. This allows study of nuclear-organelle compatibility and interaction. Some combinations, such as organelles of *S. chacoense* in *S. tuberosum* background gave fertile plants with potato morphology, but an interesting result was obtained with *S. tuberosum* as the organelle donor. In this case male sterile cybrid plants resulted, and such plants could be used as female parents for the production of hybrid true-potato seeds.

3. Direct Gene Transfer

Potato protoplasts have been used both to study transient gene expression of introduced gene constructs, and to obtain stable transformants after introduction of selectable genes (Jones et al., 1989b). Both chemical (PEG based) and electroporation methods can be used to introduce gene constructs into potato mesophyll, cell suspension and tuber protoplasts, and into protoplasts of wild *Solanum* species. Transient gene expression is valuable for rapid analysis of promoter activity of gene constructs, and allows easy quantification of relative levels of expression, free from effects on position encountered when genes are integrated into host DNA. For example, tuber specific expression of the patatin promoter can be studied in tuber protoplasts, although there is the problem that once isolated protoplasts may rapidly loose their original tissue specific characteristics. When selectable marker genes are used (e.g. nptII conferring resistance to kanamycin, with a constitutive promoter) then transgenic plants can be selected and regenerated. This approach has been used to introduce different selectable marker genes for the production of somatic hybrids (Masson et al., 1989). Co-transformation of two unlinked plasmids (e.g., 35S-*npt*II; 35S-gus), with selection for resistance conferred by one plasmid, yields co-transformants with a frequency of about 20% (Xu et al., 1990). Although transgenic plants can be obtained by direct gene transfer into protoplasts, because of the background somaclonal variation present from protoplast culture, unless large numbers of individual transformation events are required then *Agrobacterium*-mediated approaches are usually preferable.

4. Agrobacterium Transformation

Potato and related *Solanum* species are relatively amenable to transformation by standard *Agrobacterium* procedures, for both *A. tumefaciens* and *A. rhizogenes*. Approaches used include infection of leaf discs, stem segments and tuber discs (e.g., Sheerman and Bevan, 1988; Stiekeman et al., 1988; Twell et al., 1987; Visser et al., 1989a). An efficient stem segment transformation procedure is outlined in Fig. 3. Transgenic plants regenerated from 'hairy root' cultures following transformation with *A. rhizogenes* have modified tubers with the appearance of swollen stems with pronounced eyes (Ooms et al., 1985, 1986; Visser et al., 1989a). Some applications of *in vitro* transformation are considered next.

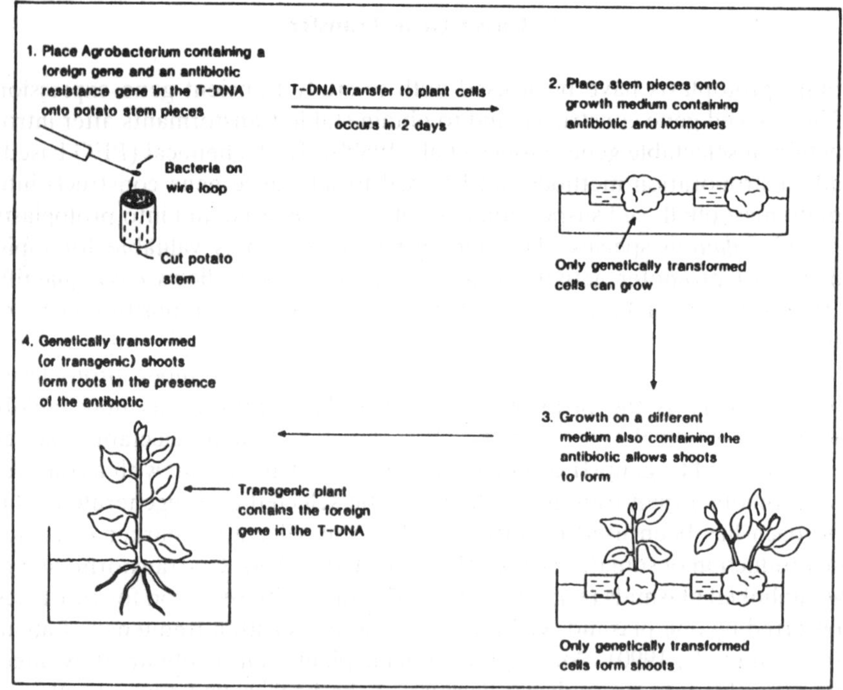

Figure 3. Stem segment transformation protocol. (Courtesy D. Twell).

5. Engineering for Potato Improvement

5.1. *Resistance to Pests and Diseases*

5.1.1. *Potato Virus Diseases*

The introduction and constitutive expression in potato plants of coat protein genes of two of the major potato viruses, PVX and PVY, has resulted in protection from those virus diseases, in the form either of delayed symptom development and drastic reduction in accumulation of virus (Hoekema et al., 1989) or complete absence of virus infection (Lawson et al., 1990). Transgenic lines expressing coat protein of both viruses also expressed resistance to both viruses (Lawson et al., 1990). This approach has also been applied to confer resistance to PLRV. An alternative approach, that is expression of an anti-sense PVX coat protein transcript was not so successful, and conferred only a low level of protection to PVX infection (Hemenway et al., 1988). Virus resistance may also be conferred by expression of defective replicase and other non-structural genes, ribozymes, satellite sequences, and expression of domed antibody genes.

5.1.2. *Bacteria, Fungi, Nematodes, Insects*

There is little information available for potato on results or proven strategies to confer resistance to bacteria, fungi and nematode pathogens. Chitinases and β1:3 glucanases may be activated in response to fungal attack, and a combination of these enzymes can inhibit fungal growth *in vitro*. Various molecular strategies can be proposed to engineer resistance to root-knot and cyst-nematodes, and rapid advances can be expected in this area (Washer et al., 1993). Approaches that have been applied to engineer resistance to insects for other crop plants are equally applicable to potato. These include production of transgenic plants expressing genes for protease inhibitors (e.g., cowpea trypsin inhibitor, CpTI, Gatehouse and Hilder, 1988) and for the protein toxin from *Bacillus thuringiensis* (Bt toxin, Fischoff et al., 1987). The latter can confer resistance to the Colorado Beetle, an important Coleopteran pest in some potato growing regions.

5.1.3. *Herbicide Resistance*

Potato plants resistant to a number of herbicides (e.g., basta, De Block et al., 1987) can be obtained using standard strategies (e.g., overproduction of the target enzyme, mutation to resistance, introduction of an insensitive target gene or of an herbicide-inactivating enzyme (Oxtoby and Hughes, 1990).

5.2. *Tuber Specific Expression*

The major tuber protein, patatin, exists as a multigene family of glycoproteins (MW 40 KDn) that can constitute up to 40% of the soluble proteins of tubers (Mignery et al., 1988). Class I patatin genes encode 98–99% of tuber patatin mRNA, and are not normally expressed in other tissues (except when explants are incubated in high concentrations of sucrose). This makes class I patatin promoters essentially tuber specific, and so they can be employed to direct tuber-specific expression of chimaeric genes (Wenzler et al., 1989).

5.3. *Wound Inducible Promoters*

When potatoes are mechanically wounded or damaged by insects or other pests, proteinase inhibitors are induced. These are serine proteinases that can accumulate to 10% of the soluble proteins; the two best characterised are inhibitor I and II. Promoters from these and other wound-induced genes (e.g., win2, Stanford et al., 1990) can be coupled to reporter genes, and shown to direct localized expression of the reporter genes at wound sites (Keil et al., 1989; Stanford et al., 1990).

5.4. *Manipulating Tuber Quality*

Two major aims are to prevent 'cold-sweetening' (the accumulation of reducing sugars on cold storage that lead to browning when tubers are processed) and to manipulate starch quality in tubers. Replacement of the cold-labile enzyme phosphofructokinase (PFK) with a heat stable PFK might prevent cold sweetening by allowing accumulated sugars to be metabolised. Considerable progress has been achieved in one aspect of the second aim, to manipulate starch. An amylose-free (amf) mutant potato has been selected from a mutagenised monohaploid line after shoot regeneration from leaf discs and staining micro-tubers for starch (normal starch, with amylopectin plus amylose, stains blue; amylose-free starch stains red, Hovenkamp-Hermelink et al., 1987). The amf mutant had a defective granule bound starch synthase (GBSS) gene. Almost the same result has been achieved by obtaining transgenic plants with constitutive expression of the anti-sense GBSS gene (Visser et al., 1989b). The amf mutation is recessive and therefore difficult to introduce into potato cultivars, whereas the anti-sense GBSS gene is dominant and can be inserted into existing cultivars, but is not expressed equally in all cell types.

6. Future Prospects

The steady progress in *in vitro* culture of potato and related *Solanum* species can be expected to continue and to be expanded to encompass more genotypes and species. There is now increased documentation of somaclonal variation of potato following culture involving a callus phase, but still a need to understand its basis and to be able to control and limit such variation. Somatic hybridization within *Solanum* spp. is set to expand, both in the production of complete hybrids and particularly of asymmetric hybrids. These approaches are being aided by associated molecular techniques, such as the use of species-specific and defined probes: the latter can be localised to specific regions of chromosome arms on the genetic map of potato. Practical applications in the introduction of useful characters from wild species, and the synthesis of desired tetraploids from selected dihaploid lines, will feed into established breeding programmes.

On the molecular front, rapid progress can be expected in the aspects previously described. In addition, potato development may be controlled for example by tuber specific expression of cytokinin genes, or increased root growth from *A. rhizogenes* transformation. Pheromones, anti-feedants and host recognition systems may also be engineered for pest protection (for example *S. berthaultii* produces the aphid alarm pheromone, E-β-farnesene (Gibson and Pickett, 1983). The first step has also been taken towards use of the potato as a bioreactor, with the expression of correctly processed human serum albumin in the potato tuber (Sijmons et al., 1990).

7. References

Austin, S., Baer, M.A. and Helgeson, J.P. (1985). Transfer of resistance to potato leaf roll virus from *Solanum brevidens* into *Solanum tuberosum* by somatic fusion. Plant Sci. 39:75–82.

Austin, S., Ehlenfeldt, M.K., Baer, M.A. and Helgeson, J.P. (1986). Somatic hybrids produced by protoplast fusion between *S. tuberosum* and *S. brevidens*: Phenotypic variation under field conditions. Theor. Appl. Genet. 71:628–690.

Austin, S., Lojkowska, E., Ehlenfeldt, M.K., Kelman, A. and Helgeson, J.P. (1988). Fertile interspecific hybrids of *Solanum*: A novel source of resistance to *Erwinia* soft rot. Phytopathology 78:1216–1220.

Barsby, T.L., Shepard, J.F., Kemble, R.J. and Wang, R. (1984). Somatic hybridization in the genus *Solanum*, *S. tuberosum* and *S. brevidens* Plant Cell Rep. 3:165–167.

de Block, M., Botterman, J., Vandeweile, M., Dokse, J., Thoen, C., Gossele, V., Rao Mowa, N., Thompson, C., van Montagu, M. and Leemans, J. (1987). Engineering herbicide resistance in plants by expression of a detoxifying enzyme. EMBO J. 6:2513–2518.

Bonierbale, M.W., Plaisted, R.L. and Tanksley, S.D. (1988). RFLP maps based on a common set of clones reveal modes of chromosomal evolution in potato and tomato. Genetics 120:1095–1103.

Carlberg, I., Glimelius, K. and Eriksson, T. (1983). Improved culture ability of potato protoplasts by use of activated charcoal. Plant Cell Rep. 2:223–225.

Ehlenfeldt, M.K. and Helgeson, J.P. (1987). Fertility of somatic hybrids from protoplast fusions of *S. brevidens* and *S. tuberosum*. Theor Appl. Genet. 73:295–402.

Evans, N.E., Foulger, D., Farrer, L., Risiott, R. and Bright, S.W.J. (1986). Variation in explant-derived potato clones over three tuber generations. Euphytica 35:353–361.

Fischoff, D.A., Bowdish, K.S., Perlack, F.J., Marrone, P.G., McCormick, S.M., Niedermeyer, J.G., Dean, D.A., Kusano-Kretzmer, K., Mayer, E.J., Rochester, D.E., Rogers, S.G. and Fraley, R.T. (1987). Insect tolerant transgenic tomato plants. Biotechnology 5:807–813.

Fish, N. and Karp, A. (1986). Improvements in regeneration from protoplasts of potato and studies on chromosome stability. Theor. Appl. Genet. 72:402–412.

Fish, N. and Jones, M.G.K. (1988). A study on the inheritance of tissue culture response from tetraploid to dihaploid *Solanum tuberosum* based on shoot regeneration from explants. Plant Cell Tissue Org. Cult. 15:201–210.

Fish, N., Karp, A. and Jones, M.G.K. (1987). Improved isolation of dihaploid *S. tuberosum* protoplasts and the production of somatic hybrids between dihaploid *S. tuberosum* and *S. brevidens*. *In vitro* 23:75–580.

Fish, N., Karp, A. and Jones, M.G.K. (1988a) Production of somatic hybrid plants of *Solanum* by electrofusion.Theor. Appl. Genet. 7:260–266.

Fish, N., Steele, S. and Jones, M.G.K. (1988b). Field characteristics of somatic hybrid plants of *S. tuberosum* and *S. brevidens*. Theor. Appl. Genet. 76:880–886.

Foulger, D. and Jones, M.G.K. (1986). Improved efficiency of genotype-dependent regeneration of important potato cultivars. Plant Cell Rep. 5:72–76.

Garner, N. and Blake, J. (1989). The induction and development of potato microtubers *in vitro* on media free of growth regulating substances. Ann. Bot. 63:663–674.

Gatehouse, A.M.R. and Hilder, V.A. (1988). Introduction of genes conferring insect resistance. Brighton Crop Protection Conference, Pests and Diseases 1988, British Crop Protection Council Thornton Heath 3:1245–1254.

Gebhardt, C., Ritter, E., Debener, T., Schachtschabel, U., Walkemeier, B., Uhrig, H. and Salamini, F. (1989). RFLP analysis and linkage mapping in *Solanum tuberosum*. Theor. Appl. Genet. 78:65–76.

Gibson, R.W. and Pickett, J.A. (1983). Wild potato repels aphids by release of aphid alarm pheromone. Nature 302:608–609.

Gibson, R.W., Jones, M.G.K. and Fish, N. (1988). Resistance to potato leaf roll virus and potato virus Y in somatic hybrids between dihaploid *S. tuberosum* and *S. brevidens*. Theor. Appl. Genet. 76:113–117.

Gibson, R.W., Pehu, E., Wood, R.D. and Jones, M.G.K. (1990). Resistance to potato virus Y and potato virus X in *Solanum brevidens*. Appl. Biol. 116:151–156.

Haberlach, G.T., Cohen, B.A., Reichert, N.A., Baer, M.A., Towiil, L.E. and Helgeson, J. (1985). Isolation, culture and regeneration of protoplasts from potato and several related *Solanum* species. Plant Sci. I 39:67–74.

Hawkes, J.G. (1990). The Potato: Evolution, Biodiversity and Genetic Resources. Belhaven, London, pp. 259.

Helgeson, J.P., Hunt, G.J., Haberlach, G.T. and Austin, S. (1986). Somatic hybrids between *Solanum brevidens* and *Solanum tuberosum*: expression of a late blight resistance gene and potato leaf roll resistance. Plant Cell Rep. 212–214.

Hemenway, C., Fang, R-X., Kaniewski, W., Chua, N-H. and Tumer, N.E. (1988). Analysis of the mechanism of protection in transgenic plants expressing the potato virus X coat protein or its antisense RNA. EMBO J 7:1273–1280.

Hoekema, A., Huisman, M.J., Molendijk, L., van den Elzen, P.J.M. and Cornelissen, B.J.C. (1989). The genetic engineering of two commercial potato cultivars for resistance to potato virus X. Bio/Technology 7:273–278.

Hovenkamp-Hermelink J.H.M., Jacobsen, E., Ponstein, A.S., Visser, R.G.F., Vos-Scheperkeuter, G.H., Bijmult, E.W., de Vries, J.N., Witholt, B. and Feenstra, W.J. (1987). Isolation of an amylase-free starch mutant of the potato (*Solanum tuberosum* L.). Theor Appl. Genet. 75:217–221.

Jones, F.G.W. and Jones, M.G. (1984). Pests of Field Crops. Arnold, London, 3rd ed., 392 pp.

Jones, M.G.K. (1988). Fusing plant protoplasts. Trends in Biotechnol 6:153–158.

Jones, R.A.C (1979). Resistance to potato leaf roll virus in *Solanum brevidens*. Potato Research 22:149–152.

Jones, H., Karp, A. and Jones, M.G.K. (1989a). Isolation, culture and regeneration of plants from potato tuber protoplasts. Plant Cell Rep. 8:307–311.

Jones, H., Ooms, G. and Jones, M.G.K. (1989b). Transient gene expression in electroporated *Solanum* protoplasts. Plant Mol. Biol. 13:503–511.

Karp, A., Nelson, R.S., Thomas, E. and Bright, S.W.J. (1982). Chromosome variation in protoplast-derived potato plants. Theor Appl. Genet. 63:265–272.

Keil, M., Sanchez-Serrano, J.J. and Willmitzer L. (1989). Both wound-inducible and tuber-specific expression are mediated by the promoter of a single member of the potato proteinase inhibitor II gene family. EMBO J. 1323–1330.

Landsman, J. and Uhrig H. (1985). Somaclonal variation in *S. tuberosum* detected at the molecular level. Theor. Appl. Genet. 71:500–505.

Larkin, P.J. and Scowcroft, W.R. (1981). Somaclonal variation – a novel source of variability from cell cultures for plant improvement. Theor. Appl. Genet. 197–214.

Lawson, C., Kaniewski, W., Haley, L., Rozman, R., Newell, C., Sanders, P. and Turner, N.E. (1990). Engineering resistance to mixed virus infection in a commercial potato cultivar: Resistance to potato virus X and potato virus Y in transgenic Russet Burbank. Biotechnol. 8:127–139.

Lentini, Z., Earle, E.D. and Plaisted, R.L. (1990). Insect-resistant plants with improved horticultural traits from interspecific potato hybrids grown *in vitro*. Theor. Appl. Genet. 80:95–104.

Masson, J., Lancelin, D., Bellini, C., Lecerf, M., Guerche, P. and Pelletier, G. (1989). Selection of somatic hybrids between diploid clones of potato (*Solanum tuberosum* L.) transformed by direct gene transfer. Theor. Appl. Genet. 78:153–159.

Mignery, G.A., Pikaard, C.S. and Park W.D. (1988). Molecular characterization of the patatin multigene family of potato. Gene 62:27–44.

Ooms, G., Karp, A., Burrell, M.M., Twell, D. and Roberts, J. (1985). Genetic modification of potato development using Ri-T-DNA. Theor. Appl. Genet. 70:440–446.

Ooms, G., Bossen, M.E., Burrell, M.M. and Karp, A. (1986). Genetic manipulation in potato with *Agrobacterium rhizogenes*. Potato Research 29:367–379.

Oxtoby, E and Hughes, M.A. (1990). Engineering herbicide tolerance into crops. Trends in Biotechnol 8:61–65.

Pehu, E., Karp, A., Moore, K., Steele, S., Dunckley, R. and Jones, M.G.K. (1989). Molecular, cytogenetic and morphological studies of somatic hybrids of dihaploid *Solanum tuberosum* and diploid *S. brevidens*. Theor. Appl. Genet. 78:696–704.

Pehu, E., Gibson, R.W., Jones, M.G.K. and Karp A. (1990a). Studies on the genetic basis of resistance to PLRV, PVX and PVY in *Solanum brevidens* using somatic hybrids of *S. brevidens* and *S. tuberosum*. Plant Sci. 69:95–101.

Pehu, E., Thomas, M., Poutala, T., Karp, A. and Jones, M.G.K. (1990b) Species-specific sequences in the genus *Solanum*: identification, characterization and application to study somatic hybrids of *S. brevidens* and *S. tuberosum*. Theor. Appl. Genet. 80:693–698.

Perl, A., Aviv, D. and Galun, E. (1988). Ethylene and in vitro culture of potato: suppression of ethylene generation vastly improves protoplast yield, plating efficiency and transient expression of an alien gene. Plant Cell Rep. 7:403–406.

Perl, A., Aviv, D. and Galun, E. (1990). Protoplast-fusion-derived *Solanum* cybrids: application and phytogenetic limitations. Theor. Appl. Genet. 79:632–640.

Potter, R.H. and Jones, M.G.K. (1991). An assessment of genetic stability of potato *in vitro* by molecular and phenotypic analysis. Plant Sci. 76:239–248.

Salaman, R.N. (1949/1985). The History and Social Influence of the Potato. Cambridge University Press, 685 pp.

Jsecor, G.A. and Shepard, J.F. (1981). Variability of protoplast-derived potato clones. Crop Sci. 21:102–105.

Sheerman, S. and Bevan, M.W. (1988). A rapid transformation method for *Solanum tuberosum* using binary *Agrobacterium tumefaciens* vectors. Plant Cell Rep. 7:13–16.

Shepard, J.F., Bidney, D. and Shahin, E. (1980). Potato protoplasts in crop improvement. Science 208:17–24.

Stanford, A.C., Northcote, D.H., Bevan, M.W. (1990). Spacial and temporal patterns of transcription of a wound-induced gene in potato. EMBO J. 9:593–603.

Stiekema, W.J., Heidekamp, F., Louwerse, J.D., Verhoeven, H.A. and Dijkhans, P. (1988). Introduction of foreign genes into potato cultivars Bintje and Desiree using an *Agrobacterium tumefaciens* binary vector. Plant Cell Rep. 7:47–50.

Sijmons, P.C., Dekker, B.M.M., Schrammeijer, B., Verwoerd, T.C., van den Elzen, P.J.M. and Hockema, A. (1990). Production of correctly processes human serum albumin in transgenic plants. Biotechnology 8:217–221.

Tempelaar, M.J. and Jones, M.G.K. (1985a). Fusion characteristics of plant protoplasts in electric fields. Planta 165:205–216.

Tempelaar, M.J. and Jones, M.G.K. (1985b). Directed electrofusion between protoplasts with different responses in a mass fusion system. Plant Cell Rep. 4:92–95.

Tempelaar, M.J., Duyst, A., de Vlas, S.Y., Krol, G., Symonds, C and Jones M.G.K. (1987). Modulation and direction of the electrofusion response in plant protoplasts. Plant Sci. 48:99–105.

Twell, D. and Ooms, G. (1987) The 5′ flanking DNA of a patatin gene directs tuber specific expression of a chimaeric gene in potato. Plant Mol. Biol. 9:345–375.

Valkonen, J.P.T., Pehu, E., Jones, M.G.K. and Gibson, R.W. (1991). Resistance in *Solanum brevidens* to both potato virus Y and potato virus X may be associated with slow cell-to-cell spread. J. Gen Virol. 72:231–236.

Visser, R.G.F., Jacobsen, E., Witholt, B and Feenstra, W.J. (1989a). Efficient transformation of potato (*Solanum tuberosum* L.) using a binary vector in *Agrobacterium rhizogenes*. Theor. Appl. Genet. 78:594–600.

Visser, R.G.F., Hergersberg, M., van der Leij, F.R., Jacobsen, E., Witholt, B. and Feenstra, W.J. (1989b). Molecular cloning and partial characterization of the gene for granule-bound starch synthase from a wild type and an amylase-free potato (*Solanum tuberosum* L.) Plant Sci. 64:185–192.

Wang, P-J. and Hu, C-Y. (1985). Potato tissue culture and its applications in agriculture. In: Li, A.P. (ed.), Potato Physiology Academic Press, Orlando.

Washer, S., Potter, R. and Jones, M.G.K. (1993). A molecular approach to engineer host resistance to root-knot nematodes (*Meloidogyne* spp.). Proc. 11th Australian Biotechnology conference, Perth, p. 279.

Wenzler, H., Mignery, G., May, G. and Park, W. (1989). A rapid and efficient transformation method for the production of large numbers of transgenic potato plants. Plant Sci. 63:39–85.

Wheeler, V.A., Evans, N.E., Foulger, D., Webb, K.J., Karp, A., Franklin, J. and Bright, S.W.J. (1985). Shoot formation from explant cultures of potato and an examination of chromosome and morphological variation in the regenerated plants. Ann. Bot. 55:309–320.

Williams, C.E., Hunt, G. and Helgeson, J.P. (1990). Fertile somatic hybrids of *Solanum* species: RFLP analysis of a hybrid and its sexual progeny from crosses with potato. Theor. Appl. Genet. 80:545–551.

Xu, Y-S., Jones, H. and Jones M.G.K. (1990). Stable transformahon of *Solanum* protoplasts by direct gene transfer. Proc. 11th Triennial Conf. Eur. Assoc. Potato Res. 439.

Xu, Y-S., Pehu, E., Malone, R. and Jones, M.G.K. (1991). Plant regeneration from protoplasts of *Solanum* species with potential agricultural value (*S. hjertingii*, *S. polyadenium*, *S. capsicibaccatum*). Plant Cell Rep. 9:520–522.

16. *In vitro* Culture of Root and Tuber Crops

Abraham D. Krikorian

Contents

Introduction

'Root' crops traditionally include plants with swollen hypocotyls (e.g. radish, beets), rhizomes, corms, cormels (e.g. Aroids) and stem tubers (e.g. true yams) as well as true roots (e.g. carrot) and root tubers (e.g. cassava, sweet potato). Because of its importance, potato is covered separately in Chapter 15, but some of the other major 'root' and tuber crops will be covered here. A few practical reasons as to why one might wish to carry out aseptic culture studies on these plants are provided under the specific entries. Two extremely valuable references with considerable insight for both broad and specific research needs that readers should be aware of, but might easily be overlooked, are Kay (1973) and Brücher (1989). For ideas on priorities of plant biotechnology research for developing countries reference may be made to National Research Council (1990) and Sasson and Costarini (1990).

It should be stated at the outset that there is little evidence to support the idea that root and tuber bearing plants present problems to the *in vitro* specialist that are very different from those encountered in any other group of plants. A number of these grow in hot humid environments, even mud (e.g. the Aroids), and special precautions may have to be taken to get 'clean' primary explants, especially if the storage organ is being used as starting

Indra K. Vasil and Trevor A. Thorpe (eds.), Plant Cell and Tissue Culture, 379–411

material. In some instances, obtaining 'clean' primary explants is antithetic to the very fact that many are vegetatively multiplied and hence may serve as reservoirs of endophytic contamination and, especially viruses. The obtaining of clean propagules for subsequent conventional multiplication may thus become an important objective in itself. In many cases, materials are 'clean' when the storage organ is sound and uninjured. Many more temperate storage organs are relatively easy to get clean. For instance, *Daucus carota*, increasingly viewed as the model species in which to study somatic embryogenesis using excised petiole or zygotic embryo segments (Krikorian and Smith, 1992), was originally almost exclusively worked with using explants from the fleshy roots since these, when healthy, provided a readily surface-sterilizable organ from which many primary explants could be taken (see Steward et al., 1975 and below). Similarly, primary explants from the tuberous stem of the topinambour of France (better known as the Jerusalem artichoke, *Helianthus tuberosus*) also have provided considerable understanding of a number of important principles in tissue culture now taken largely for granted although its use as an experimental test system has, unfortunately, declined. See for instance the importance of the time of year for explants to be taken for optimal performance (see E. Robinson in Yeoman and Davidson, 1971) and the inhibitory effect of light on callus initiation in Jerusalem artichoke explants (Yeoman and Davidson, 1971). It will be interesting as well to note that the first really successful culture of a monocotyledonous genus was achieved with an aroid (*Amorphophallus*) using a combination of coconut water (CW) and NAA (Morel and Wetmore, 1951).

A generalization that has its exceptions, of course, is that explants from generally healthy organs sink in water, whereas unsound ones float. (This does not apply to virus infected roots, however).

The now very extensive literature indicates that virtually all species and cultivars can respond to and regenerate from *in vitro* strategies. Moreover, many similar results can be obtained using a variety of approaches and media formulations. Indeed, it is a mistake to view any given protocol as settled unless it has been demonstrated to have high potential for one's perceived needs. The process chosen should be matched to the situation. In addition to genotypic differences in germplasm, subtle differences in laboratory procedures may have significant impact on success. Assuming the system is responsive, the level of control achievable in various stages in the culture process, and uniformity and speed of response then become primary objectives. For studying developmental control mechanisms, any part of the culture process becomes significant if key control points suggestive of gene activation etc. can be identified. For more practical objectives, the needs are closely identified with a specific crop or even locality. Table 1 provides a general summary of the kinds of advantages offered via aseptic culture techniques.

Table 1. Range of activities that currently available tissue culture techniques can facilitate

Research with potential for near term impact
Culture techniques can facilitate
Rapid multiplication of select specimens
Elimination of virus and specific pathogens
Virus indexing
Germplasm introduction and evaluation
Germplasm collection, preservation and management
Production of polyploids, haploids, somaclonal variants for new crop production and use in breeding, etc.
Elimination of certain breeding barriers
– *in vitro* fertilization *in ovulo*
– embryo rescue and/or storage
– androgenesis
– gynogenesis
Research with intermediate impact
All the above in more recalcitrant species, plus
Selection for complex traits such as tolerance to stress
– biotic – diseases and pests
– abiotic – temperature
salt
herbicides
In vitro mutation breeding
Cryopreservation
Research with long range implications
All the above in still more recalcitrant species, plus
Genetic engineering
– transformation by selectable genes etc.
– organelle transfer
– wide crosses-somatic hybridization
.Understanding controls in developmental and physiological processes

2. Objectives and Strategies

2.1. *Sweet Potato* (*Ipomoea batatas*)

Sweet potato, a tuberous root producing perennial cultivated as an annual with a growing period of from on average 3.5 to 8 months, is an important food crop of Neotropical origin (Fig. 1). It is a hexaploid (2n = 90), frequently self-incompatible or shows inbreeding depression and normally shows poor seed set, especially outside the tropics. Most cultivars are highly sterile but the causes of sterility are not fully understood although it can occur as failure of gametes or zygotes at any of several stages. Hybridization, even using carefully selected parents, generates tremendous new variability, and demands arduous selection among large numbers of seedling clones. It is rare indeed when a new clone combines very many desired traits. Apical

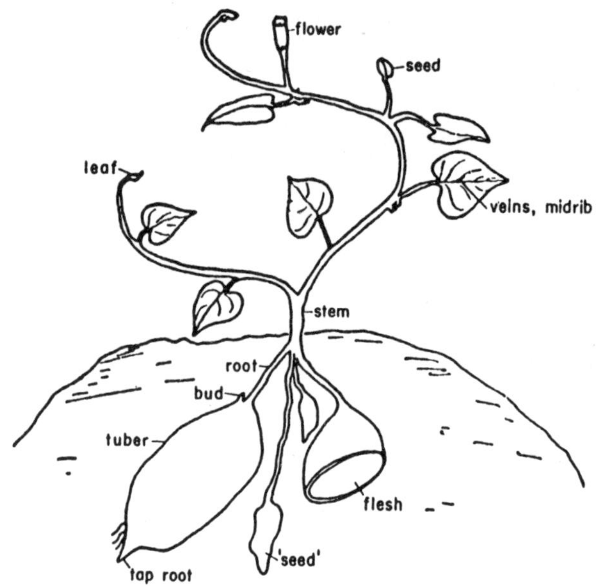

Figure 1. Diagrammatic representation of a mature sweet potato plant.

cuttings from vigorously growing plants, rather than basal or middle cuttings, have been traditionally used to multiply desired clones from thousands of variant plants. Sprouting of cut root pieces is sometimes possible, as is sprout production from detached single leaves, but these are as yet of no commercial importance. It will be apparent that the effort to generate or to maintain disease-free clones is time-consuming and costly (Martin and Jones, 1986).

Improvement in overall vigor and stress resistance, plant habit, gross morphology and keeping qualities of the tuberous root, virus, black and soft rots disease resistance, insect (e.g. sweet potato weevil) resistance, yield, nutritional value, quality and cooking characteristics are major objectives. While mentioned here under sweet potato, the points are equally applicable to a wide range of plant materials. In the tropics, it is especially difficult to maintain disease-free parental stocks and meristem culture is increasingly being appreciated as a potential means of germplasm preservation and for the production of planting materials. It has also been suggested that somatic embryos embedded in a fluid gel matrix for direct drilling could provide a major advantage for planting since a major drawback to establishment of new propagules is the tendency to desiccate.

Elimination of viruses (Henderson et al., 1984; Love et al., 1987) via meristem culture has been possible for some time as has the indexing of some viruses by ELISA (Frison and Ng, 1981; Esbenshade and Moyer, 1982) but reinfection of field grown material remains a problem of course.

Tissue culture of sweet potato early emphasized meristem culture (Elliott, 1969) and regeneration from stem explants (cf. Gunkel et al., 1972; Scaramuzzi, 1986). Litz and Conover (1978) propagated two white-fleshed varieties from shoot tips and axillary buds grown on M and S medium containing BAP, IAA, KIN and activated charcoal. Charcoal minimized callus production (most probably by adsorbing the level of immediately available growth regulators) and thus nominally sustained genotypic stability of the propagules. Small rooted plantlets that could eventually be established in the field formed from apical explants 8 weeks after primary culture initiation. Nodal segments can also readily be subcultured in a shallow liquid medium on slowly moving rotary shakers.

Adventitious shoots can be produced from callus from excised leaves *in vitro* (Sehgal, 1975) and anther callus (Sehgal, 1978) but plantlets from the latter are not haploid. Adventitious shoots form from 1–2 cm long petiole segments and root tuber plugs on M and S medium supplemented with BAP and NAA but this performance is genotype dependent (Hwang et al., 1983). Roots, of course, need extensive surface sterilization with hypochlorite or mercuric chloride to get them clean. Peeling of the periderm is important as well. (See below under carrot for a strategy to deal with fleshy roots). Ultimately, surface sterilization protocols need to be worked out systematically since it will depend on source material. Ovary-derived and shoot tip callus has been shown to be embryogenic on 2,4-D containing media, as has that from root and internodal stem segments (Liu and Cantliffe, 1984). Plantlets developed flowers and storage roots when transplanted to soil. Maintenance and proliferation of embryogenic callus in liquid with 2,4-D appears routine (Chee and Cantliffe, 1988).

Wu and Ma (1979) showed that protoplasts capable of regeneration into callus could be released within 5 hours or so from a callus initially derived from stem. A cellulase (2%) containing mineral medium supplemented with 20 mM calcium chloride (for membrane stabilization) and an osmoticum of 0.8 M mannitol was required for successful isolation. Protoplasts from petioles and from leaf mesophyll cells produced with 0.2% Macerozyme R-10 and 0.4% Cellulase 'Onozuka' R-10 were shown to be able to yield callus when plated by Bidney and Shepard (1980). A cell-layer reservoir system for gradually lowering osmoticum and nutrient environment was helpful in permitting callus regeneration at low densities. Sporadic roots were obtained but not shoots. Sihachakr and Ducreux (1987) have since regenerated plants from protoplasts prepared from petioles and stems of *in vitro* multiplied plants. Cells from leaf blades resisted digestion under their conditions.

2.2. *Taro (Colocasia esculenta) and other Aroids*

The Araceae (Monocotyledonae) comprises a large family (over 100 genera), several members of which are grown as annual or perennial herbs and yield starchy corms and cormels which are important as food. In select cases leaves and stems as well may be eaten as greens. Taxonomically the group is inadequately studied and there is considerable potential for confusion even among the edible aroids, which are highly polymorphic, and as a result variable and conflicting regional and local names frequently make matters worse. The wetland or lowland taro (*Colocasia*), which is frequently grown in a system of gently but continuously moving water, is perhaps the most popular and best studied (cf. Wang, 1983). *Colocasia* cultivars are normally broken down into two categories: those that produce a large corm with few cormels, and those that produce several small cormels rather than a large corm. Dasheens, for example, fit into the first category and have been classified as *Colocasia esculenta* var. *esculenta*, whereas eddoe(s) belong to the latter category and have been classified as *C. esculenta* var. *antiquorum*. There are intergradations, however, and there are varieties where there is a highly developed corm, considerable stolon development but there are few or no cormels. Chromosome studies indicate that large corm types, *C. esculenta* var. *esculenta*. generally have 28 chromosomes (2n = 28), whereas small cormel types, *C. esculenta* var. *antiquorum*, generally have 42 (3n = 42) but this is by no means certain and there is no general agreement on the basic chromosome number. Some have accepted 14 but even 7 has been proposed. Clearly much work remains to be done. Certainly, because it is a vegetatively multiplied plant, there is potential for tremendous variation among the many hundreds (thousands?) of clones. Mixoploidy has been especially well documented in Indian varieties of *Colocasia* and the fact that a single organ can have significantly varying cytotypes indicates that the potential for genotypic variability in this group of plants at the cell level must not be overlooked (Sharma and Sarkar, 1963). *Alocasia macrorhiza* (giant taro), *Xanthosoma* spp. (known commonly as 'new' cocoyam, tannia, yautia etc.), *Alocasia* (ta'amu, keladi etc.), *Cyrtosperma* (swamp taro, giant swamp taro), *Amorphophallus campanulatus* (elephant yam etc.), and *Amorphophallus konjac* (*A. rivieri*) (konjac etc.) are other edible aroids and provide foods of varying importance (O'Hair and Asokan, 1986; O'Hair, 1990). Plant tops, underground stem or corm (whole or cut into sections of about 4 'eyes'), offshoots or even rootstock divisions are variously used. Wet, humid tropical environments favor the growth of the edible aroids. Details of handling propagules vary but in *Colocasia* vegetative material generally consists of about 10 inches of petiole with the leaves trimmed, attached to 1/4 or 1/2 inches (ca. 15 to 25 cm and 1–1.5 cm or so respectively) of the "crown" of the corm (cf. Fig. 2 for a diagrammatic representation of a corm stripped of its leaves). These propagules intended to generate the next crop are prepared at the time of harvesting. Harvesting begins some 8 to 15 months after planting and is dependent on growth patterns and needs, rather than a specific period that results in a discrete event.

Seed propagation and breeding programs have been limited by the frequent inability of many cultivars to flower reliably, if at all. Moreover, seed viability

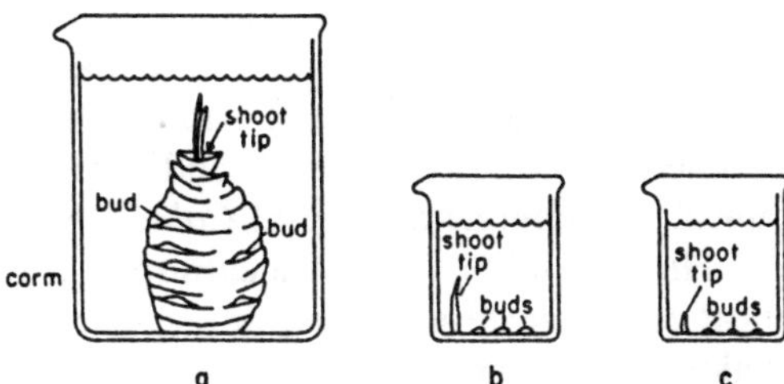

Figure 2. Schematic serial disinfestation of primary explants from taro. (a) a 10 min exposure to a 10% v/v aqueous dilution of a commercial sodium hypochlorite bleach such as Clorox; (b) a subsequent 5 min exposure to a 5% dilution of bleach; (c) and a final 1 min exposure to a 1% dilution. Lowered concentration and duration of exposure to surface sterilant accompanies removal of outer leaves at each step. The final explant has not been directly exposed to bleach. Drawn from a scheme provided the author by the late Marion O. Mapes, University of Hawaii, Manoa.

of some varieties is said to be relatively short (order of 4 to 8 weeks storage but in others longevity may be quite long) and field harvested seeds are generally fungally infested (cf. Nyman and Arditti, 1985). Edible aroids have been often viewed as seed sterile but careful observations indicate that there is considerable opportunity in this area for crop improvement via hand pollination while stigmas are still receptive (Shaw, 1975). Indeed, embryo culture from mature seeds was shown to be feasible long ago (see Abraham and Ramachandran, 1960 cited in Nyman et al., 1986).

It was early appreciated that dividing of mature plants was not only time-consuming but any of the numerous pathogens associated with the plants could be spread by this means. In addition to a range of insect pests like taro beetle, many bacterial, fungal and viral diseases affect the growth of taro or the storage of the corms once harvested (Jackson and Firman, 1982). The most serious and widespread diseases are perhaps leaf blight caused by *Phytophthora colocasiae* and soil borne *Pythium* spp. can cause major soft rot losses. Dasheen mosaic virus (DMV) is the most common of the virus diseases (Shaw et al., 1979). Varying degrees of tolerance or resistance have been noted but in other cases, growing the crop has been abandoned (cf. Wang, 1983). Crop improvement has thus far been restricted primarily to collecting, preserving and evaluating germ plasm (O'Hair, 1990) and production of resistant cultivars but flowering problems and generally slow multiplication has hampered the work.

Hartman (1974) was able to eliminate DMV by culturing apical meristems of taro comprising the dome and 1 or 2 leaf primordia. A semi-solid M and S medium supplemented with KIN (1.0 mg/l) and IAA (10 mg/l) was used. Abo El-Nil and Zettler (1976) induced protocorm-like structures from excised terminal shoots which included 2 leaf primordia using NAA and KIN. Callus was fostered by combined use of 2,4-D, p-chlorophenoxyacetic acid (p-CPA) and KIN. Darkness and a slightly modified Schenk and Hildebrandt medium seems to have been an important feature of the culture environment in terms of getting the system started but subsequent development was achieved using modified KIN and NAA levels and regular light/dark cycles. Stem tips (both apical and lateral) of several *Colocasia esculenta* var. *esculenta* cultivars from Hawaii and Samoa were cultured by Mapes in the early 1970s and multiplication was fostered by use of coconut water-supplemented M and S medium along with adenine (cf. Mapes and Cable, 1977) or by use of BAP and 2iP. The most difficult part of the procedure frequently is to get clean primary explants (Fig. 2). Taro may even be exposed to a weak solution of disinfestant along with growth regulator(s) for a few days (personal communication). One wonders whether the stress of hypochlorite activates organogenesis (Kiyosue et al., 1990)? Arditti and Strauss (1979) have produced a detailed manual of instructions for micropropagation of taro. A prominent feature of that booklet is a summary of all published work on taro carried out up to that date. Staritsky et al. (1986) have used half strength M and S medium supplemented with 1 g/l of KNO_3, 10 mg/l thiamine, 50 mg/l cysteine HCl, 100 mg/l each of inositol and casein hydrolysate, 30 g/l of sucrose, 0.5% agar (Oxoid), 5 mg/l BAP and 0.5 mg/l IBA. A 12 hour photoperiod with a day/night temperature of 28 °C day/24 °C night very successfully for a number of aroids (cf. Pardales and Dalion, 1986 for taro and Strauss and Arditti (1980), Quynh and Uyen (1987) and Liu et al. (1988) for *Xanthosoma*).

According to some (e.g. Arditti and coworkers), *antiquorum* varieties have been generally more responsive than *esculenta* varieties to aseptic culture procedures (Yam et al., 1990, 1991). Note that Mapes and Cable (1977) and Irawati and Webb (1983) both worked with var. *esculenta* (cf. Fig. 3). It has even been stated that methods for *C. esculenta* var. *antiquorum* do not work for *C. esculenta* var. *esculenta*. Use of lateral buds (rather than terminal ones) of *esculenta*, as well as the rather unconventional procedure of adding an extract made by boiling diced fresh taro corm (ca. 600 g/l of water) and adding 20 ml/l to M and S medium (the microelements of which have reduced to 1/10th normal level) along with 0.2–2.0 mg/l NAA and 0.2–2.0 mg/l BAP has been reported as being so dramatic in *esculenta* as to 'warrant establishment of two separate species, *C. esculenta* and *C. antiquorum*' (Yam et al., 1990). It should be noted that no quantitative data have yet been provided on responses *in vitro*. Neither has there been any work reported on addition of taro extract (TE) to cultures of *antiquorum* varieties to see if the two varieties respond differently. Moreover, no effort was apparently made to ash the extract and add it as a 'control' to a test medium lacking the TE, and it must remain a matter of speculation as to what the TE is doing. It could be a matter of adding mineral elements or of other heat stable constituents or growth regulators, or, conversely even the break-down products of heat labile constituents.

Taro plants showing some levels of salt tolerance have been generated by

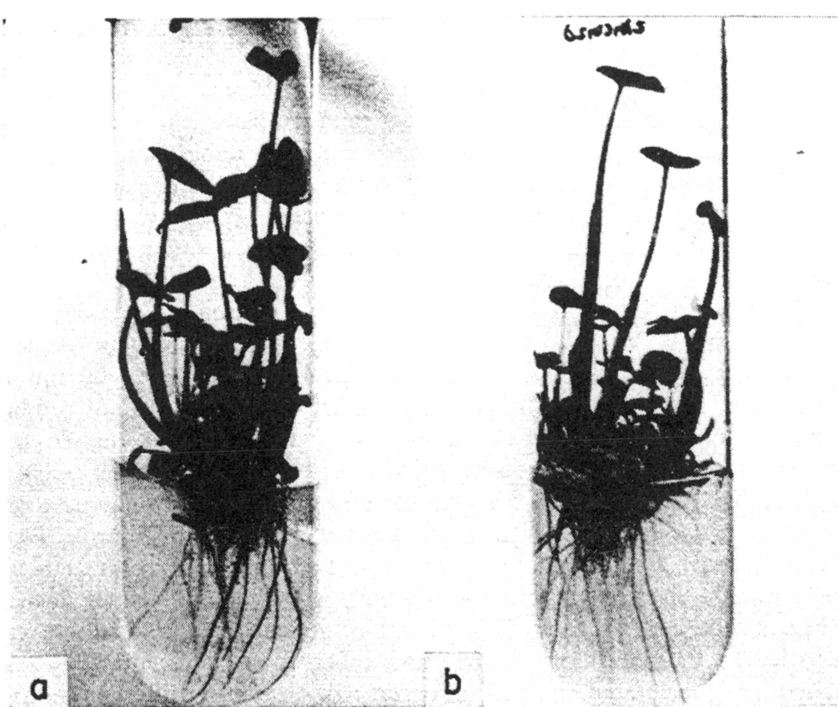

Figure 3. Micropropagated taro (*Colocasia esculenta* var. *esculenta*) *in vitro*. (a) tube with multiple shoots of Taro Niue (a preferred Samoan clone) generated from lateral buds on a shoot initially derived from a stem tip. Release of axillary buds was fostered after removal by 'decapitation' of the shoot tip of the primary axis; (b) Eleele Naioea cultivar (a clone with some resistance to *Phytophthora*) showing multiple shoots. Photographs supplied by the late Marion O. Mapes, University of Hawaii, Manoa.

exposing callus from shoot tip explants of var. *antiquorum* to selection pressure with varying levels (ca. 40% by volume) of sea water. There is no convincing evidence that these are genetically stable (cf. Nyman and Arditti, 1984). Protoplasts have been produced of taro from sections of aseptically produced leaves but no wall regeneration or cell division has been reported. Calcium oxalate crystals required removal by Ficoll density gradient separation (Nyman et al., 1984).

2.3. *Cassava* (*Manihot esculenta*)

Cassava, also known as *yuca* (Spanish), *manioc*, *tapioca* and *mandioca* (Portuguese) produces enlarged starch filled roots that are important as a source of calories for a very many people and animals in the tropics. Young shoots and leaves are also cooked and used as a vegetable in some areas of the world (O'Hair, 1990). Some variegated clones are attractive as ornamentals (Vasudevan et al., 1967). Cassava is generally high yielding, tolerant of poor soils and relatively free from pests and diseases. Once harvested, cassava roots are highly perishable and on storage rapid physiological and/or microbiological deterioration can occur. Cassava contains cyanide and cyanogenic glycosides (linamarin and lotaustralin) which can, if not processed properly, lead to toxicity. It has been shown, for instance, that cassava consumption is one of the several factors that lead to neurological disorders and goiter in populations in a poor nutritional state (Ermans et al., 1980).

Manihot is a Neotropical genus of the Euphorbiaceae comprising some hundred species dispersed from northern Argentina to northern Mexico/-southwestern United States. All the wild species are 2n = 4x = 36 and are interfertile. *M. glazovii*, a common Brazilian tree that does not produce tuberous roots, is a source of the once very popular Ceara rubber. It is immune to mosaic and brown streak virus diseases and bacterial blight as well, and has been used as a source of resistance in cassava breeding programs. Cassava is a short-lived, highly heterozygous perennial shrub. Tendency to flower is a clonal characteristic but when it flowers, it generally behaves as a short day plant and normally has male and female flowers, albeit in separate structures. But the female flowers open and close several days before the male flowers open and pollen is available, thus self-pollination is not facilitated but fruit set occurs as clones with different flowering times are grown together. Because the species is strongly outcrossing, each seedling becomes a potential variety or cultigen. This enormous variability is maintained with varying degrees of ease by vegetatively propagating clones from relatively long (ca. 30–40 cm) stem cuttings ('stakes') frequently taken from near the base of the plants. It is a relatively slow process for on average, only 10–20 cuttings can be achieved from a single mature plant per year. (Nevertheless, a method for rapid propagation of cassava by rooting of leaf cuttings under mist exists that can potentially yield 4 million plants from a single stock plant with 500 leaves (cf. Pateña and Barba, 1979). For a host of good biological reasons, internationally co-ordinated breeding efforts to increase root biomass/productivity, protein content and to improve tolerance to stresses of soil acidity, lower than optimal temperatures, drought, phosphorus-poor soils (the plant has a very high P requirement) have been met with limited success (Jennings and Hershey, 1985). Greatest success has derived from international distribution of elite clones which have been rendered virus and specific pathogen-free by means of combined meristem culture and heat treatment. Access to these clean clones has permitted dramatic improvement in yields where substantial decline had existed (see Chapters 3 and 9). Cassava common mosaic virus (CCMV) and African cassava mosaic virus (ACMV) and other virus diseases ('frogskin', brown streak virus etc.) are readily transmitted through cuttings (Lozano and Nolt, 1989) and even latent virus infections can have significant negative impact on yield and quality. Ultimately, breeding for virus resistance, which is costly and time consuming, or genetic transformation with virus coat protein genes, will have to provide the solution.

Some have noted that cassava was a relatively healthy crop as long as it was planted on a small scale basis. More recent and extensive monocultures exacerbated distribution of infected clones. While the viral diseases are the most destructive, a number of fungus diseases have worldwide distribution and have serious impact. *Cercospora*, *Colletotrichum* (causing stem anthracnosis), *Fomes* (which acts as a root fungus), *Sphaceloma* (cause of so-called superelongation disease because internodes elongate dramatically), *Diplodia* (provokes stem rot and destroys the roots) are a few examples. *Xanthomonas*-caused bacterial wilt can be another major problem (Lozano and Nolt, 1989). Clones have been identified that show varying degrees of tolerance-resistance but these may lack other desirable characteristics. Insect pests such as spider mites present their own challenges (O'Hair, 1990; Jennings and Hershey, 1985).

Elimination of cassava mosaic disease by meristem culture (explants up to 0.4 mm in length) after treatment of 'stakes' with heat (35 °C) for 30 days was reported in detail in 1975. Only explants exceeding 0.2 mm yielded plantlets; smaller explants yielded callus or callus with roots (cf. also Kartha et al., 1975; Chapter 3). The presence of leaf primordia was not a requirement for growth into a plantlet. The culture medium was M and S with the vitamins of Gamborg's B_5 supplemented with BAP, NAA and GA. Since then, the method has been perfected and made routine and reduced to a step-wise process by Roca and his co-workers (see Roca, 1980, 1982, 1984 for detailed instructions and protocols). The process is straightforward but takes some skill to execute, especially at the meristem excision stage since the meristem with or without two youngest leaf primordia are about 0.4–0.6 mm (cf. Fig. 4). Vegetative cuttings provide the starting material. Stakes (10–15 cm long and preferably from the middle of the plant) with dormant buds are surface disinfested (including a paraffin sealing of the upper cut end) and planted and treated with heat if necessary to get rid of viruses (cf. Roca and Jayasinghe, 1982). After a couple of weeks the buds sprout and will have grown enough to yield meristems appropriate for culture. Buds are collected (but not submerged in water), then exposed to disinfestation by a quick dip in 70% ethanol, followed by 2–3 min in a 0.5% hypochlorite solution. Buds are dissected under a stereoscopic microscope at 10–15 × mag using at different stages No. 11 scalpels and microscalpels made from cut double edge razor blades (see Arditti and Strauss, 1979 taro manual for an example of how to make one). Dissecting needles (3–5 mm in length) made from No. 23 hypodermic needles are used as well (see Roca publications for details with schematics). The so-called culture initiation stage is carried out on a semi-solid (0.6% agar) M and S basal salts with 2% sucrose and 0.05 mg/l BAP, 0.02 mg/l NAA and GA 0.05 mg/l (pH 5.5). (GA is apparently an important constituent for plantlet regeneration, Nair et al., 1979). Conditions include a day temperature of about 28–30 °C and a night tempera-

Figure 4. Multiplication of *Cassava* from stem tips. (a) shows a vegetative stem tip growing point used for culture. Primary explants comprise the apical meristem with 1 or 2 leaf primordia; (b) median longitudinal section of a vegetative meristem tip; (c) stage at which lateral buds are promoted on semi-solid medium in response to elevated BAP; (d) multiple shoots shown proliferated from lateral buds as at (c) in response to lowered BAP and high GA_3 in liquid shake medium (multiplication rate of 1:10); (e) rooted cultures, after 4 weeks, prior to potting; (f) potted plants in the greenhouse after 1 week. Photographs supplied by Dr. W.M. Roca, CIAT, International Center for Tropical Agriculture, Cali, Colombia. Work done by Roca, J. Narváez, J. Rodríguez, J. Beltrán, J. Roa and G. Mafla.

ture of 24–26 °C, 14 hours photoperiod, and yields from the meristem within 2–4 weeks (depending on cultivar) a small 1–2 cm tall shoot with a basal callus. The tip of the shoot so-produced is removed and rooted on semi-solid medium which is 1/3 strength M and S, 2% sucrose, and only 0.01 mg/l NAA (pH 5.7–5.8). This takes 1–2 weeks. The next step involves micropropagation, i.e. node cuttings are taken from aseptic plantlets, planted on the rooting medium just described and these yield in a few weeks a complete rooted plantlet that is ready for acclimatization and transferral to soil. An extension of the micropropagation technique involves media modification by addition of elevated (2.5–5.0 micromolar) BAP and 0.05 micromolar NAA to the primary explant to stimulate growth and emergence of multiple shoots;

these, in turn are allowed to undergo extension growth in a liquid medium in which the BAP is lowered to 0.25 micromolar and GA is maintained at 0.1 micromolar. This, in effect, yields more "nodes" for plantlet regeneration when planted on "rooting" medium (cf. Fig. 4). Both techniques are well established; world-wide germplasm distribution is routinely carried out and minimum growth storage of germplasm is standard as a supplement to conventional field germplasm maintenance. Cryogenesis still presents problems, however.

Stamp and Henshaw (1986) have summarized data on callus production from cassava derived mainly from cotyledon or leaf lobe explants and its subsequent potential to yield organized structures including somatic embryos. Indeed, high level somatic embryogenesis has been achieved (Szabados et al., 1987) using 2,4-D at a level of 4–16 mg/l on immature leaves and shoot tips (the former responding better than the latter), and plantlets have been recovered. Since there has been some interest in true seeds of cassava as planting material, this route may well have a role to play. In any case, somatic embryogenesis is an important step in achieving genetic engineering goals (Roca, 1990). Similarly, since protoplasts have been prepared from mesophyll cells and have been shown to grow into callus, their preparation from embryogenic cells and subsequent regeneration into plantlets is logical (Mabanza and Jonard, 1984). In addition to genetic engineering for such things as virus resistance and protein content (Roca, 1990), one should recall potential for mutation breeding (Vasudevan et al., 1967).

Meristem-derived plantlets have been observed to flower *in vitro* and produce viable pollen. Interestingly, this normally monoecious species behaved as if it was dioecious (cf. Tang et al., 1983).

2.4. *Yams* (*Dioscorea* spp.)

The monocotyledonous family Dioscoreaceae comprises over 600 species concentrated in but not limited to the tropics and consisting mainly of vines or trailing plants with underground tubers or axillary/aerial bulbils of varying shape and nature. The foliage of some is extremely attractive and various clones have been pointed out as having ornamental potential (Hawley, 1956). The so-called "true" yams provide some of the most popular and cheap, starch-rich staple foods of the hot, humid tropics. Over 50 have been cultivated as a food or collected/cultivated as a source of the pharmaceutically important steroidal sapogenin diosgenin (which is used as a precursor of sex hormones, corticosteroids and oral contraceptives) (Coursey, 1967; Onwueme, 1978).

There are about 12 species of *Dioscorea* that have edible tubers and 6 of these are the principal ones now grown worldwide. Many of these also produce bulbils as well. The group is inadequately studied, taxonomically complex and confusing. Among those most widely grown are *D. alata* (the so-called Water Yam or Greater Yam) which is the most geographically widespread species and hence what is generally meant when the word 'yam' is indiscriminately used), *D. esculenta* (the Chinese Yam), and *D. bulbifera* (the Air Potato). These are Asiatic species that have found world-wide distribution. *D. cayenensis* (the Yellow or Yellow Guinea Yam), *D. rotundata* (the White or White Guinea Yam – which accounts for the very largest proportion of edible yam grown in the world today), *D. bulbifera* (the Air Potato, thought to have evolved independently of the Asian forms), and *D. dumetorum* are examples of African species. *D. trifida* (the American Edible Yam, also widely called 'cush-cush') is a popular central and tropical American species but does not enter trade except locally. Examples from central America and Mexico which are important as sources of steroid precursors include *D. floribunda*, *D. composita* and *D. mexicana*. *D. deltoidea* from India is another example of an important sapogenin species. The alkaloid dioscorine is variously distributed but popular edible cultivars contain little. It is, moreover, heat labile and is destroyed on boiling. Wild species can be extremely toxic and hence have been used as poisons; elaborate detoxification procedures have been used to render toxic species edible.

Dioscorea flowers are nearly always unisexual and plants are normally dioecious. Cultivated yams have a low rate of flowering, rarely set fruit and, when they do, they frequently produce infertile seed (Coursey, 1967). From time immemorial, cultivars have been propagated vegetatively from tubers, either underground or aerial. As a result, there is considerable variation in chromosome numbers, even cultivars within the same species. The basic chromosome number of the Old World yams and African yams is $n = 10$ (the edible species being tetraploid), but higher levels of polyploidy such as hexaploid, octoploid, even decaploid and higher have been reported. In the yams of the New World the complement is based on $n = 9$, with most being tetraploids and hexaploids (cf. Pal and Sharma, 1980).

Morphologically the underground structure is complex and there remain many questions as to derivation of the edible parts. Some have viewed tubers as of root origin but work with seedlings has shown that tubers are a lateral development from the hypocotyl region. Hence it is more likely a stem tuber but there are no eyes or buds or identifiable nodes or internodes. Moreover, there are features of the tuber which are not typically stem-like, e.g. it is positively gravitropic. The proximal tip of the mature tuber (the end at which the stem is connected and the roots arise), called the yam 'head', does have visible buds, however.(Some have viewed the yam 'head' as a vestigial rhizome (Coursey, 1967; Onwueme, 1978)). 'Seed' tubers or 'setts' (about 4 oz. or more and with periderm present and treated with lime on the cut surfaces to foster suberization), generally but not invariably, from the head end are used to multiply the plant vegetatively. 'Tail' pieces can be used as well and are usually preferred to middle pieces. Onwueme (1973) studied the sprouting process from tuber pieces and showed that pieces with preformed buds sprouted fastest but shoot buds differentiated in all cases even from budless tuber pieces. These form from a meristematic layer only a very few mm from the surface of the tuber. Non-head pieces readily developed sprouting loci (described as callus-like pustules by Onwueme). The disadvantage of using storage organ setts is that material that could normally be used for food etc. is thus 'lost'. Yams are frequently harvested as early and late crops depending on variety etc. The first harvesting is often taken by cutting tubers from the head, taking care not to disturb the roots or the stem, replacing the earth so that the crown regenerates new adventitious tubers (called seed yams). Yams so removed may be used to plant another crop. A later (main) harvest is carried out a few months later. The procedure of early harvesting of mother tubers at the correct time (called milking) is clearly a clever management technique.

From all the above it becomes clear that identification of alternative sources of new propagating material becomes an important priority.

D. rotundata, *D. alata*, *D. dumetorum* are examples of commercially important yams that can be propagated by vine cuttings (Njoku, 1963). The age of the parent plant, and the presence of a node and a leaf is critical to regeneration. Axillary buds which are present do not develop in these cuttings, neither do they produce roots from the stem directly but (at least in the case of *D. alata* where it has been studied in detail) a bulbil-like organ from which roots and buds develop is formed from a bulbil primordium which is present in the node (cf. Okonkwo in Osuji, 1985). The value of clonal material generated in this way (although the rooting response is about 60%) is clear and more effort is being placed on evaluating the method although so far, it seems that the tubers produced from vine cuttings remain smaller than those regenerated from tubers. While the main means of multiplication has always been vegetative, work first carried out with *D. rotundata* has shown that with special attention seeds can be germinated and even yield monoecious plants and when planted directly in the field can produce a reasonable but still relatively small tubered crop (Sadik and Okereke, 1975a, b).

Yams have been generally regarded as relatively free of pests and diseases (Coursey, 1967; Onwueme, 1978; O'Hair, 1990 and refs. there cited) but the term relatively is significant and there are special problems in specific areas. Viruses, yam beetles, and tuber rot fungi all take their toll. Post harvest diseases of the tubers substantially reduce availability of the crop once it has been produced. An overall problem therefore is to improve productivity (overall yam yields are relatively low when compared to many other tuber crops), disease and pest tolerance. Also, reliable availability of clean seed material for planting and evaluation remains a high priority for small farmers. Storage protein levels and quality need to be improved as well. While yams remain a major source of calories for those who eat them most, they are also a main source of protein for the rural poor. Crude protein content of yams as eaten is about 1–3% despite the dry weight values being some 6–13%, a value that approximates that of cereals (Coursey, 1967) but it is still too low.

It has generally been held that prospects for yam improvement by conventional breeding are very poor (cf. e.g. Akoroda, 1983). No surprise therefore that yam improvement has been slow in coming for the same reasons stated for virtually all vegetatively propagated crops (cf. Ene and Okoli in Osuji, 1985). In yams this is especially so since there is too little research information on basic aspects of the various crops. Even in those cases where flowering occurs it remains inconsistent and the problems of hand pollinating present major obstacles to breeding. Even with special precautions, pollen viability is very limited (order of 8–14 days). There has been very little karyological work carried out on the edible species of yams, and certainly no data on linkage groups. Yam cultivation has traditionally involved staking and this increases costs of production but recent work in Puerto Rico indicates that excellent results can be achieved in that environment without staking provided clonal selection is carefully carried out. Leaf spot diseases such as caused by *Cercospora* spp. are a major problem in some clones. Where yams are grown intensively, yam die back caused by the anthracnose fungus *Colleotrichum gloeosporioides* can be a major problem but varieties differ considerably in their resistance to the disease. A soil-borne root rot fungus (*Rhizoctonia*) is the source of another problem in certain areas. Nematodes, especially ones like *Scutellonema bradys*, which preferentially destroy the meristem of the 'head' region of tubers and hence prevent sprouting of setts may also be problems.

2.4.1. *Tissue Culture-Related Biotechnologies*

Degras (in Sasson and Costarini, 1990) has provided an optimistic summary of the potential value of biotechnologies for yam improvement (cf. also Table 1). Management of genetic resources *in vitro* either as propagules or minisetts or bulbils provides the basis of a convenient way to generate planting material at will, especially since storage life of harvested yams is poor. *In vitro* storage certainly takes up little room in contrast to field or even nursery plantings. Germplasm exchange is also facilitated in this way and much needed studies on adaptation to various climatic regions can thus be fostered. Somaclonal variant generation by various tissue culture methods such as callus culture, somatic embryogenesis and protoplast fusion and regeneration is seen as a major novel way to get variation. This may also be combined with mutagenesis of *in vitro*-generated systems. Selection of clones for cultivation based on consumer preference or desired tuber characteristics is a major field trial activity where new markets are sought and the idea is that new genotypes might well be achievable.

2.4.2. *Multiplication from Nodes and Stem Tip Cultures*

Having seen the various modes of yam regeneration *in situ*, it will be appreciated that *in vitro* approaches have high potential for success because of the availability of preformed meristems, or alternatively, tissues that are capable of generating meristems *de novo*.

For *D. alata* and *D. rotundata* nodal explants, some 5–15 mm in vine length and comprising one or two axillary buds and 2–8 mm length of petiole have been shown to be excellent starting material provided the material is actively growing as evidenced by young, vigorous, succulent leaves. This means that time of year can be important.

Essentially, the objective is to render inactive meristematic tissue to be activated by the procedures. It has been recommended that mother plants be grown under 16 hours daylength to precondition them so as to be more uniformly responsive (Mantell and Haque, 1979; Mantell et al., 1979). A quick immersion of the primary explant in 70% ethyl alcohol, followed by exposure to 0.5% (w/v) solution of hypochlorite for 5 min and rinsing is usually sufficient to get clean material. This is dependent upon season however and more extreme treatments with prolonged agitation as on a shaker and with detergent may be needed. Nodal segments are planted upright in an agar based M and S medium (albeit with 2% sucrose) without auxin or cytokinin supplementation. Within a few weeks (20–30 days), shoots and roots have emerged from the axils. Removal of month-old plantlets and excision of their nodes and re-inoculation *in vitro* as single leaf segments at intervals of 2–3 weeks provides a ready means of multiplication. It has been estimated that 65,000 plants could be obtained from a single node in a six month period.

Excision of stem tip meristems from mother plants of *D. alata* maintained at 36 °C with hot air treatment for at least 14 days have yielded plants (about 19% success rate) that have been shown to be free of flexous rod viruses but not pinwheel 'inclusions' (Mantell et al., 1980). Explant size was 0.2–0.5 mm long and included 1–2 leaf primordia on the apical dome. Best recovery of plantlets from these small explants was achieved with M and S medium supplemented with 1.0 or 0.5 mg/l NAA and 0.2 or 0.1 mg/l BAP.

2.4.3. *Callus Cultures*

Early tissue studies focussed on callus production. Five to seven mm cube portions excised from *D. sansibarensis* bulbils could be grown on White's medium with 2,4-D and coconut water (CW) but showed a tendency to suberize; nevertheless some produced bud initials and roots (Rao, 1969). In 1973 Mapes and Urata reported callus production and plantlet regeneration from nodal segments grown on agar. Presence of the node was crucial to callus formation and subsequent production of plantlets. In the absence of cytokinin but with NAA, roots formed. M and S medium with CW and 2,4-D or NAA 5 mg/l did the best in terms of providing a steady supply of callus and plantlets. Removal of auxin from the medium but presence of CW helped plantlets to be established and permit transfer to soil. A key point in that paper is that more or less open-ended proliferating meristematic regions were the progenitors of the organized growth. Callus by itself was not maintainable. Similarly, Mascarenhas et al. (1976) were able to get callus production from a number of sites of *D. deltoidea* on a number of media but the tuberous region was the most responsive (recall the superficial meristem and its proliferative capability). High auxin medium generated and maintained in darkness followed by subculture to low or no auxin-containing medium with high cytokinin (BAP or kinetin) in the light fostered regeneration of shoots. Asokan et al. (1983) used darkness to initiate callus on small (6 mm) cubic explants from bulbils of *D. bulbifera* and *D. alata* using a range of growth regulators. Best callus formation and regeneration of plants from callus occurred when kinetin was used at a concentration of 10–14 mg/l in combination with 0.01 to 0.02 mg/l. Some three plantlets were produced per explant within a few months. The origin of the regenerants was not established, however, and there is a good chance that the superficial meristem present in tubers yielded the organized growth.

2.4.4. *Embryogenic Cultures*

Starting in the mid-1970s Ammirato (1984 for a comprehensive coverage of his work and the topic of yams *in vitro*) initiated studies on *D. floribunda* from the perspective of initiating embryogenic suspensions. Initially, he started cultures from zygotic embryos excised from mature seeds that had to be rigorously surface-sterilized. In darkness at somewhat elevated temperatures 25–27 °C (generally a good idea with all tropicals) M and S medium (semi-solid with 0.8% agar) with 4 mg/l 2,4-D was sufficient to yield embryogenic callus which, when transferred to liquid, permitted suspensions to be established. Maintaining a high density inoculum on subculture was especially important. Sieved suspensions (through a 250 μm mesh), washed of 2,4-D but exposed to 0.02 mg/l zeatin yielded small, conical embryos within 4–6 weeks. Further growth of somatic embryos was achieved by transfer to semi-solid medium with 0.02 mg/l zeatin and 500 mg/l glutamine. Abscisic acid (0.27 mg/l) was helpful in promoting normal embryo maturation while still in the liquid medium stage. Transfer to soil was effected using the usual precautions of thorough washing off of media and prevention of desiccation.

Chaturvedi and co-workers (Chaturvedi et al., 1977; Chaturvedi and Sinha, 1979; Sharma and Chaturvedi, 1989) have also done pioneering work on the medicinal yam species *D. floribunda* and *D. deltoidea* and have developed methods that are particularly attractive since they have taken regenerants to the field level and have considered time and financial constraints in their operations (Figs. 5–7). Moreover, the work done so far emphasizes considerable variability of response among species. Their somatic embryogenesis studies have been of special interest first because they use leaf tissue on a modified Schenk and Hildebrandt medium with BAP and 2,4-D, or BAP and 2,4-D and IAA as a starting point, and also because they note that incorporation of ammonium nitrate (500 mg/l) and reduction of levels of

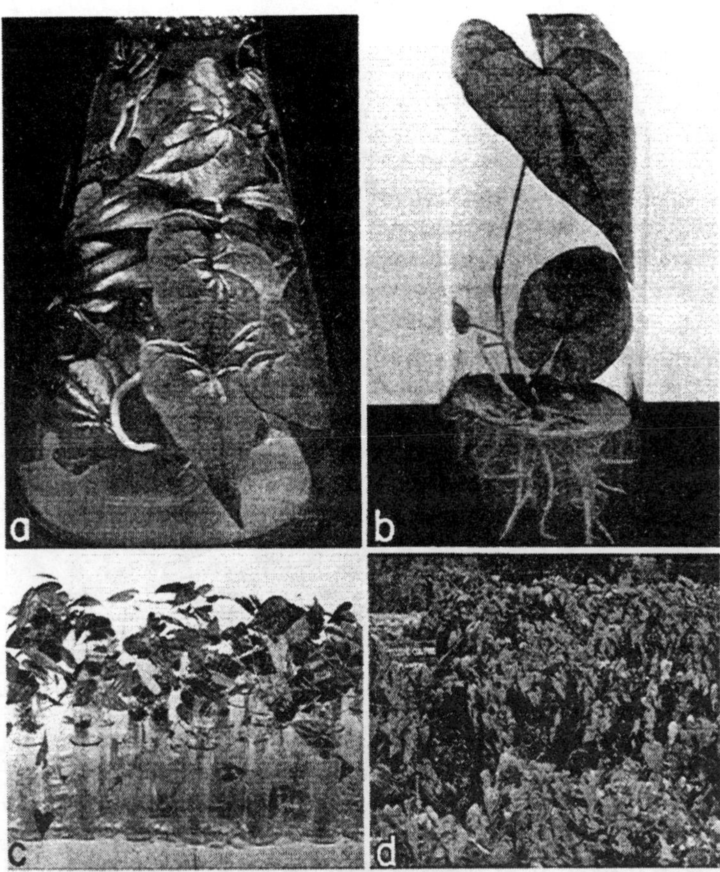

Figure 5. Clonal multiplication of *Dioscorea floribunda*, a diosgenin-producing species of Mexican origin. (a) a multi-branched plant established in aseptic culture; (b) a rooted, single-node leaf cutting taken from an aseptically grown plantlet; (c) for acclimatization, plantlets are washed free of semi-solid medium and placed in an inorganic salt solution for a period of a couple of weeks prior to establishment in soil; (d) growth at the field level is preceded by a period of growth in the greenhouse. Field cultivation here was photographed 1 year after transplantation. Photographs supplied by Dr H.C. Chaturvedi, National Botanical Research Institute, Lucknow, UP, India.

thiamine HCl, pyridoxine HCl, and nicotinic acid was essential for production of plantlets. The plantlets produced were transferable to soil at 100% efficiency and phenotypic and ploidy level differences were noted.

Zygotic embryos of *D. composita* (a medicinal yam) and the edible *D. cayenensis* (two species that can produce embryos) served as starting materials for callus induction on M and S medium with 2,4-D (Viana and Mantell, 1989). Plantlets were regenerated when transferred to medium with zeatin and glutamine following Ammirato's protocol. *D. rotundata* likewise

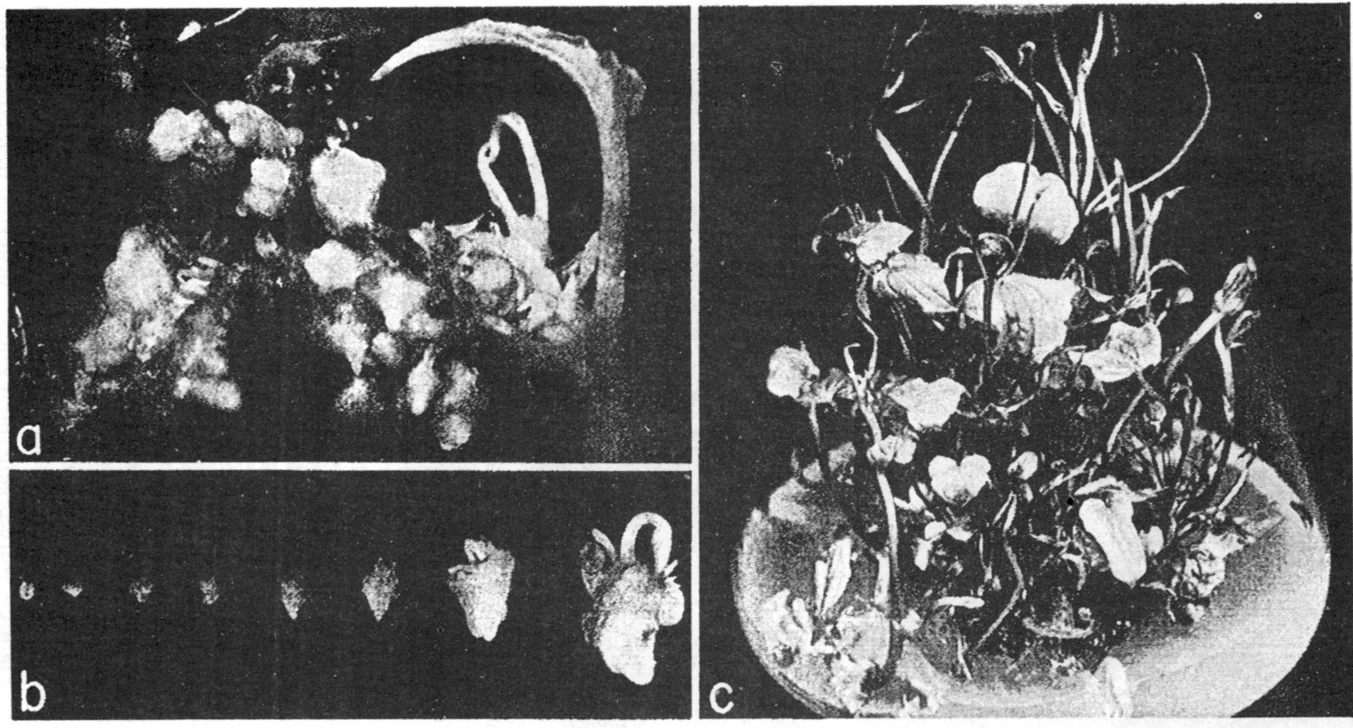

Figure 6. Production of somatic embryos of *Dioscorea deltoidea*, a diosgenin-producing species of Indian origin. (a) somatic embryogenesis from tuber-derived callus. Embryos are visible on the tissue surface; some have 'germinated' *in situ*; (b) somatic embryos at different stages of development. The mature ones (at the right) have begun to germinate; (c) erlenmeyer flask with proliferated and precociously germinated somatic embryos. Some of the plantlets so derived are multiple. Photographs supplied by Dr H.C. Chaturvedi, National Botanical Research Institute, Lucknow, UP, India.

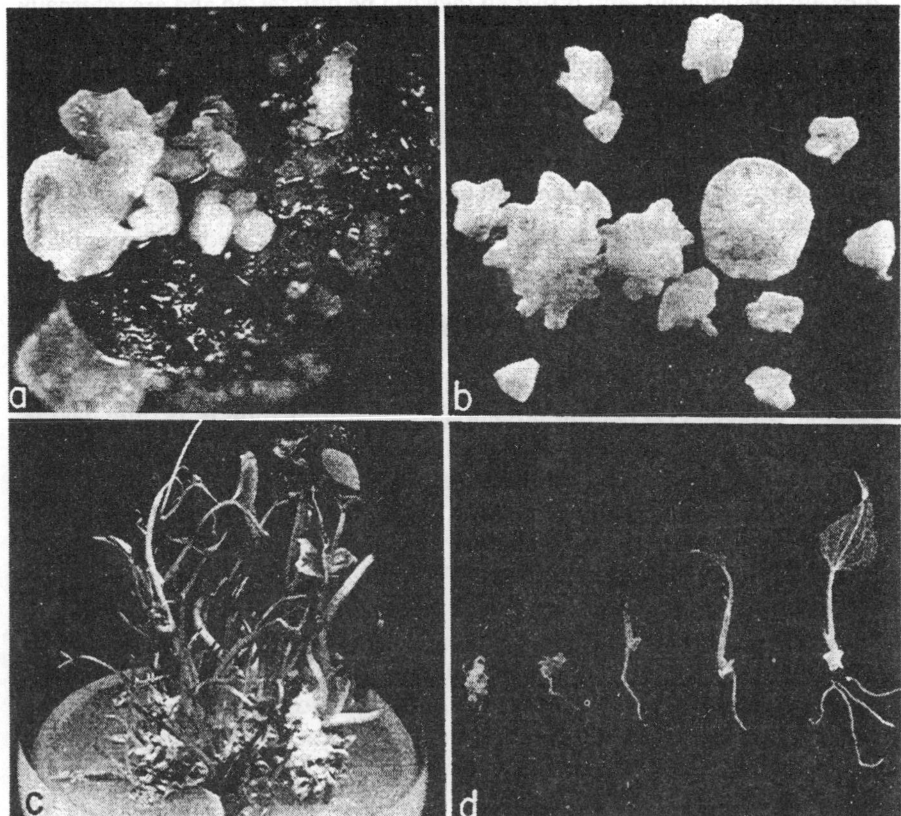

Figure 7. Production of somatic embryos of Dioscorea floribunda from leaf callus. (a) Close-up showing embryogenic tissues and embryos at different degrees of development (white areas); (b) somatic embryos isolated from a single culture showing different shapes and sizes (some twinned/multiple); (c) *in situ* of somatic embryos resulting in the formation of multiple plantlets; (d) different stages of plantlet formation from 'germinating' somatic embryos. Photographs supplied by Dr H.C. Chaturvedi, National Botanical Research Institute, Lucknow, UP, India.

has yielded somatic embryos from zygotic embryo-derived callus initiated with NAA (Osifo, 1988). Interestingly, callus generated and grown in light was embryogenic whereas dark reared material only produced roots. All this suggests that cell level differences may well be obtainable for use in improvement programs.

Protoplast work, when perfected, will have significant impact on this as well and one can be encouraged that regeneration will be achievable since embryogenic cell cultures are generally an excellent source of starting material.

Tubers also serve as an excellent starting material for embryogenic and organogenic callus. Both edible and medicinal yam species tubers, provided they are not dormant, and provided they can be disinfested and are internally clean, can yield explants that can give rise to callus. Esenowo (1986) worked with *D. rotundata* and showed that with Linsmaier and Skoog medium and with very high levels of cytokinin and auxin (order of 50–100 mg/l of KIN) along with 6% v/v CW one can sustain good growth.

A better way to start cultures and to avoid the frustrations of working with soil grown tubers is to generate tubers *in vitro* from single node segments or excised apical meristems and to use these as a source of primary explants (cf. Ammirato, 1984).

2.4.5. *Tuberization in vitro*

Asahira and Nitsch (1968) first reported that tuber formation could occur in culture. Uduebo (1971) drew attention to tuber formation in *D. bulbifera* in the presence of low auxin levels. Since then, nodal cultures and callus cultures from various species and cultivars (e.g. *D. bulbifera*, *D. floribunda*, *D. rotundata*, *D. alata*) have been shown to produce microtubers, although of greatly varying sizes (Forsyth and van Staden, 1984; Sengupta et al., 1984; Ng, 1988) under varying conditions. The value of proper photoperiod (generally short day conditions) and 'night' temperatures is drawn attention to in most of these cases but the duration of light exposure can be altered by the amount of sugar present in the medium and the growth regulator complement (cf. Forsyth and van Staden, 1984). In most cases a simple basal medium like that of M and S supplemented (or not) with 0.5 mg/l KIN and a daylength of 12 or 16 hours is sufficient to generate tubers in a matter of several weeks. An 8 hour daylength is insufficient to foster tubers in *D. rotundata* (Ng, 1988).

Microtuber production from the nodes, either aerial or at the level of the agar surface, has great potential for storage and distribution of elite germplasm of the genus. Although the problem remains of not being able to use the microtubers directly as planting material to generate edible yams because of their small size (just like the tubers from stem cuttings and seeds), they grow well if gotten into the soil relatively promptly, and can, in time, yield appropriately sized materials for use as setts. *In vitro* plantlets do not tolerate shipping conditions well (e.g. darkness) so microtubers are an excellent alternative. It is anticipated that they will play an increasingly important role in germplasm distribution since virus can be eliminated or indexed (Ng, 1988).

2.5. Beet (*Beta vulgaris*)

Beta vulgaris spp. *vulgaris* (Chenopodiaceae) comprises many edible cultivars, including sugarbeet, red beetroot, mangolds, and fodder beets. The plant is an outbreeder, hence highly heterozygous and a biennial. Diploid and triploid forms are important to agriculture. Production of beets is generally restricted to temperate, even cool climates.

The diploid chromosome numbers within the genus are 2n = 2x = 18. Both haploid and diploids have important applications in breeding. Many varieties that have long been available are being replaced with so-called monogerm varieties. Interest in monogerm varieties derives from the ability to completely mechanize crop production with these. A monogerm plant has flowers singly at most nodes of the inflorescence whereas the usual forms bear them in clusters. Fruits developed at a single node form an indehiscent cluster. This is the 'seed' used for planting. Since each fruit is single-seeded, aggregate fruits of a multigerm variety are multi-seeded; most of those from a monogerm are single-seeded. Multigerm varieties must be therefore be 'singled' and require hand operation. This is done in the field after emergence and is obviously a very laborious process. Direct seed drilling of monogerm varieties would obviate this need. Since the monogerm character is an inflorescence character, and is determined by the genotype of the seed parent, it is possible to use a multigerm pollinator on a male-sterile monogerm.

In addition to generation of valuable monogerm and other haploids and dihaploids etc. for breeding, enhancement of disease (especially curly-top etc. viral) resistance, and resistance to premature bolting or flowering which reduces the sugar yield are but a few important objectives in the breeding of sugar beets.

Tissue culture, therefore, has potential to make contributions in this area via androgenesis, parasexual hybridization, somaclonal variation, chromosome and gene transfer operations, rescue of interspecific and intergeneric crosses related embryo failures, germplasm conservation *in vitro* etc.

Hussey and Hepher (1978) reported that about 10% monogerm diploid seed could be aseptically germinated on a half-strength M and S medium with 2% sucrose and agar (20 °C, 16 hr light) until the first pair of leaves of seedlings were about 10–15 mm high. They were then surgically deprived of their cotyledons and roots and all but 4 mm of the hypocotyl and planted on a medium containing BAP. After about 5 weeks at 0.12 and 0.25 mg/l BAP, precocious axillary shoots emerged from the axils of the older leaves. Roots did not form. Axillary branches so produced could be serially propagated and roots formed when the BAP levels were reduced to below 0.03 mg/l. Adventitious shoots could form when cut ends of petioles were placed in BAP-containing medium. Axillary shoots and leaves treated with 0.3% colchicine to induce polyploidy resulted in the production polyploids and some chimeric forms. Eventually some solid tetraploids were formed. Moreover, it was shown that lower temperatures (order of 12 °C) permitted cultures to be subcultured every 18 weeks. The significance of this for minimal growth maintenance of germplasm was recognized. Since then potential for successful long term cryopreservation has been reported using aseptically isolated apical meristems (Braun, 1988).

Callus growth is very variable but a wide range of tissues has been shown to respond (Atanassov, 1986). De Greef and Jacobs (1979) have shown that selection of appropriate cell type, while important in all culture systems, can be especially important in sugarbeet. Also, it is significant that tissues can be habituated for both auxin and cytokinin within a relatively short time (more readily for cytokinin than for auxin) after initiation of a culture and still retain the ability to regenerate. These workers isolated a habituated cell line with very high regenerative capacity when they cultured leaf segments of sugarbeet at 24 °C in darkness followed by stress at 4 °C. Of special significance is that the line maintains its habituated nature and ability to regenerate, even when plants generated from the habituated lines are put through a second culture cycle at 20 °C. Indeed, the process has been regularized and used routinely for research purposes (van Geyt and Jacobs, 1985). Production of haploid sugarbeets via ovule culture has been achieved using young ovules (ca. 0.3–0.6 mm in length) on a simple medium with activated charcoal and supplemented with 0.3 mg/1 BAP and 0.05 mg/1 2, 4-D (D'Halluin and Keimer, 1986). Somatic embryos produced could make secondary embryos and showed polyembryony, but the response was highly genotype dependent. An interesting additional requirement seemed to be that donor plants be grown in light supplemented with incandescent bulbs. Greenhouse-grown plants were more responsive that growth chamber ones. Ninety percent of the somatic embryo-derived plants were haploids, 9% were spontaneous doubled haploids and 0.3% were tetraploids.

Protoplasts will have potential for additional improvement schemes and there are indications that selection of the right genotypes will help resolve some of the problems in generating regenerable materials (Bhat et al., 1986).

Braun (1988) carried out studies on cryopreservation of meristems that had been precultured on M and S medium supplemented with BAP, and the anti-auxin triiodobenzoic acid and dimethylsulfoxide (DMSO). Meristems were frozen to minus 40 °C and stored in liquid nitrogen. Cultures could be reestablished when they were quick thawed at 39 °C.

2.6. *Carrot* (*Daucus carota*)

So much has been written about carrot in terms of virtually all aspects of tissue culture that little needs to be added here. Cultivated carrots (*D. carota* var. *sativa*) and wild carrots (*D. carota* var. *carota*) are all 2n = 2x = 18 in terms of their cytotaxonomic background. Breeding of carrot has been a relatively recent activity and production of hybrids via the F1 approach using cytoplasmically male-sterile lines has been an integral part of the work. There is lack of uniformity of seed behavior, however, and there is inbreeding depression which leads to lack of vigor. Sterile or 'maintainer' lines have been difficult to keep as well. Even so, there are quite a few hybrids on the market.

Breeding has concerned itself with yield, shape, color (carotene content), earliness, top-root ratio, resistance to early flowering and other quality characteristics such as taste and tenderness etc.

From the biotechnology perspective, the uses of tissue culture methods on carrot is justified by the elements listed in Table 1. For example, use of somatic embryos for the still seminal artificial seed industry is but one of the applications that has been speculated upon. Indeed, the 'tarnished plant bug' *Lygus* spp. is an insect that frequently causes problems in the field since it can selectively remove developing embryos leaving an embryo-less 'seed'. With increasing interest in pesticide-free or pesticide-reduced agriculture, there may be a real application here. Others have speculated on the possible use of somatic embryogenesis-generated variant lines for specialty uses in the snack food industry.

Perhaps the greatest value of the carrot system is that because it has been worked on for a very long time it can serve as a model for the study of basic principles.

2.6.1. *Culture of Carrot Tap Root Phloem*

It has frequently been overlooked that carrot tap root phloem was the first source of material used for the generation of embryogenic cell suspensions. It was, moreover, the first material used on semi-solid medium that resulted in somatic embryo production (cf. Steward et al., 1975). More recently, greater attention has been placed on the use of petiole explants or zygotic embryos as a source of embryogenic cell lines (Ammirato, 1989). Because access to protocols for storage root are less readily available, a few guidelines are presented. While the procedures may seem considerably more labor-intensive than those using petioles or embryos, it has its own merits. As in all cases, the objectives sought determine the method(s) adopted.

A feature of the procedure as given below is that it entails use of a special cutter or 'vegetable slicer'. One can easily be fabricated using a rotary microtome blade as the cutter and some heavy plate stainless steel (cf. Fig. 8a). Alternatively, various commercial slicers that are sterilizable by any of several chemical means are available on the market. Also, a special cutting or explantation instrument is needed (cf. Fig. 8b and c). For instance, one can be made from a straight end needle of about 18 gauge and fitted with a rod or 'plunger' that has a close, yet freely movable tolerance. This is inserted so that extrusion of cut explants or 'plugs' can be achieved when it is depressed. When used in the cutting mode, the plunger is retracted (Fig. 8b). A crucial component of the cannula is an external 'collar' (Fig. 8b and c) made of glass tubing or stainless steel that prevents handling of the canula on surfaces that could be in contact with the aseptic plant material as explantation is being carried out. It is also necessary when making explants from turgid slices so as to exert a counterforce against the tissue being explanted and thus prevent carrot slices from being lifted by the implanted cannula.

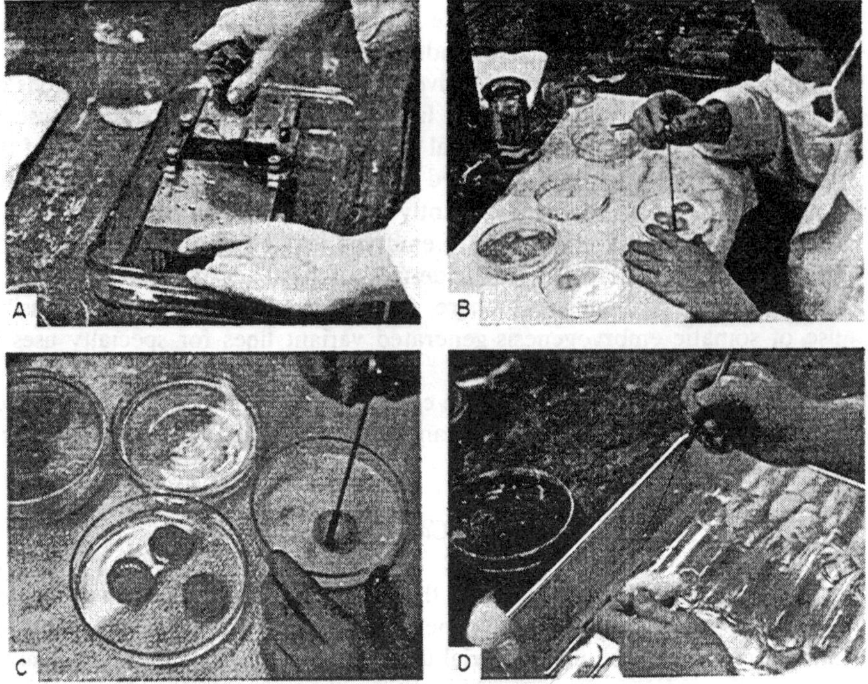

Figure 8. Demonstration of the use of cutter and cannula to make slices and carry out explantation of plugs from storage root of carrot. The technique is applicable to virtually any fleshy tuber. See text for details.

Use of a simple basal medium such as that of M and S with 2 mg/l 2,4-D or NAA is sufficient to generate an embryogenic callus from plugs removed from carrot slicers (avoid vascular cambium. White's medium will work as well (Krikorian and Smith, 1992). If the explants are placed in liquid medium, embryogenic cells are sloughed off into the medium. These may be sieved from the primary explant and subcultured serially into fresh medium of identical composition. Selective filtration as to size accompanied by visual monitoring will give a better embryogenic suspension than mere transferral. It will be easier, although not faster, to initiate on semi-solid medium and select globular stage embryogenic masses from the periphery of the explants. These masses, which are 'pearly white' can be built up through subculture, and then dispersed in liquid media of identical composition for further multiplication and build-up of supplies.

Figure 9 provides a summary of sequences in the generation and management of embryogenic somatic cells of carrot. While provided here in the context of carrot, the principles apply to virtually any plant.

A more recent procedure that has been developed in this laboratory (cf. Smith and Krikorian, 1990; Krikorian and Smith, 1992) to permit selection

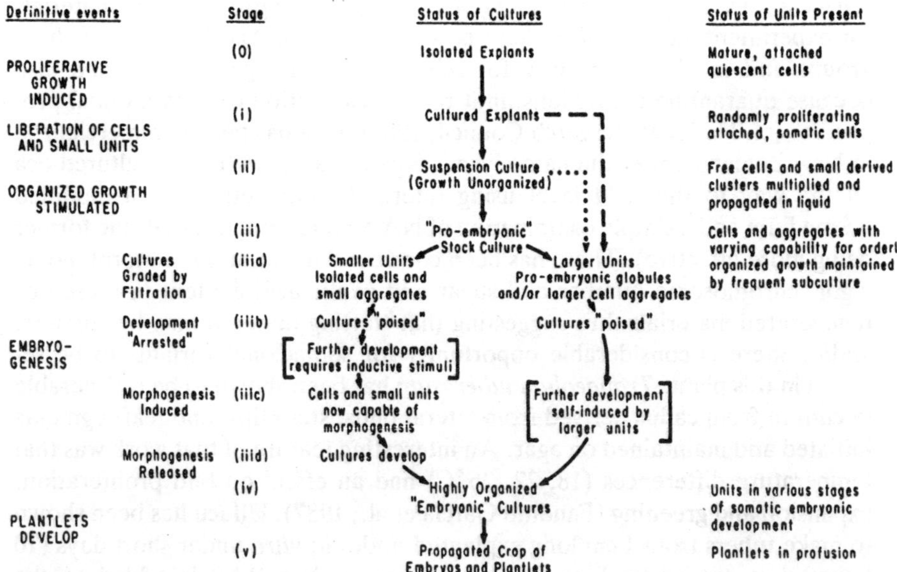

Figure 9. Stages associated with somatic embryogenesis from cultured cells of carrot grown in suspension. An attempt has been made to outline the sequences of definitive stages and events in culture. The areas of special significance from the perspective of exogenous growth regulators are indicated as Stage 0 to Stage i (Proliferated, i.e. Embryogenic Growth Induced) and Stage ii to Stage iii (Organized Growth Stimulated, i.e. Continued Development of Suspensions of Minimally-Progressed ('Unorganized') Somatic Embryos). Nevertheless, at each point along the way, growth regulators and other factors can exert their influence. The features outlined here are applicable to many embryogenic systems besides carrot. See text and Steward et al. (1975) and Krikorian and Smith (1992) for additional details.

of cells that show high embryo forming potential has been to transfer very early stage embryogenic cells initiated on hormone-containing medium to a medium lacking hormone and lowering the pH to the acid side (pH 4.5). In this procedure, all non-embryogenic cells succumb.

2.7. *Oca* (*Oxalis tuberosa*) *and Other 'Root' Crop Plants*

There are a number of 'underexploited' root crops of the New World that are increasingly being appreciated outside their hitherto very limited range (cf. O'Hair, 1990). In some cases they are being appreciated as dietary substitutes for the weight conscious or diabetics. Achira (*Canna edulis*, Cannanaceae), ahipa (*Pachyrhizus ahipa*, Leguminosae), arracacha (*Arracacia xanthorrhiza*, Umbellifereae), maca (*Lepidium meyenii*, Cruciferae), mashua (*Tropaeolum tuberosum*, Tropaeolaceae), mauka (*Mirabilis expansa*, Nyctaginaceae), oca (*Oxalis tuberosa*, Oxalidaceae), ulluco (*Ullucus tuberosus*, Basellaceae), yacon (*Polymnia sonchifolia*, Compositae) provide good materials for experimentation because there is considerable diversity in each of these groups and much opportunity for selection of elite germplasm, and also because quarantine restrictions limit ready distribution of conventional propagules (cf. National Research Council, 1989). Oca has been commercialized in New Zealand under the name 'yam'. Khan et al. (1988) have cultured oca all the way to the field level using internode and petiole on a modified B5 medium and a simple combination of NAA and zeatin or BAP (the former being more effective). There has been considerable variation in morphology, vigor, chromosome number and sugar and oxalic acid content in these oca regenerated materials thus suggesting that starting material may be chimeric and/or there is considerable opportunity for somaclonal variants to be selected in this plant. *Tropaeolum tuberosum* has been shown to be regenerable in culture from callus derived from lateral buds, stem tips, and leaf segments initiated and maintained on agar. An interesting feature of that work was that temperature differences (18, 22, 26 °C) had an effect on bud proliferation, expansion and greening (Fandino Garcia et al., 1987). Ullucu has been shown to make tubers from 1 cm long explanted nodes *in vitro* under short days (10 hours) but not under long (16 hours) days unless BAP is added to the medium (Asahira and Nitsch, 1968).

Another example of many that could be given is *Coleus barbatus* (*C. forskholii*, Labiatae), a native of India, which produces a root tuber that has long been used as a condiment, and has become increasingly attractive because it is a source of the diterpenoid forskolin (also called coleonol), a compound that stimulates adenylate cyclase in such a way as to enable study of cyclic-AMP. It is also viewed as a valuable medicinal agent for use in hypertension, glaucoma, asthma and certain cancers (Valdés et al., 1987).

3. References

Abo El-Nil, M.M. and Zettler, F.W. (1976). Callus initiation and organ differentiation from shoot tip cultures of *Colocasia esculenta*. Plant Sci. Lett. 6:401–408.

Akoroda, M.O.(1983). Variation, heritability and genetic advance of eight characters in white yam. Theor. Appl. Genet. 66:51–54.

Ammirato, P.V. (1984). Yams. In: Handbook of Plant Cell Culture W.R. Sharp, P.V. Ammirato and Y. Yamada (eds.), Vol. 2. Macmillan, New York, pp. 327–354.

Ammirato, P.V. (1989). Carrot. In: Handbook of Plant Cell Culture. D.A. Evans, W.R. Sharp and P.V. Ammirato (eds.), Vol. 4, Macmillan, New York, pp. 457–499.

Arditti, J. and Strauss, M.S. (1979). Taro Tissue Culture Manual. South Pacific Commission, Noumea, New Caledonia.

Asahira, T.L. and Nitsch, J.P. (1968) Tubérisation *in vitro*: *Ullucus tuberosus* et *Dioscorea*. Bull. Soc. Bot. France 115:345–352.

Asokan, M.P., O'Hair, S.K. and Litz, R. E. (1983). *In vitro* plant development from bulbil explants of two *Dioscorea* species. Hort. Sc. 18:702–703.

Asokan, M.P., O'Hair, S.K. and Litz, R.E. (1984). In vitro plant regeneration from corm callus of *Amorphophallus rivieri* Durieu. Sci. Hort. 24:251–256.

Atanassov, A.I. (1986). Sugar beet (*Beta vulgaris* L.). In: Biotechnology in Agriculture and Forestry 2. Crops I. Y.P.S. Bajaj (ed.), Springer Verlag, Berlin, pp. 462–470.

Bhat, S., Ford-Lloyd, B.V. and Callow, J.A. (1986). Tissue and protoplast culture in cultivated beets. In: Genetic manipulation in plant Breeding. Proceedings of the International Symposium Organized by EUCARPIA September 8–13, 1985, Berlin. W. Horn, C.J. Jensen, C.J. Jensen, W. Odenbach and O. Schieder (eds.), pp. 453–455.

Bidney, D.L. and Shepard, J.F. (1980). Colony development from sweet potato petiole protoplasts and mesophyll cells. Plant Sci. Letters 18:335–342.

Braun, A. (1988). Cryopreservation of sugarbeet germplasm. Plant Cell Tissue Org. Cult. 14:161–168.

Brücher, H. (1989). Useful Plants of Neotropical Origin and their Wild Relatives. Springer-Verlag, Berlin.

Chaturvedi, H.C. and Sinha, M. (1979). Mass Propagation of *Dioscorea floribunda* by Tissue Culture. Extension Bulletin No. 6. National Botanical Research Institute, Lucknow, India.

Chaturvedi, H.C., Sinha, M. and Sharma, A.K. (1977). Clonal propagation of *Dioscorea deltoidea* Wall. through *in vitro* culture of shoot apices and single-node leaf cuttings. In: Cultivation & Utilisation of Medicinal and Aromatic Plants. C.K. Atal and B.M. Kapur (eds.), Regional Research Laboratory, Jammu (India), pp. 500–505.

Chée, R.P. and Cantliffe, D.J. (1988). Selective enhancement of *Ipomoea batatas* Poir. embryogenic and non-embryogenic callus growth and production of embryos in liquid culture. Plant Cell Tissue Org. Cult. 15:149–159.

Chée, R.P. and Cantliffe, D.J. (1988). Somatic embryony patterns and plant regeneration in *Ipomoe batatas* Poir. *In Vitro* Cellular & Developmental Biology 24:955–958.

Coursey, D.G. (1967). Yams. An Account of the Nature, Origins, Cultivation, and Utilisation of the Useful Members of the Dioscoreaceae. Longmans, London.

De Greef, W. and Jacobs, M. (1979). *In vitro* culture of the sugarbeet: Description of a cell line with high regeneration capacity. Plant Sci. Letters 17:55–61.

D'Halluin, K., and Keimer, B. (1986). Production of haploid sugarbeets (*Beta vulgaris* L.) by ovule culture. In: Genetic manipulation in Plant Breeding. Proceedings of the International Symposium Organized by EUCARPIA September 8–13, 1985, Berlin. W. Horn, C.J. Jensen, C.J. Jensen, W. Odenbach and O. Schieder (eds.), pp. 307–309.

Elliot, R.F. (1969). Growth of excised meristem tip of kumara, *Ipomoea batatas* (L.) Poir., in axenic culture. N.Z. J. Botany 7:158–166.

Ermans, A.M., Mbulamoko, N.M., Delange, F. and Ahluwalia, R. (eds.) (1980). Role of Cassava in the Etiology of Endemic Goitre and Cretinism. IDRC-136e. International Development Research Centre, Ottawa.

Esbenshade, P.R. and Moyer, J.W. (1982). Indexing system for sweet potato feathery mottle virus in sweet potato using enzyme-linked immunosorbent assay. Plant Dis. 10:761–768.

Esenowo, G.J. (1986). The initiation and maintenance of callus from tubers of *Colocasia esculenta* and *Dioscorea rotundata*. Environmental and Experimental Botany 26:159–162.

Fandiño Garcia, T.J., Fernandez O. and Perea Dallos, M. (1987). Morfogenesis y micropropagacion de *Tropaeolum tuberosum* (Ruiz y Pavon). ACEVIV. Boletin Cientifico (Asociacion Colombiana de Estudios Vegetales *in vitro* (Bogotá) No. 2, 29–35.

Forsyth, C. and van Staden, J. (1984). Tuberization of *Dioscorea bulbifera* stem nodes in culture. J. Plant Physiol. 115:79–83.

Frison, E.A. and Ng, S.Y. (1981). Elimination of sweetpotato virus disease agents using enzyme-linked immunosorbent assay. Trop. Pest Mgt. 27:452–454.

Gunkel, J.E., Sharp, W.R., Williams, B.W., West, W.C. and Drinkwater, W.U. (1972). Root and shoot initiation in sweet potato explants as related to polarity and nutrient media variations. Bot. Gaz. 133:254–262.

Hartman, R.D. (1974). Dasheen mosaic virus and other phytopathogens eliminated from caladium, taro, and cocoyam by culture of shoot tips. Phytopathology 64:237–240.

Hawley, W.O. (1956). Dioscoreas as ornamental foliage plants. American Horticulture Magazine 35:23–29.

Henderson, J.H.M., Philis, B.R. and Whatley, B.T. (1984). Sweet potato. In: Handbook of Plant Çell Culture. W.R. Sharp, P.V. Ammirato and Y. Yamada (eds.), Vol. 2. Macmillan, New York, pp. 302–326.

Hussey, G. and Hepher, A. (1978). Clonal propagation of sugar beet plants and the formation of polyploids by tissue culture. Ann. Bot. 42:477–479.

Hwang, L.S., Skirvin, R.M., Casyao, J. and Boukamp, J. (1983). Adventitious shoot formation from sections of sweet potato grown *in vitro*. Sci. Hort. 20:119–129.

Irawati and Webb, K.J. (1983). Callus production and organogenesis from shoot tip and petiole explants of six Indonesian cultivars of *Colocasia esculenta* var. *esculenta*. Annales Bogoriensis 8:13–22.

Jackson, G.V.H. and Firman, I. (1982). Guidelines for the movement of germplasm of taro and other aroids within the Pacific. In: Regional Meeting on Edible Aroids. Suva, Fiji, 1981. International Foundation for Science, Stockholm, pp. 290–323.

Jennings, D.L. and Hershey, C.H. (1985). Cassava breeding: a decade of progress from international programmes. In: Progress in Plant Breeding. G.E. Russell (ed.), Butterworths, London, pp. 89–116.

Kartha, K.K. and Gamborg, O.L. (1975). Elimination of cassava mosaic disease by meristem culture. Phytopathology 65:826–828.

Kartha, K.K., Gamborg, O.L., Constabel, F. and Shyluk, J.P. (1974). Regeneration of cassava plants from apical meristems. Plant Sci. Letters 2:107–113.

Kay, D.E. (1973). Root Crops. Tropical Products Institute, Crop and Product Digest 2. Tropical Products Institute, London.

Khan, M.R.I., Heyes, J.K. and Cohen, D. (1988). Plant regeneration from oca (*Oxalis tuberosa* M.): the effect of explant type and culture media. Plant Cell Tissue Org. Cult. 14:41–50.

Kiyosue, T., Takano, K., Kamada, H. and Harada, H. (1990). Induction of embryogenesis in carrot by heavy metal ions. Can. J. Bot. 68:2301–2303.

Krikorian, A.D. and Smith, D.L. (1992). Somatic embryogenesis in carrot (*Daucus carota*). Plant Tissue Culture Manual: Fundamentals and Applications. K. Lindsey (ed.), Kluwer Academic Publishers, Dordrecht, pp. PCAM-A9-1-32.

Litz, R.E. and Conover, R.A. (1978). *In vitro* propagation of sweet potato. Hort. Sc. 13:659–660.

Liu, J.R. and Cantliffe, D.J. (1984). Somatic embryogenesis and plant regeneration in tissues of sweet potato (*Ipomoea batatas* Poir.). Plant Cell Rep. 3:112–115.

Liu, L.-J., Rosa-Márquez, E., Licha, M. and Biascoechea, M.L. (1988). Tanier (*Xanthosoma* spp.) propagation *in vitro*. J. Agric. Puerto Rico 72:413–426.

Love, S.L., Rhodes, B.B. and Moyer, J.W. (1987). Meristem-Tip Culture and Virus Indexing of Sweet Potatoes. Department of Horticulture, Clemson University, Clemson, S.C. and IBPGR, Rome.

Lozano, J.C. and Nolt, B.L. (1989). Pests and pathogens of cassava. In: Plant Protection and Quarantine. R.P. Kahn (ed.), Vol. II. Selected Pests and Pathogens of Quarantine Significance. CRC Press, Boca Raton, Florida, pp. 169–182.

Mabanza, J. and Jonard, R. (1984). Sur l'isolement et le développement *in vitro* des protoplastes de *Manihot glazovii* Muell. C.R. Acad. Sci. (Paris) 298:563–566.

Mantell, S.H. and Haque, S.Q. (1979). Disease-free Yams: Their Production, Maintenance and Performance. Yam Virus Bulletin No. 2. Caribbean Agricultural Research and Development Institute, University Campus, St. Augustine, Trinidad.

Mantell, S.H., Haque, S.Q. and Whitehall, A.P. (1979). A Rapid Propagation System for Yam. Yam Virus Project Bulletin No. 1. Caribbean Agricultural Research and Development Institute, University Campus, St. Augustine, Trinidad.

Mantell, S.H., Haque, S.Q. and Whitehall, A.P. (1980). Apical meristem tip culture for eradication of flexous rod viruses in yams (*Dioscorea alata*). Trop. Pest Management 26:170–179.

Mapes, M.O. and Cable, W.J. (1977). "Mericloning" of taro (*Colocasia esculenta*). Proceedings of the Third International Society for Tropical Root Crops. Ibadan. Nigeria. December 1973.

International Society for Tropical Root Crops in / collaboration with the International Institute of Tropical Agriculture, Ibadan, pp 123–126.

Mapes, M.O. and Urata, U. (1973). Aseptic stem culture of a *Dioscorea* clone. In: Tropical Root and Tuber Crops Tomorrow, Vol. II. Proceedings of the Second International Symposium on Tropical Root and Tuber Crops, pp. 25–27.

Martin, F.W. and Jones, A. (1986). Breeding sweet potatoes. Plant Breeding Revs. 4:313–345.

Mascarenhas, A.F., Hendre, R.R., Nadgir, A.L., Ghugale, D.D., Godbole, D.A., Prabhu, R.A. and Jagannathan, V. (1976). Development of plantlets from cultured tissues of *Dioscorea deltoidea* Wall. Indian J. Experimental Biol. 14:604–606.

Morel, G. and Wetmore, R.H. (1951). Tissue culture of monocotyledons. Am. J. Botan. 38:138–140.

Nair, N.G., Kartha, K.K. and Gamborg, O.L. (1979). Effect of growth regulators on plant regeneration from shoot apical meristems of cassava (*Manihot esculenta* Crantz) and on the culture of internodes *in vitro*. Z. Pflanzenphysiol. 95:51–56.

National Research Council (1989). Lost Crops of the Incas. Little-Known Plants of the Andes with Promise for Worldwide Cultivation. National Academy Press, Washington, DC.

National Research Council (1990). Plant Biotechnology Research for Developing Countries. National Academy Press, Washington, DC.

Ng, S.Y.C. (1988). *In vitro* tuberization in white yam (*Dioscorea rotundata*). Plant Cell Tissue Org. Cult. 14:121–128.

Njoku, E. (1963). The propagation of yams (*Dioscorea* spp.) by vine cuttings. W. African Science Assoc. J. 8:29–32.

Nyman, L.P. and Arditti, J. (1984). Production of salt-tolerant tissue and plantlets of taro. In: Edible Aroids. S. Chandra (ed.), Clarendon Press, Oxford, pp. 178–185.

Nyman, L.P. and Arditti, J. (1985). Germination of taro seeds. Aroideana 8:83–88.

Nyman, L.P., Strauss, M.S. and Arditti, J. (1984). Preparation and culture of protoplasts of taro. In: Edible Aroids. S. Chandra (ed.), Clarendon Press, Oxford, pp. 186–193.

Nyman, L.P., Webb, E.L., Gu, Z. and Arditti, J. (1986). Structure and *in vitro* growth of zygotic embryos of taro (*Colocasia esculenta* var. *antiquorum*). 57:623–630.

O'Hair, S.K. (1990). Tropical root and tuber crops. Hort. Revs. 12:157–196.

O'Hair, S.K. and Asokan, M.P. (1986). Edible aroids: Botany and horticulture. Hort. Revs. 8:43–99.

Onwueme, I.C. (1973). The sprouting process in yam (*Dioscorea* spp.) tuber pieces. J. Agric. Sci. 81:375–379.

Onwueme, I.C. (1978). The Tropical Tuber Crops. John Wiley, New York.

Osifo, E.O. (1988). Somatic embryogenesis in *Dioscorea*. J. Plant. Physiol. 133:378–380.

Osuji, G. (ed.) (1985). Advances in Yam Research. Anambra State University, Nsukka, Nigeria.

Pal, A. and Sharma, A.K. (1980). Analysis of cytotypes of *Dioscorea* and scope of increasing the diosgenin content. La Cellule 73:117–134.

Pardales, J.R. and Dalion, S.S. (1986). Methods for rapid vegetative propagation of taro. Trop. Agric. (Trinidad) 63:278–280.

Quynh, N.T. and Uyen N.V. (1987). Aroids propagation by tissue culture: I. Shoot tip culture and propagation of *Xanthosoma violaceum*. Hort. Sc. 22:671–672.

Pateña, L.F. and Barba, R.C. (1979). Rapid propagation of cassava by leaf-bud cuttings. Phil. J. Crop Sci. 4:53–62.

Rao, A.N. (1969). Tissue culture from bulbil of *Dioscorea sansibarensis*. Can. J. Bot. 47:565–566.

Roca, W.M. (1980). El Cultivo de Meristemas de Yuca. Guia de Estudio. CIAT serie 04SC-02.02. Cali, Colombia.

Roca, W.M. (1982). Intercambio Internacional de Clones de Yuca *in vitro*. Guia de Estudio. CIAT serie 04SC-05-02.

Roca, W.M. (1984). Cassava. In: Handbook of Plant Cell Culture. W.R. Sharp, P.V. Ammirato and Y. Yamada (eds.), Vol. 2. Macmillan, New York, pp. 269–301.

Roca, W.M. (1990). Cassava production and utilization problems and their biotechnological solutions. In: Plant Biotechnologies for Developing Countries. Proceedings of an International Symposium Organized by CTA and FAO, 26–30 June 1989, Luxembourg. A. Sasson and V. Costarini (eds.), Technical Centre for Agricultural and Rural Co-operation (CTA) and Food and Agriculture Organization of the United Nations (FAO), pp. 215–219.

Roca, W.M. and Jayasinghe, U. (1982). El Cultivo de Meristemas para el Saneamiento de Clones de Yuca. Guia de Estudio. CIAT serie 04SC-02.05. Cali, Colombia.

Sadik, S. and Okereke, O.U. (1975a). A new approach to improvement of yam *Dioscorea rotundata*. Nature 254:134–135.

Sadik, S. and Okereke, O.U. (1975b) Flowering, pollen grain germination, fruiting, germination and seedling development of white yam, *Dioscorea rotundata* Poir. Ann. Bot. 39:597–604.

Sasson, A. and Costarini, V. (eds.) (1990). Plant Biotechnologies for Developing Countries. Proceedings of an International Symposium Organized by CTA and FAO, 26–30 June 1989, Luxembourg. Technical Centre for Agricultural and Rural Co-operation (CTA) and Food and Agriculture Organization of the United Nations (FAO).

Scaramuzzi, F. (1986). Sweet potato (*Ipomoea batatas* Poir.). In: Biotechnology in Agriculture an Forestry, Vol. 2. Crops I. Y.P.S. Bajaj (ed.), Springer Verlag, Berlin, pp. 454–461.

Seghal, C.B. (1975). Hormonal control of differentiation in leaf cultures of *Ipomoea batatas* Poir. Beitr. Biol. Pflanzen 51:47–52.

Seghal, C.B. (1978) Regeneration of plants from anthers of sweet potato (*Ipomoea batatas* Poir.). Z. f. Pflanzenphysiol. 88:349–352.

Sengupta, J., Mitra, G.C. and Sharma, A.K. (1984). Organogenesis and tuberization of *Dioscorea floribunda*. Plant Cell Tissue Org. Cult. 3:325–331.

Sharma, M. and Chaturvedi, H.C. (1989). Somatic embryogenesis in callus tissue of *Dioscorea floribunda* and *D. deltoidea*. In: Tissue Culture and Biotechnology of Medicinal and Aromatic Plants. A.K. Kukreja, A.K. Mathur, P.S. Ahuja and R.S. Thakur (eds.), Central Institute of Medicinal and Aromatic Plants, Lucknow (India), pp. 29–38.

Sharma, A. and Sarkar, A.K. (1963). Cytological analysis of cytotypes of *Colocasia antiquorum*. Botanical Society of Bengal, Bulletin. Calcutta 17:16–22.

Shaw, D.E. (1975). Illustrated notes on flowering, flowers, seed and germination in taro (*Colocasia esculenta*). Papua New Guinea. Dept. of Agric., Stock and Fisheries. Research Bull. 13:39–59.

Shaw, D.E., Plumb, R.T. and Jackson, G.V.H. (1979). Virus diseases of taro (*Colocasia esculenta*) and *Xanthosoma* sp. in Papua New Guinea. Papua New Guinea Agric. J. 30:71–97.

Sihachakr, D. and Ducreux, C. (1987). Plant regeneration from protoplast culture of sweet potato (*Ipomoea batatas* Lam.). Plant Cell Rep. 6:326–328.

Smith, D.L. and Krikorian, A.D. (1990). Low external pH replaces 2,4-D in maintaining and multiplying 2,4-D initiated embryogenic cells of carrot. Physiol. Plant 80:329–336.

Stamp, J.A. and Henshaw, G.G. (1986). Adventitious regeneration in cassava. In: Plant Tissue Culture and its Agricultural Applications. L.A. Withers and P.G. Alderson (eds.), pp 149–157.

Staritsky, G., Dekkers, A.J., Louwaars, N.P. and Zandvoort, E.A. (1986). *In vitro* conservation of aroid germplasm at reduced temperatures and under osmotic stress. In: Plant Tissue Culture and its Agricultural Applications. L.A. Withers and P.G. Alderson (eds.), Butterworths, London, pp. 277–283.

Steward, F.C., Israel, H.W., Mott, R.L., Wilson, H.J. and Krikorian, A.D. (1975). Observations on growth and morphogenesis in cultured cells of carrot (*Daucus carota* L.). Phil. Trans. Roy. Soc. (London) 273:33–53.

Strauss, M.S. and Arditti, J. (1980). Plantlet regeneration from shoot tip cultures of *Xanthosoma caracu*. Ann. Bot. 45:209–212.

Szabados, L., Hoyos, R. and Roca, W.M. (1987). *In vitro* somatic embryogenesis and plant regeneration of cassava. Plant Cell Rep. 6:248–251.

Tang, A.F., Cappadocia, M. and Byrne, D. (1983). *In vitro* flowering in cassava (*Manihot esculenta* Crantz). Plant Cell Tissue Org. Cult. 2:199–206.

Uduebo, A.E. (1971). Effect of external supply of growth substances on axillary proliferation and development in *Dioscorea bulbifera*. Ann. Bot. 35:159–163.

Valdés, L.J., Mislankar, S.G. and Paul, A.G. (1987). *Coleus barbatus* (*C. forskohlii*) (Lamiaceae) and the potential new drug forskolin (coleonol). Economic Botany 41:474–483.

van Geyt, J.P.C. and Jacobs, M. (1985). Suspension culture of sugarbeet (*Beta vulgaris* L.). Induction and habituation of dedifferentiated and self-regenerating cell lines. Plant Cell Rep. 4:66–69.

Vasudevan, K.N., Nair, S.G. and Magoon, M.L. (1967). Radiation-induced mutations in cassava. Indian J. Hort. 24:95–98.

Viana, A.M. and Mantell, S.H. (1989). Callus induction and plant regeneration from excised zygotic embryos of the seed-propagated yams *Dioscorea composita* and *D. cayenensis* Lam. Plant Cell Tissue Org. Cult. 16:113–122.

Wang, J.-K. (ed.) (1983). Taro: A Review of *Colocasia esculenta* and its Potentials. University of Hawaii Press, Honolulu.

Wu, Y.-W. and Ma, C.-p. (1979). Isolation, culture, and callus formation of *Ipomoea batatas* protoplasts. Acta Bot. Sinica 21:334–338.

Yam, T.W., Webb, E.L. and Arditti, J. (1990). Callus formation and plantlet development from axillary buds of taro. Planta 180:458–460.

Yam, T.W., Ichihashi, S. and Arditti, J. (1991). Callus growth and plantlet regeneration in taro, *Colocasia esculenta* var. *esculenta* (L) Schott (Araceae). Ann. Bot. 67:317–323.

Yeoman, M.M. and Davidson, A.W. (1971). Effect of light on cell division in developing callus cultures. Ann. Bot. 35:1085–1100.

17. *In vitro* Culture of Oilseeds

C.E. Palmer and W.A. Keller

Contents

1. Introduction

This chapter provides a current overview of *in vitro* methods applicable to oilseeds. It will focus on those oilseeds with which substantial *in vitro* culture research has been undertaken. These include: oilseed *Brassicas* (i.e., *B. napus*, *B. campestris*, *B. juncea*, *B. carinata*), sunflower, flax and safflower. Reference will also be made to a number of other oilseed including *Cuphea*, Limnanthes, jojoba, Eruca, sesame, nigerseed and borage. Limited information on *in vitro* culture techniques is available for these crops.

Indra K. Vasil and Trevor A. Thorpe (eds.), Plant Cell and Tissue Culture, 413–455.

Among oilseeds, rapeseed and sunflower are exceeded only by soybean and cottonseed in terms of world production (see Robbelen et al., 1989). Significant amounts of flax, safflower and sesame are also produced world wide. In addition some oilseeds such as nigerseed, sesame, Eruca and poppy are of importance in certain geographic locations. Others, for example jojoba, Cuphea, borage and Limnanthes are sources of specialty oils and are being domesticated.

In cultivation of these oilseeds the major agronomic considerations are oil quality and quantity and meal quality. As a consequence breeding objectives are designed to improve or modify these components and ensure their availability for industrial use, animal feed or human consumption. Improvements in disease and pest resistance, stress tolerance and adaptation to environmental conditions are essential to sustained productivity of these oilseeds. Although many breeding objectives can be achieved through conventional procedures (e.g. the development of canola as described by Robbelen, 1991) the application of biotechnology and *in vitro* culture methods as adjuncts to genetic improvement programs will enhance the achievement of these objectives (Salisbury and Kadkol, 1989).

Totipotent cell culture systems have been developed for many oilseed crops. These systems include multicellular explants derived from plant/seedling organs, isolated protoplasts and gametophytic cells, particularly microspores. The availability of such systems is essential for application of genetic modification techniques. Embryo culture techniques have been developed to produce interspecific hybrids. Embryo culture offers the possibility of clonal multiplication of elite genetic lines. Interspecific and intergeneric hybrids can be obtained through the regeneration of plants from hybrid cells produced via the fusion of protoplasts from different species. Protoplast fusion techniques have also been employed to manipulate cytoplasmic traits in oilseed *Brassicas*.

One of the most useful techniques applicable to crop improvement programs is the recovery of doubled haploids through anther, microspore and ovule culture. Homozygosity can be quickly achieved, thus facilitating selection of stable genetic recombinants. The availability of totipotent gametophytic culture systems creates the opportunity for undertaking research with haploid cells. The advantage of such systems in mutagenesis and transformation studies are obvious. The induction of embryogenesis from microspores also offers the possibility of studying embryo specific metabolic pathways such as seed oil biosynthesis.

Genetic variants can be selected through cell culture as a result of mutagenesis or culture-induced genetic alterations. This approach may be useful for the recovery of disease resistance, salt tolerance and modified seed composition in some oilseeds. The use of haploid cell lines or microspores is particularly suitable for this process as recessive mutations can be efficiently detected.

Agrobacterium-mediated gene transfer technology has been developed for a number of oilseed crops particularly *Brassica napus*, flax and to some extent, sunflower. Studies to enhance the efficiency of this technique as well as to develop alternative methodology for direct transfer of genes into oilseed crops are being actively pursued by many investigators.

In a general consideration of the development of *in vitro* techniques for oilseed improvement, it is apparent that *Brassica* species and in particular Brassica napus have achieved the most advanced status (Friedt et al., 1991; Friedt, 1988). However, a significant body of information has been developed for sunflower and flax. Practical applications of *in vitro* techniques for crop improvement have been announced in recent years and as further technical advances are made greater economic benefits can be anticipated.

2. Oilseed *Brassicas*

2.1. *Plant Regeneration from Cultured Explants*

2.1.1. *Organogenesis*

Plant regeneration can be achieved with a wide variety of explants including cotyledons, stem segments, hypocotyl, leaf discs and root segments (Table 1). Response varies with species, genotype, explant source and culture medium composition (Dietert et al., 1982). Cotyledon and hypocotyls, generally isolated from aseptically-germinated seedlings, are the most commonly used source of explants with cotyledons often exhibiting a higher regenerative capacity. The age of seedling used as an explant source can influence the subsequent *in vitro* response. For example, *Brassica napus* cotyledons older than 8 days seldom responded (Sharma et al., 1990) whereas 12-day-old seedlings were a satisfactory source of hypocotyl explants (DeBlock et al., 1989). In *B. campestris* 4–5 day old cotyledons gave maximum response (Hachey et al., 1991), while in hypocotyl explants of *B. campestris* maximum shoot regeneration was achieved with 6–7 day-old explants (Yang et al., 1991). In *B. napus* shoot regeneration was obtained on root explants of 11–14 day old seedlings (Sharma and Thorpe, 1989; Lazzeri and Dunwell, 1984). Shoot formation has been Induced on leaf explants from *B. campestris* and *B. napus* (Dunwell, 1981; Newell et al., 1984; Stringam, 1979) and in each case young fully expanded leaves were employed. Regeneration of shoots can also be obtained with stem discs and stem epidermial strips of flowering plants of *B. napus* (Dale and Ball, 1991; Klimaszewska and Keller, 1985).

The composition of the culture medium is another important factor influencing organogenic response. Murashige and Skoog medium is most commonly employed with modifications depending on explant source (see Sharma and Thorpe, 1989). Auxins and cytokinins are required for shoot regeneration from many explants although cytokinin alone elicited shoot bud regeneration on cotyledon explants of *B. napus* and *B. juncea* (Moloney et

Table 1. Plant regeneration from cultured explants of oilseed *Brassicas*

Species	Explant	Developmental response	References
B. campestris	Cotyledon	Shoots	Jain et al., 1988b
			Narasimhulu and Chopra, 1988
			Narasimhulu et al., 1988
			Dale and Ball, 1991
			Hachey et al., 1991
	Hypocotyl and hypocotyl-derived cell cultures	Shoots, embryos	Singh and Chandra, 1984
	Stem	Shoots	Nanda Kumar et al., 1991
	Petiole	Embryos	Bhattacharya and Sen, 1980
	Root	Shoots	Lazzeri and Dunwell, 1984
	Leaf	Shoots	Dunwell, 1981
	Immature embryo	Embryos	Maheswaran and Williams, 1986
B. carinata	Cotyledon	Shoots	Jain et al., 1988a
			Narasimhulu and Chopra, 1987, 1988
			Narasimhulu et al., 1989
	Hypocotyl	Shoots	Bajaj and Mohapatra, 1987
			Yang et al., 1991
B. juncea	Cotyledon	Shoots	Fazakas et al., 1986
			Murata and Orton, 1987
			Jain et al., 1988a
			Narasimhulu and Chopra, 1988
			Sharma et al., 1990
			Sharma and Bhojwani, 1990
	Hypocotyl	Shoots	Sharma and Bhojwani, 1990
		Embryos	Singh et al., 1985
			Kirti and Chopra, 1989
			Chi et al., 1990
B. napus	Cotyledon	Shoots	Newell et al., 1984
			Narashimhulu and Chopra, 1988
			Moloney et al., 1989
	Hypocotyl	Shoots	Murata and Orton, 1987
			DeBlock et al., 1989
			Shukla and Sawhney, 1991
	Stem	Shoots	Klimaszewska and Keller, 1985
			Julliard et al., 1992
			Dale and Ball, 1991
	Leaf	Shoots	Stringam, 1979
			Dunwell, 1981
			Newell et al., 1984
	Microspore-derived cell cultures	Embryos	Simmonds et al., 1991

al., 1989; Sharma et al., 1991). Ethylene appears to be a critical factor in shoot morphogenesis and inhibitors of ethylene action or biosynthesis enhanced regeneration in a number of cases (Chi et al., 1990; DeBlock et al., 1989).

2.1.2. *Embryogenesis*

Although the primary mode of regeneration from *Brassica* spp. is organogenesis, somatic embryogenesis has been reported in a number of cases (Maheswaran and Williams, 1986; Kirti and Chopra, 1989; Bhattacharya and Sen, 1980). Plants have also been obtained through somatic embryogenesis in suspension cultures (Simmonds et al., 1991, Singh et al., 1981; 1985). Experiments with *B. campestris* var ToriaT15, revealed that 12–18 day-old immature zygotic embryo explants, proliferated somatic embryos on a medium containing cytokinin and yeast extract (Maheswaran and Williams, 1986). Somatic embryogenesis was obtained with several genotypes of *B. juncea* when 5–7 day-old hypocotyl segments were cultivated on a medium with auxin and cytokinin (Kirti and Chopra, 1989).

2.2. *Protoplast Culture Technology*

Methodology for protoplast isolation and culture including the recovery of regenerated plants has been well established for *Brassica napus* (Table 2). Plants have also been regenerated from *B. juncea* and *B. carinata* protoplast cultures (Jaswal et al., 1990; Barsby et al., 1986), whereas *B. campestris* has been relatively recalcitrant, with shoot and plant regeneration rarely observed (Table 2). Young fully expanded leaves from *in vitro*-grown shoots and hypocotyl of 2 to 6 day-old seedlings are the most frequently used tissue sources for protoplasts in oilseed *Brassicas*.

A number of factors have been shown to influence plating efficiency in protoplast culture (for examples see references in Table 2). An osmolarity of 500–600 m osmol. appears to be optimal and glucose has often been found to be a suitable osmoticum. Cell division and colony formation can be obtained in media supplemented with auxin and cytokinin. Protoplasts have been successfully cultured in liquid media or in liquid with agarose underlayer as well as by embedding in solid agarose media. Inclusion of Ficoll in the liquid medium enhanced cell division and colony formation in *B. napus* protoplast culture (Klimaszewska and Keller, 1987; Millam et al., 1988). After 2 to 3 weeks of culture microcalli can generally be transferred to solid regeneration media.

Plant regeneration occurs by organogenesis or embryogenesis depending on the genotype and culture medium composition (Table 2). In several genotypes of *B. napus* more than 70% of the microcalli regenerated shoots (Poulsen and Nielsen, 1989; Glimelius, 1984). However, in *B. campestris*

Table 2. Plant regeneration from protoplast cultures in oilseed *Brassicas*

Species	Source	Developmental response	References
B. campestris	Mesophyll	Shoots	Glimelius, 1984 Loudon et al., 1989
	Hypocotyl	Shoots	Glimelius, 1984
B. carinata	Hypocotyl	Shoots	Narasimuhulu et al., 1992
B. juncea	Cotyledon	Shoots	Choung et al., 1987b
	Mesophyll	Shoots	Chatterjee et al., 1985 Kao and Seguin-Swartz, 1987 Eapen et al., 1989
	Hypocotyl	Embryos, shoots	Kirti and Chopra, 1989, 1990 Pua, 1990
B. napus	Mesophyll	Embryos, shoots	Kohlenbach et al., 1992 Li and Kohlenbach, 1982 Glimelius, 1984 Kao and Seguin-Swartz, 1987 Loudon et al., 1989
	Hypocotyl	Embryos	Glimelius, 1984 Barsby et al., 1986 Kirti, 1988 Poulsen and Nielsen, 1989
	Stem layers	Shoots	Choung et al., 1987a Klimaszewska and Keller, 1987
	Roots	Shoots	Xu et al., 1982
	Microspore-derived cell culture	Embryos	Simmonds et al., 1991

less than 1% of the protoplast-derived colonies regenerated shoots. Addition of both auxin and cytokinin appears to be necessary for induction of shoot organogenesis although in some cases cytokinin alone is effective. Somatic embryogenesis has been reported in both mesophyll and hypocotyl protoplast cultures (Pua, 1990; Kirti and Chopra, 1990; Li and Kohlenbach, 1982; Eapen et al., 1989; Kranz, 1988). With hypocotyl protoplasts of *B. napus* and *B. juncea* 2,4-D was required for induction of embryogenesis (Kirti, 1988; Pua, 1990), whereas organogenesis occurred in the absence of auxin.

2.3. *Protoplast Fusion Technology*

Protoplast fusion techniques have been employed in the production of a number of interspecific (somatic) hybrid combinations involving oilseed *Brassica* species as parental lines (see Table 3). Polyethylene glycol has been the most commonly used fusion agent. A major consideration in somatic hybridization experiments relates to the development of effective strategies for selecting hybrid cells. A number of approaches have been successfully employed with oilseed *Brassicas* including: distinct morphological feature of

Table 3. Somatic hybridization of oilseed *Brassicas*

Parental combination	Hybrid selection strategy	References
B. campestris	Fluorescent dyes+flow cytometry	Schenck and Robbelen, 1982
+ *B. oleracea*	Metabolic inactivation	Glimelius et al., 1986
	Shoot/plant morphology	Robertson et al., 1987
		Sundburg et al., 1987
		Terada et al., 1987
		Terada et al., 1987
		Rosen et al., 1988
		Jourdan et al., 1989
		Heather and Earle, 1991
		Itoh et al., 1991
		Sundburg and Glimelius, 1991
B. napus	Fluorescent dyes+flow cytometry	Pelletier et al., 1983
+ *B. napus*	Micromanipulation	Yarrow et al., 1986
		Pauls and Coung, 1987
		Choung et al., 1988
		Kao et al., 1991
B. napus	Shoot/plant morphology	Primard et al, 1988
+ *B. hirta*		
B. napus	Fluorescent dyes/micromanipulation	Fahleson et al., 1988
+ *Eruca sativa*		Sundburg and Glimelius, 1991
B. napus	Flow cytometry	Sundburg and Glimelius, 1991
+ *B. nigra*	Shoot/plant morphology	Sacristan et al., 1989
		Sjodin and Glimelius, 1989a
B. napus	Shoot/plant morphology	Klimaszewska and Keller, 1988
+ *Diplotaxis harra*		
B. napus	Nuclear inactivation +	Bauer-Weston, 1990
+ *Arabidopsis thaliana*	selectable markers	
B. juncea	Shoot/plant morphology	Kirti et al., 1992
+ *Rachystoma ballii*		
B. juncea	Shoot/plant morphology	Kirti et al., 1991
+ *B. spinescens*		
B. napus	Fluorescent dyes+flow cytometry	Sundburg and Glimelius, 1991
+ *B. oleracea*		
B. nigra	Nuclear inactivation +	Sjodin and Glimelius, 1989b
B. carinata	toxin selection	
B. juncea		
+ *B. napus*		
B. napus	Fluorescent dye+flow cytometry	Sundburg and Glimelius, 1991
+ *B. juncea*		
B. napus	Fluorescent dye+flow cytometry	Sundburg and Glimelius, 1991
+ *Raphanus sativa*		

hybrid cells and/or regenerated plants (Klimaszewska and Keller 1988; Kirti et al., 1991; Kao et al., 1991; Sikdar et al., 1990), inactivation of one or both parental lines through irradiation or chemical treatment (Yamashita et al., 1989; Bauer-Weston, 1990), use of herbicide resistance as a selectable marker (Bauer-Weston, 1990) and the application of flow cytometry (Pauls

Table 4. Cybridization in oilseed *Brassicas*

Fusion partners	Trait manipulated	Selection strategy	References
B. napus + *B. napus*	CMS, triazine resistance	Micromanipulation; mass selection	Kemble et al., 1988a, b
B. napus + *B. napus*	CMS, triazine resistance	Metabolic inactivation + irradiation; mass selection	Barsby et al., 1987a, b
B. napus + *B. napus*	CMS	Irradiation; Mass selection	Menczel et al., 1987
B. napus + *B. napus*	CMS	Metabolic inactivation + irradiation; manual selection	Morgan and Maliza, 1987
B. napus + *B. napus*	Triazine resistance	Cytoplast-karyoplast fusion; manual selection	Spangenberg and Schweiger, 1986
B. napus + *B. napus*	Triazine resistance	Irradiation; micromanipulation	Thomzik and Hain, 1988
B. napus + *B. oleracea*	CMS	Irradiation + metabolid inactivation; mass selection	Yarrow et al., 1990
B. napus + *B. napus*	CMS, triazine resistance	Mass selection	Kao et al., 1991
B. napus + *B. oleracea*	CMS, triazine resistance	Metabolic inactivation; mass selection	Kao et al., 1992
B. napus + *Raphanus sativus*	CMS transfer	Cyroplast-protoplast fusion; manual selection	Sakai and Imamura, 1990

and Chuong, 1986; Glimeluis et al., 1986). Regenerated hybrids have been characterized through analysis of nuclear and organelle DNA content, chromosome size and number, isozyme profiles, and morphological traits (see Table 3 for references).

Protoplast fusion techniques have been effectively used to transfer and/or combine organelle-specific traits including cytoplasmic male sterility and triazine herbicide resistance (see Table 4). Such cytoplasmic hybrids or "cybrids" have been selected through the use of nuclear-inactivating treatments, as well as treatment with fluorescent dyes followed by micromanipulation or sorting by flow cytometry (Sundberg et al., 1991). The regenerated cybrids are generally assessed by analysis of organelle DNA (eg., Morgan and Maliga, 1987; Landgren and Glimelius, 1990).

2.4. *In vitro Culture of Embryos*

In oilseed *Brassicas*, the technique of embryo rescue has been used to recover hybrids from a number of crosses involving *Brassicas* and related genera (Table 5). This is achieved by culturing excised ovaries, ovules or immature embryos. The recovery of viable plantlets is dependent on parental genotype, culture conditions and developmental stage of the embryo at excision. Nishiyama et al. (1991) and Nanda-Kumar et al. (1988) have listed a number of crosses with success, ranging from 0–94%. In most cases, endosperm degeneration was the cause of incompatibility.

Embryo culture media have generally been based on modifications of media originally defined by White (1963), Murashige and Skoog (1962) and Nitsch and Nitsch (1969). Sucrose is usually required at relatively high levels (4–7%) along with casein hydrolysate and yeast extract as sources of reduced nitrogen. Solidified medium is preferable to liquid and there are no specific requirements for plant growth regulators, although addition of coconut milk and gibberellins has been reported to be beneficial to endosperm development (Inomata, 1990). Cultures are usually maintained at low light intensity of 300–1,000 lux under 16 hours photoperiods or continuous light, at 20–25 °C. Ovary culture is preferable to isolated embryo culture because of the ease of excision with minimum embryo damage and less demanding culture conditions.

2.5. *In vitro Production of Haploids*

Anther and isolated microspore culture techniques can be employed to produce haploid plants in oilseed *Brassicas* (see Tables 6 and 7). In both culture systems, microspores are induced to undergo embryogenesis without intervening callus formation. Reliable and efficient protocols for achieving embryogenesis and plant regeneration from isolated microspore cultures have been reported for *B. napus* (Lichter, 1982; Chuong and Beversdorf, 1985; Keller et al., 1987; Polsoni et al., 1988; Coventry et al., 1988). Under optimum conditions the frequency of embryogenesis can be as high as 10% (Swanson et al., 1987; Pechan and Keller, 1988; Swanson, 1990). High frequencies of embryogenesis have been achieved in microspore cultures of several *B. napus* genotypes and in fact microspore culture is more efficient than anther culture in this species (Siebel and Pauls, 1989; Chuong et al., 1986b). *B. carinata* also appears to be very responsive (Chuong and Beversdorf, 1985) but *B. juncea* and particularly *B. campestris* have generally given lower yields of embryos, although improved techniques for *B. campestris* microspore cultures have recently been reported (Baillie et al., 1992; Burnett et al., 1992).

Conditions which affect haploid embryogenesis in both anther and micros-

Table 5. Interspecific hybridization through embryo rescue in oilseed *Brassicas*

Parental combination	Embryo age (DAP)	References
B. campestris + *B. spinescens*	7–14	Agnihotri et al., 1990a, b
B. juncea + *B. napus*	7–12 10–14	Bajaj et al., 1986
B. oleracea + *B. campestris*	1–22	Hossain et al., 1988, 1989 Inomata, 1982, 1990
B. napus + *B. oleracea*	22	Quazi 1988
B. napus + *B. erocoides*	4	Delorme et al., 1990
B. napus + *Diplotaxis siifolia*	4–6	Batra et al., 1990
B. juncea + *B. hirta*	10–15 4–9	Mohapatra and Bajaj, 1984, 1987
B. juncea + *B. campestris*	4–9	Mohapatra and Bajaj, 1988
B. oxyrrhina + *B. campestris*		
B. juncea + *B. oxyrrhina*		Prakash and Chopra 1990
B. napus + *Sinapis alba*	10–15	Ripley and Arnisson, 1990
B. campestris + *Moricanda arvensis*	3–7	Takahata and Takeda, 1990
B. campestris + *Eruca sativa*		Agnihotri et al., 1990
B. oleracea + *Eruca sativa*	4–6	Plumper, 1991
B. juncea + *B. carinata*	4–6	Sacristan and Gerdmann, 1986
B. campestris + *B. nigra*	4–6	Campbell et al., 1991
Enarthrocarpus lyratus + *Brassica* spp.	5–15	Gundimeda et al., 1992

pore culture systems include genotype, donor plant growth conditions, stage of microspore development and *in vitro* environmental conditions (see Tables 6 and 7 for references). Sucrose concentration of greater than 10% (w/v) and elevated culture temperature (30–35 °C) during the initial culture phase are critical requirements for induction of embryogenesis.

Table 6. Embryogenesis and plant regeneration in anther cultures of oilseed *Brassicas*

Species	Significant factors affecting induction of embryogenesis	References
B. campestris	Sucrose content, elevated temperature: genotype	Keller and Armstrong, 1979; Seguin-Swartz et al., 1983 Sorvari, 1985 Aslam et al., 1990a
B. carinata	Mix of glucose and sucrose	Arora and Bhojwani, 1988
B. juncea	Combination of cold pretreatment and high temperature	George and Rao, 1982
	Elevated or low temps.	Sharma and Bhojwani, 1985
B. napus	Sequential reduction of sucrose level	Chen and Chen, 1983 Dunwell and Thurling, 1985
	Donor growth temperature	Dunwell et al., 1983, 1985 Keller et al., 1987
	Elevated culture temperature	Keller and Armstrong, 1978 Mandal, 1987
	Genotypic differences	Siebel and Pauls, 1989 Aslam et al., 1990b
	Anther development stage	Loh et al., 1982
	Liquid medium	Lichter, 1981
	Stimulation by low doses of gamma irradiation	MacDonald et al., 1988

Although the major application of microspore culture technology is the production of homozygous doubled-haploid plants primarily for plant breeding programs (Hoffmann et al., 1982), isolated micropsore and microspore-derived embryos have been employed successfully in mutant isolation and gene transfer studies (Swanson et al., 1989; Swanson and Erickson, 1989). The microspore embryos are also useful for studies of lipid biosynthesis and storage (Wiberg et al., 1991; Pomeroy et al., 1991).

2.6. *In vitro Selection of Somaclonal Variants and Mutants*

Totipotent explants and cell culture systems have been employed to produce somaclonal variants in oilseed *Brassicas* (see Table 8). Variants for seed color, oil content and agronomic characteristics were isolated from plants regenerated from cotyledon explants of *B. juncea* (George and Rao, 1983). Stable sodium chloride tolerant lines of *B. juncea* were also selected from shoots regenerating from cotyledon explants or from somatic embryos developing from hypocotyl explants (Jain et al., 1989, 1990, 1991; Kirti et al., 1991). Regenerated plants apparently retained the salt tolerant property, but convincing genetic evidence for a stable genetic mutations lacking.

Table 7. Embryogenesis and plant regeneration in isolated microspore cultures of oilseed *Brassicas*

Species	Significant factors influencing embryogenesis	References
B. campestris	Medium composition; genotype	Lichter, 1989 Baillie et al., 1992 Burnett et al., 1992
B. carinata	Culture temperature	Chuong and Beversdorf, 1985
B. napus	Benzyladine concentration	Charne and Beversdorf, 1988
	Culture temperature	Chuong and Beversdorf, 1985 Keller et al., 1987
	Genotype	Chuong et al., 1988a Gland et al., 1988 Siebels and Pauls, 1989
	Microspore developmental stage	Fan et al., 1988a, b Kott et al., 1988a, b Telmer et al., 1992
	pH	Gland et al., 1988
	Microspore density	Huang et al., 1990
	Donor growth temperature	Keller et al., 1987
	Sucrose concentration	Lichter, 1982
	Medium composition	Lichter, 1982 Gland et al., 1988 Keller et al., 1988

Table 8. In vitro selection of somaclonal variants and mutants in oilseed *Brassicas*

Species	Type of culture/selection strategy	Characteristics of selected regenerants	References
B. juncea	Hypocotyl and cotyledon callus + salt stress	Salt tolerant variants	Jain et al., 1989, 1990, 1991 Kirti et al., 1991
B. juncea	Cotyledon explants	Yellow-seeded variants	George and Rao, 1983
B. napus	Callus + Phoma ligam culture filtrate	Phoma resistant plants	Sacristan, 1982, 1985
	Mutagenized microspores/ protoplasts + herbicide	Herbicide tolerant plants	Swanson et al., 1988, 1989
	Embryos + sirodesmin	Phoma-resistant plants	Newsholme et al., 1989
	Embryos + culture filtrate	Phoma-resistant plants	MacDonald and Ingram, 1986
	Mutagenized microspore + chemical analysis of seeds of regenerants	Elevated oleic acid	Huang et al., 1991

Stable genetic tolerance to sulfonylurea and imidazolinone herbicides was obtained through mutagenesis of microspores and haploid protoplasts *B. napus* (Swanson et al., 1988, 1989). The chemical ethylnitrosourea (ENU) was equally effective as gamma irradiation as a mutagen. All colonies/embryos derived from mutagenized protoplasts/microspores were screened on lethal concentrations of the herbicide. Following screening of mutagenisized cells on media with herbicide of increasing concentration, plants were regenerated, diploidized with colchicine and self-pollinated for seed production. Mutants with elevated levels of oleic acid have also been isolated from mutagenized microspore cultures (Huang et al., 1991).

The generation of putative resistance in *B. napus* to the pathogens *Phoma lingam* and *Alternaria brassicicola* (Schw.) Wilts. has been reported (Sacristan, 1982; MacDonald and Ingram, 1986; Ingram et al., 1984; Newsholme et al., 1989). Cells were exposed to fungal culture filtrate or partially purified toxin incorporated in the culture medium. Surviving cells were then cultured on regeneration medium for plant recovery. Ethylmethylsulfonic acid (EMS) treatment of haploid callus or cells has been employed in some studies in order to induce mutants with toxin resistance. The genetic stability of the resistance traits has not been confirmed in these studies.

2.7. *Gene Transfer Technology*

Oilseed Brassicas can be transformed through *in vitro* co-cultivation of cells or tissues with *Agrobacterium tumefaciens* possessing modified Ti plasmids (See Table 9) *B. napus* has also been transformed via co-cultivation with *A. rhizogenes* (Guerche et al., 1987b; Boulter et al., 1990). The most commonly employed protocol involves the cocultivation of totipotent cotyledonary petiole and hypocotyl explants with *A. tumefaciens* possessing a bacterial kanamycin resistance gene as a selectable marker. *B. campestris*, *B. juncea* and *B. napus* have been transformed through this approach with the greatest degree of success in terms of transformation efficiency being obtained with *B. napus*. *B. napus* has also been transformed through cultivation of stem disks (Fry et al., 1987); stem peels (Pua et al., 1987; Charest et al., 1988), explants derived from microspore-derived embryos (Swanson and Erickson, 1989; Oelck et al., 1991) and isolated protoplasts (Thomzik and Hain,1990).

Frequency of stable transformants obtained varies with species and between protocols indicating the opportunity for further research to enhance the efficiency of Agrobacterium-mediated gene transfer technology in oilseed *Brassicas*. An example of protocol modification involves the *in vitro* addition of ethylene inhibitors to improve the recovery of transformants (DeBlock et al., 1989; Mehra Palta et al., 1991).

Non-biological methods of gene transfer have also been successfully employed with some oilseed crops. Direct DNA uptake through polyethylene glycol treatment or electroporation has been used to achieve transformation

Table 9. Gene transfer technology in oilseed *Brassicas*

Species	Method	Recipient cell/tissue system	Reference
B. campestris	*Agrobacterium tumefaciens*	Cotyledon, Hypocotyl	Mehra-Palta et al., 1991
B. juncea	*A. tumefaciens*	Cotyledon, Hypocotyl	Barfield and Pua, 1991 Mathews et al., 1990
B. napus	*A. tumefaciens*	Cotyledon, Hypocotyl	Radke et al., 1988 DeBlock et al., 1989 Moloney et al., 1989 DeBlock and Debrouwer, 1991 Lee et al., 1991 Altenbach et al., 1992
		Stem disks and stem peels	Fry et al., 1987 Pua et al., 1987 Miki et al., 1990 Charest et al., 1988 Misra and Gedamu, 1989
		Protoplasts	Thomzik and Hain, 1990
		Embryo segments	Swanson and Erickson, 1989 Oelck et al., 1991
	A. rhizogenes	Hypocotyls	Damgaard and Rasmussen, 1991
		Stem segments/Roots	Jouanin et al., 1987 Guerche et al., 1987b
		Roots	Jouanin et al., 1987
	Electroporation	Rototplasts	Chapel and Gimelius, 1990
	PEG mediated uptake	Protoplasts	Köhler et al., 1990
	Microinjection	Micropore-derived embryos	Neuhaus et al., 1987

of *B. napus* and *B. campestris* but plants were recovered only in *B. napus* (Chapel and Glimelius, 1990; Guerche et al., 1987, 1990; Paskowski et al., 1986). Delayed cell wall regeneration and irradiation pretreatments of protoplasts have been shown to be effective in enhancing transformation frequencies (Kohler et al., 1990; Chapel and Glimelius, 1990). Other factors influencing protoplast-mediated transformation techniques have been reviewed by Negrutiu et al. (1990). Microinjection of DNA into microspore-derived embryos of *B. napus* resulted in recovery of transgenics (Neuhaus et al., 1987), however, this technically-demanding procedures has not been widely adopted. Although successfully employed with several crops, the successful employment of biolistic techniques for the transformation of oilseed *Brassicas* has not yet been published.

Table 10. Plant regeneration from explants in sunflower *Helianthus annuus* L.

Explant	Developmental response	References
Leaf	Shoots, embryos	Greco et al., 1984
		Patterson and Everett, 1985
		Lupi et al., 1987
		Witrzens et al., 1988
Hypocotyl	Shoot, roots, embryo	Greco et al., 1984
		Lupi et al., 1987
		Robinson and Adams, 1987
		Robinson et al., 1987
		Witrzens et al., 1988
		Prado and Berville, 1990
Cell layers (hypocotyl)	Embryos	Pellissier et al., 1990
Cotyledons	Shoots	Greco et al., 1984
		Lupi et al., 1987
		Power, 1987
		Witrzens et al., 1988
		Chraibi et al., 1991, 1992
		Knittel et al., 1991
		Pugliesi et al., 1991
		Ceriani et al., 1992
Immature embryo	Embryos, shoots	Finer, 1987
		Power, 1987
		Freyssinet and Freyssinet, 1988
		McCan-Wilcox et al., 1988
		Witrzens et al., 1988, 1989
		Fspinasse et al., 1989
		Jeannin and Hahne, 1991
Mature embryo	Shoots	Power, 1987
	Shoots	Krauter and Friedt, 1991
Shoot apices	Shoots	Greco et. al., 1984
	Callus to shoots	Cavallini and Luipi, 1988
		Lupi et al., 1987
		Robinson et al., 1987
Stem segment	Shoots	Witrzens et al., 1988
Epicotyl	Shoots	Lupi et al., 1987
Suspension cells	Shoots, roots	Prado and Berville, 1990

3. Sunflower (*Helianthus annuus*)

3.1. *Plant Regeneration from Cultured Explants*

Plant regeneration from cultured sunflower explants occurs by both organogenesis and embryogenesis depending on explant source, medium composition and genotype (Table 10). Shoot apices, hypocotyl, cotyledons, leaves, stem and mature and immature embryos are all suitable explants, with immature embryos, hypocotyls and cotyledons being most commonly used.

Explant age significantly influences the *in vitro* morphogenic response. Cotyledons from 4–5 day old seedlings produced maximum response (Knittel et al., 1991) while 11–12 day old hypocotyls, from seedlings maintained in light, were the most responsive (Patterson and Everett, 1985). Immature embryos can be successfully cultured at various stages but in several genotype maximum shoot regeneration was obtained from embryos isolated 10 DAP (Wilcox-McCann et al., 1988; Witrzens et al., 1988; Espinasse and Lay, 1989). Usually these are globular or torpedo embryos, 0.2 to 1.2 mm in length. Shoot apices and epicotyls from 7–10 day old seedlings were also responsive (Greco et al., 1984; Lupi et al., 1987).

Media requirements vary with explant source but generally a medium rich in nitrate favoured shoot regeneration from hypocotyl and cotyledon explants (Patterson and Everett 1985; Knittel et al., 1991). In some cases a combination of nitrogen sources, NH_4^+, NO_3^-, and amino acids proved optimal. Cytokinin and auxin are generally required for shoot organogenesis and gibberellin (GA_3) has been reported to stimulate normal shoot development (Patterson and Everett, 1985; Lupi et al., 1987). However, in one case, shoot multiplication from shoot tip explants was more efficient with a combination of cytokinin and GA_3 while addition of auxin favoured callusing and reduced shoot multiplication (Lupi et al., 1987). For immature embryos, low auxin with high cytokinin favoured shoot organogenesis in several genotypes (Espinasse and Lay, 1989). In 7 genotypes, plant regeneration was achieved on an initial medium with auxin, ABA and BAP followed by exposure to low auxin with high cytokinin (Wilcox-McCann et al., 1988). With several genotypes examined by Witrzens et al. (1988) 3.0 mg l^{-1} BAP was optimum, low sucrose level and absence of auxin in the initial culture favoured normal shoot development. Inhibitors of ethylene action or biosynthesis have also been shown to enhance shoot regeneration (Chraibi et al., 1991).

Somatic embryogenesis occurred on hypocotyl and cotyledon derived calli (Pelissier et al., 1990; Finer, 1987, Prado and Berville, 1990). In one case (Prado and Berville, 1990), the critical factor appeared to be seedling germination on enriched MS medium, before explanting and callus induction with 2,4-D and BAP. Upon transfer to a medium with ABA and 2,4-D, 7 genotypes produced embryos. Auxin and sucrose concentrations are the major factors in induction of somatic embryogenesis on immature zygotic embryos (Finer, 1987).

A potentially promising procedure for embryogenesis was described by Pelissier et al. (1990) in which 7–8 day old hypocotyl strips containing epidermal and subepidermal layers were precultured for 5 days on enriched MS medium with auxin and cytokinin. Embryos were induced when explants were transferred to B_5 medium (Gamburg et al., 1968) with 9% sucrose and maintained in the dark. Secondary embryos were produced when these embryos were cultured on MS medium with 12% sucrose, low BAP and maintained in the dark. Sequential reduction of sucrose and addition of low auxin levels favoured embryo germination. This two step procedure was obligatory for embryogenesis and B_5 medium with high sucrose was required as the response did not occur on MS medium with high sucrose.

In most cases, whether by embryogenesis or organogenesis, plant regeneration is genotype dependent (Pelissier et al., 1990; Power, 1987). In a study of 59 genotypes, immature embryo explants were cultured on 3 different hormone combinations; several failed to respond and even the extent of callusing was genotype dependent (Espinasse et al., 1989). Shoot regeneration from cotyledon explants of 8 genotypes showed that both the percentage response and the number of shoots per explant were genotype dependent (Knittel et al., 1991).

There is little critical information on environmental factors affecting morphogenetic response of sunflower explants. Cultures are usually maintained in light of 20–100 $\mu Em^{-2} sec^{-1}$ for 16 h and a temperature of 24–28 °C. Light grown hypocotyl and cotyledon explants were more responsive than those grown in darkness, (Patterson and Everett, 1985). Development of embryos from cell layers is one example where light is critical as no embryos were formed if cultures were exposed to light during the initial stages (Pelissier et al., 1990). It should also be noted that the growth conditions of the donor plants used for zygotic embryo isolation did not affect in vitro embryogenesis (Jeannin and Hahne, 1991).

3.1. *Protoplast Culture Technology*

Viable protoplasts can be isolated from a variety of tissues including hypocotyls, cotyledons, stems, roots and leaf mesophyll. Source of protoplasts will influence response in culture as shown by Lenee and Chupeau (1986) who observed that colony formation occurred only with hypocotyl protoplast cultures. With mesophyll protoplasts, the physiological state of the donor plant affected both their yield and viability (Guilley and Hahne, 1989). A critical requirement for cell division and colony formation is the N content of the culture medium (Lenee and Chupeau, 1986). Organic N is preferable to inorganic N and therefore glutamine and casein hydrolysate are usually included in culture media. Leaf mesophyll protoplasts produced macrocalli only with glutamine or glutamine plus ammonium nitrate and KNO_3 (Guilley and Hahne, 1989). Sustained cell division can be obtained in liquid culture, but agarose bead culture enhanced cell division and colony formation (Dupuis et al., 1990; Bohorova et al., 1986; Chanabe et al., 1991). Cultures are usually maintained in light 30–50 $\mu Em^{-2} sec^{-1}$ at 25–28 °C, and in 4 genotypes better plating efficiency was obtained compared to dark culture (Bohorova et al., 1986). Reduction of auxin and sucrose levels enhanced colony development (Chanabe et al., 1989).

Sunflower is generally considered to be recalcitrant in terms of protoplast culture, however plant regeneration has been reported in two cases (see Table 11). Plant regeneration has been achieved with related species (Table

Table 11. Plant regeneration from protoplast cultures in *Helianthus* spp.

Proplast source	Species	Developmental response	References
Hypocotyl	*H. annuus*	Callus, shoots, embryos	Chanabe et al., 1989, 1991
	H. petiolaris		Burrus et al., 1991
Mesophyll	*H. annuus*	Callus, shoots, embryos	Binding et al., 1981
	H. giganteus		Krasnyanski et al., 1992

Table 12. Interspecific hybridization via *in vitro* embryo rescue in *Helianthus annuus* L.

Number of genotypes and crosses	Stage of embryo development	References
53	3–7 DAP	Chandler and Beard, 1983
1	3 DAP (100–280 μm)	Denat et al., 1991
5	Ovules 1–4 DAP	Espinasse et al., 1991
8	Ovary	Mix, 1985
3	Embryo 3–7 DAP	Assad et al., 1986
2	Mature 45–50 days	Bohorova et al., 1985
33	5–10 DAP	Kraüter et al., 1991
6	3–14 DAP	Espinasse et al., 1985

11) indicating that somatic hybridization approaches are feasible. Regeneration capacity is genotype dependent. For example, fertile plants were recovered from 2 of 6 genotypes with a frequency of 1.8–3.5% when microcalli were cultured on MS medium with NAA, BAP at 1.0 mg l^{-1} each and GA_3 0.1 mg l^{-1} (Burrus et al., 1991). Shoot recovery did not occur in the absence of GA_3. The most common morphogenic response in totipotent cultures is shoot organogenesis but embryo development has been reported in a few cases (Dupuis et al., 1988; Schmitz and Schnabl, 1989; Moyne et al., 1988).

3.3. *In vitro Culture of Embryos*

Embryo culture techniques have been employed to rescue a number of interspecific hybrids involving sunflower as one of the parents (see Table 12). Methods are now available for achieving plant recovery from the culture of very young embryos (Denat et al., 1991). Critical considerations for successful culture are stage of embryo development and media composition. Embryos 0.2–1.5 mm long (5–10 DAP) were successfully cultured on B5 medium with 9% sucrose (Kraüter et al., 1991) and a plant regeneration rate of 41% was obtained. Similarly young embryos 0.1–0.15 mm (3 DAP) were

Table 13. In vitro production of haploids in *Helianthus annus*

Number of genotypes	Explant	Developmental response	References
51	Anthers	Callus/shoot organogenesis	Alissa et al., 1985 Bohorova et al., 1985 Mix, 1985 Assad et al., 1986 Mezzarobba and Jonard, 1986 Jonard Mezzarobba, 1990 Gürel et al., 1991
8	Ovule	Embryogenesis	Cai and Zhou, 1984
8	Ovary	Embryogenesis	Gelebart and San, 1987 Yang et al., 1990

cultured and plants regenerated on a medium with 12% sucrose (Denat et al., 1991). Basal medium formulation is critical as Espinasse et al. (1988) observed that young embryos failed to develop on MS medium but did so on B_5 supplemented with amino acids and inositol. Similarly, Denat et al. (1991) observed that the mineral content of the culture medium influenced successful culture of young embryos. The culture of embryos *in ovulo* on a simpler medium prior to embryo isolation and culture has also resulted in improved frequencies of plant recovery (Espinasse et al., 1991).

3.4. *In vitro Production of Haploids*

Sunflower haploids can be obtained through culture of either male or female gametophytic cells (see Table 13). Haploids have been obtained from sunflower anther culture in a number of studies (Alissa et al., 1985; Mix, 1985; Bohorova et al., 1985; Mezzarobba and Jonard, 1986; Gürel et al., 1991a). In all cases regeneration occurred through callus proliferation followed by shoot organogenesis. Callus proliferation has also been achieved in isolated microspore culture, but plants could not be regenerated (Gürel et al., 1991b). The *in vitro* response of cultured anthers is strongly dependent on genotype (for example see Gürel et al., 1991b). Anther culture media are generally based on modified MS formulations supplemented with auxin and cytokinin. Sucrose is required at elevated concentrations, generally in the range of 6–12% (Gürel et al., 1991b). Initial culture at 35 °C for 12 h in darkness prior to transfer to 25 °C for 4–6 weeks in light, has been reported to be beneficial in terms of anther culture response (Gürel et al., 1991b). The same study also indicated that donor plant growth conditions were critical with highest anther response obtained from plants grown under.high light intensity and a day/night temperature of 20 °C/10 °C or 12 °C/8 °C.

Haploid plants have been recovered through the culture of unfertilized

Table 14. Variant and mutant selection in sunflower *Helianthus annuus* L.

Culture system and selection strategy	Response	References
Suspension cells/ metabolite analysis	Secondary Metabolite Accumulation	Rower and Mabry, 1988
Hypocotyl explants/ fungal culture filtrates	Resistant calli	Masirevec et al., 1988 Hartman et al., 1988
Immature zygotic embryo explants/ metabolite analysis	Somaclonal variants for coumarin accumulation	Roseland et al., 1991

ovaries isolated 12–24 hours before anthesis (Gelebart and San, 1987; Yang et al., 1990). Pretreatment of ovaries at 4 °C enhanced subsequent development of embryos. Optimal culture temperature was 27 °C (Yang et al., 1990). A medium with 12% sucrose, free of auxin, favored haploid development. This regime appears to suppress callus development. Further embryo development and plant recovery occurred if newly formed embryos were cultured on a medium with lower sucrose, increased nitrate, auxin, cytokinin and GA_3. Haploid plants have been recovered from 7 of 8 genotypes examined (Gelebart and San, 1987). This technique appears to be a viable alternative to anther or microspore culture and the main requirement is manipulation of culture conditions to favor development of haploid embryos at the expense of diploid callus.

3.5. *In vitro Selection of Somaclonal Variants and Mutants*

Research on generation of somaclonal variants and *in vitro* induction of mutants in sunflower has been limited, with only a few reports available in the scientific literature (see Table 14). Calli resistant to culture filtrates of *Phoma macdonaldii* and *Phomopsis* spp. have been isolated but plants were not recovered (Hartman et al., 1988; Masirevec et al., 1988). Somaclonal variants for accumulation of coumarin have been identified in plant populations regenerated from embryo-derived explants (Roseland et al., 1991). The usefulness of this approach for variant isolation will depend on selection of highly regenerable genotypes and culture conditions for efficient plant recovery, preferably at the haploid level.

Table 15. Gene transfer technology in *Helianthus annuus* L.

Target	Transformation method/ selection strategy	Response	References
Hypocotyl explants	Agrobacterium cocultivation/NPT II	Transformed plants	Everett et al., 1987
Hypocotyl explants	Agrobacterium cocultivation/PPT, NPT II, paramomycin	Transformed callus	Escandon and Hahne, 1991
Hypocotyl-derived protoplasts	PEG-mediated DNA uptake/NPT II	Transformed callus	Moyne et al., 1989
Apical meristems	Agrobacterium cocultivation/GUS, NPT II	Chimeric plants	Schrammeijer et al., 1990
Apical meristems	Agrobacterium cocultivation and/or particle bombardment/NPT II, GUS	GUS^+ sectors on plants; transformed plants	Bidney et al., 1991

3.6. *Gene Transfer Technology*

Sunflower cells have been genetically transformed through *Agrobacterium* cocultivation, particle bombardment, and polyethylene glycol-mediated DNA uptake (see Table 15). In a series of experiments on sunflower hypocotyl transformation, Escandon and Hahne (1991) concluded that medium composition, genotype, reporter gene and plant growth regulators affected the efficiency of selection following bacterial cocultivation. Manipulation of the experimental system should therefore increase the efficiency of transformation.

Transgenic plants have been obtained through cocultivation of isolated meristem and hypocotyl tissues (Everett et al., 1987; Schrammeijer et al., 1990) but the frequency of plant recovery was low and the selectable marker, kanamycin, was inhibitory to shoot regeneration (Everett et al., 1987).

Particle bombardment of apical meristems resulted in the development of tissue sectors with transformed cells (Bidney et al., 1992). However, the same study indicated that transgenic plants could be recovered if meristems wounded by bombardment were subsequently cocultured with *A. tumefaciens*. Since the regeneration capacity of this tissue is high, refinement of this method could lead to efficient recovery of transformants.

PEG-mediated uptake of DNA by hypocotyl protoplasts resulted in 4 transformed calli per 10^6 cells (Moyne et al., 1989). Transformed calli were subcultured on regeneration medium, but plants were not recovered. However, cell colony growth was observed on kanamycin, indicating that the callus was composed of transformed cells.

Table 16. Plant regeneration from cultured explants of *Linum usitatissimum*

Explant source	Development response	References
Hypocotyl	Shoots	McHughen and Swartz, 1984 Sarkar et al., 1985 Kaul and Williams, 1987 Zine et al., 1988 Zhan et al., 1989a
Cotyledon	Shoots	Sarkar et al., 1985 Zhan et al., 1989a
Root	Shoots	Zhan et al., 1989a
Leaf	Shoots	McHughen and Jordan, 1989 Zhan et al., 1989
Stem segments	Shoots	Sarkar et al., 1985
Immature embryos	Embryo shoots	Kaul and Williams, 1987

4. Oilseed Flax (*Linum usitatissimum*)

4.1. *Plant Regeneration from Cultured Explants*

In flax, hypocotyl segments, cotyledons, mature stem and immature zygotic embryo explants have been used in regeneration studies (Table 16). Explants taken from germinated seedlings and stems generally undergo shoot organogenesis. Shoot bud proliferation occurs on a number of media formulations (McHughen, 1990) in the presence of suitable levels of cytokinin and auxin. For example, hypocotyl segments 5–10 mm long proliferated shoots on MS medium with $1.0\,mg\,l^{-1}$ cytokinin and $0.02\,mg\,l^{-1}$ auxin (Jordan and McHughen, 1988a). The carbohydrate status of the culture medium has been shown to influence organogenic response (Millam et al., 1992). Addition of glutamine, yeast extract and low levels of BAP enhanced shoot proliferation from mature embryos cultured on Monniefs medium (Kaul and Williams, 1987). Somatic embryogenesis has been reported in explants derived from immature embryos (Pretova and Williams, 1986). Embryo proliferation occurred from the surface cells of the cotyledons and hypocotyl. BAP and yeast extract along with glutamine were required for induction of embryogenesis. Embryos readily matured into plantlets on a medium free of plant growth regulators.

4.2. *Protoplast Culture Technology*

Viable protoplasts have been isolated from shoot, root and cotyledon explants of flax and plants have been regenerated via organogenesis in some cases (Ling and Binding 1987; Barakat and Cocking, 1983). In a number of cases successful plant regeneration through protoplast culture has been reported in wild *Linum* species under culture conditions similar to those for flax (Ling and Binding, 1992, 1987; Barakat and Cocking 1985; Zhan et al., 1989b). A number of factors have been shown to influence cell division and colony formation in protoplast culture. Enriched media such as that developed by Kao and Michayluk (1975) can enhance plating efficiencies. Light also had a positive effect on plating efficiency (Barakat and Cocking, 1983). Plant regeneration is genotype dependent and occurred best on modified MS medium with cytokinin (Ling and Binding, 1987).

4.3. *In vitro Production of haploids*

In spite of the potential usefulness of doubled haploids in flax improvement there is relatively little information on the generation of haploid plants from anther, microspores or unfertilized embryo culture. Nichterlein et al. (1989a, b, 1991) have summarized the available information on anther culture in flax. Anthers, with microspores at the uninucleate stage, from plants grown at 16/14 °C day/night temperature and high light intensity could be cultured on modified MS medium with 6% sucrose, auxin and cytokinin (Nichterlein et al., 1989a, b). Three of six genotypes yielded callus from which haploid plants could be regenerated.

4.4. *In vitro Culture of Embryos*

Embryos isolated at the globular stage were cultured to maturation on Monnier's medium (Monnier, 1978) with 12% sucrose, glutamine and low levels of cytokinin (Pretova, 1986, 1990). Addition of cytokinin was essential for maturation, and agar solidified medium proved superior to liquid medium. Light of 3000 lux for 16 hours enhanced embryo development. Embryo culture techniques could be potentially useful for generating interspecific hybrids of flax and some of its wild relatives.

4.5. *In vitro Selection of Somaclonal Variants and Mutants*

Salt-tolerant calli were recovered when hypocotyl-derived callus was cultured on a selection medium with high levels of salts (McHughen, 1987; McHughen and Swartz, 1984). Plants were regenerated and initially shown to be salt tolerant but were not genetically characterized over several generations. A similar approach was employed for the isolation of sulfonylurea herbicide resistance (Jordan and McHughen, 1987). The mutagen, sodium azide, was included in this procedure and a number of chlorsulfuron tolerant plants were recovered.

4.6. *Gene Transfer Technology*

Hypocotyl and cotyledon explants have been used in the transformation of flax employing *Agrobacterium tumefaciens* and *A. rhizogenes* as vectors (McHughen and Dong, 1991; Jordan and McHughen 1988a, b; Basiran et al., 1987; Zhan et al., 1988a, b). A differential transformation response using both vectors, with cotyledon and hypocotyl explants, was demonstrated by Zhan et al. (1988). Cocultivation strategy was important for the efficient recovery of transformants. Long term bacterial cocultivation and removal of epidermal cell layers from hypocotyl explants improved transformation (Dong and McHughen, 1991; Jordan and McHughen, 1988a). Plant recovery was also enhanced by several subcultures of transformed tissue on selection medium (Jordan and McHughen, 1988a). This probably allowed continued proliferation of transformed cells followed by meristem initiation.

5. Safflower (*Carthamus tinctorius*)

Although Safflower is an important oilseed, limited information on its *in vitro* culture is available. Callus cultures have been established from cotyledon and hypocotyl explants on a medium containing auxin and cytokinin (Saito et al., 1989; Yoshikawa et al., 1986). Sodium chloride tolerant callus was isolated and characterized although plants were not regenerated (Nikam and Shitole, 1991). Plants were regenerated from cotyledon, hypocotyl and shoot tip explants. (Seeta and Anwar, 1991; Tejovathi and Anwar, 1989; George and Rao, 1982). Genotype, explant age and pretreatment influenced regeneration response.

The recovery of androgenetic haploid plants from several genotypes of safflower was reported (Prasad et al., 1991). Genotype and donor plant growth conditions influence anther culture response. Exposure of isolated capitula to 5 °C for 1–5 days prior to culture of anthers on media containing auxin and cytokinin was essential for achieving haploid callus proliferation. Plant regeneration was achieved through induction of shoot bud formation from proliferating haploid callus.

6. Other Oilseeds

6.1. *Cuphea*

Plants have been regenerated from explant cultures of a number of Cuphea species (Janick et al., 1986; Tagliani and Knapp, 1990). Suitable explants include, immature embryos, hypocotyls and epicotyls. Germinating seedlings are frequent sources of explants but organogenic callus can be obtained from culture of mature leaf and shoot tip explants (Janick et al., 1986). Auxin and cytokinin are required for callus proliferation and shoot regeneration. Genetic transformation of embryo explants with *Agrobacterium tumefaciens* has been reported (Tagliani and Knapp, 1990).

6.2. *Limnanthes*

Shoot multiplication was achieved by culturing nodal segments of meadowfoam *Limnanthes alba*, on medium with auxin and cytokinin (Reed, 1991). Several genotypes were examined and increasing cytokinin levels in the medium increased the number of shoots per node. Root production and shoot elongation occurred in a hormone-free medium and viable plants were recovered. Somatic embryos were obtained from the culture of immature zygotic embryos, 0.1 to 0.9 mm long, (Southworth and Kwiatkowski, 1991). High levels of sucrose and low levels of auxin and cytokinin were required. Zeatin was the most effective cytokinin. Removal of auxin allowed embryo germination but plant recovery was not reported.

6.3. *Borage*

Clonal multiplication of Borage, *Borago officinalis* L. from shoot tip cultures occurred on MS medium with coconut water, casein hydrolysate and cytokinin (Janick et al., 1987). Shoot proliferation was also obtained when immature embryos, 12–14 DAP, were cultured on MS medium with low 2,4-D and cytokinin. Addition of coconut water increased shoot proliferation several fold and was beneficial to *in vitro* rooting. However, rooting required high levels of auxin and such treatments reduced shoot elongation. Therefore a protocol for reliable plant recovery is required. Somatic embryos developed directly on cotyledons and from callus when immature zygotic embryos (12–14 DAP) were cultured on MS medium (Quinn et al., 1989a, Janick et al., 1987). Auxin, specifically 2,4-D (4.5 μM) was required for embryogenesis and this was enhanced by the inclusion of coconut water. Culture on auxin-free medium allowed development of the cotyledons but the shoot meristems were abnormal and plants were not recovered.

6.4. *Jojoba*

Clonal propagation of jojoba (*Simmondsia chinensis* linke Schneid) was achieved by culturing axillary buds on modified SH medium (Schenk and Hildebrandt, 1972), with cytokinin and auxin (Chaturvedi and Sharma, 1989). Lowering the level of cytokinin allowed continued shoot proliferation during subculture. A major problem is the low rooting efficiency of *in vitro* generated shoots. Modification of salt concentration and sugars along with auxin pulses proved effective in root development without excessive callusing (Chaturvedi and Sharma, 1989). Rooting was enhanced by the inclusion of PVP, caffeic acid and chlorogenic acid in the rooting medium (Rost et al., 1980).

Zygotic embryos 3–4 mm long (1–2 months post anthesis) produced somatic embryos when cultured on a medium with casein hydrolysate, 2,4-D and BAP (Wang and Janick, 1986a). Embryos longer than 7 mm did not respond. Optimum conditions for embryo proliferation were provided by MS medium with 3–6% sucrose 1–2 mg l^{-1} 2,4-D and 0.1 mg l^{-1} BAP. Embryogenesis occurred through a callus stage and embryos remained at the globular stage in the presence of 2,4-D. A useful aspect of embryo culture in jojoba is the comparative analysis of storage lipid accumulation in both somatic and zygotic embryos *in vitro*. Using MS medium with high sucrose levels Wang and Janick (1986b) determined that the pattern of storage lipid deposition was similar in both types of embryos.

6.5. *Eruca*

Plant regeneration through somatic embryogenesis occurred when immature zygotic embryos of *Eruca sativa* L. were cultured (Ahloowalia, 1987). Five genotypes were examined and embryo explants 20–30 DAP, torpedo stage, yielded the best results. Auxin and cytokinin were required for somatic embryo proliferation which occurred mostly on the cotyledons. A combination of asparagine and high sucrose in MS medium resulted in secondary embryo formation. Plants were recovered when embryos were cultured on hormone-free medium. Mesophyll protoplasts derived from *in vitro* grown shoots were cultured and plants regenerated (Sikdar et al., 1987). Regeneration was by both organogenesis and embryogenesis, depending on medium components. Shoot regeneration occurred when colonies were cultured in light on a medium with low sucrose, GA_3, BAP, AA and 2,4-D. Embryos developed if colonies were maintained in dark with lower levels of these hormones. Gibberellin as well as low levels of sucrose are required for high frequency regeneration. Plants were recovered on hormone-free medium.

6.6. *Sesame*

Callus can be readily obtained from seedling explants of *Sesamum indicum* L. and somatic embryos have been obtained (George et al., 1987). However, shoot proliferation was only reported from shoot tip explants obtained from seedlings germinated on high levels of cytokinins. Shoot multiplication was greatest on MS medium with high levels of cytokinin.

6.7. *Nigerseed*

Shoot organogenesis and plant regeneration has been achieved through the culture of leaf explants of several nigerseed genotypes on modified MS medium supplemented with cytokinin and auxin (Simmonds and Keller, 1987). This culture system would appear to be amenable to *Agrobacterium*-mediated gene transfer.

7. Assesment and Future Prospects

To be of value in the improvement of oilseeds, *in vitro* methods for gene manipulation and plant regeneration must be efficient and have wide applicability. Techniques such as clonal multiplication, embryo and microspore culture and protoplast culture have been applied to oilseeds. In some, such as Cuphea, meadowfoam and sesame, regeneration from explants is the only *in vitro* method reported. This method of regeneration is very efficient in certain oilseeds but, this is still dependent on species, genotype, explant source and age and media conditions. The most successful explants are cotyledon, hypocotyl, mature and immature zygotic embryos. Plant regeneration is regulated by exogenous hormones and improvements in efficiency will depend on knowledge of their role in morphogenesis.

With the availability of sensitive methods for plant hormone analysis, the relationship of endogenous levels to morphogenic competence of explants should be examined. Method are available for regeneration of plants through somatic embryogenesis in sunflower, flax and oilseed *Brassicas*. This has implications for artificial seed production if these procedures can be refined and extended to a large number of genotypes. Somatic embryos are potential sources of *in vitro* storage lipid production. Accumulation of lipids by somatic and microspore-derived embryos of several oilseed species has been reported (Janick, 1991; Pomeroy et al., 1991). To exploit the potential of this technique, research on the biochemistry of lipid synthesis and the storage capacities of somatic embryos, as well as on factors influencing lipid accumulation should be pursued.

Plant regeneration from protoplasts is advanced in the *Brassicas*, although *B. campestris* remains recalcitrant. Techniques for recovery of plants from protoplasts of sunflower and flax require further development to identify and explain species and genotypic variations. Efficient regeneration has been reported in several *Linum* species. One advantage of a reliable protoplast culture system would be production of somatic hybrids between cultivars and wild relatives to transfer useful characteristics.

In the oilseed *Brassicas* where efficient regeneration systems are available somatic hybrids have been produced between distant species such as *Trachystoma ballii* and *B. juncea*; *B. napus* and *Eruca sativa*. So far somatic hybrids have not been reported in flax and sunflower largely because of inefficient regeneration systems.

Interspecific hybridization through embryo rescue is another available method to access genes in divergent species. This has been successfully used in *Brassica* and *Helianthus*. Several media formulations have been developed for culturing young embryos to maturation and plant development. This technique is of value where the barrier to hybridization is postzygotic. *In vitro* pollination is an alternative where the barrier is prefertilization, which appears to be the case in some *Linum* species (Nichterlein et al., 1989).

Production of doubled haploid plants through microspore and anther culture is advanced in some oilseed *Brassicas*. However, *B. campestris* remains largely unresponsive. Basic information on the physiology and biochemistry of genotype specifity, high temperature and sucrose requirements and nutritional status of the donor plant is necessary to the advancement of this technique. In sunflower and flax, androgenetic haploids are limited to anther culture of a few genotypes and further research is required to develop methods for microspore culture (Gürel et al., 1991b). In one report androgenetic haploids were produced from anthers of several genotypes of safflower, without specific donor plant growth conditions. An alternative to androgenesis is gynogenesis and in sunflower conditions have been outlined for plant recovery from unfertilized ovules (Yang et al., 1990).

A haploid cell culture system offers a distinct advantage for mutagenesis and isolation of mutants, especially if the cells remain morphogenetic. Even where plants cannot be directly regenerated, haploid mutants can be rescued by cell fusion with highly regenerable cells. In some oilseeds variants have been selected for stress tolerance, herbicide resistance and fungal toxin resistance. In only very few cases were haploid cultures used. An area of deficiency with regard to *in vitro* variant selection in oilseeds is the absence of good suspension cultures for single cell cloning. In most cases mutagenesis and isolation of variants are done with callus cultures. As a result all cells are not exposed to the same selection pressure and cell-cell interaction may limit the efficiency of variant isolation. Another limitation is the non-specificity of the pathotoxins often used for variant selection. In the many oil seeds, variant selection will only be successful if the efficiency of *in vitro* plant regeneration from a number of explants is improved.

Both *Agrobacterium*-based and non-biological gene transfer techniques have been employed in the transformation of oilseeds and transgenic plants have been obtained with *Brassica*, flax and sunflower. Efficiency is limited by the absence of reliable regeneration protocol in sunflower and some Brassicas. Improvement may be obtained by selection of appropriate explants such as immature zygotic embryos and shoot meristems.

Co-cultivation of explants with *Agrobacterium* is the most frequently used technique while electroporation and PEG methods require protoplasts culture and are therefore limited. The particle gun technique has not been used extensively in oilseed transformation. Development of this method would facilitate the use of highly regenerable explants. Gene transfer technologies are the most desirable for modification of oilseeds as specific genes can be introduced without the need to eliminate unwanted background. The identification of useful genes, development of constructs with a high degree of expression and the availability of efficient regeneration systems will be critical for the advancement of the use of this technology.

8. References

Agnihotri A., Lakshmikumaran, M., Shivanna, K.R. and Jagannathan, V. (1990a). Embryo rescue of interspecific hybrids of *Brassica spinescens* × *Brassica campestris* and DNA analysis. In: Progress in Plant Cellular and Molecular Biology. H.J.J. Nijkamp, L.H.W. Vander Plas and J. Van Aartvijk (eds.), pp. 270–274.

Agnihotri A., Lakshmikumaran, M., Shivanna, K.R. and Jagannathan, V. (1990b). Production of *Eruca* and *Brassica* hybrids by embryo rescue. Plant Breeding 104:281–289.

Ahloowalia, B.S. (1987). Somatic embryogenesis and plant regeneration in *Eruca sativa*. Crop. Sci. 27:813–814.

Alissa, A., Serieys, H. and R. Jonard (1985). Sur les possibilités de régénération d'especes sauvages et d'hybrides interspecifique du genre *Hélianthus* par androgenese *in vitro*, C.R. Acad Sci. Paris t 300 Ser. III No. 1, 25–30.

Altenbach, S.B., Kuo, C.C, Staraci, L.C., Pearson, K.W., Wainwright, C., Georgescu, A. and Townsend, J. (1992). Accumulation of a brazil nut albumin in seeds of transgenic canola results in enhanced levels of protein methionine. Plant Mol. Biol. 18:235–245.

Arora, R. and Bhojwani, S.S. (1988). Production of androgenetic plants through pollen embryogenesis in anther cultures of *Brassica carinata* A. Braun. Biol. Plant. 30:25–29.

Aslam, F.M., MacDonald, M.V. and Ingram, D.S. (1990a). Rapid cycling *Brassica* species, anther culture potential of *B. campestris* L. and *B. napus* L. New Phytol. 115:1–9.

Aslam, F.N., MacDonald, M.V., Loudon, P.T. and Ingram, D.S. (1990b). Rapid cycling *Brassica* species. Inbreeding and selection of *Brassica napus* for anther culture ability and an assessment of its potential for microspore culture. Ann. Bot. 66:331–339.

Assad, A., Jonard, R., Berleys, H. and Vincourt, P. (1986). La culture d'embryons isolés *in vitro* dans un programme d' amclioration du tournesol. C.R. Acad Sci. Paris t 302 Ser. III:161–164.

Baillie, A., Epp, D., Hutcheson, D. and Keller, W.A. (1992). *In vitro* culture of isolated microspore and regeneration of plants in *Brassica campestris*. Plant Cell Rep. (in press).

Bajaj, Y.P.S., Mohajan, S.K. and Labana, K.S. (1986). Interspecific hybridization of *Brassica napus* and *Brassica juncea* through ovary ovule and embryo culture. Euphytica 35:103–109.

Bajaj, Y.P.S. and Mohapatra, D. (1987). *In vitro* plant regeneration in *Brassica carinata* A. Br. an oilseed crop. Indian J. Exptl. Biol 25:161–163.

Barfield, D.B. and Pua, E.C. (1991). Gene transfer in plants of *Brassica juncea* using *Agrobacterium tumefaciens*-mediated transformation. Plant Cell Rep. 10:308–314.

Barsby, T.L., Choung, P.V., Yarrow, S.A., Wu, S.C., Coumans, M., Kemble, R.J., Powell, A.D., Beversdorf, W.D. and Pauls K.P. (1987a). The combination of polima cms and cytoplasmic triazine resistance in *Brassica napus*. Theor. Appl. Genet. 73:809–814.

Barsby, T.L., Yarrow, S.A. and Shepard, J.F. (1986). A rapid and efficient alternative proced,.ure for the regeneration of plants from hypocotyl protoplasts of *Brassica napus* L. Plant Cell Rep. 5:101–103.

Barsby, T.L., Yarrow, S.A., Kemble, R.J. and Grant, I. (1987b). The transfer of cytoplasmic male sterility to winter type oilseed rape (*Brassica napus* L.) by protoplast fusion. Plant Sci. 53:243–248.

Barakat, M.N. and Cocking, E.C. (1983). Plant regeneration from protoplasts derived tissues of *Linum usitatissimum* L. (flax). Plant Cell Rep. 2:314–317.

Barakat, M.N. and Cocking, E.C. (1985). An assessment of the culture capabilities of protoplasts of some wild species of *Linum*. Plant Cell Rep. 4:164–167.

Basiran, H., Armitage, P., Scott, R.J. and Draper J. (1987). Genetic transformation of flax (*Linum usitatissimum* L.) by *Agrobacterium tumefaciens*, regeneration of transformed plants via a callus phase. Plant Cell Rep. 6:396–399.

Batra, V., Prakash, S. and Shivanna, K.R. (1990). Intergeneric hybridization between *Diplotaxis slifolia* a wild species and crop *Brassicas*. Theor. Appl. Genet. 80:537–541.

Bauer-Weston, B. (1990). The production and characterization of asymmetric somatic hybrids between *Arabidopsis thaliana* and *Brassica napus*. M.Sc. Thesis. Dept. Biology, Carleton University, Ottawa, Ontario, Canada.

Bhattacharya, N.M. and Sen, S.K. (1980). Production of plantlets through somatic embryogenesis in *Brassica campestris*. Z. Pflanzenphysiol. 99:357–365.

Binding, A., Nehls, R., Kock, R., Finger, J. and Mordhorst, G. (1981). Comparative studies on protoplast regeneration in herbaceous species of the dicotyledonae class. Z. Pflanzenphysiol. 101:119–130.

Bidney D., Scelonige, C., Martich, J., Burrus, M. and Sims, L. (1992). Mircoprojectile bombardment of plant tissues increases transformation frequency by *Agrobacterium tumefaciens*. Plant Mol. Biol 18:301–313.

Bohorova, N.E., Atanassov, A. and Georgieva-Todorova, J.D. (1985). *In vitro* organogenesis, androgenesis and embryo culture in the genus *Helianthus* L. Z. Pflanzenzucht. 95:35–44.

Bohorova, N.E., Cocking, E.C. and Power, J.B. (1986). Isolation, culture and callus regeneration of protoplasts of wild and cultivated *Helianthus* spp. Plant Cell Rep. 5:256–258.

Boulter, M.E., Croy, E., Simpson, P., Shields, R., Croy, R.R.D. and Shirsat, A.H. (1990). Transformation of *Brassica napus* L. (oilseed rape) using *Agrobacterium tumefaciens* and *Agrobacterium rhizogenes*, A comparison. Plant Sci. 70:91–99.

Burnett, L., Yarrow, S. and Huang, B. (1992). Embryogenesis and plant regeneration from isolated microspores of *Brassica rapa* L. ssp. *oleifera*. Plant Cell Rep. 11:215–218.

Burrus, M., Chanabe, C., Alibert, G. and Bidney, D. (1991). Regeneration of fertile plants from protoplasts of sunflower (*Helianthus annuus* L.). Plant Cell Rep. 10:161–166.

Cai, D. and Zhou, C. (1984). *In vitro* induction of haploid embryoids and plantlets from unpollinated florets and ovules of *Helianthus annuus* L. Kexue Tongbao 29:680–682.

Campbell, C.T., Seguin-Swartz, G. and Prakash, S. (1991). Artificial synthesis of *Brassica juncea* with *B. nigra* cytoplasm. Proceedings of the GCIRC Rapeseed Congress, July 9–11. D.I. McGregor (ed.), pp. 1040–1043.

Cavallini, A. and Lupi, M.C. (1987). Cytological studies of callus and regenerated plants of sunflower, *Helianthus annuus* L. Plant Breeding 99:203–208.

Ceriani, M.F., Hopp, H.E., Hahne, G. and Escandon, A.S. (1992) Cotyledons: An explant for routine regeneration of Sunflower plants. Plant Cell Physiol. 33:157–164.

Chanabe C., Burrus, M. and Alibert, G. (1989). Factors affecting the improvement of colony formation from sunflower protoplasts. Plant Sci. 64:125–132.

Chanabe, C., Burrus, M., Bidney, D. and Alibert, G. (1991). Studies on plant regeneration from protoplasts in the genus *Helianthus*. Plant Cell Rep. 9:635–638.

Chandler, J.M. and Beard, B.H. (1983). Embryo culture of *Helianthus* hybrids. Crop Sci. 23:1004–1007.

Chapel, M. and Glimelius, K. (1990). Temporary inhibition of cell wall synthesis improves the transient expression of the GUS gene in *Brassica napus* mesophyll protoplasts. Plant Cell Rep. 9:105–108.

Charest, P.J., Holbrook, L.A., Guband, J., Iyer, U.N. and Miki, B.L. (1988). Agrobacterium mediated transformation of thin cell layer explants from *Brassica napus* L. Theor. Appl. Genet. 75:438–445.

Charne, D.G. and Beversdofl, W.D. (1988). Improving microspore culture as a rapeseed breeding tool, the use of auxin and cytokinin in an induction medium. Can. J. Bot. 66:1671–1675.

Chatterjee, G., Sikdar, S.R., Das, S. and Sen, S.K. (1985). Regeneration of plantlets from mesophyll protoplasts of *Brassica juncea* L. Czern. Plant Cell Rep. 4:245–247.

Chaturvedi, A.C. and Sharma, M. (1989). *In vitro* production of cloned plants of Jojoba (*Simmondsia chinensis* (Linke) (Schneid) through shoot proliferation in long term culture. Plant Sci. 63:199–207.

Chen, Z. and Chen, Z. (1983). High frequency induction of pollen derived embryoids from anther cultures of rape (*Brassica napus* L.). Kexue Tongbao 28:1690–1694.

Chi, G.L., Barfield, D.G., Sim and Pua, E.C. (1990). Effect of $AgNO_3$ and aminoethoxyvinylglycine on *in vitro* shoot and root organogenesis from seedling explants of recalcitrant *Brassica* genotypes. Plant Cell Rep. 9:195–198.

Choung, P.V. and Beversdorf, W.D. (1985). High frequency embryogenesis through isolated microspore culture in *Brassica napus* L. and *B. carinata* Braun. Plant Sci. 39:219–229.

Choung, P.V., Pauls, K.P. and Beversdorf, W.D. (1985). A simple culture method for *Brassica* hypocotyl protoplasts. Plant Cell Rep. 4:4–6.

Choung, P.V., Pauls, K.P. and Beversdorf, W.D. (1986b). High frequency embryogenesis in male sterile plants of *Brassica napus* through microspore culture. Can. J. Bot. 66:1676–1680.

Choung, P.V., Beversdorf, W.D. and Pauls, K.P. (1987a). Plant regeneration from haploid stem peel protoplasts of *Brassica napus* L. J. Plant Physiol. 130:57–65.

Choung, P.V., Pauls, K.P. and Beversdorf, W.D. (1987b). Protoplast culture and plant regeneration from *Brassica carinata* Braun. Plant Cell Rep. 6:67–69.

Choung, P.V., Beversdorf, W.D., Powell, A.D. and Pauls, K. (1988a). The use of haploid protoplasts fusion to combine cytoplasmic atrazine resistance and cytoplasmic male sterility in *Brassica napus*. Plant Cell Tissue Org. Cult. 12:181–184.

Choung, P.V., Beversdorf, W.D., Powell, A.D. and Pauls, K.P. (1988b). Somatic transfer of cytoplasmic traits in *Brassica napus* L. by haploid protoplast fusion. Mol. Gen. Genet. 211:197–201.

Choung, P.V., Deslauriers, C., Kott, L.S. and Beversdorf, W.D. (1988a). Effect of donor genotype and bud sampling on microspore culture of *Brassica napus*. Can. J. Bot. 66:1653–1657.

Chraibi, K.M., Castelle, J.C., Latche, A., Rouston, J.P. and Fallot, J. (1992). Enhancements of shoot regeneration potential by liquid medium culture from mature cotyledons of sunflower (*Helianthus annuus* L.). Plant Cell Rep. 10:617–620.

Chraibi, K.M., Latche, A., Rouston, J.P. and Fallot, J. (1991). Stimulation of shoot regeneration from cotyledons of *Helianthus annuus* by ethylene inhibitors silver and cobolt. Plant Cell Rep. 10:204–207.

Coventry, J., Kott, L. and Beversdorf, W.D. (1988). Manual for microspore culture technique for *Brassica napus*. Dept. of Crop Sci. Techn. Bull., Univ. of Guelph. OAC Publ. 0489.

Dale, P.J. and Ball, L.F. (1991). Plant regeneration from cotyledonary explants in a range of *Brassica* species and genotypes. Proceedings of the GCIRC Rapeseed Congress, July 9–11. D.I. McGregor (ed.), pp. 1122–127.

Damgaard, O. and Rasmussen O. (1991). Direct regeneration of transformed shoots in *Brassica napus* from hypocotyl injection with *Agrobacterium rhizogenes*. Plant Mol. Biol 17:1–8.

DeBlock, M. and DeBrouwer, D. (1991). Two T-DNA's cotransformed into *Brassica napus* by a double *Agrobacterium tumefaciens* infection are mainly integrated at the same locus. Theor. Appl. Genet. 82:257–260.

DeBlock, M., DeBrouwer, D. and Tenning, P. (1990). Transformation of *Brassica napus* and *Brassica oleracea* using *Agrobacterium tumefaciens* and the expression of the bar and neo genes in the transgenic plants. Plant Physiol. 91:694–701.

Delourme, R., Eber, F. and Chevre, A.M. (1989). Intergeneric hybridization of *Diplotaxis erucoides* with *Brassica napus* I. cytogenetic analysis of F, and BC progeny. Euphytica 41:123–128.

Denat, J.F. Coumans, M., Serieys, H. and Jonard, R. (1991). Definition d'un millieu de culture permettant le development d'embryos immature de tournesol de – application a l'obtention d'hybrides interspecifiques. C.R. Acad. Sci. Paris, t 313 series III:145–151.

Dietert, M.G., Barron, S.A. and Yoder, O.C. (1982). Effects of genotype on *in vitro* culture in the genus *Brassica*. Plant Sci. Lett. 26:233–240.

Dong, J.Z. and McHughen, A. (1991). Patterns of transformation intensity on flax hypocotyls inoculated with *Agrobacterium tumefaciens*. Plant Cell Rep. 10:555–560.

Dunwell, J.M. (1981). *In vitro* regeneration from excised leaf discs of three *Brassica* species. J. Exptl. Bot. 32:789–799.

Dunwell, J.M. and Cornish, M. (1985). Influence of preculture variables on microspore embryo production in *Brassica napus* ssp. *oleifera* cv Duplo. Ann. Bot. 56:281–289.

Dunwell, J.M. and Thurling, M. (1985). Role of sucrose in microspore embryo production in *Brassica napus* ssp. *oleifera*. J. Exptal. Bot. 36:1478–1491.

Dunwell, J.M., Cornish, M. and Decourcel, A.G.L. (1985). Influence of genotype, plant growth temperature and anther incubation temperature on microspore embryo production in *Brassica napus* ssp. *oleifera*. J. Exptl. Bot. 36:679–689.

Dunwell, J.M., Cornish, M., Decourcel, A.G.L. and Middlefell-Williams, J.E. (1983). Induction and growth of microspore-derived embryos of *Brassica napus* ssp. *oleifera*. J. Exptl. Bot. 34:1–11.

Dupuis, J.M., Gomiero, M., Plantevin, C. and Chagvardieff, P. (1988). Differenciation de formes embryoides issues de protoplastes de tournesol (*Helianthus annuus* L.) C.R. Acad. Sci. Paris t 307 Serie III: 465–468.

Dupuis, J.M., Pean, M. and Chagvardieff, P. (1990). Plant donor tissue and isolation procedure effect on early formation of embryoids from protoplasts of *Helianthus annuus* L. Plant Cell Tissue Org. Cult. 22:183–189.

Eapen, S., Abraham, V., Gerdemann, M. and Shieder, O. (1989). Direct somatic embryogenesis, plant regeneration and evaluation of plants obtained from mesophyll protoplasts of *Brassica juncea*. Ann. Bot. 63:369–372.

Escandon, A.S. and Hahne, G. (1991). Genotype and composition of culture medium are important factors in the selection for transformed sunflower (*Helianthus annuus* L.) callus. Physiol. Plant 81:367–376.

Espinasse, A., Lay, C. and Dybing, D. (1985). Factors controlling *in vitro* development of sunflower embryos. Agronomie 5:825–832.

Espinasse, A. and Lay, C. (1989). Shoot regeneration of callus derived from globular to torpedo embryos from 59 sunflower genotypes. Crop Sci. 29:201–205.

Espinasse, A., Lay C. and Volin, J. (1989). Effects of growth regulator concentrations and explant size on shoot organogenesis from callus derived from zygotic embryos of sunflower. *Helianthus annuus* L. Plant Cell Tissue Org. Cult. 17:171–181.

Espinasse, A., Volin, J., Dybing, C.D. and Lay, C. (1991). Embryo rescue through *in ovulo* culture in *Helianthus*. Crop Sci. 31:102–108.

Everett, N.P., Robinson, K.E.P. and Mascharenhas, D. (1987). Genetic engineering of sunflower, *Helianthus annuus* L. Biotechnology 5:1201–1204.

Fahleson, J., Rahlen, L. and Glimelius, K. (1988). Analysis of plants regenerated from protoplast fusion between *Brassica napus* and *Eruca sativa*. Theor. Appl. Genet. 76:507–512.

Fan, Z., Armstrong, K.C. and Keller, W.A. (1988). Development of microspores *in vivo* and *in vitro* in *Brassica napus* L. Protoplasma 147:191–199.

Fazekas, G.A., Sedmach, P.A. and Palmer, M.V. (1986). Genetic and environmental effects on *in vitro* shoot regeneration from cotyledon explants of *Brassica juncea*. Plant Cell Tissue Org. Cult. 6:177–180.

Finer, J.J. (1987). Direct somatic embryogenesis and plant regeneration from immature embryos of hybrid sunflower (*Helianthus annuus* L.) on high sucrose containing medium. Plant Cell Rep. 6:372–374.

Freyssinet M. and Freyssinet G. (1988). Fertile plant regeneration from sunflower (*Helianthus annuus* L.) immature embryos. Plant Sci. 56:177–181.

Friedt, W., Nichterlein, K., Dahlhoff, M., Köhler, H. and Gürel, A. (1991). Recent progress and prospects of Biotechnology in Sunflower breeding. Fat Sci. Technol. 93:368–374.

Friedt, W. (1988). Biotechnology in breeding of industrial oil crops, Present status and future prospects. Fat Sci. Technol. 90:51–55.

Fry, J., Barnason, A. and Horsch, R.B. (1987). Transformation of *Brassica napus* with *Agrobacterium tumefaciens* based vectors. Plant Cell Rep. 6:321–325.

Gamborg, O.L., Miller, R.A. and Ojima, K. (1968). Nutrient requirements of suspension cultures of soybean root cells. Exptal. Cell Res. 50:151–158.

Gelebart, P. and San, L.H. (1987). Obtention de plants haploides par culture *in vitro* d'ovaries et d'ovules non tecondes de tournesol (*Helianthus annuus* L.) Agronomie 7:81–86.

George, L. and Rao, P.S. (1982). *In vitro* induction of pollen embryos and plantlets in *Brassica juncea* through anther culture. Plant Sci. Lett. 26:111–116.

George, L. and Rao, P.S. (1983). Yellow-seeded variants in *in vitro* regeneration of mustard (*Brassica juncea*) (Var Rai-5) Plant Sci. Lett. 30:327–330.

George, L. and Rao, P.S. (1982). *In vitro multiplication of safflower* (*Carthamus tinctorius* L.) through tissue culture. Proc. Indian Natl. Sci. Acad. 6:791–794.

George, L., Bapat, V.A. and Rao, P.S. (1987). *In vitro* multiplication of sesame (*Sesamum indicum*) through tissue culture. Ann. Bot. 60:17–21.

Gland, B., Lichter, R. and Schweiger, H.G. (1988). Genetic and exogenous factors affecting embryogenesis in isolated microspore cultures of *Brassica napus* L. J. Plant Physiol. 132:613–617.

Glimelius, K. (1984). High growth rate and regeneration capacity of hypocotyl protoplasts in some *Brassiceae*. Physiol. Plant 61:38–44.

Glimelius, K., Djupsjobacka, M. and Fellner-Feldegg, H. (1986). Selection and enrichment of plant protoplasts heterokaryons of *Brassiceae* by flow sorting. Plant Sci. 45:133–141.

Greco, B., Tanzarella, O.A., Carrozzo, G. and Blanco, A. (1984). Callus induction and shoot regeneration in sunflower, *Helianthus annuus* L. Plant Sci. Lett. 36:73–77.

Guerche, P.M., Charbonnier, L., Jouanin, C., Tourneur, J., Paszkowski and Pellitier, T. (1987). Direct gene transfer by electroporation in *Brassica napus*. Plant Sci. 52:114–116.

Guerche, P., Jouanin, L., Tepper, D. and Pelletier, G. (1987). Genetic transformation of oilseed rape (*Brassica napus*) by the RIT-DNA of *Agrobacterium rhizogenes* and analysis of inheritance of the transformed phenotype. Mol. Gen. Genet. 206:382–386.

Guerche, P., DeAlmeida, E.R.P., Schwantztein, M.A., Gander, E., Krebbers, E. and Pelletier, G. (1990). Expression of the 2S albumin from *Berthalletia excelsa* in *Brassica napus*. Mol. Gen. Genet. 221:306–314.

Guilley, E. and Hahne, G. (1989). Callus formation from isolated sunflower (*Helianthus annuus* L.) mesophyll protoplasts. Plant Cell Rep. 8:226–229.

Gundimeda, H.R. Prakash, S. and Shivanna, K.R. (1992). Intergeneric hybrids between *Enarthrocrapus lyratus*, a wild species and crop brassicas. Theor. Appl. Genet. 83:655–662.

Gürel, A. Kontowski, S., Nichterlein K. and Friedt, W. (1991a). Embryogenesis in microspore cultures of sunflower (*Helianthus annuus* L.) Helia. 14:123–128.

Gürel, A., Nichterlein, K. and Friedt, W. (1991b). Shoot regeneration from anther culture of sunflower (*Helianthus annuus* L.) and some interspecific hybrids as affected by genotype and culture procedure. Plant Breeding 106:68–76.

Hachey, J.G., Sharma, K.K. and Moloney, M.M. (1991). Efficient shoot regeneration of *Brassica campestris* using cotyledon explants cultured *in vitro*. Plant Cell Rep. 9:549–554.

Hartmann, C.L., Donald, P.A., Secor, G.A. and Miller, J.E. (1988). Sunflower tissue culture and use in selection for resistance to *Phoma macdonaldii* and white mold (*Sclerotinia sclerotiorum*). Proc. 12th Intl. Sunflower Confer., July 25–29. Novi Sad, Yugoslavia.

Heather, W.D. and Earle, E.D. (1991). Resynthesis of novel rapeseed via protoplast fusion. Proceedings of the GCIRC Rapeseed Congress, July 9–11. D.I. McGregor (ed.), pp. 1090–1095.

Hoffman, F., Thomas, F. and Wenzel, G. (1982). Anther culture as a breeding tool in rape. Theor. Appl. Genet. 61:225–232.

Hossain, M.M., Inden, H. and Asahira, T. (1990). Seed vernalized interspecific hybrid through *in vitro* ovule culture in *Brassica*. Plant Sci. 68:95–102.

Hossain, M.M., Inden, H. and Asahina, T. (1988). Intergeneric and interspecific hybrids through *in vitro* ovule culture in the *Cruciferae*. Plant Sci. 58:121–128.

Huang, B., Bird, S., Kemble, K., Simmonds, D., Keller, W.A. and Miki, B. (1990). Effects of culture density, conditioned medium and feeder cultures on microspore embryogenesis in *Brassica napus* L. cv Topaz. Plant Cell Rep. 8:594–597.

Huang, B., Swanson, E.B., Baszcynski, C.L., Macrae, W.D., Bardour, E., Armavil, V., Cooke, L., Arnaldo, M., Rozakis, S., Wescott, M., Keats, R.F. and Kemble, R. (1991). Application of microspore culture to canola improvement. Proceedings of the GCIRC Rapeseed Congress, July 9–12, 1991. D.I. McGregor (ed.), pp. 298–303.

Ingram, D.S., Loh, C.S., MacDonald, M.V. and Newsholme, D.M. (1984). Secondary embryogenesis in *Brassica*: a tool for research and crop improvement. Proceedings of the Phytochemical Society, 23:219–242.

Inomata, M. (1990). Interspecific hybridization in *Brassica* through ovary culture. In: Biotechnol. in Forestry and Agric., Vol. 10. Legumes and Oilseeds 1. Y.P.S. Bajaj (ed.), pp. 367–385.

Itoh, K., Iwabuchi, M. and Shimamoto, K. (1991). *In situ* hybridization with species specific DNA probes of genes, evidence for asymmetric nature of *Brassica* hybrids obtained by x-ray fusion. Theor. Appl. Genet. 81:356–362.

Jain, R.K., Sharma, D.R. and Chowdhory, J.B. (1989). High frequency regeneration and heritable somaclonal variation in *Brassica juncea*. Euphytica 40:75–81.

Jain, R.K., Jain, S., Nainawatee, H.S. and Chowdhory, J.B. (1990). Salt tolerance in *Brassica juncea* L. *In vitro* selection, agronomic evaluation and genetic stability. Euphytica 48:141–152.

Jain, R.K., Chowdhory, J.B. and Friedt, W. (1988a). Organogenesis and plant regeneration from cotyledon explant cultures of wild turnip (*Brassica tournefontii* L.). Plant Cell Tissue Org. Cult. 15:107–111.

Jain, R.K., Chowdhory, J.B. and Friedt, W. (1988). Genotype and media effects on plant regeneration from cotyledon explant cultures of some *Brassica* species. Plant Cell Tissue Org. Cult. 14:197–206.

Jain, S., Nainawatee, H.S., Jain, R.K. and Chowdhory, J.B. (1991). Proline status of genetically stable salt tolerant *Brassica juncea* L. Somaclones and their parent cv Prakash. Plant Cell Rep. 9:684–687.

Janick, J. (1991). Production of seed lipids via culture of somatic embryos. In Biotechnology of Plant Fats and Oils. J. Rattray (ed.), AOCS, Champaign, III, pp. 76–104.

Janick, J., Simons, J.E. and Whipkey, A. (1986). *In vitro* propagation of *Cuphea wrightii*. Hort. Sc. 21:135–137.

Janick, J., Simons, J.E. and Whipkey, A. (1987). *In vitro* propagation of borage. Hort Sci. 22:493–496.

Jaswal, S.K., Hammatt, M., Bhojwani, S.S., Cocking, E.C. and Davey, M.R. (1990). Plant regeneration from cotyledon protoplasts of *Brassica carinata*. Plant Cell Tissue Org. Cult. 22:109–116.

Jeannin, G. and Hahne, G. (1991). Donor plant growth conditions and regeneration of fertile plants from somatic embryos induced on immature zygotic embryos of sunflower *Helianthus annuus* L. Plant Breeding 107:280–287.

Jonard, R. and Mezzarobba, A. (1990). Sunflower (*Helianthus* spp.), anther culture and field studies on haploids. In: Biotechnol. in Agric. and Fores., Vol. 10. Legumes and Oilseed Crops I. Y.P.S. Bajaj (ed.), pp. 485–501.

Jordan, M.C. and McHughen, A. (1987). Selection for chlorsulfuron in flax (*Linum usitatissimum* L.) cell cultures. J. Plant Physiol. 131:333–338.

Jordan, M.C. and McHughen, A. (1988a). Glyphosate tolerant flax plants from *Agrobacterium* mediated gene transfer. Plant Cell Rep. 7:281–284.

Jordan, M.C. and McHughen, A. (1988b). Transformed callus does not necessarily regenerate transformed shoots. Plant Cell Rep. 7:285–287.

Jouanin, L., Guerche, P., Pamboubdjian, N., Tournier, C., Case-Delbart, E. and Tournier, J.

(1987). Structure of T-DNA in plants regenerated from roots transformed by *Agrobacterium rhizomes* strain A4. Mol. Gen. Genet. 206:387–392.

Jourdan, P.S., Earle, E.D. and Mutschler, M.A. (1989). Synthesis of male sterile, triazine-resistant *B. campestris*. Theor. Appl. Genet. 78:63–72.

Julliard, J., Sossountzou, L., Habricot, Y. and Pelletier, G. (1992). Hormonal requirement and tissue competency for shoot organogenesis in two cultivars of *Brassica napus*. Physiol. Plant. 84:521–530.

Kao, H.M., Brown, G.G., Scoles, G. and Seguin-Swartz, G. (1991). Ogura cytoplasmic male sterility and triazine tolerant *Brassica napus* cv Westar produced by protoplast fusion. Plant Sci. 75:63–72.

Kao, H.M., Keller, W.A., Gleddie, S. and Brown, G.G. (1992). Synthesis of *Brassica oleracea*, *Brassica napus* somatic hybrid plant with novel organelle DNA composition. Theor. Appl. Genet. 83:313–320.

Kao, K. N. and Michayluk, M. R. (1975). Nutritional requirements for growth of *Vicia hajastana* cells and protoplasts at very low population density in liquid medium. Planta 126:105–110.

Kao, H.M. and Seguin-Swartz, G. (1987). Study of factors affecting the culture of *Brassica napus* L. and *Brassica juncea* leaf mesophyll protoplasts. Plant Cell Tissue Org. Cult. 10:79–90.

Kaul, V. and Williams, E.G. (1987). Multiple shoot formation *in vitro* from hypocotyl of germinating embryos of flax *Linum usitatissimum* L. J. Plant Physiol. 131:441–448.

Keller, W.A. (1984). Anther culture of *Brassica*. In Cell Culture and Somatic Cell Genetics of Plants Vol. 1. I.K. Vasil (ed.), Acad. Press NY, pp. 302–308.

Keller, W.A. and Armstrong, K.C. (1978). High frequency production of microspore-derived plants from *Brassica napus* anther cultures. Z. Pflanzenzucht. 80:100–108.

Keller, W.A. and Armstrong, K.C. (1979). Stimulation of embryogenesis and haploid production in *Brassica campestris* anther cultures by elevated temperature treatments. Theor. Appl. Genet. 55:65–67.

Keller, W.A., Arnison, P.G. and Cardy, B.J. (1987). Haploids from gametophytic cells – recent developments and future prospects. In. Plant tissue and cell culture. Proc. 6th Intl. Tissue Cult. Congr. C.E. Green, D.A. Somers, W.P. Hackett, D.D. Biesboer (eds.), Alan. R. Liss. Inc. NY, pp. 223–241.

Kemble, R.J., Yarrow, S.A., Wu, S.C. and Barsby, T.L. (1988). Absence of mitochondrial and chloroplast DNA recombination in *Brassica napus* plants regenerated from protoplasts, protoplast fusions and anther culture. Theor. Appl. Genet. 75:875–881.

Kemble, R.J., Barsby, T.L. and Yarrow, S.A. (1988). Transformation of plant mitochondria with mitochondrial DNA plasmids via protoplast fusion. Mol. Gen. Genet. 213:202–205.

Kirti, P.B. (1988). Somatic embryogenesis in hypocotyl protoplast culture of rapeseed (*Brassica napus* L.) Plant Breeding 100:222–224.

Kirti, P.B. and Chopra, V.L. (1989). A simple method of inducing somatic embryogenesis in *Brassica juncea* L. Czern and Coss. Plant Breeding 102:73–78.

Kirti, P.B., Prakash, S. and Chopra, V.L. (1991). Interspecific hybridization between *Brassica juncea* and *B. spinescens* through protoplast fusion. Plant Cell Rep. 9:639–643.

Kirti, P.B. and Chopra, V.L. (1990). Rapid plant regeneration through organogenesis and somatic embryogenesis from cultured protoplasts of *Brassica juncea*. Plant Cell Tissue Org. Cult. 20:465–67.

Kirti, P.B. and Chopra, V.L. (1989). Plant Regeneration from hypocotyl-derived protoplasts of *Brassica juncea* L. Czern. and Coss. Plant Cell Rep. 7:708–710.

Kirti, P.B., Hadi, S., Kumar, P.A. and Chopra, V.L. (1991). Production of sodium chloride-tolerant *Brassica juncea* plants by *in vitro* selection at the somatic embryo level. Theor. Appl. Genet. 83:233–237.

Klimaszewska, K. and Keller, W.A. (1985). High frequency plant regeneration from thin cell layer explants of *Brassica napus* L. Plant Cell Tissue Org. Cult. 4:183–197.

Klimaszewska, K. and Keller, W.A. (1987). Plant regeneration from stem cortex protoplasts of *Brassica napus*. Plant Cell Tissue Org. Cult. 8:225–233.

Klimaszewska, K. and Keller, W.A. (1988). Somatic hybridization between *Brassica napus* and *Diplotaxis harra*. Plant Sci. 58:211–222.

Knittel, M., Escandon, A.S. and Hahne, G. (1991). Plant regeneration at high frequency from mature sunflower cotyledons. Plant Sci. 73:219–226.

Kohlenbach, H.W., Wenzel, G. and Hoffmann, F. (1982). Regeneration of *Brassica* plantlets in cultures from isolated protoplasts of haploid stem embryos as compared to leaf protoplasts. Z. Pflanzenphysiol. 105:131–142.

Kohler, E., Benediktsson, I., Gordon, D., Andreo, C.S. and Shieder, O. (1990). Effect of various irradiation treatments of plant protoplasts on the transformation rates after direct gene transfer. Theor. Appl. Genet. 79:679–685.

Kott, L.S., Polsoni, L. and Beversdorf, W.D. (1988a). Cytological aspects of isolated microspore culture of *Brassica napus*. Can. J. Bot. 66:1658–1664.

Kott, L.S., Ellis, B. and Beversdorf, W.D. (1988b). Autotoxicity in isolated microspore culture of *Brassica napus*. Can. J. Bot. 66:1665–1670.

Kranz, E. (1988). Somatic embryogenesis in stationary phase suspension cultures derived from hypocotyl protoplasts of *Brassica napus* L. Plant Cell Tissue Org. Cult. 12:141–146.

Krasnyanski, S., Polgar, Z., Nemeth, G. and Menczel, L. (1992). Plant regeneration from callus and protoplast cultures of *Helianthus giganteus* L. Plant Cell Rep. 11:7–10.

Kraüter, R. and Friedt, W. (1991). Propgation and multiplication of sunflower lines (*Helianthus annuus* L.) by tissue culture *in vitro*. Helia 14:123–128.

Kraüter, R., Steinmetz, A. and Friedt, W. (1991). Efficient interspecific hybridization in the genus Helianthus via embryo rescue and characterization of hybrids. Theor. Appl. Genet. 82:521–525.

Landgren, M. and Glimelius, K. (1990). Analysis of chloroplasts and mitochondrial segregation in three different combinations of somatic hybrids produced within Brassicaceae. Theor. Appl. Genet. 80:776–784.

Lazzeri, P.A. and Dunwell, J.M. (1984). *In vitro* shoot regeneration from seedling root segments of *Brassica oleracea* and *Brassica napus* cultivars. Ann. Bot. 54:341–350.

Lee, W.S., Tzen, J.T.C., Kridl, J.C., Raake, S.E. and Haung, A.H.C. (1991). Oleosin is correctly targeted to see oil bodies in *Brassica napus* transformed with maize oleosin gene. Proc. Natl. Acad. Sci. USA 88:6181–6158.

Lenee, P. and Chupeau, Y. (1986). Isolation and culture of sunflower protoplasts (*Helianthus annuus* L.) Factors influencing the viability of cell colonies derived from protoplasts. Plant Sci. 43:69–75.

Li, L.C. and Kohlenbach, H.W. (1982). Somatic embryogenesis in quite a direct way in cultures of mesophyll protoplastss of *Brassica napus* L. Plant Cell Rep. 1:209–211.

Lichter, R. (1981). Anther culture of *Brassica napus* in a liquid medium. Z. Pflanzenphysiol. 103:229–237.

Lichter, R. (1982). Induction of haploid plants from isolated pollen of *Brassica napus*. Z. Pflanzenphysiol. 105:427–434.

Lichter, R. (1989). Efficient yield of embryoids by culture of isolated microspores of different *Brassicaceae* species. Plant Breeding 103: 119–123.

Ling, H.Q. and Binding, H. (1987). Plant regeneration from protoplasts in *Linum*. Plant Breeding 98:312–317.

Ling, H.Q. and Binding, H. (1992). Improvement of plant regeneration from *Linum* protoplasts by the induction of somatic embryogenesis. J. Plant Physiol. 139:422–426.

Loh, C.S. and Ingram, D.S. (1982). Production of haploid plants from anther culture and secondary embryoids of winter oilseed rape, *B. napus* ssp. *oleifera*. New Phytol. 91:507–516.

Loudon, P.T., Nelson, R.S. and Ingram, D.S. (1989). Studies of protoplast culture and plant regeneration from commercial and rapid cycling *Brassica* species. Plant Cell Tissue Org. Cult. 19:213–224.

Lupi, M.C., Bennici, A., Locci, F., Gennai, D. (1987). Plantlet formation from callus and shoot tip culture of *Helianthus annuus* L. Plant Cell Tissue Org. Cult. 11:47–55.

MacDonald, M.V. and Ingram, D.S. (1984). Secondary embryogenesis and selection for resistance to disease in oilseed rape. British Crop Protect. Confer. Plant Diseases 38-4:223–230.

MacDonald, M.V. and Ingram, D.S. (1986). Towards the selection *in vitro* for resistance to *Alternaria brassicicola* (Schw.) Wilts. in *Brassica napus* ssp *oleifera* (Metzg) Sinsk; winter oilseed rape. New Phytol. 104:621–629.

MacDonald, M.V., Hadwiger, M.A. and Aslam, F.N. (1988). The enhancement of anther culture efficiency in *Brassica napus* ssp. *oleifera* Metzg (Sinsk) using low doses of gamma irradiation. New Phytol. 110:101–107.

Maheswaran, G. and Williams, E.G. (1986). Primary and secondary direct somatic embryogenesis from immature zygotic embryos of *Brassica campestris*. J. Plant Physiol. 124:455–463.

Mandal, A. (1987). Induction of androgenetic haploids in the breeding materials of winter rape (*Brassica napus*). Hereditas 106:189–193.

Masirevec, S.M., Secor, G.A. and Gulyn, T.J. (1988). Use of cell culture to screen sunflower germplasm for resistance to Phomopsis brown/gray stem spot. Plant Cell Rep. 7:528–530.

McCann-Wilcox, A., Cooley, G. and Van Dresser, J. (1988). A system for routine plantlet regeneration of sunflower (*Helianthus annuus* L.) from immature embryo-derived callus. Plant Cell Tissue Org. Cult. 14:103–110.

Mathews, H., Bharathan, N., Litz, R.E., Naraynnan, K.R., Rao, P.S. and Bhatia, C.R. (1990). Transgenic plants of mustard *Brassica juncea* (L) Czern Coss. Plant Sci. 72:245–252.

McHughen, A. (1987). Salt tolerance through increased vigor in a flax line (STSII) selected for salt tolerance *in vitro*. Theor. Appl. Genet. 74:727–732.

McHughen, A. (1989). Agrobacterium-mediated transfer of chlorsulfuron resistance to commercial flax cultivars. Plant Cell Rep. 8:445–449.

McHughen, A. (1990). Flax (*Linum utitatissimum* L.): *In vitro* studies. In: Biotechnology in Agriculture and Forestry 10. Legumes and Oilseed Crops I. Bajaj, Y.P.S. Ed. pp. 502–514.

McHughen, A. and Jordan, M.Ç. (1989). Recovery of transgenic plants from "escape" shoots. Plant Cell Rep. 7:611–614.

McHughen, A. and Swartz, M. (1984). A tissue-culture derived salt tolerant line of flax (*Linum usitatissimum* L.). J. Plant Physiol. 117:109–117.

Mehra-Palta, A., Graves, A., Kraemer, M., Hill, K. and Eagan, N. (1991). Genetic transformation of *Brassica napus* and *Brassica rapa*. GCIRC July 9–11. D.I. McGregor (ed.), pp. 1108–1115.

Menczel, L., Morgan, A., Brown, S. and Maliga, P. (1987). Fusion-mediated combination of Ogura-type cytoplasmic male sterility with *Brassica napus* plastids using x-irradiated CMS protoplasts. Plant Cell Rep. 6:98–101.

Mezzarobba, A. and Jonard, K. (1986). Effect du stade de prevelement et des pretraitements sur le developement cultive *Helianthus annuus* L. C.R. Acad. Sci. Paris t 303 Ser. III:181–186.

Miki, B.L., Labbe, H., Hattori, J., Ouellet, T., Gabard, J., Sanohora, G., Charest, P.J. and Iyer, V.M. (1990). Transformation of *Brassica napus* canola cultivars with *Arabidopsis thaliana* acetohydroxy acid synthase genes and analysis of herbicide resistance. Theor. Appl. Genet. 80:449–458.

Millam, S., Davidson D. and Powell W. (1992). The use of flax (*Linum usitatissimum*) as a model system for studies on organogenesis *in vitro*, The effect of different carbohydrates. Plant Cell Tissue Org. Cult. 28:163–166.

Millam, S.R., Burns, A.T.H., Hocking, T.J. and Cattell, K.J. (1988). Effects of a F1 coll-agarose system on early division and development of mesophyll protoplasts of winter oilseed rape (*Brassica napus* L.) Plant Cell Tissue Org. Cult. 12:285–290.

Misra, S. and Gedmau, L. (1989). Heavy metal tolerant transgenic *Brassica napus* L. and *Nicotiana tabacum* L. plants. Theor. Appl. Genet. 78:161–168.

Mix, G. (1985). Antheren und avarien cultur von Sonnenblumen (*Helianthus annuus* L.) Landbanforschung valkensode 35:153–156.

Mohapatra, D. and Bajaj, Y.P.S. (1984). *In vitro* hybridization in an incompatible cross – *Brassica juncea* × *Brassica hirta*. Corr. Sci. 53:489–490.

Mohapatra, D. and Bajaj, Y.P.S. (1987). Interspecific hybridization in *Brassica juncea* and *Brassica hirta* using embryo rescue. Euphytica 36:321–326.

Mohapatra, P. and Bajaj, Y.P.S. (1988). Interspecific hybridization in *Brassica juncea* and *Brassica campestris* through ovary culture. Euphytica 37:83–88.

Moloney, M.M., Walker, J.M. and Sharma, K.K. (1989). High frequency transformation of *Brassica napus* using *Agrobacterium vectors*. Plant Cell Rep. 8:238–242.

Monnier, M. (1978). Culture of zygotic embryos. In: Frontiers of Plant Tissue Culture. Proc. 4th Intl. Congr. Plant Tissue Cult. T.A. Thorpe (ed.), pp. 277–286.

Morgan, A. and Maliga, P. (1987). Rapid chloroplast segregation and recombination of mitochondrial DNA in *Brassica* cybrids. Mol. Gen. Genet. 209:240–246.

Moyne, A.L., Thor, V., Pelissier, B., Bergounious, G., Freyssinet, G. and Gadal, P. (1988). Callus and embryoid formation from protoplasts of *Helianthus annuus* L. Plant Cell Rep. 7:437–440.

Moyne, A.L., Tagu, D., Thor, V., Bergounioux, G., Freyssinet, G. and Gadal, P. (1989). Transformed calli obtained by direct gene transfer into sunflower protoplasts. Plant Cell Rep. 8:97–100.

Murashige, T. and Skoog, F. (1962). A revised medium for rapid growth and bioassays with tobacco tissue cultures. Physiol. Plant. 15:477–479.

Murata, M. and Orton, T.J. (1987). Callus initiation and regeneration capacities in *Brassica* species. Plant Cell Tissue Org. Cult. 11:111–123.

Nanda-Kumar, P.B.A. Shivanna K.R. and Prakash S. (1988). Wide hybridization in *Brassica*, crossability barriers and studies on the F_1 hybrid and synthetic amphidiploids of *B. frucuculosa* × *B. campestris*. Sexual Plant Rep. 1:234–239.

Narasimhulu, S.B. and Chopra, V.L. (1987). Plant regeneration from callus cultures of *Brassica carinata* A.Br. and its implications to improvement of oilseed *Brassicas*. Plant Breeding 99:49–55.

Narasimhulu, S.B. and Kirti, P.B., Prakash, S. and Chopra, V.L. (1992). Rapid and efficient plant regeneration from hypocotyl protoplasts of *Brassica carinata*. Plant Cell Rep. 11:159–162.

Narasimhulu, S.B. and Chopra, V.L. (1988). Species specific shoot regeneration response of cotyledonary explants of *Brassicas*. Plant Cell Rep. 7:104–106.

Narasimhulu, S.B., Prakash, S. and Chopra, V.L. (1988). Comparative shoot regeneration response of diploid *Brassicas* and their synthetic amphidiploid products., Plant Cell Rep. 7:525–527.

Narasimhulu, S.B., Chopra, V.L. and Prakash, S. (1989). The influence of cytoplasmic differences on shoot morphogenesis in *Brassica carinata* A.Br. Euphytica. 40:241–243.

Negrutiu, I., DeWulf, J., Pietrzak, M., Botterman, J., Rietveld, E., Wurger Figmrelli, E.M., Ye, D. and Jacobs, M. (1990). Hybrid genes in the analysis of transformation conditions, II: Transient expression vs stable transformation – analysis of parameters influencing gene expression levels and transformation efficiency. Physiol. Plant 79:197–205.

Neuhaus, G., Spangenberg, G., Mittelstein Shieder, O. and Schewiger, H.G. (1987). Transgenic rapeseed plants obtained by the microinjection of DNA into microspere derived embryos. Theor. Appl. Genet. 75:30–36.

Newell, C.A., Rhoads, M.L. and Bidney, D.L. (1984). Cytogenetic analysis of plants regenerated from tissue explants and mesophyll protoplasts of winter rape, *Brassica napus* L. Can. J. Genet. Cytol. 26:752–761.

Newsholme, D.M., MacDonald, M.V., Ingram, D.S. (1989). Studies of selection *in vitro* for novel resistance to phytotoxic products of *Leptosphaeria maculans* (Desm.) Ces. and de Not. in secondary embryogenic lines of *Brassica napus* ssp. *oleifera* (Metzg) Sinsk., winter oilseed rape. New Phytol. 11:117–126.

Nichterlein, K., Nickel, M., Umback, H. and Freidt, W. (1989a). Recent progress and prospects of biotechnology in breeding of linseed (*Linum usitatissimum* L.). Fat Sci. Technol. 91:272–275.

Nichterlein, K., Umbach, H. and Freidt, W. (1989b). Investigations on androgenesis in breeding of linseed (*Linum usitatissimum* L.) In Vortr. f. Pflanzenzucht. 15 II. Proc. 12th Eucarpia Congr. Gottingen 25–13.

Nichterlein, K., Umbach, H. and Freidt, W. (1991). Genotype and exogenous factors affecting shoot regeneration and anther callus of linseed (*Linum usitatissimum* L.). Euphytica. 58:157–164.

Nikam, T.O. and Shitole, M.G. (1991). Selection and characterization of NaCl-Tolerant callus lines of safflower (*Carthamssus trinctorius* L.). Proc. Intl. Soc. Plant Mol. Biol. 3rd Intl. Congr. Abst. #874.

Nishiyama, 1., Sarashima, M. and Matsuzawa, Y. (1991). Critical discussion on abortive interspecific crosses in *Brassica*. Plant Breeding 107:88–302.

Nitsch, C. and Nitsch, J.P. (1967). The induction of flowering *in vitro* in stem segments of *Plumbago indica* L. The production of vegetative buds. Planta 72:355–370.

Oelck, M.M., Phan, C.V., Eckes, P., Donn, A., Rakow, G. and Keller, W.A. (1991). Field resistance of canola transformants (*Brassica napus* L.) to ignite (Phosphinothricin). Proceedings of the GCIRC Rapeseed Congress, July 9–11. D.I. McGregor (ed.), pp. 292–297.

Paszkowski, J., Pisan, B., Shillito, R.D., Hohn, T., Hohn, B. and Potrykus (1986). Genetic transformation of *Brassica campestris* var protoplasts with an engineered cauliflower mosaic virus genome. Plant Mol. Biol. 6:303–312.

Patterson, K.E. and Everett, M.E. (1985). Regeneration of *Helianthus annuus* inbred plants from callus. Plant Sci. 42:125–132.

Pauls, K.P. and Choung, P.V. (1986). Flow cytometric identification of *Brassica napus* protoplast fusion products. Can. J. Bot. 65:834–838.

Pechan, P.M. and Keller, W.A. (1988). Identification of potentially embryogenic microspores in *Brassica napus*. Physiol. Plant 74:377–384.

Pelissier, B., Bouchefra, O., Pepin, R. and Freyssinet, G. (1990). Production of isolated somatic embryos from sunflower thin cell layers. Plant Cell Rep. 9:47–50.

Pelletier, C., Primard, C., Vedel, F., Chetrit, P., Remy, R., Rousselle, P. and Renard, M. (1983). Intergeneric cytoplasmic hybridization in cruciferae by protoplast fusion. Mol. Gen. Genet. 191:244–250.

Plumper, B. (1991). Recent progress in *Brassica napus* resynthesis by means of *in ovulo* embryo culture. GCIRC congress. D.I. McGregor (ed.), pp. 1034–1039.

Polsoni, L., Kott, L.S. and Beversdorf, W.D. (1988). Large scale microspore culture technique for mutation selection studies in *Brassica napus*. Can. J. Bot. 66:1681–1685.

Pomeroy, M.K., Kramer, J.K.G., Hunt, D.J. and Keller, W.A. (1991). Fatty acid changes during development of zygotic and microspore-derived embryos of *Brassica napus*. Physiol. Plant 81:447–454.

Potrykus, I. (1991). Gene transfer to plants, assessment of published approaches and results. Ann. Rev. Pl. Physiol. 42:205–225.

Poulsen, G.B. and Nielsen, S.V.S. (1989). Regeneration of plants from hypocotyl protoplasts of rapeseed (*Brassica napus* L. var. *oleifera*) cultivars. Plant Cell Tissue Org. Cult. 17:153–158.

Power, C.J. (1987). Organogenesis from Helianthus annuus inbreds and hybrids from cotyledons of zygotic embryos. Am. J. Bot. 74:497–503.

Prado, E. and Berville, A. (1990). Induction of somatic embryo development by liquid culture in sunflower (*Helianthus annuus* L.) Plant Sci. 67:73–82.

Prakash, S. and Chopra, V.L. (1990). Male sterility caused by cytoplasm of *Brassica* oxyrrhina in *B. campestris* × *B. juncea*. Theor. Appl. Genet. 79:285–287.

Prasad, B.R., Khadeer, M.A., Seeta, P. and Anwar, S.Y. (1991). *In vitro* induction of androgenic haploids in safflower (*Carthamus tinctorius* L.). Plant Cell Rep. 10:48–51.

Pretova, A. (1986). Growth of zygotic flax embryos *in vitro* and influence of kinetin. Plant Cell Rep. 3:210–211.

Pretova, A. and Williams, E.G. (1986). Direct somatic embryogenesis from immature zygotic embryos of flax (*Linum usitatissimum* L.). J. Plant Physiol. 126:155–161.

Pretova, A. (1990). Flax (*Linum usitatissimum* L.) embryo culture studies. In: Biotech. in Agric. and Forestry 10. Legumes and Oilseed Crops I. Y.P.S. Bajaj (ed.), pp. 515–537.

Primard, C., Vedel, F. Mathieu, C., Pelletier, G. and Chitine, A.M. (1988). Interspecific somatic hybridization between *Brassica napus* and *Brassica hirta* (*Sinapis alba* L.) Theor. Appl. Genet. 75:546–552.

Pua, E.C. (1990). Somatic embryogenesis and plant regeneration from hypocotyl protoplasts of *Brassica juncea* L. czern and coss. Plant Sci. 68:231–238.

Pua, E.C., Mehra-Palta, A., Nagy, F. and Chua, N.H. (1987). Transgenic plants of *Brassica napus* L. Biotechnol. 5:815–817.

Pugliesi, C., Cecconi, F., Mandolfo, A. and Baroncelli, S. (1991). Plant regeneration and genetic variability from tissue cultures of sunflower (*Helianthus annuus* L.) Plant Breeding 106:114–121.

Quazi, M.H. (1988). Interspecific hybrids between *Brassica napus* and *B. oleracea* developed by embryo culture. Theor. Appl. Genet. 75:309–318.

Quinn, J., Simons, J.E. and Janick, J. (1989a). Histology of zygotic and somatic embryogenesis in borage. J. Am. Soc. Hort. Sci. 114:516–520.

Quinn, J., Simons, J.E. and Janick, J. (1989a). Recovery of c-linolenic acid from somatic embryos of Borage. J. Am. Soc. Hort. Sci. 114:511–515.

Radke, S.E., Andrews, B.M., Moloney, M.M., Crouch, M.L., Kridl, J.C. and Knauf, V.C. (1988). Transformation of *Brassica napus* using *Agrobacterium tumefaciens*, Developmentally regulated expression of a reintroduced gene. Theor. Appl. Genet. 75:685–694.

Radke, S.E., Turner, J.C., Facciotti, D. and Knauf, V.C. (1990). Expression of chimeric genes in developing embryos derived from isolated microspores from transgenic *Brassica napus* plants. In: Progress in Plant Cellular and Molecular Biology. H.J.J. Nijkamp, L.H.W. Van der Plas and Van Aartvijk J. (eds.), pp. 410–415.

Reed, B.M. (1991). Micropropagation of meadowfoam (*Limnanthes* spp.). Plant Cell Rep. 10:94–96.

Ripley, V.L. and Arnison, P.G. (1990). Hybridization of *Sinapis alba* L. and *Brassica napus* L. via embryo rescue. Plant Breeding 104:26–33.

Röbbelen, G. (1991). Rapeseed in a changing world. Plant production potential. GCIRC congress July 9–11, D.I. McGregor (ed.), pp. 29–38.

Röbbelen, G., Downey, R.K., Ashri, A. (1989). Oil Crops of the World, their Breeding and Utilization. McGraw-Hill.

Robertson, D., Palmer, J.D., Earle, E.D. and Mutschler, M.A. (1987). Analysis of organelle genomes of a somatic hybrid derived from cytoplasmic male sterile *Brassica oleracea* and atrazine resistant *B. campestris*. Theor. Appl. Genet. 80:421–427.

Robinson, K.E.P. and Adams, D.O. (1987). The role of ethylene in the regeneration of *Helianthus annuus* (Sunflower) plants from callus. Physiol. Plant. 71:151–156.

Robinson, K.E.P., Adams, D.O. and Lee, R.Y. (1987). Differential physiological and morphological responses of inbred lines to the ethylene precursor 1-aminocyclopropane-1-carboxylic acid by cultured *Helianthus annuus* (Sunflower). Shoot tips. Plant Cell Rep. 6:405–409.

Robinson, K.E.P. and Everett, M.P. (1990). Sunflower (*Helianthus annuus* L.) Establishment of cultures, transformation and the regeneration of Plants. In: Biotechn. in Agric. and Forestry, Vol. 10. Legumes and Oilseed Crops. Y.P.S. Bajaj (ed.), pp. 434–452.

Roewer, I. and Mabry, T.J. (1988). Cell cultures of the wild sunflower (*Helianthus maximilliani schrader*) growth and secondary metabolite synthesis. Plant Cell Rep. 7:197–199.

Roseland, C.R., Espinasse, A. and Grosz, T.J. (1991). Somaclonal variants of sunflower with modified coumarin expression under stress. Euphytica 54:183–190.

Rosen, B., Hallden, C. and Heneen, W.K. (1988). Diploid *Brassica napus* somatic hybrids, characterization of nuclear and organellar DNA. Theor. Appl. Genet. 76:197–203.

Rost, T.L. and Hinchee, M.A.W. (1980). Preliminary report of the production of callus, organogenesis and regeneration of Jojoba (*Simmondsia chinensis* (Link Schneid)) in tissue culture. J. Hort. Sci. 55:299–305.

Sacristan, M.D. (1982). Resistance responses to *Phoma lingam* of plants regenerated from selected cell and embryogenic cultures of haploid *Brassica napus*. Theor. Appl. Genet. 61:193–200.

Sacristan, M.D. (1985). Selection for disease resistance in *Brassica* cultures. Hereditas. Suppl. 3:57–63.

Sacristan, M. and Gerdmann, M. (1986). Different behavior of *Brassica juncea* and *Brassica carinata* as sources of *Phoma lingam* resistance in experiments of interspecific transfer to *B. napus*. Plant Breeding 97:304–314.

Sacristan, M.D., Gerdmann-Knörch, M. and Schieder, O. (1989). Incorporation of hygromycin resistance in *Brassica nigra* and its transfer to *Brassica napus* through asymmetric protoplast fusion. Theor. Appl. Genet. 78:194–200.

Saito, K., Fukushima, A., Sasamoto, H. and Ashihora, H. (1989). Variation in activities of phenol oxidizing enzymes in tissue cultures of *Carthamus tinctorius* L. Acta Physiol. Plant 11: 233–240.

Sakai, T. and Imamura, J. (1990). Intergeneric transfer of cytoplasmic male sterility between *Raphanus sativus* (CMS line) and *Brassica napus* through cytoplast-protoplast fusion. Theor. Appl. Genet. 80:421–427.

Salisbury, P.A. and Kadkol, G.P. (1989). Utilization of alien genes in oilseed *Brassica* breeding. Genet. (Life Sci. Adv.) 8:65–87.

Sarkar, S., Biswas, A.K. and Bhattacharya, M.K. (1985). *In vitro* callus culture and plantlet regeneration in *Linum usitatissimum* L. Cell and Chrom. Res. 8:87–89.

Schenck, H.R. and Robbelen, G. (1982). Somatic hybrids by fusion of protoplasts from *Brassica oleracea* and *Brassica campestris*. Z. Pflanzenzucht. 89:278–288.

Schenk, R. and Hilderbrandt, A.C. (1972). Medium and techniques for induction and growth of monocotyledonous and dicotyledonous plant cell cultures. Can J. Bot. 50:199–204.

Schmitz, P. and Schnabe, H. (1989). Regeneration and evacuolation of protoplasts from mesophyll, hypocotyl and petiole from *Helianthus annuus* L. J. Plant Physiol. 135:223–227.

Schrammeijer, B., Sijmons, P.C., Van den Elzen, P.J.M. and Hockoma, A. (1990). Meristem transformation of sunflower via *Agrobacterium*. Plant Cell Rep. 9:55–60.

Seeta, P. and Anwar, S.Y. (1991). Plantlet regeneration in Safflower (*Carthamus tinctorius*). Proc. Intl. Soc. Plant Mol. Biol. 3rd Intl. Congr. Abst. #475.

Seguin-Swartz, G., Hutcheson, D.S. and Downey, R.K. (1983). Anther culture studies in *Brassica campestris*. Proc. 6th Intl. Rapeseed Conf. Paris.

Sharma, K.K. and Bhojwani, S.S. (1985). Microspore embryogenesis in anther cultures of two Indian cultivars of *Brassica juncea* (L.) Czern. Plant Cell Tissue Org. Cult. 4:235–239.

Sharma, K.K. and Bhojwani, S.S. (1990). Histological aspects of *in vitro* root and shoot differentiation from cotyledon explants of *Brassica juncea* L. Czern. Plant Sci. 69:207–214.

Sharma, K.K., Bhojwani, S.S. and Thorpe, T.A. (1991a). The role of cotyledonary tissue in the differentiation of shoots and roots from cotyledon explants of *Brassica juncea* L. Czern. Plant Cell Tissue Org. Cult. 24:55–59.

Sharma, K.K., Bhojwani, S.S. and Thorpe, T.A. (1991b). Factors affecting high frequency differentiation of shoots and roots from cotyledon explants of *Brassica juncea* L. Czern. Plant Sci. 66:247–253.

Sharma, K.K. and Thorpe, T.A. (1989). *In vitro* regeneration of shoot buds and plantlets from seedling root segments of *Brassica napus* L. Plant Cell Tissue Org. Cult. 18:129–141.

Shukla, A. and Sawhney, V.K. (1991). Comparative regeneration ability of internodal segments of wild type and genic male sterile line of rapeseed (*Brassica napus*) cultured *in vitro*. Plant Sci. 79:95–98.

Siebel, J., Pauls, K.P. (1989). A comparison of anther and microspore culture as a breeding tool in *Brassica napus*. Theor. Appl. Genet. 78:473–479.

Sikdar, S.R., Chatterjee, G., Das, S. and Sen, S.K. (1987). Regeneration of plants from mesophyll protoplasts of the wild crucifer *Eruca sativa* Lam. Plant Cell Rep. 6:486–489.

Sikdar, S.R., Chatterjee, G., Das, S. and Sen, S.K. (1990). Eurassica, the intergeneric fertile somatic hybrid developed through protoplast fusion between *Eruca sativa* Lam. and *Brassica juncea* L. Czern. Theor. Appl. Genet. 79:561–567.

Simmonds, J.A. and Keller, W.A. (1987). Proc. oilseed network meeting, IDRC-Agric. Canada haploid project. Report No. 153t, 220–224 (IDRC-Ottawa, Canada).

Simmonds, D.H., Long, N.E. and Keller, W.A. (1991). High plating efficiency and plant regeneration in low density protoplast cultures from an embryogenic *Brassica napus* cell suspension. Plant Cell Tissue Org. Cult. 27:231–241.

Singh, S. and Chandra, M. (1984). Plant regeneration in callus and suspension cultures of *Brassica campestris* cv. Yellow Sarson. Plant Cell Rep. 3: 1–4.

Singh, S., Banu S., Pireck, L.K. and Chandra, M. (1981). Induced organ differentiation in hypocotyl segments, callus and cell suspension cultures of rape, *Brassica campestris L.* Indian J. Exp. Biol. 19:658–660.

Singh, S., Gang, K. and Chandra, M. (1985). Growth and differentiation in internode callus of *Brassica campestris* cv. Yellow Sarson. Acta Botanica Indica 13:45–50.

Sjodin C. and Glimelius K. (1989). Transfer of resistance against *Phoma lingam* to *Brassica napus* by asymmetric somatic hybridization combined with toxin selection. Theor. Appl. Genet. 78:515–520.

Sorvari, S. (1985). Production of haploids from anther culture in agriculturally valuable *Brassica campestris* L.). Ann. Agric. Finnae. 24:149–160.

Southworth, D. and Kwiatkowski S. (1991). Somatic embryogenesis from immature embryos in meadowfoam *Limnanthes alba*. Plant Cell Tissue Org. Cult. 24:193–198.

Spangenberg, G., Neuhaus, G. and Schweiger, H.G. (1986a). Expression of foreign genes in a higher plant cell after electrofusion-mediated cell reconstitution of a micro injected karyoplast and a cytoplast. Europ. J. Cell Biol. 42:236–238.

Stringam, G.R. (1979). Regeneration in leaf-callus cultures of haploid rapeseed (*Brassica napus* L.). Z. Pflanzenphysiol. 92: 459–462.

Sundberg, E. and Glimelius, K. (1991). Effects of parental ploidy level and genetic divergence on chromosone elimination and chloroplast segregation in somatic hybrids within *Brassicaceae*. Theor. Appl. Genet. 83:81–88.

Sundberg, E., Landgren, M. and Glimelius, K. (1987). Fertility and chromosome stability in *Brassica napus* resynthesized by protoplast fusion. Theor. Appl. Genet. 75:96–104.

Sundberg, E., Lagererantz, U. and Glimelius, K. (1991). Effects of cell type used for fusion on chromosome elimination and chloroplast segregation in *Brassica oleracea* and *Brassica napus* hybrids. Plant Sci. 78:89–98.

Swanson, E.B. (1990). Microspore culture in *Brassica*. In: Methods in Molecular Biology, Vol. 6. Plant Cell and Tissue Culture. J.W. Pullord and J.M. Walker (eds.), Humana Press, pp. 159–169.

Swanson, E.B. and Erickson, L.R. (1989). Haploid transformation in *Brassica napus* using an octopine producing strain of *Agrobacterium tumefaciens*. Theor. Appl. Genet. 78:831–835.

Swanson, E.L., Coumans, M.P., Wu, S.C., Barsby, T.L. and Beversdorf, W.D. (1987). Efficient isolation of microspores and the production of microspore derived embryos from *Brassica napus*. Plant Cell Rep. 6:94–97.

Swanson, E.B., Coumans, M.P., Brown, G.L., Petal, J.D. and Beversdorf, W.D. (1988). The characterization of herbicide tolerance in *Brassica napus* L. after *in vitro* selection of microspores and protoplasts. Plant Cell Rep. 7: 83–87.

Swanson, E.B., Herrgesell, M.J., Orlando, M., Sippell, D.W. and Wong, R.S.C. (1989). Microspore mutagenesis and selection, Canola plants with field tolerance to the imidazolinones. Theor. Appl. Genet. 78:525–530.

Tagliani, L.H. and Knapp, S.J. (1990). Tissue culture and transformation of *Cuphea niscossima* using *Agrobacterium tumefaciens*. Agron. Abst. p. 203.

Takahata, Y. and Takeda T. (1990). Intergeneric (inter sub tribe) hybridization between *Moricanda arvensis* and *Brassica A* and *B* genome species by ovary culture. Theor. Appl. Genet. 80:38–42.

Telmer, C.A., Simmonds, D.H. and Newcomb, W. (1992). Determination of developmental stage to obtain high frequencies of embryogenic microspores in *Brassica napus*. Physiol. Plant. 84:417–424.

Terada, R., Yamashita Y., Mishibayashi S. and Shimamoto K. (1987). Somatic hybrids between *Brassica oleracea* and *B. campestris*, selection by the use of iodoacetamide inactivation and regeneration ability. Theor. Appl. Genet 73:379–384.

Tejovathi, G. and Anwar S.Y. (1986). *In vitro* plantlet regeneration and capitula induction in safflower (*Carthamus tinctorius* L.). Abst. 6th Intl. Congr. Plant Tissue Culture Minneapolis, p. 28.

Thomzik, J.G. and Hain, B. (1988). Transfer and segregation of triazine tolerant chloroplasts in *Brassica napus* L. Theor. Appl. Genet. 76:165–171.

Thomzik, J.G. and Hain, B. (1990). Transgenic *Brassica napus* plants obtained by cocultivation of protoplasts with *Agrobacterium tumefaceiens*. Plant Cell Rep. 9:213–216.

Wang, Y-C. and Janick, J. (1986). Somatic embryogenesis in jojoba. J. Am. Soc. Hort. Sci. 111:281–287.

Wang, Y-C. and Janick, J. (1986a). *In vitro* production of jojoba liquid wax by zygotic and somatic embryos. J. Am. Soc. Hort. Sci. 111:798–807.

Wang, Y-C. and Janick, J. (1986b). Sucrose concentration and osmolarity as factors affecting *in vitro* wax accumulation in jojoba embryos. Hort. Sc. 21:1048–1049.

White, P.R. (1963). The Cultivation of Animal and Plant Cells. Ronald Press, New York.

Wiberg E., Rahlen L., Hellman, M., Tillberg, E., Glimelius, K. and Shijnine S. (1991). The microspore derived embryo of *Brassica napus* L. as a tool for studying embryo specific lipid biogenesis and regulation of oil quality. Theor. Appl. Genet. 82: 515–520.

Witrzens, B., Scowcroft, W.R., Downes, R.W. and Larkin, P.J. (1988). Tissue culture and plant regeneration from sunflower *Helianthus annuus* L. and interspecific hybrids, *H. tuberosus* and *H. annuus*. Plant Cell Tissue Org. Cult. 13:61–76.

Xu, Z.H., Davey, M.R. and Cocking, E.C. (1982). Plant regeneration from root protoplasts of *Brassica*. Plant Sci. Lett. 24:117–121.

Yamashita, Y., Terada, R., Mishabayashi, S. and Shimamota, K. (1989). Asymmetric somatic hybrids of *Brassica*, partial transfer of *B. campestris* genome into *B. olercea* by cell fusion. Theor. Appl. Genet. 77:189–194.

Yarrow, S.A., Wu, S.C., Barsby, T.L., Kemble, R.J. and Shepard, J.F. (1986). The introduction of CMS mitochondria to triazine tolerant *Brassica napus* L. var. 'Regent' by micromanipulation of individual heterokaryons. Plant Cell Rep. 5: 415–418.

Yarrow, S.A., Burnett, L.A., Wildeman, R.P.and Kemble, R.J. (1990). The transfer of 'Polmia' cytoplasmic male sterility from oilseed rape (*Brassica napus*) to broccoli (*B. oleracea*) by protoplast fusion. Plant Cell Rep. 9:185–188.

Yang, M.Z., Jia, S.R. and Pua, E.C. (1991). High frequency of plant regeneration from hypocotyl explants of *Brassica carinata*. A. Br. Plant Tissue Org. Cult. 24:79–82.

Yoshihawa, T., Kaneko, H. and Kimera, T. (1986). Production of vitamin E by cell cultures of Safflower (*Carthamus tinctorius* L.) Abst. 6th Intl. Confr. Plant Tissue Cult. Minneapolis, p. 67.

Zhan, X-C., Jones, D.A. and Kerr, A. (1988). Regeneration of flax plants transformed by Agrobacterium rhizogenes. Plant Mol. Biol. 11:551–559.

Zhan, X-C., Jones, D.A. and Kerr, A. (1989a). Regeneration of shoots on explants of flax. Ann. Bot. 63:297–299.

Zhan, X-C., Jones, D.A. and Kerr, A. (1989b). *In vitro* plantlet formation *Linum marginale* from cotyledons, hypocotyls, leaves, roots and protoplasts. Aus. J. Plant Physiol. 16:315–320.

Zine, K., Chlyah, Chlyah, A. and Jacques, R. (1988). Action de la lumiere blanche et de quelques radiations monochromaliques sur la morphogenese *in vitro* du *Linum usitatissimum* L. Can. J. Bot. 66:1856–1861.

18. *In vitro* Culture of Temperate Fruits

Richard H. Zimmerman and Harry J. Swartz

Contents

1. Introduction

In vitro culture techniques have had numerous applications to fruit crops beginning nearly 60 years ago with embryo rescue techniques for stone fruits (Tukey, 1933; see Ramming, 1990). From this beginning, the method has been applied successfully to produce commercially acceptable early-ripening peach and nectarine cultivars. The methods have been adapted to other crops as well, e.g. in breeding programs to produce both early-ripening and seedless grapes (Ramming, 1990).

Historically, the second application of *in vitro* culture methods to temperate fruit crops was to eliminate disease-causing viruses from strawberries (Belkengren and Miller, 1962). Meristem-tip culture has since become an integral part of virus-indexing programs for a number of fruit crops, usually in conjunction with thermotherapy (see Zimmerman, 1983; Hutchinson and Zimmerman, 1987). In some cases, it has been necessary to micrograft the meristem tip onto an *in vitro*-grown seedling as is done for citrus (see Chapter 19, pp. 475–496).

Within the past 20 years, micropropagation of fruit crops has become an important application of *in vitro* technology. Again strawberry was the first temperate fruit crop for which the method was developed (Boxus, 1974). Now many different fruit crops are being micropropagated commercially and this usage is likely to increase in the future (Zimmerman, 1991).

More recent uses of *in vitro* culture emphasize applications for genetic

Indra K. Vasil and Trevor A. Thorpe (eds.), Plant Cell and Tissue Culture, 457–474.

improvement of fruit crops. These applications include production of hybrid plants from fused protoplasts, somaclonal variation and mutagen-driven changes in ploidy in regenerated plants, haploid plants from anther culture, and transfer of specific genes via *Agrobacterium*-mediated transformation.

2. Methods and their Applications

2.1. *Micropropagation*

The primary use of tissue culture in temperate fruit crops is for plant propagation. While the general principles are the same for all the crops, details of the techniques for specific crops often differ. Preliminary research studies have often used seedlings or embryonic tissue for method development. Practical applications have required that mature clones be used and suitable methods are now generally available for adult plants. These methods usually have evolved gradually from those originally developed for seedlings or for easily cultured clonal material. Modifications for certain genotypes are still necessary since a large genotypic variation in response exists in most crops (Brennan et al., 1989; Marcotrigiano et al., 1984; Reed, 1990; Simpson and Bell, 1989).

Some generalizations can be made about the media used for tissue culture of fruit crops. The MS medium (Murashige and Skoog, 1962) has proved to be widely adaptable and has been used for micropropagating apple, pear, peach, cherry, plum and other stone fruits, raspberry, blackberry, currant, gooseberry and strawberry (Table 1). Woody Plant Medium (WPM; Lloyd and McCown, 1980), originally developed for ericaceous plants, has been used for blueberry and also for a number of the non-ericaceous fruit crops as well, but is not as commonly used as MS for this latter group. Anderson's medium (Anderson, 1975), also developed for ericaceous plants, has been used for blueberry and also for apple and raspberry. For strawberry, a commonly used medium has been Knop's (Boxus et al., 1977). Lepoivre medium has been used for many tree fruits (Quoirin and Lepoivre, 1977), particularly in Europe. Selected media protocols for micropropagating temperate zone fruit crops are summarized in Table 1.

2.1.1. *Tree Fruits*

Micropropagation has been used to produce both rootstocks and self-rooted cultivars of most major temperate tree fruits (Hutchinson and Zimmerman, 1987). Commercial production has been chiefly of rootstocks for peach, although rootstocks for apple, pear, cherry, apricot and plum are also produced (Zimmerman, 1991). Production of self-rooted cultivars has been limited because extensive long-term field evaluations of tree performance are needed before recommendations can be made. Until then, essentially no market for such trees will exist, even though they can be readily produced.

2.1.2. *Small Fruits and Grapes*

The use of micropropagation for berry crops has been surveyed by Swartz and Lindstrom (1986) and Zimmerman (1991). Once micropropagation of strawberry had been described (Boxus, 1974), application of the technique on a commercial scale occurred rapidly because of the vastly improved rates of field propagation of tissue culture-propagated plants. With time, it became evident that certain safeguards in the procedure were essential to prevent variant phenotypes from being released for commercial production. When adequate safeguards were not used, variant plants were inadvertently distributed to commercial growers, often with disastrous financial results for growers, production laboratories or both. *Rubus* production has been greatly aided by tissue culture production of nursery plants (Swartz and Lindstrom, 1986). Virus sanitation is a major concern and the rate of field propagation from certified stock plants is unsatisfactory. Some *Vaccinium*, specifically highbush blueberry, are mass propagated *in vitro* to ease propagation and improve plant growth. Tissue culture propagation of lowbush blueberry may make clonal plantings possible because of improved growth habit. Grapes (*Vitis*) have not been mass propagated *in vitro* because of concerns about plant performance in the field and about possible changes in varietal characteristics. Tissue culture is of benefit in eliminating grape viruses and systemic *Agrobacterium vitis* (Burr et al., 1988).

2.2. *Virus Elimination*

Fruit plants are often infected with viruses that reduce plant vigor, yield and fruit quality. Since virus distribution and concentration is not uniform within a plant and virus is often not detectable in the apical meristem, the practice of isolating and culturing the meristem-tip to obtain virus-free plants was developed. This practice is usually used in conjunction with heat therapy (see Quak, 1977; Walkey, 1978; Kartha, 1984).

Elimination of viruses from a temperate fruit crop by meristem-tip culture was first demonstrated with strawberry (Belkengren and Miller, 1962). Since the *in vitro* technique does not always eliminate the virus, individual plants should be grown from each meristem-tip and then indexed for virus content. Virus-negative plants can then be re-established *in vitro* for purposes of micropropagation or germplasm conservation. McGrew (1980) pointed out that size of the meristem-tip was directly related to recovery of rooted plants and inversely related to freedom of the rooted plants from pallidosis virus.

Virus elimination from tree fruits is usually accomplished by thermotherapy (Fridlund, 1980). When thermotherapy is not capable of eliminating the

Table 1. Selected methods for micropropagating temperature zone fruit crops

Crop	Initiation	Proliferation	Rooting	References
Apple (*Malus* × *domestica* Borkh)	MS[1] + 4.4 μM BA + 0.5 μM IBA + 1.3 μM GA_3	MS + 4.4 μM BA + 0.5 μM IBA + 1.3 μM GA_3	44 mM sucrose + 1.5 μM IBA, 7 days in dark @ 30 °C; peat plugs	Zimmerman and Broome, 1981; Zimmerman and Fordham, 1985
Pear *Pyrus communis* L.)	MS + 4.4 μM BA	MS + 8.8 μM BA	MS + 2.7 μM NAA	Singha, 1980, 1982
Peach (*Prunus persica* (L.) Batsch)	Liquid $\frac{1}{2}$X MS + 0.88 μM BA + 0.05 μM IBA	MS + 8.8 μM BA + 0.05 μM IBA	$\frac{1}{2}$X MS + 5.8 μM NAA *or* 28.5 μM IAA + 100 μM chlorogenic acid *or* phloroglucinol *or* rutin	Hammerschlag, 1982; Hammerschlag et al., 1987
Cherry, sweet (*Prunus avium* (L.)	Modified Knop's[2] + MS micro-elements + 4.4 μM BA + 4.9 μM IBA	MS + 4.4 μM BA +4..9 μM IBA + 0.3 μM GA_3	$\frac{1}{2}$X MS + 2.7 μM NAA + wounding	Snir, 1982
Cherry, sour (*Prunus cerasus* L.)	Boxus and Quoirin + MS micro-elements	MS + 4.4 μM BA + 4.9 μM IBA + 0.3 μM GA_3	$\frac{1}{2}$X MS + 4.9 μM IBA for 1 week, then to $\frac{1}{2}$X MS	Snir, 1983
Strawberry (*Fragaria* × *ananassa* Duch.)	Knop's or MS 0.05 μM BA	Knop's or MS 5 μM BA	Knop's or MS 0.05 μM IBA	Boxus, 1975; Boxus et al., 1977

Table 1. Continued

Crop	Initiation	Proliferation	Rooting	References
Blueberry (*Vaccinium* sp.)	WPM or Anderson 15 μM 2iP	WPM or Anderson 15 μM 2iP	IBA dip *ex vitro*	Anderson, 1975; Lloyd and McCown, 1980; Wolfe et al., 1983
Blackberry (*Rubus* sp.)	MS + 5 μM BA + 0.5 μM IBA + 0.5 μM GA_3	MS + 5 μM BA + 0.5 μM IBA + 0.5 μM GA_3	MS + 1 μM IBA	Reed, 1990
Raspberry (*Rubus* sp.)	Andersons + + 5 μM BA + 0.5 μM IBA + 0.5 μM GA_3	Andersons + 5 to 10 μM BA + 0.5 μM IBA + 0.5 μM GA_3	Andersons + 1 μM IBA	Reed, 1990
Currants Gooseberries (*Ribes* sp.)	MS × 0.1 NH_4NO_3 + 1 μM BA + 0.1 μM NAA	MS + 4 μM BA + 0.1 μM NAA	MS + 2 μM NAA	Brennan et al., 1989
Grape (*Vitis* sp.)	MS[3] + 5 μM BA + 0.5 μM IBA	MS[3] + 5 μM BA	MS[3] + 60 mM sucrose +0.4 μM NAA	Chée and Pool, 1983

Abbreviations: MS = Murashige and Skoog (1962) medium; WPM = Woody Plant Medium of Lloyd and McCown (1980); BA = benzyladenine; 2iP = isopentenyl adenine; IAA = indoleacetic acid; IBA = indolebutyric acid; NAA = naphthaleneacetic acid; GA_3 = gibberellic acid.
[1]MS for apple, pear, sweet cherry and sour cherry contained the minimal organics of Linsmaier and Skoog (1965).
[2]See Tabachnik and Kester (1977).
[3]Medium contained 0.05 mM $MnSO_4$, $Ca(NO_3)_2$ was substituted for $CaCl_2$, and KI was eliminated.

virus, meristem-tip culture has been used successfully. *In vitro* grafting has been used also, especially where it has not been possible to root shoots developed from the isolated meristems (Huang and Millikan, 1980).

2.3. *Genetic Improvement*

Use of *in vitro* methods for genetic improvement of fruit crops has developed rapidly in the last few years as new techniques have become available. Results are still too limited to have previously merited a general review.

2.3.1 *Protoplasts*

Since the last reviews of protoplast culture of tree and small fruit genera (Hidano and Niizeki, 1988; Krul, 1988), several successful protoplast-to-plant protocols have been developed, including *Malus* (Wallin and Johansson, 1989), *Pyrus* (Ochatt and Caso, 1986), *Prunus* (Ochatt, 1990), *Fragaria* (Nyman and Wallin, 1988) and *Actinidia* (Oliveira and Pais, 1991). A recent review lists successful regeneration from protoplasts of 17 woody fruit genotypes (Ochatt, 1993). Other genera have been more recalcitrant; of note, *Vitis* protoplasts have only produced a small number of cell divisions (Krul, 1988; Theodoropoulos and Roubelakis-Angelakis, 1990). Although plants of the genera listed above that were regenerated from protoplasts have been reported normal in at least their initial growth, selection for salinity and fireblight resistance using protoclonal variation may be possible (Brisset et al., 1990; Ochatt and Power, 1989). In addition, Ochatt et al. (1989) have fused protoplasts of a pear (*Pyrus communis*) and a cherry rootstock (*Prunus avium* × *P. pseudocerasus*). Heterokaryons were selected on a medium that excluded both parental types. Regenerated plants from this work exhibited hybrid morphology and isozyme pattern; however, one off-type plant was found to be an aneuploid.

Leaves (most commonly), callus and roots have been used as sources for protoplasts. The choice of tissue can affect the success of protoplast survival and regeneration, the enzymes used for digestion [commonly cellulase (Cellulysin), hemicellulase, pectinase (pectolyase), macerase (macerozyme) or driselase], the media used, or the type of organogenesis obtained from microcalluses (e.g. see Brisset et al., 1990; Ochatt et al., 1989).

Plasmolysis or dark pretreatments are used in several species, but not consistently. Cell wall digestion media can include reducing agents and PVP, especially in pome fruits (Ochatt and Caso, 1986; Wallin and Welander, 1985). Sieving, pelleting and high osmolarity 'float' centrifugation are used to remove enzymes and cell debris and concentrate protoplasts. High pectolyase concentrations and MES, used to buffer media, have been found to be detrimental to protoplast growth (Nyman and Wallin, 1988; Wallin and Johansson, 1989). MS-based media with complex organic addenda and 0.35

to 0.7 M carbohydrates (e.g. mannitol, sucrose, glucose) or low ammonium MS salts with casein hydrolysate added as a reduced nitrogen source have proven satisfactory (Hidano and Niizeki, 1988; Ochatt, 1990; Ochatt and Caso, 1986) . Semi-solid agarose stimulates early cell division and is often covered by liquid, although initial liquid culture followed by agarose culture has also been used (Nyman and Wallin, 1988; Ochatt et al., 1989). Glass containers have been found to be more suitable than plastic ones in at least one study (Oliveira and Pais, 1991). After cell wall formation, media osmoticum (usually sucrose) is diluted to <0.2 M. MS-based media have been consistently used at the microcallus-growth stage with an occasional exception (Wallin and Johansson, 1989). The ratio of plant growth regulators used for callus growth is 3 to 10 parts auxin, usually 2,4-D, to one part cytokinin, usually BA. Microcalluses are used for regeneration at around 100 days after isolation. Approximately a 10:1 ratio of cytokinin (usually BA but often including zeatin) to auxin (often NAA but also IAA) has been used for shoot regeneration. Thidiazuron has been used to replace BA (Wallin and Johansson, 1989). Calcium pantothenate and casein hydrolysate have been found to have organogenic activity (Ochatt and Power, 1988; Ochatt, 1990). Adventitious roots derived from protoplasts have been cultured with cytokinin to produce intact plants (Ochatt, 1990). Finally, cell division and/or regeneration has been found to be stimulated by electroporation (Ochatt et al., 1988a, b).

2.3.2. *Haploids and Anther Culture*

Fruit breeders have been interested in the production of haploid plants because of the highly heterozygous nature of most fruit species.

Three approaches have been used. First, wide crosses or chemically or radiation-arrested pollen have been used to produce or attempt to produce haploid embryos (James et al., 1986; Lespinasse et al., 1983; Rajasekaran and Mullins, 1985). Heteromorphic cotyledons were used to select haploid seedlings of pear (Braniste et al., 1984). Second, mutagens, like p-fluorophenylalanine, applied to regenerating strawberry anthers and petioles have resulted in production of plants with wide variation in chromosome number, including haploids and aneuploids (Niizeki and Fukui, 1983). Third, haploids and other ploidy levels have been obtained from anther-derived apple and strawberry plants in some studies (Guangrong and Jianzhe, 1984; Niemirowicz-Szczytt et al., 1983). However, in at least one case, close anatomical study indicated new growth from strawberry anthers did not occur from microspores (Maeda et al., 1978). In addition, only a few divisions were obtained from microspore culture (Jordan, 1975). Accordingly, several authors have noticed no haploids from anther-derived plants, although this may be due to production of dihaploid plants since some variation in plant performance was obtained (Maeda et al., 1978; Oosawa and Takayanagi, 1982; Rosati et al., 1975). Haploid production has been reported occasionally from cultured grape anthers (reviewed by Cao, 1986). Anther-derived *Actinidia* plants were diploid (Fraser and Harvey, 1986). Methodology for plant formation from anthers is discussed in Section 2.3.5.

The production of polyploid plants through the use of colchicine can also be aided by tissue culture. James et al. (1987) produced a hexaploid from a triploid cherry rootstock using shoot regeneration from colchicine-treated roots. For *Rubus*, *in vitro* explants can be submerged in 5 mM colchicine for 20 h, after which treated shoots are subcultured twice in three weeks. New leaves from these treated shoots are placed on organogenesis media and shoots regenerating from these leaves have produced more than a dozen tetraploid *Rubus* clones (H.J. Swartz, unpublished data). Colchicine treatment followed by culture on a high cytokinin medium was also successfully used in apple (Lespinasse et al., 1983). A few p-fluorophenylalanine-treated plants had increased ploidy levels (Niizeki and Fukui, 1983), as did seedlings from diploid and triploid apples treated with N-nitroso-N-methylurea before pollination (Shidakow, 1986).

2.3.3 *Gene Transfer*

Disarmed *Agrobacterium tumefaciens* strains have recently been used to putatively transform several fruit genera; however, many of these reports are incomplete, lacking proper DNA fragment length hybridization or regeneration of transformed plantlets. Of the successfully documented cases of transformation in apple (James et al., 1989), strawberry (James et al., 1990; Nehra et al., 1990a, b), grape (Mullins et al., 1990), plum (Mante et al., 1991) and blackberry (Hassan, 1991), the ratio of transformed plants to number of explants inoculated is quite small, usually one to a few plants in several hundred inoculated organs. In strawberry, the transferred nopaline synthase gene was dominant (James et al., 1990).

Grapes have been particularly difficult to transform; however, these difficulties are not due to use of biovar 1 *Agrobacterium tumefaciens* when biovar 3 (now called *Agrobacterium vitis*) is the one most commonly isolated from grapevines (Baribault et al., 1989). Rather, two studies demonstrated that transformed cells and the cells that gave rise to adventitious shoots are spatially separated (Colby et al., 1991a, b). By using β-glucuronidase activity to histologically mark the tissues transformed after inoculation of tissues, the cut surface of co-incubated petioles, the wounded tissue most susceptible to transformation, was found to develop a necrotic layer. In this case, transformation frequencies would be expected to be very low unless a selection system was used to allow transformed cells to dominate growth during organogenesis. Unfortunately, grape is very susceptible to kanamycin, the antibiotic used in the most common selection system based on kanamycin resistance from the *npt*II gene (Colby and Meredith, 1990; Mullins et al., 1990). Two attempts have been made to overcome this problem. First, *Agrobacterium rhizogenes* was used to stimulate rhizogenesis from leaf petioles. Although the roots were transformed and morphologically normal, i.e. not of the 'hairy root' morphology caused by auxin genes, no plants were recovered (Guellec et al., 1990). In a second report, transformation and plant formation was obtained from secondary embryos of the rootstock *Vitis rupestris* 'du Lot' (Mullins et al., 1990). Secondary embryos commonly arise from epidermal or subepidermal cells without the development of any necrotic tissues as found in leaf-based regeneration. Southern blot analysis confirmed transformation of this clone; however, putatively transformed tissue from *V. vinifera* leaves developed buds but were unable to develop shoots. The development of tissue over the transformation/wound site does not seem to be a problem in apple or *Rubus*. In strawberry and apple, regeneration of tissues can be accompanied by an intervening callus stage (James et al., 1989; Nehra et al., 1990b). Relatively high transformation frequencies have been obtained from plum embryos (Mante et al., 1991).

The use of kanamycin and the *npt*II gene for kanamycin resistance in transformed tissues has not been as easy to adapt in fruit genera as it has in dicotyledonous herbaceous plants. Fruit plant genera are very sensitive to kanamycin and researchers have had to use low levels of kanamycin (5–25 μg/ml) during and after regeneration (Guellec et al., 1990; Hassan, 1991; Mullins et al., 1990; Nehra et al., 1990a, b) or have had to limit exposure to kanamycin to several days (James et al., 1989, 1990; Nehra et al., 1990a, b). In addition, even the kanamycin resistance of transformed plants is not sufficient for high rates of rooting on moderate kanamycin concentrations (James et al., 1989; Nehra et al., 1990b). Transformed plum embryos were selected in relatively high concentrations of kanamycin; however, only a low percentage of transformed plants rooted on kanamycin-containing medium (Mante et al., 1991). This may have been due primarily to the low percentage root initiation of these genotypes in kanamycin-free medium. In the grape studies cited above, regeneration of shoots evidently involved coordination of a large group of cells, both in the formation of a primordium and as 'nurse' cells. A group of transformed cells might not be able to produce an adventitious shoot if it is surrounded by necrotic non-transformed, kanamycin-susceptible cells. It may take a long period of callus growth on transformed nurse callus to allow regeneration of transformed shoots on high levels of kanamycin. Thus, kanamycin concentration or exposure to kanamycin should be limited to a point where callusing and chlorosis occurs but necrosis is minimal.

Transformation protocol, except for the above mentioned cases for grape, is relatively uniform. An overnight culture of a disarmed strain of *Agrobacterium tumefaciens* (commonly strains LBA 4404 or of C58 strain lineage) is centrifuged at around 10,000 × g for 5–15 min to remove NaCl or antibiotics. After resuspension in MS medium at 10^8 cells/ml, *in vitro* leaves are inoculated and coincubated on medium-soaked filter paper. Cytokinins added to this medium can enhance transformation (D.J. James, personal communication; H.J. Swartz, unpublished data). Two to three days are commonly used (e.g. James et al., 1989), but 4 days was more efficient for *Rubus* (Hassan, 1991). Embryo tissue does not need to be pregerminated prior to inoculation (Mante et al., 1991). A cefotaxime or carbenicillin post-coincubation dip is used to reduce the *Agrobacterium* population and explants are placed on organogenesis medium containing 25 to 100 μg/ml of the same antibiotic, a suitable cytokinin and an auxin, and a reduced or delayed kanamycin dosage (if the *npt*II gene is used). Cefotaxime can improve organogenesis and, under these conditions, *Agrobacterium* species do not reduce regeneration (Hassan, 1991; James et al., 1990). After regeneration, which may take more time than during model shoot organogenesis experiments, green shoots are screened for their ability to root in 10 to 25 μg/ml kanamycin. Rooted shoots are assayed for marker gene (e.g. GUS or nopaline synthase) activity and DNA analysis.

2.3.4. *Embryo Rescue*

The most important applications of this technique have been in breeding stone fruits, particularly early season peaches and nectarines, and seedless grapes. One of the important achievements of tree fruit breeding has been the development of commercially acceptable peaches and nectarines that have extended the ripening season both earlier and later than the older commercial cultivars. Breeding early ripening peaches (and other stone fruits) was difficult because the embryos either often aborted or were so underdeveloped that they could not be induced to grow when the seeds were extracted from the fruit. Embryo rescue techniques, originally shown to be feasible by Tukey (1933), permitted development of the early ripening cultivars that are now so important commercially. The historical development of the technique as applied to *Prunus* as well as specific procedures and results obtained have been thoroughly documented (Ramming, 1990). The method has been extended to *in ovulo* embryos (less than 1 mm long) for early-maturing peach, plum and nectarine hybrids (Ramming, 1990).

Seedless (stenospermic) grape cultivars are becoming increasingly important for commercial production. Breeding new cultivars has been difficult since the seedless cultivars could only be used as pollen parents. Development of ovule culture techniques has permitted the use of the seedless types as seed parents also, greatly increasing the breeding possibilities (see Ramming, 1990; Gray et al., 1990). Embryos have been recovered from a wide range of crosses; viable plants have developed from these and have been established in soil. Seedlings from crosses of seedless × seedless grapes, produced using ovule culture, have shown as high as 82% seedlessness (Ramming, 1990). A large majority, if not all, of the embryos derived from these rescue efforts have been found to be derived from zygotic tissues, i.e. they were not nucellar in origin (Durham et al., 1989).

Embryo rescue techniques have also been used in breeding of raspberries and other brambles to ensure recovery of plants from crosses with poor seed set.

2.3.5. *Adventitious Regeneration and Embryogenesis*

A prerequisite for application of genetic engineering techniques is the ability to regenerate plants from adventitious buds induced either from callus or from plant organs such as leaves or stem segments. Such a capability now exists for many, if not all, of the temperate zone fruit crops. In some cases, the techniques have been successful only with embryonic or seedling tissue, but in others regeneration from mature clones has been possible. These differences have influenced and been influenced by the strategy used for gene transfer. For some crops, transfer of a novel gene to a particular clone is the first step in a broader program of improvement emphasizing breeding. In other cases, the need is to modify a particular clone without utilizing a breeding program for further clonal improvement. Examples of the first approach would be regeneration of transformed tissue of peach and strawberry while apple would illustrate the second approach.

Recent reviews of progress on regeneration of fruit crops have been published for *Prunus* (Scorza and Hammerschlag, 1992), *Malus* (Korban, 1992) and *Pyrus* (Chevreau and Skirvin, 1992).

All major small fruit genera and many grape species have been successfully regenerated from leaves, and in some cases from anthers, internodes and roots. Regeneration from grapes was found to involve several cells and perhaps the participation of neighboring tissues, particularly young leafy tissue (Barlass et al., 1981; Colby et al., 1991a, b; Stamp et al., 1990). In contrast, strawberry shoot meristem formation seems to be initiated by one cell (Nehra et al., 1989); however, chimeric plants have been obtained from strawberry callus (Nehra et al., 1992). This indicates either a multicellular origin of shoots or that post-regeneration mutations occurred. Wound tissue often develops at the cut surface of grape petioles; however, regeneration directly from cut surfaces of *Ribes*, *Rubus*, *Fragaria* and *Vaccinium* occurs with little callus formation (Billings et al., 1988; Dweikat and Lyrene, 1988; Fiola et al., 1990; Graham and McNicol, 1991; Liu and Sanford, 1988; Nehra et al., 1989). In general, the epidermis or subepidermal tissues were the sites of organogenesis. Shoots have also been regenerated from strawberry (Jones et al., 1988; Nehra et al., 1990c) and *Actinidia* calluses (Mii and Ohashi, 1988; Revilla and Power, 1988). Some variation, mostly detrimental, occurred on callus-derived strawberry (Jones et al., 1988), especially at high 2,4-D and BA concentrations (Nehra et al., 1991).

Except for *Vaccinium*, regeneration occurs on MS-based medium with approximately 5 μM BA. In *Vaccinium* 15 to 25 μM 2iP and WPM or half-strength MS are used (Billings et al., 1988; Dweikat and Lyrene, 1988). In strawberry, more nearly equimolar auxin and cytokinin are used; however, on occasion, explants are transferred to medium containing 10 μM BA and 1 μM auxin (Liu and Sanford, 1988; Jones et al., 1988; Nehra et al., 1989; Nehra et al., 1990c). In *Rubus*, thidiazuron was found to be more active than BA on a wide range of species (Fiola et al., 1990).

Embryogenesis from grape somatic tissues has been reported on a number of occasions (Gray and Mortensen, 1987; Krul and Worley, 1977; Matsuta and Hirabayashi, 1989; Stamp and Meredith, 1988; Takeno et al., 1983). 2,4-D, or less commonly NOA, with a cytokinin (BA) concentration lower than that of auxin was used for induction of embryos. Vitamins are deleted from the medium if the new growth is overly vigorous, since excess vigor stimulates production of non-morphogenic callus (Matsuta and Hirabayashi, 1989). Continuous embryogenesis was maintained indefinitely on either initiation medium with 2,4-D or growth regulator-free medium by selection of embryogenic callus or embryos that produced secondary embryos. Embryogenic calluses will occasionally produce embryos on medium with no growth regulators. Embryos can be germinated on BA-containing medium. Regenerants have generally been reported to be uniform (Gray and Mortensen, 1987; Krul and Worley, 1977; Matsuta and Hirabayashi, 1989). Anthers of *Actinidia* (Fraser and Harvey, 1986), apple (Guangrong and Jianzhe, 1984) and pear (Mehra and Jaidka, 1985) have also been used to produce embryogenic callus. *Prunus* roots have also produced embryos (Druart, 1985; James et al., 1987).

2.4. *Germplasm Conservation*

Two approaches to germplasm conservation have been used, slow growth and cryopreservation (Withers, 1985). Slow growth is useful for short- to medium-term storage and can be considered for active collections of germplasm, whereas cryopreservation is useful for long-term storage needed for base collections of germplasm (Kartha, 1985).

Slow growth entails maintaining cultures on standard or minimal media at temperatures slightly above freezing (2–4 °C), often in darkness, with transfers to fresh medium at intervals of 6–24 months. Growth regulators or osmotic inhibitors, or a combination of factors, may also be used to retard growth (Withers, 1985). The slow growth method was first reported for strawberry (Mullin and Schlegel, 1976) and has been widely adopted, sometimes with minor modifications, for a number of fruit crops. A recent innovation has been to use gas-permeable plastic bags as the containers for growing, storing and shipping cultures. Storage times of up to 6 years without transferring to fresh medium have been reported (Mullin and Schlegel, 1976).

Cryopreservation, usually in liquid nitrogen, has been successfully demonstrated for strawberry (Kartha, 1985; Sakai, 1985) and apple (Sakai, 1985). While this method has great potential for long-term storage of plant material, it will be used infrequently by those who wish or need the rapid access to plant material provided by slow growth storage.

Few studies have been done to determine the stability of plants from long-term storage. In grape, some ploidy variation was found in root squashes from cultures stored at 9.5 °C for up to 1 year (Skene et al., 1988). However, these variations were not observed in shoot apices and were also observed in plants grown exclusively in the greenhouse.

3. References

Anderson, W.C. (1975). Propagation of rhododendrons by tissue culture, Part 1. Development of a culture medium for multiplication of shoots. Proc. Inter. Plant Prop. Soc. 25:129–135.

Baribault, T.J., Skene, K.G.M. and Scott, N.S. (1989). Genetic transformation of grapevine cells. Plant Cell Rep. 8:137–140.

Barlass, M., Skene, K.G.M. and Clingeleffer, P.R. (1981). Studies on the fragmented shoot apex of grapevine, III: A scanning electron microscope study of adventitious bud formation *in vitro*. J. Exp. Bot. 32:1079–1083.

Belkengren, R. O. and Miller, P.W. (1962). Culture of apical meristems of *Fragaria vesca* strawberry plants as a method of excluding latent A virus. Plant Dis. Rptr. 46:119–121.

Billings, S.G., Chin, C.K. and Jelenkovic, G. (1988). Regeneration of blueberry plantlets from leaf segments. Hort Sc. 23:763–766.

Boxus, P. (1974). The production of strawberry plants by *in vitro* micro-propagation. J. Hort. Sci. 49:209–210.

Boxus, P., Quoirin, M. and Laine, J.M. (1977). Large scale propagation of strawberry plants from tissue culture. In: Applied and Fundamental Aspects of Plant Cell, Tissue, and Organ Culture. J. Reinert and Y.P.S. Bajaj (eds.), Springer Verlag, Berlin, pp. 130–143.

Braniste, N., Popescu, A. and Coman, T. (1984). Producing and multiplication of *Pyrus communis* haploid plants. Acta Hort. 161:147–150.

Brennan, R., Davidson, D., Wilshin, A. and Millam, S. (1989). An assessment of the *in vitro* multiplication rates of fourteen black currant cultivars. J. Hort. Sci. 64:679–681.

Brisset, M.N., Ochatt, S.J. and Paulin, J.P. (1990). Evidence for quantitative responses during co-culture of *Pyrus communis* protoplasts and *Erwinia amylovora*. Plant Cell Rep. 9:272–275.

Burr, T.J., Katz, B.H., Bishop, A.L., Meyers, C.A. and Mittak, V.L. (1988). Effect of shoot age and tip culture propagation of grapes on systemic infestations by *Agrobacterium tumefaciens* biovar 3. Am. J. Enol. Vitic 39:67–70.

Cao, Z. (1986). Grape micropropagation. In: Handbook of Plant Cell Culture, Vol. 6. Z. Chen, D.A. Evans, W.R. Sharp, P.V. Ammirato and M.R. Sondahl (eds.), McGraw-Hill, New York, pp. 312–326.

Chée, R. and Pool, R.M. (1983). *In vitro* vegetative propagation of *Vitis*: Application of previously defined culture conditions to a selection of genotypes. Vitis 22:363–374.

Chevreau, E. and Skirvin, R.M. (1992). Pear. In: Biotechnology of Perennial Fruit Crops. F.A. Hammerschlag and R.G. Litz (eds.), CAB International, Wallingford, pp. 263–276.

Colby, S.M. and Meredith, C.P. (1990). Kanamycin sensitivity of cultured tissues of *Vitis*. Plant Cell Rep. 9:237–240.

Colby, S.M., Juncosa, A.M., Stamp, J.A. and Meredith, C.P. (1991a). Developmental anatomy of direct shoot organogenesis from leaf petioles of *Vitis vinifera* (Vitaceae). Am. J. Bot. 78:260–269.

Colby, S.M., Juncosa, A.M. and Meredith, C.P. (1991b). Cellular differences in *Agrobacterium* susceptibility and regenerative capacity restrict the development of transgenic grapevines. J. Am. Soc. Hort. Sci. 116:356–361.

Druart, P. (1985). Organogenesis and somatic embryogenesis in callus culture: a tool for genetic improvement of cherry trees. In: *In vitro* Techniques, Propagation and Long Term Storage. A. Schaefer-Menuhr (ed.), Martinus Nijhoff/Dr. W Junk Publishers, Dordrecht, pp. 167–171.

Durham, R.E., Moore, G.A., Gray, D.J. and Mortensen, J.A. (1989). The use of leaf GPI and IDGH isozymes to examine the origin of polyembryony in cultured ovules of seedless grape. Plant Cell Rep. 7: 669–672.

Dweikat, I.M. and Lyrene, P.M. (1988). Adventitious shoot production from leaves of blueberry cultured *in vitro*. Hort. Sc. 23:629.

Fiola, J.A., Hassan, M.A., Swartz, H.J., Bors, R.H. and McNicols, R. (1990). Effect of thidiazuron, light fluence rates and kanamycin on *in vitro* shoot organogenesis from excised *Rubus* cotyledons and leaves. Plant Cell Tissue Org. Cult. 20:223–229.

Fraser, L.G. and Harvey, C.F. (1986). Somatic embryogenesis from anther-derived callus in two *Actinidia* species. Sci. Hort. 29:335–346.

Fridlund, P.R. (1980). Maintenance and distribution of virus-free fruit trees. In: Proceedings of the Conference on Nursery Production of Fruit Plants Through Tissue Culture – Applications and Feasibility. R.H. Zimmerman (ed.), US Dep. Agric., Sci. Educ. Admin., ARR-NE-11, pp. 86–92.

Graham, J. and McNicol, R.J. (1991). *Ribes* regeneration and transformation. Plant Cell Tissue Org. Cult. 24:91–95.

Gray, D.J. and Mortensen, J.A. (1987). Initiation and maintenance of long term somatic embryogenesis from anthers and ovaries of *Vitis longii* 'Microsperma'. Plant Cell Tissue Org. Cult. 9:73–80.

Gray, D.J., Mortensen, J.A., Benton, C.M., Durham, R.E. and Moore, G.A. (1990). Ovule culture to obtain progeny from hybrid seedless bunch grapes. J. Am. Soc. Hort. Sci. 115:1019–1024.

Guangrong, X. and Jianzhe, N. (1984). A study on the induction of apple pollen plants. Acta Hort. Sinica 11:161–164.

Guellec, V., David, C., Branchard, M. and Tempé, J. (1990). *Agrobacterium rhizogenes* mediated transformation of grapevine (*Vitis vinifera* L.). Plant Cell Tissue Org. Cult. 20:211–215.

Hammerschlag, F. (1982). Factors affecting establishment and growth of peach shoots *in vitro*. Hort. Sc. 17:85–86.

Hammerschlag, F.A., Bauchan, G.R. and Scorza, R. (1987). Factors influencing *in vitro* multiplication and rooting of peach cultivars. Plant Cell Tissue Org. Cult. 8:235–242.

Hassan, M.A.M. (1991). The biology of *Agrobacterium* vectored *Rubus* transformation and selection of transformed plants. PhD Thesis, University of Maryland, College Park, MD USA.

Hidano, Y. and Niizeki, M. (1988). Protoplast culture of deciduous fruit trees. Sci. Hort. 37:201–216.

Huang, S.-C. and Millikan, D.F. (1980). *In vitro* micrografting of apple shoot tips. Hort. Sc. 15:741–743.

Hutchinson, J.F. and Zimmerman, R.H. (1987). Tissue culture of temperate fruit and nut trees. Hort. Rev. 9:273–349.

James, D.J., Passey, A.J., Mackenzie, K.A.D. and Menhinick, E.C. (1986). Protoplast culture and the use of apomixis *in vitro* for the regeneration of fruit plants. In: Plant Tissue Culture and its Agricultural Applications. L.A. Withers and P.G. Alderson (eds.), Butterworths, London, pp. 349–357.

James, D.J., Mackenzie, K.A.D. and Malhotra, S.B. (1987). The induction of hexaploidy in cherry rootstocks using *in vitro* regeneration techniques. Theor. Appl. Genet. 73:589–594.

James, D.J., Passey, A.J., Barbara, D.J. and Bevan, M. (1989). Genetic transformation of apple (*Malus pumila* Mill.) using a disarmed Ti-binary vector. Plant Cell Rep. 7:658–661.

James, D.J., Passey, A.J. and Barbara, D.J. (1990). *Agrobacterium*-mediated transformation of the cultivated strawberry (*Fragaria* × *ananassa* Duch.). using disarmed binary vectors. Plant Sci. 69:79–94.

Jones, O.P., Waller, B.J. and Beech, M.G. (1988). The production of strawberry plants from callus cultures. Plant Cell Tissue Org. Cult. 12:235–241.

Jordan, M. (1975). *In vitro* culture of anthers from *Prunus*, *Pyrus* and *Ribes*. Planta Medica Suppl. 1975, pp. 59–65.

Kartha, K.K. (1984). Elimination of viruses. In: Cell Culture and Somatic Cell Genetics of Plants, Vol. 1. I.K. Vasil (ed.), Academic Press, Orlando, pp. 577–585.

Kartha, K.K. (1985). Meristem culture and germplasm preservation. In: Cryopreservation of Plant Cells and Organs. K.K. Kartha (ed.), CRC Press, Boca Raton, pp. 115–134.

Korban, S.S. (1992). Apple. In: Biotechnology of Perennial Fruit Crops. F.A. Hammerschlag and R.G. Litz (eds.), CAB International, Wallingford, pp. 203–227.

Krul, W.R. (1988). Recent advances in protoplast culture of horticultural crops: Small fruits. Sci. Hort. 37:231–245.

Krul, W.R. and Worley, J.F. (1977). Formation of adventitious embryos in callus cultures of 'Seyval', a French hybrid grape. J. Am. Soc. Hort. Sci. 102:360–363

Lespinasse, Y., Godicheau, M. and Duron, M. (1983). Potential value and method of producing haploids in the apple tree, *Malus pumila* (Mill.). Acta Hort. 131:223–230.

Linsmaier, E.M. and Skoog, F. (1965). Organic growth factor requirements of tobacco tissue cultures. Physiol. Plant. 18:100–127.

Liu, Z.R. and Sanford, J.C. (1988). Plant regeneration by organogenesis from strawberry leaf and runner tissue. Hort Sc. 23:1057–1059.

Lloyd, G. and McCown B. (1980). Commercially-feasible micropropagation of mountain laurel, *Kalmia latifolia*, by use of shoot-tip culture. Comb. Proc. Int. Plant Prop Soc. 30:421–427.

Maeda, E., Villalobos, V.M. and Sugiura, T. (1978). Fine structure of the regenerating cells in strawberry (*Fragaria*) anther cultures. Japan. J. Breeding 28:143–146.

Mante, S., Morgens, P.H., Scorza, R., Cordts, J.M. and Callahan, A.M. (1991). *Agrobacterium*-mediated transformation of plum (*Prunus domestica* L.) hypocotyl slices and regeneration of transgenic plants. Bio/Technology 9:853–857.

Marcotrigiano, M., Swartz, H.J., Gray, S.E., Tokarcik, D. and Popenoe, J. (1984). The effect of benzylaminopurine on the *in vitro* multiplication rate and subsequent field performance of tissue culture propagated strawberry plants. Adv. Strawberry Prod. 3:23–25.

Matsuta, N. and Hirabayashi, T. (1989). Embryogenic cell lines from somatic embryos of grape (*Vitis vinifera* L.). Plant Cell Rep. 7:684–687.

McGrew, J.R. (1980). Meristem culture for production of virus-free strawberries. In: Proceedings of the Conference on Nursery Production of Fruit Plants Through Tissue Culture – Applications and Feasibility. R.H. Zimmerman (ed.), US Dep. Agric., Sci. Educ. Admin., Agric. Res.Results ARR-NE-11, pp. 80–85.

Mehra, P.N. and Jaidka, K. (1985). Experimental induction of embryogenesis in pear. Phytomorphology 35:1–10.

Mii, M. and Ohashi, H. (1988). *Actinidia* from cell suspensions. Acta Hort. 230:167–170.

Mullin, R.H. and Schlegel, D.E. (1976). Cold storage maintenance of strawberry meristem plantlets. Hort Sc. 11:100–101.

Mullins, M.G., Tang, F.C.A. and Facciotti, D. (1990). *Agrobacterium*-mediated genetic transformation of grapevines: transgenic plants of *Vitis rupestris* Scheele. and buds of *Vitis vinifera* L. Bio/Technology 8:1041–1045.

Murashige, T. and Skoog, F. (1962). A revised medium for rapid growth and bioassays with tobacco tissue cultures. Physiol Plant. 15:473–497.

Nehra, N.S., Stushnoff, C. and Kartha, K.K. (1989). Direct shoot regeneration from strawberry leaf disks. J. Am. Soc. Hort. Sci. 114:1014–1018.

Nehra, N.S., Chibbar, R.N., Kartha, K.K., Datla, R.S.S., Crosby, W.L. and Stushnoff, C. (1990a). Genetic transformation of strawberry by *Agrobacterium tumefaciens* using a leaf disk regeneration system. Plant Cell Rep. 9:293–298.

Nehra, N.S., Chibbar, R.N., Kartha, K.K., Datla, R.S.S., Crosby, W.L. and Stushnoff, C.

(1990b). *Agrobacterium*-mediated transformation of strawberry calli and recovery of transgenic plants. Plant Cell Rep. 9:10–13.

Nehra, N.S., Stushnoff, C. and Kartha, K.K. (1990c). Regeneration of plants from immature leaf-derived callus of strawberry (*Fragaria* × *ananassa*). Plant Sci. 66:119–126.

Nehra, N.S., Kartha, K.K., Stushnoff, C. and Giles, K.L. (1992). The influence of plant growth regulator concentrations and callus age on somaclonal variation in callus culture regenerants of strawberry. Plant Cell Tissue Org. Cult. 29:257–268.

Niemirowicz-Szczytt, K., Zakrzewska, Z., Malepszy, S. and Kubicki, B. (1983). Characteristics of plants obtained from *Fragaria* × *ananassa* in anther culture. Acta Hort. 131:231–237.

Niizeki, H. and Fukui, K. (1983). Elimination of somatic chromosomes in strawberry plants by treatment with para-fluorophenylalanine. Japan. J. Breeding 33:55–61.

Nyman, M. and Wallin, A. (1988). Plant regeneration from strawberry (*Fragaria* × *ananassa*). mesophyll protoplasts. J. Plant Physiol. 133:375–377.

Ochatt, S.J. (1990). Plant regeneration from root callus protoplasts of sour cherry (*Prunus cerasus* L.). Plant Cell Rep. 9:268–271.,

Ochatt, S.J. (1993). Woody plant protoplast technology revisited. Acta Hort. 336:285–296.

Ochatt, S.J. and Caso, O.H. (1986). Shoot regeneration from leaf mesophyll protoplasts of wild pear (*Pyrus communis* var. *Pyraster* L.). J. Plant Physiol. 1222:243–249.

Ochatt, S.J. and Power, J.B. (1988). Plant regeneration from mesophyll protoplasts of Williams' Bon Chretien (syn. Bartlett). pear (*Pyrus communis* L.). Plant Cell Rep. 7:587–589.

Ochatt, S.J. and Power, J.B. (1989). Cell wall synthesis and salt (saline) sensitivity in cultured colt cherry (*Prunus avium* × *pseudocerasus*). protoplasts. Plant Cell Rep. 8: 365–367.

Ochatt, S.J., Chand, P.K., Rech, E.L., Davey, M.R. and Power, J.B. (1988a). Electroporation-mediated improvement of plant regeneration from Colt cherry (*Prunus avium* × *psuedocerasus*). Plant Sci. 54:165–169.

Ochatt, S.J., Rech, E.L., Davey, M.R. and Power, J.". (1988b). Long-term effect of electroporation on enhancement of growth and plant regeneration of Colt cherry (*Prunus avium* × *pseudocerasus*) protoplasts. Plant Cell Rep. 7:393–395.

Ochatt, S.J., Patat-Ochatt, E.M., Rech, E.L., Davey, M.R. and Power, J.B. (1989). Somatic hybridization of sexually incompatible top-fruit tree rootstocks, wild pear (*Pyrus communis* var. *pyraster* L. and Colt cherry (*Prunus avium* × *pseudocerasus*). Theor. Appl. Genet. 78:35–41.

Oliveira, M.M. and Pais, M.S.S. (1991). Plant regeneration from protoplasts of long-term callus cultures of *Actinidia deliciosa* var. *deliciosa* cv. Hayward (kiwifruit). Plant Cell Rep. 9:643–646.

Oosawa, K. and Takayanagi, K. (1982). High yielding variants in strawberry derived from anther culture. In: Proceedings 5th International Congress of Plant Tissue and Cell Culture, Tokyo, Japan. A Fujiwara (ed.), Japan Assoc. Plant Tissue Culture, pp. 765–766.

Quak, F. (1977). Meristem culture and virus-free plants. In: Applied and Fundamental Aspects of Plant Cell, Tissue, and Organ Culture. J. Reinert and Y.P.S. Bajaj (eds.), Springer Verlag, Berlin, pp. 598–615.

Quoirin, M. and Lepoivre, P. (1977). Etude de milieux adaptes aux cultures *in vitro* de *Prunus*. Acta Hort. 78:437–442.

Rajasekaran, K. and Mullins, M. G. (1985). Somatic embryo formation by cultured ovules of 'Cabernet Sauvignon' grape: Effects of fertilization and of the male gametocide toluidine blue. Vitis 24:151–157.

Ramming, D.W. (1990). The use of embryo culture in fruit breeding. Hort. Sc. 25:393–398.

Reed, B. (1990). Multiplication of *Rubus* germplasm *in vitro*: A screen of 256 accessions. Fruit Var. J. 44:141–148.

Revilla, M.A. and Power, J.B. (1988). Morphogenic potential of long-term callus cultures of *Actinidia deliciosa*. J. Hort. Sc. 63:541–545.

Rosati, P., Devreux, M. and Laneri, U. (1975). Anther culture of strawberry (*Fragaria* × *ananassa*). Hort. Sc. 10:119–120.

Sakai, A. (1985). Cryopreservation of shoot-tips of fruit trees and herbaceous plants. In:

Cryopreservation of Plant Cells and Organs. K.K. Kartha (ed.). CRC Press, Boca Raton, pp. 135–158.

Scorza, R. and Hammerschlag, F. (1992). Stone fruits. In: Biotechnology of Perennial Fruit Crops. F.A. Hammerschlag and R.G. Litz (eds.), CAB International, Wallingford, pp. 277–301.

Shidakow, R.S. (1986). Effect of a chemical mutagen on cytoembryological processes and possibilities of obtaining polyploid apple forms. Sov. Genet. 21:1444–1451.

Simpson, D.W. and Bell, J.A. (1989). The response of different genotypes of *Fragaria* × *ananassa* and their seedling progenies to *in vitro* micropropagation and the effects of varying the concentration of 6-benzylaminopurine in the proliferation medium. Plant Cell Tissue Org. Cult. 17:225–234.

Singha, S. (1980). *In vitro* propagation of 'Seckel' pear. In: Proceedings Conference on Nursery Production of Fruit Plants Through Tissue Culture – Applications and Feasibility. R.H. Zimmerman (ed.), US Dep. Agric., Sci. and Educ. Admin., Agr Res. Results ARR-NE-11, pp. 59–63.

Singha, S. (1982). Influence of agar concentration on *in vitro* shoot proliferation of *Malus* sp. 'Almey' and *Pyrus communis* 'Seckel'. J. Am. Soc. Hort. Sci. 107:657–660.

Skene, K.G.M., Goodwins, D.R. and Barlass, M. (1988). Ploidy stability in grapevines following long term storage *in vitro*. Vitis 27:41–46.

Snir, I. (1982). *In vitro* propagation of sweet cherry cultivars. Hort. Sc. 17:192–193.

Snir, I. (1983). A micropropagation system for sour cherry. Sci. Hort. 19:85–90.

Stamp, J.A. and Meredith, C.P. (1988). Somatic embryogenesis from leaves and anthers of grapevine. Sci. Hort. 35:235–250.

Stamp, J.A., Colby, S.M.. and Meredith, C.P. (1990). Improved shoot organogenesis from leaves of grape. J. Am. Soc. Hort. Sci. 115:1038–1042.

Swartz, H.J. and Lindstrom, J.T. (1986). Small fruit and grape tissue culture from 1980 to 1985: commercialization of the technique. In: Tissue Culture as a Plant Production System for Horticultural Crops. R.H. Zimmerman, R.J. Griesbach, F.A. Hammerschlag and R.H. Lawson (eds.), Martinus Nijhoff Publishers, Dordrecht, pp. 201–220.

Tabachnik, L. and Kester, D.E. (1977). Shoot culture for almond and almond-peach hybrid clones *in vitro*. Hort. Sc. 12:545–547.

Takeno, K., Koshioka, M., Pharis, R.P., Rajasekaran, K. and Mullins, M.G. (1983). Endogenous gibberellin-like substances in somatic embryos of grape (*Vitis vinifera* × *Vitis rupestris*). in relation to embryogenesis and chilling requirement for subsequent development of mature embryos. Plant Physiol. 73:803–808.

Theodoropoulos, P.A. and Roubelakis-Angelakis, K.A. (1990). Progress in leaf protoplast isolation and culture from virus-free axenic shoot cultures of *Vitis vinifera* L. Plant Cell Tissue Org. Cult. 20:15–23.

Tukey, H.B. (1933). Artificial culture of sweet cherry embryos. J. Hered. 24:7–12.

Walkey, D.G.A. (1978). *In vitro* methods for virus elimination. In: Frontiers of Plant Tissue Culture 1978. T.A. Thorpe (ed.), University of Calgary Press, Calgary, pp. 245–254.

Wallin, A. and Johansson, L.B. (1989). Plant regeneration from leaf mesophyll protoplasts of *in vitro* cultured shoots of a columnar apple. J. Plant Physiol. 135:565–570.

Wallin, A. and Welander, M. (1985). Improved yield of apple leaf protoplasts from *in vitro* cultured shoots by using very young leaves and adding methionine to the shoot medium. Plant Cell Tissue Org. Cult. 5:69–72.

Withers, L.A. (1985). Cryopreservation and storage of germplasm. In: Plant Cell Culture: A Practical Approach. R.A. Dixon (ed.). IRL Press, Oxford, pp. 169–191.

Wolfe, D.E., Eck, P. and Chin, C.-K. (1983). Evaluation of seven media for micropropagation of highbush blueberry. Hort. Sc. 18:703–705.

Zimmerman, R.H. (1983). Tissue culture. In: Methods in Fruit Breeding. J.N. Moore and J. Janick (eds.), Purdue University Press, West Lafayette, pp. 124–135.

Zimmerman, R.H. (1991). Micropropagation of temperate zone fruit and nut crops. In: Micropropagation: Technology and Application. P.C. Debergh and R.H. Zimmerman (eds.), Kluwer Academic Publishers, Dordrecht, pp. 231–246.

Zimmerman, R.H. and Broome, O.C. (1981). Phloroglucinol and *in vitro* rooting of apple cultivar cuttings. J. Am. Soc. Hort. Sci. 106:648–652.

Zimmerman, R.H. and Fordham, I. (1985). Simplified method for rooting apple cultivars *in vitro*. J. Am. Soc. Hort. Sci. 110:34–38, 461–462.

19. *In vitro* Culture of Tropical Fruits

J. W. Grosser

Contents

1. Introduction

Contributions to tropical fruit improvement by conventional plant breeding have been limited by complex genetic systems, extended juvenility, lack of knowledge, and in many cases, lack of effort. Most of today's important tropical fruit cultivars arose from the selection of mutations within existing clones, or by selections from seedling populations derived from open pollinations. Due to the increasing significance of tropical fruits for food and export income, there has been a recent expansion of effort and resources devoted to tropical fruit cultivar improvement. Many programs are attempting to integrate emerging biotechnologies with conventional breeding methods to facilitate and expedite cultivar improvement. The objective of this chapter is to provide some pertinent information regarding *in vitro* methods for tropical fruit improvement, along with a discussion of how such methods can be applied to the genetic improvement of tropical fruits, with emphasis on increasing fruit quality and resistance to disease and environmental stress.

Florida Agricultural Experiment Station Journal Series No. R-01693.

Indra K. Vasil and Trevor A. Thorpe (eds.), Plant Cell and Tissue Culture, 475–496.

2. Plant Regeneration

2.1. *Organogenesis/Somatic Embryogenesis*

The ability to regenerate complete plants from individual protoplasts, cells, or tissues is a necessary requirement that must be established prior to the application of most biotechnological approaches to crop improvement. Efficient plant regeneration for many of the important tropical fruit species is a relatively recent phenomenon, and with the exception of *Citrus*, examples of applied biotechnology are limited. Plant regeneration in tropical fruits can be achieved by somatic embryogenic (Litz, 1985) or organogenic (Litz, 1987) pathways, and a thorough review of the subject has been provided recently by Litz and Gray (1992). Although protoplasts can be routinely isolated from many important tropical fruit species, efficient protoplast to plant regeneration has been demonstrated only in *Citrus* and *Actinidia* (Kiwifruit). Continued improvement in our understanding of morphogenesis and media requirements should lead to the development of protoplast to plant regeneration systems for many important tropical fruit species. A partial list of achievements in plant regeneration of tropical fruit species, including the regeneration pathway and examples of protoplast to plant regeneration, is provided in Table 1. With reliable plant regeneration systems in place for these species, successful genome manipulation and gene transfer may follow.

Tropical fruit tissue cultures with capacity for plant recovery via organogenesis can be established from numerous explant sources including embryonic axes, cotyledons, stems (usually hypocotyl), leaves, floral tissues, roots, and rhizomes. Tropical fruit tissue cultures regenerable via somatic embryogenesis have been initiated from nucellus, ovules, somatic and zygotic embryos, cotyledons, and leaves. In most cases, MS basal medium (Murashige and Skoog, 1962), supplemented with appropriate growth regulators has been utilized. However, MS basal medium contains high levels of macro-salts and does not have good buffering capacity. The high level of ammonia contained in MS basal medium can be toxic to *in-vitro* cultured cells and particularly to protoplasts. Several modifications of MS basal medium have been used to alleviate this problem, with the most simple being the dilution of macro-salts to half strength. Basal medium containing NH_4NO_3 and KNO_3 at half strength supplemented with 1550 mg/l glutamine and 750 mg/l KCl has been quite useful for the culture of friable embryogenic *Citrus* callus (Grosser and Gmitter, 1990a). This basal medium allows for rapid callus growth with reduced starch accumulation for several *Citrus* genotypes, and thereby facilitates protoplast isolation. The addition of glutamine at 400 mg/l has also been shown to be beneficial for culture growth and permissive induction of *Carica* sp. (Litz and Conover, 1982), *Euphoria longan* (Litz, 1988), and *Mangifera indica* (DeWald et al., 1989a).

Growth regulator regimes used to achieve organogenesis and somatic embryogenesis in fruit crop species have recently been summarized by Litz and

Gray (1992). Regimes for plant regeneration via organogenesis generally rely on moderate to high levels of cytokinin along with low levels of auxin. Gibberellic acid has also been used in some cases. BA (benzyl- adenine or 6-benzyl aminopurine) has been the most frequently used cytokinin, but kinetin, zeatin, 2iP ([N^6-[Δ^2-isopentyl]-adenine]), and thidiazuron have also been successful. NAA (napthelene acetic acid) has been the preferred auxin, but IAA (indole-3-acetic acid), 2,4-D (2,4-dichlorophenoxy-acetic acid), and IBA (indole-butyric acid) have also been successful.

Auxin is generally used to stimulate embryogenic callus induction. 2,4-D is the preferred auxin for this purpose, but NAA is also frequently utilized. IAA, IBA, and picloram have been used less frequently. *Citrus* embryogenic callus can become habituated and maintained indefinitely on growth regulator free media. This simplifies growth regulator requirements for citrus protoplast culture and may reduce somaclonal variation associated with auxin exposure (Grosser and Gmitter, 1990a).

2.2. *Protoplast to Plant*

Protoplast to plant regeneration in tropical fruit species thus far has been limited to polyembryonic genotypes of *Citrus* via the somatic embryogenesis pathway and kiwifruit (*Actinidia* spp.) via organogenesis (Table 1). Historically, examples of protoplast to plant regeneration in *Citrus* were limited to genotypes where friable embryogenic callus was available (Vardi et al., 1982; Kobayashi et al., 1983; Grosser and Gmitter, 1990c). More recently, *Citrus* plant regeneration directly from seedling leaf protoplasts has been reported for several genotypes (Tusa et al., 1990; Ohgawara et al., 1991). At the present time, plant recovery from *Citrus* seedling leaf protoplasts requires co-culture with embryogenic callus-derived protoplasts. A recent protoplast fusion experiment conducted in my laboratory to produce somatic hybrids by fusing seedling leaf protoplasts of sour orange (*Citrus aurantium* L.) with embryogenic callus protoplasts of Cleopatra mandarin (*C. reticulata* Blanco) yielded 394 plants. Of these, 171 were identified to be sour orange + Cleopatra mandarin somatic hybrids and 223 were sour orange. No Cleopatra plants were recovered, indicating that this habituated Cleopatra callus line had lost its regeneration capacity. Such a line should be quite useful in co-culture experiments to achieve plant regeneration from leaf protoplasts of citrus genotypes where suitable embryogenic callus is not available. The development of co-culture methodology should expand the number of *Citrus* genotypes amenable to somatic hybridization, cybridization, and transformation experiments. Reports of plant regeneration from protoplasts of other tropical fruit species should be forthcoming, particularly for species that exhibit *in-vitro* somatic embryogenesis.

Table 1. Partial list of plant regeneration via somatic embryogenesis or organogenesis and from protoplasts in tropical fruit species

Genus/species	Common name	Regeneration pathway[a]	Plant regeneration from protoplasts[b]	References
Actinidia chinensis	Kiwifruit	OR	CO	Tsai (1988)
		OR		Barbieri and Morini (1987)
		OR		Pais et al. (1987)
A. deliciosa	Kiwifruit	OR		Revilla and Power (1988)
		OR	CO	Oliveira and Pais (1991)
Ananus comosus	Pineapple	OR		Mapes (1974)
		OR		Wakasa (1979)
Annacardium occidentale	Cashew	OR		Philip (1984)
Annona cherimola	Custard apple	OR		Jordan (188)
A. squamosa	Sugar apple	OR		Nair et al. (1984)
Averrhoa carambola	Star fruit	OR		Litz and Conover (1980)
		OR		Litz and Griffis (1989)
Bactris gasipaes	Peach palm	SE		Arrias and Huete (1983)
Carica papaya	Papaya	OR, SE		Yie and Liaw (1977)
		OR		Litz et al. (1983)
Citrus aurantifolia	Acid lime	SE		Mitra and Chaturvedi (1972)
		SE		Gmitter and Moore (1986)
		OR		Raj Bhansali and Arya (1978a)
C. aurantium	Sour orange	OR		Bouzid (1975)
		SE		Ben-Hayyim and Neumann (1983)
		SE	C	Vardi et al. (1982)
		SE	L, C	J.W. Grosser (unpublished)
C. clementina	Clementine	SE		Navarro et al. (1985)
C. grandis	Pummelo	OR		Chaturvedi and Mitra (1974)
		SE		Rangan et al. (1968)
C. jambhiri	Rough lemon	SE		Ben-Hayyim and Neumann (1983)
		SE	C	Vardi et al. (1990)
		SE	L	J.W. Grosser (unpublished)
C. ichangensis	Ichang papeda	SE	L	J.W. Grosser (unpublished)
C. limetoides	Sweet lime	OR		Raj Bhansali and Arya (1978b)

Table 1. Continued

Genus/species	Common name	Regeneration pathway[a]	Plant regeneration from protoplasts[b]	References
Citrus limon	Lemon	OR		Bouzid (1975)
		SE		Rangan et al. (1968)
		SE	C	Vardi et al. (1982)
		SE	C	P. Spiegel-Roy and S. Saad (unpubl.)
		SE	L	Tusa et al. (1990)
C. madurensis	Calamondin	OR		Grinblat (1972)
C. microcarpa	Calamondin	SE		Maheshwari and Rangaswamy (1958)
C. mitis	Calamondin	SE	C	Sim et al., 1988)
C. nobilis	Satsuma	SE		Kochba et al. (1982)
C. paradisi	Grapefruit	OR		Chaturvedi and Mitra (1974)
		SE		Kochba et al. (1972)
		SE		Gmitter and Moore (1986)
		SE	C	Vardi et al. (1982)
		SE	L	Ohgawara et al. (1989)
		SE	L	J.W. Grosser (unpublished)
C. reticulata	Mandarin	OR		Bouzid (1975)
		SE		Sabharwal (1963)
		SE	C	Vardi et al. (1982)
		SE	C	J.W. Grosser (unpublished)
		SE	C	Hidaka and Kajiura (1988)
C. sinensis	Sweet orange	OR		Chaturvedi and Mitra (1974)
		OR		Bouzid (1975)
		OR		J.W. Grosser (unpublished)
		SE		Moore (1985)
		SE		Gmitter and Moore (1986)
		SE	C	Vardi et al. (1982)
		SE	C	Tusa et al. (1990)
		SE	L	J.W. Grosser (unpublished)
		SE	C	Kobayashi et al. (1983, 1988)
C. yuko	Yuko	SE	C	Hidaka and Kajiura (1988)
CITRUS HYBRIDS:				
C. sinensis × *Poncirus trifoliata*	Citrange	OR		Barlass and Skene (1982)
		OR		Sauton et al. (1982)
		SE	L	Ohgawara et al. (1991)

Table 1. Continued

Genus/species	Common name	Regeneration pathway[a]	Plant regeneration from protoplasts[b]	References
Citrus paradisi × *Poncirus trifoliata*	Citrumelo	OR		Grosser and Chandler (1986)
C. sinensis × *C. reticulata*	Tangor (Murcott)	SE	C	Vardi et al. (1982)
		SE	L	J.W. Grosser (unpublished)
C. reticulata × *C. paradisi*	Tangelo (Orlando)	SE		Gmitter and Moore (1986)
	(Page)	SE	C, L	J.W. Grosser (unpublished)
Cocos nucifera	Coconut	SE		Gupta et al. (1984)
		SE		Karunaratne and Periyapperuma (1989)
Diospyros kaki	Persimmon	OR		Yokoyama and Takeuchi (1976)
		OR		Tao et al. (1988)
Eriobotrya japonica	Loquat	SE		Litz (1985)
Euphoria longan	Longan	SE		Litz (1988)
Ficus religiosa	Bo-tree fig	OR		Jaisawl and Narayan (1985)
Garcinia mangostana	Mangosteen	OR		Goh et al. (1988)
Mangifera indica	Mango	SE		Litz et al. (1984)
		SE		DeWald et al. (1989 a, b)
Microcitrus sp.	Australian wild lime	SE	C	Vardi et al. (1986)
Morus alba	Mulberry	OR		Ohyama and Oka (1982)
		OR		Oka and Ohyama (1981)
Musa spp.	Banana, plantain	OR		Ma and Shii (1974)
		SE		Novak et al. (1989)
Myrciaria caulifloria	Jaboticaba	SE		Litz (1984a)
Passiflora caerulea	Passion fruit	OR		Scorza and Janick (1976)
P. suberosa	Passion fruit	OR		Scorza and Janick (1976)
Phoenix dactylifera	Date palm	SE		Reynolds and Murashige (1979)
Psidium guajava	Guava	OR		Amin and Jaiswal (1987)
Punica granatum	Pomegranate	OR		Omura et al. (1987)
Solanum quitoense	Lulo	OR		Hendrix et al. (1987)
Syzygium spp.	Jambolan	SE		Litz (1984b)

[a] OR – organogenesis; SE – somatic embryogenesis.
[b] C – embryogenic callus derived protoplasts; L – leaf derived protoplasts; CO – organogenic callus derived protoplasts.

3. Micropropagation

Efficient vegetative propagation methods already exist for most important tropical fruit species. However, continued improvements in methods for rapid clonal micropropagation are needed to increase the availability of superior newly released cultivars and pathogen-free material, and to provide enough material of promising selections for thorough evaluation. Methods for the micropropagation of tropical fruit species have been reviewed by Litz and Jaiswal (1991), and include stimulation of axillary bud proliferation from shoot tip and nodal explants, somatic embryogenesis from explants from mature trees, and adventitious shooting directly from explant or callus.

4. Pathogen Elimination

Meristem tip culture, sometimes in combination with thermo-therapy, has been used to generate pathogen free material of several important banana and plantain cultivars (Berg and Bustamante, 1974; Gupta, 1986; Surga Rivas, 1988). Surga Rivas (1988) cultured 1 mm apical meristems on MS basal medium containing 1 mg/l thiamine HCl, 1 mg/l pyridoxine, 100 mg/l nicotinic acid, 30 g/l sucrose, 3 mg/l BA, 100 mg/l myo-inositol, and 7 g/l agar, to obtain plants of banana cultivars 'Poyo' and 'Giant Cavendish' that were free of cucumber mosaic virus.

Navarro et al. (1975) developed an *in vitro* shoot-tip grafting procedure that has subsequently been used to free at least 90 *Citrus* accessions from pathogens including citrus tristeza virus, psorosis virus, stubborn spiroplasma, exocortis viroid, concave gum, cristacortis pathogen, and the impietratura pathogen (Navarro et al., 1988). Rootstocks for this procedure are grown *in vitro* on MS basal medium plus 1% agar from peeled and disinfested seed. Troyer citrange is the preferred rootstock, but Etrog citron is used for all lemon material. Infested varieties are prepared by grafting to vigorous rootstocks, hand defoliation, and maintenance at 32 °C for 12–15 days. Meristems (0.1–02 mm) are then dissected from new flush and micrografted into an inverted T incision in the prepared rootstock. Micrografted plantlets are maintained on MS basal medium containing 100 mg/l i-inositol, 4% sucrose, 0.2 mg/l thiamine HCl, 1 mg/l pyridoxine HCl, and 1 mg/l nicotinic acid. High concentrations of sucrose and B vitamins and various levels of BA, IAA, and GA_3 are used to induce shoot growth from the transplanted meristems. Grafting success of 60–70% and transplanting success of 90% have become routine. This program provides a source of pathogen-free common and foreign varieties that are available for commercial or research purposes. Several other countries where citriculture is important have established similar programs, including Italy and the United States.

Shoot tip culture methods for many important tropical fruit species have been established and recently reviewed by Litz and Jaiswal (1991). Employing such methods to free important cultivars of yield-reducing pathogens is therefore feasible and more reports should be forthcoming.

5. Haploid Production

The successful production of haploid plants via anther culture in tropical fruit species has been quite limited. In general, complicated media regimes have been required to achieve low frequencies of response. Haploids are desirable for the production of homozygous breeding lines, the selection of desirable recombinants, and studies of chromosome pairing and genome evolution. Reviews of anther culture in *citrus* (Chen, 1990), *lichti* (Fu, 1990), and *longan* (Wei, 1990) have been recently provided. Anther culture is not expected to make a major contribution to tropical fruit improvement in the near future.

6. Triploid Production

6.1. *Embryo/Ovule Rescue*

Seedlessness is an important breeding objective of most *Citrus* scion improvement programs (Grosser and Gmitter, 1990a). For this reason, the production of triploids in *Citrus* has gained much attention. Triploids in *Citrus* are usually seedless, yet they can be vigorous and fruitful. Triploid *Citrus* cultivars include 'Persian' lime (Krug and Bacchi, 1943), and 'Oroblanco' and 'Melogold' pummelo × grapefruit hybrids (Soost and Cameron, 1980, 1985). Triploid hybrids can be produced by crossing diploid and tetraploid parents. To avoid problems with nucellar polyembryony, seedy diploid monoembryonic genotypes have often been used as female parents in crosses with autotetraploid pollen parents. However, disturbed embryo:endosperm ploidy ratios associated with crosses in this direction interferes with embryo and seed development. *In vitro* rescue and culture of embryos or undeveloped ovules can be! utilized to recover triploid hybrid seedlings from such crosses. Starrantino and Recupero (1981) used embryo rescue to recover more than 300 triploid hybrids from crosses of three diploid monoembryonic seed parents with eight autotetraploid pollen parents. Embryos were excised 3–4 months after pollination and cultured on MS basal medium supplemented with 500 mg/l malt extract and 25 mg/l adenine sulfate. Several of these hybrids are in the advanced stages of field evaluation, and the release of improved seedless triploid scion cultivars from this program should follow.

More recently, allotetraploid somatic hybrids are being developed for use as pollen parents in crosses with diploids. Oiyama et al. (1991) cultured underdeveloped ovules from mature fruit of a cross between 'Clementine' mandarin seed parent and the 'Trovita' sweet orange + *Poncirus trifoliata* somatic hybrid on MT basal medium containing 1 mg/l GA_3 and recovered 85 triploid plants. Many more crosses of this nature are expected as new somatic hybrids, particularly interspecific hybrids that combine elite complementary scion cultivars, overcome juvenility, and begin to flower.

6.2. *Endosperm Culture*

Endosperm culture can be utilized to overcome barriers to sexual hybridization resulting from apomixis to generate triploid citrus hybrids. Generally, endosperm is genetically unique triploid hybrid tissue that arises from the fusion of one male gamete with two polar nuclei (Fahn, 1974). Wang and Chang (1978) first reported triploid plantlet regeneration from *Citrus* endosperm, but they were unable to establish plants in soil. More recently, Gmitter et al. (1990) established four triploid hybrid *Citrus* plants in soil that were recovered from the culture of Ridge Pineapple sweet orange (*C. sinensis* L. Osbeck) endosperm. Uninjured cellular endosperm excised 12–14 days weeks post-anthesis were cultured on a sequence of media to achieve primary callus induction and proliferation, somatic embryo induction, embryo development, and subsequent plant recovery. *In vitro* minigrafts onto Carrizo citrange seedling rootstocks were utilized to establish complete plants in soil. The most efficient callus induction medium was MT basal (Murashige and Tucker, 1969) containing (in mg/l) 5.0 BA, 2.0 2,4-D, 5.0 kinetin, 1000 casein hydrolysate, and 500 malt extract. Embryo production leading to plant regeneration was achieved on double strength MT basal medium supplemented with (in mg/l) 0.25 BA, 2.0 GA_3, 2 adenine, and 500 casein hydrolysate. These results suggest that endosperm culture research should be expanded to include other *Citrus* genotypes and apomictic tropical fruit species where seedlessness is an important breeding objective.

7. Autotetraploid Production

In vitro techniques are also being utilized in *Citrus* to facilitate the production of autotetraploids from elite diploid scions that should be useful as parents in interploid hybridizations. Gmitter and Ling (1991) recovered non-chimeric autotetraploid plants of 'Valencia' sweet orange (*Citrus sinensis*) and 'Orlando' tangelo (*Citrus reticulata* × *Citrus paradisi* Macf.) from the culture of undeveloped ovules on MT basal medium containing 500 mg/l malt extract and exposed to 0.01 or 0.10% colchicine for 21 days. Gmitter et al. (1991) recovered seven autotetraploid 'Hamlin' sweet orange (*C. sinensis*) and three 'Ridge Pineapple' sweet orange (*C. sinensis*) following the culture of habituated embryogenic callus on MT basal medium containing 0.10% colchicine for a period of 4–12 weeks. Resulting somatic embryos were germinated on

MT basal medium containing 5 mg/l GA_3. Fertility of these colchicine derived tetraploid *Citrus* plants will be assessed as they reach maturity.

8. Somatic Hybridization

Protoplast technology in *Citrus* is quite advanced, and novel somatic hybrid plants created by protoplast fusion are being reported at a rapid rate. Since the first report of somatic hybrid plants by Ohgawara et al. (1985), somatic hybrid plants of 26 different parental combinations have been reported (Table 2). The general method used to generate somatic *Citrus* hybrids is polyethylene glycol (PEG) induced chemical fusion of competent protoplasts isolated from nucellar-derived friable embryogenic callus or suspension cultures of one parent with protoplasts of the second parent isolated from seedling leaves. The capacity to recover embryogenic callus and subsequently somatic hybrid plants via somatic embryogenesis from heterokaryons is usually provided by the embryogenic parent, but also seems to be influenced by the leaf parent genotype, and stimulated by the fusion procedure itself. Although a high level of sucrose has been used in protoplast culture medium to inhibit plant regeneration from unfused protoplasts isolated from embryogenic cultures of sweet orange (*Citrus sinensis*) (Ohgawara et al., 1985), sophisticated selection schemes are not necessary for most parental combinations because somatic hybrid plant recovery is generally of comparable or greater efficiency than that of plant recovery from unfused parental protoplasts, and resulting tetraploid somatic hybrid plants are easily distinguished from diploid parental type plants based on morphology. Somatic hybrid plants generally are morphologically intermediate to their parents and exhibit increased leaf thickness due to their tetraploid nature. Further verification of somatic hybridity has been provided by determination of chromosome number in combination with electrophoretic analyses of either isozymes or DNA banding patterns (Grosser and Gmitter, 1990a).

Several strategies are being employed to apply somatic hybridization technology to *Citrus* cultivar improvement (Grosser and Gmitter, 1990a, b, c). Germplasm enhancement can be achieved by successful wide somatic hybridization of *Citrus* with species from related genera that possess traits of interest (Table 2a, b). *Citrus* relatives offer a wealth of genetic variability (Swingle and Reece, 1967; Sykes, 1988), including potential genetic variation for desirable traits that have not been identified within the genus *Citrus* (e.g. resistance or immunity to citrus blight; Timmer, 1988)). Such wide somatic hybrids, including those created from sexually incompatible parents, can be tested directly as improved citrus rootstocks without the worry of altering the genetic integrity of corresponding elite scions and, if fertile, could contribute to further gene transfer by conventional breeding at the tetraploid level.

Another promising strategy is to produce and test somatic hybrids that combine complementary rootstock genotypes. Many of the somatic hybrids identified in Table 2 with potential for rootstock improvement were created in efforts to package the following traits in various combinations as required for particular geographic locations: cold hardiness, salt tolerance, adaptation to calcareous soils, vigor, tree size control, and tolerance/resistance/immunity to citrus blight, citrus tristeza virus induced quick decline, *Phytophthora* induced diseases, and nematode diseases. Many of these hybrids are being evaluated for the expression of important parental traits and entered into commercial field trials to determine their potential as citrus rootstocks (Grosser, 1990).

The major application of somatic hybridization to *Citrus* scion improvement may be the production of interspecific allotetraploid somatic hybrids that combine complementary parents (Table 2c) for use as parents in interploid crosses to generate seedless triploid zygotic progeny. Use of such hybrids should maximize heterozygosity and variation in resulting zygotic progeny. Adequate fertility, based on pollen staining with acetocarmine, pollen germination on medium containing 20% sucrose, and viable seed production, has been observed for the following somatic hybrids: 'Trovita' sweet orange + *Poncirus trifoliata* (Kobayashi et al., 1991); Navel orange + Troyer citrange (Ohgawara et al., 1991) and 'Key' lime + 'Valencia' sweet orange (X.X. Deng, F.G. Gmitter, Jr. and J.W. Grosser, unpublished data). Interspecific somatic hybridization also allows for contributions to scion improvement from cultivars where genetic exchange by sexual hybridization has been inhibited by male sterility, polyembryony, and low seed production (Kobayashi et al., 1988). Increasing the number of unique tetraploid hybrids and autotetraploids produced by other methods will expand the gene pool that can be used to generate the diverse seedless triploid cultivars needed to meet existing and future demands of the fresh fruit citrus industry.

9. Somatic Cybridization

Protoplast fusion has also been used to create five genetically unique *Citrus* cybrids, using a donor-recipient method by Vardi et al. (1987; 1989; Table 3). With this method, donor protoplasts are irradiated to arrest nuclear division, and the cytoplasms of recipient protoplasts are rendered inactive by treatment with the antimetabolite iodoacetate. Following protoplast fusion, protoplasts were cultured using the feeder layer technique (Vardi et al., 1987). Complementation allows for plant recovery via somatic embryogenesis, and resulting cybrid plants contain the nuclear genome of the recipient parent and the organelle genomes of the donor parent. Plants with recombinant chondriomes were also recovered, and Vardi et al. (1989) suggest that the chondriome components from *Microcitrus* could induce male sterility with potential parthenocarpy in *Citrus* fresh fruit cultivars, leading to seedless fruit production.

Table 2. Allotetraploid somatic hybrid citrus plants produced by protoplast fusion.

Parents and protoplast source	Number of plants	Cultivar improvement potential	References
(a) *Intergeneric plants produced from sexually incompatible parents*			
C. sinensis 'Hamlin' (embryogenic suspension) + *Severinia disticha* (non-embryogenic callus)	>150	R,S	Grosser et al. (1988b)
C. sinensis 'Hamlin' (embryogenic suspension) + *Severinia buxifolia**	>50	R,S	J.W. Grosser and F.G. Gmitter (unpublished)
C. sinensis 'Hamlin' (embryogenic callus) + *Citropsis gilletiana* (leaf/non-embryogenic callus)	>150	R,S	Grosser and Gmitter (1990c)
C. reticulata 'Cleopatra' mandarin (embryogenic suspension) + *Citropsis gilletiana* (leaf)	>35	R	Grosser et al. (1990)
(b) *Intergeneric plants produced from sexually compatible parents*			
C. sinensis 'Trovita' (embryogenic suspension + Poncirus trifoliata (leaf)	>10	R	Ohgawara et al. (1985)
C. sinensis 'Trovita' (embryogenic suspension) + *C. sinensis* × *P. trifoliata* 'Troyer' citrange (leaf)	17	R,S	Kobayashi and Ohgawara (1988)
C. sinensis 'Hamlin' (embroygenic suspension) + *Poncirus trifoliata* 'Dragon' (leaf)	>300	R	Grosser et al. (1988a)
C. reticulata 'Cleopatra' (embryogenic callus) + *Poncirus* trifoliata 'Flying Dragon' (leaf)	>150	R	J.W. Grosser and F.G. Gmitter (unpublished)
C. reticulata 'Cleopatra' (embryogenic callus) + 'Swingle' Citrumelo (*C. paradisi* × *P. trifoliata*	>200	R	F. Sesto, J.W. Grosser and F.G. Gmitter (unpublished)
C. sinensis 'Valencia' (embryogenic callus) + *Fortunella crassifolia* 'Meiwa' kumquat	>40	S,R	Deng et al. (1992)
C. sinensis 'Valencia' (embryogenic suspension) + 'Carrizo' citrange	7	R,S	X.X. Deng, J.W. Grosser F.G. Gmitter (unpublished)
(c) *Interspecific plants produced from sexually compatible parents*			
C. aurantifolia 'Key' lime (embryogenic suspension) + *C. sinensis* 'Valencia' (leaf)	>40	R,S	Grosser et al. (1989)
C. sinensis 'Washington' navel (embryogenic suspension) + *C. unshiu* 'Hayashi' Satsuma mandarin (leaf)	1	S	Kobayashi et al. (1988)
C. sinensis 'Washington' navel (embryogenic suspension) + *C. reticulata* × *C. sinensis* (probable) 'Murcott' tangor (leaf)	?	S	Kobayashi and Ohgawara (1988)

Parents and protoplast source	Number of plants	Cultivar improvement potential	References
C. sinensis '*Trovita*' (*embryogenic suspension*) + *C. unshiu* 'Hayashi' Satsuma mandarin (leaf)	?	S	Kobayashi and Ohgawara (1988)
C. sinensis 'Valencia' (embryogenic suspension) + *C. limon* 'Femminello' (leaf)	25	S,R	Tusa et al. (1990)
C. sinensis 'Bahia' navel (embryogenic suspension) + *C. paradisi* 'Marsh' (leaf)	<10	S	Ohgawara et al. (1989)
C. paradisi 'Thompson Pink' (embryogenic suspension) + *C. reticulata* × *C. sinensis* (probable) 'Murcott' tangor (leaf)	19	S	J.W. Grosser and F.G. Gmitter (unpublished)
C. sinensis 'Hamlin' (embryogenic callus) + *C. reticulata* 'Dancy' (leaf)	1	S	J.W. Grosser and F.G. Gmitter (unpublished)
C. reticulata 'Cleopatra' (embryogenic callus) + *C. jambhiri* 'Volkameriana' (leaf)	8	R	J.W. Grosser and F.G. Gmitter (unpublished)
C. sinensis 'Hamlin' (embryogenic callus) + and *C. limon* 'Femminello' (leaf)	2	S,R	N. Tusa, J.W. Grosser F.G. Gmitter (unpublished)
C. jambhiri 'Milam' (embryogenic callus) + *C. limon* 'Femminello' (leaf)	1	S,R	N. Tusa, J.W. Grosser and F.G. Gmitter (unpublished)
C. sinensis 'Hamlin' (embryogenic callus) + *C. jambhiri* 'Rough lemon' (leaf)	11	R	X.X. Deng, J.W. Grosser and F.G. Gmitter (unpublished)
C. sinensis 'Hamlin' (embryogenic suspension) + *C. jambhiri* 'Rough lemon' (leaf)	12	R	J.W. Grosser and F.G. Gmitter (unpublished)
C. reticulata 'Cleopatra' (embryogenic callus) + *C. aurantium* 'Sour orange' (leaf)	>170	R	J.W. Grosser and F.G. Gmitter (unpublished)
C. reticulata 'Cleopatra' (embryogenic callus) + *C. jambhiri* 'Rough lemon' (leaf)	5	R	J.W. Grosser and F.G. Gmitter (unpublished)
C. sinensis 'Hamlin' (embryogenic callus) + *C. limonia* Osbeck 'Rangpur' (leaf)	>200	R	E.S. Louzada, J.W. Grosser and F.G. Gmitter (unpublished)
C. reticulata 'Cleopatra' (embryogenic callus) + *C. limonia* 'Rangpur' (leaf)	>250	R	E.S. Louzada, J.W. Grosser, and F.G. Gmitter (unpublished)
C. aurantium 'Sour orange' (embryogenic callus) + *C. jambhiri* 'Volkameriana' (leaf)	>40	R	E.S. Louzada, J.W. Grosser and F.G. Gmitter (unpublished)
C. aurantium (hybrid) 'Australian Sour orange' (embryogenic callus) + *C. jambhiri* 'Rough lemon' (leaf)	3	R	J.W. Grosser and F.G. Gmitter (unpublished)
(*C. paradisi* × *C. reticulata*) × *C. reticulata* 'Page' tangelo (embryogenic callus) + *C. sinensis* 'Succari' (leaf)	2	S	J.W. Grosser and F.G. Gmitter (unpublished)

Table 3. Unique *Citrus* cybrid plants produced by donor-recipient protoplast fusion (Vardi et al., 1987, 1989)

Donor parent	Recipient parent
'Poorman' citrus hybrid	Villafranca lemon (*C. limon*)
'Poorman' citrus hybrid	Sour orange (*C. aurantium*)
Sour orange	Villafranca lemon
Microcitrus	Rough lemon (*C. jambhiri*)
Microcitrus	Sour orange

10. Transformation

Genetic transformation of important crop species is an attractive method for adding valuable specific traits coded by one or a few genes to elite genotypes, without altering the appearance or quality of the harvestable fruit. With highly heterozygous woody fruit tree species, this is often impossible to achieve by conventional breeding methods. At present, there are only a few examples of transgenic plants that contain foreign genes contributing to crop improvement, including (see Chapters 10 and 12): transgenic tobacco plants containing bacterial toxin genes that confer insect resistance (Vaeck et al., 1987); transgenic tobacco, tomato, and potato plants containing viral coat protein genes conferring virus resistance (reviewed by Beachy et al., 1987; Grumet, 1990); and transgenic tomato plants containing an antisense polygalacturonase gene that modifies tomato ripening (Smith et al., 1988). It is expected that many more successful examples will be reported as more important single genes are identified, isolated, and cloned.

Transformation of tropical fruit species has proven to be more difficult than the herbaceous annual Solanaceous species that serve as model systems. The few examples of successful transformation in tropical fruit species are limited to *Citrus* and *Carica* (papaya), and have thus far been only for demonstration, as no genes contributing directly to crop improvement have been transferred. Pang and Sanford (1988) first reported successful *Agrobacterium* mediated transformation of papaya leaf discs. Carbenecillin at 500 μg/ml was used to kill the *Agrobacterium* following co-cultivation, and transformed leaf disc calli were selected on 1/2 strength MS medium containing 300 μg/ml kanamycin. Confirmation of transformation was based on an assay for nopaline production. No plants were regenerated and southern blot analysis was not reported. Transgenic *Citrus* callus via direct DNA uptake induced by 40% PEG (polyethylene glycol) has been reported (Kobayashi and Uchimaya, 1989). A chimeric gene containing a nopaline synthase promoter, the aminoglycoside phosphotransferase II structural gene from the bacterial transposon Tn5, and the terminator region from cauliflower mosaic virus, was transferred to 'Trovita' sweet orange (*Citrus sinensis*) protoplasts isolated from an embryogenic suspension culture maintained on MT basal medium containing 10 mg/l BA. Selection for transformants was carried out on MT medium containing 25 μg/ml kanamycin. Transformation efficiency was low (1×10^{-6}), and no plants were regenerated. Transformation was confirmed by Southern blot analysis and protein assay.

Recently, Vardi et al. (1990) reported the regeneration of two transgenic rough lemon (*Citrus jambhiri* Lush) plants containing foreign genes coding for neomycin phosphotransferase II (NPT II) and chloramphenicol acetyltransferase. Transformation was achieved by PEG induced direct DNA uptake by protoplasts isolated from a habituated embryogenic callus line. Calf-thymus DNA was used as a carrier to increase transformation efficiency. Stable transformants were selected on medium containing 0–40 μg/ml paromomycin sulfate (PAR) instead of kanamycin. The previously reported feeder layer technique was used for protoplast culture (Vardi et al., 1989). Transformation was confirmed by protein assay and Southern blot analysis.

Hidaka et al. (1990) reported the regeneration of one transgenic 'Washington' navel orange (*Citrus sinensis*) plant. Transformation was achieved by co-cultivation of embryogenic suspension cells of 'Washington' navel orange with two modified strains of *Agrobacterium*. Carbenecillin at 500 μg/ml was used to inhibit the *Agrobacterium*, and a two-step selection of transformants was carried out on medium containing either 100 μg/ml kanamycin or 20 μg/ml hygromycin followed by transfer to medium containing double-strength antibiotic. Embryoid induction was achieved on MS medium containing 0.1 M galactose and 0.1 M sorbitol. Embryo germination was achieved on the same medium supplemented with 1 μm GA_3. Confirmation of transformation was by Southern blot analysis. These researchers also reported production of transgenic 'Trovita' sweet orange callus (*Citrus sinensis*) using the same method, but subsequent plant regeneration was not achieved.

G.A. Moore (personal communication) has regenerated at least two transgenic plants of the commercially important citrus rootstock Carrizo citrange. Transformation was achieved by co-cultivation of seedling internodal stem segments with a disarmed strain of *Agrobacterium* containing a vector plasmid coding for Tn5 neomycin phosphotransferase (NPT II) and β-glucuronidase (GUS). Mefoxin (cefoxitin) at 200 μg/ml was used to inhibit the *Agrobacterium*. Kanamycin at 100 μg/ml was used to select transformed tissue. Plants were regenerated via organogenesis on MT basal medium containing 5 mg/ml BA. Confirmation of transformation was by histochemical assay for GUS, polymerase chain reaction (PCR) analysis, and Southern blot analysis.

Fitch et al. (1990) reported the production of three transgenic papaya (*Carica papaya*) isolates via microprojectible bombardment. Immature zygotic embryos and derived somatic embryos were bombarded with tungsten particles carrying chimeric NPT II and GUS genes. Embryogenic tissues were cultured on half-strength MS medium containing 10 mg/l 2,4-D, 400 mg/l glutamine, and 6% sucrose. Selection of transformants and recovery of transgenic somatic embryos and leafy shoots were achieved on 2,4-D-free medium containing 150 mg/l kanamycin sulfate. Confirmation of transformation was by histochemical analysis of GUS activity and protein assay for NPT II activity. Rooted transgenic plants in soil were not reported. The target of this research is genetically engineered coat protein mediated papaya ringspot virus resistance in commercial papaya.

It is apparent from these examples that the use of antibiotics for transformant selection and *Agrobacterium* inhibition following co-cultivation is feasible, but seems to interfere with subsequent plant regeneration via both the somatic embryogenesis and organogenesis pathways. Continued research should lead to improved methods that lessen the impact of antibiotics on the recovery of healthy normal rooted transgenic plants.

Methods of transformation that do not utilize antibiotics are also being investigated. Schell et al. (personal communication) carried out an experiment completely eliminating the use of antibiotics for the *in vitro* selection of transformants. Transformation was attempted by 40% PEG induced direct DNA uptake by protoplasts isolated from a habituated embryogenic 'Hamlin' sweet orange (*Citrus sinensis*) suspension culture. Using no selection at the cellular level, 700 healthy complete plants were regenerated via somatic embryogenesis as described by Grosser and Gmitter (1990a). Of these, 12 tested positive by histochemical analysis of GUS activity and 3 of the 12 tested positive by PCR analysis. Southern blot analysis of DNA isolated from leaves of the putative transformants is in progress.

It is evident from this discussion that transformation in tropical fruit species will not become routine in the near future. In addition to the obstacles already mentioned, efficient plant regeneration systems for many elite tropical fruit genotypes have yet to be achieved. Moreover, once transgenic plants are obtained, extensive field evaluation must be conducted to demonstrate the following: adequate expression of the foreign gene in the appropriate tissue at the appropriate time; cultivar integrity; and assurance that no other traits have been negatively altered (e.g. yield). Expansion of research effort and resources committed to transformation research in tropical fruit species should continue, but not at the expense of cultivar improvement research that utilizes other promising conventional and emerging methodologies.

11. Conclusions and Future Prospects

During the past decade, great progress has been made regarding the development of reliable *in vitro* plant regeneration systems for many important tropical fruit species. As a result, emerging biotechnologies (e.g. somatic hybridization, transformation etc.) can now be adapted and employed in cultivar improvement strategies, and future prospects appear to be excellent, particularly for citrus. Citrus somatic hybrids have potential for direct use as improved rootstocks with better resistance to disease, cold and other abiotic stresses. *In vitro* facilitated production of improved triploid seedless fresh fruit citrus varieties with expanded maturity dates should increase international fresh fruit marketing potential. Continued development of transformation systems for tropical fruit species should lead to genetically engineered coat protein mediated virus resistance in the near future. In conclusion, the application of *in vitro* techniques to tropical fruit improvement will play an important role in improving cultivar desirability and production efficiency, and thereby contribute substantially to the increasing economic significance of tropical fruit culture worldwide.

12. References

Amin, M.N. and Jaiswal, V.S. (1987). Rapid clonal propagation of guava through *in vitro* shoot proliferation and nodal explants of mature trees. Plant Cell Tissue Org. Cult. 9:235–243.

Arrias, O. and Huete, F. (1983). Propagation vegetativa *in vitro* de pejibaye (*Bactris gasipaes* H.B.K.). Turrialba 33:103–108.

Barbieri, C. and Morini, S. (1987). Plant regeneration from *Actinidia* callus cultures. J. Am. Soc. Hort. Sci. 62:107–109.

Barlass, M. and Skene, K.G.M. (1982). *In vitro* plantlet formation from *Citrus* species and hybrids. Sci. Hort. 17:333–341.

Beachy, R.N., Rogers, S.G. and Fraley, R.T. (1987). Genetic transformation to confer resistance to plant virus disease. Genetic Eng. 9:229–247.

Ben-Hayyim, G. and Neumann, H. (1983). Stimulatory effect of glycerol on growth and somatic embryogenesis in citrus callus cultures. Z. Pflanzenphysiol. 110:331–338.

Berg, L.A. and Bustamante, M. (1974). Heat treatment and meristem culture for the production of virus free bananas. Phytopathology 64:320–322.

Bouzid, S. (1975). Quelques traits du comportement de boutures de *Citrus* en culture *in vitro*. Comptes Rendus de l'Academie des Sciences Serie D 280:1689–1692.

Chaturvedi, H.C. and Mitra, G.C. (1974). Clonal propagation of *Citrus* from somatic callus cultures. Hort. Sc. 9:188–120.

Chen, Z. (1990). *Citrus*: Anther Culture, Ch. 25. In: Handbook of Plant Cell Culture, Vol. 6. Z. Chen, D.A. Evans, W.R. Sharp, P.V. Ammirato and M.R. Sondahl (eds.), Perennial Crops. McGraw-Hill, New York, pp. 365–375.

Deng, X.X., Grosser, J.W. and Gmitter, F.G., Jr. (1992). Intergeneric somatic hybrid plants from protoplast fusion of *Fortunella crassifolia* 'Meiwa' with *Citrus sinensis* 'Valencia'. Sci. Hort. 49:55–62.

DeWald, S.G., Litz, R.E. and Moore, G.A. (1989a). Optimizing somatic embryo production in mango. J. Am. Soc. Hort. Sci. 114:712–716.

DeWald, S.G., Litz, R.E. and Moore, G.A. (1989b). Maturation and germination of mango somatic embryos. J. Am. Soc. Hort. Sci. 114:837–841.

Fahn, A. (1974). Plant Anatomy. Pergamon Press, New York.

Fitch, M.M., Manshardt, R.M., Gonsalves, D., Slighton, J.L. and Sanford, J.C. (1990). Stable transformation of papaya via microprojectile bombardment. Plant Cell Rep. 9:189–194.

Fu, L. (1990). Litchi: Anther Culture, Ch. 16. In: Handbook of Plant Cell Culture. Vol. 6. Z. Chen, D.A. Evans, W.R. Sharp, P.V. Ammirato and M.R. Sondahl (eds.), Perennial Crops. McGraw-Hill, New York, pp. 376–384.

Gmitter, F.G., Jr. and Ling, X. (1991). Embryogenesis *in vitro* and nonchimeric tetraploid plant recovery from undeveloped *Citrus* ovules treated with colchicine. J. Am. Soc. Hort. Sci. 116:317–321.

Gmitter, F.G., Jr., Ling, X.B. and Deng, X.X. (1990). Induction of triploid *Citrus* plants from endosperm calli *in vitro*. Theor. Appl. Genet. 80:785–790.

Gmitter, F.G., Jr., Ling, X., Cai, C. and Grosser, J.W. (1991). Colchicine-induced polyploidy in *Citrus* embryogenic cultures, somatic embryos, and regenerated plantlets. Plant Sci. 74:135–141.

Gmitter, F.G., Jr. and Moore, G.A. (1986). Plant regeneration from undeveloped ovules and embryogenic calli of *Citrus*: Embryo production, germination, and plant survival. Plant Cell Tissue Org. Cult. 6:139–147.

Goh, H.K.L., Rao, A.N. and Loh, C.S. (1988). *In vitro* plantlet formation in mangosteen (*Garcinia mangostana* L.). Ann. Bot. 62:87–93.

Grinblat, H. (1972). Differentiation of *Citrus* stem *in vitro*. J. Am. Soc. Hort. Sci. 97:599–603.

Grosser, J.W. (1990). Citrus rootstock improvement by cell fusion. *Citrus* and Veg. Mag., September, pp. 28, 31, 32.

Grosser, J.W. and Chandler, J.L. (1986). *In vitro* multiplication of Swingle citrumelo rootstock with coumarin. Hort. Sc. 21:518–520.

Grosser, J.W. and Chandler, J.L. (1987). Aseptic isolation of leaf protoplasts from *Citrus*, *Poncirus*, *Citrus* × *Poncirus* hybrids and Severinia for use in somatic hybridization experiments. Sci. Hort. 31:253–257.

Grosser, J.W. and Gmitter F.G., Jr. (1990a). Protoplast fusion in *Citrus* improvement. Plant Breed. Rev. 8:339–374.

Grosser, J.W. and Gmitter, F.G., Jr. (1990b). Wide hybridization of *Citrus* via protoplast fusion: Progress, strategies and limitations. In: Horticultural Biotechnology, Wiley-Liss, New York, pp. 31–41.

Grosser, J.W. and Gmitter, F.G., Jr. (1990c). Somatic hybridization of *Citrus* with wild relatives for germplasm enhancement and cultivar development. Hort. Sc. 25:147–151.

Grosser, J.W., Gmitter, F.G., Jr. and Chandler, J.L. (1988a). Intergeneric somatic hybrid plants of *Citrus sinensis* cv. Hamlin and *Poncirus trifoliata* cv. Flying Dragon. Plant Cell Rep. 7:5–8.

Grosser, J.W., Gmitter, F.G., Jr. and Chandler, J.L. (1988b). Intergeneric somatic hybrid plants from sexually incompatible woody species: *Citrus sinensis* and *Severinia disticha*. Theor. Appl. Genet. 75:397–401.

Grosser, J.W., Gmitter, F.G., Jr., Tusa, N. and Chandler, J.L. (1990). Somatic hybrid plants from sexually incompatible woody species: *Citrus reticulata* and *Citropsis gilletiana*. Plant Cell Rep. 8:656–659.

Grosser, J.W., Moore, G.A. and Gmitter, F.G., Jr. (1989). Interspecific somatic hybrid plants from the fusion of 'Key' lime (*Citrus aurantifolia*) with 'Valencia' sweet orange (*Citrus sinensis*) protoplasts. Sci. Hort. 39:23–29.

Grumet, R. (1990). Genetically engineered plant virus resistance. Hort. Sc. 25:508–513.

Gupta, P.P. (1986). Eradication of mosaic disease and rapid clonal multiplication of bananas and plantains through meristem tip culture. Plant Cell Tissue Org. Cult. 6:33–39.

Gupta, P.K., Kenkurkar, S.V., Kulkarni, V M., Shirgurkar, M.V. and Mascarenhas, A.F. (1984). Somatic embryogenesis and plants from zygotic embryos of coconut (*Cocos nucifera* L.) *in vitro*. Plant Cell Rep. 3:222–225.

Hendrix, R.C., Litz, R.E. and Kirchoff, B.G. (1987). *In vitro* organogenesis and plant regeneration from leaves of *Solanum candidum* Lindl., *S. quitoense* Lam. (naranjilla) and *S. sessiflorum Dunal*. Plant Cell Tissue Org. Cult. 11:67–73.

Hidaka, T. and Kajiura, I. (1988) Plantlet differentiation from callus protoplasts induced from *Citrus* embryo. Sci. Hort. 34:85–92.

Hidaka, T., Omura, M., Ugaki, M., Tomiyama, M., Kato, A., Oshima, M. and Motoyoshi, F. (1990). *Agrobacterium*-mediated transformation and regeneration of *Citrus* spp. from suspension cells. Japan. J. Breed. 4:199–207.

Jaiswal, V.S. and Narayan, P. (1985). Regeneration of plantlets from the callus of stem segments of adult plants of *Ficus religiosa* L. Plant Cell Rep. 4:256–258.

Jordan, M. (1988). Multiple shoot formation and rhizogenesis from cherimola (*Annona cherimola* L.) hypocotyls and petiole explants. Gartenbauwissenschaften 53:234–237.

Karunaratne, S. and Periyapperuma, K. (1989). Culture of immature embryos of coconut, *Cocos nucifera* L., callus proliferation and somatic embryos. Plant Sci. 62:247–253.

Kobayashi, S. and Ohgawara, T. (1988). Production of somatic hybrid plants through protoplast fusion in *Citrus*. J. Agric. Rev. Quarterly 22:181–188.

Kobayashi, S., Ohgawara, T., Ohgawara, E., Oiyama, I. and Ishii, S. (1988). A somatic hybrid plant obtained by protoplast fusion between navel orange (*Citrus sinensis*) and satsuma mandarin. Plant Cell Tissue Org. Cult. 14:63–69.

Kobayashi, S., Oiyama, I., Yoshinaga, K., Ohgawara, T., and Ishii, S. (1991). Fertility in an intergeneric somatic hybrid plant of Rutaceae. Hort. Sc. 26:207.

Kobayashi, S. and Uchimaya, H. (1989). Expression and integration of a foreign gene in orange (*Citrus sinensis* Osb.) protoplasts by direct DNA transfer. Japan. J. Genet. 64:91–97.

Kobayashi, S., Uchimaya, H. and Ikeda, I. (1983) Plant regeneration from 'Trovita' orange protoplasts. Japan. J. Breed. 33:119–122.

Kochba, J., Spiegel-Roy, P., Neumann, H. and Saad, S. (1982). Effect of carbohydrates on somatic embryogenesis in subcultured nucellar callus of *Citrus* cultivars. Z. Pflanzenphysiol. 105:359–368.

Kochba, J., Spiegel-Roy, P. and Safran, H. (1972). Adventive plants from ovules and nucelli in citrus. Planta 106:237–245.

Krug, C.A. and Bacchi, O. (1943). Triploid varieties of *Citrus*. J. Heredity 34:277–283.

Litz, R.E. (1984a). *In vitro* somatic embryogenesis from callus of jaboticaba, *Myrciaria cauliflora*. Hort. Sc. 190:62–64.

Litz, R.E. (1984b). *In vitro* responses of adventitious embryos of two polyembryonic Eugenia species. Hort. Sc. 19:720–722.

Litz, R.E. (1985). Somatic embryogenesis in tropical fruit trees. In: Tissue Culture in Forestry and Agriculture. R.R. Henke, K.W. Hughes, M.J. Constantin and A. Hollaender (eds.), Plenum, New York, pp. 179–193.

Litz, R.E. (1987). Application of tissue culture to tropical fruits. In: Plant Tissue and Cell Culture. C.E. Green, D.A. Somers, W.A. Hackett and D.D. Biesboer (eds.), Alan R. Liss, New York, pp. 407–418.

Litz, R.E. (1988). Somatic embryogenesis from cultured leaf explants of the tropical tree *Euphoria longan* Stend. J. Plant Physiol. 132:190–193.

Litz, R.E. and Conover, R.A. (1980). Partial organogenesis in tissue cultures of *Averrhoa carambola*. Hort. Sc. 15: 735.

Litz, R.E. and Conover, R.A. (1982). *In vitro* somatic embryogenesis and plant regeneration from *Carica papaya* L. ovular callus. Plant Sci. Lett. 26:153–158.

Litz, R.E. and Conover, R.A. (1983). High frequency somatic embryogenesis from *Carica* suspension cultures. Ann. Bot. 51:683–686.

Litz, R.E. and Griffis, J.L. (1989). Carambola (*Averrhoa carambola* L.). In: Biotechnology in Agriculture and Forestry, Vol. 5. Y.P.S. Bajaj (ed.), Trees II. Springer Verlag, Berlin, pp. 59–67.

Litz, R.E. and Gray, D.J. (1992). Organogenesis and somatic embryognesis. In: Biotechnology of Perennial Fruit Crops, Chapter 1. F. Hammerschlag and R. Litz (eds.), pp. 3–34. CAB International, Wallingford, Oxon, U.K.

Litz, R.E. and Jaiswal, V.S. (1991). Micropropagation of tropical and subtropical fruits. In: Micropropagation. P.C. Debergh and R.H. Zimmerman (eds.), Kluwer Academic Publishers, Dordrecht, pp. 247–263.

Litz, R.E., Knight, R.J. and Gazit, S. (1984). *In vitro* somatic embryogenesis from *Mangifera indica* L. callus. Sci. Hort. 22:233–240.

Litz, R.E., Moore, G.A. and Srinivasan, C. (1985). *In vitro* systems for propagation and improvement of tropical fruits and palms. Hort. Rev. 7:157–200.

Litz, R.E., O'Hair, S. K. and Conover, R.A. (1983). *In vitro* growth of *Carica papaya* L. cotyledons. Sci. Hort. 19:287–293.

Ma, S.S. and Shii, C.T. (1974). Growing banana plantlets from adventitious buds. J. Chinese Soc. Hort. Sci. 20:6–12 (in Chinese).

Maheshwari, P. and Rangaswamy, N.S. (1958). Polyembryony and *in vitro* cultures of embryos of *Citrus* and *Mangifera*. Indian J. Hort. 15:272–282.

Mapes, M.O. (1974). Tissue culture of *Bromeliada*. Int. Plant Prop. Comb. Soc. 23:47–55.

Menczel, L., Nagy, F., Kiss, Z. and Maliga, P. (1981). Streptomycin resistant and sensitive hybrids of *Nicotiana tabacum* and *Nicotiana knightiana*: Correlation of resistance with *N. tabacum* plastids. Theor. Appl. Genet. 59:191–195.

Mitra, G.C. and Chaturvedi, H.C. (1972). Embryoids and complete plants from unpollinated ovaries and from ovules of *in vitro*-grown emasculated flower buds of *Citrus* spp. Bull. Torrey Bot. Club 99:184–189.

Moore, G.A. (1985). Factors affecting *in vitro* embryogenesis from undeveloped ovules of mature *Citrus* fruit. J. Am. Soc. Hort. Sci. 110:60–70.

Murashige, T. and Skoog, F. (1962). A revised medium for rapid growth and bioassays with tobacco tissue cultures. Physiol. Plant. 15:473–497.

Murashige, T. and Tucker, D.P.H. (1969). Growth factor requirements of citrus tissue culture. Proc. First Int. Citrus Symp. 3:1155–1161.

Nair, S., Gupta, P. K., Shirgurkas, M. V. and Mascarenhas, A. F. (1984). *In vitro* organogenesis from leaf explants of *Annona squamosa* Linn. Plant Cell Tissue Org. Cult. 3:29–40.

Navarro, L., Jaurez, J., Pina, J.A., Ballester, J.F. and Arregui, J.M. (1988). The citrus variety improvement program in Spain after eleven years. In: L.W. Timmer, S.M. Garnsey and L. Navarro (eds.), Proceedings 10th Conf. Intl. Organ. Cirus Virol, pp. 400–406.

Navarro, L., Ortiz, J.M. and Juarez, J. (1985). Aberrant citrus plants obtained by somatic embryogenesis of nucelli cultured *in vitro*. Hort.Sc. 20:214–215.

Navarro, L., Roistacher, C.N. and Murashige, T. (1975). Improvement of shoot-tip grafting *in vitro* for virus-free citrus. J. Am. Soc. Hort. 100:471–479.

Novak, F.J., Afza, R., Van Duren, M., Perea Dallos, M., Conger, B.V. and Xiaolang, T. (1989). Somatic embryogenesis and plant regeneration in suspension cultures of dessert (AA and AAA) and cooking (ABB) bananas (*Musa* spp.). Biotechnology 7:154–159.

Ohgawara, T., Kobayashi, S., Ishii, S., Yoshinaga, K. and Oiyama, I. (1989). Somatic hybridization in *Citrus*: Navel orange (*C. sinensis* Osb.) and grapefruit (*C. paradisi* Macf.). Theor. Appl. Genet. 71:1–4.

Ohgawara, T., Kobayashi, S., Ishii, S., Yoshinaga, K. and Oiyama, I. (1991). Fertile fruit trees obtained by somatic hybridization: Navel orange (*Citrus sinensis*) + Troyer citrange (*C. sinensis* × *Poncirus trifoliata*). Theor. Appl. Genet. 81:141–143.

Ohgawara, T., Kobayashi, S., Ohgawara, E., Uchimaya, H. and Ishii, S. (1985). Somatic hybrid plants obtained by protoplast fusion between *Citrus sinensis* and *Poncirus trifoliata*. Theor. Appl. Genet. 71:1–4.

Ohyama, K. and Oka, S. (1982). Multiple shoot formation from mulberry (*Morus alba* L.) hypocotyl by N-(2-chloro-4-pyridyl)-N′-phenylurea. 5th Intl. Cong. Plant Tissue and Cell Culture, Tokyo, pp. 149–150.

Oiyama, I., Kobayashi, S., Yoshinaga, K., Ishii, S. and Ohgawara, T. (1991). Use of pollen from a somatic hybrid between *Citrus* and *Poncirus* in the production of triploids. Hort. Sc. 26: 1082.

Oka, S. and Ohyama, K. (1981). *In vitro* initiation of adventitious buds and its modification by high concentrations of benzyladenine in leaf tissues of mulberry (*Morus alba*). Can. J. Bot. 59:68–74.

Oliveira, M.M. and Pais, M.S. (1991). Plant regeneration from protoplasts by long-term calius cultures of *Actinidia deliciosa* var. *deliciosa* cv. Hayword (Kiwifruit). Plant Cell Rep. 9; 643–646.

Omura, M., Matsuta, N., Moriguchi, T. and Kozaki, I. (1987). Adventitious shoot and plantlet formation from cultured pomegranate leaf explants. Hort. Sc. 22:133–134.

Pais, M.S., Oliveira, M.M. and Barroso, J. (1987). Use of petiole segments of *Actinidia chinensis* (Kiwi) for plant differentiation and production of friable calli for protoplast isolation. Acta Hort. 212:687–690.

Pang, S.C. and Sanford, J.C. (1988). *Agrobacterium* mediated gene transfer in papaya. J. Am. Soc. Hort. Sci. 113:287–291.

Philip, V.J. (1984). *In vitro* organogenesis and plantlet formation in cashew (*Anacardium occidentale* L.). Ann. Bot. 54:149–152.

Raj Bhansali, R. and Arya, A.C. (1978a). Shoot formation in stem and root callus of *Citrus aurantifolia* (Christm.) Swingle grown in culture. Current Sci. 47:775–776.

Raj Bhansali, R. and Arya, A.C. (1978b). Differentiation in explants of *Citrus paradisi* Macf. (grapefruit) grown in culture. Indian J. Expt. Biol. 16:409–411.

Rangan, T.S., Murashige, T. and Bitters, W.P. (1968). *In vitro* initiation of nucellar embryos in monoembryonic *Citrus*. Hort. Sc. 3:226–227.

Revilla, M.A. and Power, J.B. (1988). Morphogenetic potential of long-term callus cultures of *Actinidia deliciosa*. J. Hort. Sci. 63:541–545.

Reynolds, J.F. and Murashige, T. (1979). Asexual embryogenesis in callus cultures of palms. *In vitro* 15:383–387.

Sabharwal, P.S. (1963). *In vitro* culture of ovules, nucelli and embryos of *Citrus reticulata* Blanco var. Nagpuri. In: Plant Tissue and Organ Culture – A Symposium. P. Maheshwari and N.S. Rangaswamy (eds.), University of Delhi, Delhi, pp. 265–274.

Sauton, A., Mouras, A. and Lutz, A. (1982). Plant regeneration from citrus root meristems. J. Am. Soc. Hort. Sci. 57:227–231.

Scorza, R. and Janick, J. (1976). Tissue culture in passiflora. Proc. Trop. Region Am. Soc. Hort. Sci. 20:179–183.

Sim, G.E., Loh, C.S. and Goh, C.J. (1988). Direct somatic embryogenesis from protoplasts of *Citrus mitis* Blanco. Plant Cell Rep. 7:418–420.

Smith, C.J.S., Watson, C.F., Ray, J., Bird, C.R., Morris, P.C., Schuch, W. and Grierson, D. (1988). Antisense RNA inhibition of polygalacturonase gene expression in transgenic tomatoes. Nature 334:724–726.

Soost, R.K. and Cameron, J.W. (1980). 'Oroblanco', a triploid pummelo-grapefruit hybrid. Hort. Sc. 15:667–669.

Soost, R.K. and Cameron, J.W. (1985). 'Melogold', a triploid pummelo- grapefruit hybrid. Hort. Sc. 20:1134–1135.

Starantino, A. and Recupero, G.R. (1981). Citrus hybrids obtained *in vitro* from 2× females and 4× males. Proc. Int. Soc. Citriculture, pp. 31–32.

Surga Rivas, J.G. (1988). Obtencion de plantas libres del virus mosaico del pepino por cultivo de apices meristematicos aislados *in vitro* de dos cultivares de banano. Fitopatologia Venezolana 1:69–72.

Swingle, W.T. and Reece, P.C. (1967). The botany of citrus and its wild relatives. In: The Citrus Industry, Vol. 1. W. Reuther, L.D. Batchelor and H.J. Ueber (eds.), University of California Press, Berkeley, pp. 190–430.

Sykes, S.R. (1988). Overview of the family Rutaceae. In: R.R. Walker (ed.), Proceedings of the Breeding Workshop, Citrus Breeding in Australia and the Use of New Breeding Technologies. CSIRO, Melbourne, pp. 93–100.

Tao, R., Murayama, H., Moriguchi, K. and Sugiura, A. (1988). Plant regeneration from callus cultures derived from primordial leaves of adult Japanese persimmon. Hort. Sc. 23:1055–1056.

Timmer, L. W. (1988). Blight. In: Compendium of Citrus Diseases. J.O. Whiteside, S.M. Garnsey and L.W. Timmer (eds.), APS Press, St. Paul, pp. 66–67.

Tsai, C.K. (1988). Plant regeneration from leaf callus protoplasts of *Actinidia chinensis* Planch. var. *chinensis*. Plant Sci. 54:231–235.

Tusa, N., Grosser, J.W. and Gmitter, F.G., Jr. (1990). Plant regeneration of 'Valencia' sweet orange,'Femminello' lemon, and the interspecific somatic hybrid following protoplast fusion. J. Am. Soc. Hort. Sci. 115:1043–1046.

Vaeck, M., Reynaerts, A., Hofte, H., Jansens, S., DeBeuckeleer, M., Dean, C., Zabeau, M., Van Montagu, M. and Leemans, J. (1987). Transgenic plants protected from insect attack. Nature 328:33–37.

Vardi, A., Arzee-Gonen, P., Frydman-Shani, A., Bleichman, S. and Galun, E. (1989). Protoplast fusion mediated transfer of organelles from *Microcitrus* and *Citrus* and regeneration of novel alloplasmic trees. Theor. Appl. Genet. 78:741–747.

Vardi, A., Bleichman, S. and Aviv, D. (1990). Genetic transformation of citrus protoplasts and regeneration of transgenic plants. Plant Sci. 69:199–206.

Vardi, A., Breiman, A., and Galun, E. (1987). Citrus cybrids: Production by donor-recipient protoplast-fusion and verification by mitochondrial-DNA restriction profiles. Theor. Appl. Genet. 75:51–58.

Vardi, A. and Galun, E. (1988). Recent advances in protoplast culture of horticultural crops: *Citrus*. Sci. Hort. 37:217–230.

Vardi, A., Hutchison, D.J. and Galun, E. (1986). A protoplast to tree system in *Microcitrus* based on protoplasts derived from a sustained embryogenic callus. Plant Cell Rep. 5:412–414.

Vardi, A., Spiegel-Roy, P. and Galun, E. (1982). Plant regeneration from *Citrus* protoplasts: Variability in methodological requirements among cultivars and species. Theor. Appl. Genetics 62:171–176.

Wakasa, K. (1979. Variation in plants differentiated from tissue culture of pineapple. Japan. J. Breed. 29:13–22.

Wang, D. and Chang, C.J. (1978). Triploid citrus plantlet from endosperm culture (in Chinese). Sci. Sinica 21:822–827.

Wei, W. (1990). Longan: Anther and embryo culture, Ch. 27. In: Handbook of Plant Cell Culture. Vol. 6. Z. Chen, D.A. Evans, W.R. Sharp, P.V. Ammirato and M.R. Sondahl (eds), Perennial Crops. McGraw-Hill, New York, pp. 385–396.

Yie, S. and Liaw, S.I. (1977). Plant regeneration from shoot tips and callus of papaya. *In vitro* 13:564–567.

Yokoyama, T. and Takeuchi, M. (1976). Organ and plantlet formation from callus in Japanese persimmon (*Diospyros kaki*). Phytomorphology 26:273–275.

20. *In vitro* Culture of Plantation Crops

Abraham D. Krikorian

Contents

1. Introduction

A 'plantation' has traditionally been thought of as a large estate in a tropical or subtropical region that is generally cultivated or tended by semi-skilled or unskilled labor under the general direction of a 'manager'. Anyone who has seen large plantation operations immediately recognizes that the field workers are far from 'unskilled'. In some places, the term 'estate' crop is used synonymously with the term 'plantation' crop. Since both terms are steeped in history, and are often flavored with connotations of colonialism or unsavory exploitation (Brockway, 1979), it is perhaps better and more accurate nowadays to think of plantation crops as those wherein a very large or relatively large group of plants, usually but not always perennials, are maintained under cultivation using highly standardized or strictly controlled practices.

Indra K. Vasil and Trevor A. Thorpe (eds.), Plant Cell and Tissue Culture, 497–537.

The species grown may range from those that provide edible fruits (bananas and plantains, dates, citrus, papaya, mango, etc.) or other consumable products such as the oil from fruits of coconut or African oil palm, all the way to frequently-thought-of-as-exotic comestible commodities such as various spices, condiments and flavorings (e.g. ginger, turmeric, black pepper, cardamom, vanilla,) or the beverages coffee and tea, or, industrial commodities like rubber and the quinine-yielding *Cinchona* bark, or fibers like Manila and sisal hemp.

Even this kind of definition minimizes the artificial nature of the distinction, however, since the range of plant types grown is ever-expanding and evolving. For instance, various and 'new' forest or multi-purpose tree species are seen as providing unusual opportunities for expanding the industrial base beyond the traditional teak and eucalyptus 'plantations' (cf., e.g. Krikorian and Kann, 1987 for the Chinese tallow tree, *Sapium sebiferum*).

In short, one cannot cover in the limited space available here, in very much depth, the full range of plants that could qualify as plantation crops. As interest in biomass production increases, it is certain that plantation efforts will expand (cf. Corley, 1983). For instance, there is still interest in improving biomass production in sugarcane (which is perhaps the largest and oldest of the plantation crops and for which a considerable tissue and cell culture information base exists – cf. Maretzki, 1987). Also cotton, historically viewed as a 'plantation crop' especially in the southern USA, is not covered even though there is much *in vitro* work published on it (cf. Price and Smith, 1984; Finer et al., 1987).

From time to time considerable concern has been voiced that monocultures, i.e. plantings comprised of identical genotypes, are merely inviting disaster. The view that monocultures are 'disasters waiting to happen' is, to me, however, much exaggerated, even unjustified, and certainly very peripheral to our perspective of clonal multiplication and improvement via cell and tissue culture and related techniques. Indeed, very large, (and essentially monoclonal) plantations are managed very successfully in many parts of the world.

In fact, aseptic culture procedures offer potential to increase plantation management capability, not reduce it. Since ecotypes or élite genotypes which can be tested or eventually used in specific microenvironments can be generated in quantity, or even have potential to be generated directly or indirectly through *in vitro* and genetic engineering procedures, this should allow one to have fuller access to very specifically mixed clonal combinations of superior genotypes. Further, *in vitro* germplasm or gene bank storage capabilities will permit access *at will* to many that present seed storage problems.

It will be obvious, however, that special precautions should be taken in the propagation of disease-free or specific pathogen-free or virus-indexed or even microorganism-free materials for use in the various plantation industries. Although there may not be practical ways once in the field to prevent or control re-infection of a clean propagule, the initial superior performance of such a plant will generally more than off-set the effort exerted to generate it in the first place.

But this is far from a simple procedure in some plants and we are limited by our lack of knowledge in many instances. From the perspective of plants which look healthy but may in reality be infected with latent pathogenic viruses, this is a particularly onerous task. Figure 1 provides a summary scheme for generation of virus-indexed plants. The size of an initial explant is significant since careless use of unnecessarily large stem tip explants in clonal multiplication procedures for orchids and other plants (see Mori, 1971 for minimum sizes of some meristems needed to achieve virus elimination) has resulted in unwitting dissemination of virus-infected plants.

For example, thirty years ago when seeds were exclusively used for rearing orchids, many of the orchid virus diseases now rampant were practically unknown. Had apical meristem procedures been adopted, and combined as needed with thermotherapy, these problems would more than likely not have been encountered since the apical dome is frequently virus-free (Wang, 1990; see also Chapter 3). Dissemination of cucumber mosaic (CMV) and banana bunchy top virus (BBTV) infection has been a problem in some cases with bananas which have been casually multiplied clonally. Indeed, in the latter instance, the use of meristem culture, even with thermotherapy, is insufficient to eliminate virus (Krikorian, 1990). Happily, use of ELISA (enzyme linked immunosorbent assay) procedures to detect various known virus infections is a possibility. Indeed, we can even detect viruses that we don't know the identity of using doubled-stranded RNA techniques (Jordan, 1986).

In addition to the problem of viruses, the more usual problem of bacterial or fungal contamination of propagules needs to be addressed in all *in vitro*-generated plants but perhaps especially for those being multiplied for plantation use.

2. Objectives and Strategies

Specific objectives and the role of *in vitro* techniques will depend on the crop, of course, and it will be recognized that large, well-established plantation industries are, understandably, extremely conservative in their practices. The [US] National Research Council (1990) and Sasson and Costarini (1990) are but two examples of efforts made by international agencies to set some priorities for developing appropriate biotechnologies, especially as they affect the less economically developed countries. The specific recommendations in these sets of priorities might well be argued, but the fact remains that there is a keen interest in evaluating the status of the modern technologies (including plant tissue and cell culture) so that advantage may be taken of new developments (cf., e.g. Rao, 1982; Srivastava et al., 1986; Bhaskaran

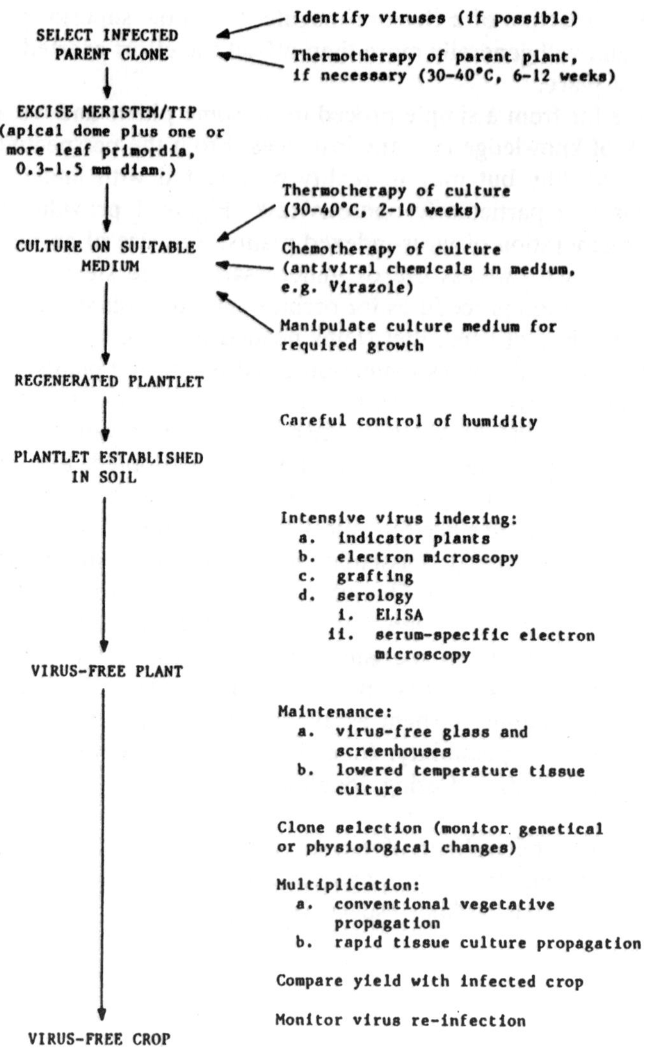

Figure 1. Schematic representation of the sequence of events involved in the production of a virus-'free' crop. Note that this is merely a guide which emphasizes the complexity of the operation (cf. Krikorian and Rao, 1990).

and Prabhudesai, 1986, 1989; Smith and Drew, 1990; Esan, 1991). Some plants have been favored by the establishment of networks that permit excellent communication regarding tissue culture technologies that are particularly relevant to them – e.g. bamboo (cf. Krikorian and Rao, 1990); bananas and plantains (Vuylsteke, 1989; Jarret, 1990; Krikorian et al., 1990).

While rapid multiplication and distribution of specific pathogen-free or disease tolerant material for field evaluation and research is commonly a high priority, as is the development of the potential for germplasm preservation and conservation, a major and continuing effort should be the development of the full range of *in vitro* procedures so that, as need eventually arises for their use, they will be in place. All these technologies are necessarily long range. For example, artificial seed via somatic embryos will probably not be a practical means for the introduction of materials into the field for some time (Redenbaugh, 1990) but in the long run, the ability to regenerate plants from free cells and protoplasts, is probably one of the most rational instruments of research and routes for improvement through genetic engineering (Krikorian, 1982; Grumet, 1990; Grierson, 1991).

3. Some Important Points in Development of Protocols

3.1. *Background Information*

It should be emphasized that a very helpful primary step in determining which *in vitro* strategies might be adopted for a given plant is to familiarize oneself with the overall biology, genetics and normal mode(s) of multiplication that are used to propagate and manage the species in question. Unfortunately, many plantation crops, despite their having been exploited for many years, generally lack an adequate scientific information base. That which is available is often out-of-date (cf. Garner et al., 1976) or accessible only with some difficulty, (e.g. Bavappa et al., 1982; Shama Bhat and Radhakrishnan Nair, 1985 for the betel nut palm, *Areca catechu*). Some crops do have a reasonably good data base that has been conveniently assembled and interpreted (e.g. banana – Stover and Simmonds, 1987; sugarcane – Heinz, 1987; coffee – Clarke and Macrae, 1988). One wishes that more species were covered in such exemplary fashion. Alvim and Kozlowski (1977) is somewhat dated but that volume contains much valuable information on a range of plantation crops including pineapple, sugarcane, coffee, cacao, rubber, tea, oil palm, coconut palm, citrus, banana, cashew and mango. Simmonds (1991) provides a lucid summary of the genetics of horizontal resistance to diseases in crops that should be read by all seeking to improve crops.

Many encyclopedic handbooks are currently available which contain chapters that summarize *in vitro* procedures for a great many plants and, in some cases provide detailed protocols. Additional and slightly different perspectives on how to approach a given tissue culture problem may be gained from information contained in Packer and Douce (1987), Krikorian (1988), Krikorian and Smith (1992) and Pollard and Walker (1990).

In my view, the further one can get away from the idea of an inflexible protocol that should be followed, the better. This is especially true in the early part of a project wherein one is seeking to render a given plant controllably responsive to *in vitro* culture. Plants are highly 'plastic' systems,

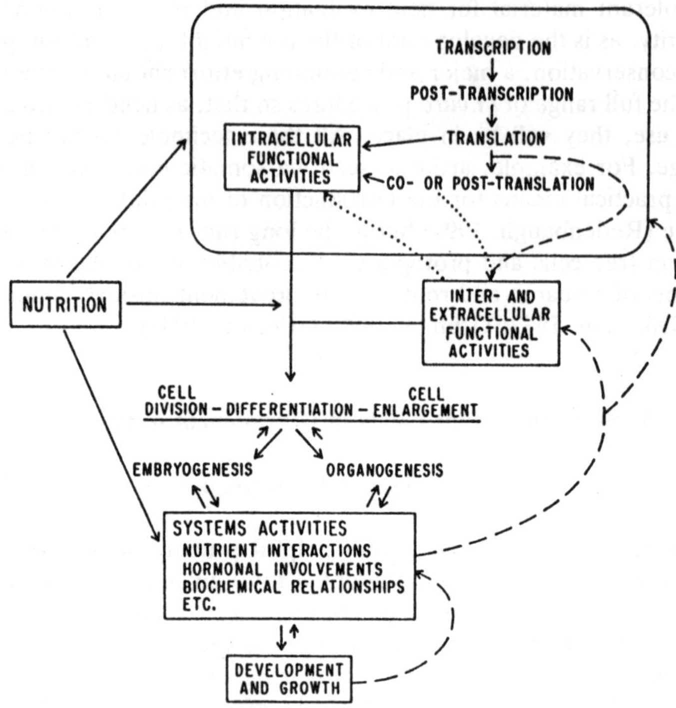

Figure 2. Some relationships between nutrition, and exogenous and endogenous controls in the establishment, maintenance and control of organs, tissues, cells and protoplasts *in vitro*. See text for details.

especially when they are freed of correlational or positional controls (cf. Barlow and Carr, 1984). Figure 2 provides a skeleton scheme of some of the relationships between nutrition and various intra-, inter- and extra-cellular activities as they relate to cell division, differentiation and enlargement. Embryogenesis and organogenesis, and their sensitivity relative to nutrition and various other system activities, are key responses to special signals (or an alteration or breakdown in signal processing). These, in turn, derive directly or indirectly from what occurs at a given place at a particular time (so-called spatial and temporal responses). The high degree of communication between intra- and extra-cellular functional activities is represented by the dotted lines. The major paths for feedback and regulation are suggested by dashed lines. Even a cursory examination of this scheme emphasizes that there are many points where complex interactions may occur. A major objective is to evaluate how well defined these are and controlled they may be rendered.

3.2. *Maintenance and Monitoring of Asepsis*

Special mention need not be made here of the absolute requirement to achieve and maintain asepsis in the *in vitro* process. What has long been appreciated by experienced workers but is in desperate need of re-statement is that maintenance of asepsis is not a trivial matter and one must be ever-vigilant to ensure that good laboratory procedures are in place and followed (cf. Knauss, 1976). This is perhaps no more important in tropical environments than in temperate ones, but it is certainly more difficult to put into practice in environments that are hot and humid and hence conducive to microbial growth.

Investigators who have not initiated their own cultures and thereby have not been able to exercise quality control over the generation and maintenance of their materials from primary explants are urged to at least test acquired cultures for cryptic bacterial and/or fungal infestation before they continue working with them.

This may be conveniently carried out by coarsely slicing tissues and incubating them at 28–30 °C in or on something like Difco Bacto Nutrient Broth (with or without a gelling agent like Gelrite–Kelco Commercial Development Division of Merck and Co., San Diego, CA) and on BBL Mycophil Agar (BBL Microbiology Systems, Becton Dickinson and Co., Cockeysville, MD) or its equivalent like a potato dextrose agar (PDA) or Sabouraud's medium. (Note: 10 gm each of phytone and dextrose in a liter of water will produce the equivalent of Mycophil broth.) Cryptic contaminants will generally be disclosed anywhere from 1–5 days after inoculation depending on the degree of contamination. It is important to test for both bacteria and fungi using an appropriate medium for the growth of each since in some cases contaminants are fastidious as to their growth requirements (Leifert and Waites, 1990). At Stony Brook we generally do not attempt to re-disinfest contaminated primary cultures unless rarity of the material makes it essential to attempt to do so. The use of antibiotics, while helpful or even essential in some molecular biological procedures, is fraught with its own set of special problems. Reference may be made, nevertheless, to the use of fungicides (Shields et al., 1984; Hauptmann et al., 1985; Haldeman et al., 1987) or antibiotics against bacteria (Falkner, 1990) in cultures.

It will be surprising to some that dipping culture instruments into 95% ethanol and flaming over an alcohol lamp for periods long enough to cause severe discomfort to the fingers of most is generally insufficient to ensure sterilization. Exposure to newer tools such as an electric incinerator (e.g. Bacti-Cinerator III®) for 20 sec or longer is necessary, for instance, in the case of bananas to prevent introduction of microbial contaminants (cf. also e.g. Singha et al., 1987 for efforts to incinerate tools contaminated with *Bacillus* species).

3.3. *Use of an Appropriate Gelling Agent*

It has long been recognized that while agar generally provides an excellent means of rendering a nutrient medium semi-solid, it has also been appreciated that there is, due to its biological origin and nature of production, considerable variation in its quality. This includes the level and nature of the number of 'contaminants' (both inorganic and organic) associated with it (cf. Marshall and Orr, 1954). All these variables, which incidentally have been very poorly characterized, can have significance for the success of the tissue culture process (cf., e.g. Debergh, 1983). The phenomenon of vitrification or hyperhydricity, or the condition in which tissues become or are glassy or excessively watery, may be controllable by increase in agar concentration (Kevers et al., 1984; Williams and Leopold, 1989).

Washing of agar used to be routine in most fastidious plant tissue culture laboratories and the practice hardly needed emphasis. As time has passed and it has become apparent to me that it does need to be re-stated, I have taken efforts to include a simple but effective protocol for washing of agar in a publication dealing with carrot somatic embryos (Smith and Krikorian, 1989).

In many instances, it does not matter what 'grade' of agar is used; in other cases it is pivotal to success. It is interesting to note that this fact is being re-discovered. Because many plantation crops are rich in phenolics etc., and tend to darken in culture, it is important to 'shock' them as little as possible. Also, because tropical crops may be difficult to disinfest, and hence may require extra-ordinary disinfestation procedures, it is well to consider as clean and 'stress-free' (e.g. 'salt-free') a substrate as possible. 'Ordinary' food or pharmaceutical grade agar (once official in the United States Pharmacopoeia, the USP, but now relegated to the National Formulary, NF) is perfectly adequate for most plant tissue culture uses and has the advantage of always being available and being much cheaper than laboratory grades. It should be washed, however.

3.4. *Gelrite®*

Gelrite, Phytagar and Gel-Gro® are but a few trade names of examples of naturally- (usually bacterially-) derived gelling polymers that are increasingly becoming popular as gelling substrates for the *in vitro* culture of plant cells and tissues. A particularly useful feature is that it forms a clear gel. A general rule of thumb is that these gellans are generally used at about a quarter concentration as that of agar (e.g. 2 gm of Gelrite/l of solution). Its gelling properties are different from those of agar and it must be recalled that one cannot get a gelled medium using distilled water; salts must be present (Baird et al., 1983).

3.5. *Some Other Alternatives to Agar*

Having made the point that substrates can have major effects on growth, reference may be made to alginate (cf., e.g. Adaoha Mbanasao and Roscoe, 1982), various other cellulose crystallite substrates (George et al., 1983), agarose (Lörz et al., 1983; Adams and Townsend, 1983), and specialized starches (Sorvari and Schieder, 1987; Henderson and Kinnersley, 1988).

3.6. *Use of Activated Charcoal*

Just as in the case of agar, there has been the idea that various brands of activated charcoal are in and of themselves far better than others for use in plant tissue culture. While this may well be true since there are a terrific number of parameters that may vary (cf., e.g. ASTM Standards, 1985), one is constantly faced with the possibility that suppliers may change their source. Key manufacturers whose major sales involve activated charcoal largely for purposes of decolorization etc. to such industry giants as sugar refiners have had little incentive to worry about what plant tissue culturists require.

Activated charcoal has been used for many years as an adsorbent of presumably toxic substances in the growing of plants. Epiphytic orchid growers have long relied on charcoal chips as an additive to their potting media or substrates (Yam et al., 1989). Proskauer and Berman (1970) and Werckmeister (1971) used activated charcoal in aseptic culture media for orchids. Activated charcoal can provide a darkened environment for the growth of roots but since it is 'activated', it does things other than merely darkening the medium. For instance, adsorption of growth regulators being supplied to the tissue by activated charcoal must be taken into account, and this in turn, is dependent not only on the kind of charcoal and degree of activation, but the amount added. Amounts much lower than 0.02% w/v have not been of much use to us at Stony Brook and levels more in the range of 0.05 to 1% have been more helpful. One may, however, have to add 100 times more growth regulator (especially adenyl cytokinins) in the presence of activated charcoal but this needs to be empirically determined (Krikorian and Kann, 1987; Krikorian et al., 1987; Arditti and Ernst, 1993).

Use of activated charcoal can make a major difference in the success or failure of a given culture attempt (cf., e.g. Krikorian and Kann, 1987; Krikorian, 1988 and refs. there cited). Activated charcoal does adsorb growth regulators but it also adsorbs substances presumed to be deleterious, like phenolics, oxidized phenolics, cell lysis products etc. Volatiles like ethylene and methane, likewise thought by some to be injurious to some tissue culture efforts, can also be adsorbed. Too little is known about controlled release of growth regulators or adsorption of metabolic byproducts to make generalizations (Loomis, 1974; Krikorian et al., 1987).

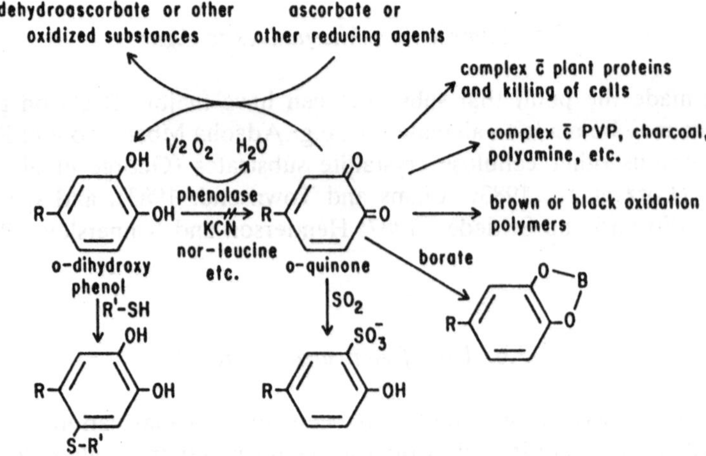

Figure 3. Scheme of some relationships between phenolics, their oxidation, and the potential for intervention in the process with various reducing and adsorbing agents (Krikorian, 1988).

3.7. *Anti-oxidants and Anti-darkening Agents*

Phenolics frequently (but unjustifiably cf. Clemetson and Andersen, 1966) have the connotation of being inhibitory substances that should be avoided or eliminated from *in vitro* environments. Primary explants, especially, of tropical species particularly but by no means exclusively woody ones, are especially likely to darken and it has more or less been routinely assumed that this darkening is due to polyphenol oxidase activity (cf., e.g. Uritani, 1978; McClure, 1979; Vaughn and Duke, 1984). Because of this, many agents have been used to counter the darkening. Some examples are: ascorbic acid (order of 10^{-1} M), sodium hydrosulfite (10^{-3} M), cysteine (10^{-1} M), diethyldithiocarbamate – DTT (also known as Cleland's Reagent) (10^{-3} M), potassium ethylxanthate (10^{-3} M), thiourea (0.1–0.2%), benzimidazole (10^{-3} M), sodium bisulfite (10^{-3} M), polyvinylpyrrolidone (PVP) soluble, say PVP 10 – average molecular weight 10,000 (1%), Polyclar AT (0.7%). To this list may be added citric acid, glutathione, bovine serum albumin etc.

One can easily find references in the literature that can substantiate or refute the nominal value of each of these. One notable example is that control of oxidative browning in *Strelitzia reginae* with 0.04% filter-sterilized DTT was pivotal to successful culture according to Ziv and Halevy (1983). Figure 3 provides a summary scheme that gives some relationships between phenolics, their oxidation to yield products at various levels of oxidation and the places where use of various reducing and adsorbing agents might play a role as countermeasures.

4. Some Specific Crops

Against this general background, attention may now be drawn to some species that have been placed in aseptic culture with the view of using it as an instrument of improvement.

4.1. *Pineapple* (*Ananas comosus*)

Pineapple (2n = 50) is a perennial, herbaceous monocotyledon which had its origins in the neotropics (Smith and Downs, 1979). The plant is normally self-incompatible and vegetatively propagated except when the objective is breeding new cultivars. While the juicy and delicious parthenocarpically developed fruit (a syncarp which develops without prior pollination) is the main product of importance (Py et al., 1984), some cultivars and closely related species are grown for the fiber which may be obtained from their leaves (cf. Leal and Amaya, 1991 for the curagua, *Ananas lucidus*) or for extraction of the proteolytic enzyme bromelain. Also, some varieties, especially variegated and chimeric ones are of considerable interest as ornamentals.

Breeding is a slow process and an F_1 plant may take some 4 years to produce fruit. It is said that in an interspecific cross, 4 back-cross generations are generally required to produce a commercial variety. All this adds up to some 25 years to develop an introduction, and that still needs to be multiplied prior to distribution (Bartholomew and Kadzimin, 1977; Madhusudanan et al., 1982).

Micropropagation for rapid increase of new varieties for early release has been seen as a major advantage of aseptic culture but it has also been recognized that these are merely extensions of the more traditional methods (see Wee, 1982). Multiplication has been mainly from axillary buds (see Krauss, 1948a, b, c for a thorough rundown of the gross morphology and anatomy of pineapple). Methods have been variations on the now-familiar theme (cf. Corlett and Caldis, 1972; Mapes, 1973; Ting and Chua, 1975; Mathews and Rao, 1980; Rangan, 1984; Cásale and García, 1987; De Wald et al., 1988 and refs. there cited). See Fig. 4a and b for an indication of the prolific multiplication that is possible.

A major problem is to get clean explants. Surface-disinfestation of stems stripped of their leaves has been achieved with a 20% v/v dilution of commercial laundry bleach for 20 minutes prior to bud excision. Or, a succession of increasingly dilute exposures to hypochlorite after leaves have been peeled off the shoot end can be tried, e.g. 10% v/v dilution, followed by 5%, 1% and then sterile distilled water.

A ratio of 4:1 auxin to cytokinin ratio seems to be helpful in breaking dormancy and then a reversal of the ratios seems helpful for bud proliferation. Pulsing the explant from liquid to semi-solid can have its benefits until

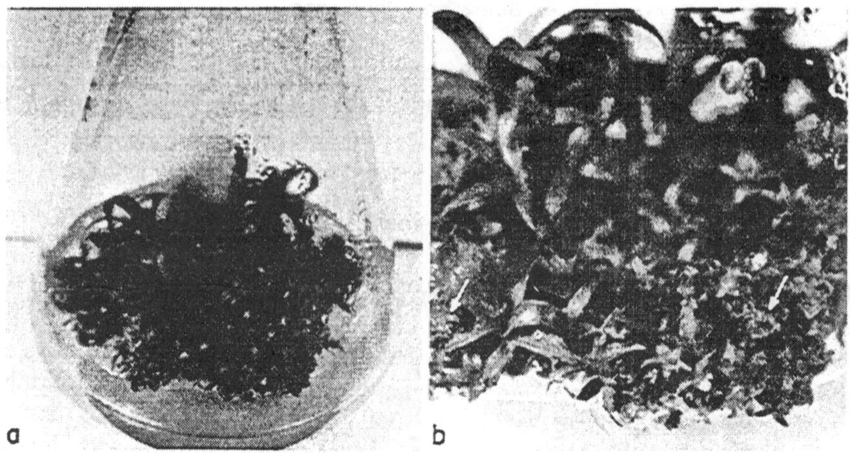

Figure 4. Proliferation in a pineapple culture. (a) overall appearance of a culture showing masses of tightly clustered shoots after 60 days growth; (b) close-up of a similar culture showing callus (arrows), nodular calloid growths, and shoots. Medium comprised of M and S mineral salts, vitamins and sucrose supplemented with 30 mg/l adenosine. Photograph provided by the late Marion O. Mapes.

multiplication is established. Callus can, and usually does form at the basal portion of excised buds (cf. Aghion and Beauchesne, 1960; Lakshmi Sita et al., 1974; Teo, 1974; Pannetier and Lanaud, 1976; Mathews and Rangan, 1979). Mathews et al. (1976) describe the production of shoots from proliferating callus. Similarly, Srinivasa Rao et al. (1981) induced regenerable callus from the radicles of excised embryos of a pineapple hybrid. Normally dormant axillary buds from the crown provided excellent primary explants in the work of Mathews and Rangan (1981), and even leaf explants (3rd–5th leaf from the apex) were shown to form callus from which plantlets could be recovered. *In vitro*-obtained shoots are also good starting materials for callus induction (Mathews and Rangan, 1981). Interestingly, 2,4-D was not a good growth regulator in their hands but NAA and coconut water (CW) were useful. Also, these investigators showed that albino plantlets could be produced in vitro. Wang and Huang (1978) obtained polyploids by exposure of young leaves which had regenerable callus to colchicine.

Ananas erectifolius, a wild pineapple, can be induced to form somatic embryos. Exposure of buds to kinetin 5 mg/l and 2,4-D 0.2 mg/l in a Schenk and Hildebrandt medium (as a liquid – 15 ml in a 50 ml erlenmeyer flask) can get things off to a good start. p-Chlorophenoxyacetic acid can take the place of 2,4-D. A callus will form in a few weeks at the cut end of the explant. This is then transferred to an MS medium (semi-solid with adenosine at 30 mg/l). Alternatively, MS liquid medium with BAP 5 mg/l and kinetin

5 mg/l will permit this callus to proliferate and grow. When transferred to half-strength MS with NAA 2.5 mg/l or 0.25 mg/l (or even 2,4-D 0.2 mg/l), somatic embryos will develop (Mapes, 1974 and unpublished personal communication).

Zepeda and Sagawa (1981) have estimated that 5,000 or so plants can be produced from an explant from a single crown in a year. They used a liquid MS medium supplemented with 25% v/v CW in the light, transferred every 2 weeks to half-strength MS with the same level of CW, and within a couple of months after initial explanting, got solitary shoots with several leaves. These, in turn, were subcultured to BAP-containing medium; this forced axillary shoots which could be rooted in growth regulator-free medium. Because of the implications of proper selection of material for the plantation and because of the length of time from *in vitro* culture to the field, there is considerable interest in developing and evaluating efficient substrates for hardening off and preparing *in vitro*-generated plantlets for the field (Folliot and Marchal, 1990).

Wakasa et al. (1978) generated large numbers of plantlets via protocorm-like structures (so-called globular bodies) from a number of morphological sites and noted that there was considerable variation in the material. At the field level, materials derived from syncarps and slips showed the most variation; crowns and axillary bud-derived cultures yield the lowest numbers of off-types (Wakasa, 1979). Changes in leaf color and morphology, waxiness, spininess, presence of albinism etc., were among the features noted.

Mutations are well known among the pineapples (Furtado, 1926/29; Collins and Kerns, 1938; Loison-Cabot, 1990). It seems inadvisable, therefore, to refer indiscriminately to tissue culture-associated changes as due to somaclonal variation. Smith and Drew (1990) relate that 5% off-types have been acceptable to the Australian growers. Roguing has long been an integral part of pineapple cultivation procedures (Wee, 1982).

4.2. *Palms – African Oil Palm* (*Elaeis guineensis*), *Date Palm* (*Phoenix dactylifera*) *and Coconut Palm* (*Cocos nucifera*)

Krikorian (1989) has given an overview of most of the relevant literature as it applies to palm tissue culture. Reference may also be made to Tisserat (1984) as it applies to date palm, Pannetier and Buffard-Morel (1986) as it applies to coconut, and Paranjothy (1984, 1987b) and Cui and Gong (1990) for oil palm.

4.2.1. *African Oil Palm*

The African oil palm (2n = 32) is an arborescent monocotyledon which is very important to the world economy. Although its cultivation is restricted to an area 10° N and S of the Equator, it is ranked in importance second only to soybean as an oilseed crop. There has been considerable interest in development of reliable and commercially viable techniques for the rapid and large scale multiplication of élite tenera palms (Paranjothy, 1984; Duval et al., 1988).

Elaeis guineensis occurs in nature in three fruit forms or varieties. One of these, *dura*, with a homozygous dominant character for the presence of a shell, is characterized by the absence of a fiber ring and the presence of a thick shell. A homozygous recessive character for shell presence, the *pisifera*, has no shell but a fiber ring around the seed. A hybrid of the *dura* and *pisifera* forms, the *tenera*, has a thin shell and fiber ring around the seed. The tenera gives the best yield of palm oil and a relatively good supply of palm kernel oil. It is, therefore, generally cultivated as the oil palm of intensive plantation agriculture. Unfortunately, since it is a hybrid, it does not breed true. It segregates on selfing and gives all three fruit forms in a 1:2:1 ratio, indicating incomplete dominance of dura over the *pisifera*. *As a result of the segregation occurring on selfing commercial oil palm, the cross has to be repeated each time with dura* and *pisifera*. The desirability of achieving large numbers of clonal plantlets in a plant that does not produce axillary buds which are vegetative, at once becomes apparent. (See Analytical Bibliography, 1987 for an extensive listing of oil palm tissue culture work).

Early enthusiasm over progress made in cloning oil palm from somatic embryos was somewhat diminished due to observations that there may be aberrations of normal floral development on both male and female inflorescences in nominally clonal palms (cf. Corley et al., 1986; Paranjothy, 1987b). The first tissue culture-derived clones tested by Unilever and its affiliates in the field showed 'bunch failure due to an inadequate number of fertilized fruits to sustain development of the bunch to ripeness, and absence of pollen' (Paranjothy, 1987b). The reasons for this problem are not clear but the fact that French workers have experienced far less serious problems (cf. Ahee et al., 1981; Noiret et al., 1985) suggests that inappropriate tissue culture procedures may have played a role. At least this is the assumption. Results from Ivory Coast indicate that some 3.1% of the plants have floral problems which make them unproductive but it is emphasized that this is at the same level of plants obtained from seed (Durand-Gasselin, 1990). (See also Krikorian and Kann, 1986 for citations to the literature on a number of off-types of oil palm that may be encountered in nature).

The significant improvement in production has established the theoretically predicted value of cloning oil palm via tissue culture (Meunier et al., 1988). It remains to be proven whether off-types can be minimized by changes in tissue culture protocols. Clearly, clone performance in the field is being examined very carefully, far more carefully than has ever been done before (cf. Krikorian, 1982, 1989a and b for a discussion of historical inability in many cases to examine clone performance because sufficient numbers have never been available in the first place). As more clones produced via *in vitro* techniques are evaluated, better quality control procedures are sure to emerge (cf. Tahardi, 1989 for an example of research work on oil palm in Indonesia).

A more immediately achievable goal than establishing highly cost-effective, problem-free cloning from élite palms that can be accomplished is the generation of clones of *pisiferas* for use as male parents in *tenera* production. *Pisiferas* are frequently male-sterile and their seeds often have a much reduced level of germination. Because their pollen is preferable for commercial *tenera* production, the poor viability of their seed limits breeding. Any means of increasing availability of those *pisifera* palms which have limited fertility, or those which were sterile or infertile and have reverted to a fertile or semi-fertile condition, either from their embryos, seedlings or mature trees would be of value in breeding. Cryopreservation shows itself to be a possibility as well for the maintenance of germ plasm banks (Engelmann et al., 1988).

Nwankwo and Krikorian (1983) showed that over 50% germination could be obtained from *pisifera* seeds merely by aseptically storing seeds, removed from fruits, in unaerated water. Since the testa of *pisifera* is thin and the fruit susceptible to microbial rot, simple aseptic storage was sufficient to 'rescue' germ plasm that would otherwise have been lost.

Mature seeds readily serve as a good model for explants from fully grown and tested palms (cf. Krikorian and Kann, 1986). Moreover, it is a good idea to have some familiarity with the behavior of the tissue before attempting to obtain primary explants from soil grown materials. Callus can be induced readily by addition of NAA or 2,4-D to a 'half-strength' basal medium such as that of Linsmaier and Skoog to which $NaH_2PO_3 \cdot H_2O$ 170 mg/l has been added. Adenine sulfate (ca. 40 mg/l), casein hydrolysate (enzymatic ca 500 mg/l) and inositol (100 mg/l) as additional supplements can be helpful. An important additive is activated charcoal (order of 500 mg/l). The callus-forming response to growth regulator is much more uniform than one might ordinarily expect, and although callus growth is not rapid, ample amounts can be generated for further manipulation. The broad range of 10–60 mg/l growth hormone, provided activated charcoal is in the medium, is sufficient to yield embryogenic callus which is regenerable into plantlets.

Embryogenic callus can also be initiated and maintained in the absence of activated charcoal; however, callus is more likely to darken. Also, the callus is more nodular and 'swollen' and tends to yield plantlets that are not as discrete and 'neat' (cf. Fig. 5). Rooting can occur in liquid (cf. Krikorian and Kann, 1986) but a clean agar semi-solid substrate can lead to a vigorous root system as well (cf. Fig. 5. These figures have been selected to emphasize that there is much room for variations on a theme and for improvement in any of the many so-called protocols that have been published (cf. Duval et al., 1988).

Indeed, it has not been established whether non-embryogenic callus can be induced controllably to develop via indirect organogenesis. Such histological examination as has been done shows that in the presence of exogenously

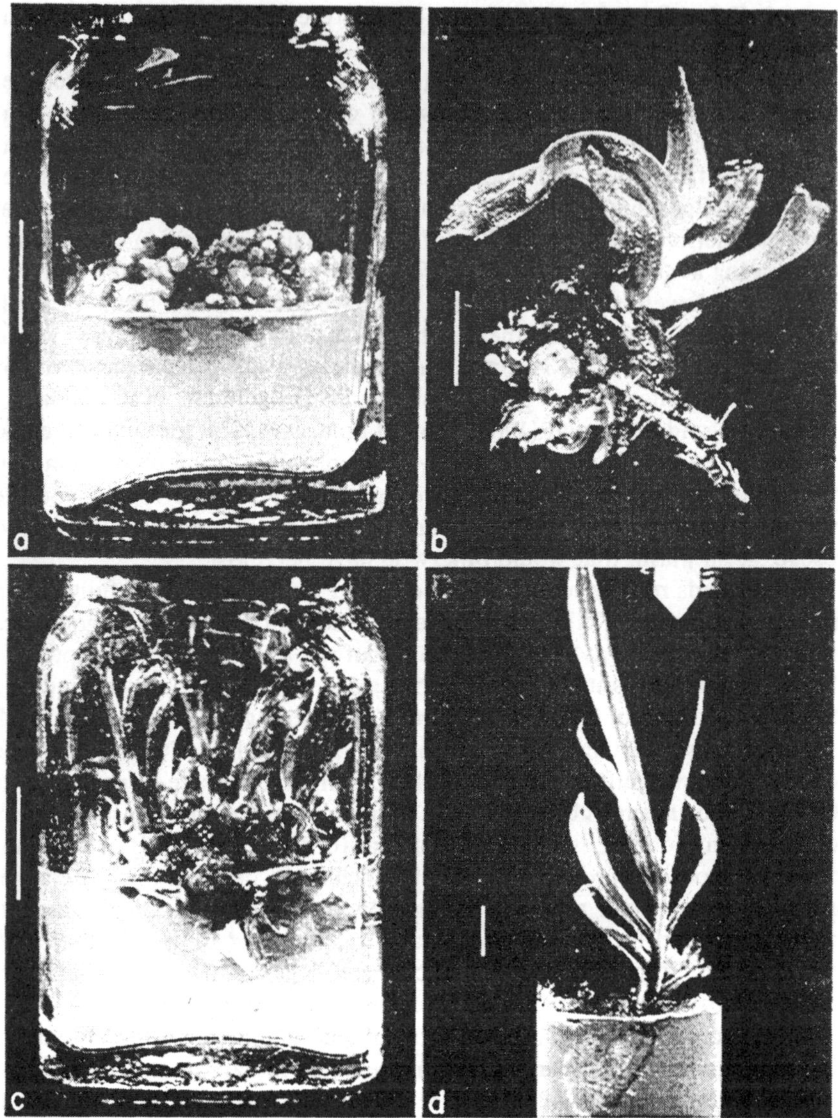

Figure 5. From callus to plantlet in *tenera* oil palm. (a) nodular, embryogenic callus grown on semi-solid medium (0.7% agar). Note that some nodules may be organized and show prominent meristematic dome-like structures. This particular culture was initiated and maintained on M and S (half-strength), 1000 mg/l casein hydrolysate and 5 mg/l NAA; (b) shoot and roots developed from a subculture of callus as grown at 'a' transferred to a half-strength M and S with kinetin (0.2 mg/l) and NAA (0.5 mg/l). Since this complex did not regenerate from a single somatic embryo, it is somewhat 'messy'; (c) shoots developed from a brownish callus mass. The plantlets were initiated on M and S (half-strength) with kinetin (0.05 mg/l) and NAA (0.1 mg/l, and then allowed to continue their development on M and S with 5 mg/l NAA; (d) plantlet in a culture tube on half-strength M and S ready for transfer to soil. Scale bars, 1 cm.

Figure 6. Root development in *tenera* oil palm plantlets. (a) jar full of shoots with minimal root development ready for subdivision and transfer to culture tubes for further root development. The medium consisted of half-strength M and S; (b) plantlet removed from culture medium and washed free of agar prior to planting in soil. Scale bar, 1 cm.

added hormone, cells within or in close proximity to the provascular tissue yield numerous meristematic centers, which in turn, give rise to somatic embryonal structures (cf. Nwankwo and Krikorian, 1983; Schwendiman et al., 1988, 1990). Bipolar structures can be removed, exposed for a short time to a basal medium containing low levels (ca. 1 mg/l) of GA_3, and reared to full size within a few weeks (cf. Krikorian and Kann, 1986 for a detailed summary of the methods published up until that time). Non-chlorophyllous immature leaf explants and root explants from more mature and élite palms

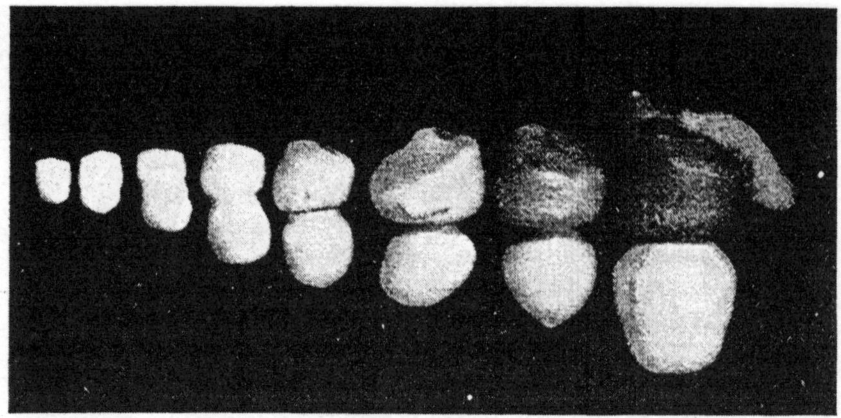

Figure 7. Zygotic embryos of coconut removed from the seed at different stages of germination. The youngest is at the far left. The cotyledonary end of the embryo is pointing upwards in this photograph; the haustorial end is downwards. Note the 'constriction' that is increasing apparent as the embryos germinates.

are equally responsive (cf. Wooi, 1984; Brackpool et al., 1986; Schwendiman et al., 1988; Tahardi, 1989).

4.2.2. *Coconut Palm*

Like the oil palm, coconut (2n = 32) does not sucker and hence must be reproduced from seed. They are essentially outbreeders and show heterosis on crossing. Breeding improved palms has become a major objective and much-improved and disease tolerant germ plasm has been developed, although it is usually in short supply (cf. Krikorian, 1989a).

Kovoor (1981) first summarized tissue culture work on coconut. Since then, appreciation of the potential has been high but relatively few groups have worked upon it. De Guzman and her colleagues (cf. Krikorian, 1989a for refs.) carried out embryo rescue on the Makapuno mutant which will not germinate (Samonte et al., 1989), and thus showed that a relatively simple procedure as embryo culture could be immensely important. The procedure involves first splitting the seed with an appropriate knife (such as a bolo or machete) so as to expose the contents. Using a large diameter cork borer, the solid endosperm in which the embryo is embedded is cored. The material so removed protects the embryo until it can be surface sterilized with something like calcium hypochlorite at 5% w/v and the final removal of the embryo can be achieved (cf. Fig. 7 for examples of normal coconut embryos removed at different stages of fruit growth and germination). White's medium with 4% dextrose in liquid for about 8 weeks, followed by MS medium with 1% activated charcoal and 4% dextrose for about 8 weeks and finally,

MS with 8% sucrose and 1% activated charcoal for another 8 weeks has been useful in rescuing Makapuno. Tahardi and Warga-Dalem (1982) essentially repeated the work done in the Philippines using the Indonesian equivalent of Makapuno, Kopyor. In semi-solid medium, MS medium supplemented with 10 mg/l IAA, 0.05 mg/l IBA, 6% dextrose and 0.5% activated charcoal fostered balanced root growth of the Kopyor seedlings.

Zygotic embryo culture has been used to collect coconut germ plasm in the field. MS medium with 2% sucrose and 2 g/l activated charcoal, kept in darkness until sprouting is observed has given a good success level. Assy-Bah et al. (1988) provide a detailed protocol with illustrations of the several steps. The process of plantlet development is slow, however (order of 5–6 months before well developed leaves are formed). Auxin helps root development. Moreover, there seems to be relationship between age of embryo and the culturability (cf. also Assy-Bah, 1986). Zygotic embryos have been recovered from a number of genotypes, and taken through various steps to the field level (cf. Assy-Bah et al., 1989). Ashburner et al. (1991) have shown that culture of embryos is best in liquid at the outset, but shoot and root growth performance is better on semi-solid medium if the period in liquid is restricted to 4 to 8 weeks. They recommend that the correct time for switching from liquid to semisolid medium may be gauged at 4–8 weeks after initial explanting for that is when visible germination has taken place.

Callus growth from coconut has been generally regarded as recalcitrant (Raju et al., 1984; Gupta et al., 1984; Prakash Kumar et al., 1985; Pannetier and Buffard-Morel, 1985) but steady progress is being made. Indeed, callus formation from coconut has been achieved using 2,4-D and activated charcoal from tissues ranging from those of young zygotic embryos to leaves and inflorescences of mature palms.

The successful procedures used nowadays largely seem akin to those used for oil palm. Field plantings have been established from somatic embryos (Verdeil et al., 1989). Androgenesis seems to be a possibility (Monfort, 1985).

Many problems remain in virtually all aspects of tissue culture of coconut, however. There is, moreover, need for a more open literature on this topic.

4.2.3. *Date Palm*

Date (which is dioecious and 2n = 36) is another of the obligate outbreeders and as such there is much variability in the off-spring of a cross. Vegetative multiplication does occur in some varieties but is slow and there seem to be limited prospects for improvement. Tisserat (1984, 1987) has done extensive work on the tissue culture of date and detailed protocols exist (cf. also Krikorian, 1989a and refs. there cited). The basal salts of Schenk and Hildebrandt as well as those of Gamborg's B_5 have been shown to be excellent for callus induction, whereas S and H is useful in producing somatic embryos. Activated charcoal is used at 1.5 to 2 g/l. High levels of NAA are active in inducing embryogenic callus and 2iP and zeatin are the more active of the cytokinins (in contrast with BAP, and kinetin). A major limitation to implementation of the process seems to be the cost (personal communication, Dr. Mostafa Abo El Nil).

4.2.4. *Other Palms and Oil-producing Plants*

Many palms are of interest because of their fruits and their 'hearts' or tender growing points (cf. Uhl and Dransfield, 1987; Tomlinson, 1990). The pejibaye (*Bactris gasipaes*, syn. *Guilielma gasipaes*) is an example where suckering may be highly variable. Arias and Huete (1983) have shown that plantlets can be regenerated from callus induced in the darkness with 2,4-D and BAP and subsequent removal of growth regulators. Somatic embryogenesis has also been induced from callus derived from *in vitro* cultured shoot tips of young field-grown plants using BAP and picloram (Valverde et al., 1987); and from zygotic embryos extracted from mature fruit (Villalobos et al., 1987) using methods virtually identical to those used for oil palm (Nwanko and Krikorian, 1983). Similarly, *Euterpe edulis*, an endangered Brazilian palm used as a source of hearts of palm, has also been shown to be responsive to 2,4-D treatment and activated charcoal, nominally leading to production of somatic embryos from excised mature zygotic embryos without intervening callus (Guerra and Handro, 1988).

4.3. *Bananas and Plantains* (*Musa clones*)

These large, perennial herbs are excellent candidates for tissue culture (Krikorian, 1990). The corms or suckers are large and inconvenient for germ plasm movement, frequently harbor pathogens and are often in short supply – especially when large new plantings are contemplated. Germ plasm maintenance in the field is expensive and space consuming. Breeding of the essentially seed-sterile clones of interest to plantation agriculture is fraught with many difficulties (Stover and Buddenhagen, 1986; Rowe and Rosales, 1994). Moreover, there are a number of diseases like *Fusarium* wilt for which control is either unavailable or extremely costly (cf. Soto, 1985; Ploetz, 1990 and refs. there cited).

The means to micropropagate *Musa* are reasonably well established, well publicized and have been widely adopted in various parts of the world. Technical guidelines for safe movement of germ plasm have been delineated and most are aware of the problems of assuming that meristems or *in vitro* cultures are necessarily 'clean' (Frison and Putter, 1989). There are too many possibilities of introducing virus diseases if adequate precautions are not taken to see that plants are virus indexed or quarantined (Lockhart, 1986). Various publications summarize in greater or less detail the methods to make primary explants and to initiate multiple shoot cultures (cf. Krikorian and

Cronauer, 1984; Cronauer and Krikorian, 1986, 1988; Vuylsteke, 1989; Rowe, 1991).

There is considerable opportunity for changes to be implemented in these procedures, and refinements, in time, will surface.

Shoot tip cultures may be initiated from corms or floral buds. Material intended for culture should be carefully examined, and determined to be sound. Prompt culture of material from the field is essential in order to avoid contamination. The initial cutting (done with a single edge razor blade) can be done in a non-clean work area such as a greenhouse. One by one the outer leaves are removed by first cutting across the top of the corm and then cutting down the length of the leaf base. The leaf may then be readily peeled away. After removal of several leaves, the leaf bases comprising the tender pseudostem appear slender and brittle. The top should be cut away (leaving about 4 inches above the corm) to prevent the entire pseudostem from snapping off (taking the apex with it). Continue removing leaves until the leaf base area is about 1/2 inch long. Using a sharp knife or scalpel, rectangular cuts should be made into the corm so as to enable the apex to be lifted out. This apex needs to be taken to a clean area (laminar flow hood or room) and further processed. First, sterilization needs to be carried out. Then, the rest of the procedure is carried out under a dissecting microscope until the appropriate sized apex is obtained (order of a couple of mm an or even smaller). 1% v/v Clorox or equivalent laundry bleach with a drop of wetting agent such as Tween 20/80 for 5 min has been satisfactory. Floral buds may be handled in a similar manner except that flowers and bracts are removed row by row.

MS liquid medium with 5 mg/l BAP and 4% sucrose provides an adequate starting point for the cultures. Pulsing the culture from liquid to semi-solid fosters growth (Cronauer and Krikorian, 1984). Multiple shoot production may be stimulated some 3 weeks after initial explantation by a longitudinal cut. Multiple shoots may be separated at 3 week intervals and placed in fresh cytokinin-containing medium. These multiple shoot cultures can be separated, transferred to a culture medium with or without auxin, or with or without activated charcoal and rooted. They are readily transferred to a soil environment with virtually no loss provided moderate care is taken to prevent drying out. Some effort is being given to the development of non-traditional substrates for the mass multiplication of bananas (Matsumoto and Yamaguchi, 1989).

Callus culture procedures with *Musa* have been steadily progressing and somatic embryos have been generated from a number of species and clones (cf. Novak et al., 1989; Krikorian et al., 1990 for a review and summary of different strategies for somatic embryogenesis). While dicamba and picloram are valuable as auxins, the more conventional ones like 2,4-D and 2,4,5-T are adequate. The finding that Gelrite was nominally crucial to successful callus culture may be explained on the basis of failure to use adequately clean agar (Huang and Chi, 1988). Even so, Gelrite® has a real place in

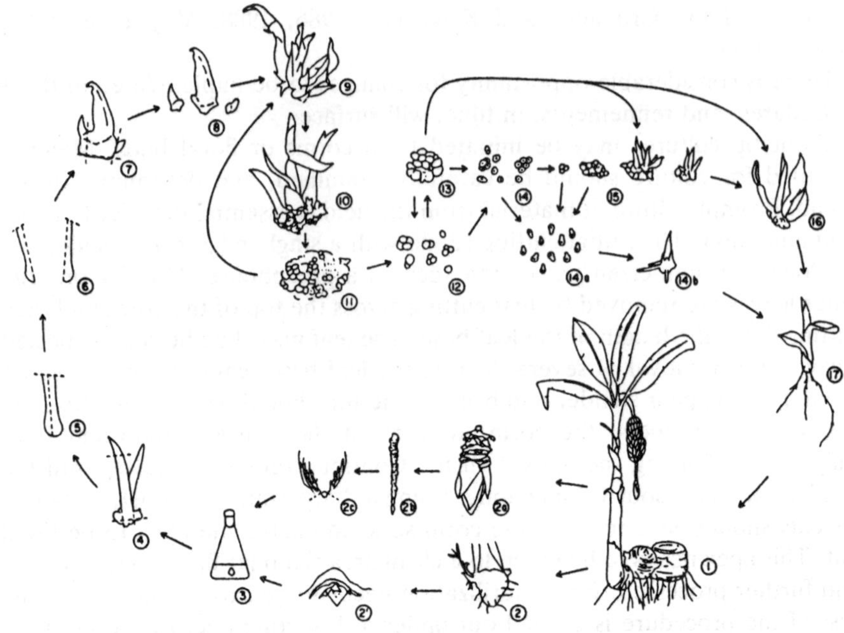

Figure 8. Diagrammatic representation of the procedures followed in the establishment of aseptic shoot and floral apex cultures. See text for details.

Musa tissue culture. It is very helpful, for instance, in detecting subtle contaminants.

Figure 8 provides a schematic view of the various tissue culture procedures for *Musa* and how they relate to one another. (1) shows a mature plant that has formed from a ratoon or sucker. Excision of the vegetative apex from a non-flowering corm (2 through 3), or alternatively an apex from the floral bud (sometimes called the 'bell') – (2a through 2c), leads to a system where leaf-opposed and adventitious buds are generated (4 through 8). Subdivision of these multiple shoots and 're-planting' in cytokinin-containing culture medium (semi-solid or shallow liquid) can keep the system going (9). Further treatment of aseptic shoot tips *in vitro* with auxins can foster the development of callus, calloid, and embryogenic tissues (10–13) (cf. Novak et al., 1989; Krikorian et al., 1990). These various-sized and variously-developed units may be further broken down as to unit size, or regenerated into true somatic embryos (14a) or organogenic callus masses (15). These in turn can be reared to full-sized propagules that may be placed in soil (14b, 16–17). Protoplast work has begun as well (Krikorian et al., 1988; Panis et al., 1993).

There is considerable evidence that off-types can pose problems for those seeking to use *Musa* shoot tip cultures for massive multiplication. The genus shows a relatively high mutation rate under normal field conditions and there are many reasons to suspect that poor selection of material and poor laboratory practice have both played a role (see Smith, 1987; Krikorian, 1989b; Drew and Smith, 1990). Whether or not prior roguing of offtypes at the laboratory or nursery level can be can be carried out has not been completely resolved but holds some promise for at least some off-types such as dwarfism (cf. Reuveni, 1990).

In other cases, nominal 'offtypes' can be valuable sources of new varieties (cf. Irizarry et al., 1991; Krikorian et al., 1993). Again, many opportunities exist to improve upon and further define the requirements for the systems to be fully effective and reliably usable for both research and practical objectives (cf., e.g. Molina and Krauz, 1989; Novak et al., 1990).

4.4. *Cacao* (*Theobroma cacao*)

Cacao is of Neotropical origin and is normally an outbreeding species which is extremely variable ($2n = 20$, Glicenstein and Fritz, 1989). Even so, it has traditionally been multiplied from seed, although rooted cutting and budding methods for clonal propagation exist (rooting was first introduced in the 1930s). These are of course, essential to the breeder. For the larger scale grower, however, the expense involved and the innate problems of dimorphic growth habit (i.e. plagiotropy and orthotropy) has been a deterrent (cf. Charrier, 1969; Greathouse et al., 1971 and refs. cited). The plant shows some unusual features and mechanisms of self-incompatibility involving both sporophytic and gametophytic control and these have consequences for breeding.

Much effort has recently been put into identifying parents with good combining qualities for use in breeding schemes. Extensive plantings throughout the world derive largely from unimproved seedlings and the vision is that increased demands for chocolate and cocoa butter will necessitate expanded plantings with improved materials that will be amenable to intensive cultivation. Diseases such as *Monilia* (the cause of water pod rot), *Phytophthora* (black pod rot), witches broom – caused by *Crinipellis perniciosa* – cf. Evans and Solórzando, 1982), and capsids (insect family Miridae) are only a few of the many serious constraints to production (Hunter, 1990). One aspect of this is that quarantine restrictions are a bottleneck to germ plasm movement for breeding purposes (Gregory, 1977).

Hybrid seeds (using upper Amazonian types in breeding) which show precocity, large bean size, appropriate flavor when processed, and disease resistance have thus far been the mainstay of modern improvement plans. Some have suggested that any earlier trend on the part of producers to move from use of seed to clonal plantings will probably be reversed, however. On this score it will be a situation going against the usual tide of 'seedlings to clones'. But this trend could change overnight with the development of reliable micropropagation techniques and indeed, prospects for biotechnology at a range of levels, including the use of somatic embryo cotyledons to produce biochemically important compounds (cf. Griffiths and Harwood, 1991; Paiva et al., 1982) seem to hold high promise (Janick, 1986; Mabbett, 1987; Fritz, 1987).

The cocoa plant is generally viewed as recalcitrant *in vitro* (Dublin, 1984, 1991) but as more work is carried out, each of the aseptic culture systems are sure to fall into place. Esan first demonstrated in 1975 that asexual embryos could be produced via direct somatic embryogenesis from the embryo axis from mature seeds (cf. Esan, 1982, 1985). Somatic embryos could also be generated from proliferating callus derived from mature embryo axis and immature embryo cotyledons. MS medium supplemented with NAA, coconut water and casein was used. Adenine sulfate and kinetin were shown to be optional additives. Whereas various stages of asynchronous embryonic development were encountered, somatic embryos became dormant and could not be induced to grow out as plantlets. Janick and his co-workers (cf. e.g. Pence et al., 1980) independently essentially encountered the same failure. Adu-Ampomah et al. (1988) initiated somatic embryos from zygotic embryos and demonstrated that the removal of cotyledons from the somatic embryos enhanced the capacity to 'germinate'. Unfortunately, this technique seems to be valuable only in certain genotypes and the situation of getting somatic embryos of cacao (however well developed they may be morphologically – cf. Dos Santos and Machado, 1989), to continue their development into whole plants has remained a problem.

Micrografting of somatic embryos onto zygotic seedlings germinated *in vitro* has recently offered some promise in recovering plantlets (Villalobos, 1990) but is sure to be labor-intensive (cf. Kadje and Ngambi Ndjama, 1982). Also, new findings using elevated carbon dioxide and increased light indicate that vigorous germination and growth can be fostered in recalcitrant somatic embryos (cf. Figueira et al., 1991; Sondahl, personal communication). It may well turn out that this procedure benefits many nominally recalcitrant 'germinating' cultures. More recently, it has been shown that nucellar and petal explants (from flower buds about 3–5 mm in length) yield vigorous somatic embryos capable of developing to plants – embryogenesis response from petal explants at a level of about 4.3% (based on 9,000 explants) and nucellar explants at about 2.0% (based on 29,000 explants). For petal cultures, additives to an otherwise conventional MS semi-solid medium for callus induction were 2,4-D (10 μM) and kinetin (1 μM). Regeneration medium included, besides half-strength MS but with 3% sucrose, BAP (5 μM). In the case of nucellus explants, Kao's medium (Kao and Michayluk, 1975) with 2,4-D (3.0 μM) was used. It was, however, supplemented with PVP-40 (5 gm/l) – presumably to prevent oxidation, coconut water (5% v/v), activated charcoal (0.2% w/v) and MES as a buffering agent (600 mg/l) was used. There has been thus far, the relatively brief interval of some 40 weeks from primary explant to establishment of plantlets in soil (cf. Sondahl et al., 1993 for details).

Axillary bud cloning has been fraught with difficulties as well and generally few or very modest numbers of plants are recovered from micropropagation efforts (cf., e.g. Passey and Jones, 1983; Paranjothy, 1987a; Legrand et al., 1984; Glicenstein et al., 1990; Flynn et al., 1990). The Woody Plant Medium of Lloyd and McCown (1980) has found some usefulness but the fact of recalcitrance stands. More work is needed on the controls that govern shoot tip behavior (Hardwick et al., 1982).

Spontaneous haploids were first recognized in *Theobroma* in 1972 (Dublin, 1972) when polyembryonic seeds were dissociated. Hybrid heterogeneity, especially in the newer Amazonian hybrids, has emphasized the value of haploids. These have been used in research programs for breeding purposes after they have been rendered diploid and homozygous (Dublin, 1974,1991). Protoplasts from leaf tissue have been prepared but growth has not been forthcoming (Thompson et al., 1989). The potential for transformation has been recognized and *Agrobacterium* susceptibility identified (Purdy and Dickstein, 1989). Work on cloning and characterization of chloroplast DNA etc. has moved forward (Yeoh et al., 1990; Couch and Fritz, 1990).

4.5. *Coffee* (*Coffea* sp.)

The genus *Coffea* has a basic chromosome number of n = 11. *Coffea arabica* is the only polyploid (2n = 4x = 44) in the genus and is self-fertile. Virtually all the others of interest are highly self-incompatible. *Coffea canephora* is the most important of the diploid coffees. Although generally agreed as producing a poorer quality of coffee than *arabica*, it is tolerant to rust and is more tolerant to adverse growing conditions. Resistance to disease, especially rust (*Hemileia vastatrix*) and coffee berry disease (*Colletotrichum coffeanum*), have been the main objectives in breeding to date. Other features such as short stature and yield capacity, and even organoleptic qualities of the end product when brewed have also occasionally entered into breeding considerations (cf. Sondahl and Loh, 1988).

Tissue culture methods for coffee are well advanced, constantly being improved and indeed a comprehensive information base for the crop is beginning to emerge. Moreover, the rationale for using biotechnology has been carefully thought out and enunciated. Also, thanks to work done in predominantly Brazil there is an excellent body of literature that covers in considerable detail the various protocols that might be followed (cf. Sondahl et al., 1984, 1991; Sondahl and Loh, 1988).

Early studies by Sondahl and Sharp (see Sondahl references above and Sondahl et al., 1980) established a protocol for so-called high frequency somatic embryogenesis (HFSE) in two steps. The first involves callus production on a medium containing a high level of auxin relative to cytokinin; followed by transfer to a medium with the reverse ratio. Dublin (1981) showed that high cytokinin in the absence of any auxin could induce 'direct' embryogenesis on leaf discs. An interesting and labor-saving feature that some of the leaf disc or strip methods employ is that a pre-culture (actually a flotation on liquid hormone-free basal medium) is carried out for about 4 days to eliminate the necrotic, oxidized or contaminated explants before treatment in culture.

Somatic embryogenesis studies have increased over the last several years and include work on many species besides *arabica* and *canephora* that have potential in transmitting disease resistance like *C. liberica*, *C. eugenioides*, *C. congensis* etc. Besides the Brazilian work, studies have been done in Portugal (cf. Marques, 1990) and Colombia (cf. de Peña, 1984, 1990). A recent development is that a single medium may now be used to initiate somatic embryos (M and S mineral salts, sucrose 116.9 Mm, inositol 555 μM, malt extract 250–1000 mg/l, BAP (up to 44.4 μM, agar 0.8% (Marques, 1990).

Protoplasts were first regenerated from *canephora* by Schöpke et al. (1987) from embryogenic cultures. Protoplasts have recently been recovered from *arabica* (Acuña and de Peña, 1991). It would be interesting to see whether protoplasts can be regenerated from leaves or other plant parts since it is well appreciated that plantlets are best regenerated from embryogenically competent suspensions (cf. Fitter and Krikorian, 1981, 1982).

4.6. *Tea* (*Camellia* spp.)

The genus *Camellia* (Camelliaceae) includes some 82 species. *Camellia sinensis* (a diploid, 2n = 30) can be produced from seed (it is highly outbreeding), and most of the older plantations or gardens around the world are of seed origin. Newer operations are more or less exclusively derived from cuttings made from single leaf internodes, and clonal selection and multiplication by this means has played an important role in the rejuvenation of the industry.

It was recognized very early that *in vitro* regeneration could play a significant role in tea improvement because clonal selection could be facilitated (Singh, 1978; Singh and Jain, 1990). Also, because tea is self-sterile, one could attempt to use anther culture to achieve homozygous diploid breeding lines. Germ plasm preservation and a host of other objectives could be served as well, if reliable methods could be developed.

Tissue culture methods such as obtaining polyploids through somatic embryo culture and multiple shoot production from single bud explants have been developed for the genus but procedures and efficiency still needs improvement (cf., e.g. Haldeman et al., 1987; Iddagoda et al., 1988; Kato, 1989; Samartin, 1989; Haldeman and Thomas, 1989; Singh and Jain, 1990; Esan, 1992).

4.7. *Rubber* (*Hevea brasiliensis*)

South American leaf blight, caused by the ascomycete *Microcyclus ulei*, has limited plantation production of rubber in the Neotropics despite the fact that the genus had its origins in the Amazon basin (Brücher, 1989). Extensive plantation culture in Southeast and South Asia, and to a lesser extent in Africa has been possible due to its absence in those areas. Nevertheless, there is interest in breeding for horizontal resistance (cf. Simmonds, 1990) against the eventuality of accidental introduction of leaf blight to these areas.

The commercial rubber tree is an interspecific hybrid (actually an amphidiploid) that is bud grafted onto unselected seedling rootstocks. Of the nine well characterized species of *Hevea*, virtually all are diploid (2n = 36) with the exception of one triploid, *Hevea guianensis* (2n = 3x = 54). Bud grafting of superior clones has been used for many years and the procedure maintains intraclonal heterogeneity. Test procedures involved in breeding start out with hand pollination, and end with seedling evaluation. This seedling evaluation of clones is a lengthy procedure, often taking up to 30 years from pollination to practical impact. There is substantial interest in breeding for disease resistance, increased latex yield and tolerance to wind damage, this latter character of wind fastness being shown to be negatively correlated with yield.

A major interest in rubber tissue culture has been the obtaining of haploids via androgenesis (Chen, 1984; Chen et al., 1988, 1990). Detailed protocols have been published. Special emphasis is placed on using anthers with well developed microspores predominantly in the mononucleate stage and some just entering the binucleate stage. The effect of weather is emphasized as excessively hot weather adversely affects pollen performance. Gynogenesis has also been achieved (the stages of pollen development correlate well with embryo sac in the ovule development). Uninucleate to 4-nucleate embryo sacs are said to respond best.

Just as rooting of cuttings has been known to be very difficult and essentially impractical in *Hevea*, stem tip cuttings, especially of mature élite clones, have been very recalcitrant to multiplication and rooting *in vitro* (cf. Chen, 1984). Even excised axillary buds from seedlings that have been forced by dipping the stem tip in BAP have not been that easy to root (Enjalric and Carron, 1982). Somatic embryogenesis holds greater promise as a means of micropropagation of rubber (El Hadrami et al., 1989). It seems that the gaseous environment is of great importance in the generation of embryogenic callus of rubber (thin explants of internal integument of immature seed are used). Removal of ethylene, or adequate gas exchange in the culture vessels is critical and prevention of ethylene synthesis in the tissue is helpful as well (Auboiron et al., 1990). Also, the water status of the tissue undergoing somatic embryogenesis is apparently very important (cf. Etienne et al., 1991) and abscisic acid is thought to play a decisive role in the process.

Finally, mention should be made of guayule (*Parthenium argentatum*, fam. Compositae) as an example of an alternative rubber-producing plant that has had a longish and frustrating history in terms of attempts to domesticate and improve it (Miller and Backhaus, 1986). The species (2n = 18) is a native of the arid regions of northern Mexico and Texas. It is a slow-growing plant, and while cuttings can be made, it is not a practical approach for establishing large plantings. Coated seeds have permitted direct seeding. Tissue culture work has not been easy with this plant (cf. Staba and Nygaard, 1983).

4.8. *Spice, Condiment, Flavor Plants, Tropical Fruit Trees*

Work on a great many tropical fruit trees has shown the tremendous potential for tissue culture studies (cf., Litz, 1989; Hammerschlag and Litz, 1992).

Since ornamental orchid culture is well established, progress with vanilla (*Vanilla planifolia*) is not unexpected (cf. Philip and Nainar, 1976). It is of great interest, however, that root meristems of vanilla can be transformed in culture and give rise to shoot meristems, and ultimately into plants (Philip and Nainar, 1988).

Progress has been made on callus culture of cashew, *Anacardium occidentale* (cf. Philip, 1984 for use of cotyledonary explants; Timir Baran Jha, 1988 for use of immature embryos), as well as on micropropagation from nodal explants taken from 6–15 month old plants (Lievens et al., 1989). Clove and related fruit crop species (*Syzygium* spp.) are also beginning to show signs of response to *in vitro* methods although so far, the work has been done on seedling material (Mathew and Hariharan, 1990; Yadav et al., 1990).

Ginger (*Zingiber officinale*) of the Zingiberaceae suffers severe losses from a number of diseases. Excision of rhizome buds and their stimulation to form axillary and adventitious shoots has provided a major way to multiply planting material. A medium such as MS supplemented with BAP has proven adequate although slight refinements and modifications can be seen throughout the protocols (cf. Mapes and Bullock, 1975; Hosoki and Sagawa, 1977; Nadgauda et al., 1980; Pillai and Kumar, 1982; Ikeda and Tanabe, 1989; Balachandran et al., 1990).

Turmeric (*Curcuma longa*), also of the Zingiberaceae, has been multiplied by means of *in vitro* culture as well (Nadgauda et al., 1978; Balachandran et al., 1990). Multiplication is normally through rhizome pieces but this is a slow process.

Similarly, cardamom (*Elleteria cardamomum*, Zingiberaceae) a very important spice crop in India (cf. Cherian, 1986) has responded to *in vitro* methods (Nadgauda et al., 1983). The methods are commercially viable and are being used. Clonal multiplication of élite cultivars, coupled with their being rendered free of pathogens etc. have made the methods very useful. Since callus can be used to regenerate plants as well (Srinavasa Rao et al., 1982), one can see the possibility of improvement of each of these Zingiberaceous crops by mutation breeding (Micke et al., 1987). Precautions will have to be taken, however, to avoid chimeras (Bergann, 1967). Dekkers et al. (1991) have examined possibilities for germ plasm preservation at room temperatures. Philip et al. (1993) have micropropagated black pepper.

4.9. *Some Fiber crops*

Manila hemp (abacá) or *Musa textilis* can be handled much like other *Musa* species. Members of the Agavaceae (e.g. *Agave sisalana*, the source of sisal hemp, *A. fourcroydes*, the source of hennequin fiber are also amenable to *in vitro* culture. These species normally multiply slowly to reasonably well vegetatively depending on a number of factors (e.g. those that are multiplied by bulbils in the axils of the inflorescence may take as long as 30 years to get because of the monocarpic flowering habit). Binh et al. (1990) have described a fairly straightforward protocol using explants from stolon-derived plants that were established in a greenhouse. Callus may also be established from various *Agave* sp. and these can yield plantlets (Groenewald et al., 1977; Villalobos et al., 1991). Jute (*Corchorus* sp., Tiliaceae) has likewise been subjected to tissue culture (Das et al., 1987).

5. Acknowledgments

Help which enabled me to present as up-to-date coverage as possible is acknowledged. I especially thank Drs. E.B. Esan (Cocoa Research Institute of Nigeria, Ibadan), D.V. Marques (Centro de Investigaçao das Ferrugens do Cafeeiro, Quinta do Marquês, Oeiras, Portugal), Myriam de Penã (Laboratorio Investigaciones Quimica del Cafe-LIQC Federacafé, Bogotá, Colombia), H.C. Chaturvedi (National Botanical Research Institute, Lucknow, India) and J.S. Tahardi (Bogor Research Institute for Estate Crops, Bogor, Indonesia).

References

ASTM Standards (1985). 1985 Annual Book of ASTM Standards. Section 15.01 Refractories; Carbon and Graphite Products; Activated carbon. ASTM, Philadelphia.

Acuña, J.R. and de Peña, M. (1991). Plant regeneration from protoplasts of embryogenic cell suspensions of *Coffea arabica* L. cv. Caturra. Plant Cell Rep. 10:345–348.

Adams, T.L. and Townsend, J.A. (1983). A new procedure for increasing efficiency of protoplast plating and clone selection. Plant Cell Rep. 2:165–168.

Adaoha Mbanasao, E.N. and Roscoe, D.H. (1982). Alginate: An alternative to agar in plant protoplast culture. Plant Sci. Letters 25:61–66.

Adu-Ampomah, Y., Novak, F.J., Afza, R., Van Duren, M. and Perea-Dallos, M. (1988), Initiation and growth of somatic embryos of cocoa (Theobroma Cacao L.). Café Cacao Thé 32 (No. 3), 187–200.

Aghion, D. and Beauchesne, G.. (1960). Utilisation de technique de culture stérile d'organes pour obtenir des clones d'ananas. Fruits 15:444–446.

Ahee, J., Arthuis, P., P., Cas, G., Duval, Y., Guénin, G., Hanower, J., Hanower, P., Lievoux,

D., Lioret, C., Malaurie, B., Pannetier, C., Raillot, D., Varechon, C. and Zuckerman, L. (1981). La multiplication végétative *in vitro* du palmier á huile par embryogénèse somatique. *Oléagineux* 36:113–116.

Alvim, P. de T. and Kozlowski, T.T. (1977). Ecophysiology of Tropical Crops. Academic Press, New York.

Analytical Bibliography (1987). Bibliographie analytique. La culture *in vitro* du palmier à huile (*Elaeis guineensis* Jacq.). Oléagineux 42: DA101–DA111.

Anonymous (1988). Proceedings of the Workshop for the Improvement of Oil- and Steroid-Producing Plants, Jakarta, 5–6 April, 1988. The Bogor Research Institute for Estate Crops, Jakarta, Indonesia.

Arditti, J. and Ernst, R. (1993). Micropropagation of Orchids. John Wiley & Sons, New York.

Arias, O. and Huete V., F. (1983). Propagacion vegetativa *in vitro* de pejibaye (*Bactris gasipaes* H.B.K.). Turrialba 33:103–108.

Ashburner, G.R., Thompson, W.K., Maheshwaran, G. and Burch, J.M. (1991). The effect of solid and liquid phase in the basal medium of coconut (*Cocos nucifera* L.) embryo cultures. Oléagineux 46:149–152.

Assy-Bah, B. (1986). Culture *in vitro* d'embryons zygotiques de cocotiers. Oléagineux 41:321–328.

Assy-Bah, B. Durand-Gasselin, T. and Pannetier, C. (1988). Use of zygotic embryo culture to collect germplasm of coconut (*Cocos nucifera* L.). FAO/IBPGR Plant Genetic Resources Newsletter (Rome, Italy) 71:4–10.

Assy-Bah, B., Durand-Gasselin, T., Engelmann, F. and Pannetier, C. (1989). Culture *in vitro* d'embryons zygotiques de cocotier (*Cocos nucifera* L.). Méthode révisée et simplifiée, d'obtention de plants de cocotiers transférables au champ. Oléagineux 44:515–523.

Auboiron, E., Carron, M.-P. and Michaux-Ferrière, N. (1990) Influence of atmospheric gases, particularly ethylene, on somatic embryogenesis of *Hevea brasiliensis*. Plant Cell Tissue Org. Cult. 21:31–37.

Baird, J.K., Sandford, P.A. and Cottrell, I.W. (1983). Industrial applications of some new microbial polysaccharides. Bio/Technology 1:778–783.

Balachandran, S.M., Bhat, S.R. and Chandel, K.P.S. (1990). *In vitro* clonal multiplication of tumeric (*Curcuma* spp.) and ginger (*Zingiber officinale* Rosc.). Plant Cell Rep. 8: 521–524.

Barlow, P.W. and Carr, D.J. (1984). Positional Controls in Plant Development. Cambridge University Press, Cambridge and London.

Bartholomew, D.P. and Kadzimin, S.B. (1977). Pineapple. In: Ecophysiology of Tropical Crops. P. de T. Alvim and T.T. Kozlowski (eds.), Academic Press, New York, pp. 113–156.

Bavappa, K.V.A., Nair, M.K. and Prem Kumar, T. (eds.) (1982). The Arecanut Palm. Central Plantation Crops Research Institute, Kasaragod, India.

Bergann, F. (1967). The relative instability of chimeral clones – the basis for further breeding. Abh. Dtsch. Akad. Wiss. Berlin. Kl. Med. Jahrgang 1967 Nr. 2, 287–300.

Bhaskaran, S. and Prabhudesai, V.R. (1986). Applications of tissue culture to plantation crops. In: Plantation Crops. Opportunities and Constraints: H.C. Srivastava, Bhartendu Vatsya and K.K.G. Menon (eds.), Vol. 2. Oxford & IBH Publishing, New Delhi, pp. 191–205.

Bhaskaran, S. and Prabhudesai, V.R. (1989). Tissue culture of plantation crops. In: Applications of Biotechnology in Forestry and Horticulture. V. Dhawan (ed.), Plenum Press, New York, pp. 87–107.

Binh, L.T., Muoi, L.T., Oanh, H.T.K., Thang, T.D. and Phong, D.T. (1990). Rapid propagation of Agave by *in vitro* tissue culture. Plant Cell, Tissue Org. Cult. 23:67–70.

Brackpool, A.L., Branton, R.L. and Blake, J. (1986). Regeneration in palms. In: Cell Culture and Somatic Cell Genetics of Plants, Vol. 3. Academic Press, Orlando, pp. 207–222.

Brockway, L.H. (1979). Science and Colonial Expansion. The Role of the British Royal Botanical Gardens. Academic Press, New York, London.

Brücher, H. (1989). Useful Plants of Neotropical Origin and their Wild Relatives. Springer Verlag, Berlin.

Casale, O., I.M. and García, E.C. (1987). Multiplicacion clonal acelerada de tres variedades de piña. AVEVIV. Asociacion Colombiana de estudios Vegetales *In vitro* 2:3–18.

Charrier, A. (1969). Contribution à l'étude de la morphogenèse et de la multiplication végétative du cacoyer (*Theobroma cacao* L.). Café Cacao Thé 13:97–115.

Chen, Z. (1984). Rubber (*Hevea*). In: Handbook of Plant Cell Culture. W.R. Sharp, D.A. Evans, P.V. Ammirato, and Y. Yamada (eds.), Vol. 2. Macmillan, New York, pp. 505–545.

Chen, Z., Li, W. and Zhang, L. (1988). Production of haploid plantlets in cultures of unpollinated ovules of *Hevea brasiliensis* Muell.-Arg. In: Somatic Cell Genetics of Woody Plants. M.R. Ahuja (ed.), Kluwer Academic Publishers, Dordrecht, pp. 39–44.

Chen, Z., Xu, X. and Pan, R. (1990). Rubber tree: anther and ovule culture. In: Handbook of Plant Cell Culture. Z. Chen, D.A. Evans, W.R. Sharp, P.V. Ammirato, and M.R. Sondahl (eds.), Vol. 6. McGraw-Hill, New York, pp. 440–452.

Cherian, A. (1986). Cardamom plantation industry in India. In: Plantation Crops. Opportunities and Constraints. 2 Vols. Oxford & IBH Publishing, New Delhi. I, 245–261.

Clarke, R.J. and Macrae, R. (1988). Coffee. 4 Vols. Elsevier, London, New York.

Clemetson, C.A.B. and Andersen, L. (1966). Plant polyphenols as antioxidants for ascorbic acid. Ann. NY Acad. Sci. 136 (Art. 14): 339–378.

Collins, J.L. and Kerns, K.R. (1938). Mutations in the pineapple. J. Heredity 29:163–164.

Corlett, Jr., D.A. and Caldis, P.D. (1972). Plant culture method for product. United States Patent 3,683,550. 15 August 1972.

Corley, R.H. (1983). Potential productivity of tropical perennial crops. Exptl. Agric. 19:217–237.

Corley, R.H.V., Lee, C.V., Law, I.H. and Wong, C.Y. (1986). Abnormal flower development in oil palm clones. The Planter (Kuala Lumpur) 62:233–340.

Couch, J.A. and Fritz, P.J. (1990). Isolation of DNA from plants high in polyphenols. Plant Mol. Biol. Reporter 8:8–12.

Cronauer, S.S. and Krikorian, A.D. (1984). Multiplication of *Musa* from excised stem tips. Ann. Bot. 53:321–328.

Cronauer, S.S. and Krikorian, A.D. (1986). Banana (*Musa* spp.). In: Biotechnology in Agriculture and Forestry 1. Trees I. Y.P.S. Bajaj (ed.), Springer Verlag, Berlin, pp. 233–252.

Cronauer, S.S. and Krikorian, A.D. (1988). Temporal, spatial and morphological aspects of multiplication in aseptically cultured *Musa* clones. In: Progress and Prospects in Crop and Forest Biotechnology. State University of New York College of Environmental Science and Forestry, Colloquium April 18–20, 1985, Syracuse. Springer Verlag, New York, pp. 45–57.

Cui, Y. and Gong, Z. (1990). Oil palm. In: Handbook of Plant Cell Culture. Z. Chen, D.A. Evans, W.R. Sharp, P.V. Ammirato, and M.R. Sondahl (eds.), Vol. 6. McGraw-Hill, New York, pp. 440–452.

Das, B., Haque, M.M., Islam, A S., Rahman, M.H. and Imdadul Haque, M, (1987). *In vitro* plantlet development in jute. In: Agricultural Applications of Biotechnology. A.N. Rao and H.Y. Mohan Ram (eds.), COSTED, Madras, pp. 83–95.

Debergh, P.C. (1983). Effects of agar brand and concentration on the tissue culture medium. Physiol. Plant. 59:270–276.

Dekkers, A.J., Rao, A.N. and Goh, C.J. (1991). *In vitro* storage of multiple shoot cultures of gingers at ambient temperatures of 24–29 °C. Sci. Hort. 47:157–167.

de Peña, M. (1984). Somatic embryo induction and plant regeneration from *Coffea canephora* and *Coffea arabica*. In: Simposio sobre Ferrugens do Cafeeiro. Centro de Investigaçao das Ferrugens do Cafeeiro. Oeiras, Portugal, pp. 495–512.

de Peña, M. (1990). El cultivo *in vitro* de tejidos de cafe en Colombia. In: 50 Años de CENICAFE 1938–188. Conferencias Conmemorativas. Federacion Nacional de Cafeteros de Colombia, Chinchiná, Caldas, Colombia, pp. 194–198.

DeWald, M.G., Moore, G.A., Sherman, W.B. and Evans, M.H. (1988). Production of pineapple plants *in vitro*. Plant Cell Rep. 7:535–537.

Dos Santos, A.V.P. and Machado, R.D. (1989). A scanning electron microscope study of *Theobroma cacao* somatic embryogenesis. Ann. Bot. 64:293–296.

Drew, R.A. and Smith, M.K. (1990). Field evaluation of tissue-cultured bananas in southeastern Queensland. Australian J. Exptl. Agric. 30:569–574.

Dublin, P. (1972). Polyembronie et haploidie chez *Theobroma cacao*. Café Cacao Thé 16:295–311.

Dublin, P. (1974). Les haploides de *Theobroma cacao* L.: Diploidisation et obtention d'indivus homozygotes. Café Cacao Thé 18:83–96.

Dublin, P. (1981). Embryogenèse somatique directe sur fragments de feuilles de caféier Arabusta. Café Cacao Thé 25:237–242.

Dublin, P. (1984) Cacao. In: Handbook of Plant Cell Culture. P.V. Ammirato, D.A. Evans, W.R. Sharp and Y. Yamada (eds.), Vol. 3. Macmillan, New York, pp. 541–563.

Dublin, P. (1991). Multiplicación vegetativa de café, hevea y cacao. In: Cultivo de Tejidos en la Agricultura. Fundamentos y Aplicaciones. W.M. Roca and L.A. Mroginski (eds.), CIAT, Centro Internacional de Agricultura Tropical, Cali, Colombia, pp. 577–642.

Durand-Gasselin, T., Le Guen, V., Konan, K. and Duval, Y. (1990). Plantations en Côte-d'Ivoire de palmiers à huile (Elaeis guineensis Jacq.), obtenus par culture *in vitro*: Premiers résultats. Oléagineux 45:1–11.

Duval, Y., Durand-Gaselin, T., Konan, K. and Pannetier, C. (1988). Multiplication végétative du palmier à huile (*Elaeis guineensis* Jacq.) par culture *in vitro*. Oléagineux 43:39–47.

El Hadrami, I., Carron, M.-P. and d'Auzac, J. (1989). Variabilité clonale du potential embryogène chez *Hevea brasiliensis*; relations avec les polyamines (PA) et les peroxydases (PO) des cals. C.R. Acad. Sci. (Paris) sér. III:299–305.

Engelmann, F., Duval, Y. and Pannetier, C. (1988) Utilisation des techniques de cryoconservation pour la création de banques d'embryons somatiques de palmier à huile (*Elaeis guineensis* Jacq.). Oléaqineux 43:323–328.

Enjalric, F. and Carron, M.-P. (1982) Microbouturage *in vitro* de jeunes plants d'Hevea brasiliensis (Kunth., Mull. Arg.). C.R. Acad. Sci. (Paris) 295:259–264.

Esan, E.B. (1982). *In vitro* studies on the morphogenesis and plant regeneration in tissue cultures of cacao and *Herrania* species – A progress report. In: 8th Intl. Cocoa Research Conference. Proceedings. Cartagena, Colombia. Cocoa Producers' Alliance, Lagos, Nigeria, pp. 245–252.

Esan, E.B. (1985). Clonal multiplication of *Theobroma cacao* L. zygotic embryos *in vitro* – asexual embryogenesis. Nigerian Agricult. J. 19/20:23–38.

Esan, E.B. (1992). Status of *in vitro* regeneration of tropical plantation crops. In: Biotechnology: Enhancing Research on tropical Crops in Africa. International Institute of Tropical Agriculture (IITA), Ibadan, Nigeria, pp. 157–164.

Etienne, H., Berger, A. and Carron, M.P. (1991). Water status of callus from *Hevea brasiliensis* during induction of somatic embryogenesis. Physiol. Plant. 82:213–218.

Evans, H.C. and Solórzando, G.R. (1982). 'Witches' broom' disease: Wrong experiment right results. In: 8th Intl. Cocoa Research Conference. Proceedings. Cartagena, Colombia. Cocoa Producers' Alliance, Lagos, Nigeria, pp. 415–418.

Falkner, F.R. (1990). The criteria for choosing an antibiotic for control of bacteria in plant tissue culture. International Association for Plant Tissue Culture, Newsletter 60:13–23.

Figueira, A., Whipkey, A. and Janick, J. (1991). Increased CO_2 and light promote growth and development of *Theobroma cacao*. J. Am. Soc. Hort. Sci. 116:585–589.

Finer, J.J., Reilly, A.A. and Smith, R.H. (1987). Establishment of embryogenic suspension cultures of a wild relative of cotton (*Gossypium klotzschianum* Anderss.). *In vitro* Cell. Devel. Biol. 23:717–722.

Fitter, M.S. and Krikorian, A.D. (1981). Recovery of totipotent cells and plantlet production from daylily protoplasts. Ann. Bot. 48:591–597.

Fitter, M.S. and Krikorian, A.D. (1982). Plant Protoplasts. Some Guidelines for their Preparation and Manipulation in Culture. Calbiochem-Behring, LaJolla, California.

Flynn, W.P., Glicenstein, L.J. and Fritz, P.J. (1990). *Theobroma cacao* L.: An axillary bud in vitro propagation procedure. Plant Cell Tissue Org. Cult. 20:111–117.

Folliot, M. and Marchal, J. (1990). Influence du support de culture sur la vitesse de croissance des vitroplants d'ananas en phase d'acclimation. Fruits 45:367–376.

Frison, E.A. and Putter, C.A.J. (eds.) (1989). FAO/IBPGR Technical Guidelines for the Safe Movement of *Musa* Germplasm. FAO, Rome/International Board for Plant Genetic Resources, Rome.

Fritz, P.J. (1987). Cacao biotechnology – An introduction. In: 41st Pennsylvania Manufacturing Confectioners' Association Annual Production Conference. Proceedings, pp.49–54.

Furtado, C.X. (1926–29). *Ananas sativa*. Gardens' Bull. (Straits Settlements. Singapore) ser. 3, 4:85–86.

Garner, R.J., Chaudhri, S.A. et al. (1976). The Propagation of Tropical Fruit Trees. Horticultural Review No. 4. FAO, Rome. Commonwealth Agricultural Bureaux, Farnham Royal, Slough, UK.

George, J., Höfer, E., Leike, H., Dautzenberg, H. and Loth, F. (1983). The use of novel stabilizing substrate for the *in vitro* cultivation of cauliflower (*Brassica olearacea* L. var. *botrytis*). Plant Cell Reports 2:189–191.

Glicenstein, L.J. and Fritz, P.J. (1989). Ploidy level in *Theobroma cacao* L. J. Heredity 80:464–467.

Glicenstein, L.J., Flynn, W.P. and Fritz, P.J. (1990). Clonal propagation of cacao. Cocoa Growers' Bulletin 43:7–10.

Greathouse, D.C., Laetsch, W.M. and Phinney, B.O. (1971). The shoot growth rhythm of a tropical tree, *Theobroma cacao*. Am. J. Bot. 58:281–286.

Gregory, P.H. (1977). Cocoa (*Theobroma cacao* L.). In: Pant Health and Quarantine in International Transfer of Genetic Resources. W.B. Hewitt and L. Chiarappa (eds.), CRC Press, Cleveland, pp 119–124.

Grierson, D. (ed., 1991). Plant Genetic Engineering. Blackie: Glasgow and London.

Griffiths, G. and Harwood, J.L. (1991). The regulation of triacylglycerol biosynthesis in cocoa (*Theobroma cacao*) L. Planta 184:279–284.

Groenewald, E.G., Wessels, D.C.J. and Koelman, A. (1977) Callus formation and subsequent plant regeneration from seed tissue of an *Agave* species (Agavaceae). Z. Pflanzenphysiol. 81:369–373.

Grumet, R. (1990) Genetically engineered plant virus resistance. Hort. Sc. 25:508–513.

Guerra, M.P. and Handro, W. (1988). Somatic embryogenesis and plant regeneration in embryo cultures of *Euterpe edulis* Mart. (Palmae). Plant Cell Rep. 7:550–552.

Gupta. P.K., Kendurkar, S.V., Kulkarni, V.M., Shirgurkar, M.V. and Mascarenhas, A.F. (1984). Somatic embryogenesis and plants from zygotic embryos of coconut (*Cocos nucifera* L.) *in vitro*. Plant Cell Rep. 3:222–225.

Haldeman, H.H. and Thomas, R.L. (1989), Multiple shoot production from single buds of tea, *Camellia sinensis*. The *Camellia* J. 44(2):23–26.

Haldemann, J.B., Thomas, R.L. and McKamy, D.L. (1987). Use of benomyl and rifampicin for *in vitro* shoot tip culture of *Camellia sinensis* and *Camellia japonica*. Hort. Sc. 22:306–307.

Hammerschlag, F.A. and Litz, R.E. (1992, eds.). Biotechnology of perennial fruit crops. CAB International, Wallingford, UK.

Hannigan, K.J. (1985). Biotechnology may improve natural flavors: Vanilla tech. Food Engineering, Sept., pp. 58–59.

Hardwick, K., Abo-Hamed S., and Collin, H.A. (1982). Hormonal control of shoot apex activity in *Theobroma cacao* L. In: 8th Intl. Cocoa Research Conference. Proceedings. Cartagena, Colombia. Cocoa Producers' Alliance, Lagos, Nigeria, pp. 253–257.

Hauptmann, R.M., Widholm, J.M. and Paxton, J.D. (1985). Benomyl: A broad spectrum fungicide for use in plant cell and protoplast culture. Plant Cell Rep. 4:129–132.

Heinz, D. (ed.) (1987). Sugarcane Improvement through Breeding. Elsevier, Amsterdam.

Henderson, W.E. and Kinnersley, A.M. (1988). Corn starch as an alternative gelling agent for plant tissue culture. Plant Cell Tissue Org. Cult. 5:17–22.

Hosoki, T. and Sagawa, Y. (1977). Clonal propagation of ginger (*Zingiber officinale* Roscoe) through tissue culture. Hort. Sc. 12:451–452.

Huang, L.-C. and Chi, D.-L. (1988) Pivotal roles of picloram and gelrite in banana callus culture. Environ. Exptl. Botany 28:249–258.

Hunter, J.R. (1990) The status of cacao (*Theobroma cacao*, Sterculiaceae) in the Western hemisphere. Economic Botany 44:425–439.

Iddagoda, N., Kateva, N.N. and Butenko, R.G. (1988) *In vitro* clonal micropropagation of tea (*Camellia sinensis*). I. Defining the optimum conditions for culturing by means of mathematical design technique. Indian J. Plant Physiol. 31:1–10.

Ikeda, L.R. and Tanabe, M.J. (1989). *In vitro* subculture applications for ginger. Hort. Sc. 24:142–143.

Irizarry, H., Rivera, E., Krikorian, A.D. and Rodriguez, J.A. (1991). Proper bunch management of the French-type 'Superplantain' (*Musa acuminata* × *M. balbisiana*, AAB) in Puerto Rico. J. Agric. Univ P.R. 75:163–171.

Janick, J. (1986). Embryogenics: The technology of obtaining useful products from the culture of asexual embryos. In: Biotechnology of Plants and Microorganisms. O.J. Crocomo, W.R. Sharp, D.A. Evans, J.E. Bravo, F.C.A. Tavares and E.F. Paddock (eds.), pp. 97–117.

Jarret, R.L. (1990). Identification of Genetic Diversity in the Genus *Musa*. Proceedings of an International Workshop held at Los Baños, Philippines 5–10 September 1988. INIBAP, International Network for the Improvement of Banana and Plantain, Montferrier sur-Lez, France.

Jordan, R.L. (1986). Diagnosis of plant viruses using double-stranded RNA. In: Tissue Culture as a Plant Production System for Horticultural Crops. R.H. Zimmerman, R.J. Griesbach, F.A. Hammerschlag and R.H. Lawson (eds.), Martinus Nijhoff Publishers, Dordrecht, pp. 125–134.

Kadje, A. and Ngambi Ndjama, J.-M. (1982). Essai de mise au point d'une methodologie de greffage chez le cacoyer (*Theobroma cacao* L.). In: 8th International Cocoa Research Conference. Proceedings. Cartagena, Colombia. Cocoa Producers' Alliance, Lagos, Nigeria, pp. 49–55.

Kao, K.N. and Michayluk, M.R. (1975). Nutrient requirements for growth of *Vicia hajastana* cells and protoplasts at a very low population density in liquid media. Planta 126:105–110.

Kato, M. (1989). Polyploids of *Camellia* through culture of somatic embryos. Hort. Sc. 24:1023–1025.

Kevers, C., Coumans, M.,Coumans-Gillès, M.-F. and Gaspar, Th. (1984). Physiological and biochemical events leading to vitrification of plants cultured *in vitro*. Physiol. Plant. 61:69–74.

Knauss, J.F. (1976). A tissue culture method for producing *Dieffenbachia picta* cv. 'Perfection' free of fungi and bacteria. Proc. Fla. State Hort. Soc. 89:293–296.

Kotalawala, J. (1971). Mass production of pineapple planting material. Tropical Agriculturist (Ceylon) 127:199–202.

Kovoor, A. (1981). Palm Tissue Culture: State of the Art and its Application to the Coconut. FAO Plant Production paper 30. FAO, Rome.

Krauss, B.H. (1948a). Anatomy of the vegetative organs of the pineapple, *Ananas comosus* (L.) Merr. I. Introduction, organography, the stem, and the lateral branch or axillary buds. Bot. Gaz. 110:159–217.

Krauss, B.H. (1948b). Anatomy of the vegetative organs of the pineapple, *Ananas comosus* (L.) Merr. II. The leaf. Bot. Gaz. 110:333–404.

Krauss, B.H. (1948c). Anatomy of the vegetative organs of the pineapple, *Ananas comosus* (L.) Merr. III. The root and the cork. Bot. Gaz. 110:550–587.

Krikorian, A.D. (1982). Cloning higher plants from aseptically cultured tissues and cells. Biol. Rev. 57:151–218.

Krikorian, A.D. (1988). Plant tissue culture: Perceptions and realities. Proc. Indian Acad. Sci. (Plant Sci.) 98:425–464.

Krikorian, A.D. (1989a). The context and strategies for tissue culture of date, African oil and coconut palms. In: Applications of Biotechnology in Forestry and Horticulture. V. Dhawan (ed.), Plenum Press, New York, pp. 119–144.

Krikorian, A.D. (1989b). Bananas and plantains from tissue culture: Background, update and call for information. Tropical Agriculture (Trinidad) 66:194–200.

Krikorian, A.D. (1990). Baseline tissue and cell culture studies for use in banana improvement schemes. In: *Fusarium* Wilt of Banana. R.C. Ploetz (ed.), APS Press, American Phytopathological Society, St. Paul, Minnesota, pp. 127–133.

Krikorian, A.D., Irizarry, H., Cronauer-Mitra, S.S. and Rivera, E. (1993). Clonal fidelity and variation in plantain (*Musa* AAB) regenerated from vegetative and floral apices *in vitro*. Ann. Bot. 71:519–535.

Krikorian, A.D. and Cronauer, S.S. (1984). Banana. In: Handbook of Plant Cell Culture. W.R. Sharp, D.A. Evans, P.V. Ammirato and Y. Yamadat (eds.), Vol. 2. Macmillan, New York, pp. 327–348.

Krikorian, A.D. and Kann, R.P. (1986). Oil palm improvement via tissue culture. Plant Breeding Rev. 4:175–201.

Krikorian, A.D. and Kann, R.P. (1987). Multiplication of Chinese tallow, *Sapium sebiferum*. In: Tissue Culture in Forestry case Histories: Gymnosperms, Angiosperms and Palms. J. Bonga and D.J. Durzan (eds.), Vol. 3. pp. 187–205.

Krikorian, A.D. and Rao, A.N. (1990). Bamboo production: Imperatives and research strategies. In: Bamboos: Current Research. Proceedings of the International Bamboo Workshop Held in Cochin, India from 14–18 November 1988. I.V. Ramanuja Rao, R. Gnanaharan and C.B. Sastry (eds.), The Kerala Forest Research Institute, India and International Development Research Centre, Canada, pp. 353–370.

Krikorian, A.D. and Smith, D.L. (1992). Somatic embryogenesis in carrot (*Daucus carota*). In: Plant Tissue Culture Manual: Fundamentals and Applications. K. Lindsey (ed.), Kluwer Academic Publishers, Dordrecht, pp. PTCM-A9-1-32..

Krikorian, A.D., Cronauer-Mitra, S.S. and Fitter Corbin, M.S. (1988). Protoplast culture of perennials. Sci. Hort. 37:277–293.

Krikorian, A.D., Singh, M. and Quinn, C.E. (1982). Aseptic micropropagation of Cinchona: prospects and problems. In: Tissue Culture of Economically Important Plants. Proceedings of the International Symposium Held at the Botany Department, National University of Singapore, Singapore, 28–30, April, 1981. A.N. Rao (ed.), COSTED and ANBS, Singapore, pp. 167–174.

Krikorian, A.D., Scott, M.E., Cronauer-Mitra, S.S. and Smith, D.L. (1990). *Musa* callus and cell culture: Strategies, achievements and directions. In: *In-vitro* Mutation Breeding of Bananas and Plantains I. Report of the First Research Coordination Meeting for the FAO/-IAEA, 29 May–2 June 1989. Technical Document, International Atomic Energy Agency, Vienna, 1989. IAEA 312.D2.RC.411, pp. 9–23.

Krikorian, A.D., Smith, D.L. and Kelly, K. (1987). Hormones in plant tissue culture and propagation. In: Plant Hormones and their Role in Plant Growth and Development. P.J. Davies, (ed.), Martinus Nijhoff/Dr. W. Junk Publishers, Dordrecht, pp. 592–613.

Lakshmi Sita, G.R., Singh, R. and Iyer, C.D.A. (1974). Plantlets through shoot-tip cultures in pineapple. Curr. Sci. 44:724–725.

Leal, F. and Amaya, L. (1991). The curagua (*Ananas lucidus*, Bromeliaceae) crop in Venezuela. Economic Botany 45:216–224.

Legrand, B., Cilas, C. and Mississo, E. (1984). Comportement des tissus de *Theobroma cacao* L. var. *amelonado* cultivés *in vitro*. Café Cacao Thé 28:245–250.

Leifert, C. and Waites, W.M. (1990). Contaminants of plant tissue culture. International Association for Plant Tissue Culture Newsletter 60:2–13.

Lievens, C., Pylyser, M. and Boxus, Ph. (1989). First results about micropropagation of *Anacardium occidentale* by tissue culture. Fruits 44:553–557.

Litz, R.E. (1989). *In vitro* strategies for tropical tree improvement. In: Applications of Biotechnology in Forestry and Horticulture. V. Dhaw (ed.), Plenum Press, New York, pp. 109–118.

Lloyd, G. and McCown, B. (1980). Commercially feasible-micropropagation of Mountain Laurel, *Kalmia latifolia*, by use of shoot-tip culture. Intl. Plant Prop. Soc. Proc. 30:421–427.

Lockhart, B.E.L. (1986). Purification and serology of a bacilliform virus associated with banana streak disease. Phytopathology 76:995–999.

Loison-Cabot, C. (1990). Génétique de l'ananas: Hérédité de certains caractères, leur stabilité au cours des cycles végétatifs. Fruits 45:447–455.

Loomis, W.D. (1974). Overcoming problems of phenolics and quinones in the isolation of plant enzymes and organelles. Methods in Enzymology 31A:528–544.

Lörz, H., Larkin, P.J., Thomson, J. and Snowcroft, W.R. (1983). Improved protoplast culture and agarose media. Plant Cell Tissue Org. Cult. 2:217–226.

Mabbett, T. (1987) How biotechnology can improve cocoa quality. Coffee & Cocoa International 3:57, 59.

Madhusudanan, K.N., Nabeesa, E., Umadevi, V. and Nandakumar, S. (1982). Differentiation of propagules and crop growth pattern in 20 pineapple cultivars. Sci. Hort. 18:215–224.

Mapes, M.O. (1973). Tissue culture of bromeliads. Proc. Int. Plant Prop. Soc. 23:47–55.

Mapes, M.O. (1974). Aseptic propagation and cloning of bromeliads. In: 3rd Intl. Congress of Plant Tissue and Cell Culture. Leicester, England. Abstracts: Demonstrations. Abstract 264.

Mapes, M.O. and Bullock, R.M. (1975). Tissue culture of edible ginger (*Zingiber officinale* Roscoe). Hort. Sc. 10:53.

Maretzki, A. (1987). Tissue culture: Its prospects and problems. In: Sugarcane Improvement through Breeding. D.J. Heinz, (ed.), Elsevier, Amsterdam, pp. 343–384.

Marques, D.V. (1990). One-step high frequency somatic embryos on Coffea spp. explants cultured *in vitro*. In: Abstracts 7th Intl. Congress on plant Tissue and Cell Culture. Amsterdam, June 24–29, 1990, p. 259.

Marshall, S.M. and Orr, A.P. (1954) Effect of different ions on gel strength of red seaweed extracts. Advan. Chem. Ser. 11:101–103.

Mathew, M. and Hariharan, M. (1990) In vitro shoot regeneration in *Syzygium aromaticum*. Ann. Bot. 65:277–279.

Mathews, V.H. and Rangan, T.S. (1979) Multiple plantlets in lateral bud and leaf explant *in vitro* cultures of pineapple. Sci. Hort. 11:319–328.

Mathews, V.H. and Rangan, T.S. (1981) Growth and regeneration of plantlets in callus cultures of pineapple. Sci. Hort. 14:227–234.

Mathews, V.H. and Rao, P.S. (1980) Morphogenetic investigations on members of Bromeliaceae (*Ananas comosus* and *Cryptanthus bromtricolor*). In: Proceedings of a National Symposium on Plant Tissue Culture, Genetic Manipulation and Somatic Hybridization of Plant Cells. P.S. Rao, M.R. Heble and M.S. Chadha (eds.), Department of Atomic Energy, Government of India, Bombay, pp. 192–196.

Mathews, V.H., Rangan, T.S. and Narayanaswamy, S. (1976). Micro-Propagation of *Ananas sativus in vitro*. Z. Pflanzenphysiol. 79:450–454.

Matsumoto, K. and Yamaguchi, H. (1989). Nonwoven materials as a supporting agent for *in vitro* culture of banana protocorm-like bodies. Trop. Agric. (Trinidad) 66:8–10.

McClure, J.W. (1979). The physiology of phenolic compounds in plants. Rec. Adv. Phytochem. 12:525–536.

Meunier, J., Baudoin, L., Nouy, B. and Noiret, J.M. (1988). Estimation de la valeur des clones de palmier à huile. Oléagineux 43:195–200.

Micke, A., Donini, B. and Maluszynski, M. (1987). Induced mutations for crop improvement – a review. Trop. Agric. (Trinidad) 64:259–278.

Miller, J.M. and Backhaus, R.A. (1986). Rubber content in diploid guayule (*Parthenium argentatum*): Chromosomes, rubber variation, and implications for economic use. Economic Botany 40:366–374.

Molina, G.C. and Krausz, J.P. (1989). A phytotoxic activity in extracts of broth cultures of *Mycosphaerella fiiensis* var. *fijiensis*. Plant Dis. 73:142–144.

Monfort, S. (1985). Androgenesis of coconut: Embryos from anther culture. Z. Pflanzenzucht. 94:251–254.

Mori, K. (1971). Production of virus-free plants by means of meristem culture. Japan Agric. Res. Ouart. 6:1–7.

Nadgauda, R.S., Kulkarni, D.D., Mascarenhas, A.F. and Jagannathan, V. (1980). Development of plantlets from cultured tissues of ginger (*Zingiber officinale* Roscoe). In: Proceedings of a National Symposium on Plant Tissue Culture, Genetic Manipulation and Somatic Hybridization of Plant Cells. P.S. Rao, M.R. Heble and M.S. Chadha, (eds.), Department of Atomic Energy, Government of India, Bombay, pp. 358–365.

Nadguada, R.S., Mascarenhas, A.P. and Madhusoodanan, K.J. (1983). Clonal multiplication of cardamom (*Elleteria cardamomum* Maton) by tissue culture. J. Plantation Crops 11:60–64.

Nadgauda, R.S., Mascarenhas, A.F., Hendre, R.R. and Jagannathan, V. (1978). Rapid multiplication of turmeric (*Curcuma longa* Linn.) plants by tissue culture. Indian J. Exptl. Biol. 16:120–122.

National Research Council (1990). Plant Biotechnology Research for Developing Countries. National Academy Press, Washington, DC.

Noiret, J.M., Gascon, J.P. and Pannetier, C. (1985) La production de palmier à huile par huile. Oléagineux 40:365–372.

Novak, F.J., Afza, R., van Duren, M., Perea-Dallos, M., Conger, B.V. and Xiolang, T. (1989). Somatic embryogenesis and plant regeneration in suspension cultures of dessert (AA and AAA) and cooking (ABB) bananas (*Musa* spp.). Biotechnology 7:154–159.

Novak, F., Afza, R., van Duren, M. and Omar, M.S. (1990). Mutation induction of *in vitro* cultured shoot-tips of banana and plantain (*Musa* cvs.). Trop. Agr. (Trin.) 67: 21–28.

Nwankwo, B.A. and Krikorian, A.D. (1983). Morphogenetic potential of embryo- and seedling-derived callus of *Elaeis guineensis* Jacq. var. *pisifera* Becc. Ann. Bot. 51:65–76.

Packer, L. and Douce, R. (eds.) (1987). Plant cell membranes. Methods in Enzymology 148:1–762.

Paiva, M., Wright, D.C. and Janick, J. (1982). Alkaloid variation in *Theobroma cacao* L. In: 8th Intl. Cocoa Research Conference. Proceedings. Cartagena, Colombia. Cocoa Producers' Alliance, Lagos, Nigeria, pp. 229–233.

Panis, B., Van Wauve, A. and Swennen, R. (1993). Plant regeneration through direct somatic embryogenesis from protoplasts of banana (*Musa* spp.). Plant Cell Rep. 12:403–407.

Pannetier, C. and Lanaud, C. (1976). Divers aspects de l'utilisation possible des cultures *in vitro* pour la multiplication végétative de l'*Ananas comosus* L. Merr., variété 'Cayenne lisse'. Fruits 31:739–750.

Pannetier, C. and Buffard-Morel, J. (1986). Coconut palm (*Cocos nucifera* L.). In: Biotechnology in Agriculture and Forestry 1. Trees I. Y.P.S. Bajaj (ed.), Springer-Verlag, Berlin, pp. 430–450.

Paranjothy, K. (1984). Oil palm. In: Handbook of Plant Cell Culture. P.V. Ammirato, D.A. Evans, W.R. Sharp and Y. Yamada (eds.), Vol. 3. Macmillan Publishing, New York, pp. 591–605.

Paranjothy, K. (1987a). *Hevea* tissue culture. In: Cell and Tissue Culture in Forestry. J.M. Bonga and D.J. Durzan (eds.), Vol. 3. Martinus Nijhoff Publishers, Dordrecht, pp. 326–337.

Paranjothy, K. (1987b). Recent developments in cell and tissue culture of oil palm. In: Agricultural Applications of Biotechnology. A.N. Rao and H.Y. Mohan Ram (eds.), COSTED, Madras, pp. 83–95.

Passey, A.J. and Jones, O.P. (1983). Shoot proliferation and rooting *in vitro* of *Theobroma cacao* L. type Amelonado. J. Hort. Sci. 58:589–592.

Pence, V.C., Hasegawa, P.M. and Janick, J. (1980). Initiation and development of asexual embryos of *Theobroma cacao* L. *in vitro*. Z. Pflanzenphysiol. 98:1–14.

Philip, V.J. (1984). *In vitro* organogenesis and plantlet formation in cashew (*Anacardium occidentale* L.). Ann. Bot. 54:149–152.

Philip, V.J., Joseph, D., Triggs, G.S. and Dickinson, N.M. (1992). Micropropagation of black pepper (*Piper nigrum* L.) through shoot tip cultures. Plant Cell Rep. 12:41–44.

Philip, V.J. and Nainar, S.A.Z. (1976). Clonal propagation of *Vanilla planifolia* (Salisb.) Ames using tissue culture. J. Plant Physiol. 122:211–215.

Philip, V.J. and Nainar, S.A.Z. (1988). *In vitro* transformation of root meristem to shoot and plantlets in *Vanilla planifolia*. Ann. Bot. 61:193–199.

Pillai, S.K. and Kumar, K.B. (1982). Note on the clonal multiplication of ginger *in vitro*. Indian J. Agric. Sci. 52:397–399.

Ploetz, R.C. (ed.) (1990). *Fusarium* Wilt of Banana. APS. The American Phytopathological Society, St. Paul.

Pollard, J.W. and Walker, J.M. (eds.) (1990). Plant Cell and Tissue Culture. Methods in Molecular Biology 6. Humana Press, Clifton, New Jersey.

Prakash Kumar, P., Raju, C.R., Chandramohan, M. and Iyer, R.D. (1985). Induction and maintenance of friable callus from the cellular endosperm of *Cocos nucifera* L. Plant Sci. 40:203–207.

Price, H.J. and Smith, R.H. (1984). Cotton. In: Handbook of Plant Cell Culture. P.V. Ammirato, D.A. Evans, W.R. Sharp and Y. Yamadaf (eds.), Vol. 3. Macmillan, New York, pp. 486–510.

Proskauer, J. and Berman, R. (1970). Agar culture medium modified to approximate soil conditions. Nature (London) 227:1161.

Purdy, L.H. and Dickstein, E.R. (1989). *Theobroma cacao*, a host for *Agrobacterium tumefaciens*. Plant Disease 73:638–639.

Py, C., Lacoeuilhe, J.J. and Teisson, C. (1984). L'Ananas. Sa Culture, Ses Produits. Editions Maissoneuve & Larouse, Paris.

Raju, C.R., Prakash Kumar, Chandramohan, M. and Iyer, R.D. (1984). Coconut plantlets from leaf cultures. J. Plantation Crops 12: 75–79.

Rangan, T.S. (1984). Pineapple. In: Handbook of Plant Cell Culture. P.V. Ammirato, D.A. Evans, W.R. Sharp, Y. Yamada (eds.), Vol. 3. Macmillan, New York, pp. 373–382.

Rao, A.N. (ed.) (1982). Tissue Culture of Economically Important Plants. Proceedings of the International Symposium Held at the Botany Department, National University of Singapore, Singapore, 28–30, April, 1981. COSTED and ANBS, Singapore.

Redenbaugh, K. (1990). Application of artificial seed to tropical crops. Hort. Sc. 25:251–255.

Reuveni, O. (1990). Methods for detecting somaclonal variants in 'Williams' bananas. In: Identification of Genetic Diversity in the Genus *Musa*. Proceedings of an International Workshop held at Los Baños, Philippines 5–10 September 1988. R.L. Jarret 7 (ed.), INIBAP, International Network for the Improvement of Banana and Plantain, Montferrier sur-Lez, France, pp. 108–113.

Rowe, P. (1991). Breeding bananas and plantains. Plant Breed. Rev. 2:135–155.

Rowe, P. and Rosales, F.E. (1994). Bananas and plantains. In: Fruit Breeding. 2nd ed., J. Janick (ed.), John Wiley & Sons, New York (in press).

Samonte, J.L., Mendoza, E.M.T., Ilag, L.L., de la Cruz, N.B. and Ramirez, D.A. (1989). Galactomannan degrading enzymes in maturing normal and Makapuno and germinating normal coconut endosperm. Phytochemistry 28:2269–2273.

Samartin, A. (1989). A comparative study of effects of nutrient media and cultural conditions on shoot multiplication of *in-vitro* cultures of *Camellia japonica* explants. J. Hort. Sc. 64:73–79.

Sasson, A. and Costarini, V. (eds.) (1990). Plant Biotechnologies for Developing Countries. Proceedings of an International Symposium Organized by CTA and FAO, 26–30 June 1989, Luxembourg. Technical Centre for Agricultural and Rural Co-operation (CTA) and FAO, Rome.

Schöpke, C., Müller, L.E. and Kohlenbach, H.W. (1987). Somatic embryogenesis and regeneration of plantlets in protoplast cultures from somatic embryos of coffee (*Coffea canephora* P. ex Fr.). Plant Cell Tissue Org. Cult. 8:243–248.

Schwendiman, J., Pannetier, C. and Michaux-Ferrière, N. (1988). Histology of somatic embryogenesis from leaf explants of the oil palm *Elaeis guineensis*. Ann. Bot. 62:43–52.

Schwendiman, J., Pannetier, C. and Michaux-Ferrière, N. (1990). Histology of embryogenic formations during *in vitro* culture of oil palm *Elaeis guineensis* Jacq. Oléagineux 45:409–418.

Singh, I.D. (1978). Application of plant cell, tissue and organ cultures in tea research and development. Two and A Bud 25:65–69.

Shama Bhat, K. and Radhakrishnan Nair, C.P. (1985). Arecanut. Research and Development. Central Plantation Crops Research Institute, Kasaragod, India.

Shields, R., Robinson, S.J. and Anslow, P.A. (1984). Use of fungicides in plant tissue cultures. Plant Cell Rep. 3:33–36.

Simmonds, N.W. (1990). Breeding horizontal resistance to South American leaf blight of rubber. J. Nat. Rubb. Res. 5:102–113.

Simmonds, N.W. (1991). Genetics of horizontal resistance to diseases of crops. Biol. Rev. 66:189–241.

Singh, R. and Jain, S.M. (1990). The application of biotechnologies to tea. In: Plant Biotechnologies for Developing Countries. Proceedings of an International Symposium Organized by CTA and FAO, 26–30 June 1989, Luxembourg. Technical Centre for Agricultural and Rural Co-operation (CTA) and FAO, Rome.

Singha, S., Bissonnette, G.K. and Double, M.L. (1987). Methods for sterilizing instruments with *Bacillus* sp. from plant tissue cultures. Hort. Sc. 22:659.

Smith, D.L. and Krikorian, A.D. (1989). Release of somatic embryogenic potential from excised zygotic embryos of carrot and maintenance of proembryonic cultures in hormone-free medium. Am. J. Bot. 76:1832–1843.

Smith, L.B. and Downs, R.J. (1979). Bromelioideae (Bromeliaceae). Flora Neotropica. Monogr. 14(3):1493–2142.

Smith, M.K. (1987). A review of factors influencing the genetic stability of micropropagated bananas. Fruits 43:219–223.

Smith, M.K. and Drew, R.A. (1990). Current applications of tissue culture in plant propagation and improvement. Austral. Physiol. 17:267–289.

Sondahl, M.R. and Loh, W.-H.T. (1988). Coffee biotechnology. In: Coffee. R.J. Clarke and R. Macrae (eds.), Vol. 4. Elsevier, London and New York, pp. 235–262.

Sondahl, M.R., Caldas, L.S., Maraffa, S.B. and Sharp, W.R. (1980). The physiology of asexual embryogenesis. Hort. Revs. 2:268–310.

Sondahl, M., Liu, S., Bellato, C. and Bragin, A. (1993). Cacao somatic embryogenesis (in press).

Sondahl, M.R., Nakamura, T. and Sharp, W.R. (1991). Propagación in vitro del café. In: Cultivo de Tejidos en la Agricultura. Fundamentos y Aplicaciones. W.M. Roca and L.A. Mroginski (eds.), CIAT, Centro Internacional de Agricultura Tropical, Cali, Colombia, pp. 621–642.

Sondahl, M.R., Nakamura, T., Medina-Filho, H.P., Carvalho, A., Fazuoli, L.C. and Costa, W.M. (1984). Coffee. In Handbook of Plant Cell Culture. P.V. Ammirato, D.A. Evans, W.R. Sharp and Y. Yamada (eds.), Vol. 3. Macmillan, New York, pp. 486–510.

Sorvari, S. and Schieder, O. (1987). Influence of sucrose and melibiose on barley anther cultures in starch media. Plant Breeding 99:164–171.

Soto, B.M. (1985). Bananos; Cultivo y Comercialización. Litografía e Imprenta LIL, San José, Costa Rica.

Srinivasa Rao, N.K., Dore Swamy, R. and Chacko, E.K. (1981). Differentiation of plantlets in hybrid embryo callus of pineapple. Sci. Hort. 15:235–238.

Srivastava, H.C., Vatsya, B. and Menon, K.K.G. (eds.) (1986). Plantation Crops. Opportunities and Constraints. 2 Vol. Oxford & IBH Publishing Co., New Delhi.

Srinivasa Rao, N.K., Narayanaswamy, S., Chacko, E.K. and Dore Swamy, R. (1982). Regeneration of plantlets from callus of *Elleteria cardamomum* Maton. Proc. Indian Acad. Sci. (Plnt Sci.) 91:37–41.

Staba, E.J. and Nygaard, B.G. (1983). *In vitro* culture of guayule. Z. Pflanzenphysiol. 109:371–378.

Stover, R.H. and Buddenhagen, I. (1986). Banana breeding: polyploidy, disease resistance and productivity. Fruits 45:175–191.

Stover, R.H. and Simmonds, N.W. (1987). Bananas. 3rd ed. Longman, New York.

Tahardi, S. (1989). *In vitro* approaches to oil palm improvement. In: Proceedings on the Workshop on Biotechnology for the Improvement of Oil- and Steroid-Producing Plants, Jakarta, 5–6 April 1988. The Bogor Research Institute for Estate Crops, Bogor, Indonesia, pp. 49–55.

Tahardi, S. and Warga-Dalem, K. (1982). Kultur embrio kelapa kopyor *in vitro*. Menara Perkebunan 50:127–130.

Teo, C.K.H. (1974). Clonal propagation of pineapple (*Ananas comosus*) by tissue culture. Planter (Kuala Lumpur) 50:58–59.

Thompson, W., Collin, H.A., Isaac, S. and Hardwick, K. (1987). Isolation of protoplasts from cocoa (*Theobroma cacao* L.) leaves. Café Cacao Thé 31: 115–120.

Timir Baran Jha (1988). *In vitro* morphogenesis in cashewnut *Anacardium occidentale* L. Indian J. Exptl. Biol. 26:505–507.

Ting, T.W. and Chua, B.K. (1975). Tissue culture of pineapple. In: Proceedings of the National Plant Tissue Culture Symposium, 21–22 February 1975. J.C. Rajarao and K. Paranjothy (eds.), Organised by the Rubber Research Institute of Malaysia and the Malaysian Agricultural Research and Development Institute, Kuala Lumpur, pp. 51–58.

Tisserat, B. (1984). Date palm. In: Handbook of Plant Cell Culture. W.R. Sharp, D.A. Evans, P.V. Ammirato and Y. Yamada (eds.), Vol. 2. Macmillan, New York, pp. 505–545.

Tisserat, B. (1987). Palms. In: Cell and Tissue Culture in Forestry. J.M. Bonga and D.J. Durzan (eds.), Vol. 3. Martinus Nijhoff Publishers, Dordrecht, pp. 338–356.

Tomlinson, P.B. (1990). The Structural Biology of Palms. Clarendon Press, Oxford.

Uhl, N.W. and Dransfield, J. (1987). Genera Palmarum. A Classification of Palms Based on the Work of Harold E. Moore, Jr. The L.H. Bailey Hortorium and The International Palm Society, Allen Press, Lawrence, Kansas.

Uritani, I. (1978). Biochemistry of host response to infection. Prog. Phytochem. 5:29–63.

Valverde, R., Arias, O. and Thorpe, T.A. (1987). Picloram-induced somantic embryogenesis in pejibaye palm (*Bactris gasipaes* H.B.K.) Plant Cell Tissue Org. Cult. 10:149–156.

Vaughn, K.C. and Duke, S.O. (1984). Function of polyphenol oxidase in higher plants. Physiol. Plant. 60:106–112.

Verdeil, J.L., Buffard-Morel, J. and Pannetier, C. (1989). Somatic embryogenesis of coconut (*Cocos nucifera* L.) from leaf and inflorescence tissue. Oléagineux 44:403–411.

Villalobos A., V.M., Mejía Muñoz, J.M. and Escobar Araya, H.A. (1991). Micropropagación de opuntias y agaves. In: Cultivo de Tejidos en la Agricultura. Fundamentos y Aplicaciones. W.M. Roca and L.A. Mroginski (eds.), CIAT, Centro Internacional de Agricultura Tropical, Cali, Colombia, pp. 643–650.

Villalobos, I., Gámez, A. and Arias, O. (1987). Morphogenetic response of Pejibaye (*Guillielma gasipaes* (H.B.K.) *in vitro*. In: International Congress of Plant Tissue Culture Tropical Species. Abstracts. A. Angarita Z. (ed.), Bogota, Colombia, p. 16.

Villalobos, V.M. (1990). Advances in tissue culture methods applied to coffee and cocoa. In: Plant Biotechnologies for Developing Countries. Proceedings of an International Symposium Organized by CTA and FAO, 26–30 June 1989, Luxembourg. Technical Centre for Agricultural and Rural Co-operation (CTA) and FAO, Rome, pp. 247–250.

Vuylsteke, D.R. (1989). Shoot-tip Culture for the Propagation, Conservation and Exchange of *Musa* Germplasm. International Board for Plant Genetic Resources, Rome.

Wakasa, K. (1979). Variation in the plants differentiated from the tissue culture of pineapple. Ikushu-gaku Zasshi (Japan J. Breed.) 29:13–22.

Wakasa, K., Koga, Y. and Kudo, M. (1978). Differentiation from *in vitro* of *Ananas comosus*. Ikushu-gaku Zasshi (Japan. J. Breed.) 28:113–121.

Wang, P.J. and Huang, H.W. (1978). The induction of polyploid of pineapple through tissue culture. Ann. Report. Institute of Botany. Academia Sinica. TaiPei, R.O.C. July 1977–June 1978, pp. 18–19.

Wang, P.J. (1990). Regeneration of virus-free plants through tissue culture. Adv. in Biochem. Engin. 18:61–99.

Wee, Y.C. (1982). Vegetative propagation, tissue culture and the Malaysian pineapple industry. In: Tissue Culture of Economically Important Plants. Proceedings of the International Symposium Held at the Botany Department, National University of Singapore, Singapore, 28–30, April, 1981. A.N. Rao (ed.), COSTED and ANBS, Singapore, pp. 121–123.

Werckmeister, P. (1971). Light induction of geotropism, and the control of proliferation and growth of *Cymbidium*. Bot. Gaz. 132:346–350.

Williams, R.J. and Leopold, A.C. (1989). The glassy state in corn embryos. Plant Physiol. 89:977–981.

Wooi, K.C. (1984). Palm tissue culture. In: Micropropagation of Selected Rootcrops, Palms, Citrus, and Ornamental Species. FAO Plant Protection Paper 59. FAO, Rome, pp. 88–112.

Yadav. U., Lai, M. and Jaiswal, V.S. (1990). *In vitro* micropropagation of the tropical fruit tree *Syzygium cuminii* L. Plant Cell Tissue Org. Cult. 21: 87–92.

Yam, T.W., Arditti, J. and Weatherhead, M.A. (1989). The use of darkening agents in orchid seed germination and tissue culture media for orchids: A review. J. Orchid Soc. India 3:35–39.

Yeoh, H.H., Chung, D.K. and Fritz, P.J. (1990). *Theobroma cacao* chloroplast DNA: Isolation, molecular cloning, and characterization. Café, Cacao, Thé 34:173–178.

Zepeda, C. and Sagawa, Y. (1981). *In vitro* propagation of pineapple. Hort. Sc. 16:495.

Ziv, M. and Halevy, A. (1983). Control of oxidative browning and *in vitro* propagation of *Strelitzia reginae*. Hort. Sc. 18: 434–436.

21. *In vitro* Culture of Forest Trees

Indra S. Harry and Trevor A. Thorpe

Contents

1. Introduction

Forest resources are disappearing at an unprecedented rate. An important component of forest ecosystems are trees, which provide food, fuel, construction and industrial products. In addition, trees are recognized as the critical elements in maintaining stability in the world's atmosphere. Currently, a major concern is the erosion of genetic variability for a given species. Therefore, both *in situ* and *ex situ* management is necessary for biodiversity, conservation and consequently for tree crop improvement (NRC, 1991).

When grown as a crop, forest trees represent a renewable resource and although lagging behind agricultural crops, attempts have been made to improve several economic species by conventional breeding. Tree improvement involves managing genetic resources, and includes conservation, selection and breeding, and propagation (Cheliak and Rogers, 1990). Biotechnology has the potential to accelerate the genetic improvement of forest yield (Timmis et al., 1987) and in the short term, this impact will revolve around the cloning of selected phenotypes. As *in vitro* methods become more integrated with breeding and selection programs, other aspects such as engineering trees for disease and herbivore resistance, fiber quality, drought tolerance and so on will be added to tree improvement programs. However, the applications of tissue culture technology to tree improvement, other than for micropropagation, have remained largely unfulfilled (Thorpe, 1983; Haissig et al., 1987).

Indra K. Vasil and Trevor A. Thorpe (eds.), Plant Cell and Tissue Culture, 539–560.

The traditional methods for propagation of trees are by seeds and rooted cuttings or grafting. Vegetative propagation for *Cryptomeria* has been used for centuries in Japan (Isikawa, 1986). This method remains popular since it retains the genetic integrity of the donor trees. Currently, *in vitro* propagation is being viewed as an alternative method of clonal propagation. Considerable progress has been made since true plantlets were obtained via organogenesis from *Populus tremuloides* (Winton, 1970). Until 1975, hardwood micropropagation involved the production of callus, and the subsequent regeneration of plantlets from these cultures, but more recently, organ cultures are being used (Chalupa, 1987).

Gymnosperms or softwoods represent a division of cone bearing plants which includes conifers. Most species are evergreen trees or shrubs, and some tree species form extensive temperate forests. Economically, this group is extremely important because they are the chief source of softwood lumber, pulp and paper (Wenger, 1984). Shoots from *Sequoia* callus were obtained in the 1950s (Ball, 1950), but it was not until the 1970s that complete plantlets were obtained from *Pinus palustris* (Sommer et al., 1975). Since then, rapid progress has been made in developing *in vitro* methods for forest trees. This chapter describes the factors involved in clonal propagation using both organogenic and somatic embryogenic routes, as well as various aspects of protoplast culture and the implications for genetic engineering, and problems and prospects for future applications of these methods.

2. Clonal Propagation

There are two main objectives for developing clonal techniques for forest trees. Firstly, viable large-scale clonal systems can be an asset for selected trees and secondly, reliable protocols are necessary for further genetic manipulation. Some of the requirements for a large-scale system are simplicity, reproducibility, cost effectiveness and adaptability to nursery practices (Ajuha, 1987).

Asexual multiplication of both hardwoods and softwoods using tissue culture methods can be achieved by axillary bud breaking, production of adventitious buds and somatic embryogenesis. The first two approaches lead to the formation of unipolar shoots which are subsequently rooted, and somatic embryogenesis leads to the development of a bipolar embryo, through stages similar to zygotic embryogenesis. To date, plantlets have been obtained from over 70 angiosperms and 30 gymnosperms (see Table 1, Thorpe et al., 1991). Plantlets via axillary buds were produced from about 45% of the angiosperms, 13% from adventitious buds, 25% indirectly from callus and 17% from somatic embryogenesis; for gymnosperms, the percentages were 28, 55, 5, and 12%, respectively. Therefore, for angiosperms the most successful method was axillary bud multiplication, and for gymnosperms, direct adventitious bud formation.

2.1. *Organogenesis*

2.1.1. *Plant Material*

A major advantage of manipulating angiosperm trees *in vitro* is that a wide range of material can be used as explants. These include seeds, seedling parts, leaves, shoots and buds from both mature and juvenile trees, floral parts and in some cases even roots. This extensive source of explants is crucial since the viable period of seeds from many hardwood species is quite short. For gymnosperms, whole embryos or excised parts are the most succesful explants. Needle fascicles have been used with various degrees of success, but in general explants used for clonal propagation of conifers depend on the species, research objectives, and to the recalcitrant nature of conifers in general.

When seeds are used as explants, they must be decontaminated before removal of the embryos or for aseptic germination. For tropical species, decontamination procedures are fairly stringent because of the high degree of natural contamination, and a combination of $HgCl_2$, NaOCl and ethanol is often used (Le Roux, 1991; Mascarenhas et al., 1987). The procedures for gymnosperms are not as stringent, and a combination of NaOCl, H_2O_2 and sometimes ethanol usually is sufficient for decontamination. However, several steps like scarification, imbibition, sterilization, and sometimes stratification and re-sterilization e.g., Canary Island pine (Martinez Pulido et al., 1990) are required. In general, the morphogenic response and germination percentage of most conifers improves with an empirically determined stratification period e.g., radiata pine (Aitken-Christie and Thorpe, 1984). These steps are modified according to the species, and smaller seeds like lodgepole pine can be imbibed first and then sterilized, or decontaminated in a dry state and imbibed overnight in sterile water like the procedure established by Patel and Thorpe (1984). Decontamination procedures are modified according to the size of the seeds, the thickness of the seed coats, how the seeds were collected and subsequently stored, the presence or absence of seed coat appendages, and the geographical region of origin.

Depending on the species, several types of shoots can be used as explants. These include coppice, scion or epicormic shoots. Again, these shoots have to be contaminant-free, and the objective of sterilization is to obtain disinfected material without damaging or killing the tissues. The age of the material and when and where it was harvested, determine which treatment should be used. Also, several preharvest techniques can be employed to reduce surface contaminants. For example, shoots can be treated *in vivo* with fungicides before explants are removed. Conversely, an intermediary stage can be added and shoots or specimens can be removed from field conditions and grown under controlled environmental conditions like greenhouses or plant growth chambers. Explants from these sources will require less stringent, decontamination procedures like those used for chestnut (Vieitez et al., 1986).

Buds are also used as explants for several woody angiosperm species like *Populus*, *Quercus* and *Betula* (McCown and McCown, 1987). Usually branches are collected in the spring and the intact terminal or axillary buds are sterilized and the shoot apices (the apical meristem and about two leaf primordia) excised and cultured. Unlike angiosperms, shoots and buds are used rarely as explants for initiating cultures of conifers. Some exceptions include redwood (*Sequoiadendron giganteum*) where apical meristems were used to clone mature trees (Monteuuis, 1991), and Douglas fir (*Pseudotsuga menziesii*) where dormant buds from 17 to 20-year-old trees were used as explants (Dunstan et al., 1986).

Flowers, anthers and leaves are also used for hardwood culture. An important consideration for the use of anthers is the stage of microspore development. Therefore, a common practice when using poplar anther culture for example, is to harvest branches with catkin buds from selected trees in early spring, and keep these in water for further bud development (Stoehr and Zsuffa, 1990). Alternatively, branches can be collected in late winter and stored at low temperatures. These can be forced to flower and flush with leaves as required. In addition, branches can be rooted, and grown under controlled conditions to lessen the contamination load and provide a readily available source of explants. Lignotubers can also be used for *in vitro* shoot generation. These are produced at the bases of young eucalyptus trees, and both *in situ* and *in vitro* treatments are necessary for decontamination before any culture manipulation can be done (Le Roux, 1991).

2.1.2. *Media and Culture Conditions*

A variety of media and salt concentrations have been used, but Murashige and Skoog or MS salts (1962) still remain the most widely used formulation for angiosperms. Several researchers have used modifications of this medium or other formulations like Lloyd and McCown's woody plant medium or WPM (1980), Gresshoff and Doy's GD formulation (1972), or Gamborg et al. B5 salts (1968). Various formulations used for conifers but rarely for hardwoods include von Arnold and Eriksson's AE or LP (1981), Schenk and Hildebrandt's SH (1972), Bornman's MCM (1983), Quoirin and LePoivre's QP (1977), Litvay et al. (1981), Gupta and Durzan's DCR (1985) and to a lesser extent MS salts. These salt combinations are used either at full-strength or with modifications based on experimental results. Several researchers have combined the macro salts or organic components of one formulation with that of another. For example, the medium for larch induction using whole excised embryos was half-strength QP salts with SH organics (Harry et al., 1991). These formulations show differences in nitrogen, potassium and phosphate levels, and in the overall ionic strength of the medium itself (McCown and Sellmer, 1987).

Numerous additional components such as coconut milk, casein hydrolysate, glycine, vitamins, sucrose, phytohormones and various kinds of agar are added. The single most important medium additive are the cytokinins. Most commonly used cytokinins are N^6-benzyladenine (BA), 2iP, kinetin or zeatin, and auxins include α-naphthaleneacetic acid (NAA) and 2,4-dichlorophenoxyacetic acid (2,4-D) for callus induction, and indoleacetic acid (IAA) and indolebutyric (IBA) and sometimes β-indolepropionic acid (IPA) for rooting. The most widely used cytokinin is BA and concentrations vary from 0.1 to 1.0 μM for hardwoods (McCown and McCown, 1987) and up to 25 μM for softwoods (Thorpe and Harry, 1991). However, with some species, mixed cytokinins have proven to be beneficial especially in the quality of shoots produced. Kinetin and zeatin in combination with BA produced more vigorous shoots in larch than BA used singly (Harry et al., 1991). This factor is very important because conifer shoots elongate very slowly. The addition of auxins, while sometimes used in hardwood culture, is rarely used for conifers because even in low concentrations, callus formation increases and organogenesis is reduced.

Incubation conditions are similar for many cultures but in some instances, based on whether the original material is tropical or temperate, environmental conditions have to be determined. For tropical species like teak, for example, 28 °C was used (Mascarenhas et al. 1987), and for temperate poplar species, a range of 22 to 26 °C was appropriate depending on the explants being cultured (Park and Son, 1988). Although conifers grow naturally under a wide range of climatic conditions, the *in vitro* culture requirements are surprisingly similar for most species. For example, both lodgepole pine and Canary Island pine, a temperate and semi-tropical species, were cultured and maintained at 27 ± 1 °C, 16 h photoperiod, and light intensity of ca. 80–100 $\mu E.m^{-2} s^{-1}$ from Gro-Lux® WS lamps (Patel and Thorpe, 1984; Martinez Pulido et al., 1990). However, several species respond better at lower temperatures, e.g., 24 °C for larch (Harry et al., 1991).

2.1.3. *Culture Establishment*

After sterilization, explants are transferred to pre-determined media with the necessary supplements and phytohormones added for multiplication of axillary buds, *de novo* production of adventitious buds, or for the induction of somatic embryogenic tissue. The starting point for several species like teak and eucalyptus is the germination of decontaminated seeds on water agar or in sterile soilless mixes for a source of axenic juvenile explants. Several species are propagated utilizing pre-formed buds on shoot tips and nodal cuttings. This method has formed the basis for a rapid multiplication system for eucalyptus yielding about one million microshoots per year from one shoot (Le Roux, 1991), and is also successful with poplar, a rapidly growing forestry species used for pulp, paper and plywood proliferation. From 120 to 220 shoots were obtained from each original bud explant after 6–8 weeks and about one million poplar plantlets could be produced annually from one bud (Ajuha, 1987). Examples of established cultures of birch and

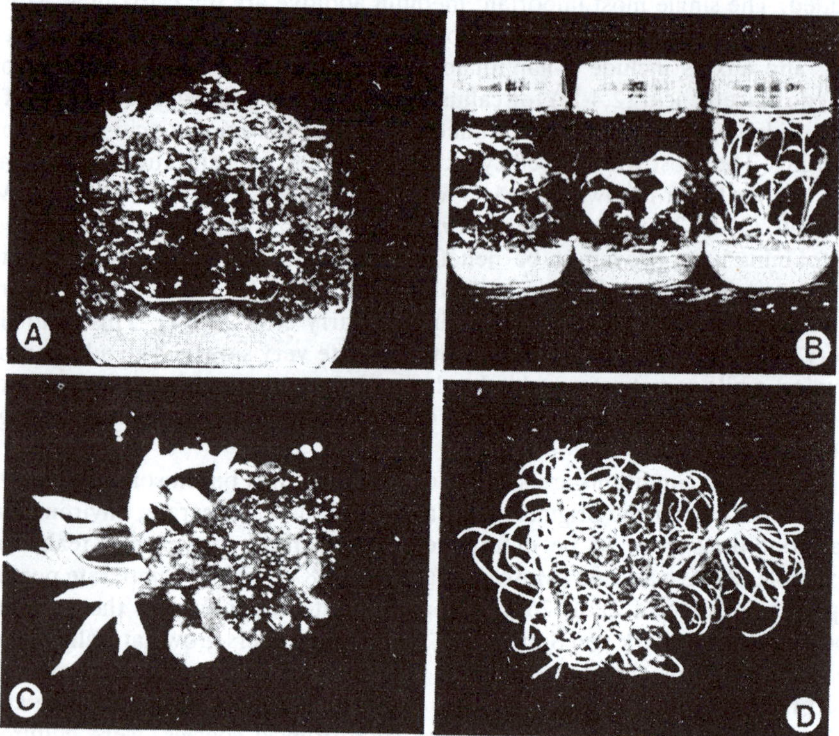

Figure 1. (A) Birch shoot cultures obtained from buds (photograph courtesy Brent McCown); (B) Poplar shoot cultures obtained from buds (photograph courtesy Jim Sellmer); (C) Embryonic explant of larch showing developing buds after 3 weeks in culture; (D) Embryonic explant of larch showing elongated shoots after 4 months in culture.

poplar shoots are shown in Fig. 1A and 1B, respectively. Both types of shoots were generated from buds from mature trees.

Leaves of several hardwood species can be used as a source of explants for producing adventitious buds. Park and Son (1988), for example, used punctured leaves from *in vitro* stock cultures of *Populus nigra* × *P. maximowiczii* to generate shoot buds. For other species like elm, no wounding was necessary and shoot primordia appeared directly from the margins of elm leaves cultured on MS medium with 2–4 μM BA. After a critical size, shoot clusters can be removed from the original explant and cultured separately for microshoot development (Fink et al., 1986).

To date, most of the clonal propagation of conifers is through *de novo* production of adventitious buds directly on the explants. Two stages in this process using whole excised larch embryos as explants are illustrated in Fig. 1C and 1D. The former shows developing buds after a 14-day induction period, and the latter, elongating shoots after 3 months in culture. After seed sterilization, whole embryos are excised and plated on induction medium or germinated (on semi-solid media or soilless mixes) for seedling parts. The most responsive explants, phytohormone treatments and time of exposure all have to be determined empirically. For many temperate species with smaller seeds, whole embryos are commonly used, but with larger seeds like radiata or Canary Island pine, cotyledons are the most effective explants providing a higher multiplication rate per seed. Culture material from mature conifer trees have not been as successful as more juvenile explants. However, when utilized, the apical dome with the youngest leaf primordia was excised from sterile shoot tips of redwood (Monteuuis, 1991) and only the bud-bases, i.e., without the meristematic dome and several needle primordia, were cultured for Douglas fir (Dunstan et al., 1986), and white spruce (Dunstan et al., 1987).

After a two to three-week induction period, treated explants are transferred to hormone-free basal media for the formation of functional shoot apices with juvenile leaf primordia. This phase involves the development of the nodular tissue formed during bud induction into shoots with primary needles (Thorpe and Hasnain, 1988). In general, this process may require transfers to medium with altered nutritional or hormone levels (Thorpe and Biondi, 1984). For example, bud growth on Canary Island pine cotyledonary explants was promoted by transferring the cotyledons after 2 weeks on induction medium to hormone-free half-strength medium. Several manipulations like cutting the explants into pieces, the addition of activated charcoal and lowering the sucrose level to 1–2% enhance shoot elongation. Various phytohormones can also be used at low concentrations, but this method is avoided because it can interfere with rooting and eventual topophysis (growth morphology) of the plantlet. Hyperhydricity of buds and shoots can occur with a few species e.g., larch (Karnosky and Veerville, 1988), and Canary Island pine (Martinez Pulido et al., 1990). Factors such as lower agar concentrations (Debergh et al., 1981), and free water and high humidity in the culture vessel also contribute to this problem (Aitken et al., 1981).

Indirect adventitious budding can also occur via callus formation. This process is fairly common among angiosperms and has been reported for at least 28 forestry species (Thorpe et al., 1991). A variety of plant organs can be used, and in general, meristematic tissue like cambium responds effectively. Callus can be induced effectively with 2,4-D or NAA, and with angiosperms it is fairly straightforward to maintain the callus by serial subculture. However, with few exceptions, it has been difficult to maintain conifer callus through several subcultures. In addition to the rapid loss of regeneration capacity, the chances of producing abnormal plantlets are increased (Patel and Berlin, 1982). Only one case of plantlet regeneration, *Pinus eldarica*, from long-term subcultured (true) callus has been reported (Gladfelter and Phillips, 1987). Laliberté and Lalonde (1988) have also obtained adventitious shoots via callus cultures of *Larix* × *eurolepsis* initiated from short shoot buds from a 12-year-old tree. They have subcultured this tissue for up to 32 months with sustained caulogenesis. Another interesting development is the use of sub-culturable meristematic nodular callus (not a true callus) in *Pinus radiata* for plantlet regeneration (Aitken-Christie et al., 1988).

2.1.4. *Shoot Multiplication*

The ability to rapidly multiply shoot cultures *in vitro* is essential for the establishment of economically feasible micropropagation systems. Usually, once proliferating shoot cultures of hardwoods are stabilized, they can provide a continuous source of microshoots. As larger shoots are harvested, the basal portions of the proliferating clusters are divided and placed on fresh multiplication medium thus ensuring a continuous supply of rootable shoots every month (Chalupa, 1987). This process is particularly important for conifers since the number of shoots produced by adventitious budding (per explant) is relatively low for most species. The approaches used with softwood adventitious shoots include low phytohormone pulses, or the removal of the shoot apices. Since there is a direct relationship between shoot size and the number of axillary buds, taller shoots have a larger multiplication ratio. This method has been used successfully with several species. However, careful monitoring of plantlet growth patterns should be made to ensure that a 'normal' orthotropic response is obtained under field conditions. Axillary multiplication is particularly successful with the Cupressaceae and in particular with the genera *Cupressus*, *Thuja* and *Juniperus*. In our experience, shoot cultures of *Cupressus sempervirens*, *Thuja occidentalis* and *Juniperus cedrus* can be propagated indefinitely without many culture manipulations or transfers. Using these species, hundreds of plantlets can be produced per year with little additional labour and material costs.

2.1.5. *Rooting*

When shoots are about 30 mm tall, they can be rooted and acclimatized before transfer to greenhouse or field conditions. For temperate hardwoods, in addition to height, leaves should be well developed, since shoots deteriorated during the rooting stage if the leaves were not fully formed (McCown and McCown, 1987). For rooting, an auxin may or may not be used, and these are generally applied through an agar or liquid pulse. For example, both agar and commercial soilless mixes were used successfully and about 98% rooting was obtained for birch and poplar shoots when commercial mixes, and no hormone treatments, were used (McCown and McCown, 1987). For other species like teak and eucalyptus, auxin treatments were required (Mascarenhas et al., 1987; Gupta and Mascarenhas, 1987). A major concern in rooting is to reduce the number of steps thus lessening the chances of obtaining morphological variation, while at the same time reducing the production costs.

In contrast to angiosperms, rooting of coniferous shoots is a slow and difficult process probably because unlike easy-to-root woody species, these shoots do not usually possess preformed root initials (Mohammed and Vidaver, 1988). Shoot vigor, juvenility, a reduction in minerals and sucrose concentrations, increases in type as well as agar concentrations, and the use of soilless mixes are some of the factors affecting rooting. For conifers, auxins are necessary for rooting, and NAA, IBA and IAA are generally used. Their mode of application and duration of exposure are all critical for the induction of root primordia. These can be applied in three ways: continuous exposure, short duration pulses and in a rooting powder (Mohammed and Vidaver, 1988). Root primordia appear in a shorter time (about 4–6 weeks) in semi-solid media, but unlike rooting in angiosperm species, these are often thicker with few (if any) lateral roots. When a liquid pulse is used, an effective auxin or combination of phytohormones, concentration of the rooting solution and time of exposure have to be determined. For example, the best treatment was a 4-h liquid pulse in a 100 μM IBA solution, followed by primordia development in peat : vermiculite (1 : 1) for *Pinus canariensis* (Martinez Pulido et al., 1990). *Ex vitro* rooting is becoming more common, even though dessication and disease infestations still occur. For some species like eastern white cedar, shoots can be rooted ex vitro using a commercial rooting powder. Reduced shade and high humidity were critical at this stage, since the rooting and hardening phases were carried out together (Harry et al., 1987).

2.1.6. *Field Trials*

Plantlets of several species have been transferred to field conditions either through a gradual acclimatization period under a controlled environment or directly to field conditions. In general, plantlets exhibited early maturation, flowering and in some species, increased biomass production when compared to seedling growth.

Before field transfer, special pretreatments to harden plantlets and improve root growth, are necessary. For example, eucalyptus plantlets were grown in pots containing a 3 : 3 : 1 mix of soil : sand : compost, and after 15 to 20 days they were transferred to polybags for further root development. Soil was drenched with fungicide before plantlets were transferred to field conditions (Mascarenhas et al., 1987). Initially, these plantlets grew faster than seedlings, but by the fifth year, there were no significant differences in growth rates. Also, flowers were observed on plantlets by the third year, and in seedlings (of the same species) after 7 to 10 years. For temperate species, McCown and Amos (1979), recommended that plantlets be handled like seedlings, and that roots be air pruned, fertilized and plantlets be vigorous before field trials. Again, they found that plantlets were slower than seedlings, but showed a more uniform growth pattern.

Information on field performance of gymnosperm species is limited since comparative studies on plantlets and seedlings began less than 10 years ago. The quality and acclimatization of the *in vitro* generated shoots have a direct correlation to field survival, growth rate, plantlet morphology and disease susceptibility. A crucial concern has been the difference in quality between the seedling and plantlet roots. While the plantlets of loblolly pine had thick, unbranched roots, the seedlings had thinner roots and more lateral root development (McKeand and Allen, 1984). Also, the nutrient uptake (nitrogen and phosphorus) was related to the root morphology: seedlings were more efficient based on per gram dry weight of root, but there was no difference when uptake was calculated as a function of root area. In contrast, uptake and translocation of ^{32}P and ^{86}Rb were similar in plantlets and seedlings of *Thuja occidentalis* (Bender et al., 1987).

Field trials by the Weyerhaeuser Company (USA) showed a survival rate of 91% for plantlets, 99% for seedlings and 49% for rooted cuttings of Douglas fir (Ritchie and Long, 1986). The plantlets and seedlings were of different sizes at the time of planting out, and by the end of five years, plantlets were still 11% shorter than seedlings. However, the annual growth increments were similar. Although some plantlets were plagiotrophic initially, by the third year, they had become orthotropic. Similarly, four years of field evaluation showed that plantlets of loblolly pine lagged behind seedlings, although growth increments were comparable by the fourth year (Amerson et al., 1988). Another important finding of this study was that 5-year-old plantlets seemed to be less susceptible to fusiform rust, a devastating fungal disease of loblolly pine. Genetic differences cannot account for this resistance since both test materials came from the same half-sib families.

Differences in plantlet and seedling topophysis were subtle. Thus, two years after field transplanting, plantlets of loblolly pine had numerous growth and morphological characteristics of mature trees (McKeand, 1985), and in Douglas fir, some of the plantlets had shorter and fewer branches, darker foliage and coarser needles (Ritchie and Long, 1986). Since gymnosperms live for a long time and grow very slowly, long term monitoring is crucial before definitive assessments can be made.

2.2. *Somatic Embryogenesis*

To date, shoot cultures are the most successful method of propagation for hardwood trees. However, somatic embryogenesis is being viewed as an alternative method for several species. Once an *in vitro* protocol is defined, it can prove to be a more effective method for the rapid propagation of selected trees. Other advantages include the lower costs associated with a rapid multiplication system based on somatic embryos, the potential for generating artificial seeds and eventually for using an automated system, and suspensions systems can provide embryogenic protoplasts which potentially can be used for genetic engineering (Thorpe et al., 1991). Somatic embryogenic tissue can be generated directly on a variety of treated explants or obtained indirectly by manipulating non-embryogenic callus *in vitro*. The sequence of steps which leads to plantlets include induction, tissue maintenance, maturation and germination of somatic embryos, and acclimatization of plantlets before field trials.

In conifers, success has been achieved mainly with immature and mature embryos, and in rare cases with megagametophytic tissue, e.g., *Larix decidua* (Nagmani and Bonga, 1985). The first evidence for conifer somatic embryogenesis was presented by Hakman et al. (1985) for *Picea abies* (Norway spruce) using immature zygotic embryos, but currently, protocols have been established for about 19 species, with success using mature embryos entirely restricted to the genus *Picea* (Tautorus et al., 1991). In general, precotyledonary zygotic embryos are the most successful explants for pines, and cotyledonary zygotic embryos for spruces (Becwar et al, 1988). In addition, excised cotyledons of *Picea abies* (Krogstrup, 1986), and more recently, of *P. glauca* and *P. mariana* (Lelu et al., 1990; Lelu and Bornman, 1990), have also yielded embryogenic tissue. Of significance also, is that Attree et al. (1990) were able to use 12- to 30-day-old seedlings of *Picea mariana* and *P. glauca* as sources of explants.

A variety of media formulations can be used, and both NAA and 2,4-D in combination with BA have been used successfully. Factors such as light regimes, concentrations of the basal medium, sucrose, nitrogen level and composition, mineral elements, agar, plant growth regulators and pH affect induction (Tautorus, 1991). Once the characteristic mucilaginous embryogenic tissue is formed (Fig. 2A) it can be removed from the original explant and multiplied on induction medium. This tissue is relatively stable, and can be subcultured (every 2–3 weeks) for an extended period. How time in culture affects maturation has yet to be determined. Overall, regular transfers are not the only effective method of maintaining embryogenic tissue. Kartha et al. (1988) were able to cryopreserve tissue from *Picea glauca*, and obtain viable cultures after one year. Joy et al. (1991) were able to store similar tissue of white spruce for up to a year in serum-capped flasks at ambient temperatures and still obtain regeneration.

The removal of phytohormones (particularly auxins used for induction and maintenance) in hardwood somatic embryogenic tissue usually leads to embryo development. However, for conifers, the production of cotyledonary stage embryos usually requires transfer to medium containing abscisic acid (ABA). Cotyledonary stage embryos developed after 4–6 weeks (Fig. 2B). So far, this phytohormone has proven to be critical for development with all conifer species, and high exogenous levels appear to suppress precocious germination (Roberts et al., 1990), allow the accumulation of storage products (Joy et al., 1991), and reduce cleavage polyembryony (Durzan and

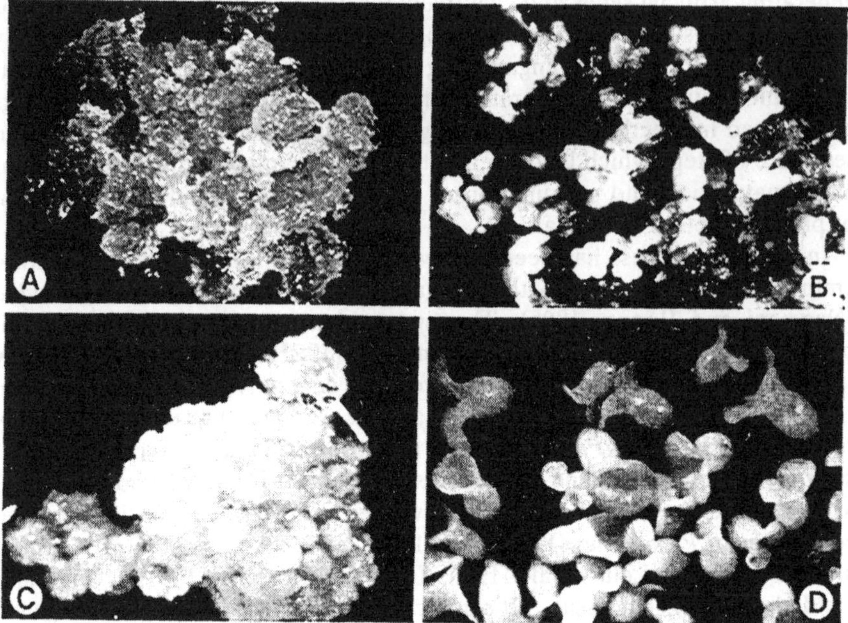

Figure 2. (A) Embryogenic tissue obtained from mature embryos of red spruce showing mucilaginous characteristic and early filamentous stage embryos; (B) Cotyledonary stage somatic embryos of red spruce after 4 weeks on medium containing abscisic acid; (C) Nodular embryogenic tissue obtained from immature zygotic embryos of yellow-poplar (photograph courtesy of Scott Merkle); (D) Torpedo-stage somatic embryos of yellow-poplar with developing cotyledons (photograph courtesy of Scott Merkle).

Gupta, 1988). A range of concentrations has been used, but levels seem to vary with species. For *Picea rubens* (Harry and Thorpe, 1991) and *P. engelmanii* × *glauca* (Roberts et al., 1990) a range of 40–60 μM ABA was best. A partial drying procedure enhanced germination of the cotyledonary embryos for both species (see Roberts et al., 1990).

To date, embryogenesis has been obtained from several hardwood species, and in the genus *Eucalyptus* alone, seven species have been propagated via this route (Le Roux, 1991). For some species like sandalwood, an important forestry species in southeast Asia, both direct and indirect protocols have been established (Bapat and Rao, 1989). Embryogenic protocols have been achieved with several temperate species including walnut, poplar, oak and birch. Using walnut for example, Tulecke and McGranahan (1985) were able to obtain induction on immature cotyledonary tissue (6–11 weeks after pollination) on medium containing BA, IBA, kinetin, and L-glutamine. Culture manipulation included a 2–4 week induction period, followed by transfer to hormone-free medium and globular embryos were visible after 6 weeks. Typically, embryogenic tissue obtained from angiosperm forestry species is different in morphology from that of coniferous trees. The tissue obtained from immature zygotic embryos of yellow poplar, for example, was nodular, friable and pale yellow in appearence (Fig. 2C); globular and cotyledonary stage somatic embryos (Fig. 2D) were obtained by subculturing on hormone-free medium (Merkle and Sommer, 1991). Germination of mature somatic embryos is sometimes problematic but can be improved with pre-treatments e.g., a cold period of 8–10 weeks (Tulecke and McGranahan, 1985), or dessication for 3–5 days with relative humidity maintained at 60–95% (Deng and Cornu, 1992) improved germination of walnut embryos.

In addition to immature embryos, embryogenic material has also been obtained from other plant tissue. For example, embryos were obtained from punctured poplar leaves on BA and 2,4-D (Park and Son, 1992) and from seeds and one-week-old seedlings and leaves of birch, and catkins (pedicle/-peduncle regions) of *Quercus bicolor* (Gingas, 1991). Haploid plantlets have also been obtained from *Populus maximowiczii* using anther microspores at the uninucleate stage (Stoehr and Zsuffa, 1990). Although embryo development was incomplete, and no cotyledons differentiated, some embryoids germinated precociously, and formed only main and secondary roots; these roots were used as explants for the production of adventitious shoots.

While the number of species which can be potentially propagated by this method has increased, the conversion rate to plantlets still remains relatively low. To date, a high percentage conversion has been obtained from only two species. About 1600 somatic embryos (71 genotypes) of white spruce (Webster and Roberts, 1989), and about 4000 Norway spruce embryos (17 genotypes) (Gupta et al., 1991) were converted to plantlets. Maturation and conversion is particularly difficult for pines, where additional steps like hormone removal, suspensor and proembryo development, and osmoticm treatments are critical. Plantlets have been established in the soil for *Larix decidua* × *L. leptolepsis* and *L. leptolepsis* × *L. decidua*, *Picea abies*, *P. glauca* × *engelmannii*, *P. mariana*, *P. sitchensis*, *Pinus taeda* and *Pseudotsuga menziesii* (Tautorus et al., 1991). The most extensive field trial is being conducted with interior spruce where 1200 plants (71 different genotypes) are being observed (Webster et al. 1990).

Artificial seed technology and scaling-up of embryogenic suspension cultures in bioreactors is progressing. The labor costs for sorting and selecting the germinative somatic embryos are still very high since these are done manually (Gupta et al., 1991). Artificial seeds, obtained through the encapsulation of somatic embryos after dessication or hydration, provide a means of lowering the cost of producing clonal material, increasing the long-term storage of somatic embryos and a mechanism through which the field planting process can be mechanized. The encapsulation medium and environment should mimic the endosperm or megagametophytic tissue in conifers. However, the role of this tissue on the growth and development of the embryo is not fully understood (Tautorus et al., 1991). This factor in combination with the low conversion rate of somatic embryos makes the production of artificial seeds extremely difficult. Cotyledonary stage somatic embryos of Norway spruce and loblolly pine also have been encapsulated in hydrated calcium alginate gel but only 1–2% germination has been achieved (Durzan and Gupta, 1988). Sandalwood is one of the few hardwood forestry species used for the production of artificial seeds; somatic embryos were encapsulated in sodium alginate, but after storage at 4 °C for 45 days, and plating on MS basal medium, only about 10% germination was recorded (Bapat and Rao, 1989).

Success with scaling-up the production of somatic embryos is limited. One of the few reports on the use of bioreactor technology projected a yield of one million plants per year from each 2L units based on current conversion rates (Smith et al., 1991). Several problems such as slow growth rates, shear sensitivity and cell aggregation have to be resolved with plant cells grown in suspensions (Tatick et al., 1991).

3. Protoplast Culture and Genetic Engineering

Protoplast culture provides an alternative method of improving plants, and is one of the few systems which can utilize somaclonal variation, somatic hybridization, cybridization and genetic engineering (Yarrow and Barsby, 1988). In addition, the latter can potentially overcome obstacles like size, long life cycles, and the lack of basic genetic information for improving woody perennials (McCown et al., 1991).

Protoplasts have been isolated from both tropical and temperate species of hardwoods. The determination of source material and preculture conditions, plating densities, culture procedures and conditions are all important in obtaining viable and regenerable protoplasts. For sandalwood, for example, four types of source material namely leaf mesophyll, hypocotyl and stem callus, and embryogenic suspension cultures, were used (Bapat and Rao, 1989). Efforts are underway to develop a viable somatic hybridization system for several species of eucalyptus. Dos Santos et al. (1990) working with *Eucalyptus grandis*, *E. urophylla* and *E. saligna* were able to establish cell cultures for protoplast isolation and culture. Of interest also was that eucalyptus protoplasts, unlike whole cells, showed variable resistance to low temperatures, and viability was correlated to the frost resistance of the donor trees (Le Roux and Van Staden, 1991). Similar procedures are used for protoplast isolation in temperate deciduous hardwood species. Park and Son (1992) using fully expanded leaves from *in vitro* cultures of *Populus nigra* × *P. maximowiczii* were able to obtain 10^4 to 10^6 protoplasts per gram fresh weight, and to regenerate shoots from these cultures.

Transformation of forest species has been successful with only a few genera like *Populus* and *Juglans* mainly using the *Agrobacterium* vector system.

Pythoud et al. (1987) using *Agrobacterium rhizogenes* were able to transfer the neomycin phosphotransferase II gene (resistance to kanamycin) to poplar cells (*Populus trichorpa* × *P. deltoides*). Juvenile axenic shoots were used as explants and the resulting transformation proved to be stable. For walnut, a repetitive somatic embryogenic system and *Agrobacterium tumefaciens* were co-cultured. Screening involved two phases: first, the vector-inoculated embryos were cloned, and secondly, these clones were screened on 75 μg/mL kanamycin over a three month period to obtain transformed tissue (McGranahan et al., 1988). More recently, McCown et al. (1991) were able to use protoplast-derived cells, nodules and stems for transformation. This is the first report of the incorporation of pest resistance genes into a forest tree species using electric discharge particle acceleration. Of particular importance is that both pests, the gypsy moth and the tent caterpillar are major defoliators of North American hardwoods.

The application of traditional plant breeding methods to conifers has been limited by their long generation times (Sederoff et al., 1986) and protoplast technology allows genetic manipulation in a shorter time period. Unlike angiosperms, there are few reports of protoplast isolation from gymnosperms. Early work included protoplast isolation from non-embryogenic callus of *Pseudotsuga menziesii*, *Pinus echinata*, *Pinus taeda*, *Pinus contorta* and *Tsuga heterophylla* (Winton et al., 1975). More recently, emphasis has shifted to using embryogenic cell suspension systems, and regeneration has been successful for about 8 species, two of which have been carried to the plantlet stage (Tautorus et al., 1991).

Successful regeneration depends on species and on the cell line being used. Protoplast isolation for white spruce (Attree et al., 1987) involved collecting cell aggregates from the suspension medium, pre-plasmolysis, digestion, followed by protoplast separation. About 75–85% viability was obtained, and first divisions occurred after 24 h; somatic embryos were recovered after 23 days. Other species respond more slowly, with the first division appearing after 7–10 days, and somatic embryos after 8–10 weeks for *Pinus taeda* (Gupta and Durzan, 1987).

The regeneration of somatic embryos from protoplasts provides a suitable system for direct gene transfer experiments with conifers (Tautorus et al., 1991). Some of the current methods used are electroporation, polyethlene glycol (PEG), and *Agrobacterium tumefaciens*. Particle bombardment (or biolistics) is also being viewed as a novel way of transferring genes to intact plant tissues and organelles (Klein et al., 1992). Recently, Robertson et al. (1992) reported stable transformation of Norway spruce using mature somatic embryo explants and microprojectile bombardment. Recombinant DNA technology adds new dimensions to a conifer tree improvement program, e.g., changing lignin chemistry for better pulping properties and lower costs, and manipulating the above ground carbon allocation of conifers are now possibilities (Timmis and Trotter, 1989).

4. Problems and Prospects

One of the major problems preventing a wider use of micropropagation technology is the difficulty in using explants from mature trees. Most success involves the use of excised embryos and seedling parts, and by the time trees are old enough for evaluation, they are often recalcitrant in culture (Thorpe and Harry, 1990). Tree propagation for the most part is carried out with unproven material. Since it is easier to propagate juvenile than mature material two options are available. These include selecting the most juvenile tissues from within the tree, or rejuvenating parts of the donor tree (Bonga, 1987). In some conifers hedging or repeated pruning is useful e.g., radiata pine (Libby et al., 1972). Rejuvenation or reactivation treatments include repeated spraying of branches intended for excision with BA (Abo El-Nil, 1982), or by serial grafting (Franclet et al., 1987). Also, serial rooting of mature cuttings have produced scions with more juvenile characteristics in both hardwoods and softwoods, and repeated subculture of shoot apices in cytokinin-containing medium has resulted in reactivated meristems in many species (Thorpe and Harry, 1990).

Woody plants, especially angiosperms often secrete substances into the medium in response to wounding. These polyphenols and tannins can inhibit development of the explant and often cause necrosis. Several approaches such as the addition of activated charcoal, PVPP, antioxidants and some amino acids to the medium can alleviate these problems (Pierik, 1987). However, in some cases, the only successful method is the very labour intensive process of frequent transfer of the explant to fresh medium. Cultured tissues also produce volatile substances which can accumulate and become auto-intoxicating. These include carbon dioxide, ethylene, ethane, acetaldehyde and ethanol; the latter two are characteristic of callus cultures and woody nodal explants (Thomas and Murashige, 1979). The manipulation of the gaseous environment is essential since the importance of some of these gases in *de novo* organogenesis has been observed (Kumar et al., 1987).

Contamination can become a severe problem if explants have systemic infections. These organisms are not removed by surface decontamination, and can spread quickly when tissues are cultured. This problem has appeared in cultures of black and English walnut (Boulay, 1985), and although the bacteria did not seem to affect the growth of shoots, they caused severe problems during the rooting phase. Internal infestations can be partially overcome by antibiotics (Young et al., 1984; Boulay, 1985), but more research on types, concentration, duration and timing of applications is needed. To date, the most successful approach has been to elongate infected material rapidly in culture, and then subculture meristem-tips to fresh media (Zimmerman, 1986). Another problem often encountered is vitrification or hyperhydric malformation which affect growth, rooting and transfer to field conditions. The leaves of vitrified shoots are often dark green, translucent, thick, curled, elongated in conifers, and the needles are often stuck together;

the stems of these shoots are often watery and brittle (Boulay, 1985; Gasparet al., 1987). Vitrification is caused by various factors, and an increase in the agar concentration, lower cytokinin and ammonia nitrogen levels have normalized vitrified tissues in some species (Gaspar et al., 1987).

Finally, the cost of generating plantlets is very high. The most advanced micropropagation technology is still labour and cost intensive (Thorpe and Hasnain, 1988). Therefore, this process can only be profitable if species cannot be multiplied conventionally and a high return is expected from genetic gain, yields, disease-free stock, etc. (George and Sherrington, 1984).

5. References

Abo, El-Nil, M.M. (1982). Method for asexual reproduction of coniferous trees. US Patent No. 4,353,184.

Aitken, J., Horgan, K.J. and Thorpe, T.A. (1981). Influence of explant selection on the shoot-forming capacity of juvenile tissue of *Pinus radiata*. Can. J. For. Res. 11:112–117.

Aitken-Christie, J. and Thorpe, T.A. (1984). Clonal propagation: Gymnosperms. In: Cell Culture and Somatic Cell Genetics of Plants, Vol. 1. I.K. Vasil (ed.), Academic Press, Orlando, pp. 82–95.

Aitken-Christie, J., Singh, A.P., Davies, H. (1988). Multiplication of meristematic tissue: A new tissue culture system for radiata pine. In: Genetic Manipulation of Woody Plants. J.W. Hanover and D.E. Keathley (eds.), Plenum Press, New York, pp. 413–432.

Ajuha, M.R. (1987). *In vitro* propagation of poplar and aspen. In: Cell and Tissue Culture in Forestry. Vol. 3. J.M. Bonga and D.J. Durzan (eds.), Martinus Nijhoff Publishers, Dordrecht, pp. 207–223.

Amerson, H. V., Frampton Jr., L.J., Mott, R.L. and Spaine, P.C. (1988). Tissue culture of conifers using loblolly pine as a model. In: Genetic Manipulation of Woody Plants. J.W. Hanover and D. E. Keathley (eds.), Plenum Press, New York, pp. 117–137.

Attree, S. M., Bekkaoui, Dunstan, D.I. and Fowke, L.C. (1987). Regeneration of somatic embryos from protoplasts isolated from an embryogenic suspension culture of white spruce (*Picea glauca*). Plant Cell Rep. 6:480–483.

Attree, S.M., Tautorus, T.E., Dunstan, D.I. and Fowke, L.C. (1990). Somatic embryo maturation, germination, and soil establishment of plants of black and white spruce (*Picea mariana* and *Picea glauca*). Can. J. Bot. 68:2583- 2589.

Ball, E.A. (1950). Differentiation in a callus culture of *Sequoia sempervirens*. Growth 16:295–325.

Bapat, V.A. and Rao, P.S. (1989). *In vitro* strategies for sandalwood propagation. In: Applications of Biotechnology in Forestry and Horticulture, V: Dhawan (ed.), Plenum Press, New York, pp. 145–156.

Becwar, M.R., Wann, S.R., Johnson, M.A., Verhagen, S.A., Feirer, R.P. and Nagmani, R. (1988). Development and characterization of *in vitro* embryogenic systems in conifers. In: Somatic Cell Genetics of Woody Plants. M.R. Ahuja (eds.), Kluwer Academic Publishers, Dordrecht, pp. 1–18.

Bender, L., Harry, I.S., Yeung, E.C. and Thorpe, T.A. (1987). Root histology, and nutrient uptake and translocation in tissue culture plantlets and seedlings of *Thuja occidentalis* L. Trees 1:232–237.

Bonga, J.M. (1987). Clonal propagation of mature trees: Problems and possible solutions. In: Cell and Tissue Culture in Forestry. J.M. Bonga and D.J. Durzan (eds.), Martinus Nijhoff Publishers, Dordrecht, pp. 249–271.

Bornman, C.H. (1983). Possibilities and constraints in the regeneration of trees from cotyledonary needles of *Picea abies in vitro*. Physiol. Plant. 57: 5–16.

Boulay, M. (1985). Some practical aspects and applications of the micropropagation of forest trees. In: *In vitro* Propagation of Forest Tree Species. Proc. Intl. Symp, May 1984, Bologna, Italy, pp. 51–81.

Bourgkard, F. and Favre, J.M. (1988). Somatic embryos from callus of *Sequoia sempervirens*. Plant Cell Rep. 7:445–448.

Chalupa, V. (1987). European hardwoods. In: Cell and Tissue Culture in Forestry, Vol. 3. J.M. Bonga and D.J. Durzan (eds.), Martinus Nijhoff Publishers, Dordrecht, pp. 224–246.

Cheliak, W.M. and Rogers, D.L. (1990). Integrating biotechnology into tree improvement programs. Can. J. For. Res. 20:452–463.

Debergh, P., Harbaoui, Y. and Lemeur, R. (1981). Mass propagation of globe artichoke (*Cynara scolymus*): Evaluation of different hypotheses to overcome vitrification with special reference to water potential. Physiol. Plant. 53:181–187.

Deng, M.-D. and Cornu, D. (1992). Maturation and germination of walnut somatic embryos. Plant Cell Tissue Org. Cult. 28:195–202.

Dos Santos, A.V.P., D'Utra Vaz and Power, J.B. (1990). Towards somatic hybridization in the genus *Eucalyptus*. A8-17, p. 208.

Dunstan, D.I. 1988. Prospects and progress in conifer biotechnology. Can. J. For. Res. 18: 1497–1506.

Dunstan, D.I., Mohammed, G.H. and Thorpe, T.A. (1987). Morphogenetic response of vegetative bud explants of adolescent and mature *Picea glauca* (Moench) Voss *in vitro*. New Phytol. 106:225–236.

Dunstan, D.I., Mohammed, G.H. and Thorpe, T.A. (1986). Shoot production and elongation on explants from vegetative buds excised from 17- to 20-year-old *Pseudotsuga menziesii*. New Zealand J. For. Sci. 16:269–282.

Durzan, D.J. and Gupta, P. (1988). Somatic embryogenesis and polyembryogenesis in conifers. Adv. Biotech. Process 9:53–81.

Fink, C.V.M., Sticklen, M.B., Lineberger, R., Daniel and Domir, S.C. (1986). *In vitro* organogenesis from shoot tip, internode, and leaf explants of *Ulmus* × 'Pioneer'. Plant Cell Tissue Org. Cult. 7:237–245.

Franclet, A., Boulay, M., Bekkaoui, F., Fouret, Y., Verschoore-Martouzet, B. and Walker, N. (1987). Rejuvenation. In: Cell and Tissue Culture in Forestry, Vol.1. J.M. Bonga and D.J. Durzan (eds.), Martinus Nijhoff Publishers, Dordrecht, pp. 232–248.

Gamborg, O.L., Miller, R.A., and Ojima, K. (1968). Nutrient requirements of suspension cultures of soybean root cells. Exp. Cell. Res. 50:151–158.

Gaspar, Th., Kevers, C., DeBergh, P., Maene, L., Paques, M. and Boxus, Ph. (1987). Vitrification: Morphological, physiological and ecological aspects. In: Cell and Tissue Culture in Forestry, Vol. 1. J.M. Bonga and D.J. Durzan (eds.), Martinus Nijhoff Publishers, Dordrecht, pp. 232–248.

George, E.F. and Sherrington, P.D. (1984). Plant Propagation by Tissue Culture. Exegetics Ltd., England.

Gingas, V.M. (1991). Asexual embryogenesis and plant regeneration from male catkins of *Quercus*. Hort. Sc. 26:1217–1218.

Gladfelter, H.J. and Phillips, G.C. (1987). *De novo* shoot organogenesis of *Pinus eldarica* Medw. *in vitro*. 1. Reproducible regeneration from long-term callus cultures. Plant Cell Rep. 6:163–166.

Gresshoff, P.M. and Doy, C.H. (1972). Development and differentiation of haploid *Lycopersicon esculentum* (tomato). Planta 107:473–497.

Gupta, P.K. and Durzan, D.J. (1987). Somatic embryos from protoplasts of loblolly pine proembryonal cells. Bio/Technology 5:710–712.

Gupta, P.K. and Durzan, D.J. (1985). Shoot multiplication from mature trees of Douglas fir, *Pseudotsuga menziesii*, and sugar pine, *Pinus lambertiana*. Plant Cell Rep. 4:177–179.

Gupta, P.K. and Mascarenhas, A.F. (1987). Eucalyptus. In: Cell and Tissue Culture in Forestry,

Vol. 3. J.M. Bonga and D.J. Durzan (eds.), Martinus Nijhoff Publishers, Dordrecht, pp. 385–399.

Gupta, P.K., Timmis, R. and Mascarenhas, A.F. (1991). Field performance of micropropagated forestry species. *In vitro* Cell. Dev. Biol. 27P:159–164.

Gupta, P.K., Timmis, R., Pullman, G., Yancey, M., Kreitinger, M., Carlson, and Carpenter, C. (1991). Development of an embryogenic system for automated propagation of forest trees. In: Cell Culture and Somatic Cell Genetics of Plants, Vol. 8. I.K. Vasil (ed.), Academic Press, New York, pp. 75–93.

Haissig, B.E., Nelson, N.D. and Kidd, G.H. (1987). Trends in the use of tissue culture in forest improvement. Bio/Technology 5:52–57.

Hakman, I., Fowke, L.C., von Arnold, S. and Eriksson, T. (1985). The development of somatic embryos in tissue cultures initiated from immature embryos of *Picea abies* (Norway spruce). Plant Sci. 38:53–59.

Harry, I.S. and Thorpe, T.A. (1991). Somatic embryogenesis and plant regeneration from mature zygotic embryos of red spruce. Bot. Gaz. 152:446–452.

Harry, I.S., Thompson, M.R. and Thorpe, T.A. (1991). Regeneration of plantlets from mature embryos of western larch. *In vitro* Cell. Dev. Biol. 27P:89–98.

Harry, I.S., Thompson, M.R., Lu, C.-Y. and Thorpe, T.A. (1987). *In vitro* plantlet formation from embryonic explants of eastern white cedar (*Thuja occidentalis* L.). Tree Physiol. 3:273–283.

Isikawa, H. (1986). Cryptomeria (*Cryptomeria japonica* Don.). In: Biotechnology in Agriculture and Forestry, Vol. 1. Trees I. Y.P.S. Bajaj (ed.), Springer Verlag, Berlin, pp. 316–321.

Joy, R.W. IV., Kumar, P.P. and Thorpe, T.A. (1991). Long-term storage of somatic embryogenic white spruce tissue at ambient temperature. Plant Cell Tissue Org. Cult. 25:53–60.

Joy, R.W. IV., Yeung, E.C., Kong, L. and Thorpe, T.A. (1991). Development of white spruce somatic embryos: I. Storage product deposition. *In vitro* Cell. Devel. Biol. 27P:32–41.

Karnosky, D.F. and Verville, J.D. (1988). Economics of larch micropropagation. Paper presented at the 4th Intl. Tissue Culture Work Group, August 8–12, 1988, Sask., Canada.

Kartha, K.K., Fowke, L.C., Leung, N.L., Caswell, K.L. and Hakman, I. (1988). Induction of somatic embryos and plantlets from cryopreserved cell cultures of white spruce (*Picea glauca*). J. Plant Physiol. 132:529–539.

Klein, T.M., Arentzen, R., Lewis, P.A. and Fitzpatrick-McElligott, S. (1992). Transformation of microbes, plants and animals by particle bombardment. Bio/Technology 10:286–291.

Krogstrup, P. (1986). Embryolike structures from cotyledons and ripe embryos of Norway spruce (*Picea abies*). Can J. For. Res. 16:664–668.

Kumar, P.P., Reid, D.M. and Thorpe, T.A. (1987). The role of ethylene and carbon dioxide in differentiation of shoot buds in excised cotyledons of *Pinus radiata in vitro*. Physiol. Plant. 69:244–252.

Laliberté, S. and Lalonde, M. (1988). Sustained caulogenesis in callus cultures of *Larix* × *eurolepsis* initiated from short shoot buds of a 12-year-old tree. Am. J. Bot. 75:767–777.

Le Roux, J.J. and Van Staden, J. (1991). Micropropagation and tissue culture of *Eucalyptus*: A review. Tree Physiol. 9:435–477.

Lelu, M.-A. and Bornman, C.H. (1990). Induction of somatic embryogenesis in excised cotyledons of *Picea glauca* and *Picea mariana*. Plant Physiol. Biochem. 28:785–791.

Lelu, M.-A., Boulay, M.P. and Bornman, C.H. (1990). Somatic embryogenesis in cotyledons of *Picea abies* is enhanced by an adventitious bud-inducing treatment. New For. 4:125–135.

Libby, W.J., Brown, A.G. and Fielding, J.M. (1972). Effects of hedging radiata pine on production, rooting and early growth of cuttings. New Zealand J. For. Sci. 2:263–283.

Litvay, J.D., Johnson, M.A., Verma, D., Einspahr, D. and Weyrauch, K. (1981). Conifer suspension culture medium development using analytical data from developing seeds. Institute Paper Chemistry Tech. Paper Series No. 115, Appleton, WI.

Lloyd, G. and McCown, B. (1980). Commercially-feasible micropropagation of mountain laurel, *Kalmia latifolia*, by the use of shoot-tip culture. Comb. Proc. Intl. Plant Prop. Soc. 30:421–437.

Martinez Pulido, C., Harry, I.S. and Thorpe, T.A. (1990). *In vitro* regeneration of plantlets of Canary Island pine (*Pinus canariensis*). Can. J. For. Res. 20:1200–1211.

Mascarenhas, A.F., Kendurkar, S.V., Gupta, P.K., Khuspe, S.S. and Agrawal, D.C. (1987). Teak. In: Cell and Tissue Culture in Forestry, Vol. 3. J.M. Bonga and D.J. Durzan (eds.), Martinus Nijhoff Publishers, Dordrecht, pp. 300–315.

McCown, B.H. and Amos, R. (1979). Initial trials with commercial micropropagation of birch selections. Comb. Proc. Intl. Plant Prop. Soc. 29:387–393.

McCown, D.D. and McCown, B.H. (1987). North American hardwoods. In: Cell and Tissue Culture in Forestry, Vol. 3. Martinus Nijhoff Publishers, Dordrecht, pp. 247–260.

McCown, B.H. and Sellmer, J.C. (1987). General media and vessels suitable for woody plant culture. In: Cell and Tissue Culture in Forestry, Vol. 1. J.M. Bonga and D.J. Durzan (eds.), Martinus Nijhoff Publishers, Dordrecht, pp. 4–16.

McCown, B.H., McCabe, D.E., Russell, D.R., Robison, D.J., Barton, K.A. and Raffa, K.F. (1991). Stable transformation of *Populus* and incorporation of pest resistance by electric discharge particle acceleration. Plant Cell Rep. 9:590–594.

McGranahan, G.H., Leslie, C.A., Uratsu, S.L., Martin, L.A. and Dandekar, A.M. (1988). *Agrobacterium*-mediated transformation of walnut somatic embryos and regeneration of transgenic plants. Bio/Technology 6:800–804.

McKeand, S.E. (1985). Expression of mature characteristics by tissue culture plantlets of loblolly pine. J. Am. Soc. Hort. Sci. 110:619–623.

McKeand, S.E. and Allen, H.L. (1984). Nutritional and root development factors affecting growth of tissue culture plantlets of loblolly pine. Physiol. Plant. 61:523–528.

Merkle, S. A. and Sommer, H.E. (1991). Yellow-poplar. In: Biotechnology in Agriculture and Forestry, Vol. 16. Trees III. Y.P.S. Bajaj (ed.), Springer Verlag, Berlin.

Mohammed, G.H. and Vidaver, W.E. (1988). Root production and plantlet development in tissue cultured conifers. Plant Cell Tissue Org. Cult. 14:137–160.

Mohammed, G.H. and Vidaver, W.E. (1991). Early development of Douglas-fir plantlets following transfer to the greenhouse. Plant Sci. 76:259–265.

Monteuuis, O. (1991). Rejuvenation of a 100-year-old *Sequoiadendron giganteum* through *in vitro* meristem culture. I. Organogenic and morphological arguments. Physiol. Plant. 81:111–115.

Murashige, T. and Skoog, F. (1962). A revised medium for rapid growth and bioassays with tobacco tissue cultures. Physiol. Plant. 15:473–497.

Nagmani, R. and Bonga, J.M. 1985. Embryogenesis in subcultured callus of *Larix decidua*. Can. J. For. Res. 15:1088–1091.

National Research Council (1991). Managing Global Genetic Resources: Forest trees. National Academy Press, Washington, DC.

Park, Y.G. and Son, S.H. (1988). *In vitro* organogenesis and somatic embryogenesis from punctured leaf of *Populus nigra* × *P. maximowiczii*. Plant Cell Tissue Org. Cult. 15:95–105.

Park, Y.G. and Son, S.H. (1992). *In vitro* shoot regeneration from leaf mesophyll protoplasts of hybrid poplar (*Populus nigra* × *P. maximowiczii*. Plant Cell Rep. 11:2–6.

Patel, K.R. and Berlyn, G.P. (1982). Genetic instability of multiple buds of *Pinus coulteri* regenerated from tissue culture. Can. J. For. Res. 12:93–101.

Patel, K.R. and Thorpe, T.A. (1984). *In vitro* differentiation of plantlets from embryonic explants of lodgepole pine (*Pinus contorta* Dougl. ex Loud). Plant Cell Tissue Org. Cult. 3:131–142.

Pierik, R.L.M. (1987). *In vitro* culture of higher plants. Martinus Nijhoff Publishers, Dordrecht.

Pythoud, F., Sinkar, V.P., Nester, E.W. and Gordon, M.P. (1987). Increased virulence of *Agrobacterium rhizogenes* conferred by the vir region of pTiBo542: Application to genetic engineering of poplar. Bio/Technology 5:1323–1327.

Quoirin, M. and LePoivre, P. (1977). Études de milieux adaptés aux cultures *in vitro* de *Prunus*. Acta Hort. 78:437–442.

Ritchie, G.A. and Long, A.J. (1986). Field performance of micropropagated Douglas fir. New Zealand J. For. Sci. 16:343–356.

Roberts, D.R., Sutton, B.C.S., and Flinn, B.S. (1990). Synchronous and high frequency germination of interior spruce somatic embryos following partial drying at high relative humidity. Can. J. Bot. 68:1086–1090.

Schenk, R.V. and Hildebrandt A.C. (1972). Medium and techniques for induction and growth of monocotyledonous and dicotyledonous plant cell cultures. Can J. Bot. 50:199–204.

Sederoff, R., Stomp, A.-M., Chilton, W. Scottt and Moore, L.W. (1986). Gene transfer into loblolly pine by *Agrobacterium tumefaciens*. Bio/Technology 4:647–649

Smith, D.R., Donaldson, L.A. and Warr, A. (1991). Bioreactor technology for mass propagation of Pinus radiata by somatic embryogenesis. *In vitro* Cell. Dev. Biol. 27: Abstract 29, p. 43A.

Sommer, H.E., Brown, C.L. and Kormanik, P.D. (1975). Differentiation of plantlets in longleaf pine (*Pinus palustris* Mill.) tissue cultured *in vitro*. Bot. Gaz. 136:196–200.

Stoehr, M.U. and Zsuffa, L. (1990). Induction of haploids in *Populus maximowiczii* via embryogenic callus. Plant Cell Tissue Org. Cult. 23:49–58.

Taticek, R.A., Moo-Young, M. and Legge, R.L. (1991). The scale-up of plant cell culture: Engineering considerations. Plant Cell Tissue Org. Cult. 24:139–158.

Tautorus, T.E., Fowke, L. C. and Dunstan, D.I. (1991). Somatic embryogenesis in conifers. Can. J. Bot. 69:1873–1899.

Thomas, D. and Murashige, T. (1979). Volatile emissions of plant tissue cultures, I: Identification of the major components. *In vitro* 15:654–658.

Thorpe, T.A. (1983). Biotechnological applications of tissue culture to forest tree improvement. Biotech. Advs. 1:263–278.

Thorpe, T.A. and Biondi, S. (1984). Conifers. In: Handbook of Plant Cell Culture, Vol. 22. W.R. Sharp, D.A. Evans, P.V. Ammirato and Y. Yamada (eds.), Macmillan, New York, pp. 435–470.

Thorpe, T.A. and Harry, I.S. (1991). Clonal propagation of conifers. In: Plant Tissue Culture Manual. K. Lindsey (ed.), Kluwer Academic Publishers, Dordrecht, pp. C3: 1–16.

Thorpe, T.A. and Harry, I.S. (1990). Special problems and prospects in the propagation of woody species. In: Plant Aging: Basic and Applied Approaches. R. Rodríguez, R. Sanchez Tames and D.J. Durzan (eds.), Plenum Press, New York, pp. 67–74.

Thorpe, T.A. and Hasnain, S. (1988). Micropropagation of conifers: Methods, opportunities and costs. In: Tree Improvement – Progressing Together, Proceedings 21st Meeting of the Canadian Tree Improvement Association, Truro, NS, August 1987. E.K. Morgenstern and J.B. Boyle (eds.), Can. Fors. Serv., Ont., pp. 68–84.

Thorpe, T.A., Harry, I.S. and Kumar, P.P. (1991). Application of micropropagation to forestry. In: Micropropagation. P.C. Debergh and R.H. Zimmerman (eds.), Kluwer Academic Publishers, Dordrecht, pp. 311–336.

Timmis, R. and Trotter, P.C. (1989). Biological and economic feasibility of genetically engineered trees for lignin properties and carbon allocation. In: Applications of Biotechnology in Forestry and Horticulture. V. Dhawan (ed.), Plenum Press, New York, pp. 349–367.

Timmis, R., Abo El-Nil, M.M. and Stonecypher, R.W. (1987). Potential genetic gain through tissue culture. In: Cell and Tissue Culture in Forestry, Vol. 1. J.M. Bonga and D.J. Durzan (eds.), Martinus Nijhoff Publishers, Dordrecht, pp. 198–215.

Tulecke, W. and McGranahan, G. (1985). Somatic embryogenesis and plant regeneration from cotyledons of walnut, *Juglans regia* L. Plant Sci. 40:57–63.

Vieitez, A.M., Vieitez, M.L. and Vieitez, E. (1986). Chestnut (*Castanea* spp.) In: Biotechnology in Agriculture and Forestry, Vol. 1. Trees I. Y.P.S. Bajaj (ed.), Springer Verlag, Berlin, pp. 393–414.

von Arnold S. and Eriksson, T. (1981). *In vitro* studies of adventitious shoot formation in *Pinus contorta*. Can. J. Bot. 59:870–874.

Webster, F.B. and Roberts, D.R. (1989). Optimization of somatic embryogenesis and germination of interior spruce for clonal propagation. Abst. IUFRO Workshop on Somatic Cell Genetics, Placerville, CA, Oct. 15–19, p. 29.

Webster, F.B., Roberts, D.R., McInnis, S.M. and Sutton, B.C.S. (1990). Propagation of interior spruce by somatic embryogenesis. Can. J. For. Res. 20:1759–1765.

Wenger, K.F. (1984). Forestry Handbook. John Wiley, New York.

Winton, L. L. (1970). Shoot and tree production from aspen cultures. Am. J. Bot. 57:904–909.

Winton, L.L., Parham, R.A. and Kaustinen, H.M. (1975). Isolation of conifer protoplasts. Inst. Paper Chem., Appleton, pp. 1–7.

Yarrow, S.A. and Barsby, T.L. (1988). Recent advances in protoplast culture of horticultural crops: Vegetable crops. Sci. Hort. 37:179–200.

Young, P.M., Hutchins, A.S. and Canfield, M.L. (1984). Use of antibiotics to control bacteria in shoot cultures of woody plants. Plant Sci. Lett. 34:203–209.

Zimmerman, R.H. (1986). Regeneration in woody ornamentals and fruit trees. In: Cell Culture and Somatic Cell Genetics of Plants, Vol. 3. I.K. Vasil (ed.), Academic Press, New York, pp. 243–258.

22. *In vitro* Culture of Ornamentals

Pierre Debergh

Contents

1. History – Economic Importance

The history and economic importance of tissue culture for ornamental plants has recently been reviewed (Capellades et al., 1990; Debergh and Read, 1991). From the data presented in these reviews it can be concluded that world-wide more than 500 million plants are micropropagated annually, and the majority are ornamentals. Since this information was gathered some new developments have occurred. Eastern Europe, especially Poland and Hungary, have become very important producers, generally at the expense of the laboratories in Western Europe. Also, Far East Asia plays a more and more predominant role in the world of micropropagation. This is particularly important because the total market size is limited and rather static.

Most often micropropagation is associated with ornamentals and vice versa. Apart from micropropagation, tissue culture has much more to offer to the ornamentals. An overview of possible applications is given hereafter.

Indra K. Vasil and Trevor A. Thorpe (eds.), Plant Cell and Tissue Culture, 561–573.

2. Major Applications

2.1. *Germination in vitro*

This procedure for obtaining plants from seeds or spores has a major impact in orchids and ferns, but also some bromeliads (Pierik et al., 1984; Mekers, 1977) and cacti (Zimmer and Buttner, 1982).

2.1.1. *Orchids*

Since Knudson (1925) showed that germination of orchid seeds was possible on a simple medium it became a routine business for most tropical and subtropical orchids, but also for many Western European terrestrial orchids (Van Waes and Debergh, 1986a, b).

Key protocol

- The duration of seed pretreatment with $Ca(OCl)_2$ + Tween 80 is one of the most important factors controlling the germination in Western European orchids (Van Waes and Debergh, 1986a). For tropical and subtropical orchids it is far easier to sterilize a closed seed capsule. Lucke (1971) claimed that, in principle orchid seeds can be sown when the seed is half way between fertilization (takes place very late after pollination) and ripening of the capsule.
- Most often agar solidified media are used. The defined media for orchids are very simple. Frequently used mineral media are: Knudson C, Vacin and Went, Thomale, Burgeff-3f (Pierik, 1987). Fairly often non-defined media are used, especially in the numerous laboratories in the Far East. These latter media contain coconut milk, banana pulp, potato extract, pineapple juice, etc. Sucrose (1–3%) is generally added. Vitamins are sometimes added, but growth regulators are most often not necessary.
- Depending on the species, light (12–16 hours, low irradiance) or darkness are required. Usually temperatures between 20 and 25 °C are maintained.

2.1.2. *Ferns*

Germination of fern spores *in vitro* allows one to avoid phytosanitary problems during this very susceptible early period in the growth cycle of these plants. Indeed, a major problem under normal cultural conditions is infection with bacteria, fungi, algae and mosses.

Key protocol (after Pierik et al., 1986)

- Spores are first cleaned and put in a filter paper bag. Sterilization steps are: a few seconds in alcohol, 45 min in commercial bleach (12.5%), followed by several rinses in sterile water.

- An agar solidified Murashige and Skoog (1962) medium, with half strength major elements, is used, with or without charcoal.
- Culture conditions are: low intensity light and 25 °C.

2.2. *Micropropagation*

Micropropagation of ornamentals is the most widely used tissue culture technique. In some cases their impact is tremendous; e.g. in Western Europe 100% of the propagation material of *Gerbera jamesonii* is tissue cultured 90% for *Spathiphyllum*, and 75% for *Anthurium* spp.

More and more examples are available where micropropagated material is not used as the final product, but rather as micro mother plants for cutting production. This is the case for greenhouse ornamentals (many *Ficus* spp.), but also for most ornamental shrubs (e.g. *Rhododendron*; Marks, 1991).

Routinely the following stages are identified: Stage 0 (preparation of the stock plants), Stage I (initiation of the culture), Stage II (rapid multiplication) and Stage III (preparation of the propagules for rooting and establishment in the greenhouse).

Axillary budding and single node explants are the most popular ways of propagation (Sagawa and Kunisaki, 1990).

As the list of micropropagated ornamentals is extremely long it is impossible to formulate a general key procedure for all of them. Therefore, hereafter we only enumerate a few general considerations and list only selected references, review articles or books.

General considerations about a key protocol for micropropagation

- Stage 0: This stage is now generally accepted as an intrinsic part of a micropropagation scheme. It was originally introduced (Debergh and Maene, 1981) as a remedy for contamination problems. However, stage 0 also includes many kinds of interventions that can make an explant more suitable or more reliable as the starting material (Debergh, 1987). The most often manipulated parameters are light, temperature and growth regulators (Debergh and Read, 1991).
- Stage I: Nodes, buds, and shoot apices are the most reliable explants for starting a micropropagation scheme. However, in several instances leaf explants (*Ficus lyrata*, *Anthurium andreanum*, *A. scherzerianum*) or explants from flowers (*Rhododendron* spp., *Gerbera jamesonii*) have been used, especially in those cases where other types of explants do not provide enough guarantees to yield 'disease free' explants.
- For sterilization procedures standard products are used, and appropriate protocols are provided in the literature (George and Sherrington, 1984).
- Media for stages I and II are most often solidified with, and derived, from the Murashige and Skoog (1962) or Woody Plant medium (Lloyd and McCown, 1980). With the axillary bud technique a hormonal balance

in favor of cytokinins will release apical dominance and stimulate the development of axillary buds. With the single node culture the original bud explant will elongate; cytokinins are not always required in this case. Details about media and hormonal balances used for different ornamentals are provided by George et al. (1987). The highest propagation ratio is not always the best. Indeed, too high a cytokinin concentration, the use of very strong cytokinins, and too frequent subcultures on high cytokinin containing media can be responsible for physiological aberrations during culture *in vitro*, but also during subsequent culture in the greenhouse or in the field. When the original buds or shoots are obtained adventitiously, the subsequent propagation step will usually be by axillary buds.

- Stage III. Rooting can be performed *in vitro* and *in vitro*. Media with a low salt concentration and supplemented with an auxin are favoured by many researchers, although rooting can also occur on hormone free or charcoal-containing media. For many herbaceous and woody ornamentals, the double layer technique (Maene and Debergh, 1985) has proven to be appropriate to induce elongation and/or rooting.
- The quality of the micropropagated material is a determining factor for success in acclimatization. Hyperhydricity and other types of physiological disorders should be avoided (Preece and Sutter, 1991; Ziv, 1991).

Key protocol for micropropagation with flower parts as explants

- Select flower buds at different developmental stages, but always prior to anthesis.
- Standard sterilization procedures are used.
- For *Rhododendron*: individual florets with as much pedicel as possible, Anderson's medium, Z or 2iP (12–15 mg/l) (Meyer, 1983) is used.
- for *Gerbera* fragments of the capitulum are cultured in the dark on a Murashige and Skoog medium with BA (up to 20 mg/l) (Pierik et al., 1982).

2.3. *Meristem Culture*

In ornamentals, meristem culture was one of the first and is still a most widely used procedure in commercial applications of tissue culture. Morel and Martin (1952) were to first to develop this technique for *Dahlia*. Later it became a routine operation for many ornamentals, especially cut flowers. The most important applications are in the eradication of viruses in *Dianthus caryophyllus* (carnation), *Dendranthema morifolium* (former name *Chrysanthemum morifolium*), *Lilium* spp. and *Pelargonium* spp. Nowadays, more than ever, growers are aware of the importance of the phytosanitary quality of propagation material and of the final product. This explains the renewed interest in real meristem culture.

Indeed, the term meristem culture is rather often abused to identify material that has been micropropagated in one way or another, but without taking the precautions of thoroughly testing the progeny. The consequences can be dramatic and disastrous. Instead of eradicating viruses, micropropagation can result in the systematic spread of diseased plants (Capellades et al., 1991).

The importance of meristem culture is demonstrated by the fact that it saved the Dutch *Lilium* industry about 20 years ago. Virus diseases were jeopardizing the subsistance of this crop. Nowadays 95% of all certified lily bulbs originate in the test tube.

Meristem culture is not only effective in eradicating viruses. *Pelargonium* meristem culture, combined with appropriate testing by immunofluorescence, has been effective in the production of mother plants freed of *Xanthomonas pelargonii* (Reuther, 1983).

Meristem culture has no value when it is not combined with thorough testing of the plant material afterwards.

Key protocol

- Meristem culture must be combined with the following experimental treatments: chemotherapy, thermotherapy or cryotherapy.
- When real meristem culture is practised sterilization procedures are most often not required.
- Most of the commonly used culture media are fairly rich in essential elements. The suitability of MS medium has been emphasized by many workers. There has been no critical assessment of the role of various vitamins and amino acids (Bhojwani and Razdan, 1983).
- Plant growth regulators are generally used, although is has not always been essential. In some cases the addition of N^6- substituted adenines has proven to be detrimental, due to severe browning reactions (Nitsch and Strain, 1969).

2.4. *Ovary and Ovule Culture*

Ovary and ovule culture in ornamentals have been used for different purposes:

- Overcoming incompatibility problems in breeding programs: e.g. production of F_1-hybrids in *Chrysanthemum morifolium* (Watanabe, 1977); interspecific hybridization between *Hibiscus syriacus* × *H. rosa-sinensis* (Paek et al., 1989).
- Gynogenesis: success in ovary culture, to yield haploids, came in the middle of the 1970s, approximately 10 years later than that of anther culture. Ornamentals for which haploids have been obtained by this technique include: *Gerbera jamesonii* (Cappadocia et al., 1988; Meynet and Sibi, 1984; Sitbon, 1981); *Lilium davidii* (Gu and Cheng, 1983), *Petunia*

axillaris (DeVerma and Collins, 1984) and *Petunia hybrida* after pollination with irradiated pollen grains (Raquin, 1985, 1986).
- Micropropagation: *Hosta sieboldiana* (Hill et al., 1989), *Hyacinthus orientalis* (Saniewski, 1975), *Rhododendron prinophyllum* (Dai et al., 1987), *Helianthus annuus* (Mix, 1985). In the latter case the idea was to yield haploids.

Guidelines for a key protocol
- Practically always unfertilized or dormant ovaries are used as primary explant.
- Most often the medium is a modification of the medium of Murashige and Skoog (1962).
- Generally speaking the pathway is callus induction followed by shoot differentiation.
- The average success rate in haploid production is 5%, in some cases the cultivar influence is predominant over medium composition (*Helianthus*).

2.5. *Embryo Rescue*

Embryo rescue techniques are routinely used in breeding programs to overcome embryo abortion and to recover an offspring that would be otherwise lost *in vivo*. These techniques have proven to be very helpful in solving breeding problems in many ornamental crops.

In *Pelargonium* × *hortorum* direct embryo culture only succeeded when embryos were removed at least 13 days after pollination. With the process of ovary culture, embryo development and germination could be achieved as early as 5–6 days after pollination. Embryo rescue effectiveness increased with the age of cultivated ovaries (Scemama and Raquin, 1990).

Embryo culture has also been effective in rescuing embryos, 60 days after pollination, from interspecific crosses between *Iris laevigata* and *I. ensata* (Yabuya and Yamagata, 1981).

Especially in *Lilium*, embryo rescue has played a predominant role in the development of interspecific crosses. Emsweller and Uhring (1962) were the first to apply this technique in *Lilium*.

2.6. *Anther and Microspore Culture*

So far, the impact of anther culture in the production of haploids from ornamentals is rather limited. Different reports mention the production of haploid callus without regeneration, and/or the production of plants from somatic tissues. Haploids have been obtained in the following species: *Lilium longiflorum* (Qu et al., 1988; Sharp and Rasquin, 1971), *Mamillaria* (Cheema and Mehra, 1977), *Pelargonium hortorum* (Abo El-Nil and Hildebrandt,

1973), *Pelargonium roseum* (Kato et al., 1980), *Petunia hybrida* (Straub, 1973), *Primula obconica* (Bajaj, 1981), and *Saintpaulia ionantha* (Hughes et al., 1975).

Key protocol

- In most cases successful buds contain anthers that are in or near the first mitotic division.
- Rather often callus is formed, followed by plantlet development.
- Different types of salt formulations and adjuvants are used.

2.7. *Polyploids*

Only a few examples exist. Somatic embryos of *Camellia japonica* differentiated from single cells of hypocotyl explants placed on MS-medium containing GA_3. When these somatic embryos were treated with 0.1% colchicine for 1 week secondary polyploid embryos formed on the hypocotyl of the primary embryos, and they developed into plantlets that exhibited autopolyploid morphological characteristics (Kato, 1989).

To overcome the sterility of the F_1-hybrids between *Iris laevigata* × *I. ensata* (2n = 24), embryos were treated with colchicine (0.05%) for 1, 2 and 4 days. This resulted in the production of fertile amphidiploids (2n = 56), with a high pollen and seed fertility. Such material appeared to be suitable for use as breeding material for interspecific cross breeding of *I. ensata* (Yabuya, 1985).

Amphidiploids have also been induced from the hybrid between *Lilium longiflorum* × *L. henryi* by 0.1% colchicine solution treatment of the scales one week after scaling (Kunishige and Hirata, 1975).

2.8. *Embryogenesis*

In *Ranunculus asiaticus* direct embryogenesis has been induced from superficial anther tissues (Meynet and Duclos, 1990) and from callus derived from thalamus tissue (M. Beruto, San Remo, personal communication). Anther-derived plants manifested somaclonal variation, which was not the case with the thalamus derived plants. Recently, embryogenesis has been described in *Rosa* (Arif and Khatamian, 1990; Noriega et al., 1990). Other ornamentals where somatic embryogenesis has been reported are: *Camellia japonica* (Vieitez and Barciela, 1990); *Cyclamen persicum* (Wicart et al., 1984); *Hibiscus acetosella* (Reynolds and Blackmon, 1983); *Hoya carnosa* (Maraffa et al., 1981); *Iris* (Reuther, 1977); *Petunia inflata* (Handro et al., 1973).

Key protocol

- Embryogenesis has been obtained from embryo and leaf explants.

- Most often callus is induced on auxin containing media (2,4-D, but also other auxins) and successively embryo regeneration occurs on media devoid of auxins.

2.9. *Mutation Breeding*

Mutation breeding is of considerable importance in ornamentals. Combining *in vitro* propagation procedures and mutation techniques may be a tool for further improvement of higher plants. Under normal cultural conditions the diplontic selection between mutated cells and the surrounding non-mutated cells is often lost by the mutated cells. Tissue culture techniques offer possibilities to increase the expression of genetic variability (somaclonal variants) and to induce numerous mutants by a combined protocol using tissue culture and gamma or X-rays, or chemical mutagens.

Key protocols

- Gamma irradiation of tissues already in culture: meristem cultures (Johnson, 1980); plantlets (Dubuc-Lebreux and Vieth, 1987); seeds, seedlings and callus (Bapat and Rao, 1976).
- Gamma irradiation of stock plants, followed by tissue culture: pedicels (De Jong and Custer, 1986), bulb scales (Wang et al., 1989); irradiated pollen grains were used for pollination, and later on the embryos were rescued and cultured *in vitro* (Raquin, 1985, 1986).
- Treat nodal explants or regenerating organs with a solution of ethyl methyl sulfonate (10–1000 μg/ml), nitrosomethylurea (5 mM) or others, for a few hours or days and subculture on appropriate regeneration medium.

2.10. *Protoplast Culture*

Little work has been done directly on ornamental genotypes, and the available technology is not ready for exploitation, it is still at the research level. In his review paper on protoplasts of ornamental plants, Griesbach (1988) mentioned 46 species that had been regenerated into whole plants. Approximately 60% of the regenerated species were either from the *Petunia* or *Nicotiana* genus. Since 1988 not many new ornamental species have been regenerated from protophasts. One of the most well established systems has been developed by Fitter and Krikorian (1981), who regenerated complete plants from protoplasts prepared from morphogenic *Haemerocallis* suspension cultures; no other source of protoplasts was successful. The plating efficiency was greater than 50%.

Key protocols
In general two strategies for the development of complete protoplasts systems were utilized (McCown, 1988):
(1) the isolation of leaf protoplasts from *in vitro* shoot cultures with subsequent shoot regeneration via organogenesis, often from nodular callus.
(2) isolation of protoplasts from embryonic tissue sources (cotyledons or embryogenic cultures) with subsequent shoot regeneration via somatic embryogenesis.

2.11. *Genetic Engineering*

Recent advances in molecular biology, especially gene isolation, manipulation and transfer between species, make it possible to alter the aesthetic and commercial properties of plants in a highly directed fashion. As a result commercial enterprises have begun molecular flower-breeding programs which use recombinant DNA techniques (Mol et al., 1989).

By manipulating endogenous genes involved in pigmentation and senescence, new, longer lasting varieties will be bred soon. The large scale production of ornamental plants will also benefit from disease- and herbicide-resistance genes that are already being tested in field crops. However, more fundamental knowledge is required on the effects of hormones, other molecules and regulatory genes on gene expression, and phenotypes such as floral shape, plant form and habit and time of flowering (Mol et al., 1989).

Petunia, an interesting experimental plant, is one of the only representatives among the ornamentals that has been genetically engineered. Interspecific and intergeneric somatic hybrids and cybrids have been obtained (Boeshore et al., 1985; Clark et al., 1986) by protoplast fusion. Vector mediated DNA transfer has been reported by Meyer et al. (1987). This was the first case of flower color manipulation, whereby the maize gene for dihydroflavonol was transferred to *Petunia*.

It has been reported that the Dutch company Florigene has transferred some unspecified genes to some *Dendranthema* cultivars (Anonym 1989).

2.12. *Cryopreservation*

Little information is available on the cryopreservation of ornamentals *in vitro*. Examples are *Lilium* (Bouman and De Klerk, 1990), *Chrysanthemum* (Fukai, 1990).

3. Future Prospects

Although tissue culture techniques have already had a tremendous impact on the production of ornamentals, much more research is required to overcome many problems and to make the techniques more reliable in order to exploit some of the other possibilities to a larger extent.

Major problems are: mastering contaminations and variation. All kinds of contaminations are still hampering tissue culture laboratories, and especially bacterial infections are a real burden in most facilities (Cassells, 1988). Somaclonal variation and physiological abnormalities (epigenetic variation) are quite often responsible for serious problems in commercial production.

The head space of the containers is still largely terra incognita. More work is required to understand photosynthesis (De Riek et al., 1991), and to develop autotrophy (Kozai, 1991).

The future will probably allow us to witness not only a larger impact from embryogenesis and artificial seeds, but also the large scale use of fermentors for cultures (Preil, 1991).

4. References

Abo El-Nil, M.M. and Hildebrandt, A.C. (1973). Origin of androgenetic callus and haploid geranium plants. Can. J. Bot. 51:2107–2109.

Anonym (1989). Succesvol overplanten genen bij chrysant. Vakblad Bloemisterij 47:19.

Arif, M.B. and Khatamian, H. (1990). *In vitro* embryogenesis derived from leaf callus of 'Tiffany' Rose. Hort. Sc. 25:1085–1086.

Bajaj, Y.P.S. (1981). Regeneration of plants from ultra-low frozen anthers of *Primula obconica*. Sci. Hort. 14:93–95.

Bapat, V.A. and Rao, P.S. (1976). Differential radiosensitivity of seeds, seedlings and callus cultures of *Petunia inflata*. Plant Sci. Lett. 6:291–298.

Bhojwani, S.S. and Razdan, M.K. (1983). Plant Tissue Culture: Theory and Practice. Elsevier, Amsterdam.

Boeshore, M.L., Hanson, M.R. and Izhar, S. (1985). A variant mithochondrial DNA arrangement specific to *Petunia* stable sterile somatic hybrids. Plant Mol. Biol. 4:125–132.

Bouman, H. and De Klerk, G.J. (1990). Leliemeristemen bewaren door invriezen. Vakblad Bloemisterij 12:52–53.

Cappadocia, M., Chrétien, L. and Laublin, G. (1988). Production of haploids in *Gerbera jamesonii* via ovule culture: Influence of fall versus spring sampling on callus formation and shoot regeneration. Can. J. Bot. 66:1107–1110.

Capellades Queralt, M., Beruto, M., Vanderschaeghe, A. and Debergh, P. (1991). Ornamentals. In: Micropropagation: Technology and Application. P.C. Debergh and R.H. Zimmerman (eds.), Kluwer Academic Publishers, Dordrecht, pp. 215–229.

Cassells, A.C. (1988). Symposium on bacterial and bacteria-like contaminants of plant tissue cultures. Acta Hort. 225. I.S.H.S, The Netherlands.

Cheema, G.S. and Mehra, P.N. (1977). Anther culture of a Cactus. Nat. Cact. Succ. J. 36:8 - 11.

Clark, E., Schnabelrauch, L., Hanson, M.R. and Sink, K.C. (1986). Differential fate of plastid and mitochondrial genomes in *Petunia* somatic hybrids. Theor. Appl. Genet. 72:748–755.

Dai, C., Lambeth, V.N., Taven, R. and Mertz, D. (1987). Micropropagation of *Rhododendron prinophyllum* by ovary culture. Hort. Sc. 22:491–493.

Debergh, P. (1987). Recent trends in the application of tissue culture to ornamentals. In: Plant tissue and cell culture. C.E. Green, D.A. Somers, W.P. Hackett and D.D. Biesboer (eds.), Alan R. Liss, New York, pp. 383–393.

Debergh, P.C. and Maene, L.J. (1981) A scheme for commercial propagation of ornamental plants by tissue culture. Sci. Hort. 14:335–345.

Debergh, P.C. and Read, P.E. (1991) Micropropagation. In: Micropropagation: technology and application. P.C. Debergh and R.H. Zimmerman (eds.), Kluwer Academic Publishers, Dordrecht, pp. 1–13.

Debergh, P., Roggemans, J. and Standaert-De Metsenaere, R. (1990). Tissue culture in relation to ornamental plants. In: Plant Tissue Culture: Applications and Limitations. S.S. Bhojwani (ed.), Elsevier, Amsterdam, pp. 161–189.

De Jong, J. and Custer, B.M. (1986). Induced changes in growth and flowering of *Chrysanthemum* after irradiation and *in vitro* culture of pedicles and petal epidermis. Euphytica 35:137–148.

De Riek, J., Van Cleemput, O. and Debergh, P. (1991). Carbon metabolism of micropropagated *Rosa multiflora* L. *In vitro* cellular developmental biology. Plant 27:57–63.

DeVerma, J.W. and Collins, G.B. (1984). Maternal haploids of *Petunia axillaris* (Lam.) B.S.P. via culture of placenta attached ovules. Theor. Appl. Genet. 69:187–192.

Dubuc-Lebreux, M.A. and Vieth, J. (1987) Effets des rayons gamma ^{60}Co sur des tigelles de *Gerbera jamesonii* irradiées *in vitro*. Can. J. Bot. 65:261–267.

Emsweller, S.L. and Uhring, J. (1962). Endosperm-embryo incompatibility in *Lilium* species. In: Adv. Hort. Sci. & Applications, Vol. II. Pergamon Press, Oxford, pp. 360–367.

Fitter, M.S. and Krikorian, A.D. (1981). Recovery of totipotent cells and plantlet production from daily protoplasts. Ann. Bot. 48:591–597.

Fukai, S. (1990). Cryopreservation of chrysanthemum shoot tips. Sci. Hort. 45:167–174.

George, E.F., Puttock, D.J.M. and George, H. (1987). Plant Culture Media; Vol. 1. Formulations and Uses, Vol. 2. Commentary and Analysis, Exegetics Ltd., Everseley.

George, E.F. and Sherrington, P.D. (1984). Plant Propagation by Tissue Culture. Exegetics Ltd., Eversley.

Griesbach, R.J. (1988). Recent advances in the protoplast biology of flower crops. Sci. Hort. 37:247–256.

Gu, Z.P. and Cheng, K.C. (1983). *In vitro* induction of haploid plantlets from unpollinated ovaries of lily and its embryological observations. Acta Bot. Sin. 24:24–28.

Handro, W., Rao, P.S. and Harada, H. (1973) A histological study of the development of buds, roots, and embryos in organ cultures of *Petunia inflata* R. Fries. Ann. Bot. 37:817–821.

Hill, R.A., Tuskan, G.A. and Boe, A.A. (1989). *In vitro* propagation of *Hosta sieboldiana* using excised ovaries from immature florets. Plant Cell Tissue Org. Cult. 17:71–76.

Hughes, K.W., Bell, S.L. and Caponetti, J.D. (1975). Anther-derived haploids of the African violet. Can. J. Bot. 53:1442–1444.

Johnson, R.T. (1980). Gamma irradiation and *in vitro* induced separation of chimeral genotypes in carnation. Hort. Sc. 15:605–606.

Kato, M. (1989). Polyploids of *Camellia* through culture of somatic embryos. Hort. Sc. 24:1023–1025.

Kato, M., Suga, T. and Tokumasu, S. (1980). Effect of 2,4-D and NAA on callus formation and haploid production in anther culture of *Pelargonium roseum*. Mem. Coll. Agr., Ehime Univ. 24:199–207.

Knudson, L. (1925). Physiological study of the symbiotic germination of orchid seeds. Bot. Gaz. 29:345–379.

Kozai, T. (1991) Micropropagation under photoautotrophic conditions. In: Micropropagation: Technology and Application. P.C. Debergh and R.H. Zimmerman (eds.), Kluwer Academic Publishers, Dordrecht, pp. 447–470

Kunishige, M. and Hirata, Y. (1975). Induction of polyploidy in *Lilium*. Bull. Hort. Res. Stat. (Kurume) pp. 5–8.

Lloyd, G. and McCown, B. (1980). Commercially-feasible micropropagation of Mountain Laurel, *Kalmia latifolia*, by use of shoot tip culture. Comb. Proc. Intl. Plant Prop. Soc. 30:421–427.

Lucke, E. (1971). Zur Ernte unreifer Samenkapseln. Dies Orchidee 22:146–147.

Maene, L.J. and Debergh, P.C. (1985). Liquid medium additions to established tissue cultures to improve elongation and or rooting *in vitro*. Plant Cell Tissue Org. Cult. 5:23–33.

Maraffa, S.B., Sharp, W.R., Tayama, H.K. and Fretz, T.A. (1981). Apparent asexual embryogenesis in cultured leaf sections of *Hoya*. Z. Pflanzenphysiol. 102:45–55.

Marks, T.R. (1991). *Rhododendron* cuttings. I. Improved rooting following 'rejuvenation' *in vitro*. J. Hort. Sci. 66:103–111.

McCown, B.H. (1988). Recent advances in protoplast culture of horticultural crops: Ornamental trees and shrubs. Sci. Hort. 37:257–265.

Mekers, O. (1977). *In vitro* propagation of some Tillandsioideae (Bromeliaceae). Acta Hort. 78:311–320.

Meyer, M.M. (1983). Clonal propagation of perennial plants from flowers by tissue culture. Proc. Intl. Plant Prop. Soc. 33: 402–407.

Meyer, P., Heidmann, I., Forkmann, G. and Saedler, H. (1987). A new petunia flower colour generated by transformation of a mutant with a maize gene. Nature (London) 33: 677–678.

Meynet, J. and Duclos, A. (1990). Culture *in vitro* de la renoncule des fleuristes (*Ranunculus asiaticus* L.), III: Etude des plantes produites par embryogenèse somatique à partir des tissus superficiels de l'anthère. Agronomie 10:285–290.

Meynet, J.L and Sibi, M. (1984). Haploid plants from *in vitro* culture of unfertilized ovules in *Gerbera jamesonii*. Z. Pflanzenzücht. 93:78–85.

Mix, G. (1985). Antheren- und Ovarienkultur von Sonnenblumen (*Helianthus annuus* L.). Landbauforschung Volkenrode 35:153–156.

Mol, J., Stuitje, A., Gerats, A., Van der Krol, A. and Jorgensen, R. (1989). Saying it with genes: molecular flower breeding. Trends in Biotechnology 7:148–153.

Morel, G. and Martin, C. (1952). Guérison des dahlias atteints d'une maladie à virus. C.R. Acad. Sci. Paris 235: 1324–1325.

Murashige, T. and Skoog, F. (1962). A revised medium for rapid growth and bioassays with tobacco tissue culture. Physiol. Plant. 15:473–497.

Nitsch, J.P. and Strain, G.C. (1969). Effet de diverses cytokinines sur le brunissement d'explantats de canne à sucre. C.R. Acad. Sci. Paris 268:806–809.

Noriega, C., Sondahl, M.R. and Whitaker, R. (1990). Studies on plant regeneration in hybrid tea roses. In: Abstracts 7th Intl. Congress on Plant Tissue and Cell Culture. IAPTC (ed.), p. 133.

Paek, K.Y., Hwang, J.K., Jong, S.K. and Park, S.I. (1989). *In vitro* pollination, ovule culture of *Hibiscus syriacus* and differentiation of adventitious bud and somatic embryo from callus. J. Korean Soc. Hort. Sci. 30: 145–156.

Pierik, R.L.M. (1987). *In Vitro* Culture of Higher Plants. Martinus Nijhoff Publishers, Dordrecht.

Pierik, R.L.M., Oerlemans, A.P.C., Verhoeven, M.A.P., Mentjox, M..A. and Molenberg, H. (1986). Zaai van varens in kweekbuizen. Vakblad Bloemisterij 10:48–49.

Pierik, R.L.M., Steegmans, H.H.M. and Hendriks, J. (1984). The influence of NAA on the growth of *in vitro* cultivated seedlings of *Bromeliaceae*. Sci. Hort. 24:193–199.

Pierik, R.L.M., Steegmans, H.H.M., Verhaegh, J.A.M. and Wouters, A.N. (1982). Effect of cytokinin and cultivar on shoot formation in *Gerbera jamesonii in vitro*. Neth. J. Agric. Sci. 30:341–346.

Preece, J.E. and Sutter, E.G. (1991). Acclimatization of micropropagated plants to the greenhouse and the field. In: Micropropagation: Technology and Application. P.C., Debergh and R.H. Zimmerman (eds.), Kluwer Academic Publishers, Dordrecht, pp. 71–94.

Preil, W. (1991) Application of bioreactors in plant propagation. In: Micropropagation: Technology and application. P.C. Debergh and R.H. Zimmerman (eds), Kluwer Academic Publishers, Dordrecht, pp. 425–445.

Qu, Y., Mok, M.C., Mok, D.W.S. and Stang, J.R. (1988). Phenotypic and cytological variation among plants derived from anther cultures of *Lilium longiflorum*. *In vitro* Cellular, Developmental Biology 24:471–476.

Raquin, C. (1985). Induction of haploid plants by *in vitro* culture of *Petunia* ovaries pollinated with irradiated pollen. Z. Pflanzenzücht. 94:166–169.

J Raquin, C. (1986). Etude des conditions d'obtention de Petunias gynogénétiques par culture *in vitro* d'ovaires de plantes pollinisées par du pollen irradié. In: Nuclear Techniques and *in vitro* Culture for Plant Improvement. Intl. Atomic Energy Agency (ed.), pp. 207–211.

Reuther, G. (1977). Embryoide Differenzierungsmuster im Kallus der Gattungen *Iris* und *Asparagus*. Ber. Deutsch. Bot. Ges. 90:417–437.

Reuther, G. (1983). Propagation of disease-free *Pelargonium* cultivars by tissue culture. Acta Hort. 131:311–319.

Reynolds, B.D. and Blackmon, W.J. (1983). Embryogenesis and plantlet regeneration from callus of *Hibiscus acetosella*. J. Am. Soc. Hort. Sci. 108:307–310.

Sagawa, Y. and Kunisaki, J.T. (1990). Micropropagation of floriculture crops. In: Handbook of Plant Cell Culture. Vol. 5. P.V. Ammirato, D.A. Evans, W.R. Sharp and Y.P.S. Bajaj (eds.), McGraw-Hill, New York, pp. 25–56.

Saniewski, M. (1975). Studies on the physiology of Hyacinth bulbs, VII: Root and bulblet-like regenerations from the ovary wall of *Hyacinthus orientalis* L. Experientia 31:302.

Scemama, C. and Raquin, C. (1990). An improved method for rescuing zygotic embryos of *Pelargonium* × *hortorum* Bailey. J. Plant Physiol. 135:763–765.

Sharp, W.R. and Raskin, R.S. (1971). Haploidy in *Lilium*. Phytomorphology 21:334–337.

Sitbon, M. (1981). Production of haploid *Gerbera jamesonii* plants by *in vitro* culture of unfertilized ovules. Agronomie 1:807–812.

Straub, J. (1973). Die genetische Variabilität haploider Petunien. Z. Pflanzenzücht. 70:265–274.

Van Waes, J.M. and Debergh, P.C. (1986a). *In vitro* germination of some Western European orchids. Plant Physiol. 67:253–261.

Van Waes, J.M. and Debergh, P.C. (1986b). Adaptation of the tetrazolium method for testing the seed viability, and scanning electron microscopy study of some Western European orchids. Plant Physiol. 66:435–442.

Vieitez, A.M. and Barciela, J. (1990). Somatic embryogenesis and plant regeneration from embryonic tissues of *Camellia japonica*. Plant Cell Tissue Org. Cult. 21:267–274.

Wang, Y., Xi, Y., Wei, Z. and Lu, W. (1989). Effects of gamma rays and colchicine on mutagenesis in somaclones of *Lilium davidii* var. Willmottiae. Jiangsu J. Agr. Sci. 5:31–37.

Watanabe, K. (1977). Successful ovary culture and production of F_1 hybrids and androgenic haploids in Japanese *Chrysanthemum* species. J. Heridity 68:317–320.

Wicart, G., Mouras, A. and Lutz, A. (1984). Histological study of organogenesis and embryogenesis in *Cyclamen persicum* Mill. tissue cultures: Evidence for a single organogenetic pattern. Protoplasma 119:159–167.

Yabuya, T. (1985). Amphidiploids between *Iris laevigata* Fisch. and *I. ensata* Thunb. induced through *in vitro* culture of embryos treated with colchicine. Japan. J. Breed. 35:136–144.

Yabuya, T. and Yamagata, H. (1981). Embryo growth and cultural conditions in *Iris ensata* Thunb. Japan J. Breed. 31:377–382.

Zimmer, K. and Buttner, P. (1982). Ersatz des Lichtbedürfnisses bei der Keimung von Kakteensamen durch Gibberellinesäure. Gartenbauwissenschaft 47:97–101.

Ziv, M. (1991) Vitrification: Morphological and physiological disorders of *in vitro* plants. In: Micropropagation: Technology and Application. P.C., Debergh and R.H. Zimmerman (eds.), Kluwer Academic Publishers, Dordrecht, pp. 45–70.

Subject Index

Species Index